Materials Horizons: From Nature to Nanomaterials

Series Editor

Vijay Kumar Thakur, School of Aerospace, Transport and Manufacturing, Cranfield University, Cranfield, UK

Materials are an indispensable part of human civilization since the inception of life on earth. With the passage of time, innumerable new materials have been explored as well as developed and the search for new innovative materials continues briskly. Keeping in mind the immense perspectives of various classes of materials, this series aims at providing a comprehensive collection of works across the breadth of materials research at cutting-edge interface of materials science with physics, chemistry, biology and engineering.

This series covers a galaxy of materials ranging from natural materials to nanomaterials. Some of the topics include but not limited to: biological materials, biomimetic materials, ceramics, composites, coatings, functional materials, glasses, inorganic materials, inorganic-organic hybrids, metals, membranes, magnetic materials, manufacturing of materials, nanomaterials, organic materials and pigments to name a few. The series provides most timely and comprehensive information on advanced synthesis, processing, characterization, manufacturing and applications in a broad range of interdisciplinary fields in science, engineering and technology.

This series accepts both authored and edited works, including textbooks, monographs, reference works, and professional books. The books in this series will provide a deep insight into the state-of-art of Materials Horizons and serve students, academic, government and industrial scientists involved in all aspects of materials research.

More information about this series at http://www.springer.com/series/16122

Neethu T. M. Balakrishnan · Raghavan Prasanth
Editors

Electrospinning for Advanced Energy Storage Applications

Springer

Editors
Neethu T. M. Balakrishnan
Department of Polymer Science
and Rubber Technology (PSRT)
Cochin University of Science
and Technology (CUSAT)
Cochin, Kerala, India

Raghavan Prasanth
Department of Polymer Science
and Rubber Technology (PSRT)
Cochin University of Science
and Technology (CUSAT)
Cochin, Kerala, India

ISSN 2524-5384 ISSN 2524-5392 (electronic)
Materials Horizons: From Nature to Nanomaterials
ISBN 978-981-15-8843-3 ISBN 978-981-15-8844-0 (eBook)
https://doi.org/10.1007/978-981-15-8844-0

This Springer imprint is published by the registered company Springer Nature Singapore Pte Ltd.
The registered company address is: 152 Beach Road, #21-01/04 Gateway East, Singapore 189721, Singapore

Dedicated to our
naughty little prince,

Aaron,

who were no help
in writing this book,
but without his
tolerance this book
could not be
completed on time

Foreword

The energy crisis that was happencd by the in-depth uses of non-renewable sources such as fossil fuels, which not only results in the extinction of resources but also leads to severe environmental issues. For a reliable energy supply and for fuel diversification, the substitution of non-renewables is important since it always results in better and safer impacts environmentally and on a socio-economic level. For exploring the renewables at its extreme as well as to make it completely flexible with the desire, energy storage systems are introduced that can efficiently store the energy produced. Different storage devices such as batteries, capacitors, supercapacitors, and fuel cells were introduced for the accomplishment of the energy desire. Even though these different storage devices have their own significant impact, batteries are considered to be the best and are extensively used. Being an energy storage device, batteries are the important one that is capable to power the everyday appliances. The era of the batteries started even before the advancement of technology. Ancient technological traces of these storage systems were found to be used in the Parthian period and later were stored in Bagdad Museum. Even after that different batteries were developed from rechargeable to non-rechargeable. Heavier batteries such as lead acids were ruling the automobiles, and it was believed that the larger size is required for the accomplishment of better storage. This concept was misspelled after the discovery of flexible, lightweight, and thin rechargeable systems. A green signal for a reliable rechargeable system was got in 1985 when Akira Yoshino developed the first prototype of lithium-ion battery after completing the safety test. After that the commercialization was made with the C/LCO system which was the contribution of John Goodenough that is LCO cathode. For this remarkable invention that greatly influenced the advancement of public life, John Goodenough, Akira Yoshino, and Stanley Wittingham, who introduced the concept of rechargeable system, shared the prestigious Nobel Prize in Chemistry in 2019. Lithium-ion battery seeks much attention owing to its ability to provide high energy as well as power density which helps to acquire a rechargeable system that can charge and discharge at a high rate with a shorter time period. Ever since then lithium-ion battery technology was in the path of development. For meeting the best performance of the system, different variations were done within the battery. Even though the system was commercialized with the carbonaceous anode and intercalating transition metal compounds along with liquid electrolytes, different lithium metal alloys, as well as transition metal oxides cathodes without lithium were also examined for this battery system. One of the best choices always preferred for this best performing system was the modification of the liquid counterpart. Different organic liquids were employed as the electrolyte in lithium-ion batteries. Even though they can provide high ionic conducting systems, the safety risk caused by them was very severe. This issue was mitigated by the introduction of polymer electrolytes, which is currently employed in lithium-ion batteries that are commercially used. On the path of developing polymer electrolytes, a variety of methods were practiced in order to obtain the best system with enhanced electrochemical performance that ranges from the basic solvent casting technique to the advanced electrospinning technique. For the synthesis of the best porous membrane, electrospinning was chosen as the best as well as electrospun porous free-standing electrodes are also of prime importance for the development of an electrochemically well-performing system. Electrospinning seeks much attention in lithium-ion batteries, which can result in the formation of porous membranes comprising of continuous fibers having nanometer dimensions that are capable to exhibit excellent electrochemical performance. In view of this, this book lights up to the basics and current trends in electrospinning techniques for the development of different components in lithium-ion batteries. Currently, different polymers are employed as the matrix for electrolytes, for each of such systems; the performance enhancement can be done with the introduction of electrospinning techniques.

This book entitled *Electrospinning for Advanced Energy Storage Devices* is comprised of 20 chapters distributed among three portions, viz. electrolytes, anodes, and cathodes. Chapter 1 deals with the detailed history of this commercially irreplaceable device and the milestones which paved the way for its commercialization and eventually the Nobel Prize. The basics, consisting of the working principle and their operating mechanisms, are clarified along with its history. Chapter 2 introduces the electrospinning method that is widely used in research currently despite its early introduction. The first section, electrolytes, covers the role of electrospun polymer electrolytes for the lithium-ion battery application. The application of polymer matrices such as polyvinylidene difluoride, polyvinylidene difluoride-*co*-hexafluoro propylene, and polyacrylonitrile as electrospun electrolytes in lithium-ion batteries is elaborated in Chaps. 3–5. Chapter 6 covers the composite polymer electrolytes in general, whereas Chap. 7 focuses on silica-based composite polymer electrolytes for battery application. The final chapter of the section, Chap. 8, portraits the electrospun blend polymer electrolytes.

The second section comprises of anodes in lithium-ion battery applications that are prepared using electrospinning. Electrospun iron oxides nanocomposites as anodes in the form of nanofibers, nanotubes, and nanorods are distributed in Chaps. 9 and 10. The use of other nanocomposites based on cobalt, manganese oxide, tin, carbon, and silicon by electrospinning as anodes is the content of Chaps. 11–15, respectively, in the section.

The final section of the book which covers the electrospun cathodes in lithium-ion battery comprises the rest of the five chapters. The electrospinning of the conventional and non-conventional lithium-based cathodes, namely lithium cobaltate and lithium iron phosphate constituting nanocomposites, is discussed in Chaps. 16 and 17. Chapters 18 and 19 are composed of non-lithium nanocomposite cathodes such as vanadium pentoxide and manganese oxide using electrospinning. The final chapter deals with electrospun nanocomposites made of mixed metal oxides as cathodes for the lithium-ion battery application.

In a nutshell, this book is one of a kind since it deals with the elaborative review on the application of the electrospinning in lithium-ion battery realm. This book has tried its best to be clear on this topic by proper classification and by addressing almost all the materials related to it without fail. This book will be a useful link for the academic and industrial researchers, who want to get a clear idea on the application of electrospinning in the Nobel Prize-winning energy storage device, the lithium-ion battery.

Singapore, Singapore

Prof. Rachid Yazami
(Draper Prize 2014)

Contents

Editors and Contributors

About the Editors

Neethu T. M. Balakrishnan is currently a graduate student pursuing her Ph.D. thesis with Prof. Raghavan Prasanth at Material Science and NanoEngineering Laboratory, Department of Polymer Science and Rubber Technology, Cochin University of Science and Technology (CUSAT), India. She completed her Master's degree in Polymer Science from the Central Institute of Plastic Engineering and Technology (CIPET), Ahmedabad, Gujarat, after obtaining her B.S. degree in Chemistry from the University of Calicut. She has research experience in different fields viz, polymer engineering, materials science, and energy storage devices. Before joining CUSAT, she worked at the Indian Rubber Manufacturers Research Association (IRMRA), Mumbai. Currently, she is working in the field of energy storage devices with more focus on the development of electrospun based polymer gel electrolytes and electrodes for high performance lithium ion batteries in collaboration with Gyeongsang National University, Republic of Korea. She has 25 peer-reviewed publications on lithium ion batteries and several international conference publications to her credits. Her current area of interest includes electrospinning of polymers, carbon nanomaterials, ceramics, metal oxides, and green synthesis of ionic liquids from renewable resources for flexible and bendable energy storage solutions.

Dr. Raghavan Prasanth, Professor, at the Department of Polymer Science and Rubber Technology, Cochin University of Science and Technology (CUSAT), Kerala, India, and Visiting Professor at the Department of Materials Engineering and Convergence Technology, Gyeongsang National University, Republic of Korea. Dr. Prasanth also holds the Associate/Adjunct Faculty position at the Inter-University Centre for Nanomaterials and Devices (IUCND), CUSAT. He earned his Ph.D. in Engineering under the guidance of Prof. Jou-Hyeon Ahn, from the Department of Chemical and Biological Engineering, Gyeongsang National University, Republic of Korea, in 2009, supported by a prestigious Brain Korea (BK21) Fellowship. Dr. Prasanth earned his B.Tech and M.Tech from CUSAT, India. After a couple of years of attachment as Project Scientist at the Indian Institute of Technology (IIT-D), New Delhi, he moved to the Republic of Korea for his Ph.D. studies in 2007. His Ph.D. research was focused on fabrication and investigation of nanoscale fibrous electrolytes for high-performance energy storage devices. He completed his Engineering doctoral degree in less than three years, still a record in the Republic of Korea. After his Ph.D., Dr. Prasanth joined the Nanyang Technological University (NTU), Singapore, as a Research Scientist, in collaboration with the Energy Research Institute at NTU (ERI@N) and TUM CREATE, a joint electromobility research centre between Germany's Technische Universität München (TUM) and NTU, where he was working with Prof. (Dr.) Rachid Yazami, who has successfully introduced graphitic carbon as an anode for commercial lithium-ion batteries, and received the Draper Prize, along with the Nobel laureates, Prof. (Dr.) John B. Goodenough and Prof. (Dr.) Akira Yoshino. After four years in Singapore, Dr. Prasanth moved to Rice University, USA as Research Scientist, where he worked with Prof. (Dr.) Pulickal M Ajayan, the co-inventor of carbon nanotubes, and was fortunate to work with 2019 Chemistry Nobel Prize laureate, Prof. (Dr.) John B. Goodenough. Dr. Prasanth was selected for Brain Korea Fellowship (2007), SAGE Research Foundation Fellowship, Brazil (2009), Estonian Science Foundation Fellowship, European Science Foundation Fellowship (2010),

Faculty Recharge, University Grants Commission (UGC), Ministry of Higher Education, India (2015), etc. Dr. Prasanth has received many international awards, including Young Scientist award, Korean Electrochemical Society (2009), and was selected for the Bharat Vikas Yuva Ratna Award (2016) etc. He developed many products such as a high-performance breaking parachute, flex wheels for space shuttles, high-performance lithium-ion batteries for leading companies for portable electronic device and electric automobiles etc. Dr. Prasanth has a general research interest in polymer synthesis and processing, nanomaterials, green/nanocomposites, and electrospinning. His current research focuses on nanoscale materials and polymer composites for printed and lightweight charge storage solutions, including high-temperature supercapacitors and batteries. Dr. Prasanth has published many research papers in high-impact factor journals, and a number of books/book chapters, and has more than 7000 citation and a h-index of 50 plus. Apart from science and technology, Dr. Prasanth is a poet, activist and a columnist in online portals and printed media.

Contributors

Jou-Hyeon Ahn Department of Materials Engineering and Convergence Technology, Gyeongsang National University, Jinju-daero, Jinju, Republic of Korea

Wazed Ali Department of Textile and Fiber Engineering, Indian Institute of Technology (IIT-D), New Delhi, India

Neethu T. M. Balakrishnan Department of Polymer Science and Rubber Technology (PSRT), Cochin University of Science and Technology (CUSAT), Cochin, India

Aiswarya Bhaskar Electrochemical Power Sources Division, CSIR-Central Electrochemical Research Institute, Karaikudi, Tamil Nadu, India

Jayesh Cherusseri NanoScience Technology Centre, University of Central Florida, Orlando, USA

Akhila Das Department of Polymer Science and Rubber Technology (PSRT), Cochin University of Science and Technology (CUSAT), Cochin, India

Ditty Dixon Electrochemical Power Sources Division, CSIR-Central Electrochemical Research Institute, Karaikudi, Tamil Nadu, India

Harsha Enale Electrochemical Power Sources Division, CSIR-Central Electrochemical Research Institute, Karaikudi, Tamil Nadu, India

M. J. Jabeen Fatima Department of Polymer Science and Rubber Technology (PSRT), Cochin University of Science and Technology (CUSAT), Cochin, India

N. S. Jishnu Rubber Technology Centre, Indian Institute of Technology-Kharagpur (IIT-KGP), Kharagpur, West Bengal, India

Jarin D. Joyner Department of Chemistry, Rice University, Houston, TX, USA; Department of Materials Science and Nano Engineering, Rice University, Houston, TX, USA

Indrapal Karbhal B101, Polymer and Advanced Materials Laboratory Physical & Material's Chemistry Division, CSIR-National Chemical Laboratory, Pune, India;
Academy of Scientific and Innovative Research (AcSIR), Ghaziabad, UP, India

Mahadevappa Y. Kariduraganavar Department of Chemistry, Karnatak University, Dharwad, India

Nizao Kong College of Materials Science and Engineering, Beijing University of Chemical Technology, Chaoyang District, Beijing, People's Republic of China

M. A. Krishnan Department of Mechanical Engineering, Amrita Vishwa Vidyapeetham, Amritapuri, India;
Department of Electrical Engineering, Pennsylvania State University, University Park, Pennsylvania, USA

R. Krishnapriya Sustainable Materials and Catalysis Research Laboratory (SMCRL), Department of Chemistry, Indian Institute of Technology-Jodhpur, Jodhpur, Rajasthan, India

Devika Laishram Sustainable Materials and Catalysis Research Laboratory (SMCRL), Department of Chemistry, Indian Institute of Technology-Jodhpur, Jodhpur, Rajasthan, India

O. Manaf Materials Research Laboratory, Department of Chemistry, National Institute of Technology (NIT-C), Calicut, India

Nikhil Medhavi Department of Polymer Science and Rubber Technology (PSRT), Cochin University of Science and Technology (CUSAT), Cochin, India

Satishkumar R. Naik Department of Chemistry, Karnatak University, Dharwad, India

Golu Parte B101, Polymer and Advanced Materials Laboratory Physical & Material's Chemistry Division, CSIR-National Chemical Laboratory, Pune, India; Academy of Scientific and Innovative Research (AcSIR), Ghaziabad, UP, India

Apurva Patrike B101, Polymer and Advanced Materials Laboratory Physical & Material's Chemistry Division, CSIR-National Chemical Laboratory, Pune, India; Academy of Scientific and Innovative Research (AcSIR), Ghaziabad, UP, India

Raghavan Prasanth Department of Polymer Science and Rubber Technology (PSRT), Cochin University of Science and Technology (CUSAT), Cochin, India; Department of Materials Engineering and Convergence Technology, Gyeongsang National University, Jinju-daero, Jinju, Republic of Korea

Leya Rose Raphael Department of Polymer Science and Rubber Technology, Cochin University of Science and Technology (CUSAT), Cochin, Kerala, India

Bhagirath Saini Sustainable Materials and Catalysis Research Laboratory (SMCRL), Department of Chemistry, Indian Institute of Technology-Jodhpur, Jodhpur, Rajasthan, India

Rakesh K. Sharma Sustainable Materials and Catalysis Research Laboratory (SMCRL), Department of Chemistry, Indian Institute of Technology-Jodhpur, Jodhpur, Rajasthan, India

Manjusha Shelke B101, Polymer and Advanced Materials Laboratory Physical & Material's Chemistry Division, CSIR-National Chemical Laboratory, Pune, India; Academy of Scientific and Innovative Research (AcSIR), Ghaziabad, UP, India

Sreekanth J. Varma Department of Physics, Sanatana Dharma College, Alappuzha, Kerala, India

Ammu Surendran Electrochemical Power Sources Division, CSIR-Central Electrochemical Research Institute, Karaikudi, Tamil Nadu, India

Anjumole P. Thomas Department of Polymer Science and Rubber Technology (PSRT), Cochin University of Science and Technology (CUSAT), Cochin, India

Sabu Thomas School of Chemical Sciences, Mahatma Gandhi University, Athirampuzha, Kerala, India

M. Thripuranthaka B101, Polymer and Advanced Materials Laboratory Physical & Material's Chemistry Division, CSIR-National Chemical Laboratory, Pune, India;
Academy of Scientific and Innovative Research (AcSIR), Ghaziabad, UP, India

Anand I. Torvi Department of Chemistry, Karnatak University, Dharwad, India

S. K. Vineeth Department of Polymer and Surface Engineering, Institute of Chemical Technology, Matunga, Mumbai, India

Xiaoping Yang College of Materials Science and Engineering, Beijing University of Chemical Technology, Chaoyang District, Beijing, People's Republic of China

Yunhua Yu College of Materials Science and Engineering, Beijing University of Chemical Technology, Chaoyang District, Beijing, People's Republic of China

Abbreviations

1D	1 Dimensional
2D	2 Dimensional
AFD	Average fiber diameter
BPCNF	Black phosphorus-doped carbon nanofibers
CA	Cellulose acetate
CD	Charge–discharge
CF	Carbon fibers
CNFs	Carbon nanofibers
CNFW	Carbon Nanofibers Web
CNT	Carbon nanotubes
CV	Cyclic Voltammetry
CVD	Chemical vapor deposition
DEC	Diethyl carbonate
DMAc	Dimethyl acetamide
DMC	Dimethyl carbonate
DMF	Dimethyl formamide
DSSC	Dye-sensitized solar cell
EtOH	Ethanol
EV	Electric vehicles
FESEM	Field Emission Scanning Electron Microscope
FSiGCNF	Flexible Si and graphene/carbon nanofibers
G	Graphene
GCNF	Graphene/carbon nanofiber
GO	Graphene oxide
HCl	Hydrochloric acid
HCNF	Hollow carbon nanofibers
HCTs	Hollow carbon tubes
HEPAF	High-efficiency particulate air filters
HEV	Hybrid electric vehicles
HF	Hydrofluoric acid

HNT	Halloysite nanotubes
HR-TEM	High-resolution TEM
IEEE	Institute of Electrical and Electronics Engineers
JSPS	Japan Society for the Promotion of Science
LCO	Lithium cobalt oxide
LFP	Lithium iron phosphate
LiBOB	Lithium bisoxalate borate
LIBs	Lithium-ion batteries
LiPVAOB	Lithium polyvinyl alcohol oxalate borate
LLTO	Lanthanum titanate oxide
LMO	Lithium manganese oxide
MA	Maleic anhydride
m-SiNFs	Mesoporous silicon nanofibers
NASA	National Aeronautics and Space Administration
NATO	North Atlantic Treaty Organization
NCNF	Nitrogen-doped carbon nanofibers
NFs	Nanofibers
Ni	Nickel
Ni-Cd	Nickel–cadmium
Ni-MH	Nickel metal hydride
NiS	Nickel sulfide
NPCNFs	Nitrogen-rich porous carbon nanofibers
NPs	Nanoparticle
NTs	Nanotubes
NWs	Nanowires
PAA	Poly(acrylic) acid
PAN	Poly(acrylonitrile)
PCNF	Phosphorus-doped carbon nanofibers
PDLA	Poly(DL-lactic acid)
PE	Polymer electrolyte
PEO	Polyethylene oxide
pHCNFs	Porous hollow carbon nanofibers
PLGA	Poly(lactic-*co*-glycolic acid)
PLLA	Poly(L-lactic acid)
PMIA	Poly(m-phenylene isophthalamide)
PMMA	Polymethyl methacrylate
PS	Polystyrene
PSA	Polysulfone amide
PU	Polyurethane
PVA	Polyvinyl alcohol
PVdF	Polyvinylidene diflouride
PVdF-*co*-HFP	Poly(vinylidene flouride-co-hexaflouropropylene)
PVP	Polyvinyl pyrrolidone
RAM	Random access memory
rGO	Reduced graphene oxide

$ROCO_2Li$	Lithium alkyl carbonates
RPCNF	Red phosphorus-doped carbon nanofibers`
SAED	Selected area electron diffraction
SAN	Styrene-co-Acrylonitrile
SEI	Solid electrolyte interphase
SEM	Scanning electron microscopy
Si@MC	Silicon at multichannel carbon fibers
TEM	Transmission electron microscopy
TEOS	Tetraethoxysilane
THF	Tetrahydrofuran
T_m	Melting temperature
TMOs	Transition metal oxides
TPU	Thermoplastic polyurethane

Symbols and Formulas

$(NH_4)_2MoS_4$	Ammonium molybdenum sulfide
Ag	Silver
AgCl	Silver chloride
Al_2O_3	Aluminium oxide
Ar	Argon
$BaTiO_3$	Barium titanate
$C_2H_3LiO_2$	Lithium acetate
$CaCO_3$	Calcium carbonate
$CaMnO_3$	Calcium manganese oxide
CeO_2	Selenium oxide
$CH_3COOLi.2H_2O$	Lithium acetate dihydrate
CH_3OH	Methanol
ClO_4^-	Perchlorate ion
Co	Cobalt
$Co(Ac)_2$	Cobalt acetate
$Co(CH_3COO)_2.4H_2O$	Cobalt acetate tetrahydrate
CO_2	Carbon dioxide
Co_3O_4	Cobalt oxide
$CoFe_2O_4$	Cobalt ferrite
CoO	Cobalt oxide
CoO_2	Cobalt oxide
CoS	Cobalt sulfide
Cr	Chromium
CrS_2	Chromium sulfide
Cu	Copper
$CuFe_2O_4$	Copper ferrite
CuO	Copper oxide
CuS	Copper sulfide
Fe	Iron

$Fe\ (C_2H_3O_2)_2$	Iron acetate
$Fe(acac)_2$	Iron acetylacetonate
$Fe(NO_3)_3.9H_2O$	Iron nitrate
$Fe(NO_3)_3.9H_2O$	Iron(II) nitrate nonahydrate
Fe_2O_3	Iron(II) oxide
Fe_3O_4	Iron(III) oxide
$FeC_6H_5O_7$	Iron citrate
$FeNO_3$	Ferric nitrate
FeOOH	Iron oxide
$FePO_4$	Iron phosphate
FeS_2	Ferrous sulfide
Ge	Germanium
H_2	Hydrogen
H_2SO_4	Sulfuric acid
H_3PO_4	Phosphoric acid
HNO_3	Nitric acid
$K_2V_8O_{21}$	Potassium vanadate
KCl	Potassium chloride
KH_2PO_4	Potassium hydrogen phosphate
Li	Lithium
Li^+	Lithium ion
$Li_{1.5}MnTiO_{4+\delta}$	Lithium manganese titanium oxide
Li_2O	Lithium oxide
Li_2S	Lithium disulfide
Li_2TiO_3	Lithium titanate
Li_2ZrO_3	Lithium zirconate
Li_3N	Lithium nitride
$Li_4Ti_5O_{12}$	Lithium titanium oxide
$LiAlO_2$	Lithium aluminate
$LiBF_4$	Lithium tetrafluoroborate
$LiCF_3SO_3$	Lithium trifluoromethanesulfonate
LiCl	Lithium chloride
$LiClO_4$	Lithium perchlorate
$LiCoO_2$	Lithium cobalt oxide
$LiCoPO_4$	Lithium cobalt phosphate
$LiCuO_2$	Lithium copper oxide
LiF	Lithium fluoride
$LiFePO_4$	Lithium iron phosphate
LiH_2PO_4	Lithium
$LiMn_2O_2$	Lithium manganese oxide
$LiMnO_4$	Lithium manganese oxide
$LiNi_{1/3}Co_{1/3}Mn_{1/3}O_2$	Lithium nickel cobalt manganese oxide
$LiNiO_2$	Lithium nickel oxide
$LiNiPO_4$	Lithium nickel phosphate
$LiNO_3$	Lithium nitrate

LiOH H_2O	Lithium hydroxide monohydrate
$LiPF_6$	Lithium hexafluorophosphate
$LiTiS_2$	Lithium titanium disulfide
LiV_2O_5	Lithium vanadium oxide
$MgAl_2O_4$	Magnesium aluminate
MgO	Magnesium oxide
Mn	Manganese
$Mn(CH_3COO)_2$	Manganous acetate
$Mn(CH_3COO)_2.4H_2O$	Manganous acetate tetra hydrate
Mn_3O_4	Manganese oxide
$MnCl_2$	Manganese chloride
MnO_2	Manganese oxide
Mo	Molybdenum
MoO_3	Molybdenum trioxide
MoS_2	Molybdenum disulfide
N_2	Nitrogen
Na	Sodium
$Na_{0.7}Fe_{0.7}Mn_{0.3}O_2$	Sodium iron manganese oxide
Na_2SO_4	Sodium sulfate
NaCl	Sodium chloride
NaH_2PO_4	Sodium hydrogen phosphate
$NaNO_3$	Sodium nitrate
NaOH	Sodium hydroxide
Nb	Niobium
NbO_5	Niobium pentoxide
$NH_4H_2PO_4$	Ammonium dihydrogen phosphate
Ni	Nickel
$Ni(CH_3COO)_2.4H_2O$	Nickel acetate tetrahydrate
Ni_3S_4	Nickel sulfide
$NiCo_2O_4$	Nickel cobaltite
NiO	Nickel oxide
$NiPS_3$	Nickel phosphorus trisulfide
P	Phosphorus
PF^{6-}	Hexaflourophosphate ion
RuO_2	Rubidium oxide
Si	Silicon
SiO_2	Silicon oxide
SnO_2	Tin oxide
SnS	Stannous sulfide
SnSe	Stannic selenide
Ta	Thallium
Tb	Terbium
Ti	Titanium
TiO_2	Titanium dioxide
TiS_2	Titanium disulfide

V	Vanadium
V_2O_3	Vanadium trioxide
V_2O_5	Vanadium pentoxide
VO	Vanadium oxide
VO_2	Vanadium dioxide
WS_2	Tungsten disulfide
Zn	Zinc
$ZnAl_2O_4$	Zinc aluminate
$ZnFe_2O_4$	Zinc ferrite
ZnO	Zinc oxide
ZrO_2	Zirconia

Units

$S\ cm^{-1}$	Siemens per centimeter
$mAh\ g^{-1}$	Milliampere hour per gram
$mV\ s^{-1}$	Millivolt per second
$mS\ cm^{-1}$	Milli-Siemens per centimeter
kV	Kilovolt
$kg\ cm^{-1}$	Kilogram per centimeter
μm	Micrometer
$S\ \mu m^{-1}$	Semen per micrometer
°C	Degree Celsius
cm	Centimeter
V	Volt
%	Percentage
M	Molar
$ml\ h^{-1}$	Milliliter per hour
nm	Nanometer
MPa	Mega pascal
μm	Micrometer
$ml\ min^{-1}$	Milliliter per minute
Å	Armstrong units
$kJ\ mol^{-1}$	Kilojoule per mol
wt.%	Weight percentage
GW	Gigawatt
TWh	Terawatt hour
$Wh\ L^{-1}$	Watt hour per liter
$Wh\ kg^{-1}$	Watt hour per kilogram
Ω	Ohm
$J\ g^{-1}$	Joule per gram
$mA\ g^{-1}$	Milliampere per gram

$A\ g^{-1}$	Ampere per gram
h	Hours
$cm^2\ S^{-1}$	Square centimeter per semen

Chapter 1
The Great History of Lithium-Ion Batteries and an Overview on Energy Storage Devices

Neethu T. M. Balakrishnan, Akhila Das, N. S. Jishnu, Leya Rose Raphael, Jarin D. Joyner, Jou-Hyeon Ahn, M. J. Jabeen Fatima, and Raghavan Prasanth

1.1 Introduction

In the current scenario, the energy demand is increasing annually. The overall rate of increase in the global energy demand is around 2% p.a. [1]. Currently, the energy generated is mostly relied on the fossil fuels (coal). The depletion in fossil fuel and increase in the environmental pollution by the combustion products of these non-renewable energy sources have led to establishment of alternative sources for energy generation. The energy source is partially being replaced by renewable energy sources such as solar energy, wind energy. This energy needs to be stored for the uninterrupted usage. Hence, the need for efficient and reliable energy storage device has been aroused as prime requirement. The global warming and climatic changes have been a vital concern. The replacement for fossil fuels with electrical energy for

N. T. M. Balakrishnan · A. Das · L. R. Raphael · M. J. Jabeen Fatima (✉) · R. Prasanth (✉)
Department of Polymer Science and Rubber Technology, Cochin University of Science and Technology (CUSAT), Cochin, Kerala 682022, India
e-mail: jabeen@cusat.ac.in

R. Prasanth
e-mail: prasanth@cusat.ac.in

N. S. Jishnu
Rubber Technology Centre, Indian Institute of Technology-Kharagpur (IIT-KGP), Kharagpur, West Bengal 721302, India

J. D. Joyner
Department of Materials Science and Nano Engineering, Rice University, 6100 Main Street, MS 364, Houston, TX 77005, USA

J.-H. Ahn (✉) · R. Prasanth
Department of Materials Engineering and Convergence Technology, Gyeongsang National University, 501 Jinju-Daero, Jinju 52828, Republic of Korea
e-mail: jhahn@gnu.ac.kr

N. T. M. Balakrishnan and R. Prasanth (eds.), *Electrospinning for Advanced Energy Storage Applications*, Materials Horizons: From Nature to Nanomaterials,
https://doi.org/10.1007/978-981-15-8844-0_1

the transportation is another most vibrant area of research. The electric vehicles are being commercialized worldwide; hence, there is requirement for efficient energy storage device. Energy storage is again the necessity for these transport systems. Many energy storage systems are being widely used and explored such as lithium ion battery (LIB), fuel cells, supercapacitors.

1.2 Emergence of Energy Storage Devices

The importance of portable energy storage devices was put forward with the introduction of batteries. Batteries are broadly classified as primary and secondary batteries. Primary batteries are irreversible batteries that should be disposed after once it completely used, while the secondary batteries are the renewable and reversible battery systems that can be charged and discharged a number of times and can be effectively used. During Parthian period (248 BC), a battery was been used and later stored in Baghdad Museum. Wilhelm Konig investigated the details of this battery and was termed as "Baghdad battery." The battery was a kind of primary voltaic cell with a copper compartment and a pointed iron rod. On addition of acidic electrolyte such as vinegar, lime juice would initiate the cell reaction [2]. Later in 1749, Benjamin Franklin coined the name "battery" since his experiments on capacitors. But a discovery of the battery was done by Alessandro Volta, by year 1800. The experiment was conducted with aqueous salt membrane sandwiched between copper and zinc disks and the connection produced a voltage of 0.76 V. The experiment is considered to be the basic experiment of electrochemistry. The cell so formed was known as "galvanic cell" with two half cells (zinc and copper). During 1836, John Frederic Daniel developed another type of electrochemical cell which produced about 1 V [3]. In 1859, French Physicist Gaston Planté introduced the first rechargeable battery "lead acid battery" [4]. The battery was fabricated by rolling sandwiched lead foils with rubber strips. Georges-Lionel Leclanché introduced a new type of electrochemical cell in 1866, with carbon as anode and zinc as cathode in an electrolyte of ammonium chloride with a voltage of 1.4 V. The cell was termed as "Leclanché cell" [5]. A modified version of Leclanché cell was later on commercialized as dry cell with carbon anode and zinc as cathode. These cells were widely used in early telecommunication systems. A thermoelectric battery was introduced by Ernst Waldmar Jungner in 1869 [6]. In 1908, he also introduced "electrodes for reversible galvanic batteries." Nickel iron, nickel cadmium, etc. were used as electrodes for the fabrication of rechargeable batteries in alkaline electrolytic medium.

Lead acid battery is known to be the first commercial battery. These batteries possess comparatively huge efficiency (80–90%) but the capacity available decreases on removal of input power [7]. A part from decrease in efficiency, huge size, weight of these batteries, leakage, etc. is some common disadvantages diminishing the application of these batteries as energy storage devices. Ignoring these failures, the system is commercially used as household energy storage combined with inverters as well as in fossil fuel powered transportation systems. Usually, aircrafts powered by lead acid battery contain 6–12 batteries connected in series to produce a voltage ranging from 12 to 24 V [8]. Another major commercial battery is nickel cadmium battery. Until past three decades, Ni-Cd batteries were the mostly marketed energy storage device. These rechargeable energy storage systems were widely seen in portable electronic devices like toys, AA-type battery, AAA-type battery, etc. Usually, the power generated is less; hence, to enhance the power, two or more systems are connected in series with each other and sealed in a stainless-steel pack. The major drawback of these batteries is the "memory effect" [9]. The memory effect can be explained as a memory of initial point of the charging cycle owing to which a sudden potential drop is experienced at the same point. This drop affects the performance battery for potential applications. During initial stages of battery commercialization, alkaline batteries were used as AA and AAA batteries. But since these showed leakage issues, basic components were replaced by nickel cadmium, nickel metal hydride and lithium ion batteries. The current energy storage is leaned on lithium ion batteries.

1.3 Next Generation Energy Storage Devices

Among energy storage devices known, lithium ion batteries (LIB) have arisen as an inevitable part of the day-to-day life. The introduction of the portable devices has paved a revolution of LIBs. In the current era, even days without these portable devices are unimaginable for even ordinary peoples.

1.3.1 Supercapacitors

Supercapacitors are otherwise known as ultracapacitors. These devices show the intermediate properties of that of conventional capacitors (electrochemical capacitors) and batteries. The important prominent properties of the batteries are their lower power density and high energy density, but the supercapacitors show a contradictory character with high power density and lower energy density. The structural fabrication of supercapacitor is quite similar to that of batteries. It also contains a cathode, anode and electrolyte sandwiched between these electrodes separated with a semipermeable membrane. A typical schematic representation of a supercapacitor is given in Fig. 1.1.

1.3.2 Fuel Cells

Fuel cells are electrochemical energy storage devices which converts chemical energy in to electrical energy. Hence, works similar to that of a battery. The fuels like hydrogen and oxygen are passed on the electrodes which are converted to water.

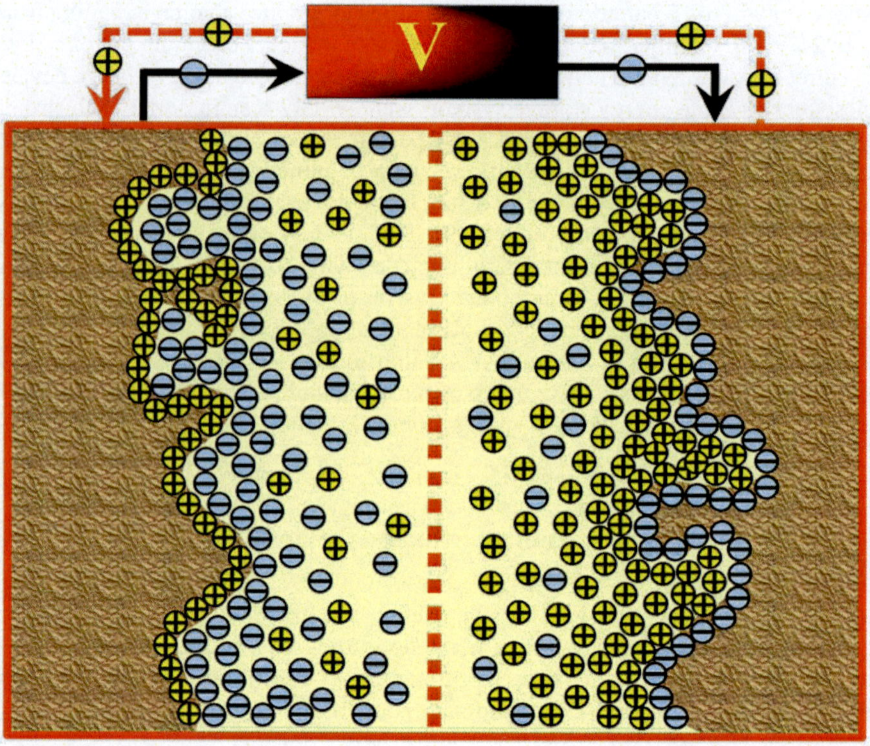

Fig. 1.1 Schematic representation of a supercapacitor

Varieties of fuel cells are known such as alkaline fuel cells, solid oxide fuel cell, direct methanol fuel cell, phosphoric acid fuel cell, molten carbonate fuel cell, proton exchange membrane fuel cell. This fuel cell varies in the components and working but the basic principle remains the same. A schematic representation of a fuel cell is given in Fig. 1.2. William Robert Grove is known as the "father of fuel cell technology" [10]. He was a physical scientist and he explained the thermal dissociation of molecules. He also found that platinum electrode can dissociate acidic solution into hydrogen and oxygen which aroused as the basic principle of fuel cell to generate energy [10].

1.4 Lithium Ion Batteries: History to the Present

Lithium ion battery (LIB) is the most advanced battery technology that empowers mobile powers used in most of the portable electronics. The importance of lithium as a battery is that it is the lightest among metals. It shows greatest electrochemical potential and largest specific energy per weight [11, 12]. The general structure of a LIB consists of an anode cathode and an electrolyte sandwiched in between these two electrodes separated by a semipermeable membrane. Even though LIB is commercially marketed in a large scale, the efficiency and safety have not yet achieved up to

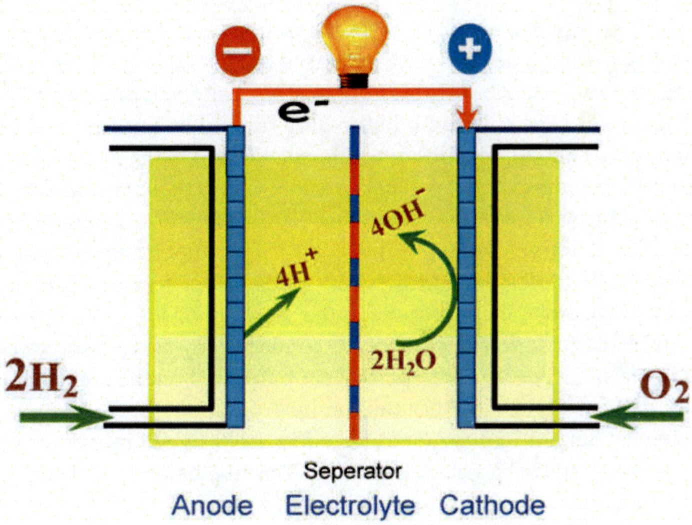

Fig. 1.2 Schematic representation of a fuel cell

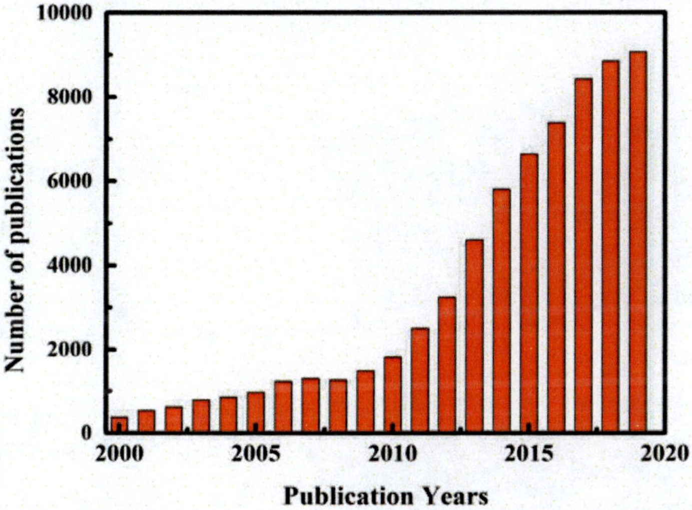

Fig. 1.3 Statistics of publications on the subject of lithium ion batteries from 2000 to 2020 (search engine: Web of Science and search key word lithium ion batteries). For 2020, there are already 650 publications reported by January

the target. The research statistics of LIB from the discovery to the present is being depicted in Fig. 1.3. Maximum publications were reported in the year 2018 (8697 as per the report from Scopus).

The need for energy storage devices for the military and civilians led to the investigation of energy storage devices with increased energy density. In 1964, Selis et al. [13] reported the importance of lithium on testing battery fabricated with calcium and silver electrodes. The calcium lithium alloy formed in situ from the reaction of negative electrode (calcium) and lithium chloride present in the electrolyte enhanced the energy of the proposed battery. Later in 1965, Casey et al. [14] also reported the importance of lithium ions in the electrolyte particularly at temperature >55 °C in nickel hydroxide-cadmium battery. Similarly in 1970, importance of lithium ions in the electrolyte for batteries operating at 450 °C based on copper oxide-magnesium batteries [15]. Ultimately, the initiation on the working of lithium ion batteries was introduced in 1974 by Mark, in a conference conducted by power sources at Chicago [16]. Later in 1976 and 1978, few reports were published on the enhancement in the conductivity of electrolytes with lithium ion incorporation.

The chemistry behind LIB was introduced in 1978 by Whittingham [17] which led to the discovery of rechargeable batteries. The intercalation of alkali metal like lithium into transition metal sulfides like titanium disulfide (TiS_2) was analyzed and found that they also possessed electrochemical reversibility similar to that of a battery. Crystallographic image of TiS_2 is given in Fig. 1.4 [18]. This finding laid the foundation for the current LIBs. Later in 1979, Li^+-ion ordering in $LiTiS_2$ for an intercalation battery was studied theoretically using statistical mechanics for lattice

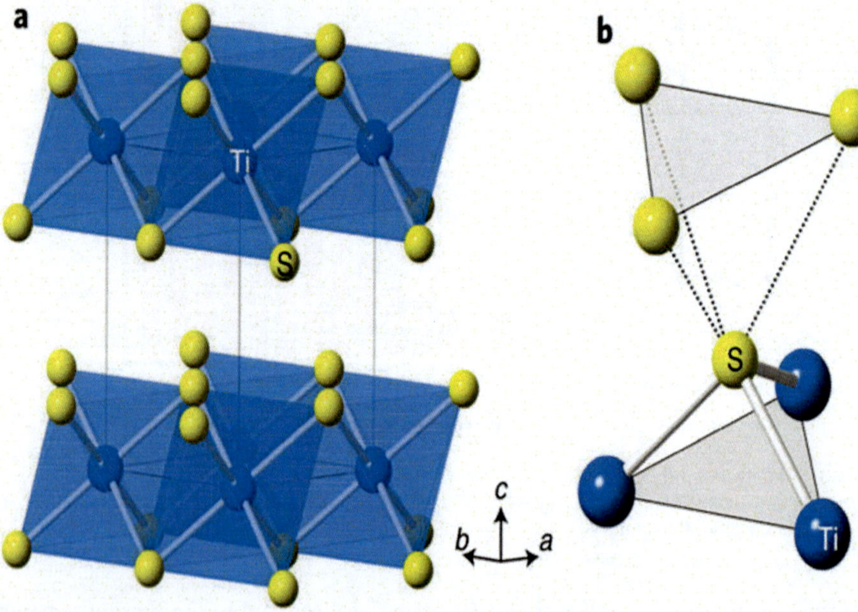

Fig. 1.4 Crystallographic image of TiS_2. Adapted and reproduced from Ref. [18], Copyright 2018 Springer Nature

gas model [19]. In the proposed work, they analyzed the effect of Li^+-ion interaction with the sites available in the electrode enhancing the voltage of battery [19]. In the same year, lithium nitride (Li_3N) was reported as one of the efficient solid Li^+-ion conducting solid electrolyte with an ionic conductivity of $10^{-3}\ \Omega^{-1}\ cm^{-1}$ [20].

Layered titanium disulfide (TiS_2) cathode material showed Li^+-ion storage mechanism which lay to a high-density energy storage device. Whittingham called the Li^+-ion storage mechanism as "intercalation." A first battery system was developed in Exxon research and Engineering Company under the supervision of Whittingham, using lithium metal as anode and TiS_2 as cathode. The battery fabrication was a studious and risky process owing to the high reactivity of metallic lithium with the atmospheric oxygen. The counter electrode, titanium disulfide was also highly reactive with atmosphere leading to the formation of hydrogen sulfide, which has a pungent smell. Hence, an inert atmosphere was required for the assembling the whole battery which was commercially not a feasible process. Another major issue faced by these batteries is their highly reactive anode (lithium metal) which readily reacts with the electrolytes and produces a passivation layer on the electrode surface, normally called solid electrolyte interface (SEI). This passivation layer is permeable to lithium ions. But at some extend, ongoing discharge process may cause an irregularity in SEI and that may cause uneven lithium ion deposition that leads to short circuit [21, 22]. Commercialization of lithium metal cathode was further restricted when there was a firing that occurred in cell phone that made use of MOLICELL batteries which operated with lithium metal anode and MoS_2 as cathode. These defects limited the further research on these material and its usage is restricted into the coin cell and it applied only in the watch batteries [23].

In 1981, Goodenough et al. [24] proposed a new class of intercalation compounds based on $LiCoO_2$ for cathodes in LIBs based on the structural similarity of dichalcogenides with the lithium cobalt oxide that can effectively intercalate the lithium ions. Even though LCO poses similar structure as that of TiS_2, it poses some superior properties over TiS_2 such as its stability at ambient conditions. In the subsequent year (1982), Gauthier et al. [25] reported the electrochemical performance of Li^+-ion intercalation on various chalcogenides and found that TiS_2 gave a cycling stability up to 300 cycles with molten salt electrolyte. In the same year, Besenhard et al. [26] investigated the role of oxides of molybdenum and chromium as cathodes for LIBs. The results revealed that these metal oxides formed irreversible chemical reaction with lithium to form stable species, hence is not a suitable candidate as cathode. In 1989, Thackeray [27] reported Li/MnO_2 cathodes for LIB system.

The development of a safer battery system required replacement of lithium metal anode and reliable electrolyte. In 1978, Armand proposed the use of solvent free polymer electrolytes. The polymer electrolyte was a combination of lithium salt that is lithium triflate along with a coordinating polymer, polyethylene oxide. But the replacement of anode was not able to achieve, and hence, the commercialization of the Lithium polymer battery was again a dream. Later during 80s, a new idea was proposed the desire of two intercalation-type electrodes became necessary that can accept lithium ion, which can act as the anode and the other that can release the same which can act as the cathode. This system was called as lithium rocking chair battery in which lithium ion rock between two electrodes. Even though this concept was actually happened in late 1970s, its practical demonstration was made possible in 1980s.

The graphite lithium intercalation was first proposed by Basu et al. [28, 29] in 1978–79. In the proposed work, Li^+-ions formed LiC_6 and it is the least two-dimensional (2D) compound formed via intercalation reaction of Li^+-ions with graphite. In 1982, Besenhard et al. [30] also investigated the role of graphite as a potential candidate for Li^+ ion intercalation and deintercalation. In 1983, Yazami and Touzain [31] proposed lithium ion intercalation in graphite electrode with solid polymer electrolyte incorporating lithium perchlorate ($LiClO_4$) in polyethylene oxide (PEO). In 1983, Akira Yoshino [32] fabricated the new class of LIB with $LiCoO_2$ as cathode, polyacetylene as cathode with non-aqueous electrolyte. During 1977, Shirakawa et al. [31] reported halogen derivative of polyacetylene with good electrical conductivity. Shirakawa was the recipient of prestigious Nobel Prize in Chemistry 2000 for his discovery of conducting polymers [33]. Akira Yoshino is the man behind the lithium ion batteries who made a safer and reliable future rechargeable system. According to Yoshino, lithium ion batteries are defined as "*non-aqueous secondary battery using transition-metal oxides containing lithium ion such as* $LiCoO_2$ *as a positive electrode and carbonaceous materials as a negative electrode.*" [32] Even though these cells was functional, the low real density and chemical stability of polyacetylene made difficulties. Vapor phase grown carbon fibers were reported 1976, which were the structural materials that effectively substitute polyacetylene in Yoshino's battery. The most challenging part of the commercialization of LIBs was its safety test. In 1986, Yoshino carried out the first safety test by colliding iron lump with battery. Violent ignition occurred in the case of battery with lithium metal anode where the Yoshino's battery does not showed any explosion [32].

Commercial lithium ion battery was established in 1990 by Sony successfully announced the first lithium ion battery [34]. Initially, LIB was commercialized with graphite anode, lithium cobalt oxide as cathode and a liquid electrolyte. The safety issues made the researchers to replace the electrolyte with polymer electrolyte. The risk of dendritic projection formation due to lithium metalation extending from anode to cathode leading to blasting of the battery lead to the development of safer and reliable electrolytic system. The volume expansion and the safety issues caused by the lithium cobalt oxide led to the replacement by more efficient and safer system like lithium iron oxide (LFP). Currently (in 2019), most of the lithium ion batteries are relied on the LFP cathode, graphite anode separated by a polymer electrolyte.

1.5 Structure of Lithium Ion Batteries

Lithium ion batteries are commercially available and mostly marketed portable batteries. Most of the new generation gadgets are relied on this energy storage system. The components may vary from battery to battery but the basic construction is the same. The size, shape and components of the batteries are varied depending on the application. A LIB consists of four major parts—an anode, cathode, electrolyte and a separator. A schematic of the detailed structure of LIBs is displayed in Fig. 1.5. Anode of a rechargeable battery is a positive electrode which receives electron while charging, whereas during discharging, the electron flows from the anode toward cathode. In the LIB available in the market, anode used is graphite rods. Another anode materials widely explored are silicon, lithium, alloys, ternary metal oxides, etc. Cathode is negative electrode while charging which gives electron to the external circuit owing to the deintercalation of lithium ions. Normally used cathode materials are lithium cobalt oxide ($LiCoO_2$ simply termed as LCO), lithium iron phosphate ($LiFePO_4$ simply termed as LFP), lithium manganese oxide ($LiMnO_4$ simply termed as LMO), etc. The chemical reactions taking place at cathode and anode in a typical LIB are given below (Eqs. 1.1 and 1.2);

$$\text{Anode (Graphite):}\ \ C_n + x\text{Li}^+ + xe^- \rightarrow C_n\text{Li}_x \tag{1.1}$$

$$\text{Cathode (LCO):}\ \ \text{LiCoO}_2 \rightleftharpoons \text{Li}_{1-x}\text{CoO}_2 + x\text{Li}^+ + xe^- \tag{1.2}$$

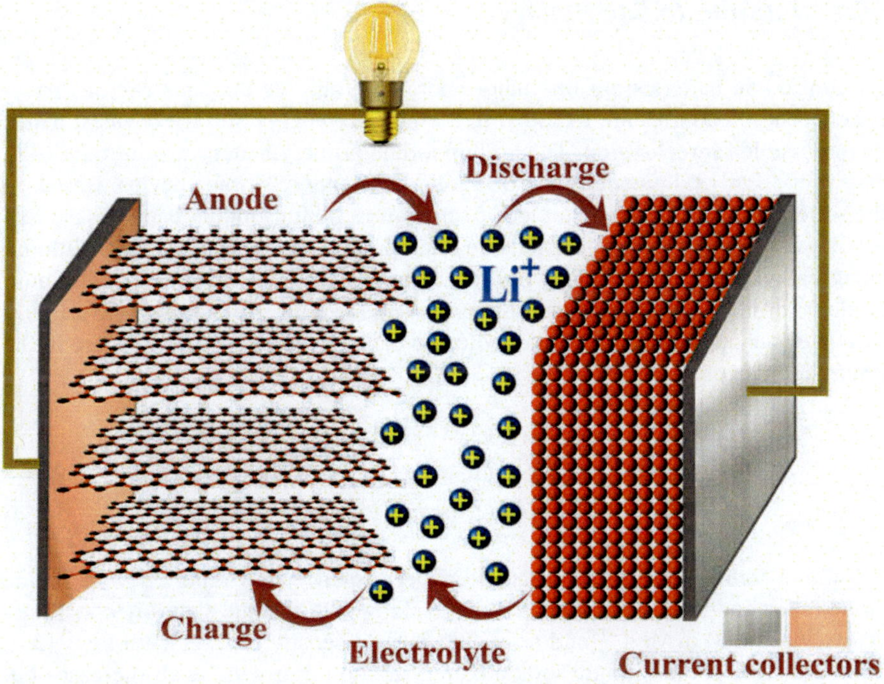

Fig. 1.5 Schematic representation on the structure of lithium ion battery

1.6 Other Types of Lithium-Based Batteries

LIB is the most marketed energy storage device in the globally. Miniature devices to e-vehicle battery vary in the size and capacity of the battery. Batteries are the key energy suppliers for most of the portable devices and equipments. Implantable devices used in the medical field like pacemakers, neurostimulators, and drug delivery systems, etc. Both primary and secondary batteries based on lithium such as lithium iodide battery, lithium manganese oxide battery have been employed chiefly as energy storage devices in these medical implants and equipments. The lithium ion batteries are main energy storage device in the laptops, palmtops and mobile phones. Normal lithium ion batteries are being widely used in these portable devices. High-density batteries are required for the electric vehicles. Lithium ion batteries with polymer electrolytes are safer and more reliable power sources, hence employed in transportation systems. Lithium iron phosphate and lithium manganese oxide are major cathode materials of commercial e-vehicle batteries. Several other energy storage devices based on lithium other than normal LIB are being explored recently such as lithium iodide battery, lithium air battery, lithium sulfur battery.

1.6.1 Lithium Iodide Battery

Lithium iodide batteries are the major energy storage for implants such as pacemakers. These batteries are included in the primary energy storage devices, hence are impossible for recharging. The lithium iodine primary battery was introduced in 1972, by Moser [35] patenting the first solid state energy storage device. Based on this solid state battery, first attempt of implanting lithium iodide battery—cardiac pacemaker was achieved in the same year [36]. The anode of the battery is lithium metal, cathode is iodine-poly-2-vinyl pyridine and at the contact layer of lithium iodide is formed which conducts lithium ions performing as electrolyte [37]. Equations (1.3) and (1.4) represent the anodic and cathodic reactions taking place in the lithium iodide cell.

$$\text{Anode: } \mathrm{Li}_x \rightarrow x\mathrm{Li}^+ + x\mathrm{e}^- \tag{1.3}$$

$$\text{Cathode: } x\mathrm{I}_2 + 2x\mathrm{e}^- \rightarrow x\mathrm{I}^- \tag{1.4}$$

Lithium iodide battery with enhanced performance was reported with iodine iodide redox couple-based aqueous cathode along with organic electrolyte to obtain a capacity of 200 mAh g^{-1} and a specific energy density of 0.33 kWh kg^{-1} [38]. Basic principle of the lithium iodide battery is the same with in all the cases but the reagents and the phase of the system differ from battery to battery. A typical schematic image of lithium iodide battery used in pacemakers is given in Fig. 1.6 [39].

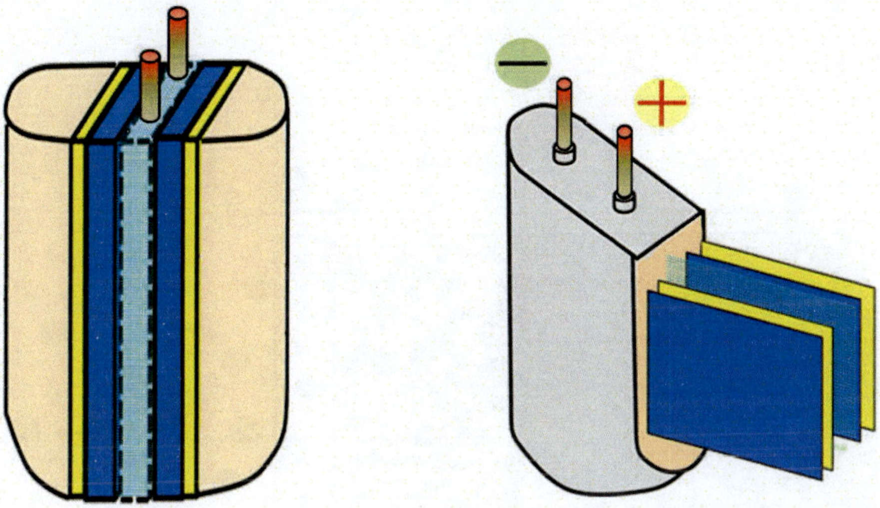

Fig. 1.6 Schematic representation of the structure of lithium iodide battery used in pacemakers. Adapted and reproduced from Ref. [39], Copyright 2016 Springer Nature

1.6.2 Lithium Air Battery

Lithium air battery utilized the electrochemical reaction of lithium and oxygen which has a high theoretical energy density of 3500 Wh kg^{-1} which can substantiate the current demand of energy. A new type of energy storage device was first introduced in 1987 by Semkow and Sammells [40] in which they used lithium alloy with general formula Li_xFeSi_2, which has been immersed in a ternary molten salt (LiF, LiCl and Li_2O), stabilized zirconia as solid electrolyte and $La_{0.89}Sr_{0.10}MnO_3$ as oxygen electrode [40]. The overall reaction occurring in the typical cell is depicted in Eq. 1.5 [40].

$$Li_xFeSi_2 + O_2 \rightleftarrows LiO_2 + FeSi_2 \tag{1.5}$$

The major cathode materials reported for the lithium oxygen batteries are activated carbon, carbon nanotubes, carbon nanofibers, graphene oxide, platinum oxide, palladium oxide, ruthenium oxide, copper, cobalt, metal alloys, transitional metal carbides, chalcogenides, etc. [41, 42]. Anodes mainly employed are based on lithium such as lithium metal, alloys. Lithium salt-based electrolytes in various organic solvents have been tried for lithium oxygen batteries [43]. A schematic of lithium air battery is displayed in Fig. 1.7.

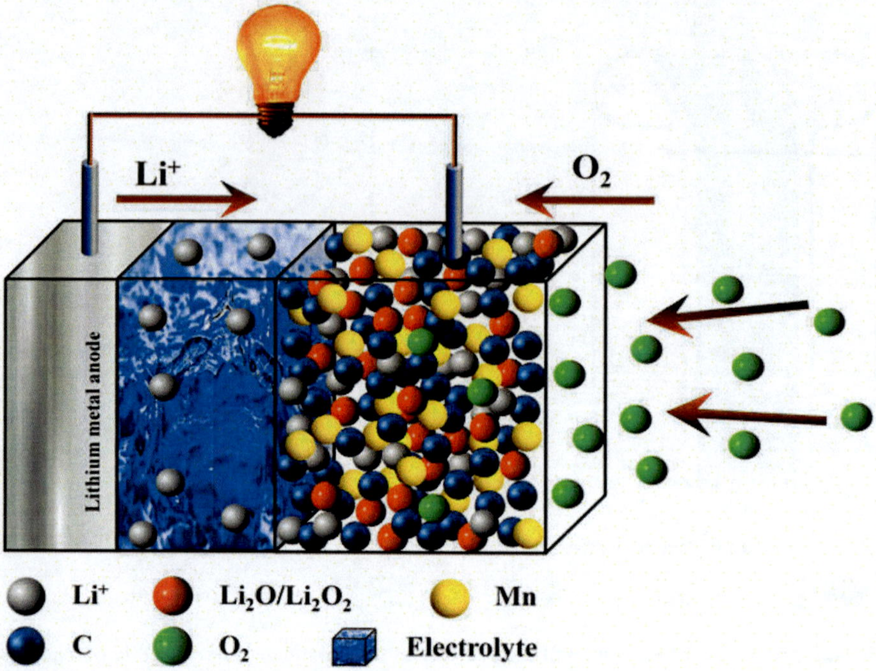

Fig. 1.7 Schematic representation of a typical lithium air battery

1.6.3 Lithium Redox Flow Battery

Lithium redox flow battery is a secondary battery similar to other rechargeable batteries. The usual batteries are working by storing chemical energy, whereas in redox flow battery, the electrodes are chemical components (anode and cathode) stored in separate reservoirs, which is circulated continuously. These circulating mediums are separated by an ion selective semipermeable membrane. Both the reservoir contain redox moiety which are active materials dispersed in aqueous electrolyte (Fig. 1.8) [44]. During the discharging process, the redox moiety present in the anolyte (anodic electrolyte) undergoes oxidation to form Li^+-ions, whereas in catholyte (cathode electrolyte), redox moiety undergoes reduction in the redox ion. In a typical charging process, the anolyte undergoes reduction from lithium ion to lithium metal, whereas catholyte undergoes oxidation reaction. The charging reaction is given in Eqs. (1.6) and (1.7).

$$\text{Cathode:}\ \ RC^{x+} + ye^- \rightarrow RC^{(x-y)+} \tag{1.6}$$

$$\text{Anode:}\ \ zLi^+ + ze^- \rightarrow zLi \tag{1.7}$$

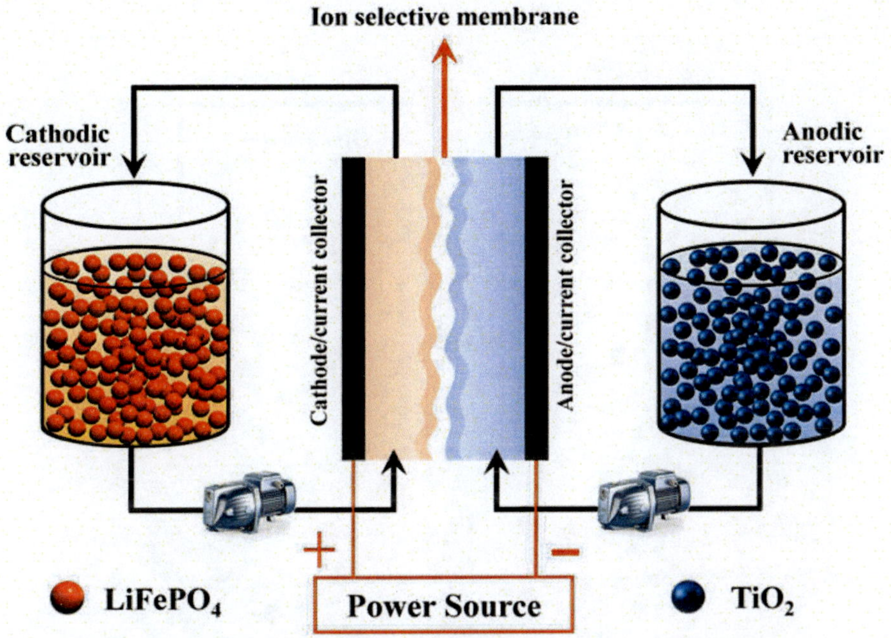

Fig. 1.8 Schematic representation of lithium redox flow battery

1.6.4 Lithium Sulfur Battery

Lithium sulfur battery is another major energy storage device under study. The high theoretical specific capacity of sulfur 1675 mAh g^{-1} makes it more attractive in the area of energy storage devices. It also possess a theoretical energy density of ~2600 Wh kg^{-1} which is much higher compared to other storage systems. In a typical Li-S battery, lithium is used as anode, sulfur as cathode and an electrolyte capable of conducting the sulfur ions. The overall reaction is being depicted in Eq. 1.8. Even though the capacity is much higher, they lack in the experimental performance owing to the poor conductivity of the sulfides used in the battery as well as volume expansion occurring the sulfur during cycling. Hence, several attempts were tried to improve the properties of cathode materials used in the Li-S battery.

$$16\mathrm{Li} + \mathrm{S}_8 \rightleftarrows 8\mathrm{Li}_2\mathrm{S} \tag{1.8}$$

The electrolyte employed in the preliminary stage of Li-S battery was liquid electrolyte which was subsequently replaced by solid state, gel electrolyte and polymer electrolytes [45]. Lithium polysulfide reactions can be catalyzed by using several metals such as platinum, cobalt and metal oxides such as MnO_2, VO_2, CeO_2. The polysulfide shuttle taking place in the lithium sulfur battery is the back bone of the

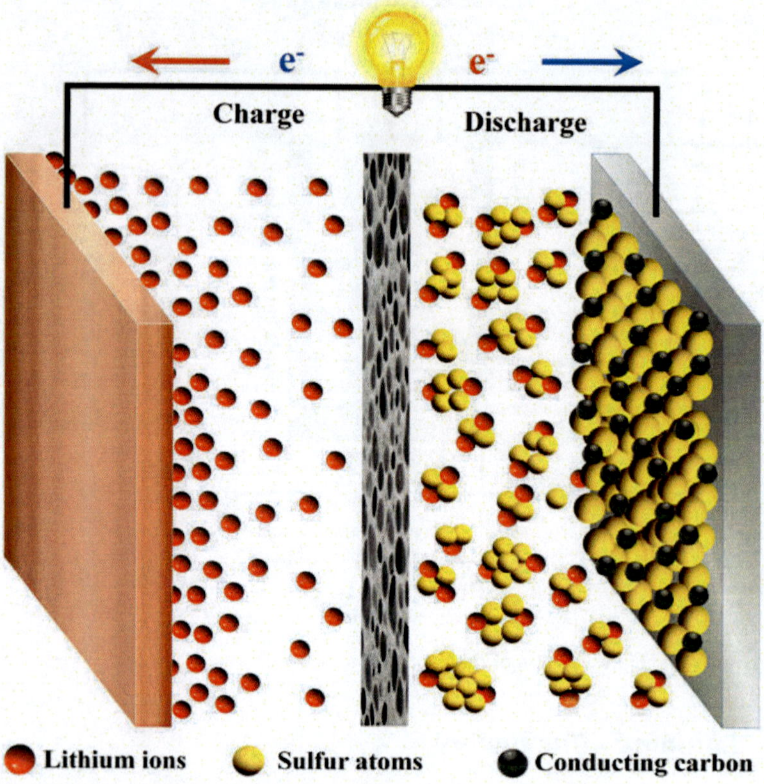

Fig. 1.9 Schematic representation of the structure of lithium sulfur battery

enhanced performance of these systems [46]. A schematic representation of lithium sulfur battery is given in Fig. 1.9.

1.7 Nobel Prize and Lithium Ion Battery 2019

Nobel Prize is the most prestigious award for the eminent scientist working under the field of physics, chemistry, medicine, peace and literature in the globe. The prize is named on Swedish scientist Alfred Nobel. He was a chemist as well as a business man. The scientist had left a testament in which majority of his wealth was kept as safe deposits and the amount obtained from the treasure was to be utilized to establish a prize, which was later on named "Nobel Prize." The testament included the lines "*the interest on which shall be annually distributed in the form of prizes to those who, during the preceding year, shall have conferred the greatest benefit on mankind*" (derived from the official website of Nobel Prize) [47]. The first Nobel Prize was

Fig. 1.10 Images of Nobel laureates 2019 (from right to left)—John B. Goodenough, Stanley M. Whittingham, Akira Yoshino. Adapted and reproduced from Ref. [47], Copyright © Nobel Media 2019. Illustration: Niklas Elmehed

awarded for peace in the year 1901. In 2019, the Nobel Prize was rewarded for the establishment of rechargeable world by introduction of lithium ion batteries. Three eminent personalities who were behind the development and establishment of lithium ion batteries (1) Prof. John. B. Goodenough (2) Prof. Stanley M. Whittingham (3) Prof. Akira Yoshino (Fig. 1.10) [47]. The most deserving eminent personalities who worked hard to make a secondary battery successful. The prize was awarded for they work that led to commercialization of portable high-density batteries which in turn has been practically made possible the usage of mobile phones, laptops, palmtops, tablets and even electric vehicles.

1.7.1 Prof. John Bannister Goodenough

Prof. John Bannister Goodenough is world famous American scientist. He was known as solid state physicist. He was graduated in mathematics. He completed his MS and Ph.D. in physics from the University of Chicago, USA. He joined in the famous Massachusetts Institute of Technology Lincoln laboratory in 1952, where he developed the concept of cooperative orbital ordering in the transition metal compounds. This invention led to the discovery of random-access memory (RAM). Goodenough was also known for explaining electron transfer (super exchange) of electrons between the overlapped orbitals and is known as "Goodenough-Kanamori rules." In 1976, he joined University of Oxford, Inorganic Chemistry Laboratory as Head. The major achievement in the battery industry, the introduction of lithium cobalt oxide as an intercalating compound of lithium was introduced in 1981 [24]. This led to the commercialization of wireless revolution with lithium ion batteries. After ten years of career in Oxford University, he moved to University of Texas, Austin. There he is positioned as Virginia H. Cockrell Centennial Chair of Engineering in the Cockrell School of Engineering [47]. As per Scopus, the publication list of J. B. Goodenough has 771 articles published in peer-reviewed journals and 27 reviews owned to his credit. The maximum publication annually reported was around 40 (43 in 2016). About 663 publications have been published from University of Texas. He has several awards and recognitions to his credit other than Charles Draper Prize in 2014. He has won Japan Prize in 2001, Enrico Fermi award in 2009, National medal of Science in 2011, the Eric and Sheila Samsun Prime Minister's Prize for Innovation in Alternative Fuels for Transportation in 2015. He also won the fellow of Electrochemistry Society award and National Academy of Inventors in 2016. The Welch Award in Chemistry was awarded in 2017 and in the subsequent year awarded with Benjamin Franklin Award and finally awarded with the Hon'ble Nobel Prize in Chemistry for the year 2019 [47].

1.7.2 Prof. Michael Stanley Whittingham

Prof. Michael Stanley Whittingham is an American scientist specialized in chemistry. He is currently serving as Professor in Binghamton University, State University of New York, USA. He was the man behind the development of intercalation of compounds. He also invented first rechargeable battery implementing the intercalation behavior of transitional metal chalcogenides. He was working for Exxon research and development company from where he developed and patented the first rechargeable lithium ion battery in 1977 [48]. In his first battery, lithium metal was used as anode and titanium disulfide as cathode. This invention laid the foundation of rechargeable batteries leading to portable battery era. Later in 1988, he joined as Professor in Binghamton University and still continuing his works in field of energy [48].

1.7.3 Prof. Akira Yoshino

Prof. Akira Yoshino is working as Meijo University in Nagoya, Japan and Honorary Fellow, Asahi Kasei Corporation. He is a Japanese chemist who completed his MS in the 1970 from Department of Petrochemistry, Graduate School of Engineering, Kyoto University. He received doctorate from Graduate School of Engineering, Osaka University in 2005. From 1972 onwards, he is working for Asahi Kasei Corporation in various posts. In 1987, he patented the secondary batteries (first rechargeable batteries) [49]. In 1999, he received Fiscal 1998 Chemical Technology Prize from the Chemical Society of Japan and the award of the electrochemical society. In 2001, he received Meritorious Achievement Prize, from the New Technology Development Foundation (Ichimura Foundation) and Encouragement Prize of Invention of the Minister of Education, Culture, Sports, Science and Technology, from the Japan Institute of Invention and Innovation. In 2014, he obtained the Medal with purple ribbon from Government of Japan. In 2011, he obtained Yamazaki-Teiichi Prize from the Foundation for Promotion of Material Science and Technology of Japan and C&C Prize from the NEC C&C Foundation. In 2012, he was awarded with Fellow of the Chemical Society of Japan and IEEE Medal for Environmental and Safety Technologies from the Institute of Electrical and Electronics Engineers. In 2013, he obtained the Global Energy Prize and the Kato Memorial Prize from the Kato Foundation for Promotion of Science. In 2016, he received the NIMS Award 2016 from the National Institute for Materials Science. In 2018, he was awarded with "The Japan Prize." In 2019, he received the Prestigious Nobel Prize in Chemistry and the European Inventor Award.

1.8 Draper Prize and Lithium Ion Battery (2014)

US National Academy for Engineering situated at Washington, USA, is a non-government organization established in 1964. They established a prize on the name of Charles Stark Draper (1901–1987) in 1988. Charles Stark Draper was a well-known American scientist/engineer, known as "Father of Inertial Navigation." He was also the founder of "Massachusetts Institute of Technology." Later, the name of Massachusetts Institute of Technology was changed over to Charles Stark Draper Laboratory. He was an expert in aerospace engineering and contributed in several sensing and space craft equipment. He also worked as part of several national institutions such as US National Academy for Engineering. The Charles Stark Draper prize was established as an endowment prize for the leading contributors in science, engineering and technology. The prize consists of a certificate (inscribed in hand), a cash prize ($500,000) and a gold medal. It is one of the prestigious awards which are included as one of three Nobel Prizes in engineering. In 2014, four distinct pioneers behind the development, establishment and commercialization of lithium ion batteries were awarded with this prestigious award. Prof. John B Goodenough, Mr. Yoshio Nishi, Prof. Rachid Yazami and Prof. Akira Yoshino were the four eminent personalities behind lithium ion battery (Fig. 1.11) [44]. Brief biography of Prof. John B Goodenough and Prof. Akira Yoshino was been discussed in Sect. 1.7 and the biography of Prof. Rachid Yazami and Mr. Yoshio Nishi is being discussed herewith.

1.8.1 Prof. Rachid Yazami

Prof. Rachid Yazami is the man behind the graphite intercalation of lithium ions. He was a native of Morocco. He completed his post-graduation in chemistry and

Fig. 1.11 Photograph of Draper Prize winners in lithium ion batteries of year 2014. From right side, Prof. Akira Yoshino, Prof. John B. Goodenough, Mr. Yoshio Nishi, and Prof. Rachid Yazami. Adapted and reprinted with permission from National Academy of Sciences for the National Academy of Engineering Ref. [44]

chemical technology in 1978. In 1983, the insertion of alkali metal ion like lithium into graphite was patented as part of his research work [50]. He shared this patent along with Philippe Touzain and Jacques Maire. From 1985 onwards, he worked as research associate at Center National de la Recherche Scientifique (CNRS), and from 1998 onwards, he joined as research director at CNRS. In year 2013, he joined as director of energy storage program, Professor and adjunct Principal Scientist at Energy Research Institute (ERI) at Nanyang Technological University (NTU). As per the Scopus list, he have 162 publications in peer-reviewed journals to his credit and above 140 patents. Recently in 2019, Rachid Yazami was received a patent for electrode fabricated with silicon, tin and a non-alloy of transition metal for battery [51]. He have received several awards and recognitions apart from the Draper Prize of 2014. He was awarded with "Royal Wissam of Intellectual Competency" and Global energy Prize in 2014. In 2016, he received Medal of Chevalier de la Légion d'Honneur; in 2018, he received TAKREEM Awards Laureate for Science and Technology Achievement. Apart from these, he have also received Institute of Electrical and Electronics Engineers (IEEE), Japan Society for the Promotion of Science (JSPS), North Atlantic Treaty Organization (NATO), National Aeronautics and Space Administration (NASA) and Marius Lavet Prize.

1.8.2 Mr. Yoshio Nishi

Mr. Yoshio Nishi was retired senior vice-president and chief technology officer of the Sony Corporation. He was a specialist in chemistry. He was graduated from the Faculty of Applied Chemistry of the Department of Technology at Keio University in 1966. He joined Sony where he was engaged in research and development on fuel cells and electrochemical cells with non-aqueous electrolytes. In 1991, Yoshio Nishi and Keizaburo Tozawa commercialized first lithium ion battery (LIB). He has several awards to his credit. Technical award was obtained from Electrochemical Society in 1994. He was also awarded with the Kato Memorial Award from Kato Foundation for Promotion of Science (Japan) in 1998. He was also awarded with Ichimura Award from the New Technology Development Foundation (Japan) in 2000 in recognition of his contributions to LIB technology.

1.9 Conclusion and Future Outlook

The patent filed by Dr. Akira Yoshino in US patent "secondary batteries" laid the foundation for establishment and commercialization of lithium ion battery as a prime energy storage device. The flexibility of these secondary energy storage devices to tune the size, shape and morphology has led to use these batteries from miniature devices to heavy systems like electric vehicles. The energy stored in these electrochemical devices is being stored as chemical energy and the chemical reaction causes the release of the electrons to the external circuit. The research in the area is still progressing to bring a system with enhanced energy density as well as power density with safe and reliable storage system. The world is trying to replace the fuel consumption from fossil fuel to renewable energy, and hence, highly efficient energy storage system is required to store the converted energy from the sources like solar power. Another major target of the globe is to minimize the pollution by reducing the use of vehicles powered by fossil fuels and establishing electric vehicles. These electric vehicles require more efficient energy storage system. Even though currently lithium ion batteries are being used in these areas, further research is advancing to improve the current system.

Acknowledgements Authors Dr. Jabeen Fatima M. J. and Dr. Prasanth Raghavan would like to acknowledge Kerala State Council for Science, Technology and Environment (KSCSTE), Government of Kerala for financial assistance.

References

1. Dudley B (2019) BP statistical review of world energy statistical review of world. BP Stat Rev World Energy 68:1–61

2. Mills AA (2001) The "Baghdad Battery". Bull Sci Instrum Soc 68:35–37
3. Martins GF (1990) Why the Daniell cell works! J Chem Educ 67:482
4. Ruftschi P (1977) Review on the lead-acid battery science and technology. J Power Sources 2:3–24
5. Leclanchè GL (1867) Improvement in combining generating and secondary or accumulating galvanic battery, pp 22–24
6. Jungner EW (1894) Thermoelectric battery, pp 2–4
7. Pinnangudi B, Kuykendal M, Bhadra S (2017) Smart grid energy storage. Elsevier Ltd, London
8. Eismin TK (2014) Study guide for aircraft electricity and electronics. McGraw Hill Professional, Portsmouth
9. Berndt D (1997) Maintenance-free batteries : lead-acid, nickel/cadmium, nickel/metal hydride. A handbook of battery technology
10. World of chemicals. William Robert Grove—father of fuel cell technology
11. Facility T, Number S (2010) Recombinant salmon calcitonin (RSCT): a 4-week oral (gavage) toxicity study in sprague dawley rats testing facility. Anim Welf 1–10
12. Lundgren CA, Xu K, Jow TR et al (2017) Lithium-ion batteries and materials. Springer handbook of electrochemical energy. Springer, Berlin, pp 449–495
13. Selis SM, Wondowski JP, Justus RF (1964) A high-rate, high-energy thermal battery system. J Electrochem Soc 111:6–13. https://doi.org/10.1149/1.2426065
14. Casey EJ, Lake PE, Dubois AR, Moroz WJ (1965) Effects of foreign ions on nickel hydroxide and cadmium electrodes. J Electrochem Soc 112:371–383. https://doi.org/10.1149/1.2423552
15. Thaller LH (1970) Catholyte studies in copper oxide-magnesium thermal cells. J Electrochem Soc 117:1590–1594
16. Salomon M (1974) Secondary lithium ion batteries. In: Power sources symposium, 26, pp 71–73
17. Whittingham MS (1978) Chemistry of intercalation compounds: metal guests in chalcogenide hosts
18. Kasai H, Tolborg K, Sist M et al (2018) X-ray electron density investigation of chemical bonding in van der Waals materials. Nat Mater 17:249–252. https://doi.org/10.1038/s41563-017-0012-2
19. Berlinsky AJ, Unruh WG, McKinnon WR, Haering RR (1979) Theory of lithium ordering in Li_xTiS_2. Solid State Commun 31:135–138. https://doi.org/10.1016/0038-1098(79)90421-6
20. Alpen UV (1979) Li_3N: a promising Li ionic conductor. J Solid State Chem 29:379–392. https://doi.org/10.1016/0022-4596(79)90195-6
21. Scrosati B (2011) History of lithium batteries. J Solid State Electrochem 15:1623–1630. https://doi.org/10.1007/s10008-011-1386-8
22. Xu W, Wang J, Ding F et al (2014) Lithium metal anodes for rechargeable batteries. Energy Environ Sci 7:513–537. https://doi.org/10.1039/c3ee40795k
23. Goodenough JB (2018) How we made the Li-ion rechargeable battery. Nat Electron 1:204. https://doi.org/10.1038/s41928-018-0048-6
24. Mizushima K, Jones PC, Wiseman PJ, Goodenough JB (1981) Li_xCoO_2 ($0 < x \sim <1$): a new cathode material for batteries of high energy density. Solid State Ionics 4:171–174
25. Gauthier M, Vassort G, Bélanger A (1982) Electrochemical studies on lithium intercalation materials in molten salts. J Electrochem Soc 129:970–978. https://doi.org/10.1149/1.2124075
26. Besenhard JO, Heydecke J, Fritz HP (1982) Characteristics of molybdenum oxide and chromium oxide cathodes in primary and secondary organic electrolyte lithium batteries. I. Morphology, structure and their changes during discharge and cycling. Solid State Ionics 6:215–224. https://doi.org/10.1016/0167-2738(82)90042-X
27. Thackeray MM (1989) Development of rechargeable MnO_2 electrodes for lithium batteries. Mater Res Soc Symp Proc 135:585–596. https://doi.org/10.1557/PROC-135-585
28. Basu S, Zeller C, Flanders PJ et al (1979) Synthesis and properties of lithium-graphite intercalation. Mater Sci Eng 38:275–283
29. Zanini M, Basu S, Fischer JE (1978) Alternate synthesis and reflectivity spectrum of stage 1 lithium-graphite intercalation compound. Carbon N Y 16:211–212. https://doi.org/10.1016/0008-6223(78)90026-X

30. Besenhard JO, Mohwald H, Nickl JJ, Theodoridou E (1982) Electrochemical applications of graphite intercalation compounds. Synth Met 4:211–223
31. Shirakawa H, Louis EJ, MacDiarmid AG, Chiang CK, Heeger AJ (1977) Synthesis of electrically conducting organic polymers: halogen derivatives of polyacetylene, (CH) x. J Chem Soc Chem Commun 16:578–580. https://doi.org/10.1039/C39770000578
32. Yoshino A (2012) The birth of the lithium-ion battery. Angew Chem Int Ed 51:5798–5800. https://doi.org/10.1002/anie.201105006
33. N media (2000) Hideki Shirakawa biographical
34. Nishi Y (2016) The dawn of lithium-ion batteries. Electrochem Soc Interface Fall 25:71–74
35. Moser JR (1972) Solid state lithium-iodine
36. Greatbatch W, Holmes CF (1992) The lithium/iodine battery: a historical perspective. PACE Pacing Clin Electrophysiol 15:2034–2036. https://doi.org/10.1111/j.1540-8159.1992.tb03016.x
37. Mallela VS, Ilankumaran V, Rao NS (2004) Technical series. Trends in cardiac pacemaker batteries. Indian Pacing Electrophysiol J 4:201–212
38. Zhao Y, Wang L, Byon HR (2013) High-performance rechargeable lithium-iodine batteries using triiodide/iodide redox couples in an aqueous cathode. Nat Commun 4:1–7. https://doi.org/10.1038/ncomms2907
39. Khanna VK (2016) Batteries for Implants. In: Implantable Medical Electronics. Springer, Cham, pp 167–183
40. Semkow KW, Sammells AF (1987) A lithium oxygen secondary battery. J Electrochem Soc 134:2084–2085. https://doi.org/10.1149/1.2100826
41. Wang C, Xie Z, Zhou Z (2019) Lithium-air batteries: challenges coexist with opportunities. APL Mater 7:040701. https://doi.org/10.1063/1.5091444
42. Jung KN, Kim J, Yamauchi Y et al (2016) Rechargeable lithium-air batteries: a perspective on the development of oxygen electrodes. J Mater Chem A 4:14050–14068. https://doi.org/10.1039/c6ta04510c
43. Abraham KM, Jiang Z (1996) A polymer electrolyte-based rechargeable lithium/oxygen battery. J Electrochem Soc 143:1–5
44. Charles Stark Draper Prize for Engineering (2014)
45. Lin Z, Liang C (2015) Lithium-sulfur batteries: from liquid to solid cells. J Mater Chem A 3:936–958. https://doi.org/10.1039/c4ta04727c
46. He J, Manthiram A (2019) A review on the status and challenges of electrocatalysts in lithium-sulfur batteries. Energy Storage Mater 20:55–70. https://doi.org/10.1016/j.ensm.2019.04.038
47. Prize N History. In: Nobel media
48. Gregersen E (2019) M. Stanley Whittingham. Encycl. Br
49. Akira Yoshino, Kenichi Sanechika TN (1987) Secondary battery
50. Touzain P, Yazami R, Maire J (1986) Insertion compounds of graphite with improved performances and electrochemical applications of those compounds, pp 2–4
51. Rachid Yazami WZ (2019) Electrode, battery cell and battery cell arrangement, p 1

Chapter 2
Electrospinning: The State of Art Technique for the Production of Nanofibers and Nanofibrous Membranes for Advanced Engineering Applications

Akhila Das, Neethu T. M. Balakrishnan, Jarin D. Joyner, Nikhil Medhavi, O. Manaf, M. J. Jabeen Fatima, Jou-Hyeon Ahn, Wazed Ali, and Raghavan Prasanth

2.1 Introduction

Nanofibers are nanometer-scale structures with theoretically unlimited length having huge surface area and a very large surface area to volume ratio [1]. This exceptionally nanosized electrospun fibers find application in various fields, such as, nanocatalysis, tissue engineering scaffolds, protective clothing, filtration, biomedical, pharmaceutical, optical, electronics, healthcare, biotechnology, defense, energy storage and environmental engineering [2–7]. Both natural [8] and synthetic polymers [1, 9]

A. Das · N. T. M. Balakrishnan · N. Medhavi · M. J. Jabeen Fatima · R. Prasanth (✉)
Department of Polymer Science and Rubber Technology, Cochin University of Science and Technology (CUSAT), Cochin 682022, India
e-mail: prasanth@cusat.ac.in

J. D. Joyner
Department of Materials Science and Nano Engineering, Rice University, 6100 Main Street MS 364, Houston, TX 77005, USA

O. Manaf
Materials Research Laboratory, Department of Chemistry, National Institute of Technology (NIT-C), Calicut 673601, India

J.-H. Ahn · R. Prasanth
Department of Materials Engineering and Convergence Technology, Gyeongsang National University, 501 Jinju-Daero, Jinju 52828, Republic of Korea

W. Ali (✉)
Department of Textile and Fiber Engineering, Indian Institute of Technology (IIT-D), HauzKhas, New Delhi 110016, India
e-mail: wazed@textile.iitd.ac.in

N. T. M. Balakrishnan and R. Prasanth (eds.), *Electrospinning for Advanced Energy Storage Applications*, Materials Horizons: From Nature to Nanomaterials,
https://doi.org/10.1007/978-981-15-8844-0_2

metal oxides [10, 11], ceramic materials [12, 13] etc. have been successfully electrospun into nanosized fibers for various applications. In electrospinning, a continuous solid fiber is generated as the electrified jet of viscous polymer/polymer composite solution is continuously stretched due to the electrostatic repulsion between the surface charges and evaporation of solvent. Only a few methods are currently known for nanofiber preparation; specifically, template synthesis [13], self-assembly [14], phase separation [15] drawing [16] and electrospinning [1]. Among the different process for nanofiber production electrospinning is a versatile and most frequently used method [17]. A wide array of researchers have investigated electrospinning as the simplest and mass production method for the preparation of micro- and nanofibers and several excellent reviews of the state-of-the-art in electrospinning have been published [18–20]. The electrospinning spinning unit consists of four components (i) spinneret (generally an injection syringe fitted with a blunt injection metal needle), (ii) a syringe pump for forcing the polymer solution from the syringe to the metal needle), (iii) a high voltage source for electrifying the polymer droplets comes out from the needle and (iv) a collection grid for collecting the micron to nano sized fibers. The simple schematic of the electrospinning set up is shown in Fig. 2.1 and their modified versions for the production of fibers having engineered morphology is displayed in Fig. 2.2. To electrospin the fibers from polymer/polymer composite, a solvent at least as volatile as water is generally chosen for preparing the solution. The uniform bubble free, viscous polymer/polymer composite solution is first fed into the syringe fitted with bund steel needle connected to the positive terminal of the high voltage source. A high voltage (typically 10–30 kV) is applied to the droplet of polymer solution forced through the injection needle using the syringe pump or air pressure. When the charge at the droplet surface prevail over the surface tension of the solution, a fine jet elongates from the droplet and is collected on a grounded target plate. The diameter and morphology of the resulting fiber has been greatly depends on all variables in the electrospinning process, which is mainly categorized in three groups: system, solution and ambient parameters. The objectives of this chapter are to briefly discuss the history and process of electrospinning in the view of important governing parameters which influence the fiber diameter and morphology, to demonstrate the potential of this technique to produce fibers with engineered secondary structures, morphologies and functionalities. The chapter also address the issues related to the organization of fibers into hierarchical assemblies.

2.2 History of Electrospinning

The history of any particular technology is always difficult to trace and tease out from ever growing science and civilization. When trying to portrait the picturesque history of electrospinning, it is found that the time line is spreaded mainly in four centuries starts from seventeen to twenty century, hence a huge number of literatures, documents and patents on electrospinning are available to date. To reduce this abstract of history on electrospinning to manageable dimensions, the authors have

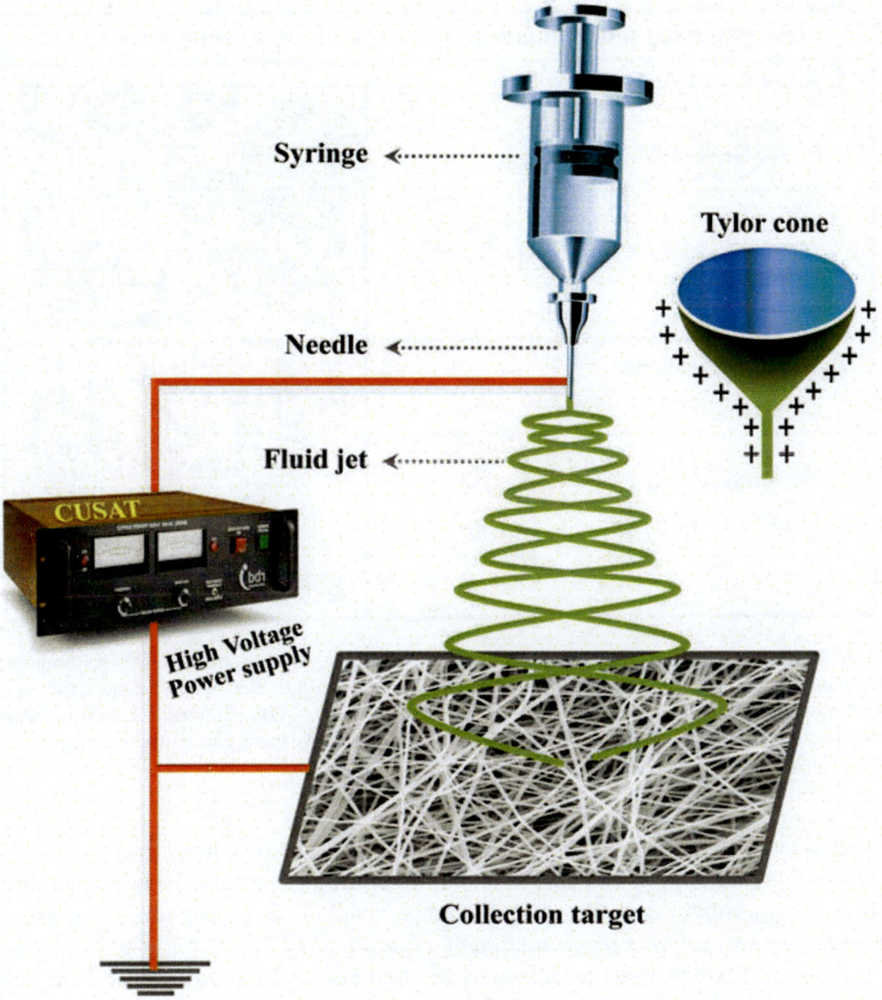

Fig. 2.1 Schematic representation of typical laboratory type electrospinning setup

limited the story to the inventions and discoveries that directly relate to the genesis and development of electrospinning and the milestones in the history of electrospinning is tabulated in Table 2.1. Even though the history of electrospinning begins in seventeenth century, it is commonly supposed that the real story of electrospinning starts with the considerable contributions of Anton Formhals in the 1930s. However, the direct history of the industrial process begins thirty years before then with Cooley [23] when he filed the first patent on electrospinning in 1900. In that filed patent he proposed four types of indirectly charged spinning heads; a conventional head, a coaxial head, an air assisted model, and a spinneret featuring a rotating distributor. Later in 1912, W. B. Wiegand and B. F. Burton published a paper, which describes

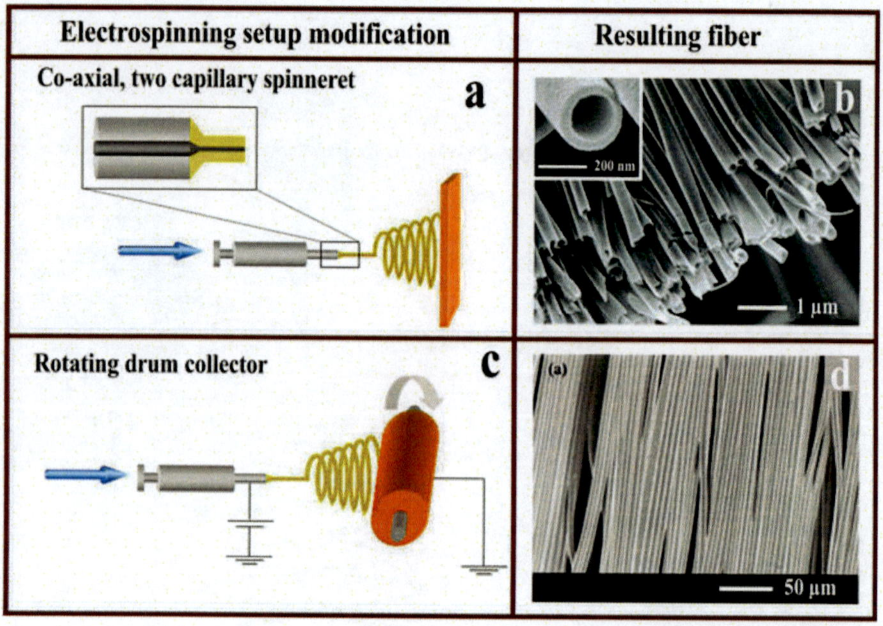

Fig. 2.2 Modifications of the typical electrospinning setup used to produce meshes with unique morphologies. **a** A co-axial, two capillary spinneret can be used to electrospin hollow nanofibers shown in (**b**). Adapted an reproduced from Ref. [21], Copyright 2008 Elsevier. **c** A rotating drum collector can be used to produce aligned fibers shown in (**d**). Adapted and reproduced from Ref. [22], Copyright 2005 American Chemical Society

the effect of electricity on streams of water drops [24] and explaining the relationship between surface tension and charge on the water drop. They surprisingly noted that when using a triboelectrically charged ebonite paddle, "*during the process of atomization, one may observe the formation of a fine jet of liquid*". Later in 1914, Zeleny [25] a physicist working at the University of Minnesota had started to concern himself with discharges from liquid surfaces. In his experiments he used a hemispherical drop of liquid on the end of a capillary tube and observed the tendency of the hemisphere to distort at high voltage. Also with similar experimental set-up he observed the flight of liquid drops from the liquid meniscus [26] and photographed a 4 μm fine stream of liquid jet are flows out from the metal capillary [27]. This interesting observation leads mathematically modeling the behavior of fluids under electrostatic forces. Zeleny [28] adapted the theories of Rayleigh [29] on the stability of an isolated charged liquid droplet to the case of an electrified droplet losing its stability when jetting begins at its vertex. In the theoretical studies, Rayleigh predicted that the charged liquid droplet becomes unstable and fission takes place when the charge becomes sufficiently large compared to the stabilizing effect of the surface tension.

Hagiwara [30], who worked on silk and artificial fibres, used electricity to orientate the molecular structure of colloidal liquid Viscose precursor prior to spinning

Table 2.1 Historical milestones engraved in golden letters in the field of electrospinning

Year	Development	Reference
1900	First patent filed on electrospinning	[23]
1902	First patent on Electrospinning	[51]
1912	Paper, which describes the effect of electricity on streams of water drops and explaining the relationship between surface tension and charge on the water drop	[24]
1914	Observed the flight of liquid drops from the liquid meniscus	[26]
1917	Photographed a 4 μm fine stream of liquid jet are flows out from the metal capillary	[27]
1917	Mathematically modeling the behavior of fluids under electrostatic forces	[27]
1927	Spinning process by using an 80 Hz oscillating current	[31]
1934	First patent filed on yarn fabrication using electrospinning by Formhals	[33]
1936	The melt electrospinning using a combined electrostatic (100 kV) and air-jet assist method	[40]
1938	Constructed a system to produce short fibers of controlled fiber lengths	[35]
1939	Proposed co-spinning of fibers with opposite charges to produce a product with no net charge	[36]
1940	Developed a method for producing composite fiber webs from multiple polymer and fiber substrate by electrostatically spinning polymer fibers on a moving base substrate	[38]
1941	Commercial application of electrospun filter materials known as "Petryanov filters" in USSR	
1955	Correlation between the electrostatic force required to the dielectric constant of the fluid and the radius of curvature of the fluid droplet	[43]
1964	Mathematically modeling the shape of the cone formed by the fluid droplet under the influence of an electric field and fundamental studies on the jet forming process	[52]
1971	Devised a method of photographing electrospun fibers in flight	[45]
1974	Proposed use of electrospun fibers as wound dressing	[53]
1978	First recorded study of electrospun fibers as implantable (vascular) graft	[54]
1981	Fabricated a melt electrospinner for the melt spinning of polymers	[28]
2001	Using electrospinning to fabricate inorganic fibers	[55]
2003	Core-shell electrospinning process	[56]
2005	First scientific journal publication of electrospun continuous yarn	[57]
2006	3D block electrospun fiber	[58]
2008	Receipt of CE Mark for Electrospun implantable graft	[59]
2012	Marketed consumer product constructed from electrospun nanofiber-based composite	[60]
2015	High-tech nanofibers could help nutrients in food hit the spot	[61]
2018	Drinking water quality was improved by nano fiber electrospun technique removing nitrates	[62]

(continued)

Table 2.1 (continued)

Year	Development	Reference
2019	New startup on nano clean by IIT Delhi alumni's on producing nanofibers in India	[63]

[31] and the resulting colloidal liquid is called as *fibrous sol* in 1927. This Hagiwara fibrous sol is very similar to the *stäbchen sol* reported in 1923 by the Andor Szegvári and Emmy Schalek in the Kaiser-Wilhelm Institut für Physikalische Chemie und Elektrochemie at Berlin-Dahlem (Kaiser Wilhelm Institute for Physical Chemistry and Electrochemistry at Berlin-Dahlem). In the series of experiments with viscose and other forms of cellulose including nitrocellulose, and cellulose acetate, gelatine, albumen and a natural silk solution, Hagiwara observed that this electrically treated Viscose produces a conventionally spun fiber "*free of irregular aggregation of the particles*" and with improved lustre due to the increased degree of molecular orientation in the Viscose. Based on this observation Hagiwara extended the spinning process by using an 80 Hz oscillating current [32] in conjunction with air pressure to increase spinning speed and fiber diameter. In this experiment, Hagiwara observed that ozone evolved as the fiber loses charge in-flight decomposes the hydrogen sulphide produced in the coagulation bath. This ozone formation provides the first evidence of loss of charge in flight and obtained American patent in 1929.

Even though until 1933, this spinning technique had been known as electrostatic spinning, and there were only a few publications are reported its use in the production of thin fibers [3, 28]. Between 1931 and 1944, Anton Formhals made significant contributions to the development of electrospinning, filed a sequence of 22 patents on aspects of the electrospinning process taken out in America, France, the United Kingdom and Germany. Formhals patented his first invention in 1934, relating to the process and the apparatus for producing artificial filaments using electric charges. His first patent [33] describe on a machine design based on a saw-toothed rotating fiber emitter. The emitter resembles a circular saw blade and dips into a trough of the spinnable liquid. Charge concentration at the wetted tooth tips causes fibers to be evolved and to fly off towards a movable thread collecting device in the form of a spinning drum or rotating disc targets similar to the one in the conventional spinning, on which the threads were collected in a stretched state. Also in the patent Formhals suggests that the as-spun fiber can be passed through a coagulating bath if required. The main disadvantage of this spinning method was obtaining the completely dried fibers due to the short distance between the spinning zone and the collection targets. However, the setup was capable of producing parallel threads onto the collection device in such a fashion that they can be unwound continuously.

By 1937, Formhals had turned his attention to nozzle design observing that nozzles made by drilling a plain parallel sided hole were prone to blocking, hence he made a new design. His nozzle design has a conical taper and can be disassembled for ease of cleaning [34]. Later in 1938 [35], in the next design, he constructed a system to produce short fibers of controlled fiber lengths, which is achieved by interrupting the current flow to the multiple spinning heads of the machine and fiber was collected on a slatted belt. In this, the individual spinnerets were fed from a manifold kept at a constant pressure by a header tank. In the same year, he further refined this design [36] for more effective removal of the as-spun fibers from the salted belt by the addition of a rotating notched disc counter electrode meshing with the slatted belt. A year later, in 1939, Formhals revised the disadvantages of the earlier setups by altering the distance between the spinneret and the collection device to get more drying time for the electrospun fibers [3] and, was seeking to control as-spun fibers in flight by manipulating the electrical field [36] and patented the methods. The later design [37] projects the fiber strand between two parallel wire electrodes that are connected to an alternating supply of up to 100 kV. The variation in electrostatic field resulting from this parallel wire set up caused the fiber to be deposited in hanks. Then inn 1940, Formhals developed a method for producing composite fiber webs from multiple polymer and fiber substrate by electrostatically spinning polymer fibers on a moving base. Formhals also proposed co-spinning of fibers with opposite charges to produce a product with no net charge [36] and made serious efforts to devise winding devices to gather up the fiber in a usable form [38, 39]. In 1936, Charles Ladd Norton patented the melt electrospinning using a combined electrostatic (100 kV) and air-jet assist method [40] to produce lofted fibers which is more suitable for electrical insulation and packaging. He used deflector plates running at about 2 kV AC, to influence the flight path of the fiber and kept the target up to 6 m away from the spinning head or the point of emission. The use for air-jet in conjunction with high voltage made possible to keep the target far away from the emission point (even up to 6 m) and assist to remove charge from the product, improving the rate of deposition.

In the late 1930s Nikolai Albertowich Fuchs [41] from Aerosol Laboratory, USSR advanced a theory of ultrafine fibrous materials and in 1938 his coworkers Igor' Vasil'evich Petryanov-Sokolov and Natalya D Rosenblum developed a filter material called "*Petryanov filters*" from electrospun fibres, which led to the establishment of a factory in Tver' for the manufacture of electrospun smoke filter elements for gas masks. For this work Igor' Vasil'evich Petryanov-Sokolov and Natalya D. Rosenblum were awarded the Stalin Prize in 1938 [41]. In 1952 Bernard Vonnegut and Raymond L. Neubauer [42] and in 1955 VadimDrozin at Columbia University, USA [43] examined the formation of fluid jets and droplets by electrostatic force. The Raymond's study [43] observed that the formation of uniform sized fluid droplets and pair of like charged droplet clouds would repel each other—the origin of the "tramline" effect often observed in multiple spinning head devices. They also made an estimate of the size of the liquid droplets (about 1 μm) from the observation of the rainbow colors of high order Tyndall spectra in the droplet cloud. In VadimDrozin's study [43], He made a correlation between the electrostatic force required to the dielectric constant of the fluid and the radius of curvature of the fluid droplet. He also measured limits of operation in terms of specific conductivity, dipole moment and refractive index [43].

Between 1964 and 1969, Sir Geoffrey Ingram Taylor significantly contributed to the theoretical underpinning and dynamic mechanism of electrospinning by mathematically modeling the shape of the cone formed by the fluid droplet under the

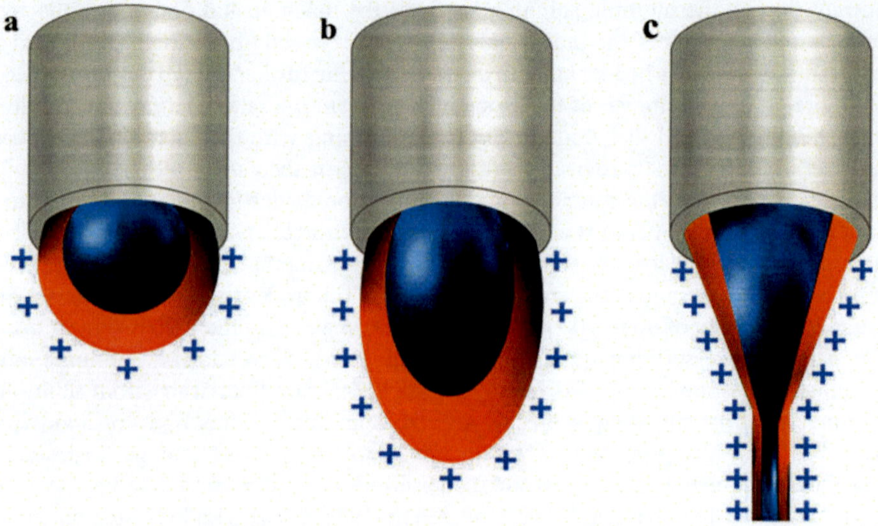

Fig. 2.3 Schematic illustration of compound Taylor cone formation. **a** Surface charges on the sheath solution, **b** viscous drag exerted on the core by the deformed sheath droplet, **c** sheath-core compound Taylor cone formed due to continuous viscous drag

influence of an electric field [16] Tylor initiated with fundamental studies on the jet forming process and Zeleny's [27] theoretical results were corrected by him in seminal work in which the spheroidal approximation of droplet shapes was proposed for the case of (i) a charged droplet sustained at a given potential F_0 relative to the ground, or (ii) an uncharged droplet in a given uniform field E. In this, Tylor studied the shape of the polymer droplet produced at the tip of the needle when an electric field is applied and showed that it is conical in shape. This characteristic droplet shape forming at the tip of the spinneret is now known as the Taylor cone and the jets were ejected from the vertices of the cone. The schematic illustration of compound Taylor cone formation is presented in Fig. 2.3. By the detailed examination of different viscous fluids, Tylor determined that an angle of 49.3° is required to balance the surface tension of the polymer fluids with the electrostatic force. Further Tylor worked with Melcher [44] to develop the "*leaky dielectric model*" also called Taylor-Melcher model for conducting fluids, which is the standard model for the electrodynamics of poorly conducting leaky dielectric fluids under an electric field. Taylor introduced the leaky dielectric model to explain the behavior of fluid droplets deformed by a steady state electric field. In 1971, Peter Karl Baumgarten [45], worked for E. I. du Pont Nemours & Co. USA, devised a method of photographing electrospun fibers in flight, and in the course of investigating the effect of solution viscosity, surrounding gas, voltage and jet radius on fiber diameter and jet length described the electrospinning process.

Then in 1981, Larrondo and St. John Manley [28] fabricated a melt electrospinner for the melt spinning of polymers similar in principle of operation to a melt flow index devices. In this electrospinner, a static load was applied to a piston running in a heated barrel to produce a drop of polymer melt at the tip of spinneret and the fine fibers were drawn from the fluid drop by electrostatic force. In subsequent years researchers studied the relationship between the structural morphology and the process parameters on the electrospinning of polymer solutions and melts [28, 45, 46]. In the early 1990s, several research groups notably that of Reneker [47, 48] revived interest in this technique by demonstrating the fabrication of ultra thin fibers from a broad range of polymers. By then, the term electrospinning has been coined starts to widely used in the literature [1, 47, 49, 50]. The timely demonstrations by Doshi and Reneker [47] gained dynamic momentum in the research on nanofiber, which triggered a lot of experimental and theoretical studies related to electrospinning. It is notable that, since then, the number of publications about electrospinning has been increasing exponentially every year (Fig. 2.4), on account of the remarkable simplicity, versatility and potential application of this technique.

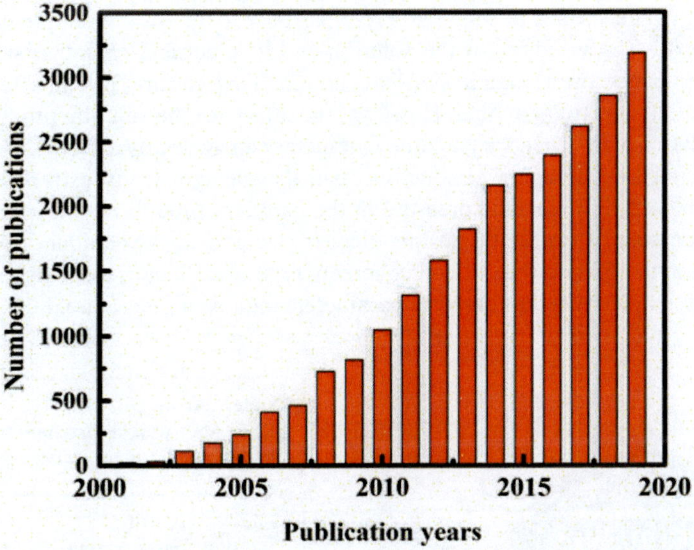

Fig. 2.4 The annual number of publications on the subject of electrospinning as provided by search engine Scopus (key word: electrospinning). For 2019, there are already publications reported by December

2.3 Parameter Affecting Electrospinning

From the previous theory and process of electrospinning, it is clear that electrospinning process can be manipulated by a number of variables. Doshi and Reneker [47] listed out the parameters, that can influence the transformation of polymer solutions into nanofibers through electrospinning. The important spinning parameters on nanofiber production by electrospinning is shown in the Fig. 2.5. The parameters affecting electrospinning of polymer solutions are broadly classified as (i) system parameters, (ii) solution properties, (iii) governing variables (process parameters), (iv) ambient parameters. The system parameters include the molecular weight and molecular weight distribution of the polymers and the architecture of the polymer, such as linear, branched or block copolymers. Polymer solution parameters include the solution concentration, solution viscosity, surface tension, temperature, electrical and thermal conductivity and dielectric constant. The solution parameters have the most significant influence in the electrospinning process and electrospun fiber morphology. The surface tension has a role partially in the formation of beads along the fiber axis. The viscosity of the solution and its electrical characteristics determine the extent of the elongation of the solution. This, in turn, has an effect on the diameter of the electrospun fiber. However, the effects of the solution properties can be difficult to isolate since varying one parameter can generally affect other solution properties (e.g., changing the conductivity can also change the viscosity). Governing variables includes hydrostatic pressure in the capillary tube, flow rate, bore size of the needle/capillary, needle tip design, electric potential at the capillary tip, and the gap between the tip and the collecting screen, type of collector, and its speed in the case of a dynamic type. The ambient parameters include temperature, humidity, and

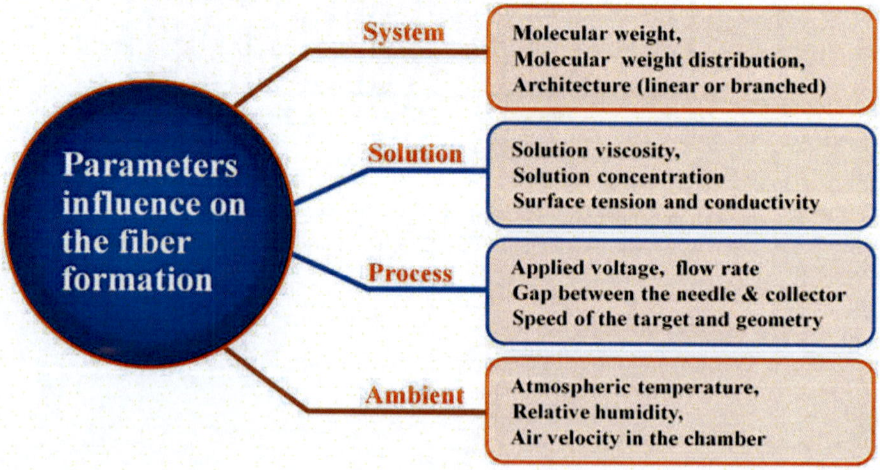

Fig. 2.5 The important spinning parameters affect the production and morphology of nanofiber by electrospinning

air velocity in the electrospinning chamber. The ambient parameters have certain degree of influence on fiber morphology although they are less significant than the effect of the solution parameters. In this section, studies that investigate the effect of each parameter on electrospun fiber morphologies and sizes are highlighted.

A lower concentration or deviation of the voltage from the optimum as well as a smaller distance between needle and target or faster flow rate resulted in beaded fibers with different diameters. The size range of PS fibers produced in this work ranges from 0.8 to 20 ($\pm$ 10) μm. An increasing solution concentration or increasing flow rate as well as a decrease in spinning voltage resulted in increased fiber size. The influence of these parameters on the pore formation is not as obvious as it is for bead formation and fiber size. The pores range in size from 20 to 350 nm. They vary in size and shape with changing PS concentration. There are even pores of differing size and shape in one PS fiber electrospun from THF (18 wt%).

2.3.1 Solution Parameters

The major factor that govern the electrospinning technique is the solution parameters. Solution parameters includes viscosity, electrical conductivity, surface tension, solution properties, vapour pressure of solution etc. This rheological properties mainly viscosity influence the electrospinning. Higher the viscosity leads to the hard ejection of the polymer solution and very less viscous polymer cannot produce fibers. Optimization of such parameters results in the generation of fine, smooth fibrous membranes. Higher electric field results in smooth fibers, higher surface tension provides smaller diameters etc. are some of the parameters that affect the electrospinning technique. A detailed description of the influence of parameters in electrospinning is given below.

2.3.1.1 Conductivity

The impact of the solution conductivity and charge density on the fiber diameters and surface morphology of electrospun fibers was also studied. The studies demonstrated that solution conductivity has important role in the fiber formation and its morphology in electrospinning. Generally the polymers are insulators, except few dielectric materials or conducting polymers. Many studies demonstrated that the charged ions in the polymer solution are highly influential in jet formation, thereby the fiber diameter and surface morphology of the fibers. The higher conductivity of the polymer solution resulted in a higher charge density on the surface of the solution jet during the electrospinning, bringing more electric charges to the jet. As the charges carried by the jet increase, higher elongation forces were imposed to the jet under the electrical field, resulting bead free fibers with thinner fiber diameters. Zong et al. [64] found that addition of ionic salts improves the conductivity of the spinning solution and the resulting nanofibers were bead-free. Addition of salts, increase charged ions in the polymer solution and charge carrying capacity of the jet, thereby subjecting the jet to higher tension with the applied electric field resulted beadless fibers with relatively smaller diameters ranging from 200 to 1000 nm. It was hypothesized that volatile salt additive would not remain in the fibers and therefore would not affect the properties of the fibers. Shin et al. [65] found a higher jet stability for PEO solution than glycerol, attributed to the higher conductivity of the PEO solution. Lee et al. [66] observed that throughput increases with increasing electric conductivity of PS solutions. Addition of alcohol, [67] pyridiumformiate [68] non-ionic surfactant such as dodecyltrimethylammonium bromide and tetrabutylammonium chloride [69] has also been used to increase conductivity of polymer solutions and demonstrated a significant reduction in bead formation with good smooth surface finish. These studied indicates, highly conducting polymer solutions may be easily collected in aligned form in electrospinning. This, however, does not imply that a higher applied electrical field could result in fewer beads and smoother nanofibers. In fact, Deitzel et al. [70] investigated the influence of electrical charge, which was applied for electrospinning, on the morphology of PEO nanofibers. They reported that with the increase of the electrical potential the resulting surface morphology of the nanofibers became more rough.

Zhang et al. studied the effect of conductivity on fiber diameter by adding different salts such as NaCl, KH_2PO_4, NaH_2PO_4 into PVA [71] and PDLA [64] polymer solution and showed that PVA fiber diameters were decreased from 214 ± 19 nm to 159 ± 21 nm when NaCl concentration was increased from 0.05 to 0.2%. The PVA solution prepared using water as solvent and electrospun at an applied voltage of 5 kV, flow rate of 0.2 mL h^{-1} and a distance of 10 cm between spinneret and collection target. The studies with PDLA found that salts with smaller ionic radii produced smaller fibers (~210 nm) while salts with larger ionic radii yielded larger ones (~1000 nm). They attributed this difference to the higher charge density, and thereby mobility, of ions with smaller radii; the higher mobility resulted in increased elongational forces exerted on the fiber jet yielding a smaller fiber. Likewise, the addition of proteins [72] and ionic surfactants [69] was found to yield fibers with smaller diameters with uniform morphology. Since the addition of proteins did not affect the solution viscosity, the variation in fiber diameter resulted from changing the solution conductivity [72]. It was hypothesized that charged surfactants increased the solution conductivity and net charge density, causing an increase in the whipping instability yielding more uniform fibers [69]. The increase in conductivity, however, leads to the formation of fibers with larger diameter for PAA [73] and polyamide [74] due to the large increase in the mass flow during spinning.

2.3.1.2 Viscosity/Solvent Effects (Solution Concentration)

Solution viscosity plays an important role in determining the fiber size and morphology during spinning of polymeric fibers. It has been found that with very low viscosity there is no continuous fiber formation and with very high viscosity there is difficulty in the ejection of jets from polymer solution, thus there is a requirement of optimal viscosity for electrospinning. Generally, the solution viscosity can be tuned by adjusting the polymer concentration of the solution. The relationship between the polymer viscosity and/or concentration on fibers obtained from electrospinning has been studied in a number of systems [64, 70, 75–83] including poly(DL-lactic acid) (PDLA) [64], poly(lactic-*co*-glycolic acid) (PLGA) [76], poly(ethylene oxide) (PEO) [70, 78], poly(vinyl alcohol) (PVA) [79–82], poly(methyl methacrylate) (PMMA) [80], polystyrene [83], poly(L-lactic acid) (PLLA) [68], gelatin [84], Dextran [72] and collagen type I-PEO [72]. There should be an optimum solution viscosity/concentration for the electrospinning process, as at low viscosity/concentrations beads are formed instead of fibers and at high concentrations the formation of continuous fibers are prohibited because of the inability to maintain the flow of the solution at the tip of the needle [85].

For solution of low polymer concentration/viscosity, surface tension is the dominant factor and just beads or beaded fiber formed [74, 86–91]. The process generally not considered as electrospinning and is referred as electrospraying [87]. The solution having suitable viscosity, continuous fibers with bead free morphology can be obtained. The viscosity of the solution can simply increase by polymer concentration. At critical concentration/viscosity, (the minimum concentration of the polymer at in the solution or the minimum viscosity of the polymer solution, at which the fibers can be form) electrospinning yield uniform fibers with few beads and junctions [70, 83]. In most of the cases, the surface tension of the polymer solution may get change with polymer concentration [70]. Jarusuwannapoom and Hongrojjanawiwat [83] demonstrated the effect of viscosity on the morphology of the electrospun membranes. In the study, polystyrene solutions with constant surface tension and varying viscosity were used. The morphological evaluation showed bead formation is decreased with increased viscosity [83]. At very high viscosity/concentration the droplet formed at the top of the spinneret dried out due to the fast evaporation of the solvent which prevent the formation of jet or spinning [64, 87, 92]. There are many attempts have been made to quantify the minimum polymer concentrations and viscosities required to form the fibers by electrospinning. It was found that for PVA, the solution with $[\eta]C > 5$ [79] and PEO $[\eta]C > 10$ [77] forms ultrafine fibers by electrospinning, where $[\eta]$ is the intrinsic viscosity and C is the concentration. The cross-section of PVA fibers was circular-shaped at $[\eta]C > 9$ and flat at $[\eta]C > 10$ [77]. Recently the minimum solution concentration was quantified in terms of entanglement concentration, Ce [93] and critical chain overlap concentration C^* [75]. A concentration greater than entanglement concentration or critical chain overlap concentration yielded uniform fibers with bead-free morphology. In both cases solutions with concentrations double than entanglement concentration or critical chain overlap concentration is required to form uniform fibers. When the solution concentration was less than critical chain overlap concentration there was insufficient chain overlap to form polymer fibers and resulted droplet formation, while in the semi-dilute region (concentrations between C^* and Ce), fibers with beads were obtained.

Increase in solution concentration/viscosity results in increased fiber diameter (Fig. 2.6) [1, 20, 47, 93–100]. Researchers have aimed to find a relationship between solution concentration and the fiber diameter. Demir et al. [87] derived a power

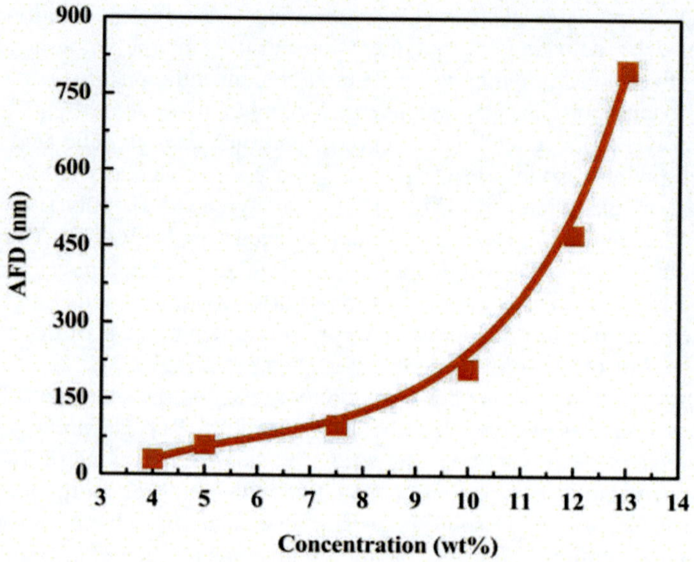

Fig. 2.6 Average fiber diameter (AFD) of electrospun fibers as a function of solution concentration. Adapted and reproduced from Ref. [20], Copyright 2012 Elsevier

relationship between the AFD and concentration.

$$\text{AFD} = (\text{Concentration})^3$$

Deitzel et al. [70] electrospun PEO; below 4 wt% concentration, they obtained a mixture of fibers and droplets while above 10 wt%, spinning was prohibited by high viscosity. They observed that the AFD is related to the solution concentration through a power law relationship with exponent of about 0.5. Near 8 wt%, second population fibers with diameters about one third of that of the primary population fibers are produced due to fiber splaying. Jun et al. [70] prepared PLLA fibers with diameters of 100–300 nm from 1 wt% solutions while 5 wt% solutions yielded fiber diameter of 800–2400 nm. Similarly, the diameter of PVA fibers increased from 87 $\pm$ 14 nm to 246 $\pm$ 50 nm by increasing the PVA concentration from 6 to 8% [81].

It was found that an increased fiber diameter correlated directly to a decrease in the surface area of electrospun mats [20, 81]. Figure 2.7 shows the FE-SEM images of PA 6.9 (Mw: 60,000 g mol^{-1}) prepared with different solution concentrations respectively [101]. PA 6.9 (Mw: 60,000 g mol^{-1}) dissolved in acetic acid:formic acid (50:50 v/v) solvent mixtures electrospun at a flow rate of 2 mL h^{-1}, applied voltage of 18 kV and tip of the needle to collector 6 cm. The ambient parameters are monitored during electrospinning, the temperature is 21 $\pm$ 2 °C and the relative humidity is 43 $\pm$ 5% RH. A higher flow rate is selected because more nanofiber material can be collected at one time and a lower tip-to-collector distance (TCD) selected, since it allows the use of lower voltages.

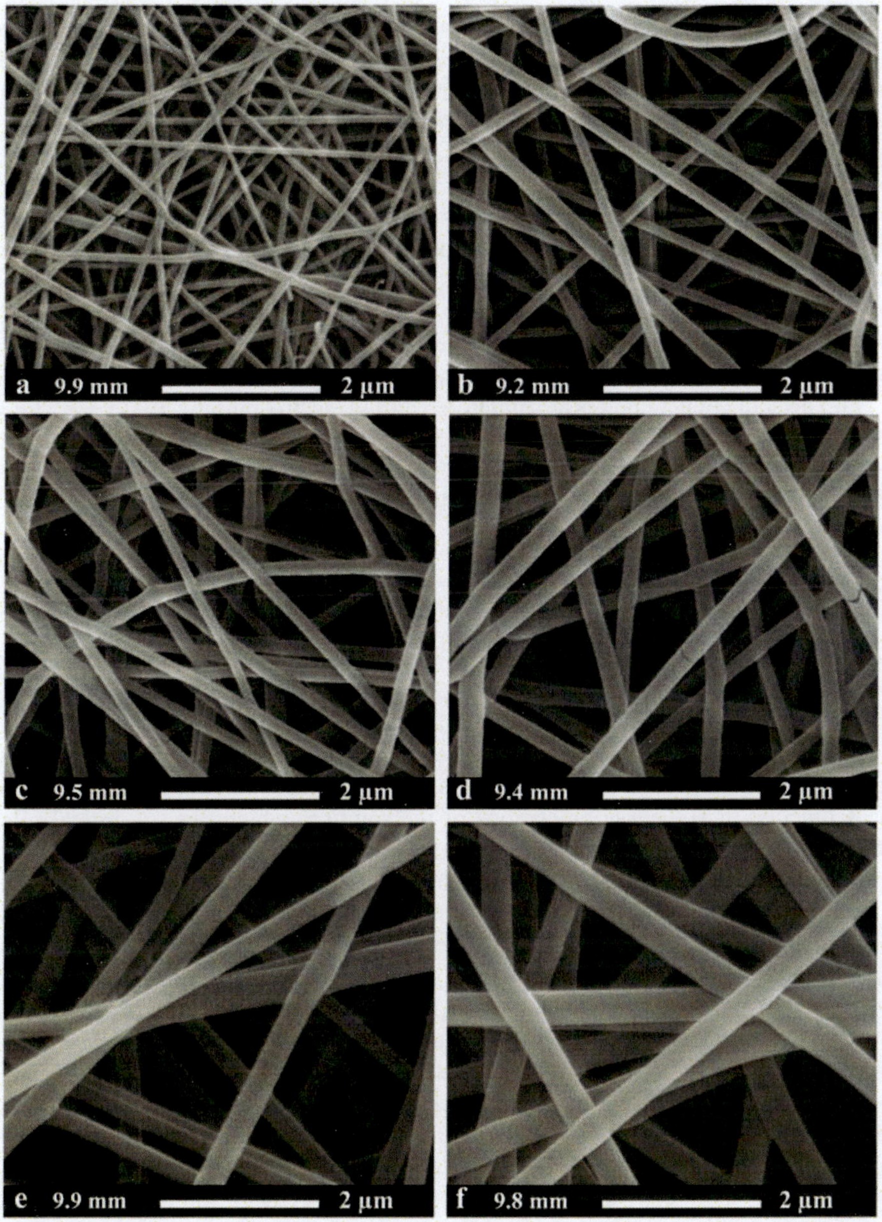

Fig. 2.7 FE-SEM images of nanofibers made with different concentrations of PA 6.9 (wt%), **a** 10, **b** 12, **c** 14, **d** 16, **e** 18, **f** 20 (magnification 50,000). Adapted and reproduced from Ref. [101], Copyright 2012 Springer Science

Brune and Bicerano [102] observed that beads are less likely to form using more viscous solutions. They also observed that lower surface tension favors bead-free filaments. Drew et al. [103] had demonstrated through a study on TiO_2 filled PEO that above a critical polymer concentration, the viscosity effect on fiber morphology was the same whether the viscosity was due to polymer or some other material. Lee et al. [66] reported increasing bead aspect ratio with increasing concentration for electrospun PS. Above 15 wt%, fibers without beads formed due to high viscosity. FESEM images of different acetic acid: formic acid ratio (Fig. 2.8) revels the effect of solvent ratio on surface tension and viscosity. Solvent ratio determines the average fiber diameter. PA 6.9 (Mw: 60,000 g mol^{-1}) dissolved in acetic acid:formic acid solvent mixtures to prepare 14 wt% polymer solution and electrospun at a flow rate of 2 mL h^{-1}, applied voltage of 23 ± 2 kV and tip of the needle to collector distance of 6 cm. The ambient parameters are monitored during electrospinning, the temperature is 21 ± 2 °C and the relative humidity is $43 \pm 5\%$ RH. A higher flow rate is selected because more nanofiber can be collected at one time and a lower tip-to-collector distance (TCD) selected, since it allows the use of lower voltages.

2.3.1.3 Surface Tension

Surface tension is a contractive tendency of the surface of a liquid that allows it to resist an external force. The cohesive forces among liquid molecules are responsible for the phenomenon of surface tension. In the bulk of the liquid, each molecule is pulled equally in every direction by neighboring liquid molecules, resulting in a net force of zero. Hence the surface tension of polymer solution, more likely to be a function of solvent compositions of the solution, which plays a critical role in the electrospinning process. Generally, in electrospinning, fibers with bead free morphology can be obtained by reducing the surface tension of the spinning solution. The polymer solution with high surface tension inhibits the electrospinning process because of the instability of the jets and results electrospraying rather than electrospinning [96]. The surface tension of the spinning solution greatly depends on the solvent used and can be controlled by using a solvent mixture, for example N,N′-dimethyl acetamide/acetone (70:30) mixture for preparing PVdF-*co*-HFP solution for electrospinning [20, 50, 104]. In some cases, increasing the concentration of a polymer solution can also affect the surface tension of the polymer solution [70].

The impact of surface tension on the average fiber diameter and surface morphology of electrospun fibers has been widely investigated. Zueo et al. [67] prepared poly-3-hydroxy butyrate-*co*-valerate (PHBV) solutions (4% solution) in different solvents, such as chloroform or its mixture with alcohol, tetrachloromethane, or N,N-dimethyl formamide, having similar conductivities and different surface tensions with/without triethylbenzyl ammonium chloride and studied its effect on fiber formation and its morphology. The study showed that surface tension is an important factor for bead formation during electrospinning. The solution of higher surface tension was favoring a bead-on-string structure. The physics of this phenomenon may be that surface tension tries to make the area per

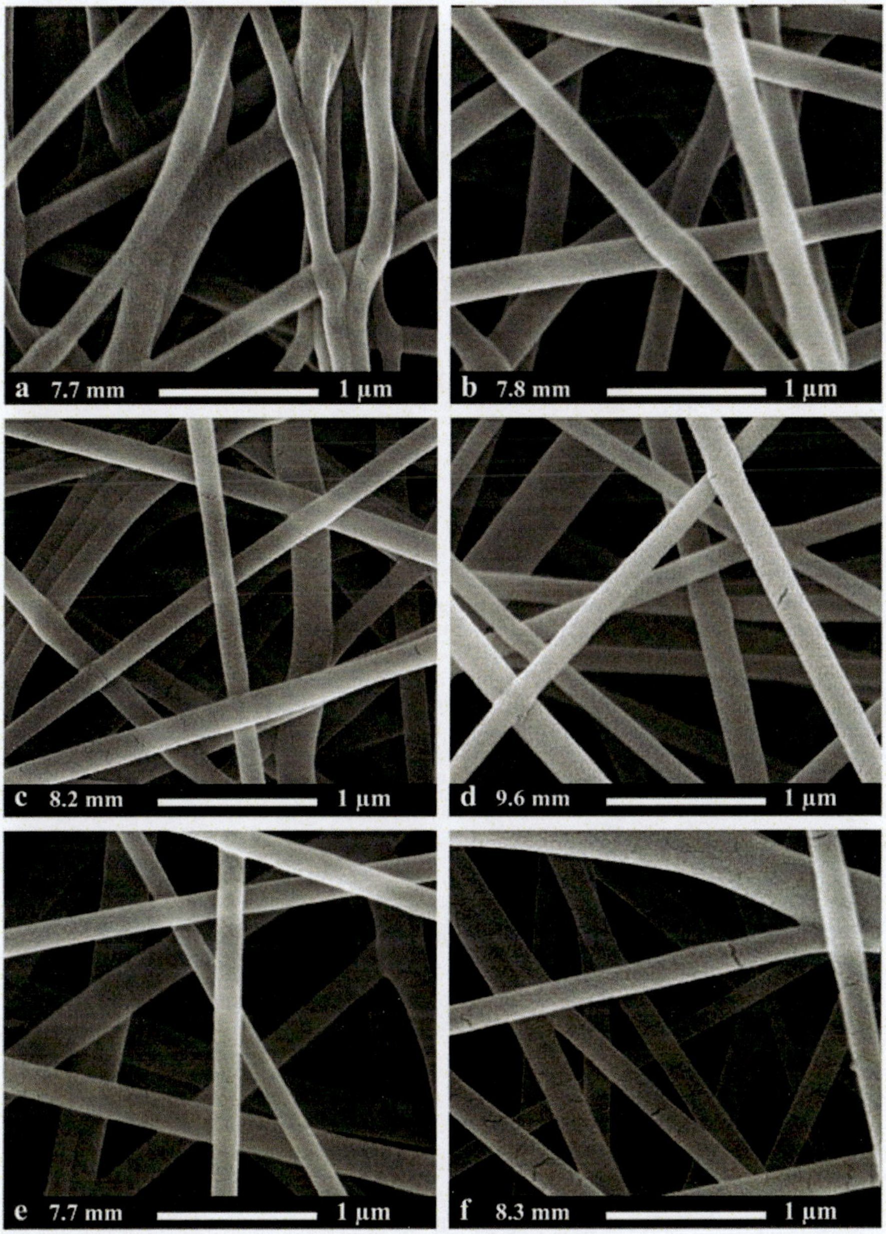

Fig. 2.8 FE-SEM-images of nanofibers made with different acetic acid:formic acid solvent ratios (v:v%), **a** 35:65, **b** 40:60, **c** 45:55, **d** 50:50, **e** 55:45, **f** 60:40 (magnification 100,000) Adapted and reproduced from Ref. [101], Copyright 2012 Springer Science

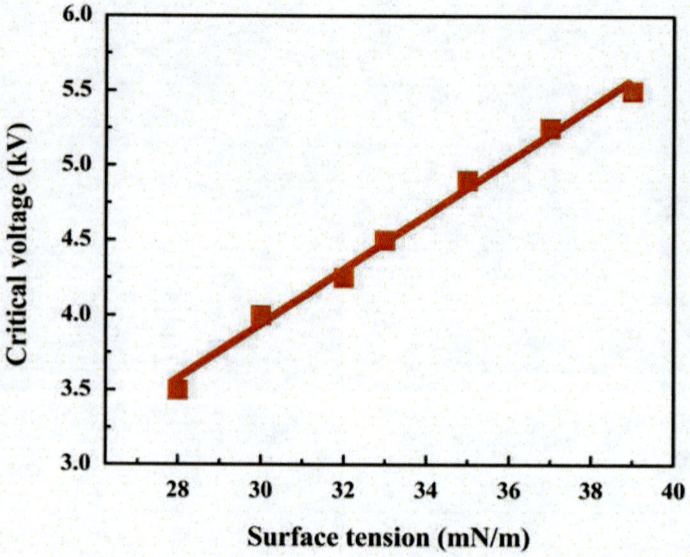

Fig. 2.9 Plot of critical voltage versus surface tension for electrospinning of PS in THF/DMF. Adapted and reproduced from [20], Copyright 2012 Elsevier

unit mass smaller, by changing the jet into spheres. The addition of ethanol to PEO [78] and PVA [81] PHBV [67] solutions lowered the surface tension. In the case of the PEO [78] and PHBV [67], the solution containing ethanol exhibited less beading; however, when ethanol was added to PVA solutions, beading was increased [81]. The difference in the effect of adding ethanol to these systems was attributed to the fact that it is a non-solvent for PVA and a solvent for PEO or PHBV. Another approach to reducing the surface tension of polyurethane urea solutions was to add polydimethylsiloxane, but no significant effect in fiber morphology was observed [87]. Lee et al. [66] found the critical voltage "Vc" increase linearly with surface tension (Fig. 2.9) [20, 50, 104] is a plot of surface critical voltage versus surface tension of PS.

2.3.1.4 System Parameters

Morphology of the electrospun fibers are greatly influenced by the system parameters, The structure of the polymer, molecular weight, molecular weight distribution, etc. are the key factors that include in the system parameters. Smoothness of an electrospun fiber depends upon whether it is low or high molecular weight polymer since the molecular weight has an effect on viscosity, conductivity, surface tension, etc. Molecular weight plays a vital role in the formation of fiber diameters. Hence a detailed discussion of system parameters is inevitable and the effect of electrospun fibers on applying such parameters is given below.

2.3.1.5 Molecular Weight

Molecular weight of the polymer also has an important effect on the morphologies of electrospun fiber. Many researchers have examined the relationship between polymer molecular weight and the morphology and average fiber diameter (AFD) of electrospun fibers. Molecular weight of the polymer has a significant effect on the rheological and electrical properties such as viscosity, surface tension, conductivity, and dielectric strength of the solution. It has been observed that too low a molecular weight solution tends to form beads rather than fibers and a high molecular weight solution gives fibers with larger average fiber diameters. In principle, the molecular weight reflects the entanglement of polymer chains in solutions, namely the solution viscosity. By keeping the concentration fixed and lowering the molecular weight of the polymer trends to form beads rather than smooth fiber. However, the studies showed that high molecular weights are not always essential for the electrospinning process if sufficient intermolecular interactions can provide a substitute for the interchain connectivity through chain entanglements. Using this principle, researchers have prepared oligomer-sized phospholipids from lecithin solutions into nonwoven membranes through electrospinning [105]. Gupta et al. [75] systematically studied to explore the scaling relationship between polymer molecular weight or molecular weight distribution and diameter of electrospun fibers electrospun at identical conditions (solvent, dimethylformamide; applied voltage, 10 kV; distance to a collector, 15 cm; flow rate, 3 mL h^{-1}) [75]. A series of linear homopolymers of poly(methylmethacrylate) with molecular weight (Mw) ranging from 12,470 to 205,800 g mol^{-1} and narrow molecular weight distribution (1.03–1.35) were synthesized and compared the effect of molecular weight on fiber diameter with commercial PMMA with varying in molecular weight (Mw) from 34,070 to 365,700 g mol^{-1} and broader molecular weight distribution (1.6–2.12). They determined critical chain overlap concentration, c^* for each of these polymers, and found that c^* decreased with increasing molecular weight. To isolate and clearly understand the effect of molecular weight, all the selected polymers are dissolved in the same organic solvent (dimethylformamide) and electrospun at identical spinning parameters, i.e., voltage = 10 kV, distance to collector = 15 cm, flow rate = 3 mL h^{-1} and ambient conditions. Upon electrospinning, they found as the molecular weight increased, the number of beads and droplets was reduced and obtained uniform fibers with high molecular weight polymer (Mw > 205,800 g mol^{-1}). Uniform fiber formation was observed at lower concentration ($c/c^* \sim 6$) for all the narrow molecular weight distribution polymers of Mw of 12,470–205,800 g mol^{-1} (Mw/Mn is 1.03–1.53) but for the broad narrow molecular weight distribution polymers of Mw of 34,070 and 95,800 g mol^{-1} (Mw/Mn is 1.6–2.12) uniform fibers were not formed until higher concentrations ($c/c^* \sim 10$). These scaling relationships were in general agreement with that observed by Mckee et al. [93].

Geng et al. [106] also studied the effect of molecular weight on the fiber formation and fiber diameter of electrospun chitosan fibers. Chitosan with three different molecular weight 30×10^3, 106×10^3, and 398×10^3 g mol^{-1} was used in the study. The results showed that among chitosans with different molecular weights and polymer concentrations, only about 7–7.5% solution of 106×10^3 g mol^{-1} chitosan (viscosity ranging from 484 to 590 cP) can produce a continuous and uniform fiber. Only chitosan solution of a molecular weight of 106×10^3 g mol^{-1} at 7% polymer concentration, dissolved in 90% aqueous acetic acid solution, formed a bead-free and more uniform nanofiber. The electrospun samples of the lower molecular weight chitosan solution (9.5–10.5% solution) contained large size beads and the fibers were fragile, while those of the higher molecular weight chitosan solution (2.5–3% solution) showed rougher and finer nanofibers with some bead defects. With 30×10^3 g mol^{-1} chitosan only fragile fibers with several droplets were obtained. Uniform nanofiber morphology was stabilized at a molecular weight of 106×10^3 g mol^{-1}, while very fine fibers (average diameter, 60 nm) with rougher surfaces were produced at 398×10^3 g mol^{-1}. The large population of smaller fiber at the highest molecular weight chitosan is attributed to the higher charge density of the chitosan molecules due to higher amino groups per molecule. Drastic morphological changes were observed when chitosan concentration increased at the same molecular weight [106]. For poly(N-isopropyl acrylamide), lowering the molecular weight produced smaller fibers that were more densely packed [74]. Varying the molecular weight of PEO from 6×10^5 to 40×10^5 g mol^{-1} during the spinning of PEO/chitosan solutions resulted in little difference in fiber diameter [106]. It was found that the spinning of polyamide at various molecular weights produced fibers with different average fiber diameters that were dependent upon the solution viscosity [74]. At a given concentration, fibers obtained from polyamide of higher molecular weights appeared to be larger in diameter, but it was observed that the average diameters of the fibers from polyamide of different molecular weights had a common relationship with the solution viscosities which could be approximated by an exponential growth equation.

2.3.2 *Process Parameters*

Process parameters play a key role in the electrospinning technique since it includes geometry of spinneret, solution feed rate or flow rate, the distance between the tip of the spinneret or needle to a collector, applied voltage hydrostatic pressure in the capillary, movement of collector plate or drum, etc. Applied voltage is a crucial element in this technique since it affects the fiber diameter since surface charges exerting on the jet depends on the voltage. Electrostatic repulsive force increases with the increase in voltage leading to a decrease in the fiber diameter. Low feeding rate is suitable for the solid nanofibers and the easy evaporation of the solvent. Thus, desirable electrospun fibers can be obtained by optimizing the process parameters and a detailed description is discussed below.

2.3.2.1 Applied Voltage/Field Strength

In the electrospinning process, the most crucial element is the application of high voltage to polymer the solution and is one of the most studied parameters among the controlled variables. Only if the applied voltage is above a given limiting value required to overcome the surface tension of the solution (threshold voltage), induces the necessary charges on the solution along with the electric field and the fibers get ejected out from the spinneret. The charge transport due to the applied voltage in the electrospinning process is mainly due to the flow of the polymer jet towards the collector. In electrospinning, the electrical field is defined as the applied voltage divided by the distance between the tip and collector grid and can be controlled by changing the distance between the tip and collector or by changing the applying voltages. When the applied voltage just above the threshold voltage, the volume of the drop at the tip of the spinneret decreases, causing the Taylor cone to recede. Then the jet originates from the liquid surface within the tip results fibers with beads. As the voltage is increased further, the jet eventually moves around the edge of the tip, with no visible Taylor cone; at these conditions, the presence of many beads can be observed [70] Using laser diffraction, it has also been shown that increased voltages produces jets with larger diameters and ultimately lead to the formation of multiple jets [87]. High voltages and high electric field strengths are also associated with higher productivity, which is mainly considered to be due to increase in electrical current and electrostatic stresses, which in turn draw more material out of the spinneret and facilitates the formation of fibers with a larger diameter [1], Raghavan et al. [1] electrospun 16% poly(acrylonitrile) solution in DMF at a different applied voltage ranging from 10 to 25 kV and observed that average fiber diameter increases and fiber distribution narrow down with increasing voltage. Similar trend was reported for poly(vinyl alcohol) electrospun at 5–15 kV [81], polyurethane electrospun at 0–35 kV [87] and poly(D,L-lactic acid) (PDLA) and semi-crystalline poly(L-lactic acid) (PLLA) electrospun at 0–35 kV [64].

However, there exists controversy in the literature as to the effect of increasing the voltage on the final diameter of the electrospun nanofibre. Experimental voltage effect analysis was performed for the Nylon-6 using an output voltage range between 12 and 18 kV showed, the average fibre diameter decreased from 1.35 to 1.21 μm. The phenomena are attributed to higher electrostatic forces on the jet induced by higher voltage or field results the higher repulsive forces and favor the formation of the thinner fibres [107]. Similar results are reported for electrospun silk-like polymer with fibronectin functionality [90] and bisphenol-A polysulfone [108]. The fiber diameter tended to decrease with increasing applied voltage. In most cases, a higher voltage causes greater stretching of the solution due to the greater columbic forces in the jet as well as a stronger electric field and these effects lead to a reduction in the fiber diameter. Larrondo et al. showed [28, 46, 109] the decrease of fiber diameter by roughly half by doubling the applied electric field. Rapid evaporation of solvent from the fibers also results in fibers with smaller fiber diameter. Thus, the significance of applied voltage on fiber diameter varies with the polymer solution concentration and on the distance between the tip and the collector [110].

At a higher voltage, there is a greater probability of beads formation, and the phenomena are demonstrated by many groups [19, 70, 80, 87, 90, 111–114]. The presence of beads and junctions at high voltages was found when spinning solutions of PEO [70, 78], polystyrene [19], PVA [80], polyurethane [87], PDLA [64], bisphenol-A polysulfone [108], chitosan [106], protein-polymer [90] and gelatin [84]. As the fluid exits from the spinneret and enters the electric field, it assumes a conical (or nearly conical) shape, from the apex of which a thin jet emerges. In inviscid fluids, the shape that the meniscus attains depends on the interplay between the electric field and the surface tension and has been studied extensively. Increasing the applied voltage to a certain level changes the shape of the pendant droplet from which the jet originates. Chowdhury and Stylios [107] studies demonstrated the stable fiber shape cannot be achieved for electrospun Nylon-6 with increasing applied voltage due to the change in the shape of the droplet formed at the tip of the spinneret. Yarin et al. [115, 116] also demonstrated that the electrified meniscus can assume a wide variety of shapes that are different from that of the Taylor cone. In the process of electrospinning, the electric current associated with the spinning process typically is measured in a microampere. The droplets or fibers transport charge across the gap between the charged needle and the electrically grounded target, i.e., closing of the circuit. In general, the researchers have suggested that when higher voltages are applied, more volume of polymer gets ejected from the charged needle and facilitates the formation of fibers having a larger diameter. On the other hand, many studies are reported that an increase in the applied voltage (i.e., increasing the electric field strength), increases the electrostatic repulsive force on the fluid jet which ultimately favors the narrowing of fiber diameter. Theoretically, it is widely accepted that a higher voltage causes greater stretching of the solution due to the greater columbic forces in the jet as well as a stronger electric field. This greater degree of stretching leads to the reduction in the fiber diameter and rapid evaporation of solvent from the fibers results in the disposition of more dried fibers on the collection target. FESEM images of nanofibers produced by applying a different voltage (Fig. 2.10) depicted on applying a voltage no significant effect was observed.

2.3.2.2 Flow Rate or Feed Rate

The flow rate of the polymer fluid from the syringe to the spinneret (blunt steel needle) where the high voltage is to be applied is an important process parameter to premine the morphology of the electrospun fibers. It has a great influence on the material transfer rate (amount of polymer solution flows out from the syringe/s) and the jet velocity. The adjustments on the solution flow rate are essential to maintain a stabilized Taylor cone during electrospinning [117]. To a certain extent, increasing the flow rate while maintaining the Taylor cone produces larger fibre diameters [118]. Many researchers have systematically investigated the relationship between solution feed or flow rate on fiber diameter and morphology. There should be a minimum flow rate of the spinning solution to form fibers during spinning, which greatly depends on the solution concentration and viscosity of the spinning solution. Fong et al. [78]

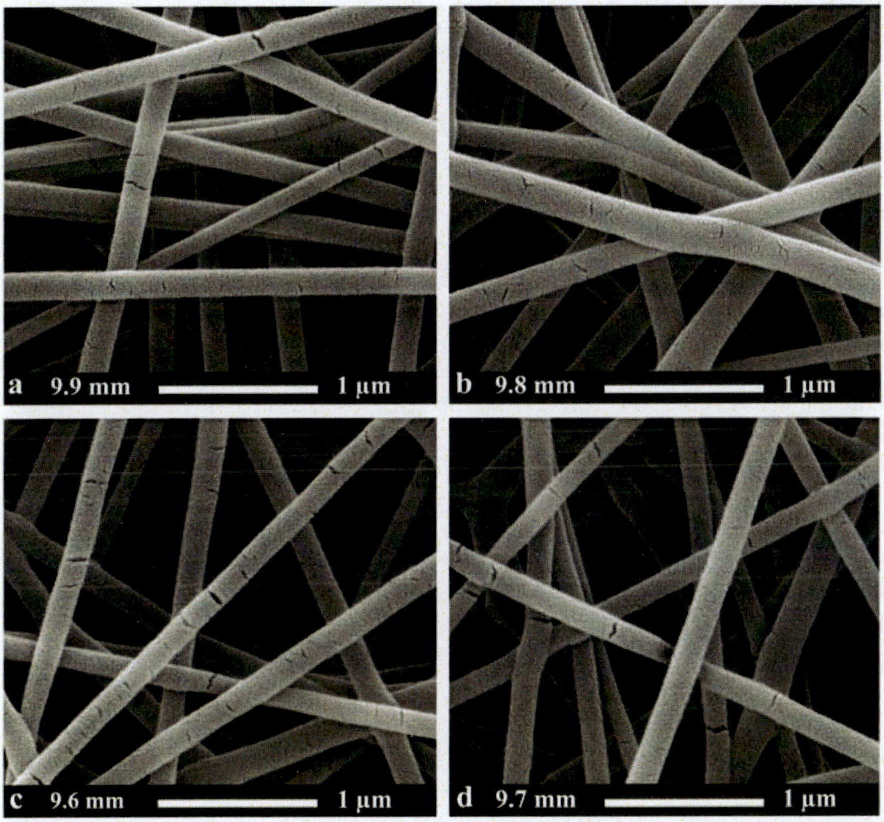

Fig. 2.10 FE-SEM-images of nanofibers made with varying applied voltage (kV), **a** 16, **b** 18, **c** 20, **d** 22 (magnification 100,000). Adapted and reproduced from Ref. [101], Copyright 2012 Springer Science

observed that a minimum solution volume at the end of the spinneret should be present for the formation of an equilibrium Taylor cone. Generally, it was found that a lower flow rate is more desirable as the polymer solution will get enough time for polarization and solvent evaporation results stalking of dried fibers with smaller diameters on the collection grid [64]. If the flow rate is too slow, smaller fibers with spindle-shaped beads are formed and with increasing flow rate, the average fiber diameter, pore size, and bead formation increases [19, 78, 108].

Increasing the flow rate too much causes fibers to be collected without sufficient solvent evaporation leading to a flattened web-like appearance [119]. If the flow rate is very high, bead fibers with thick diameter will form rather than the smooth fiber with thin diameter owing to the short drying time prior to reaching the collector and low stretching forces [67, 81, 108, 120]. Zong et al. [64] reported that in addition to the bead morphology, the higher flow rate the mean pore-size increased from 90 to 150 nm. Optimally, a lower solution flow rate, while maintaining the Taylor cone is more ideal since it permits time for solvent evaporation and produces better fibers. However, a solution flow rate below the threshold for a given voltage causes evaporation at the needle tip, thus preventing electrospinning.

It has been observed that the fiber diameter and the pore diameter increases with an increase in the polymer flow rate in the case of polystyrene (PS) fibers and by changing the flow rate from 0.004 to 0.24 mL min^{-1}, the morphological structure also changed [64]. Bead formation is observed when the flow rate is 0.10 mL min^{-1} and higher. With increasing flow rate, the ribbon-shaped fibers increase in size from about 5 to 20 μm and show the typical microtexture and nanopores on the fiber. The pore sizes range in diameter from 50 to 250 nm and the mean pore size is observed to increase from about 90 to 150 nm with increasing flow rate. Figure 2.11 shows the FESEM images of PS solution in THF, electrospun at different flow rates, by keeping other parameters constant. The PS solution of concentration 30 to 35%, electrospun at a flowrate of 0.07 mL min^{-1}, applied voltage 7.5 to 10 kV, working distance 35 cm produced beadless ribbon-shaped fibers having fiber diameter 20 ± 10 μm with a microtextured surface and densely packed nanopores of diameters in the 50 to 200 nm range. The study suggested that bead formation on the electrospun fibers is a function of the concentration as well as of the spinning voltage, the working distance, and the feed rate. The fiber diameter range of PS fibers obtained at different flow rate ranges from 0.8 to 20 (± 10) μm with increasing flow rate.

Zuo et al. [67] also found that the solution feeding rate is an important factor for bead formation during electrospinning. When 4 wt% PHBV solution was electrospun at different feed rates, while all other spinning parameters kept constant, it was observed that bead size increased from 8 to 23 μm with increasing flow rate from 2 to 9 mL h^{-1}. Surprisingly, however, the average fiber diameter was found to be unchanged (1.5 μm) with feed rate. This suggested that, for electrospinning of PHBV solution, the flow rate influence only on bead formation and its size rather than fiber diameter. As the feeding rate increases, more solutions get ejected out from the needle tip at the same time. When the electric field force is not able to stretch the jet sufficiently, the superfluous solutions form the beads, which may be caused by surface tension. [67].

Chowdhury and Stylios [107] reported at the low flow rate (about 0.20–0.25 mL h^{-1}) the Nylon 6 electrospun fiber is cylindrical and uniform. At higher flow rates (about 0.26–0.30 mL h^{-1}) the fiber surface is rougher and if the flow rate is much above 0.3 mL h^{-1}, the fluid jet bents from the outset and breaks before it progresses any further. The average fiber diameter of Nylon 6 increased from 1.0 to 1.6 μm with an increasing feed rate of 0.20 to 0.30 mL h^{-1}. The study concluded that at low feed rates 0.20 to 0.25 mL h^{-1}, the solvent has a sufficient amount of time to evaporate until the fibers are collected on the plate and thinner and uniform fibers are produced. At a high feed rate 0.26 to 0.30 mL h^{-1} which seems above the quasi-stable point, the solution is not completely carried away to the collector, which resulted in unstable jet and larger fibers [107]. SEM images of fibers with different flow rate reveals a steady increase in average fiber diameter on increasing the flow rate to 1.8 mL h^{-1} and remains constant.

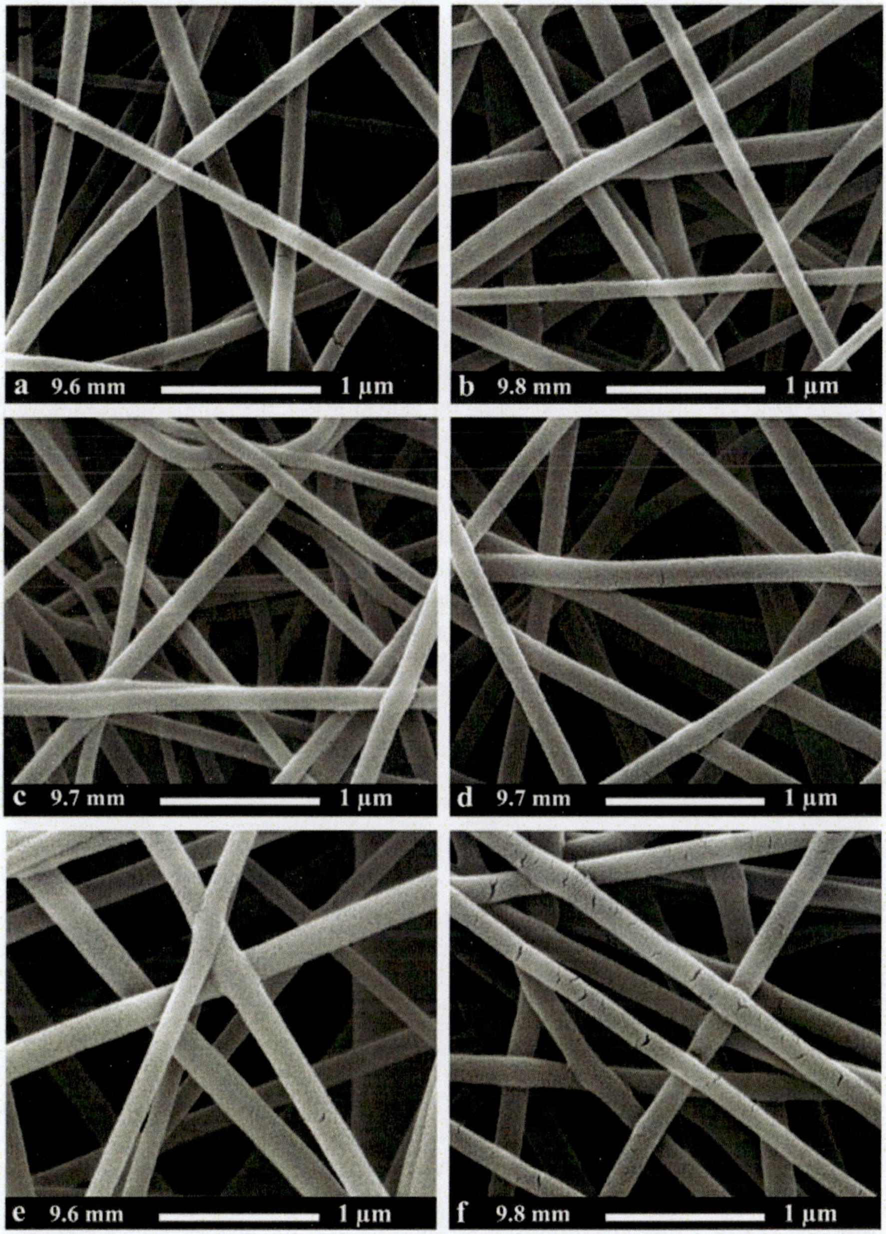

Fig. 2.11 FE-SEM-images of nanofibers produced with varying flow rate (mL h^{-1}), **a** 0.4, **b** 0.8, **c** 1.2, **d** 1.6, **e** 2.0, **f** 2.8 (magnification 100,000). Adapted and reproduced from Ref. [102], Copyright 2002 Springer Science

Zong et al. [64] studied the effect of feed rate on electrospun nanofiber from poly(D,L-lactic acid) (PDLA) solution with 1 wt% KH_2PO_4 electrospun at 20 kV. It was found that at a lower feeding rate, smaller fibers with spindle-like beads were formed. Relatively large fiber diameters and beads were seen in fibers spun from a higher solution feeding rate (75 mL/min). This behavior of the process is attributed to the velocity of the fluid jet. Since the droplet suspended at the end of the spinneret is larger with a higher feeding rate, the jet of the solution can carry the fluid away with a faster velocity. As a result, the electrospun fibers are harder to dry before they reached the collection drum. This can result in large beads and junctions in the final membrane morphology [64]. Similar characteristics were observed for the spinning of PVA solution. At the flow rate of 0.3 mL h^{-1}, a few big beads were observed on the PVA fibers. When the flow rate exceeded a critical value, the delivery rate of the solution jet to the capillary tip exceeded the rate at which the solution was removed from the tip by the electrostatic forces. This shift in the mass-balance resulted in sustained but unstable jet and fibers with big beads are use to form [81, 90]. All these studies suggest that flow rate could affect electrospinning process and a certain minimum value of the solution volume suspended at the end of the spinneret should be maintained in order to form an equilibrium Taylor cone. Therefore, different morphology can be obtained with the change in the feeding rate at a given applied voltage. Yuan et al. [108], investigated the effect of the flow rate on the morphologies of the PSF fibers from 20% PSF/DMAC solution and showed, bead fibers with thicker diameters can be obtained as the flow rate is 0.66 mL h^{-1}.

2.3.2.3 Tip to Collector Distance

The distance between the tip of the spinneret/needle and the collector has been also shown to affect the size and morphology of electrospun fibers [84]. One important physical aspect of the electrospinning nanofibers is their dryness from the solvent used to dissolve the polymer [121]. Thus, there should be an optimum distance between the tip and collector is required to dry the fibers before reaching the collector [106]. If the distance from the tip of the needle to the collector is either too close or too far, forms fibers with bead morphology [80, 84]. With an increasing tip to collector distance, the electrostatic force decreases and so the drawing stress decreases. This leads to a decrease in the aspect ratio and an increase in the size of beads, although it facilitates the deposition of more solidified or dried fibers on the collector.

A number of researchers examined the effect of varying the distance between the tip and the collector on the fiber morphology and average fiber diameter of electrospun fibers of PVA [80], chitosan [106], gelatin [84], bisphenol-A [108], and poly(vinylidene fluoride) (PVdF) [91]. Compared to other parameters, no significant effect of the distance between the tip and collector on the fiber size and morphology was observed for the spinning of PVA [81], gelatin [84], Chitosan [106], and poly(vinylidene fluoride) [91]. It has been reported that flatter fibers can be produced at closer distances but with an increase in distance rounder fibers have been observed with the spinning of silk-like polymer with fibronectin functionality

[90]. For bisphenol-A polysulfone, closer distances between the tip and collector yielded smaller fibers [108, 122]. Yuan et al. [108] also demonstrated that a little long-distance favors the thinner fiber diameter as shown in Fig. 2.12. In brief, if the distance is too short, the fiber will not have enough time to solidify before reaching the collector, whereas if the distance is too long, bead fiber can be obtained. However, the distance between needle and collector is less significant to the fiber morphology and size when compared to the other sinning parameters.

2.3.2.4 Collector Composition and Geometry

One important aspect of the electrospinning process is the type of collector used. In this process, a collector serves as a conductive substrate where the nanofibers are collected. Generally, aluminum foil is used as a collector but due to difficulty in transferring of collected fibers and with the need for aligned fibers for various applications, other collectors such as conductive paper, conductive cloth, wire mesh [123, 124], parallel or grided bar [125], rotating rod, rotating wheel [126], liquid nonsolvent such as methanol coagulation bath [127] have also been studied.

A number of materials and geometries have been studied for the collection of electrospun polymeric fibers. The fibrous membrane morphology (such as random fibers, aligned fibers, intertwined fibers, etc.) and physical properties (such as crystal morphology, molecular orientation, stretching, etc.) of electrospun fibers are affected by the type and geometry of collectors [76, 98, 128]. Kim et al. [76] collected PLLA and PLGA fibers on metal collectors, a water reservoir, and a methanol collector. They found that smooth fibers were obtained using the metal collector. Collection on the surface of the water caused the hydrophobic polymer fibers to shrink, while methanol caused swelling of the fibers [76] Cellulose acetate was collected using copper mesh, aluminum foil, water, and paper [98]. It was found that the packing density was influenced by the conductivity of the collectors: the more conductive collectors dissipated the charge of the fibers. When this charge was not dissipated (non-conductive collectors), the fibers repelled one another, yielding a more porous structure [98]. Also, porous collectors, such as paper and copper mesh, produced a less-packed structure as compared to fibers collected on aluminum foil and water [98]. Poly(ethylene-co-vinyl alcohol) was even spun directly onto a human hand [89].

Collectors with various geometries have also been designed and used. The collectors can be either stationary or dynamic. The most commonly used stationary collector is a plate collector or screw jacks wrapped with aluminum foil and a dynamic collector is a grounded rotating drum made up of stainless steel or aluminum. The common types of drum collectors are shown in Fig. 2.13. With dynamic collectors (due to drum speed), the diameter of the fiber can be controlled by controlling the speed of rotation of the collector [129, 130] A significant number of aligned fibers can be achieved by rotating drum collectors having rotating speed higher than 250 rpm. As the speed increases the degree of aligned fibers increases. Rotating disk type

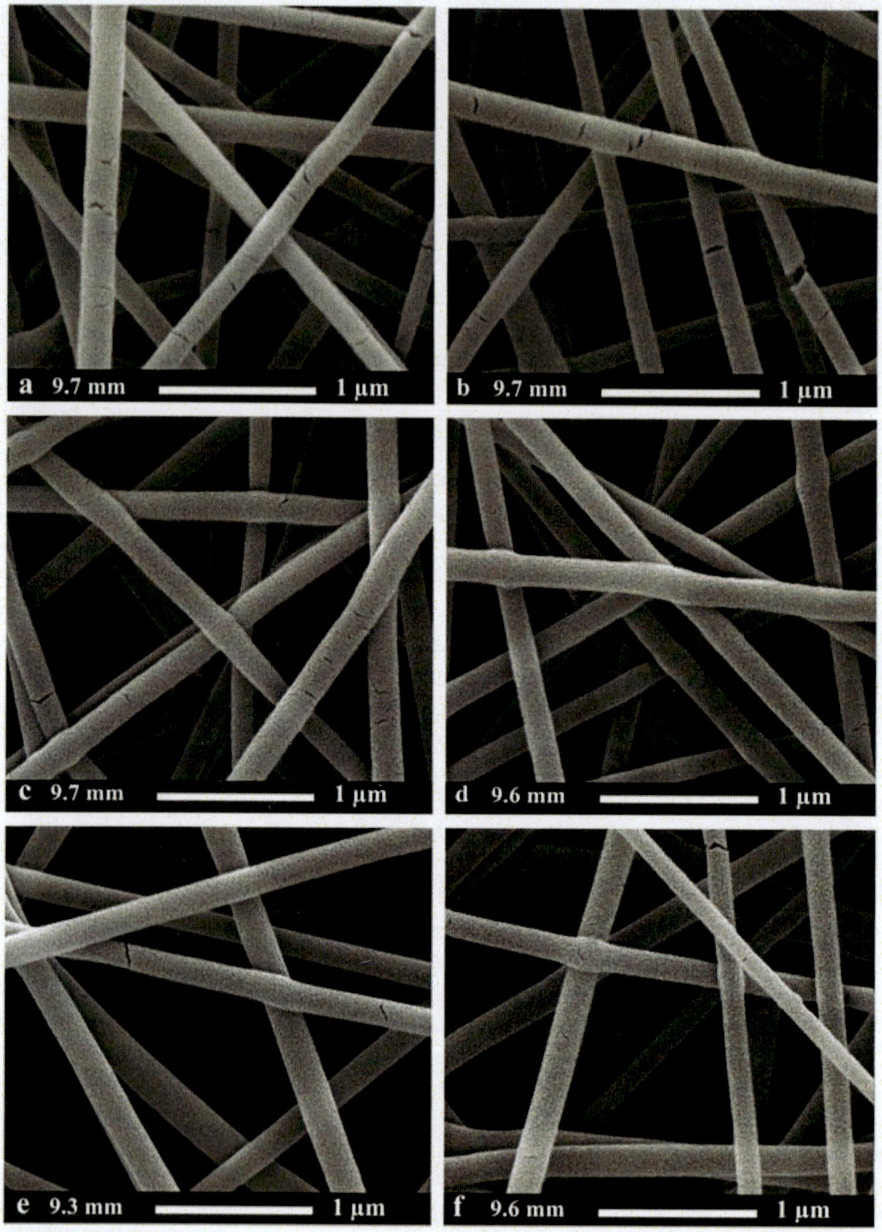

Fig. 2.12 FE-SEM-images of nanofibers made with varying tip to collector distance (TCD) (cm), **a** 6, **b** 7, **c** 8, **d** 9, **e** 10, **f** 11 (magnification 50,000). Adapted and reproduced from Ref. [102], Copyright 2002 Springer Science

Fig. 2.13 Different configurations of cylindrical or disc type collectors. Adapted and reproduced from Ref. [141]

collector is popular for creating uniaxially aligned fibers. The morphological difference between the fibers collected by using a rotating drum collector and disc collector is displayed in Fig. 2.13 (FE-SEM images). The most important benefit of using disk collector instead of drum collector is that a large number of the fibers is residue on the disk edge and the aligned fibers are gathered as nanofibers [126, 131–134]. Another type of collector parallel plate collectors using two or more parallel plates (electrodes), so that excellent fiber alignment can be obtained on each electrode based on air holes through the electrodes that cause electrostatic rebuttal between the fibers [131, 135, 136]. Rutledge and co-workers used two parallel plates when spinning their fibers in order to generate uniform electric fields [65, 137]. Frame collectors were shown to yield aligned fibers with a conductive frame producing better alignment than a non-conductive one [137]. Also, an array of electrospun fibers has been produced using two conductive, collection rings [137]. The fibers were suspended between the rings, and fibers up to 10 cm long were obtained. The rotation of one of the collection rings allowed for the production of a multi-filament yarn [137]. PEO was also spun using a multiple field method in which the polymer jet passed through three parallel rings, each connected to an independent power supply. This method produced smaller, bead-free fibers that were collected in a more focused area [137]. Fibers have also been collected using a rotating cylindrical drum collector rather than a stationary target; doing so has allowed for the alignment of the fibers [70, 138]. Better alignment was observed when this idea was extended to use a drum composed of copper wires that were grounded [130]. In another variation, a thin, steel pin was used as a counter electrode and was placed behind a rotating, non-conductive cylindrical collector; aligned fibers greater than 10 cm in length were obtained [124]. The rotating drum was also combined with the previously mentioned multiple field method to enable the collection of fibers into thin strips or yarns [139]. Theron et al. [94] employed a "tapered and grounded wheel-like bobbin" to collect and align the nanofibers of PEO [129]. Their approach yielded fibers that were several hundred microns in length with diameters from 100 to 300 nm; braided fibers were also obtained. Electrospun poly(lactide-co-glycolide) (PLGA) and poly(ε-caprolactone) (PCL), PLGA/PCL membrane fabricated by custom-designed electrospinning set-up with aligned and randomly oriented SEM images showing an average fiber diameter of 554 nm (Fig. 2.14) [140].

Fibers (isotropic or anisotropic) are affected by speed and type of collector plate or disc or drum. The rotation speed of the collector has a significant role play on the crystal orientation. Higher speed of rotation can improve the crystal orientation of fibers, due to the polymer molecular chains' alignment along the direction of fiber axis

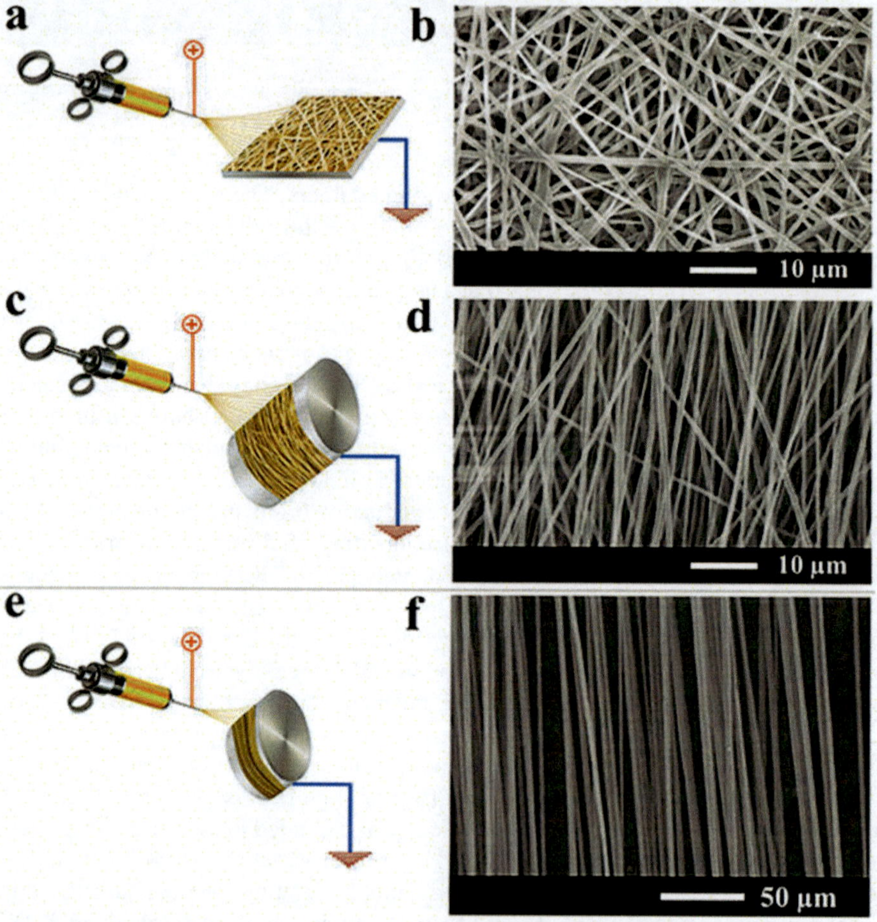

Fig. 2.14 Schematic of experimental setup diagram and SEM images of electrospun PLGA and PCL, PLGA/PCL membrane; **a** Experimental setup, **b** FE-SEM image of randomly oriented electrospun nano fiber (ENF), **c** experimental setup of aligned ENF, **d** SEM image. Adapted and reproduced from Ref. [140], Copyright 2015 Elsevier, **e** experimental setup, **f** SEM images of aligned nanofibers polyurethane fibers. Adapted and reproduced from Ref. [142], Copyright 2017 Royal Society of Chemistry

which is obtained due to the force of the rotational speed of collector [143]. It must be mentioned that using high-speed rotational collector can create a ventilator effect and evaporation of the solvent [144]. Yee et al. reported a modified rotating disc collector for producing well-aligned electrospun fibers. In the modified version o the collector is made by attaching separate, parallel electrodes onto a rotating disk collector as shown in Fig. 2.15. Using the modified rotating disc collector, well-aligned electrospun polyvinylidenedifluoride (PVdF), PVdF/carbon nanotube nanocomposite, and Poly(vinylidene fluoride–trifluoroethylene) copolymer nanofibers are directly deposited onto flat substrates forming relatively large, uniform, and compact fibrous thin films. The attachments alter the electric-field distribution on the rotating disk, which fosters the fanning of the nanofibers, while the electric field between the separate electrodes and the mechanical force exerted by the rotational disk facilitate the fiber alignment. The study also showed that the specific environment and force fields were created on the modified rotating disk because the electrospun fibers being effectively stretched to form highly oriented β-form crystallites with slightly reduced inter-chain distance. They also lead to slight increases in crystallinity and crystal size. The proposed hypothetic mechanism for the structural alteration is that: it may be related to the formation of extended-chain crystals with a more perfect planar-zigzag conformation under high tensile stress [145].

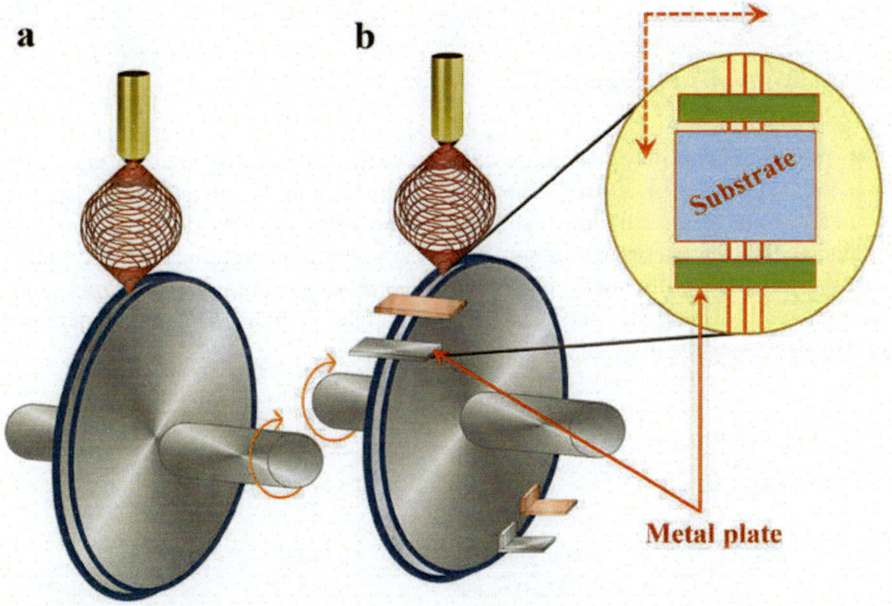

Fig. 2.15 Schematics showing the electrospinning process using **a** a conventional rotating disk collector, **b** the modified rotating disk collector. Adapted and reproduced from Ref. [142], Copyright 2017 Elsevier

2.3.2.5 Dipole Moment and Dielectric Constant

Few studies have been performed to date to investigate the effect of dipole moment and dielectric constant of the spinning solution on fiber properties. Table 2.2 shows the conductivity and dielectric constants of different solvents, commonly used for the preparation of polymer solutions. The spinning of polystyrene was studied in 18 different solvents, and the only solvents found to be used successfully had high values of dipole moment [83]. Solvents with a high dielectric constant were used to produce PMMA “cups” [146]. Also, the productivity (number of fibers produced per unit time) of spinning polystyrene fibers was found to correlate with the dipole moment and dielectric constant [147]. To date, few studies have methodically examined the effect of dipole moment and dielectric constant on the resulting fiber morphology since these parameters are difficult to isolate.

Son et al. [77] studied the effect of solution properties on the electrospinning of PEO to observe that the higher the dielectric constant of the solvent, the lower the fiber diameter. As the dielectric constant of the polymer fluid increases, the AFD decreases marginally. Solvent composition also affects fiber morphology. Lee et al. [66] studied the bead morphology of electrospun PS from a mixture of THF and DMF. It was found that bead concentration depends on the DMF content; half hollow sphere structures appeared in fibers from PS in THF only.

2.3.3 *Ambient Parameters*

Ambient parameters plays an important role in the morphology and average fiber diameter of a polymer matrix. The major studies include relative humidity, the temperature of the surroundings, solution temperature, air-flow rate, etc. Ambient parameters mainly encompass humidity and ambient temperature. Humidity influence the morphology in such a way that on increasing humidity, uniform circular pores can be obtained. Smaller fibers can be obtained by increasing the temperature thereby decreases the viscosity [49, 74].

Table 2.2 Conductivity and dielectric constants of different solvents

Solvent	Conductivity	Dielectric constant
Acetone	0.0202	20.7
Dimethyl formamide	1.09	3.67
Tetrahydrofuran (THF)	0.037	7.6
Distilled water	0.447	
Ethanol	0.0554	2.46
Methanol	0.12	
Isopropanol	0.0385	20.1
DMAc		

2.3.3.1 Ambient Temperature

Few studies have been conducted to examine the effects of ambient parameters (i.e., temperature and humidity) on the electrospinning process. Mit-Uppatham et al. [74] spun polyamide-6 fibers at temperatures ranging from 25 to 60 °C. They found that increasing the temperature yielded fibers with a decreased fiber diameter, and they attributed this decline in diameter to the decrease in the viscosity of the polymer solutions at increased temperatures.

2.3.3.2 Relative Humidity

The humidity was varied by Casper et al. [148] while spinning polystyrene solutions. Their work showed that increasing the humidity resulted in the appearance of small circular pores on the surface of the fibers; increasing the humidity further lead to the pores coalescing. Spinning has also been performed under vacuum in order to obtain higher electric fields; doing so produced fibers and yarns with larger diameters [49].

In summary, we have highlighted studies that examine the effect of manipulating each parameter in the electrospinning process and the effect of that parameter on the resulting fiber morphology. Table 2.3 summarized the general effect of processing parameters on fiber dimensions and Table 2.4 lists each parameter studied and its effect on the fiber morphology. However, it is difficult to isolate the effect of many of the parameters since they are interrelated. For example, changing the solution concentration/viscosity affects other solution properties, such as conductivity and surface tension. Additionally, though a large number of distinct polymers have been electrospun, there has been a little systematic investigation of the conditions required for successful spinning. Typically, a trial-and-error approach has been employed in which the solution properties and spinning parameters are varied until uniform, defect-free fibers are obtained. Previously, it has been shown that a minimum concentration is required to electrospin; if there is insufficient chain overlap, uniform fibers

Table 2.3 Effect of processing parameters on fiber dimensions

Parameters (increasing)	Effect of diameter
Viscosity	Increasing
Feed rate	Increasing (more beads formation)
Solution conductivity	Decreasing
Surface tension	Decreasing
Needle inner diameter	Increasing (more beads formation)
Voltage	First decrease after that continuously increasing
Distance between tip and collector (working distance)	Decreasing

Table 2.4 Effect of different spinning parameters on the fiber morphology of electrospun fibers

Process parameter	Effect on fiber morphology	References
Viscosity/concentration	• Low concentrations/viscosities yielded defects in the form of beads and junctions; increasing concentration/viscosity reduced the defects • Fiber diameters increased with increasing concentration/viscosity	[67, 68, 74, 75, 79–83, 86, 88–90, 92, 93, 98, 99, 149–153]
Conductivity/solution charge density	• Increasing the conductivity aided in the production of uniform bead-free fibers • Higher conductivities yielded smaller fibers in general (exceptions were PAA and polyamide-6)	[68, 69, 72, 73, 79, 88, 90, 99, 146, 151]
Surface tension	• No conclusive link established between surface tension and fiber morphology	[67, 78, 81, 87]
Polymer molecular weight	• Increasing molecular weight reduced the number of beads and droplets	[74, 75, 88, 92]
Dipole moment and dielectric constant	• Successful spinning occurred in solvents with a high dielectric constant	[83, 146, 147]
Flow rate	• Lower flow rates yielded fibers with smaller diameters • High flow rates produced fibers that were not dry upon reaching the collector	[67, 81, 108, 147]
Field strength/voltage	• At too high voltage, beading was observed • Correlation between voltage and fiber diameter was ambiguous	[68, 80, 83, 87, 88, 90, 150]
Distance between tip and collector	• A minimum distance was required to obtain dried fibers • At distances either too close or too far, beading was observed	[80, 81, 84, 90, 106, 108, 153]

(continued)

Table 2.4 (continued)

Process parameter	Effect on fiber morphology	References
Needle tip design	• Using a coaxial, 2-capillary spinneret, hollow fibers were produced • Multiple needle tips were employed to increase throughput	[154–157]
Collector composition and geometry	• Smoother fibers resulted from metal collectors; more porous fiber structure was obtained using porous collectors • Aligned fibers were obtained using a conductive frame, rotating drum, or a wheel-like bobbin collector • Yarns and braided fibers were also obtained	[65, 78, 82, 86, 124, 129, 131, 137, 139, 158]
Ambient parameters	• Increased temperature caused a decrease in solution viscosity, resulting in smaller fibers • Increasing humidity resulted in the appearance of circular pores on the fibers	[49, 74, 148]

will not be obtained. This minimum concentration is typically 5–10 times c^*; once this minimum concentration has been determined, varying other process parameters can be used to tune the resulting fiber morphology. Overall, these studies highlight the potential of electrospinning for producing polymeric fiber meshes. By changing the solution and processing conditions, such as the polymer concentration, the properties of the resulting mesh can be easily tuned. These meshes have tremendous potential as tissue-engineered scaffolds, and the applications to date are discussed below. FESEM images of 12 wt% polymer solution were electrospun with different relative humidity at 16, 43, 74% is shown below (Fig. 2.16). As the relative humidity increases, significant decreases in the fiber diameters were observed.

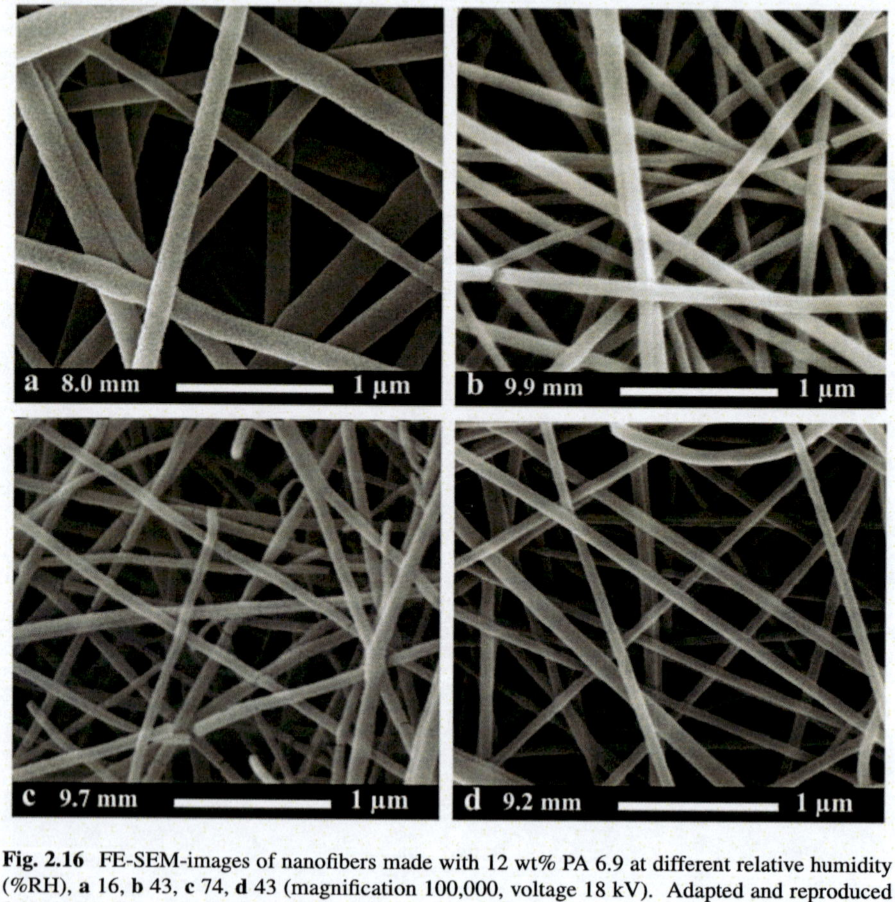

Fig. 2.16 FE-SEM-images of nanofibers made with 12 wt% PA 6.9 at different relative humidity (%RH), **a** 16, **b** 43, **c** 74, **d** 43 (magnification 100,000, voltage 18 kV). Adapted and reproduced from [102], Copyright 2002 Springer Science

2.3.3.3 Surrounding Gas

The environmental conditions around spinneret such as the surrounding air, the RH, any vacuum conditions, and the surrounding gas influence the fiber structure and morphology of the electro-spun membrane. Baumgarten [45] observed that in dry air (RH < 5%) spinning drops tended to dry out, spinning is possible for only a few minutes. In humid air (RH > 60%), the fibers do not dry properly and become tangled in the grounded screen. The breakdown voltage of the atmospheric gases influences the charge retaining capacity of the fibers. In helium, no spinning is possible as the gas undergoes electrical breakage at 2500 V. Fibers spun in Freon-12 gas have diameter 1.4–2.6 times greater than those spun in the air, possibly due to the higher breakdown voltage of Freon-12 gas than air, which allows fibers to retain electric charges for a long time. Megelski et al. [19] investigated the pore characteristics of polystyrene fibers at varied RH and evaluated the importance of the phase separation mechanisms in explaining the pre-formation of electrospun fibers.

2.4 Applications of Electrospinning

In the modern technological era, electrospun membranes conquer the technological fields owing to their higher performance. The non-woven membranes produced by the electrospinning technique attained attraction due to its higher porosity, the higher surface to volume ratio, smaller diameter, etc. Other than single polymers, blended as well as composite polymers were also introduced for the fabrication of fibrous membranes. It provides a variety of applications in textile industries, the medical field, sensors, tissue engineering, scaffolds, energy storage devices, etc. the major highlights for the demand of such technology is the ability to produce fibers from micro to nanometer size. The global interest in this file is increasing day by day. Though nanofibers can be prepared by hydrothermal synthesis vapor phase techniques, etc. these methods were handicapped by high cost, low yield, complexity in the preparation process, etc. A schematic diagram of the application of the electrospun membrane is given in Fig. 2.17.

2.4.1 Energy Storage Devices

Electrospun membranes play a vital role in energy storage devices mainly includes batteries, supercapacitors, sensors, piezoelectric devices, flexible electronics, etc. This versatile technique is attracted in this area due to its high surface-to-volume ratio, good flexibility, multiple functionalities, high length to diameter ratio, high porosity which leads to the faster conduction of salts and improves the conductivity of the material. Depending upon the necessity, the pore size can be adjusted by controlling the voltage, tip to collector distance, temperature, etc. There are many reports on

Fig. 2.17 Potential applications of electrospun fibers in different technological fields

lithium-based batteries since it increases the ionic conductivity of batteries, owing to the wide range of applied research is mainly focusing on this technique and the impact of electrospun fibers in different energy storage devices in terms of a number of publications per year is illustrated in Fig. 2.18 [159] and reports published on the electrospun lithium-ion and sodium-ion batteries in the period of 2010–2019 are shown in Fig. 2.19.

Electrospun materials were also fabricated in Dye-Sensitized Solar Cell (DSSC) in which a photoanode, counter electrodes, etc. photosensitizers plays a major role in which it absorbs photons and they dissociates into free electrons and holes. Efficiency increase with an increase in the absorption of electros and I-D fibers are of a better choice. For the easy and efficient fabrication of such fibers, electrospinning was found to be a better choice. Electrospun membranes also play an inevitable role in supercapacitors. Nanofibers based micro-supercapacitors are interesting technology providing higher efficiency with cycling stability and retention capacity. Electrospun carbon nanofibers and polymer-based materials offer higher specific capacitance compared to other techniques. Batteries consist of anodes, electrolytes, separators, and cathodes and electrospinning of such materials potentially improves the efficiency of batteries. The first commercialized cathode material was $LiCoO_4$ attained attention owing to its properties such as high specific energy density, high cyclic stability, low self-discharge, etc. but these materials were handicapped by slower state diffusion which can be rectified by electrospun nanofibrous membrane offering a fast rate of diffusion. In a similar way, a drastic change in the electrochemical properties was observed by the introduction of electrospun membranes. Carbon and

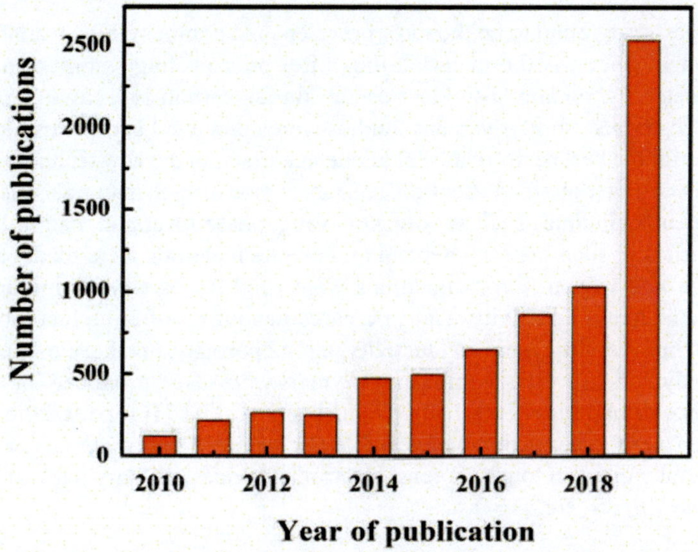

Fig. 2.18 Number of publications on electrospun fibers on different energy storage devices in terms of different year of publications in the period of 2010–2019

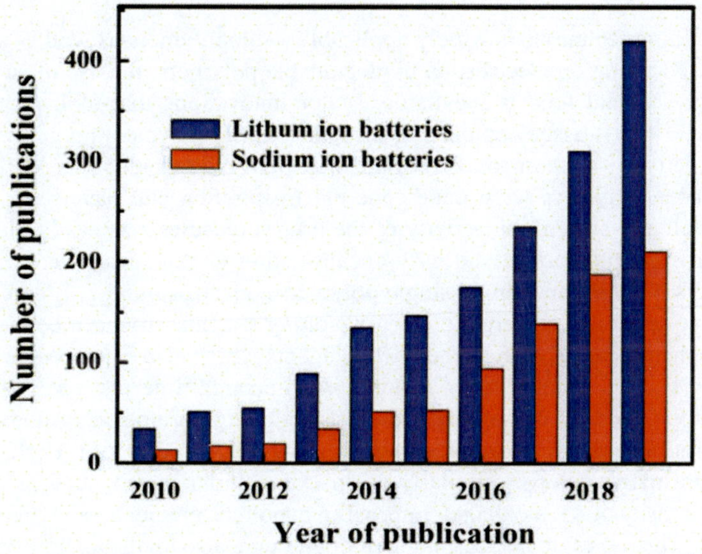

Fig. 2.19 Number of research articles published on electrospun fiber for energy applications in lithium ion batteries and sodium ion batteries published in the period of 2010–2019 (Scifinder data: keywords electrospinning + Li-ion batteries)

its allotropes were found to be the most prevalent and commercially available materials. Though Silicon is the material, they offer poor cycling stability. In contrast to these materials, doping si to electrospun fibrous carbon is a suitable candidate for anode materials. Similarly, many studies were conducted in electrospun Carbon Nanotubes for the best performance of anode materials. In the new technological era, polymer electrolytes attained attention, and fabrication of such materials was possible with different techniques such as solvent casting, phase inversion, ball milling, etc. The properties of such material depend on how the materials were made. Recently, electrospun non-woven fibrous materials were used for battery testing and found to produce higher conductivity. Thus, prepared nanofibers offer high columbic efficiency, cycling stability, ionic conductivity, higher porosity, uptake, etc. Electrospun blended polymers as well as composite polymers were also introduced for the better performance of batteries [160–162]. Chinnappan et al. [163] reported the effect of electrospun membranes in flexible/wearable electronics. These materials have a wide range of applications in touch screens, electronic devices, military garment devices, portable electronics, etc. [164].

2.4.2 Medicinal Field

Electrospinning technique is widely applicable to biopolymers as well as synthetic polymers. Focusing on electrospun biodegradable polymers, the use of such materials in the medical field is inevitable. Major innovations that took place in the medical field are (i) tissue engineering for scaffolds (ii) enzyme immobilization and (iii) drug delivery. Biocompatible polymers of both biodegradable as well as non-biodegradable polymers were undergone electrospinning and have been used. In the drug delivery system, bioactivity of the drug molecules was easily maintained and electrospun membranes aid easy encapsulation of therapeutic agents present in the fibers. Other than using a single polymer in electrospinning, Blending more than one polymer and undergone electrospinning was also found a better alternative for higher encapsulation of therapeutic agents which also enhances the number of binding sites in drug delivery systems. At the same time rate of drug, release can also controlled by controlling the electrospinning parameters. Another major field is tissue engineering for the manufacturing of Scaffolds [165]. Highly porous electrospun matrix behaves similarly as an extracellular matrix such as collagen. Scaffolds consist of such collagen improving nanofiber efficiency and distinct fiber alignment. This types of electrospun membranes care also applicable for the wound healing process. Electrospun polymers can also act as catalysts such as TiO_2, ZnO, etc. biodegradable polymers such as fibroin, chitosan, collagen, polyvinyl alcohol, etc. were used [152, 166, 167].

2.4.3 Textile Industry

In earlier times, electrospun nanofibers in textile industries were used for spinning yarns.. different types of non-woven membranes were fabricated for the use of everyday clothing, sports-type clothes, etc. the major aim of preparing such electrospun textiles include the production of heat resistant fabrics for the use of firefighters suits, fabricating anti-bacterial clothing suited for medical purpose, preparing water-resistant clothing's and suits, etc. [165, 168]. 3-D printing method was also fabricated in wearable clothing. Mold has been used in this clothing and electrospun 3-D printing was fabricated in this mold. The major challenges of this method are its durability, design, and fabrication processing time. Other than conventional colors, different other dyes were also used for obtaining colorful fibers in the clothing.

2.4.4 Environmental and Filtration

Environmental field mainly discusses two major factors i.e., water and air. Water purification and air purification are the two chief elements that research should concern about. In this contest, filters play a major platform for healthier and biomedical aspects of life. The major application of electrospun membranes infiltration was the air filtration process owing to the advantage of forming pores of nanometer scale. Air filtration thus improves its efficiency attributing its smaller fiber size with a higher pressure drop. This can be also used in water treatment bringing towards nanofiltration from macro filtration. The conventional method of using filters was high-efficiency particulate air filters (HEPAF). The quality factor of the electrospun membrane was very much higher than the conventional HEPAF [169, 170]. Liquid filtration helps to filter out proteins, viruses, emulsions, etc. these materials should have very little surface pore size (0.1 μm). Though electrospun technologies have been widely used in the environmental field these materials faced many challenges including the production of nanofibers at a larger scale, selection of polymeric matrix etc. [171, 172]. The electrospun non-woven membranes will be further explored in different fields. In the future, researchers will mainly focus on the large wide variety of applications that could enhance the efficiency of the materials.

2.5 Conclusion

Electrospinning attained its attention owing to its formation of fibrous membranes producing fibrous membranes, large surface area, etc. The tremendous progress in electrospinning of producing nanofibers revolutionalizes in the era of advanced materials. High voltage source is applied and ultrafine fibers will ejects which is collected from the grounded metallic targets. The nature of such electrospun fibers depends upon the type of polymer that has been chosen. Major factors that influence the electrospinning are the nature of polymer solution, properties of polymer solution such as surface tension, viscosity, conductivity, process parameters such as flow rate, the distance between tip to a collector, applied voltage, ambient parameters including temperature, humidity, etc. Due to this reason electrospinning technique expands in a wide variety of fields such as medical, energy storage, textile, environmental, etc. Among all these parameters type of polymer chosen is a superior parameter which is determined by the solvent that we have used for mixing, its mixing ratio, etc. Thus, by tailoring the parameters of the electrospun fibers a continuous controlled and reproducible fibers can be generated. The scientific and engineering world considers electrospinning as the most powerful tool for the high-performance applications with a broad range of functionalities.

References

1. Raghavan P, Manuel J, Zhao X et al (2011) Preparation and electrochemical characterization of gel polymer electrolyte based on electrospun polyacrylonitrile nonwoven membranes for lithium batteries. J Power Sources 196:6742–6749. https://doi.org/10.1016/j.jpowsour.2010.10.089
2. Luu YK, Kim K, Hsiao BS et al (2003) Development of a nanostructured DNA delivery scaffold via electrospinning of PLGA and PLA–PEG block copolymers. J Control Release 89:341–353
3. Subbiah T, Bhat GS, Tock RW et al (2004) Electrospinning of nanofibers. J Appl Polym Sci 96:557–569. https://doi.org/10.1002/app.21481
4. Journal I, Nanofibers E, Journal I et al (2017) Electrospinning drug-loaded poly (butylenes succinate-co-bytylene terephthalate) (PBST) with Acetylsalicylic Acid (Aspirin). Int J Electrospun Nanofibers Appl 1:1–6
5. Journal I, Nanofibers E, Journal I et al (2017) Preliminary investigation of electrospun collagen and polydioxanone for vascular tissue engineering applications. Int J Electrospun Nanofibers Appl 1:73–87
6. Welle A, Kro M, Do M et al (2007) Electrospun aliphatic polycarbonates as tailored tissue scaffold materials. Biomaterials 28:2211–2219. https://doi.org/10.1016/j.biomaterials.2007.01.024
7. Ye K, Kuang H, You Z et al (2019) Electrospun nanofibers for tissue engineering with drug loading and release. Pharmaceutics 11:182. https://doi.org/10.3390/pharmaceutics11040182
8. Bhattarai DP, Aguilar LE, Park CH, Kim CS (2018) A review on properties of natural and synthetic based electrospun fibrous materials for bone tissue engineering. Membranes 8(3):62. https://doi.org/10.3390/membranes8030062
9. Raghavan P, Choi JW, Ahn JH et al (2008) Novel electrospun poly(vinylidene fluoride-co-hexafluoropropylene)-in situ SiO_2 composite membrane-based polymer electrolyte for lithium batteries. J Power Sources 184:437–443. https://doi.org/10.1016/j.jpowsour.2008.03.027
10. Mondal K, Sharma A (2016) Recent advances in electrospun metal-oxide nanofiber based interfaces for electrochemical biosensing. RSC Adv 6:85495–85514. https://doi.org/10.1039/C6RA21477K
11. Ramakrishna S, Peng S (2012) Electrospun metal oxides for energy applications. https://doi.org/10.1007/978-4-431-54067-0
12. Nanofibers C (2006) Ceramic nanofibers and nanotubes. J Am Ceram Soc 89:1861–1869. https://doi.org/10.1111/j.1551-2916.2006.00989.x

13. Shabestari MM (2016) Fabrication of ceramic nanofibers and effect of calcination parameters on grain growth
14. Soc JAC, Whitesides GM, Grzybowski B (2002) Self-assembly at all scales. Science 295:2418–2421
15. Ma PX, Zhang R (1998) Synthetic nano-scale fibrous extracellular matrix. J Biomed Mater Res 46:60–72
16. Search H, Journals C, Contact A et al (1998) Drawing a single nanofibre over hundreds of microns. Europhys Lett 42(2):215–220
17. Bhardwaj N, Kundu SC (2010) Electrospinning: a fascinating fiber fabrication technique. Biotechnol Adv 28:325–347. https://doi.org/10.1016/j.biotechadv.2010.01.004
18. Xue J, Xie J, Liu W, Xia Y (2017) Electrospun nanofibers: new concepts, materials, and applications. Acc Chem Res 50:1976–1987. https://doi.org/10.1021/acs.accounts.7b00218
19. Megelski S, Stephens JS, Bruce Chase D, Rabolt JF (2002) Micro- and nanostructured surface morphology on electrospun polymer fibers. Macromolecules 35:8456–8466. https://doi.org/10.1021/ma020444a
20. Raghavan P, Lim DH, Ahn JH et al (2012) Electrospun polymer nanofibers: the booming cutting edge technology. React Funct Polym 72:915–930. https://doi.org/10.1016/j.reactfunctpolym.2012.08.018
21. Miao R, Liu B, Zhu Z et al (2008) PVDF-HFP-based porous polymer electrolyte membranes for lithium-ion batteries. J Power Sources 184:420–426. https://doi.org/10.1016/j.jpowsour.2008.03.045
22. Chew SY, Wen J, Yim EKF, Leong KW (2005) Sustained release of proteins from electrospun biodegradable fibers. Biomacromolecules 6:2017–2024. https://doi.org/10.1021/bm0501149
23. Cooley JF (1900) Improved methods of an apparatus for electrically separating the relatively volatile liquid component from the component of relatively fixed substances of composite fluids
24. Burton EF, Wiegand WB (2009) Effect of electricity on streams of water drops. Philos Mag J Sci 23:148–165
25. Zeleny N (1914) The electrical discharge from liquid points, and a hydrostatic method of measuring the electric intensity at their surfaces. Phys Rev 3:69
26. Hendricks CD Jr (1962) Charged droplet experiments. J Colloid Sci 17:249–259
27. Zeleny J (1917) Instability of electrified liquid surfaces. Phys Rev 10:1
28. Larrondo L, Manley RSTJ (1981) electrostatic fiber spinning from polymer melts. I. Experimental observations on fiber formation and properties. J Polym Sci Polym Phys 19:909–920
29. Rayleigh L (1881) On the equilibrium of liquid conducting masses charged with electricity. Philos Mag 14:184–186
30. Tucker N, Ph D, Stanger JJ et al (1995) The history of the science and technology of electrospinning from 1600 to 1995. J Eng Fiber Fabr 7:63–73. https://doi.org/10.1177/155892501200702S10
31. Li D, Xia Y (2004) Electrospinning of nanofibers: reinventing the wheel? Adv Mater 16:1151–1170. https://doi.org/10.1002/adma.200400719
32. Hagiwara K, Oji-machi OF (1929) Process For manufacturing artificial silk and other filaments by applying electric current, pp 1–2
33. Formhals A (1975) Process and apparatus for preparing artificial threads
34. Formhals A (1937) Production of artificial fibers, 1936–1938
35. Formhals A (1938) Artificial fiber constructions, vol 2, p 333
36. York N (1941) Method and apparatus for producing artificial fibers. Field 2:742
37. Hals AF (1939) Method of producing artificial fifers, vol 2, p 415
38. Schrei R (1940) Artificial thread and method of producing same
39. Formhals A (1943) Production of artificial fibers from fiber forming liquids, pp 2–5
40. Norton CL (1936) Method of and apparatus for producing fibrous or filamentary material, pp 1–4
41. Nicolai Albertowich Fuchs: the pioneer of aerosol science

42. Electrospray R, Luis F (1971) The production of charged monodisperse fuel droplets by electrical dispersion The production of charged monodisperse fuel droplets by electrical dispersion
43. Drozin VG, Department (1954) The electrical dispersion of liquids as aerosols
44. Melcher JR, Warren EP (1971) Electrohydrodynamics of a current-carrying semi-insulating jet. J Fluid Mech 47:127–143
45. Baumgarten PK (1971) Electrostatic spinning of acrylic microfibers. J Colloid Interface Sci 36:71–79
46. Larrondo L (1981) Electrostatic fiber spinning from polymer melts. 111. Electrostatic deformation of a pendant drop of polymer melt. J Polym Sci B Polym Phys 19:933–940
47. Process E, Doshi J, Reneker DH (1995) Electrospinning process and applications of electrospun fibers. J Electrostat 35:151–160
48. Tomczak N, Gu S, Han M (2006) Single light emitters in electrospun polymer nanofibers: effect of local confinement on radiative decay. Eur Polym J 42:2205–2210. https://doi.org/10.1016/j.eurpolymj.2006.06.017
49. Reneker DH, Chun I (1996) Nanometre diameter fibres of polymer, produced by electrospinning. Nanotechnology 7:216–223
50. Raghavan P, Zhao X, Kim J et al (2008) Ionic conductivity and electrochemical properties of nanocomposite polymer electrolytes based on electrospun poly(vinylidene fluoride-co-hexafluoropropylene) with nano-sized ceramic fillers. Electrochem Acta 54:228–234. https://doi.org/10.1016/j.electacta.2008.08.007
51. Cooley JF (1902) Apparatus for electrically dispersing fluids, p 7
52. Taylor G (1964) Disintegration of water drops in an electric field. Proc R Soc Lond A 280:383–397. https://doi.org/10.1098/rspa.1964.0151
53. Trapp PEW (1977) Fibrillar product of electrostatically spun organic material, p 331
54. Annis D et al (1978) An elastomeric vascular prosthesis. Trans Am Soc Artif Intern Organs 24:209–214
55. Wang Y, Serrano S, Santiago-Aviles JJ (2002) Electrostatic synthesis and characterization of $Pb(Zr_xTi_{1-x})O_3$ micro/nano-fibers. Mater Res Soc Symp Proc 702:359
56. Sun BZ, Zussman E, Yarin AL et al (2003) Compound core-shell polymer nanofibers by co-electrospinning. Adv Mater 15:1929–1932. https://doi.org/10.1002/adma.200305136
57. Sanderson RD, Smit E, Bu U (2005) Continuous yarns from electrospun fibers. Polymer 46:2419–2423. https://doi.org/10.1016/j.polymer.2005.02.002
58. Teo WE, Ramakrishna S (2006) A review on electrospinning design and nanofibre assemblies. Nanotechnology 17(14):R89–R106. https://doi.org/10.1088/0957-4484/17/14/R01
59. Tillman BW, Yazdani SK, Jin S et al (2009) The in vivo stability of electrospun polycaprolactone-collagen scaffolds in vascular reconstruction. Biomaterials 30:583–588. https://doi.org/10.1016/j.biomaterials.2008.10.006
60. Online VA (2013) Electrospun composite nanofiber-based transmucosal patch for anti-diabetic drug delivery. J Mater Chem B 1:3410–3418. https://doi.org/10.1039/c3tb20487a
61. Tucker N (2015) High-tech nanofibers could help nutrients in food hit the spot. Science Daily
62. Myung N (2018) Nanofiber filter improves drinking water quality. Applied Sciences from Technology Networks
63. Sharma P, Vyas T, Kewlani J (2019) IIT Delhi alumni's startup Nanoclean is now working on mass producing nanofibres in India. Yoursstory
64. Zong X, Kim K, Fang D et al (2002) Structure and process relationship of electrospun bioabsorbable nanofiber membranes. Polymer 43:4403–4412
65. Shin YM, Hohman MM, Brenner MP, Rutledge GC (2001) Experimental characterization of electrospinning: the electrically forced jet and instabilities. Polymer 42:9955–9967
66. Lee KH, Kim HY, Bang HJ et al (2003) The change of bead morphology formed on electrospun polystyrene fibers. Polymer 44:4029–4034. https://doi.org/10.1016/S0032-3861(03)00345-8
67. Zuo W, Zhu M, Yang W et al (2005) Experimental study on relationship between jet instability and formation of beaded fibers during electrospinning. Polym Eng Sci 45(5):704–709. https://doi.org/10.1002/pen.20304

68. Jun Z, Hou H, Schaper A et al (2003) Poly-L-lactide nanofibers by electrospinning—influence of solution viscosity and electrical conductivity on fiber diameter and fiber morphology. e-Polymers 3:1–9
69. Lin T, Wang H, Wang H, Wang X The charge effect of cationic surfactants on the elimination of fibre beads in the electrospinning of polystyrene. Nanotechnology 15(9):1375–1381. https://doi.org/10.1088/0957-4484/15/9/044
70. Deitzel JM, Kleinmeyer J, Harris D, Tan NCB (2001) The effect of processing variables on the morphology of electrospun nanofibers and textiles. Polymer 42:261–272
71. Park J, Ito T, Kim K et al (2010) Electrospun poly (vinyl alcohol) nanofibers: effects of degree of hydrolysis and enhanced water stability. Polym J 42:273–276. https://doi.org/10.1038/pj.2009.340
72. Jiang H, Fang D, Hsiao BS et al (2004) Optimization and characterization of dextran membranes prepared by electrospinning. Biomacromolecules 5:326–333
73. Kim B, Park H, Lee S, Sigmund WM (2005) Poly (acrylic acid) nanofibers by electrospinning. Mater Lett 59:829–832. https://doi.org/10.1016/j.matlet.2004.11.032
74. Mit-uppatham C, Nithitanakul M, Supaphol P (2004) Ultrafine electrospun polyamide-6 fibers: effect of solution conditions on morphology and average fiber diameter. Macromol Chem Phys 6:2327–2338. https://doi.org/10.1002/macp.200400225
75. Gupta P, Elkins C, Long TE, Wilkes GL (2005) Electrospinning of linear homopolymers of poly(methyl methacrylate): exploring relationships between fiber formation, viscosity, molecular weight and concentration in a good solvent. Polymer 46:4799–4810. https://doi.org/10.1016/j.polymer.2005.04.021
76. Symp M (2005) Morphological characterization of electrospun nano-fibrous membranes of biodegradable poly(L-lactide) and poly(lactide-co-glycolide). Macromol Symp 224(1):145–154. https://doi.org/10.1002/masy.200550613
77. Keun W, Ho J, Seung T, Ho W (2004) The effects of solution properties and polyelectrolyte on electrospinning of ultrafine poly (ethylene oxide) fibers. Polymer 45:2959–2966. https://doi.org/10.1016/j.polymer.2004.03.006
78. Fong H, Chun I, Reneker DH (1999) Beaded nanofibers formed during electrospinning. Polymer 40:4585–4592
79. Koski A, Yim K, Shivkumar S (2004) Effect of molecular weight on fibrous PVA produced by electrospinning. Mater Lett 58:493–497. https://doi.org/10.1016/S0167-577X(03)00532-9
80. Lee JS, Choi KH, Ghim H Do et al (2004) Role of molecular weight of atactic poly (vinyl alcohol) (PVA) in the structure and properties of PVA nanofabric prepared by electrospinning. J Appl Polym Sci. https://doi.org/10.1002/app.20602
81. Zhang C, Yuan X (2005) Study on morphology of electrospun poly (vinyl alcohol) mats. Eur Polym J 41:423–432. https://doi.org/10.1016/j.eurpolymj.2004.10.027
82. Ding BIN, Kim H, Lee S et al (2002) Preparation and characterization of a nanoscale poly (vinyl alcohol) fiber aggregate produced by an electrospinning method. J Polym Sci Part B: Polym Phys 40:1261–1268. https://doi.org/10.1002/polb.10191
83. Jarusuwannapoom T, Hongrojjanawiwat W (2005) Effect of solvents on electro-spinnability of polystyrene solutions and morphological appearance of resulting electrospun polystyrene fibers. Eur Polym J 41:409–421. https://doi.org/10.1016/j.eurpolymj.2004.10.010
84. Seok C, Hyun D, Don K et al (2005) Characterization of gelatin nanofiber prepared from gelatin–formic acid solution. Polymer 46:5094–5102. https://doi.org/10.1016/j.polymer.2005.04.040
85. Sukigara S, Gandhi M, Ayutsede J et al (2003) Regeneration of *Bombyx mori* silk by electrospinning—Part 1: Processing parameters and geometric properties. Polymer 44:5721–5727. https://doi.org/10.1016/S0032-3861(03)00532-9
86. Cha DI, Kim HY, Lee KH et al (2004) Electrospun nonwovens of shape-memory polyurethane block copolymers. Appl Polym 96:460–465. https://doi.org/10.1002/app.21467
87. Demir MM, Yilgor I, Yilgor E, Erman B (2002) Electrospinning of polyurethane fibers. Polymer 43:3303–3309

88. Chen H, Hsieh Y (2004) Ultrafine hydrogel fibers with dual temperature- and pH-responsive swelling behaviors. J Polym Sci Part A: Polym Chem 42:6331–6339. https://doi.org/10.1002/pola.20461
89. Kenawy E, Layman JM, Watkins JR et al (2003) Electrospinning of poly (ethylene-co-vinyl alcohol) fibers. Biomaterials 24:907–913
90. Buchko CJ, Chen LC, Shen Y, Martin DC (1999) Processing and microstructural characterization of porous biocompatible protein polymer thin films. Polymer 40:7397–7407
91. Zhao Z, Li J, Yuan X et al (2004) Preparation and properties of electrospun poly (vinylidene fluoride) membranes. J Appl Polym Sci 97:466–474. https://doi.org/10.1002/app.21762
92. Duan B, Dong C, Yuan X (2012) Electrospinning of chitosan solutions in acetic acid with poly(ethylene oxide). J Biomater Sci Polym Ed 15:797–811
93. Mckee MG, Wilkes GL, Colby RH, Long TE (2004) Correlations of solution rheology with electrospun fiber formation of linear and branched polyesters. Macromolecules 37:1760–1767. https://doi.org/10.1021/ma035689h
94. Theron SA, Zussman E, Yarin AL (2017) Experimental investigation of the governing parameters in the electrospinning of polymer solutions. Polymer 45:2017–2030. https://doi.org/10.1016/j.polymer.2004.01.024
95. Jung HR, Ju DH, Lee WJ et al (2009) Electrospun hydrophilic fumed silica/polyacrylonitrile nanofiber-based composite electrolyte membranes. Electrochim Acta 54:3630–3637. https://doi.org/10.1016/j.electacta.2009.01.039
96. Hohman MM, Shin M, Rutledge G et al (2001) Electrospinning and electrically forced jets. II. Applications. Phys Fluids 13:2221. https://doi.org/10.1063/1.1384013
97. Kim G, Kim W, Kim G, Kim W (2006) Formation of oriented nanofibers using electrospinning Formation of oriented nanofibers using electrospinning. Appl Phys Lett 233101:31–34. https://doi.org/10.1063/1.2210972
98. Liu H, Hsieh Y (2002) Ultrafine fibrous cellulose membranes from electrospinning of cellulose acetate. J Polym Sci Part B: Polym Phys 40:2119–2129. https://doi.org/10.1002/polb.10261
99. Jun Y, Yong H, Hyung K et al (2003) Transport properties of electrospun nylon 6 nonwoven mats. Eur Polym J 39:1883–1889. https://doi.org/10.1016/S0014-3057(03)00096-X
100. Reneker DH, Yarin AL (2008) Electrospinning jets and polymer nanofibers. Polymer 49:2387–2425. https://doi.org/10.1016/j.polymer.2008.02.002
101. De Schoenmaker B, Goethals A, Van Der Schueren L et al (2012) Polyamide 6.9 nanofibres electrospun under steady state conditions from a solvent/non-solvent solution. J Mater Sci 47:4118–4126. https://doi.org/10.1007/s10853-012-6266-9
102. Brune DA, Bicerano J (2002) Micromechanics of nanocomposites: comparison of tensile and compressive elastic moduli, and prediction of effects of incomplete exfoliation and imperfect alignment on modulus. Polymer 43:369–387
103. Taylor P, Drew C, Wang X et al (2013) The effect of viscosity and filler on electrospun fiber morphology. J Macromol Sci Part A: Pure Appl Chem 40:1415–1422. https://doi.org/10.1081/MA-120025320
104. Laudenslager MJ, Scheffler RH, Sigmund WM (2010) Electrospun materials for energy harvesting, conversion, and storage: a review. Pure Appl Chem 82:2137–2156. https://doi.org/10.1351/PAC-CON-09-11-49
105. Mckee MG (2014) Phospholipid nonwoven electrospun membranes. Science 311:353–355. https://doi.org/10.1126/science.1119790
106. Geng X, Kwon O, Jang J (2005) Electrospinning of chitosan dissolved in concentrated acetic acid solution. Biomaterials 26:5427–5432. https://doi.org/10.1016/j.biomaterials.2005.01.066
107. Chowdhury M, Stylios G (2010) Effect of experimental parameters on the morphology of electrospun nylon 6 fibres. Int J Basic Appl Sci 10:70–78
108. Yuan X, Zhang Y, Dong C, Sheng J (2004) Morphology of ultrafine polysulfone fibers prepared by electrospinning. Polym Int Polym Int 53:1704–1710. https://doi.org/10.1002/pi.1538
109. Larrondo L, Manley RSTJ (1981) Electrostatic fiber spinning from polymer melts. 11. Examination of the flow field in an electrically driven jet. J Polym Sci Polym Phys Ed 19:921–932

110. Yördem OS, Papila M, Menceloğlu YZ (2008) Effects of electrospinning parameters on polyacrylonitrile nanofiber diameter: an investigation by response surface methodology. Mater Des 29:34–44. https://doi.org/10.1016/j.matdes.2006.12.013
111. Haghi AK, Akbari M (2007) Trends in electrospinning of natural nanofibers. Phys Status Solidi Appl Res 1834:1830–1834. https://doi.org/10.1002/pssa.200675301
112. Katti DS, Robinson KW, Ko FK, Laurencin CT (2004) Bioresorbable nanofiber-based systems for wound healing and drug delivery: optimization of fabrication parameters. J Biomed Mater Res. https://doi.org/10.1002/jbm.b.30041
113. Mo XM, Xu CY, Kotaki M, Ramakrishna S (2004) Electrospun P (LLA-CL) nanofiber: a biomimetic extracellular matrix for smooth muscle cell and endothelial cell proliferation. Biomaterials 25:1883–1890. https://doi.org/10.1016/j.biomaterials.2003.08.042
114. Bb S (2004) Biomedical nanoscience: electrospinning basic concepts, applications, and classroom demonstration. Mater Res Soc Symp Proc 827:1–12
115. Yarin AL, Koombhongse S, Reneker DH (2005) Taylor cone and jetting from liquid droplets in electrospinning of nanofibers Taylor cone and jetting from liquid droplets in electrospinning of nanofibers. J Appl Phys 90:4836–4846. https://doi.org/10.1063/1.1408260
116. Yarin AL, Koombhongse S, Reneker DH (2001) Bending instability in electrospinning of nanofibers. J Appl Phys 89:3018–3026. https://doi.org/10.1063/1.1333035
117. Haider S, Al-masry WA, Al-zeghayer Y et al (2011) Fabrication of chitosan nanofibers membrane via electrospinning. TechConnect Briefs 1:810–812
118. Homayoni H, Abdolkarim S, Ravandi H, Valizadeh M (2009) Electrospinning of chitosan nanofibers: processing optimization. Carbohydr Polym 77:656–661. https://doi.org/10.1016/j.carbpol.2009.02.008
119. Norris ID, Shaker MM, Ko FK, Macdiarmid AG (2000) Electrostatic fabrication of ultrafine conducting fibers: polyaniline r polyethylene oxide blends. Synth Met 114:109–114
120. Kim K, Jeong L, Park H et al (2005) Biological efficacy of silk fibroin nanofiber membranes for guided bone regeneration. J Biotechnol 120:327–339. https://doi.org/10.1016/j.jbiotec.2005.06.033
121. Jalili R, Hosseini SA, Morshed M (2005) The effects of operating parameters on the morphology of electrospun polyacrilonitrile nanofibres. Iran Polym J 14:1074–1081
122. Pham QP, Sharma U, Mikos AG (2006) Electrospun poly(E-caprolactone) microfiber and multilayer nanofiber/ microfiber scaffolds: characterization of scaffolds and measurement of cellular infiltration. Biomacromolecules 7:2796–2805
123. Um IC, Fang D, Hsiao BS et al (2004) Electro-spinning and electro-blowing of hyaluronic acid. Biomacromolecules 5:1428–1436
124. Sundaray B, Subramanian V, Natarajan TS et al (2004) Electrospinning of continuous aligned polymer fibers. Appl Phys Lett 1222:10–13. https://doi.org/10.1063/1.1647685
125. Li BD, Wang Y, Xia Y (2004) Electrospinning nanofibers as uniaxially aligned arrays and layer-by-layer stacked films. Adv Mater 16:361–366. https://doi.org/10.1002/adma.200306226
126. Xu CY, Inai R, Kotaki M, Ramakrishna S (2004) Aligned biodegradable nanofibrous structure: a potential scaffold for blood vessel engineering. Biomaterials 25:877–886. https://doi.org/10.1016/S0142-9612(03)00593-3
127. Ki CS, Kim JW, Hyun JH et al (2007) Electrospun three-dimensional silk fibroin nanofibrous scaffold. J Appl Polym Sci. https://doi.org/10.1002/app.26914
128. Kongkhlang T, Tashiro K, Kotaki M, Chirachanchai S (2008) Electrospinning as a new technique to control the crystal morphology and molecular orientation of polyoxymethylene nanofibers. J Am Chem Soc 130:15460–15466. https://doi.org/10.1021/ja804185s
129. Search H, Journals C, Contact A et al (2001) Electrostatic field-assisted alignment of electrospun nanofibres. Nanotechnology 12(3):384
130. Katta P, Alessandro M, Ramsier RD, Chase GG (2004) Continuous electrospinning of aligned polymer nanofibers onto a wire drum collector. Nano Lett 4:2215–2218
131. Huang ZM, Zhang YZ, Kotaki M, Ramakrishna S (2003) A review on polymer nanofibers by electrospinning and their applications in nanocomposites. Compos Sci Technol 63:2223–2253. https://doi.org/10.1016/S0266-3538(03)00178-7

132. Dersch R, Liu T, Schaper AK et al (2002) Electrospun nanofibers: internal structure and intrinsic. J Polym Sci Part A: Polym Chem 41:545–553
133. Mathew G, Hong JP, Rhee JM et al (2005) Preparation and characterization of properties of electrospun poly (butylene terephthalate) nanofibers filled with carbon nanotubes. Polym Test 24:712–717. https://doi.org/10.1016/j.polymertesting.2005.05.002
134. Fennessey SF, Farris RJ (2004) Fabrication of aligned and molecularly oriented electrospun polyacrylonitrile nanofibers and the mechanical behavior of their twisted yarns. Polymer 45:4217–4225. https://doi.org/10.1016/j.polymer.2004.04.001
135. Search H, Journals C, Contact A et al (1878) Electrospun fibre bundle made of aligned nanofibres over two fixed points. Nanotechnology 16:1878. https://doi.org/10.1088/0957-4484/16/9/077
136. Wong S, Baji A, Leng S (2008) Effect of fiber diameter on tensile properties of electrospun. Polymer 49:4713–4722. https://doi.org/10.1016/j.polymer.2008.08.022
137. Fridrikh SV, Yu JH, Brenner MP, Rutledge GC (2003) Controlling the fiber diameter during electrospinning. Phys Rev Lett 90:144502. https://doi.org/10.1103/PhysRevLett.90.144502
138. Xian J, Ling L, Liau L et al (2017) Electrospun collagen nanofibers and their applications in skin tissue engineering. Tissue Eng Regen Med. https://doi.org/10.1007/s13770-017-0075-9
139. Deitzel JM, Kleinmeyer JD, Hirvonen JK, Tan NCB (2001) Controlled deposition of electrospun poly (ethylene oxide) fibers. Polymer (Guildf) 42:8163–8170
140. Zhong J, Zhang H, Yan J, Gong X (2015) Effect of nanofiber orientation of electrospun nanofibrous scaffolds on cell growth and elastin expression of muscle cells. Colloids Surf B Biointerfaces 136:772–778. https://doi.org/10.1016/j.colsurfb.2015.10.017
141. Mirjalili M, Zohoori S (2016) Review for application of electrospinning and electrospun nanofibers technology in textile industry. J Nanostruct Chem. https://doi.org/10.1007/s40097-016-0189-y
142. Batnyam O, Suye SI, Fujita S (2017) Direct cryopreservation of adherent cells on an elastic nanofiber sheet featuring a low glass-transition temperature. RSC Adv 7:51264–51271. https://doi.org/10.1039/c7ra10604a
143. Kim KW, Lee KH, Khil MS et al (2004) The effect of molecular weight and the linear velocity of drum surface on the properties of electrospun poly(ethylene terephthalate) nonwovens. Fibers Polym 5:122–127
144. Yee WA, Nguyen AC, Lee PS et al (2008) Stress-induced structural changes in electrospun polyvinylidene difluoride nanofibers collected using a modified rotating disk. Polymer (Guildf) 49:4196–4203. https://doi.org/10.1016/j.polymer.2008.07.032
145. Aik W, Chien A, See P et al (2008) Stress-induced structural changes in electrospun polyvinylidene difluoride nanofibers collected using a modified rotating disk. Polymer 49:4196–4203. https://doi.org/10.1016/j.polymer.2008.07.032
146. Liu J, Kumar S (2005) Microscopic polymer cups by electrospinning. Polymer (Guildf) 46:3211–3214. https://doi.org/10.1016/j.polymer.2004.11.116
147. Wannatong L, Sirivat A, Supaphol P (2004) Effects of solvents on electrospun polymeric fibers: preliminary study on polystyrene. Polym Int 53:1851–1859. https://doi.org/10.1002/pi.1599
148. Casper CL, Stephens JS, Tassi NG et al (2004) Controlling surface morphology of electrospun polystyrene fibers: effect of humidity and molecular weight in the electrospinning process. Macromolecules 37:573–578
149. Jeong L, Park WH (2014) Preparation and characterization of gelatin nanofibers containing silver nanoparticles. Int J Mol Sci 15:6857–6879. https://doi.org/10.3390/ijms15046857
150. Kim HS, Kim K, Chin I (2005) Morphological characterization of electrospun nanofibrous membranes of biodegradable poly(L-lactic acid) and poly(lactide-co-glycolide). Macromolecules 224:145–154. https://doi.org/10.1002/masy.200550613
151. Huang L, Nagapudi K, Apkarian RP, Chaikof EL (2001) Engineered collagen–PEO nano fibers and fabrics. J Biomater Sci Polym Ed 12:979–993
152. Chew S, Wen Y, Dzenis Y, Leong K (2006) The role of electrospinning in the emerging field of nanomedicine. Curr Pharm Des 12:4751–4770. https://doi.org/10.2174/138161206779026326

153. Zhao Z, Li J, Yuan X et al (2005) Preparation and properties of electrospun poly(vinylidene fluoride) membranes. J Appl Polym Sci 97:466–474. https://doi.org/10.1002/app.21762
154. Theron SA, Yarin AL, Zussman E, Kroll E (2005) Multiple jets in electrospinning: experiment and modeling. Polymer 46:2889–2899. https://doi.org/10.1016/j.polymer.2005.01.054
155. Li D, Xia Y (2004) Direct fabrication of composite and ceramic hollow nanofibers by electrospinning. Nano Lett 4(5):933–938
156. Kidoaki S, Kwon IK, Matsuda T (2005) Mesoscopic spatial designs of nano- and microfiber meshes for tissue-engineering matrix and scaffold based on newly devised multilayering and mixing electrospinning techniques. Biomaterials 26:37–46. https://doi.org/10.1016/j.biomaterials.2004.01.063
157. Ding B, Kimura E, Sato T et al (2004) Fabrication of blend biodegradable nanofibrous nonwoven mats via multi-jet electrospinning. Polymer 45:1895–1902. https://doi.org/10.1016/j.polymer.2004.01.026
158. Dong B, Arnoult O, Smith ME, Wnek GE Electrospinning of collagen nanofiber scaffolds from benign solvents. Macromol Rapid Commun 30:539–542. https://doi.org/10.1002/marc.200800634
159. Liu Q, Zhu J, Zhang L, Qiu Y (2018) Recent advances in energy materials by electrospinning. Renew Sustain Energy Rev 81:1825–1858. https://doi.org/10.1016/j.rser.2017.05.281
160. Dong Z, Kennedy SJ, Wu Y (2011) Electrospinning materials for energy-related applications and devices. J Power Sources 196:4886–4904. https://doi.org/10.1016/j.jpowsour.2011.01.090
161. Wang H, Yuan S, Ma D et al (2015) Environmental science electrospun materials for lithium and sodium rechargeable batteries: from structure evolution to electrochemical performance. Energy Environ Sci 8:1660–1681. https://doi.org/10.1039/C4EE03912B
162. Sun G, Sun L, Xie H, Liu J (2016) Electrospinning of nanofibers for energy applications. Nanomaterials 6:129. https://doi.org/10.3390/nano6070129
163. Chinnappan A, Baskar C, Baskar S et al (2017) An overview of electrospun nanofibers and their application in energy storage, sensors and wearable/flexible electronics. J Mater Chem C 5:12657–12673. https://doi.org/10.1039/c7tc03058d
164. Yang Y, Jia Z, Li Q, Guan Z (2006) Electrospinning and its applications. Gaodianya Jishu/High Volt Eng 32:91–95. https://doi.org/10.1088/2043-6262/1/4/043002
165. Miśkiewicz P (2018) Nanotechnology in textile industry. WSN 100:74–85
166. Manea LR, Popa A, Bertea AP (2017) Medical applications of functional electrospun nanofibers—a review. Key Eng Mater 752:132–138. https://doi.org/10.4028/www.scientific.net/KEM.752.132
167. Khan N (2012) Applications of electrospun nanofibers in the biomedical field. SURG J 5:63–73
168. Fabrics from electrospinning (2017) Electrospin tech
169. Bjorge D, Daels N, De Vrieze S et al (2009) Performance assessment of electrospun nano fibers for filter applications. Desalination 249:942–948. https://doi.org/10.1016/j.desal.2009.06.064
170. Graham K, Ouyang M, Raether T et al (2002) Polymeric nanofibers in air filtration applications. In: Fifteenth annual technical conference & expo of the American Filtration & Separations Society
171. Thavasi V, Singh G, Ramakrishna S (2008) Electrospun nanofibers in energy and environmental applications. Energy Environ Sci 1:205–221. https://doi.org/10.1039/b809074m
172. Yoon K, Hsiao BS, Chu B (2008) Functional nanofibers for environmental applications. J Mater Chem 18:5326–5334. https://doi.org/10.1039/b804128h

Chapter 3
Electrospun Polyvinylidene Difluoride Membranes for High-Performance Application in Lithium Ion Batteries

Akhila Das, Neethu T. M. Balakrishnan, Jarin D. Joyner, Nikhil Medhavi, Jou-Hyeon Ahn, M. J. Jabeen Fatima, and Raghavan Prasanth

3.1 Introduction

Lithium ion batteries (LIBs) attained a greater demand in the present technological era. Lead-acid batteries, nickel-cadmium batteries (Ni-Cd), nickel-metal hydride batteries (Ni-MH), etc., are some of the batteries that have been used for commercial application. But these batteries could not attain the demanding efficiency of the prevailing lithium ion batteries owing to its high energy density, cycling stability, shelf life, etc. Hence, LIBs are in the lime light from 1991 when the Sony commercialized it. This fast-developing lithium ion technology, almost replaced all other battery technologies and received the Nobel prize this year (2019), which is shared by John B. Goodenough, M. Stanley Whittingham, Akira Yoshino and Draper prize in 2014 shared by John Goodenough, Rachid Yazami, Akira Yoshino each made substantial contributions to its development. An era of electric vehicles is pronounced since LIBs are prominent and had wide application in consumer electronics. As a result of these, many countries are readying its roadmap for the transition from gasoline to zero emission electric mobility by exploring the opportunities in its varieties in LIBs. Batteries are the heart of electric vehicles, and hence the safety of LIBs is

A. Das · N. T. M. Balakrishnan · N. Medhavi · M. J. Jabeen Fatima (✉) · R. Prasanth (✉)
Department of Polymer Science and Rubber Technology (PSRT), Cochin University of Science and Technology (CUSAT), Cochin 682022, India
e-mail: jabeen@cusat.ac.in

R. Prasanth
e-mail: prasanth@cusat.ac.in

J. D. Joyner
Department of Materials Science and Nano Engineering, Rice University, 6100 Main Street, MS 364, Houston, TX 77005, USA

J.-H. Ahn · R. Prasanth
Department of Materials Engineering and Convergence Technology, Gyeongsang National University, 501 Jinju-Daero, Jinju 52828, Republic of Korea

N. T. M. Balakrishnan and R. Prasanth (eds.), *Electrospinning for Advanced Energy Storage Applications*, Materials Horizons: From Nature to Nanomaterials,
https://doi.org/10.1007/978-981-15-8844-0_3

a matter of discussion. LIBs are still expensive and the main concern of battery is its thermal runway. At high functioning temperature, these batteries would have negative impact on the cycle life and its electrochemical performance. The gases are produced due to the electrochemical oxidation of the electrolytes paving the way for the bulging of batteries at high temperatures. Thus, electrolytes have a major role in the electrochemical performance and safety of batteries.

Performance of conventional liquid and solid electrolytes in LIBs was hindered by its safety issues and lower ionic conductivity, respectively. Fabrication of an efficient polymer electrolyte is inevitable since polymer electrolytes can solve the boundless requirement for the energy density of LIBs. In addition to this, polymer electrolytes provide good dimensional stability, chemical inertness, flexibility and safety. These types of electrolytes are thermally stable at high temperatures and prevent dendrite formation, hence polymer electrolytes are most suitable electrolytes for the fabrication of thermally and dimensionally stable LIBs. However, for the further improvement in ionic conductivity and cycle life, polymer electrolytes are suitable matrix for the development of advanced LIBs, which could power the electric vehicles.

Apart from the different synthesis of polymer electrolyte membrane, such as solution casting [1–3], phase inversion [4–8], ball milling [9–11], electrospinning is the advanced technique that improves the battery performance. Earlier this technique was reported for in medical fields [12], textile industries [13], environmental [14], filtration [15] applications, etc. The name ‘electrospinning’ was derived from electrostatic spraying and the first patent was published in the field of textile industry. Later on, the use for electrospun membranes and fibres are successfully demonstrated in various advanced applications and it touches in most of the emerging fields such as electronic, medical and environmental applications. The major advantage of this technique is easy to prepare fibrous membranes having high porosity, surface area, aspect ratio, etc. Electrospun nanofibres are not only used as electrodes but also applicable for the preparation of efficient polymer electrolytes. There are many reports on the electrospun host polymer such as poly ethylene oxide (PEO) [16, 17], polyacrylonitrile (PAN) [18–20], polyaniline (PANI) [14, 21, 22], polyvinylidene difluoride (PVdF), copolymer of PVdF [23–26] etc. PEO is one of the promising candidates for all solid-state batteries but shows poor conductivity due to the crystalline nature of PEO. Simultaneously PAN shows good electrochemical performance due the presence of –CN group in its molecular structure, which helps to dissociate more lithium salts, and this polymer matrix was handicapped not only by the brittleness of membrane but also by its cost [27]. Hence, materials PVdF came into lime light due to its high dielectric constant (8.4), thermal stability (<300 °C), electrochemical stability (~5 V), wide voltage range (1.2–5 V), etc. There are many applications in fibrous PVdF in medicine [28], piezoelectric field [29] and acting as air filters [30] etc., other than its application in batteries. There are many reports on PVdF-based polymer electrolytes for the better battery performance [31]. But these materials could not achieve the effective electrochemical stability having the break down voltage. At the same time, electrospun PVdF-based materials provide high porosity, interconnected pore structures, high surface area, etc. Detailed investigation of the electrospun fibrous anodes, separators cum electrolytes and cathodes by choosing the advantages of 1-D materials was also reported [9–11].

3.1.1 Polyvinyledene Difluoride Polymer Gel Electrolytes

Polyvinyledene difluoride (PVdF) is a non-reactive thermoplastic fluoropolymer which is formed by replacing two hydrogen atom with fluorine atoms $[–CH_2CF_2–]_n$. It possesses good mechanical strength and is chemical-resistant polymer. It possesses high dielectric constant (8.4) offering higher dissociation of Li ions owing to the increase in the charge carriers. PVdF possesses a melting temperature between 160 and 170 °C and the presence of electron withdrawing (–CF–) group enhances the anodic stability of batteries. PVdF attained its attention in ferroelectric, piezoelectric and pyroelectricity owing to its properties such as possessibility, lightweight and flexibility. PVdF can easily form a homogeneous solution when it is dissolved in dimethyl carbonate (DMC), ethylene carbonate EC, diethyl carbonate (DEC), lithium salts, etc., in an appropriate proportions. Thermal stability is also an important factor for the development of batteries, and PVdF possesses high thermal stability (greater than 300 °C). It provides wettability in all organic solvents.

3.1.1.1 Polymorphs of Polyvinyledene Difluoride

Polyvinyledene difluoride exists in 5 crystalline polymorphs such as α, β, γ, δ and ε (Fig. 3.1). For the extensive application, α, β and γ have been used in which α-phase of PVdF will easily obtain. There are possibilities to convert from one phase to another phase. β-phase exists as zig-zag structure attributing higher dielectric constant. Among these polymorphs, β-phase provides maximum electrolytic uptake (<400%), good electrical and thermal properties. There are different techniques available for the conversion of one phase from other phases; solvent casting [32], phase transition methods [33], spin coating [34], electrospinning [35], etc., are some of the popular techniques. By the electrospinning technique, β-phases can be easily synthesized. This technique is relatively straight forward, simple and easier. By controlling the parameters of electrospinning such as voltage, temperature, tip-to-collector distance and feeding rate. β-phase of PVdF can be easily synthesized. For temperature below 70 °C, β-phase is prominent, whereas for temperature above 110 °C, α-phase is prominent and γ-phase is only observed during melting (T_m of PVdF is 160–170 °C).

Capillary tube, conducting substrate and high voltage supply are the major components in electrospinning. Electrospun PVdF was used as separators as well as electrolytes in LIBs. Definite amount of PVdF dissolved in organic solvent at room temperature and this solution is loaded in the syringe. On applying high voltage, the electric field is developed between the sample collector and the drum. At critical

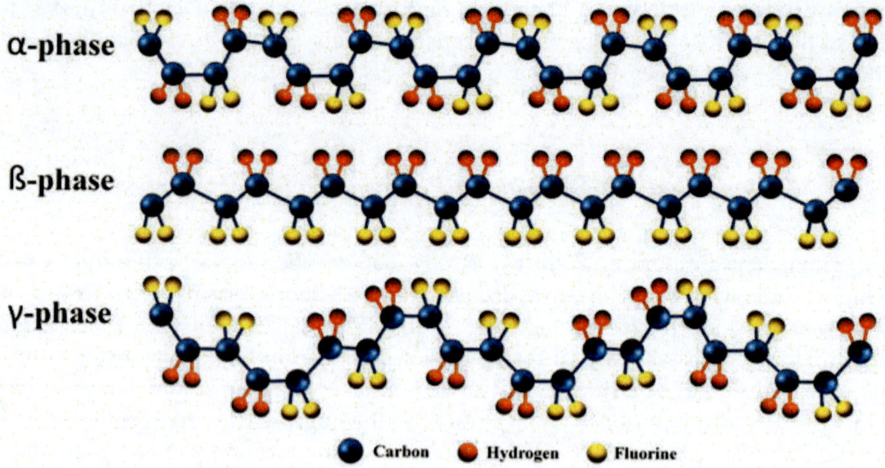

Fig. 3.1 Schematic representation of different phases of PVdF of α-, β- and γ-phases, respectively

voltage, the surface tension of the solution was suppressed by the electrostatic force leading to the initiation of the jet from the tailor cone tip to the negative current collector. The volatile solvents evaporate during its fly and thus form solid fibres. β-phase fibre was formed after electrospinning which depends on certain parameters such as distance between tip-to-sample collector, feed rate and spinning temperature. This fibrous membrane can be brought down to nanometer scale from micrometre scale. The spinnability depends on the type of solvent that has been chosen and the solubility of the polymer. The major advantage of electrospun membrane compared to the polymer membrane prepared by other techniques is its significant enhancement in porosity and improvement in mechanical strength, surface area, etc. [36].

It is reported that nanofibrous non-woven membranes obtained by electrospinning technique show greater than 400% uptake and ionic conductivity in the range of 10^{-3} S cm^{-1} at ambient temperature [37]. Optimization of average parameters by controlling the voltage range is also another highlight of this electrospinning technique. At higher voltage, jet velocity arises due to the increase in the charge density at the surface of the jet resulting a higher elongation force on the jet. Subsequently a decrease in fibre diameter as well as formation of jumbled diameters with increase in fibre boarders is observed with increase in voltage. Thus, electrospinning is a versatile technique for the preparation of polymer electrolytes. This chapter mainly focusses on the PVdF-based polymer electrolytes and its electrochemical properties. Enhancement in electrochemical properties is observed in PVdF blend electrolytes and composite PVdF-based electrolytes which is discussed in this chapter.

3.2 Polyvinyledene Difluoride-Electrospun Membranes for Lithium Ion Batteries

Electrospun membranes show better porosity and electrochemical properties. Ferroelectric β-phase of PVdF can be easily attained by electrospinning technique. The amount of β-phase that formed depends on the type of solvent and the processing parameters. Andrew and Clarke [38] dissolved PVdF in DMF solvent while Choi et al. [39] dissolved the PVdF polymer in acetone: DMAc solvent (17%) for lithium ion batteries. As the voltage increases, an increase in the electric pole arises in between the collector and syringe tip. At higher voltage range (15–20 kV), the fraction of β-phase tends to increase with an increase in fibre diameter for DMF solvent whereas 15 kV voltage applied for obtaining thinnest fibre diameter. Optimized solvent ratio of DMAc: Acetone (5/5 wt%) results higher solvent uptake, increase in pore size. DMAc will partially dissolve in the solution and remain in the surface of the polymer which enhancing the crosslinking points. Higher viscous DMAc (less volatile) suppresses the acetone from the easy evaporation. Electrospun PVdF membrane is soaked in $LiPF_6$ with 1 M solution of EC/DMC. Ethylene carbonate easily solvates to lithium salts paving the way for good conduction behaviour. Space between ultrathin fibres resulting in the formation of pore structure was the reason for the large uptake of liquid electrolytes. This polymer matrix provides higher ionic conductivity of 1 mS cm^{-1} with wider voltage range of 4.5 V (Fig. 3.2).

The same group reported [40] the electrochemical performance of electrospun PVdF membrane in lithium ion batteries. The electrospun fibrous PVdF has undergone hot press method. A decrease in electrolyte uptake (85%) and porosity was observed for the polymeric matrix after hot pressing technique compared to that of non-pressed membrane (uptake ~395%). Due to the deficiency of liquid electrolytes, a continuous fading of capacity was observed in hot-pressed membrane. Capacity fading was also noted after 2 C rate which is almost 72%. This matrix

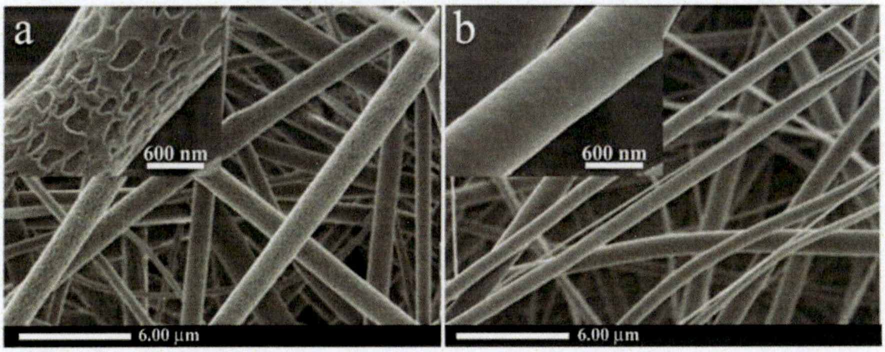

Fig. 3.2 **a** FE-SEM images on the surface morphology of electrospun PVdF membranes using **a** acetone, **b** acetone/DMAc (50/50). Adapted and reproduced from Ref. [39], Copyright 2007 American Chemical Society

exhibits different ionic conductivity towards the electrospun PVdF membrane with and without hot-pressed. The electrospun PVdF polymer membrane synthesised with hot press method has lower ionic conductivity (0.85 mS cm^{-1}) compared to that of PVdF membrane synthesised without hot press method (1.0 mS cm^{-1}). This result was confirmed by Kim et al. [41] by fabricating electrospun PVdF in the same solvent (DMAc:Acetone) that delivers an ionic conductivity of 1.0 mS cm^{-1} in 1 M $LiPF_6$ in EC/PC/DMC. This paper mainly focusses on the physical properties of PVdF membrane followed by its thermal stability. 12 wt% of PVdF solution dissolved in suitable solvent undergone electrospinning. The electrochemical characteristics depend on the amount of liquid electrolytes that are encapsulated in the polymer matrix. The highest porosity of (<90%) was obtained with those polymer matrix with average fibre diameter (AFD) of 0.45–1.38 μm providing higher electrolyte uptake of 300–350%. Fibrous polymer solution contains three parts (i) swollen phase, (ii) solid fibrous phase and (iii) liquid phase. Liquid phase enhanced the ionic conductivity of the matrix. Solution leakage was prevented by the amorphous swollen phase by holding the electrolytes tightly. Higher electrolyte uptake was obtained with lower average fibre diameter. Electrospun fibrous PVdF is tuned not only by the dissolution of solvent in the polymer but also by the addition of lithium salt. Lithium nitrate ($LiNO_3$) salt was added to polymer host for enhanced properties such as uniformity of fibres and porosity which is reported by Janakiraman et al. [42]. Ionic conductivity was raised to 1.61 mS cm^{-1} which is greater than the above reported. At first, PVdF was electrospun and different ratio in lithium nitrate ($LiNO_3$) was added and optimized. Roughness and average fibre diameter decrease with an increase in the doping salt from 1 to 2%. Charge delocalization, high interfacial resistance and high electronegativity are the major highlights of chosen lithium salt attributing its improved mobility and ionic conductivity (1.61 mS cm^{-1}). On electrospinning PVdF and β-phase were confirmed due to the drawing of PVdF polymer into longer stretched fibres. As the percentage of doping salt increases, enhancement in the ionic conductivity was observed owing to the transportation of doping salt increases from $LiNO_3$. A higher discharge capacity of 119 mAh g^{-1} and voltage stability of 4.2 V, which is higher than pristine PVdF, was observed. The electrochemical performance of prepared PVdF membrane was compared with commercial Celgard 2500® separator and revealed that PVdF membrane exhibits higher discharge capacity compared to that of commercial separator (Celgard 2500®). Encapsulation of this electrolyte solution in fibrous PVdF membrane occurs very well, and the enhancement in the surface area due to the thin fibre diameters also ensures the higher ionic conductivity of a polymeric material.

3.3 Polyvinyledene Difluoride Blend Electrolytes

Electrospun PVdF has attained greater attention in lithium ion batteries. Instead of using single polymer in a polymer matrix for the battery applications, two or more polymers are mixed together in which no such physical contact between them exists.

Blending of such materials enhances its physical properties as well as electrochemical properties in lithium ion batteries [43, 44]. There are many reports on the development of PVdF blend electrolytes such as PVdF/PMMA blend electrolytes [45, 46], PVdF/PEO blend electrolytes [47] and PVdF/HDPE [3] (high density polyethylene) electrolytes which improved electrochemical properties compared to that of single PVdF polymer membrane. Electrospun PVdF blend polymer electrolytes exhibit higher ionic conductivity as well as cycling stability than electrospun PVdF membrane alone.

3.3.1 Polymer Blend Based on Thermoplastic Polyurethane (TPU)

Thermoplastic polyurethane is an important choice of battery application since it provides an inherent morphology. TPU is a class of polyurethane plastics which are block copolymers consisting of soft and hard segments. Hard segments are responsible for its higher mechanical strength, spatial stability and the soft segment assist the dissociation of anions and cations by dissolving lithium salts which paves the enhanced ionic conductivity of the matrix [48]. The dimensional stability attributed by the hard segment is beneficial for gel polymer electrolytes for safer lithium ion batteries. TPU is thermally stable up to 305 °C and compatible with lithium salts. When TPU is mixed with PVdF polymer, the hydrogen atoms present in the amino group of TPU makes hydrogen bond with fluorine atoms of PVdF which makes the system stronger. But the organic electrolytes used in GPE in TPU are not safe. Wu et al. [49] first reported TPU blending with PVdF polymer in a solvent of DMAc:acetone by electrospinning technique. Fibrous membrane was obtained by electrospinning the polymer solution at 25 kV with tip to collector distance of 15 cm. The membrane shows an ionic conductivity of 1.6 mS cm^{-1} which is very much greater than blending PVdF with polyurethane (PU) reported by Santhosh et al. [50] (ionic conductivity ~0.32 mS cm^{-1}) An improvement in ionic conductivity was reported by Peng et al. [51] by synthesizing electrospun PVdF/thermoplastic polyurethane (TPU) gel polymer electrolyte with and without (PVdF-g-MA) PVdF–*g*-maleic anhydride. Report clearly reveals the difference between the electrochemical performance of simple PVdF/TPU and PVdF/TPU/PVdF-*g*-MA. Polymer membrane PVdF/TPU and PVdF/TPU/PVdF-g-MA of 9 wt% undergone electrospinning at an ambient temperature using acetone: dimethyl formamide (DMF) solvents. PVdF/TPU/PVdF-g-MA shows higher tensile strength and enhanced ionic conductivity (4.3 mS cm^{-1}) with wide voltage gap 5 V. The discharge capacity of PVdF/TPU/g-Ma is 166.9 mAh g^{-1} and AC impedance spectra of PVDF/TPU and PVDF/TPU/PVdF-g-MA having ionic conductivity of 1.6 and 4.7 mS cm^{-1} (Fig. 3.3).Enhancement in the electrochemical behaviour is due to the high porous morphology exhibited by PVdF-g-MA.

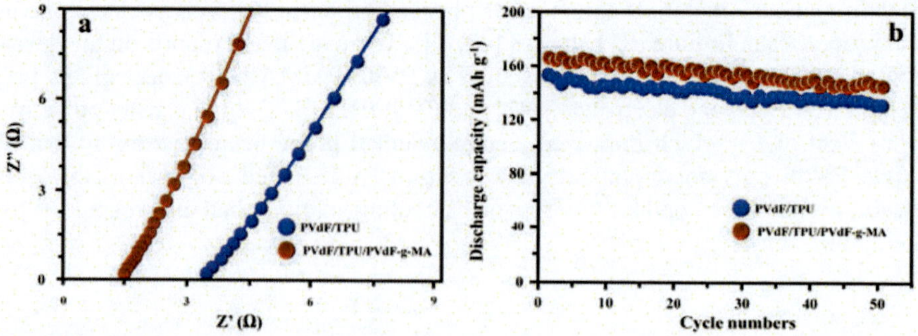

Fig. 3.3 **a** Impedance spectra and **b** first charge–discharge capacity of PVdF/TPU electrospun-based gel polymer electrolyte with an without PVdF-g-MA. Adapted and reproduced from Ref. [51], Copyright 2017 Elsevier

Reports reveals that an enhancement in ionic conductivity obtained by applying PVdF/TPU blend systems. Further improvement in electrochemical properties as well as mechanical properties were observed by Xu et al. [52, 53] in which PPC and PAN were blended along with TPU/PVdF blend systems. Polypropylene carbonates are biodegradable polymers which are environmental friendly. PPC can act as good substrate for GPE owing to its good interface effect with electrodes in batteries and its structure similar to the carbonate plasticizer. Higher electrolyte uptake is exhibited by this polymer due to the presence of polar ester group which absorb the liquid electrolytes strongly [54]. A considerable decrease in the glass transition temperature was observed when PPC was blended with PVdF/TPU polymer matrix. PVdF/TPU/poly propylene carbonate (PPC) has blended fibrous polymer that was obtained by applying a voltage of 24 kV producing average fibre diameter of 300 nm. 12 wt% of solution shows higher ionic conductivity of 5.39 mS cm^{-1} with lower crystallinity. Blended polymer shows wide electrochemical stability window (2–4.2 V) with a tensile strength of 9.9 MPa. Polyacrylonitrile (PAN) attained its attraction due to its high melting point (300 °C), high modulus of elasticity, soluble in most of the polar solvents, etc. [54]. There is a decline in the mechanical properties when plasticizers are added to this polymer hence free standing of such material is so difficult. At the same time blending PAN with PVdF/TPU shows excellent mechanical and electrochemical stability (5.5 V). Electrospun PVdF-based blend with TPU as well as with polyacrylonitrile (PAN) since PAN reduces the dendrite growth which found to be a good candidate for blending. PAN exhibits higher ionic conductivity of 6.9 mS cm^{-1} and high electrochemical stability of 5.5 V owing to the VdF group in PVdF, CN groups in PAN and ester groups in TPU which is greater than TPU/PVdF/PPC system. PVdF/TPU/PAN (2/2/1) ratio was optimized and 9 wt% solution of polymer blend was undergone electrospinning at a high voltage of 24 kV (1.0 mold m^{-3}) in $LiPF_6$ soaked in EC/DMC solvent in electrospun PVdF/TPU/PAN forms gel polymer electrolytes (GPE). The first discharge capacity of PVdF/TPU/PAN was about 167.52 mAh g^{-1} comparable to PVdF/TPU/PPC system (165.8 mAh g^{-1}).

The same group reported the effect of 0TPU blended with its copolymer of PVdF [55]. Electrospun TPU/PVdF-HFP delivered an ionic conductivity of 6.62 mS cm^{-1} which is higher than earlier reported PVdF/TPU (~0.32 mS cm^{-1}). The polymer matrix was undergone electrospinning with voltage of 24.5 kV and the polymer along with its doping salt $LiFePO_4$ shows higher initial discharge capacity (161 mAh g^{-1}) capacity.

3.3.2 *Other Blends*

Polyvinyl chloride (PVC) provides attractive properties including cost effective, good compatibility, easy of processing, durable material, etc. PVC lacks good ionic conductivity due to the immiscibility with liquid electrolytes. Thus, blending PVC with other polymer can reduce its limitations and act as good polymer matrix for the preparation of GPE. The effect of PVdF/PVC blend was reported by Rajendran et al. [56] with different salts of $LiClO_4$ and $LiBF_4$ providing good ionic conductivity than host polymers. Optimized concentration of PVdF/PVC (80/20) was prepared by solution casting technique. Even then these blend materials could not achieve a conductivity greater that 10^{-4} S cm^{-1}. This challenge was overcome by introducing electrospun PVdF/PVC (8:2 ratio) blend polymer electrolyte for advance lithium ion battery applications which was reported by Zhong et al. [57]. The polymer matrix was dissolved in DMF/acetone stirred for 6 h in 60 °C. From the morphological studies, the polymer matrix provides higher uptake and porosity than pure PVdF due to higher surface area, and polar nature of DMF and low volatile nature of THF solvent. Polymer solution has undergone electrospinning with high voltage of 20 kV and syringe tip distance of 20 cm at a flow rate of 3 ml h^{-1}. Electrospun PVdF/PVC provides a voltage stability greater than 5 V inferring the addition of PVC that enhances the amorphous nature of the material. Blending PVdF with PVC increases the interconnected pore structure as well as decrease in the crystallinity due to the increase in the PVC content to PVdF membrane. Compared with pure PVdF fibres (932 nm), average fibre diameter was less in PVdF/PVC (624 nm) fibres. A significant increase in conductivity was also reported with an increase in temperature owing to the increase in the carrier ions and expansion of the polymer matrix that pave an easy path to the carrier ions. This polymeric matrix with doping salt $LiFePO_4$ shows low ionic conductivity of 2.25 mS cm^{-1} at 25 °C compared with PVdF/PEO polymer matrix. Blending of PVdF along with PEO explores both properties to form hybrid gel polymer electrolytes (HGPE) resulting in high-performance battery applications. Though there are reports on PEO/PVdF blend electrolytes in batteries [58], electrospun PVdF/PEO blend electrolytes were first reported by Li et al. [59] which deliver an ionic conductivity of 4.8 mS cm^{-1} much higher than PVdF/PEO systems. Both PVdF/PEO and PVdF/PVC exhibit similar electrolytic uptake (~250%). Blended polymer solution was electrospinned at a voltage of 15–17 kV with drum rotation of 150 rpm having collector to drum rotation of 15 cm. An enhancement in electrochemical stability of HGPE was observed (4.8 V) (Fig. 3.4a)

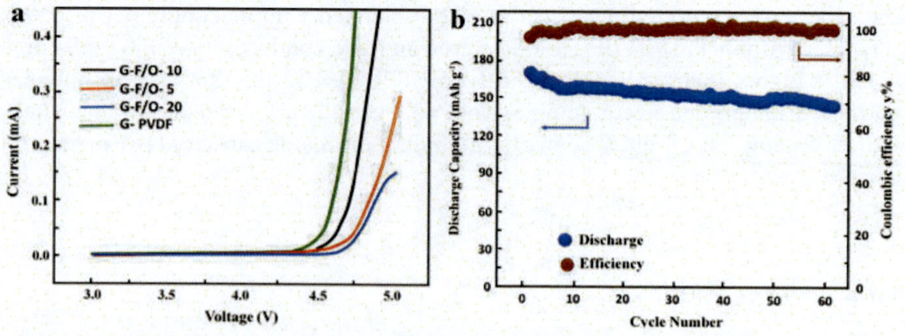

Fig. 3.4 **a** Electrochemical stability window of HGPE, **b** charge/discharge cycling performance of the hybrid gel polymer electrolyte with F/O ratio 5 (Li/HGPE/$LiCOO_2$ cell, at room temperature). Adapted and reproduced from Ref. [59], Copyright 2015 Elsevier

compared to pure PVdF polymer membrane (4.5 V) with PVdF:PEO weight ratio of 5:1. PVdF blend-based polymer electrolytic membrane can also act as separators. This membrane delivers high initial discharge capacity of 171 mAh g^{-1} (Fig. 3.4b).

PMMA is a transparent polymer having good mechanical properties. These are hard and low water absorption. The amorphous nature exhibited by PMMA improved the ionic conductivity of the polymer electrolyte and is less reactive towards lithium salt [60, 61]. PMMA is well-suited with aqueous electrolytes as well as it enhances the amorphous nature which removing the major drawback of using PVdF as polymer matrix. Liang et al. [62] reported the effect of electrospun PVdF/PMMA blend electrolytes in the performance of lithium ion batteries. Optimized PVdF/PMMA (90/10) has undergone electrospinning in a solvent of DMAc:acetone with a voltage range of 18 kV. Blended PVdF/PMMA of 90:10 offers an ionic conductivity of 2.54 mS cm^{-1} with electrochemical stability window (5 V) whereas 80:10 ratio delivers ionic conductivity of 1.95 mS cm^{-1}. From this report, it was noted that PVdF/PMMA blend offers higher ionic conductivity and can also act as best separators. Therefore, blending PVdF/PMMA along with another polymer also contributes to the function of improved electrochemical achievement of batteries. Yvonne et al. [63] reported PVdF/PMMA/cellulose acetate (CA) blend-based membrane prepared by electrospinning technique. 90:0:10 ratio (PVdF: PMMA: CA) with DMAc:acetone (7:3) was found to be best material with 99.1% porosity, highest melting temperature (161.8 °C), highest porosity (324%), low crystallinity, etc., compared to the pristine PVdF. A voltage of 60 kV was applied with 20 cm of tip-to-syringe distance having aluminium foil as a current collector. Biodegradable polymer cellulose has high absorption capability attributing higher porosity and thereby its uptake.

Other than simply blending, sandwiching polyester in between PVdF was reported by Wei et al. [64] for the application of lithium air batteries. A bifunctional $La_{0.5}Ce_{0.5}Fe_{0.5}Mn_{0.5}O_3$ (LCFMO) cathode material was prepared by wet chemical co-precipitation technique whereas PVdF/PET/PVdF was fabricated by electrospinning method with voltage range between 15 and 25 kV. These polymer blend membranes exhibit an ionic conductivity of 7.9×10^{-11} S cm^{-1} higher than glass fibre (7.1×10^{-12} S cm^{-1}). Similar studies was reported by Zhai et al. [65] in which poly(m-phenylene isophthalamide) (PMIA) was sandwiched in between PVdF. This polymer matrix mainly functions as separators owing to the easy shutdown mechanism and mechanical stability of PMIA polymer. Sandwiched structure undergone electrospinning by applying 30 kV with tip distance of 25 cm and lithium fluorophosphates ($LiPF_6$) in DMAc were utilized as doping salt for the electrochemical studies. This matrix compares electrochemical characterization including ionic conductivity with PVdF/PMIA/PVdF, pure PVdF and found to be of 0.81 and 0.61 mS cm^{-1}, respectively. This polymer membrane reached higher voltage stability greater than 5 V with discharge capacity of 131.5 Ah g^{-1} after 100 cycles. But these matrix could not attain the ionic conductivity of actual PMIA.

Prasanth et al. [66] reported the effect of polyethylene oxide on PVdF which was prepared by electrospinning technique. Uptake and porosity leakage were observed with different doping salt such as $LiClO_4$, $LiPF_6$, $LiCF_3SO_3$, $LiBF_4$, LiTFSI and highest electrochemical properties including ionic conductivity (4.9 mS cm^{-1}) which greater than PVdF/PEO blend reported earlier by Li et al. [59] (4.8 mS cm^{-1}) (Fig. 3.5) since this doping salt has highest dissociation constant.

LITFSI exhibits discharge capacity of 162 mAh g^{-1} whereas $LiClO_4$ only shows a capacity of 152 mAh g^{-1} (Fig. 3.6). Compatibility of PVdF along with PEO attributes better blending properties resulting for the dissociation of lithium salts in an effective way. 16 wt% of polymer in a solvent dimethyl acetamide (DMAc) was undergone electrospinning with a voltage range of 18–20 kV with collector-to-tip distance of 20 cm. $LiFePO_4$ is the cathode materials used since it is thermally stable, cheap, non-toxic, etc., with higher theoretical capacity. LITFSI and $LIClO_4$ were chosen owing to the compatibility and ionic conductivity of the material. Initial discharge capacity of LITFSI shows 168 mAh g^{-1} whereas for $LiClO_4$ it is 168 mAh g^{-1}.

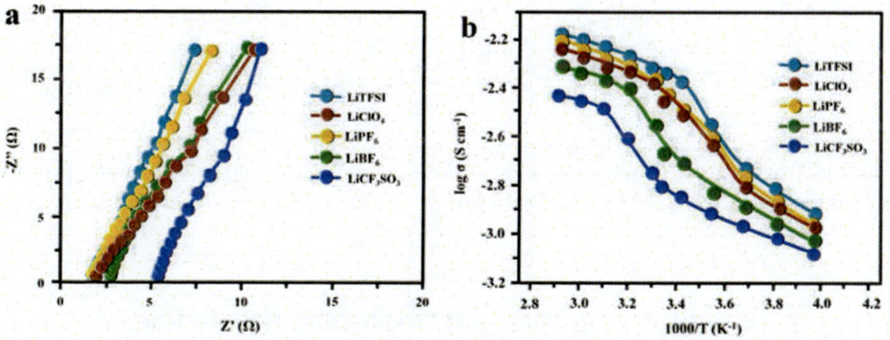

Fig. 3.5 **a** AC impedance spectra of PE based on electrospun PVdF/PEO blend membranes with electrolytes (SS/PGE/SS cells, frequency range 10 mHz to 1 MHz, amplitude 20 mV). **b** Effect of temperature on ionic conductivity of PE is based on electrospun PVdF/PEO polymer blend membranes activated with different electrolytes. Adapted and reproduced from Ref. [66], Copyright 2014 Elsevier

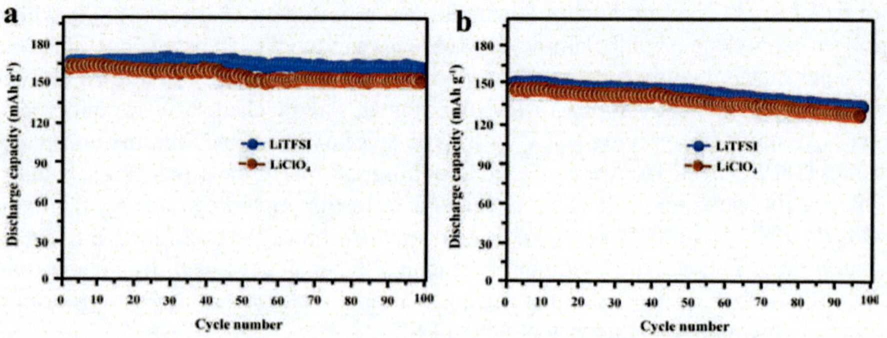

Fig. 3.6 Cycling properties of Li/PGE/$LiFePO_4$ cells with polymer electrolyte based on electrospun membrane (25 C, 0.1 C rate): **a** PVdF/PEO polymer blend and **b** PVdF. Adapted and reproduced from Ref. [66], Copyright 2014 Elsevier

Table 3.1 Electrochemical characterization of polymer blend

Polymer blend	Ionic conductivity (mS cm^{-1})	Discharge capacity (mAh g^{-1})	Cycle no.	Retention capacity	Voltage stability (V)	References
PVdF–PVC	2.25	130.8	50	90.1	5.1	[57]
PVdF/TPU/PAN	6.91	168.56	50	96.2	4.7	[53]
PVdF/TPU/PPC	5.32	165.1	50	94.7	5.4	[52]
PVdF/PMIA/PVdF	0.81	135.29	100	93.1	5.15	[65]
PVdF/TPU/PVdF-g-MA	4.7	166.9	50	98	5	[51]
PVdF/PMMA	2.54	–	–	–	5	[62]
PVdF/PEO	4.8	171.8	60	86	4.8	[59]
PVdF-*co*-HFP/TPU	6.62	163	50	–	3.5	[55]
PVdF/PEO	4.9	168	50	96	4	[66]
PVdF/PET/PVdF	6.02	100	–	–	4.5	[64]

The electrochemical properties of different electrospun blend polymer electrolyte are tabulated in Table 3.1.

3.4 Electrospun Composite Polyvinyledene Difluoride Polymer Electrolyte

Composite polymer electrolytes are electrolytes prepared by dispersing nanofillers such as silica, clay, alumina and titania into the polymers. These materials enhance the ionic conductivity, mechanical stability, thermal stability, etc. There are reports on composite polymer electrolytes for various energy storage applications such as in supercapacitors [67] and lithium ion batteries [68]. Advance study of composite polymer electrolytes mainly focussed the electrospun composite polymer electrolytes that enhance the surface area as well as porosity paving the way to higher ionic conduction. By introducing composite polymer electrolytes, both the electrochemical and morphological properties will improve followed by the increment in the electrode electrolytes stability. Thus, alumina and silica are very important fillers for the preparation of gel polymer electrolytes. Aluminosilicate clay halloysite nanotubes (HNT) are those materials which show both the properties of alumina and silica. The inner cylinder core of HNT is similar to alumina whereas outer surface shows properties similar to that of silica providing large variety of applications [69, 70]. The above properties confirm the importance of using HNT in lithium ion composite polymer electrolytes. HNT are advanced fillers which have been employed in PVdF polymer for the fabrication of nanocomposite polymer membrane that diminishes the problems of PVdF such as thermal stability, electrochemical stability and crystallinity [71]. HNT fibres enhance the mechanical strength, environmentally benign and cost effectiveness. This electrospun gel polymer composite membranes in lithium ion battery can also act as separator.

10 wt% of the PVdF/HNT fibre has undergone electrospinning with a voltage of 20 kV. From the morphological study of HNT, it was observed that a tubular structure in which its inner and outer surfaces are chemical similar to that of TiO_2 as well as Al_2O_3. From the TEM image, it is visible that HNT are aligned along the PVdF polymer because of the induced alignment (Fig. 3.7).

HNT favours electrospinning technique due to the charge potential present in the surface of HNT fibres offering easy electrostatic conduction. Lithium cobalt oxide with electrospun composite nanofibre electrolyte delivers specific capacitance of 131 mAh g^{-1}, with its high electrolyte uptake of 493%. Formation of interconnected pore structures as well as surface area improved the electrochemical stability (5.3 V) and ionic conductivity (1.77 S cm^{-1}). This enhanced cation mobility as well as

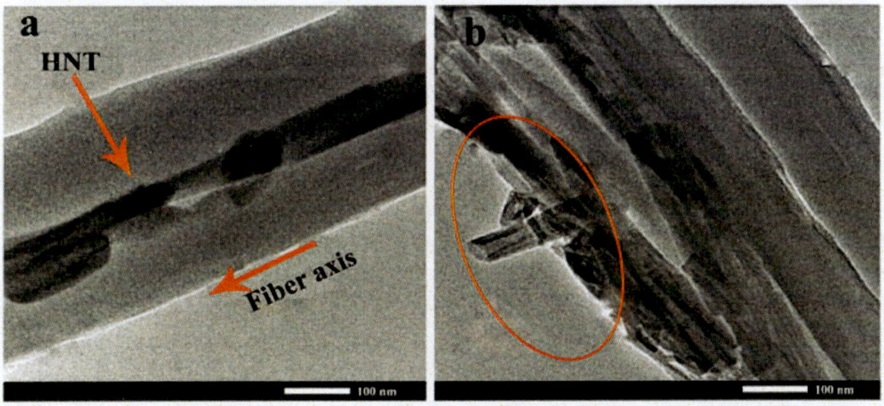

Fig. 3.7 **a, b** TEM micrographs of nanocomposite nanofibre (NCNF). Adapted and reproduced from Ref. [71], Copyright 2016 John Wiley and Sons

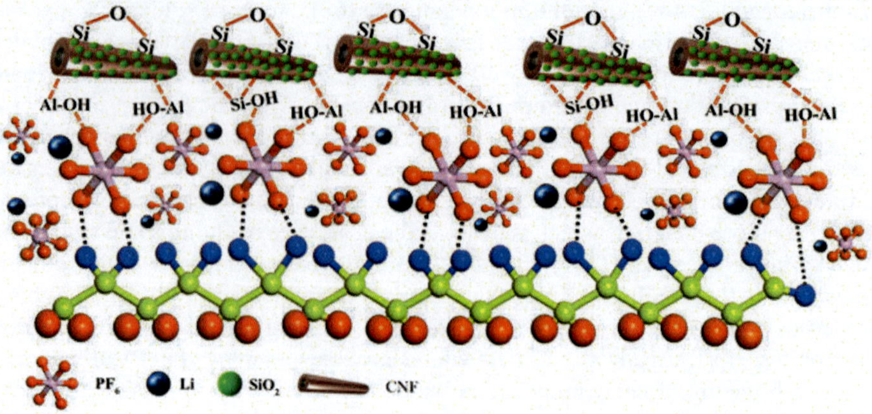

Fig. 3.8 Interactions between nanocomposite nanofibre (NCNF) and $LiPF_6$. Adapted and reproduced from Ref. [71], Copyright 2018 John Wiley and Sons

ionic conductivity inhibits the nucleation of the dendrite formation and material can show better performance than the other separators such as Celgard 2400®, PVdF separators. But it could not achieve the ionic conductivity of composite polymer electrolytes containing SiO_2 and TiO_2 as fillers [72]. PVdF/HNT fibres enhance the effective production of polymer matrix extended to large scale due to the cost effectives, environmentally benign, mechanical and thermal stabilities. From the schematic diagram, the mechanism of the interaction between the electrolytes and HNT is clear (Fig. 3.8). The interactions are mainly by (i) Formation of hydrogen bonds between OH groups in HNT and fluorine atoms in $LiPF_6$, (ii) hydrogen bonding between PVdF and fluorine atom (iii) Dipole moment existing between Li^+ ions in $LiPF_6$ and oxygen atom in the HNT group owing to the increment in the number of free Li^+ ions and enhance the ionic conductivity.

3.4.1 *Effect of SiO_2 and TiO_2 Fillers in Electrospun Polyvinylidene Difluoride*

Cao et al. [51, 52, 55] reported electrospun PVdF/TPU-based polymer matrix membrane for the application of lithium ion batteries which was described earlier. Similar reports based on the effect of ceramic fillers on PVdF/TPU were reported in the year of 2011 by the same group. Wu et al. [73] reported the effect of fillers in PVdF/thermoplastic poly urethane (TPU) blend polymer gel electrolytes. Both SiO_2 and TiO_2 was optimized and added to this polymer electrolyte blend which was then undergone electrospinning technique. In situ synthesized SiO_2 and TiO_2 take advantages of improved porosity, better ionic conductivity, etc., because of the Lewis acid–base interactions between polar group and the filler particles. In situ synthesized fillers were added to 9 wt.% polymer solution providing an electrochemical stability greater than 4.6 V. 3 wt% TiO_2 shows high electrochemical performance compared to 3 wt% SiO_2. The presence as well as absence of in situ addition of fillers was reported. Ionic conductivity of polymer electrolytes without filler was low (3.2 mS cm^{-1}) compared to the addition of fillers. Though there is an improvement in ionic conductivity observed by the addition of in situ SiO_2 filler (3.8 mS cm^{-1}), in situ addition of TiO_2 provided better dispersion with high ionic conductivity 4.8 mS cm^{-1}. Since ionic conduction is closely related to porosity and there by atomic force diameter (AFD), SiO_2 nanoparticles suffer feeble tortuosity in the submicron phase. Though TiO_2 plays a vital role as fillers, Ding et al. [72] reported the effect of electrospun composite polymer membrane as separators in which TiO_2 was synthesized by tetrabutyl titanate. Electrospun PVdF/TiO_2 membrane was fabricated by dissolving PVdF: TiO_2 (95:5) the polymer in acetone: ET solvent (7:3) and applied a 25 kV voltage having 15 cm tip to collector distance. An interconnected fibre structure was observed from its morphological study that provides better mechanical strength and less tensile and elongation modulus. An enhancement in porosity and uptake was observed compared to pure PVdF membrane. Ionic conductivity of PVdF/TiO_2 (1.4 mS cm^{-1}) was increased compared to pure PVdF membranes (0.89 mS cm^{-1}). Zhang et al. [74] reported a new class of electrospinning as well as electrospraying technique for the better performance of lithium ion battery. Electrospinning of PVdF polymer nanofibre along with the electrospraying technique of nanoparticle SiO_2 is utilized as ceramic filler which serves as separators in battery. Even though both the techniques are based on electric forces that could help to eject the polymer liquids and collected in the collector, fine droplets were produced from the jets when the electric force is greater than surface tension which takes place in electrospraying technique whereas nanofibres were collected in the collector without any breakage due to the inability of electric repulsion force to surpass intermolecular force. Adding SiO_2 nanoparticle enhances the ionic conductivity (2.6 mS cm^{-1}) compared to commercial microporous PP membrane with high uptake and porosity soaked in $LiFePO_4$. Compared to PP membrane, a slight increase in initial discharge capacity was reported (159 mAh g^{-1}) with high cycling performance. A new kind of preparation of composite gel polymer electrolytes (CGPE) by doping ion complex (SiO_2-PAALi) into PVdF polymer matrix has undergone electrospinning with high voltage of 15 kV [75]. To SiO_2-C=C bonds, acrylic acid (AA) was added followed by ultra-centrifugation and then undergone polymerization (PAA). Prepared composite membrane (PVdF-SiO_2-PAALi) is compared with pure PVdF and PVdF/SiO_2 composites. This membrane delivers high thermal stability as well as electrolyte uptake providing discharge capacity of 156 mAh g^{-1} whereas for PVdF and PVdF/SiO_2 with capacity of 149 and 152 mAh g^{-1}. A schematic illustration of lithium ion conduction pathway in PVdF/SiO_2 along with liquid electrolyte is given below (Fig. 3.9).

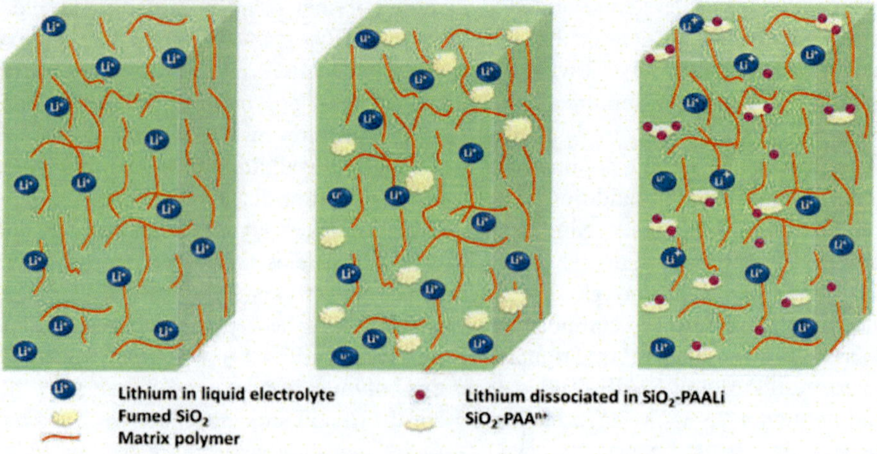

Fig. 3.9 Channel pathway of lithium ion conducting gel polymer electrolytes

3.4.2 *Other Type Fillers*

Zeng et al. [76] reported a composite gel polymer electrolyte comprising of electrospun PVdF along with lithium polyvinyl alcohol oxalate borate (LiPVAOB) exhibiting good mechanical strength and ionic conductivity. The structure of LiPVAOB was similar to that of lithium bisoxalate borate (LiBOB) which is cost effective, thermally stable, etc., but suffers from poor mechanical and ionic conductivity. One molar solution of $LiPF_6$ along with this electrospun polymer matrix provides a gel polymer membrane delivering an ionic conductivity of 0.26 mS cm^{-1}. This type of gel polymer electrolytes enhances the safety and has high thermal stability attributing a better matrix for the future lithium ion batteries. The evaporation rate of this liquid electrolyte is very less compared to the other polymer matrix (only 35 wt% loss). But these materials suffer from poor ionic conductivity. In order to overcome the thermal shrinkage as well as poor mechanical stability caused by PVdF membrane nano fillers such as Al_2O_3, SiO_2 and clay were used. Fang et al. [77] reported the addition of the class of clay nano particles called 'montmorillonite' dispersing to PVdF membrane undergone electrospinning that overcomes the barriers of PVdF membranes. MMT was dissolved in DMF:acetone in 7:3 ratio and 15 wt% of this polymer solution was dispersed in MMT and this homogeneous solution was taken in 12 ml syringe followed by electrospinning with an applied voltage of 12 kV. 7 wt% of montmorillonite (MMT) was optimized and obtained an ionic conductivity of 4.2 mS cm^{-1} which is greater than commercial separators such as Celgard®. Comparing with Celgard PP separators, stable cyclic performance was observed in composite membrane with capacitive retention even after 50 cycles at C rate of 0.2 C (Fig. 3.10). An intercalation behaviour of polymer matrix into the layered host provides high electrochemical as well as mechanical stability. Increasing the

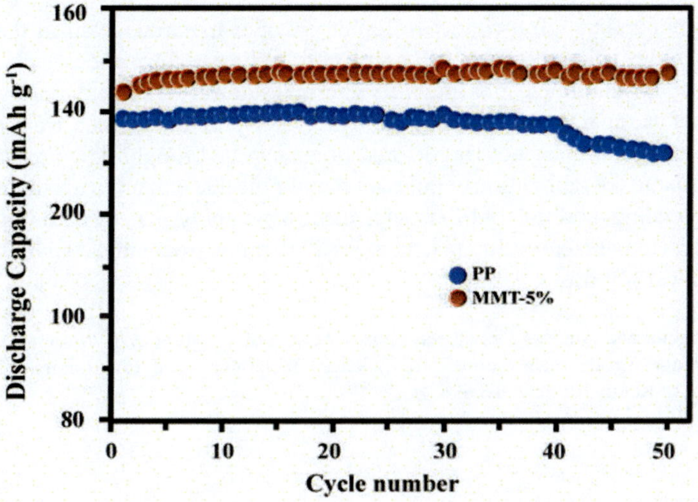

Fig. 3.10 Cycling performance of Li/$LiFePO_4$ cells containing Celgard PP and PVdF/montmorillonite-5% membranes (0.2 C rate). Adapted and reproduced from Ref. [77], Copyright 2016 Elsevier

Table 3.2 Electrochemical performance of electrospun composite PVdF-based materials

Polymer composite	Ionic conductivity (10^{-3} S cm^{-1})	Discharge capacity (mAh g^{-1})	Cycle no.	Retention capacity (%)	Voltage stability (V)	References
PVdF/HNT	1.77	138	50	97	4.2	[71]
TPU/PVdF/(SiO_2&TiO_2)	4.8	–	–	–	5.4	[73]
PVdF/TiO_2	1.4	137	20	96.6	4.1	[72]
SiO_2/PVdF	2.6	132	100	78	3.5	[74]
PVdF/SiO_2-PAALi	3.5	156.5	30	97	5.05	[75]
PVdF/MMT	4.2	138	50	96	5	[77]

MMT content in the polymer matrix enhances lithium conducting channel pathways. The electrochemical performance of PVdF-based composite polymer electrolytes is tabulated in Table 3.2.

3.5 Conclusion

Electrospun PVdF-based membrane attained lots of attraction in the fabrication of polymer electrolytes in electrochemical world. These materials offer higher ionic conductivity, uptakes, porosity, etc., due to the fibrous structure that emerges after electrospinning. Though there are different polymer materials avail in the electrochemical studies, electrospun PVdF offers outstanding performances such as higher dielectric constant, dimensional stability, thermal stability, chemical inertness which can also act as separators in lithium ion batteries. Nanofibres which were produced from electrospinning are not fragile, and fibres can be brought down to nanometre scale. Similar to the fabrication of polymer membrane that enhances the ionic conductivity by blending, addition of fillers, etc., electrospun polymers by blending or addition of fillers to the electrospun matrix offers higher electrochemical stability as well as mechanical stabilities.

Acknowledgements Authors Dr. Jabeen Fatima M. J. and Dr. Raghavan Prasanth, would like to acknowledge Kerala State Council for Science, Technology and Environment (KSCSTE), Government of Kerala for financial assistance.

References

1. Hwang YJ, Nahm KS, Kumar TP, Stephan AM (2008) Poly(vinylidene fluoride-hexafluoropropylene)-based membranes for lithium batteries. J Memb Sci 310:349–355. https://doi.org/10.1016/j.memsci.2007.11.006
2. Wang H, Huang H, Wunder SL et al (2000) Novel microporous poly (vinylidene fluoride) blend electrolytes for lithium-ion batteries novel microporous poly (vinylidene fluoride) blend electrolytes for lithium-ion batteries. J Electrochem Soc 147:2853–2861. https://doi.org/10.1149/1.1393616
3. Liu H, Liu J, Zhou Q, Wang J (2013) Novel polymer electrolyte based on PVDF/HDPE blending for lithium-ion battery. Mater Lett 99:164–167. https://doi.org/10.1016/j.matlet.2013.03.033
4. Karabelli D, Leprêtre J, Dumas L et al (2015) Crosslinking of poly (vinylene fluoride) separators by gamma-irradiation for electrochemical high power charge applications. Electrochim Acta 169:32–36. https://doi.org/10.1016/j.electacta.2015.04.034
5. Pereira AO, Miranda CR (2015) First-principles investigation of transition metal dichalcogenide nanotubes for Li and Mg Ion battery applications. J Phys Chem C 119:4302–4311. https://doi.org/10.1021/jp510182u
6. Nonlinear O, Materials O, Films OT et al (2003) An electrospun poly(vinylidene fluoride) nanofibrous membrane and its battery applications. Adv Mater 15:2027–2032. https://doi.org/10.1002/adma.200304617
7. Manuel Stephan A, Nahm KS (2006) Review on composite polymer electrolytes for lithium batteries. Polym (Guildf) 47:5952–5964. https://doi.org/10.1016/j.polymer.2006.05.069
8. Naderi R (2016) Composite gel polymer electrolyte for lithium ion batteries. Masters thesis
9. Nitta N, Wu F, Lee JT, Yushin G (2015) Li-ion battery materials: present and future. Mater Today 18:252–264. https://doi.org/10.1016/j.mattod.2014.10.040
10. Tarascon JM (2010) Key challenges in future Li-battery research. Philos Trans R Soc A Math Phys Eng Sci 368:3227–3241. https://doi.org/10.1098/rsta.2010.0112
11. Ryu KS (2013) Synthesis and electrochemical performance of graphene/metal oxide nanocomposite electrodes in lithium secondary batteries and supercapacitors. In: 8th international forum on strategic technology 2013, IFOST 2013—Proceedings, vol 1, pp 96–100. https://doi.org/10.1109/IFOST.2013.6616953
12. Myung SK, Hak YK, Young SK et al (2005) Preparation of electrospun oxidized cellulose mats and their in vitro degradation behavior. Macromol Res 13:62–67

13. Servais T, Brocke R, Fatka O (1996) The effect of processing variables on the morphology of electrospun nanofibers and textiles. Palaeontology 39:389–405. https://doi.org/10.1016/S0032-3861(00)00250-0
14. Thavasi V, Singh G, Ramakrishna S (2008) Electrospun nanofibers in energy and environmental applications. Energy Environ Sci 1(2):205–221. https://doi.org/10.1039/b809074m
15. Bjorge D, Daels N, De Vrieze S et al (2009) Performance assessment of electrospun nano fibers for filter applications. Desalination 249:942–948. https://doi.org/10.1016/j.desal.2009.06.064
16. Zhou X, Guo Y-G (2013) A PEO-assisted electrospun silicon–graphene composite as an anode material for lithium-ion batteries. J Mater Chem A Mater energy Sustain 1:9019–9023. https://doi.org/10.1039/c3ta11720k
17. Ballengee JB, Pintauro PN (2011) Morphological control of electrospun Nafion nanofiber mats. J Electrochem Soc 158:568–572. https://doi.org/10.1149/1.3561645
18. Rao M, Geng X, Liao Y et al (2012) Preparation and performance of gel polymer electrolyte based on electrospun polymer membrane and ionic liquid for lithium ion battery. J Memb Sci 399–400:37–42. https://doi.org/10.1016/j.memsci.2012.01.021
19. Ji L, Yao Y, Toprakci O et al (2010) Fabrication of carbon nanofiber-driven electrodes from electrospun polyacrylonitrile/ polypyrrole bicomponents for high-performance rechargeable lithium-ion batteries. J Power Sources 195:2050–2056. https://doi.org/10.1016/j.jpowsour.2009.10.021
20. Gopalan AI, Santhosh P, Manesh KM et al (2008) Development of electrospun PVdF-PAN membrane-based polymer electrolytes for lithium batteries. J Memb Sci 325:683–690. https://doi.org/10.1016/j.memsci.2008.08.047
21. Li C, Chartuprayoon N, Bosze W et al (2014) Electrospun polyaniline/poly (ethylene oxide) composite nanofibers based gas sensor. Electroanalysis 26:711–722. https://doi.org/10.1002/elan.201300641
22. Miao Y, Yan J, Huang Y et al (2015) Electrospun polymer nanofiber membrane electrodes and an electrolyte for highly flexible and foldable all-solid-state supercapacitors. RSC Adv 5:26189–26196. https://doi.org/10.1039/C5RA00138B
23. Lee Y, Liu Y (2017) Crosslinked electrospun poly (vinylidene difluoride) fiber mat as a matrix of gel polymer electrolyte for fast-charging lithium-ion battery. Electrochim Acta 258:1329–1335. https://doi.org/10.1016/j.electacta.2017.11.191
24. Tong Y, Que M, Su S, Chen L (2016) Design of amphiphilic poly (vinylidene fluoride-co-hexafluoropropylene)-based gel electrolytes for high-performance lithium-ion batteries. Int J Ionics. https://doi.org/10.1007/s11581-016-1662-9
25. Bicy K, Suriyakumar S, Anu AS et al (2018) Highly lithium ion conductive, Al_2O_3 decorated electrospun P(VDF-TrFE) membranes for lithium ion battery separators. New J Chem 42:19505–19520. https://doi.org/10.1039/c8nj01907j
26. Zhao H, Deng N, Ju J et al (2018) Novel configuration of heat-resistant gel polymer electrolyte with electrospun poly (vinylidene fluoride-co-hexafluoropropylene) and poly-m-phenyleneisophth-alamide composite separator for high-safety lithium-ion battery. Mater Lett. https://doi.org/10.1016/j.matlet.2018.10.067
27. Raghavan P, Manuel J, Zhao X et al (2011) Preparation and electrochemical characterization of gel polymer electrolyte based on electrospun polyacrylonitrile nonwoven membranes for lithium batteries. J Power Sources 196:6742–6749. https://doi.org/10.1016/j.jpowsour.2010.10.089
28. Guo W, Tan C, Shi K, Li J, Wang X-X, Sun B, Huang X, Long Y-Z, Jiang P (2018) Wireless piezoelectric device based on electrospun PVDF/$BaTiO_3$ NW nanocomposite fibers for human motion monitoring. Nanoscale 10(37):17751–17760. https://doi.org/10.1039/C8NR05292A
29. Structures M, Pan C, Tsai K et al (2017) Large-area piezoelectric PVDF fibers fabricated by near-field electrospinning with multi-spinneret structures. Micromachines (Basel) 8:1–16. https://doi.org/10.3390/mi8040097
30. Graham K, Ouyang M, Raether T et al (2002) Polymeric nanofibers in air filtration applications. In: Fifteenth annual technical conference & expo of the American Filtration & Separations Society

31. Muniyandi N, Kalaiselvi N, Periyasamy P et al (2001) Optimisation of PVdF-based polymer electrolytes. J Power Sources 96:14–19. https://doi.org/10.1016/S0378-7753(01)00562-6
32. Ruan L, Yao X, Chang Y et al (2018) Properties and applications of the β phase poly(vinylidene fluoride). Polymer (Guildf) 10:1–27. https://doi.org/10.3390/polym10030228
33. Doll WW, Lando JB (2016) The polymorphism of poly (vinylidene fluoride)IV. The structure of high-pressure-crystallized poly(vinylidene fluoride). J Macromol Sci Part B 2348:889–896. https://doi.org/10.1080/00222347008217130
34. Ramasundaram S, Yoon S, Kim KJ, Lee JS (2009) Crystalline structure and ferroelectric response of poly(vinylidene fluoride)/organically modified silicate thin films prepared by heat controlled spin coating. Macromol Res 210:951–960. https://doi.org/10.1002/macp.200800600
35. Aik W, Kotaki M, Liu Y, Lu X (2007) Morphology, polymorphism behavior and molecular orientation of electrospun poly (vinylidene fluoride) fibers. Polymer (Guildf) 48:512–521. https://doi.org/10.1016/j.polymer.2006.11.036
36. Sun G, Sun L, Xie H, Liu J (2016) Electrospinning of nanofibers for energy applications. Nanomaterials 6:129. https://doi.org/10.3390/nano6070129
37. Raghavan P, Zhao X, Choi H et al (2014) Electrochemical characterization of poly(vinylidene fluoride-co-hexafluoro propylene) based electrospun gel polymer electrolytes incorporating room temperature ionic liquids as green electrolytes for lithium batteries. Solid State Ionics 262:77–82. https://doi.org/10.1016/j.ssi.2013.10.044
38. Andrew JS, Clarke DR (2008) Effect of electrospinning on the ferroelectric phase content of polyvinylidene difluoride fibers. Langmuir 24:670–672. https://doi.org/10.1021/la7035407
39. Choi SW, Kim JR, Ahn YR et al (2007) Characterization of electrospun PVdF fiber-based polymer electrolytes. Chem Mater 19:104–115. https://doi.org/10.1021/cm060223+
40. Lee SW, Choi SW, Jo SM et al (2006) Electrochemical properties and cycle performance of electrospun poly(vinylidene fluoride)-based fibrous membrane electrolytes for Li-ion polymer battery. J Power Sources 163:41–46. https://doi.org/10.1016/j.jpowsour.2005.11.102
41. Kim JR, Choi SW, Jo SM et al (2004) Electrospun PVdF-based fibrous polymer electrolytes for lithium ion polymer batteries. Electrochim Acta 50:69–75. https://doi.org/10.1016/j.electacta.2004.07.014
42. Janakiraman S, Surendran A, Ghosh S, Anandhan S, Venimadhav A (2018) A new strategy of PVDF based Li-salt polymer electrolyte through electrospinning for lithium battery application. Mater Res Express 6(3):035303
43. Rajeswari N, Selvasekarapandian S, Prabu M et al (2013) Lithium ion conducting solid polymer blend electrolyte based on bio-degradable polymers. Bull Mater Sci 36:333–339. https://doi.org/10.1007/s12034-013-0463-2
44. Wang H, Huang H, Wunder SL (2002) Novel microporous poly(vinylidene fluoride) blend electrolytes for lithium-ion batteries. J Electrochem Soc 147:2853. https://doi.org/10.1149/1.1393616
45. Xiao Q, Li Z, Gao D, Zhang H (2009) A novel sandwiched membrane as polymer electrolyte for application in lithium-ion battery. J Memb Sci 326:260–264. https://doi.org/10.1016/j.memsci.2008.10.019
46. Ma T, Cui Z, Wu Y et al (2013) Preparation of PVDF based blend microporous membranes for lithium ion batteries by thermally induced phase separation: I. Effect of PMMA on the membrane formation process and the properties. J Memb Sci 444:213–222. https://doi.org/10.1016/j.memsci.2013.05.028
47. Deng F, Wang X, He D et al (2015) Microporous polymer electrolyte based on PVDF/PEO star polymer blends for lithium ion batteries. J Memb Sci 491:82–89. https://doi.org/10.1016/j.memsci.2015.05.021
48. Yu R, Bao JJ, Chen TT et al (2017) Solid polymer electrolyte based on thermoplastic polyurethane and its application in all-solid-state lithium ion batteries. Solid State Ionics 309:15–21. https://doi.org/10.1016/j.ssi.2017.06.013
49. Na Wu, Jing Bo, Cao Qi, Wang Xianyou, Hao Kuang QW (2010) A novel electrospun TPU/PVdF porous fibrous polymer electrolyte for lithium ion batteries. J Appl Polym Sci 116:2658–2667. https://doi.org/10.1002/app

50. Santhosh P, Vasudevan T, Gopalan A, Lee KP (2006) Preparation and characterization of polyurethane/poly(vinylidene fluoride) composites and evaluation as polymer electrolytes. Mater Sci Eng B Solid-State Mater Adv Technol 135:65–73. https://doi.org/10.1016/j.mseb.2006.08.033
51. Peng S, Cao Q, Yang J et al (2015) A novel electrospun poly(vinylidene fluoride)/thermoplastic polyurethane/poly(vinylidene fluoride)-g-(maleic anhydride) porous fibrous polymer electrolyte for lithium-ion batteries. Solid State Ionics 282:49–53. https://doi.org/10.1016/j.ssi.2015.09.018
52. Xu J, Liu Y, Cao QI et al (2019) A high-performance gel polymer electrolyte based on poly (vinylidene fluoride)/thermoplastic polyurethane/ poly (propylene carbonate) for lithium-ion batteries. J Chem Sci 131:1–10. https://doi.org/10.1007/s12039-019-1627-4
53. Liu Y, Peng X, Cao Q et al (2017) Gel polymer electrolyte based on poly(vinylidene fluoride)/thermoplastic polyurethane/polyacrylonitrile by electrospinning technique gel polymer electrolyte based on poly (vinylidene fluoride)/thermoplastic polyurethane/polyacrylonitrile by the electrospinning technique. J Phys Chem. https://doi.org/10.1021/acs.jpcc.7b03411
54. Li XH, Meng YZ, Wang SJ et al (2004) Completely biodegradable composites of poly(propylene carbonate) and short, lignocellulose fiber *Hildegardia populifolia*. J Polym Sci Part B: Polym Phys 42:666–675. https://doi.org/10.1002/polb.10761
55. Peng X, Zhou L, Jing B et al (2015) A high-performance electrospun thermoplastic gel polymer electrolyte for Li-ion batteries. J Solid State Electrochem. https://doi.org/10.1007/s10008-015-3030-5
56. Rajendran S, Sivakumar P, Shanker R (2007) Studies on the salt concentration of a PVdF–PVC based polymer blend electrolyte. J Power Sources 164:815–821. https://doi.org/10.1016/j.jpowsour.2006.09.011
57. Zhong Z, Cao Q, Jing B et al (2012) Electrospun PVdF–PVC nanofibrous polymer electrolytes for polymer lithium-ion batteries. Mater Sci Eng B 177:86–91. https://doi.org/10.1016/j.mseb.2011.09.008
58. Xi J, Qiu X, Li J et al (2006) PVDF-PEO blends based microporous polymer electrolyte: effect of PEO on pore configurations and ionic conductivity. J Power Sources 157:501–506. https://doi.org/10.1016/j.jpowsour.2005.08.009
59. Li W, Wu Y, Wang J et al (2015) Hybrid gel polymer electrolyte fabricated by electrospinning technology for polymer lithium-ion battery. Eur Polym J. https://doi.org/10.1016/j.eurpolymj.2015.04.014
60. Lee L, Kim IJ, Yang S, Kim S (2013) Electrochemical properties of PEO/PMMA blend-based polymer electrolytes using imidazolium salt-supported silica as a filler. Res Chem Intermed 39:3279–3290. https://doi.org/10.1007/s11164-012-0839-8
61. Mathew CM, Kesavan K, Rajendran S (2015) Structural and electrochemical analysis of PMMA based gel electrolyte membranes. Int J Electrochem 2015:1–7. https://doi.org/10.1155/2015/494308
62. Liang Y, Cheng S, Zhao J, Zhang C (2013) Preparation and characterization of electrospun PVDF/PMMA composite fibrous membranes-based separator for lithium-ion batteries. Adv Mater Res 752:1914–1918. https://doi.org/10.4028/www.scientific.net/AMR.750-752.1914
63. Yvonne T, Zhang C (2014) Properties of electrospun PVDF/PMMA/CA membrane as lithium based battery separator. Cellulose 21:2811–2818. https://doi.org/10.1007/s10570-014-0296-1
64. Wei C, Karuppiah C, Yang C et al (2019) Bifunctional perovskite electrocatalyst and PVDF/PET/PVDF separator integrated split test cell for high performance Li-O_2 battery. J Phys Chem Solids 133:67–78. https://doi.org/10.1016/j.jpcs.2019.05.012
65. Zhai Y, Wang N, Mao X et al (2014) Sandwich-structured PVdF/PMIA/PVdF nanofibrous separators with robust mechanical strength and thermal stability for lithium ion batteries. J Mater Chem A 2:14511–14518. https://doi.org/10.1039/c4ta02151g
66. Prasanth R, Shubha N, Hoon H, Srinivasan M (2014) Effect of poly(ethylene oxide) on ionic conductivity and electrochemical properties of poly(vinylidene fluoride) based polymer gel electrolytes prepared by electrospinning for lithium ion batteries. J Power Sources 245:283–291. https://doi.org/10.1016/j.jpowsour.2013.05.178

67. Wang X, Xing W, Feng X et al (2017) MoS2/polymer nanocomposites: preparation, properties, and applications. Polym Rev 57:440–466. https://doi.org/10.1080/15583724.2017.1309662
68. Scrosati B, Garche J (2010) Lithium batteries: status, prospects and future. J Power Sources 195:2419–2430. https://doi.org/10.1016/j.jpowsour.2009.11.048
69. Tarasova E, Naumenko E, Rozhina E et al (2019) Cytocompatibility and uptake of polycations-modified halloysite clay nanotubes. Appl Clay Sci 169:21–30. https://doi.org/10.1016/j.clay.2018.12.016
70. Lazzara G, Cavallaro G, Panchal A et al (2018) An assembly of organic-inorganic composites using halloysite clay nanotubes. Curr Opin Colloid Interface Sci 35:42–50. https://doi.org/10.1016/j.cocis.2018.01.002
71. Khalifa M, Janakiraman S, Ghosh S et al (2018) PVDF/halloysite nanocomposite-based non-wovens as gel polymer electrolyte for high safety lithium ion battery. Polym Compos 40:2320–2334. https://doi.org/10.1002/pc.25043
72. Wu D, Huang S, Zhao X et al (2008) Preparation of PVdF-based electrospun membranes and their application as separators. Sci Technol Adv 9:015005. https://doi.org/10.1088/1468-6996/9/1/015005
73. Wu N, Cao Q, Wang X et al (2011) In situ ceramic fillers of electrospun thermoplastic polyurethane/poly(vinylidene fluoride) based gel polymer electrolytes for Li-ion batteries. J Power Sources 196:9751–9756. https://doi.org/10.1016/j.jpowsour.2011.07.079
74. Yanilmaz M, Lu Y, Dirican M et al (2014) Nanoparticle-on-nanofiber hybrid membrane separators for lithium-ion batteries via combining electrospraying and electrospinning techniques. J Memb Sci 456:57–65. https://doi.org/10.1016/j.memsci.2014.01.022
75. Li W, Xing Y, Wu Y et al (2015) Study the effect of ion-complex on the properties of composite gel polymer electrolyte based on electrospun PVdF nano fibrous membrane. Electrochim Acta 151:289–296. https://doi.org/10.1016/j.electacta.2014.11.083
76. Zhu Y, Xiao S, Shi Y et al (2013) A composite gel polymer electrolyte with high performance based on poly(vinylidene fluoride) and polyborate for lithium ion batteries. Adv Energy Mater 4:1300647. https://doi.org/10.1002/aenm.201300647
77. Fang C, Yang S, Zhao X et al (2016) Electrospun montmorillonite modified polyvinylidene fluoride) nanocomposite separators for lithium-ion batteries. Mater Res Bull 79:1–7. https://doi.org/10.1016/j.materresbull.2016.02.015

Chapter 4
Electrospun Nanofibrous Polyvinylidene Fluoride-*co*-Hexaflouropropylene-Based Polymer Gel Electrolytes for Lithium-Ion Batteries

Jarin D. Joyner, M. J. Jabeen Fatima, Neethu T. M. Balakrishnan, N. S. Jishnu, and Raghavan Prasanth

4.1 Introduction

With the ever-increasing demand to power portable electronic devices, electric vehicles, and large industrial equipment via fossil fuel-less, alternative energy methods, lithium-ion batteries (LIBs) have remained one of the most viable energy storage options [1]. In comparison to preceding secondary battery technologies such as lead–acid, nickel–cadmium, and nickel–metal hydride systems, LIBs have exhibited superior properties that include longer lifetimes and higher energy densities as well as specific energies [2]. LIBs are comprised of three 'active' components that allow for the storage of chemical energy to then be converted into electrical energy: the negative electrode known as the anode, the positive electrode known as the cathode, and an electrolyte that is typically a non-aqueous-based system. During the charging process in battery cycling, the battery is exposed to an external electrical

J. D. Joyner (✉)
Department of Chemistry, Rice University, 6100 S Main Street, Houston, TX 77005, USA
e-mail: jarin.d.joyner@rice.edu

Department of Materials Science and Nano Engineering, Rice University, 6100 Main Street, Houston, TX 77005, USA

M. J. J. Fatima · N. T. M. Balakrishnan · R. Prasanth (✉)
Department of Polymer Science and Rubber Technology, Cochin University of Science and Technology (CUSAT), Cochin 682022, India
e-mail: prasanth@cusat.ac.in

N. S. Jishnu
Rubber Technology Centre, Indian Institute of Technology-Kharagpur (IIT-KGP), Kharagpur, West Bengal 721302, India

R. Prasanth
Department of Materials Engineering and Convergence Technology, Gyeongsang National University, 501 Jinju-Daero, Jinju 52828, Republic of Korea

N. T. M. Balakrishnan and R. Prasanth (eds.), *Electrospinning for Advanced Energy Storage Applications*, Materials Horizons: From Nature to Nanomaterials,
https://doi.org/10.1007/978-981-15-8844-0_4

power source, allowing for the migration of positively charged lithium ions (Li^+-ions) from the cathode to the anode along the electrolyte medium to then be oxidized by the reducing agent activity of the anode. Upon discharging, the battery is exposed to the respective device of interest and the Li^+-ions migrate in the reverse direction, i.e., from the anode to the cathode, and are effectively reduced by the 'oxidizing agent' nature of the cathode, carrying a current that allows for the battery to serve as the electrical power source. Typically, anodes in secondary LIBs consist of porous and/or layered carbon materials such as graphite to allow for the intercalation of lithium ions, while cathode materials consist of other intercalative materials such as lithium cobalt dioxide ($LiCoO_2$) and lithium iron phosphate ($LiFePO_4$), etc. [3]. The schematic of a typical LIB system is displayed in Fig. 4.1.

Despite not being an 'active' component in the storage of chemical energy to subsequently be converted into chemical energy in LIB systems, another constituent crucial to optimal battery performance is that of the separator. The primary function of the separator is to serve as a physical barrier and thus prevent contact and subsequent short circuiting between the interfaces of the anode and cathode components. In addition, the role of the separator is to also provide a path of free ionic transport between the anode and cathode. Because the various properties of separators nonetheless play an important role in features of battery performance such as energy

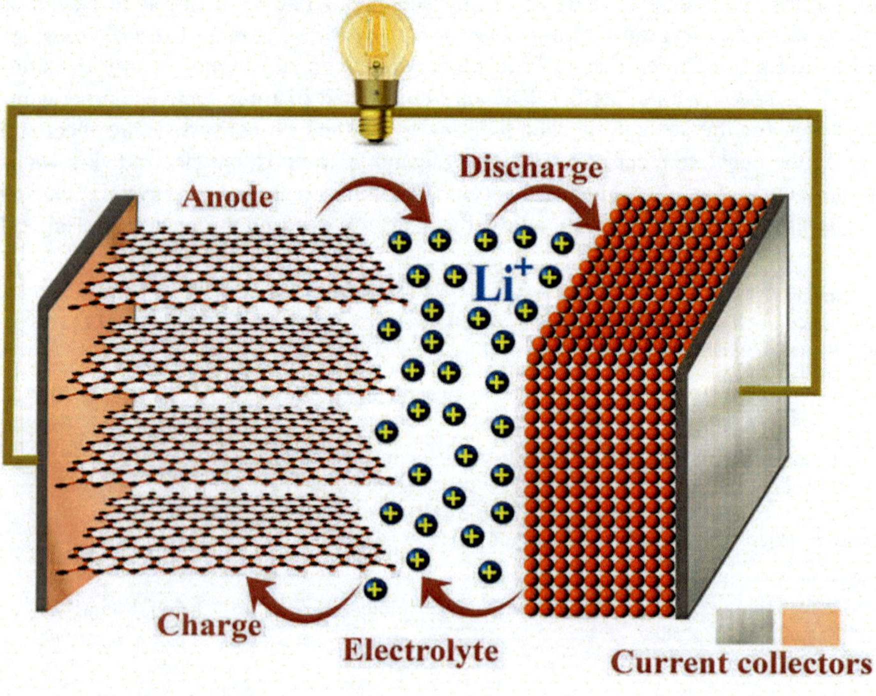

Fig. 4.1 Schematic representation of the main components of a LIB as well as the charging and discharging processes

Table 4.1 List of basic requirements for optimal separator performance in LIB systems

Property	Requirement
Mechanical property	>1000 kg·cm^{-1} (98.06 MPa)
Thickness	20–25 μm
Pore size	<1 μm
Porosity	40–60%
Permeability	<0.025 s·μm^{-1}
Thermal stability	<5% shrinkage after 60 min at 90 °C
Dimensional stability	No curl-up or shrinkage upon electrolyte wetting
Chemical and electrochemical stability	Stable over a large period of time (>250 cycles)

and power density, cycle life, and safety, there are a number of requirements (listed in Table 4.1) stipulated to ensure optimal separator utilization [4].

With respect to structural and morphological classifications, separators are typically made up of either a polymeric, micro-porous membrane (that features micropores <1 μm), non-woven fabric mats, or inorganic-based composite membranes [5]. In the case of polymeric micro-porous membranes, the processes for fabrication typically incorporates the heating of an extruded tubular film that is then stretched to form micropores (as well as other wet-phase techniques), while the development of non-woven fabric mats typically utilizes electrospinning techniques (and other techniques such as solution extrusion) that yield mats made up of randomly oriented microfibers [6, 7]. To date, the majority of separators that are considered the most reliable and utilized on a commercial basis are that of Celgard®, which has many different types of polymer specifications. These include Celgard® 2320 (a three-layer laminate made up of [PP] and [PE] layers), 2400, and 2500 (2400 and 2500 both being made up of a single PP layer but with varying pore sizes and percent porosities) [8, 9].

Despite being utilized on such a large commercial level due to their favorable porosity, electrochemical as well as chemical stability, and other mechanical properties, these aforementioned polyolefin-based separators have exhibited adverse properties mainly due to membrane shrinkage arising from processing techniques as well as characteristic softening and/or melting at unfavorably low temperatures (occurring between 120 and 170 °C) [10]. In addition, polyolefin-based separators also suffer from poor wettability and retention to polar electrolytes arising from their hydrophobic nature, leading to the search for alternative membrane technologies [11]. Thus, research involving more novel polymer-based membranes along with additional chemical and physical modifications to form polymer composites to be used as separators has been pursued. One such alternative to traditional polyolefin separators is that of co-polymerized PVdF-based separators as well as their subsequent physical and chemical modifications.

4.2 PVdF-*co*-HFP Nanofibers as Separators

4.2.1 Properties

PVdF is a semi-crystalline, non-reactive polymer consisting of a partially fluorinated backbone composed of approximately 59 wt% fluorine and 3 wt% hydrogen and is synthesized by polymerizing the vinylidene fluoride monomer [12]. PVdF exists in five crystalline polymorph types (α, β, γ, δ, and ε), with the α, β, and γ phases being of interest and the β phase being the most extensively used in electronic and energy storage applications due to its all-trans confirmation, which leads to the induction of significant dipole moments [13–15]. The structural confirmations of the α, β and γ phases can be seen in Fig. 4.2. PVdF, along with many of its co-polymeric counterparts such as trifluoroethylene [PVdF-*co*-TrFE], chlorotrifluoroethylene, [PVdF-*co*-CTFE] and hexaflouropropylene (PVdF-*co*-HFP), has been shown to have remarkable properties that are key for battery separators, which include excellent thermal and mechanical properties, chemical inertness, wettability by organic electrolytes, and relatively high dielectric constants [16]. The co-polymerization of PVdF with these additional fluorinated monomeric functional groups also decreases its crystallinity, while enhancing the ionic conductivity as well as anodic stability via the increase in fluorine content, which provides a source of strong electron withdrawing groups (-CF-) [16–18]. It has even been found that in comparison to commercially available separators, the critical surface tension of electrospun PVdF-*co*-HFP-based membranes is slightly lower ($_c$ of 25 mNm^{-2}), allowing for better wettability of non-aqueous electrolytes [19].

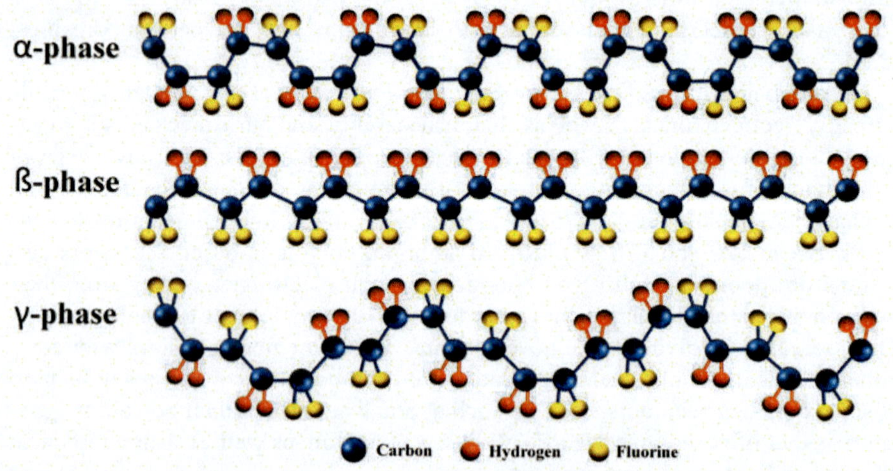

Fig. 4.2 Schematic representation of backbone chain confirmation of the α, β, and γ phases of PVdF

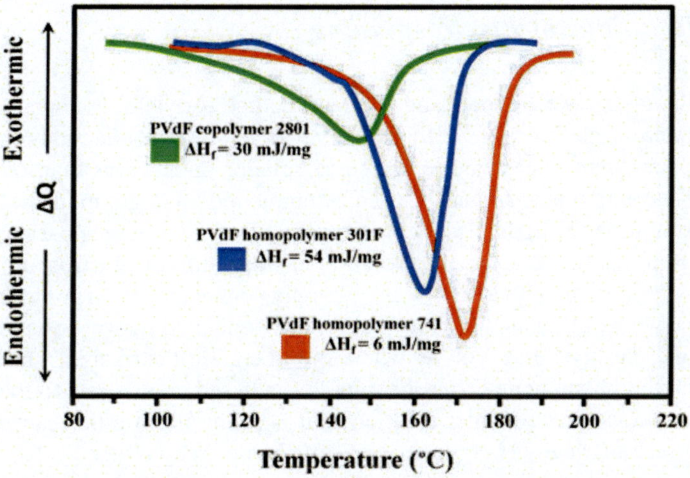

Fig. 4.3 DSC plot of the melting points of commercial PVdF homopolymers 301F and 741 as well as PVdF co-polymer 2801. Adapted and reproduced from Ref. [20]. Copyright 2000 Elsevier

When considering non-woven mat structures used as separators in LIBs, pristine PVdF is by far the most commonly studied polymer, but falls short in practical industrial use, because of its relatively higher degrees of crystallinity, which is theorized to result in lower ionic conductivities and stability (due to slower rates of Li migration) [4]. Differential scanning calorimetry (DSC) experiments carried out by Michot et al. [20] found that commercial PVdF-*co*-HFP membranes exhibit broader endothermic peaks at lower temperature ranges (100–160 °C) than its homopolymeric PVdF counterparts (145–180 °C). This observation of having a lower melting point (and thus lower heat of fusion) indicates that the respective co-polymer indeed has a lower degree of crystallinity than PVdF homopolymers (Fig. 4.3). Therefore, when considering the use of non-woven mats as separators, PVdF-*co*-HFP as an alternative to pristine PVdF has been viewed as a favorable option.

The enhanced ionic conductivity, electrochemical stability, and electrochemical performance of PVdF-*co*-HFP are attributed to the co-polymer's high dielectric constant ($\varepsilon = {\sim}8.4$) and favorable anodic stability, which can operate upwards of 5 V [20, 21]. Furthermore, previous studies have shown that when used as a separator in Li/$LiFePO_4$ cells (with Li serving as the anode and $LiFePO_4$ serving as the cathode) with $LiPF_6$ as well as other room temperature ionic liquid-based electrolytes, electrospun PVdF-*co*-HFP non-woven mats exhibit suitable electrochemical stabilities with respect to cycle performance [22]. However, to improve on this co-polymer's mechanical stability (<10 MPa) and thermal stability (due to a melting point of 172 °C), which is considered disadvantageous for LIBs that need to operate at high discharge rates as well as sustain fail-safe mechanisms [21, 23, 24], many studies have been employed in order to incorporate ceramic fillers, surface modifications, and core–shell structures.

4.2.2 Technique of Electrospinning

Electrospinning is a well-known and efficient method for yielding fibers on either a nano-, or micro-scale. Because of this, mats generated by the collection of randomly oriented electrospun fibers can be scaled to smaller thicknesses than other traditional mat-forming methods and thus can be employed on larger industrial levels to produce separators for LIBs [25]. A prototypical electrospinning apparatus consists of a syringe to contain the polymer solution of interest to be electrospun, a nozzle attached to the syringe with a high-voltage supply connected, and a grounded collection surface where upon an electrical potential is generated between that respective interface and the needle as well as where the fibers are distributed. The principle upon which the phenomena of electrospinning is based upon allots to the uniaxial elongation of a viscous polymer solution as it is spent out of a thin nozzle, which then forms an electrically charged droplet from an applied voltage [21].

This aforementioned charged droplet then forms a conically shaped jet known as a 'Taylor cone,' which arises from its interactions with electrostatic forces from both repulsive forces between the surface charges as well as Coulombic forces which originate from the external electric field [21, 26, 27]. When the applied potential reaches a certain voltage (typically >10 kV), the electrostatic forces overcome the surface tension of the polymer solution, in turn, resulting in the ejection of the jet and subsequent evaporation of the solvent leading to a non-woven fiber that is distributed on the collector [26]. This description can be seen in Fig. 4.4. As expected, many parameters and factors affect the morphology and nature of fibers that are electrospun and thus lead to a large number of related studies. These include properties of the polymer solution (such as the solution concentration and subsequent solution viscosity, relative molecular mass of the polymer, and conductivity of the polymer solution) as well as experimental parameters (such as the voltage of the applied external field, the distance from the nozzle tip to the collection interface, the flow rate of the polymer solution), and environmental factors (such as the ambient temperature and relative humidity) [28].

Studies carried out by Fridrikh et al. [29] as well as others [30] have entailed the computational and theoretical investigations of controlling the fiber diameter by analytically determining stability domains of the spinning jet based on the kinematic viscosity of the polymer solution, solution density, and intensity of the electrostatic field. Beachley et al. [31] reported that when using a parallel collection surface set up, the fibrous properties of polycaprolactone (PCL) such as average fiber diameter, uniformity, and quality were indeed affected by the distance between the parallel plates as well as the concentration of the PCL solution (in a dichloromethane: methanol, 7:3 mixture) and applied voltage. With respect to the electrospinning of fibers based upon PVdF, similar studies have also been carried out by Motamedi et al. [32]. These authors reported that increasing the PVdF solution concentration (w/v %) in *N, N'*-dimethylacetamide (DMAc)/acetone resulted in larger fiber diameters. However, increases in the distance from the spinning nozzle to the collection plate yielded smaller diameters of the PVdF fibers [32].

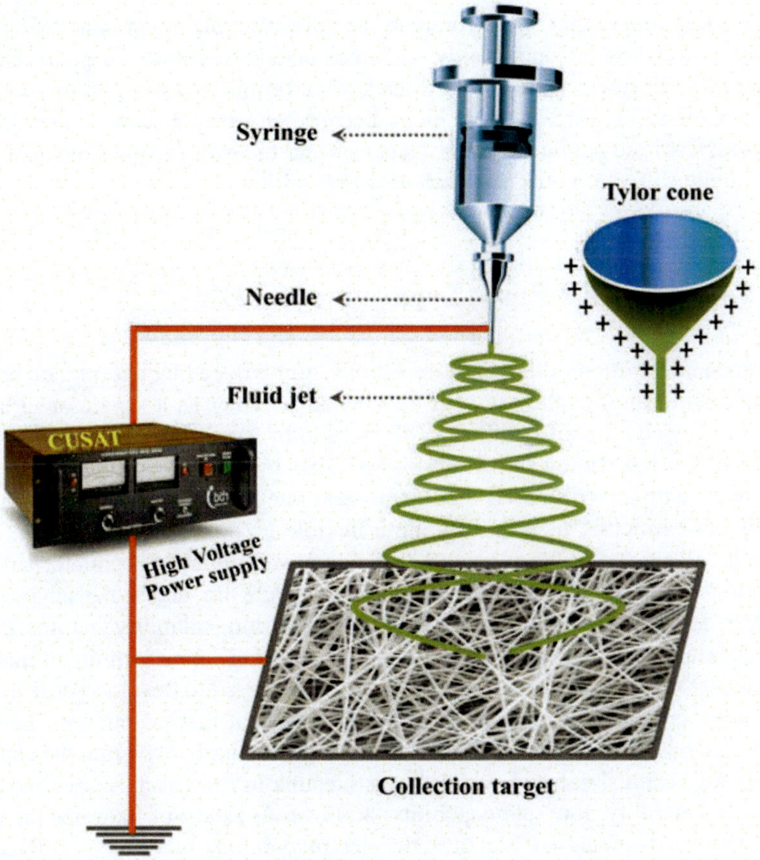

Fig. 4.4 Schematic of basic principles of electrospinning

With respect to the variation of the voltage applied during the electrospinning process, Gao et al. [33] found that with an increase in the voltage range of 8–15 kV, the diameter of PVdF nanofibers as well as crystallinity decreased. This observation of smaller diameters over higher voltages was suggested to be a result of the fact that a higher applied voltage leads to a higher charge density on the jet surface, which in turn results in a faster velocity and thus subsequent higher elongation forces of the jet as it is spent. In addition, because of the faster velocity of the charged jet, the time that it takes for the nanofibers to form and reach the grounded collector shortens, resulting in the maintenance of molecular chains in a non-crystalline state, hence the smaller degree of crystallinity in PVdF nanofibers at higher voltages. Zulifkar et al. [34] found that in order to limit the amount of bead defects on PVdF nanofibers when electrospun in *N, N'*-dimethylformamide (DMF), it was observed that flow rates of ~0.04 $mL{\cdot}min^{-1}$ lead to a minimal amount (and increase as flow rate decreases). This most likely is attributed to the fact that at higher flow rates, fibers are formed at faster rates resulting in a higher strain force on the collection plate and thus defects form at much slower rates. Concurrently, at larger nozzle-to-collection plate distances (~15 cm), PVdF fibers were found to have a more uniform morphology with less bead formation [34]. This is most likely because at larger distances, the electrical field strength of the applied voltage decreases and the evaporation time as the fiber settles on the collection interface increases [34, 35].

4.3 PVdF-*co*-HFP-Based Separators in LIBs

The first demonstration (although not a fibrous, non-woven mat but rather a polymer laminate membrane) of the use of a fluorinated polymer as a separator in an LIB system was indeed that of PVdF-*co*-HFP in 1996 by Tarascon et al. [36] This study resulted in the fabrication of an 'all-plastic' LIB by solvent casting the electrode components with a PVdF-*co*-HFP binder to generate flexible membranes along with a plasticizer extraction method forming a flexible PVdF-*co*-HFP separator. At the time, this 'all-plastic' battery proved to rival its traditional LIB counterparts with respect to gravimetric/volumetric energy density, cycle life, and self-discharge rates, while also demonstrating enhanced flexibility, safety, and scalability features. In addition to the plasticizer extraction method, a multitude of different techniques including solution casting and phase inversion have been employed to develop PVdF-*co*-HFP membranes as separators in other studies such as that carried out by Cheruvally et al. [37]. However, in the case of solvent casting and phase inversion, the challenge of removing residual solvent, which in turn results in drawbacks concerning electrochemical stability, low rate capability, and overall safety of batteries have been considered disadvantageous, and thus, the electrospinning of PVdF-*co*-HFP has been considered a preferred method [38].

One of the earliest examples of the preparation of fibrous membranes of PVdF-*co*-HFP via electrospinning was carried out by Kim et al. [39] In this study, it was found that the variation of the PVdF-*co*-HFP concentration in an acetone/DMAc (7:3 by weight) was a facile strategy to control fiber diameter, porosity, and pore size (Fig. 4.5a). These separators were able to achieve ionic conductivities above 1×10^{-3} S·cm^{-1} at room temperature, withstand an electrochemical window up to 4.5 V using a 1 M $LiPF_6$ ethylene carbonate/dimethylene carbonate/diethyl carbonate (EC/DMC/DEC, 1:1:1 by weight) electrolyte, and exhibited suitable charge/discharge properties with minimal amounts of capacity fading (Fig. 4.5b).

Li et al. [22] were able to achieve electrochemical stability potentials for PVdF-*co*-HFP-based separator membranes that exceeded 4.5 V (~5 V) using a 1 M $LiCF_3SO_3$ in tetra (ethylene glycol) dimethyl ether (TEGDME) electrolyte and demonstrated high initial discharge capacities and minimal capacity fading for longer cycles when using PVdF-*co*-HFP-based electrolytes prepared by electrospinning a 16% by weight solution in acetone/DMAC (7:3 by weight) at 18 kV.

More recent studies on these respective electrospun co-polymer-based membranes have considered the fact that porosity, uptake, and ionic conductivity can be as high

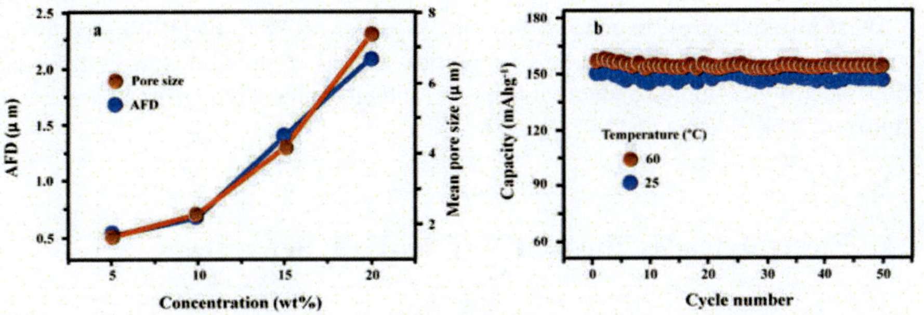

Fig. 4.5 **a** Average fiber diameter and mean pore size as a function of PVdF-*co*-HFP concentration in acetone/DMAC solvent, **b** cycle performance for a PVdF-*co*-HFP separator having an average fiber diameter of 0.5 μm (C/PE/LCO). Adapted and reproduced from Ref. [39]. Copyright 2005 Journal of Electrochemical Society

as 70%, 247%, and 3.2×10^{-3} S·cm^{-1}, respectively, using 1 M $LiPF_6$ in EC/DMC (1:2 v/v) despite poor mechanical integrity. This property was greatly improved by electrospinning a tri-layer that consisted of two PVdF-*co*-HFP layers that sandwich a polyvinyl chloride (PVC) layer. The stress–strain behavior of this tri-layer polymer electrolyte membrane is shown in Fig. 4.6 [19].

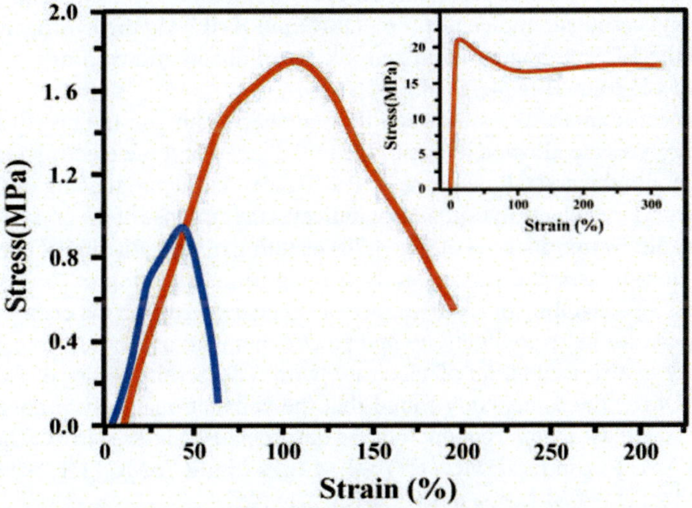

Fig. 4.6 Stress versus behavior of single layer of electrospun PVdF-*co*-HFP layer and electrospun PVdF-*co*-HFP/PVC/PVdF-*co*-HFP tri-layer. Adapted and reproduced from [19]. Copyright 2014 Elsevier

These promising results based off of the properties of PVdF-*co*-HFP-based fibers have led to further studies into how properties such as compatibility with different liquid electrolytes, ionic conductivity, uptake, and mechanical integrity can be further improved.

4.4 Electrolytes Used with PVdF-*co*-HFP Separators

The current state of both LIBs available on the market as well as research that revolves around LIBs mostly employs the use of non-aqueous (organic) aprotic, alkyl, carbonate-based electrolytes such as EC, DMC, DEC, and ethyl methyl carbonate (EMC) due to their suitable electrochemical stability, high ionic conductivity, and fairly low toxicity [40]. Lithium containing salts such as $LiPF_6$, LiTFSI, $LIClO_4$ are also typically added to the aforementioned electrolytes in order to further enhance the ionic conductivity and to assist in the formations of a solid electrolyte interface (SEI) layers [40, 41]. Due to the strong fluorinated groups present in PVdF-co-HFP separators that provide high polarity, as well as dielectric permittivity, these membranes exhibit a strong affinity for polar liquid electrolytes [42]. With respect to electrospun fibers of the pristine PVdF polymer, FT-Raman spectra studies carried out by Choi et al. [43] confirmed that strong interactions take place between the hydrogens of the methylene groups in the PVdF backbone and the carbonyl oxygen in linear alkyl carbonate molecules (i.e., DMC and EMC) via the hydrogen bonding (where as the EC component of mixed alkyl carbonates more so assists with the solvation of Li^+ ions) (Fig. 4.7a) [43, 44].

These specific interactions between the polymer chain and the alkyl carbonate-based electrolytes are allowed to occur due to the fact that these electrolytes are able to penetrate into the crystalline phase of PVdF (seen in the broadening of the PVdF fiber-based polymer electrolyte peak that indicates the collapse of the crystal structure in the X-ray diffraction analysis of Fig. 4.7b) resulting in the reduction of crystallinity of the polymer matrix and subsequently leading to the swelling of the PVdF fiber membranes. This swelling of PVdF to form a gel phase results in the enhancement of room temperature ionic conductivity and electrochemical properties. Because ionic conductivity is also a function of viscosity, with a decreasing order of EC > DEC > EMC > DMC, the same study found that the conventional polycarbonate-based electrolytes decrease in ionic conductivity in the order of 1 M $LiPF_6$ in EC/DMC/DEC (1:1:1) > 1 M $LiPF_6$ in EC/DMC (1:1) > 1 M $LiPF_6$ in EC/EMC (1:1) > 1 M $LiPF_6$ in EC/DEC (1:1) > 1 M $LiPF_6$ in EC/DMC (2:1) [45].

Solvent effects on characteristics such as uptake on PVdF and PVdF-*co*-HFP-based membranes have been summarized in Fig. 4.8, and different uptake ratios can be observed due to interactions with cations and/or anions and the polymer by the solvation process [42].

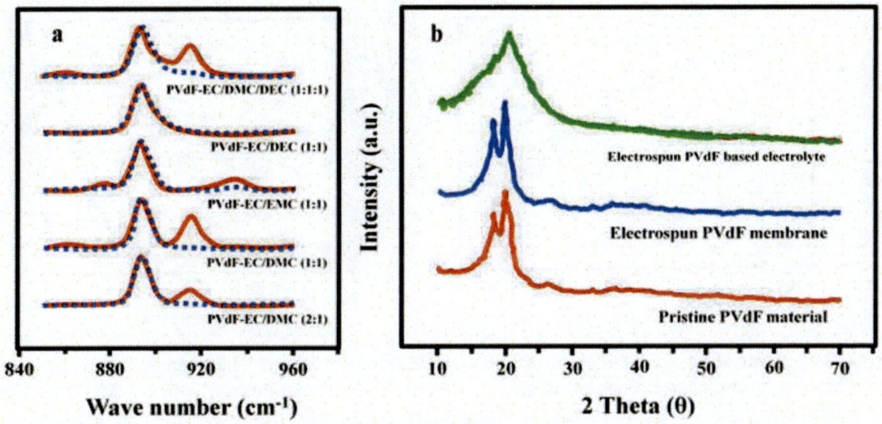

Fig. 4.7 **a** Normalized FT-Raman spectra of the EC breathing mode and C-O stretching mode of linear carbonates in mixed solvents and polymer with mixed solvents, **b** X-ray diffraction spectra of raw PVdF material along with corresponding electrospun PVdF fiber-based membrane and PVdF fiber-based membrane soaked in 1 M $LiPF_6$ in EC/DMC/DEC (1:1:1 by weight). Adapted and reproduced from [43]. Copyright 2007 American Chemical Society

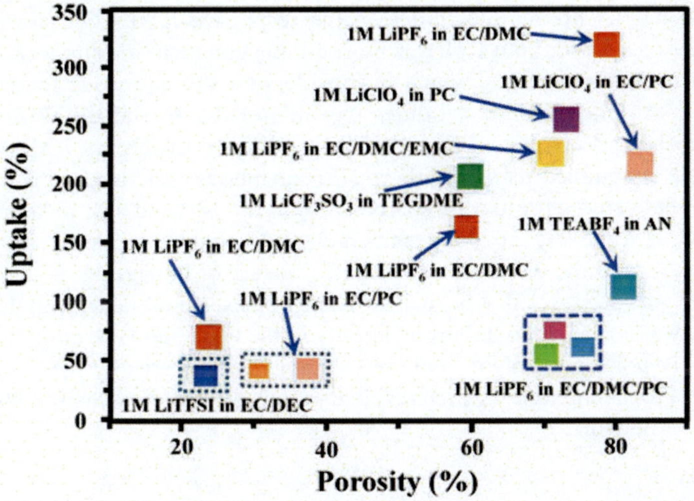

Fig. 4.8 Porosity versus percentage uptake of various electrolyte solutions incorporated into PVdF membranes. Adapted and reproduced from Ref. [42]. Copyright 2013 Royal Society of Chemistry

In addition to conventional organic-based electrolytes, room temperature ionic liquids (RTILs) have drawn much attention as electrolytes in LIBs due to physical properties such as low volatility and non-flammability and other characteristics such as thermal stability, tunability (due to selectability of cations and anions), and recyclability [45]. Cheruvally et al. [37] investigated the

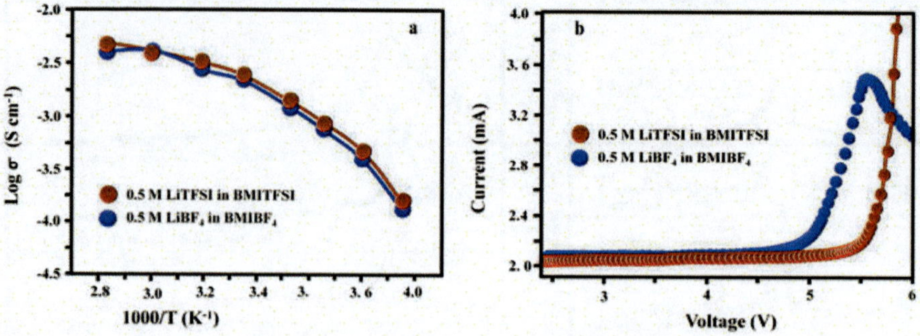

Fig. 4.9 **a** Variation of ionic conductivity with temperature of electrospun PVdF-*co*-HFP membranes, **b** anodic stabilities of electrospun PVdF-*co*-HFP membranes with 0.5 M LiTFSI in BMITFSI and 0.5 M $LiBF_4$ in $BMIBF_4$. Adapted and reproduced from Ref. [37]. Copyright 2007 Elsevier

compatibility of electrospun PVdF-*co*-HFP fibers with two RTILs, a 0.5 M solution of lithium bis(trifluoromethanesulfonyl) imide (LiTFSI) in 1-butyl-3-methylimidazolium bis(trifluoromethanesulfonyl) imide (BMITFSI) and a 0.5 M solution of $LiBF_4$ in 1-butyl-3-methylimidazolium tetrafluoroborate ($BMIBF_4$). By immersing the fibrous membrane in the respective RTILs and assembling a Li/$LiFePO_4$-based cell, both RTILs achieved ionic conductivities up to 2.3×10^{-3} $S{\cdot}cm^{-1}$ at room temperature, which is considered a suitable value for LIBs. The RTILs also exhibited anodic stabilities upwards of 4.5 V, and the PVdF-*co*-HFP membrane showed compatibility specifically with the LiTFSI in BMITFSI. The lithium cells assembled using PVdF-*co*-HFP membrane activated with LiTFSI in BMITFSI, delivered an initial discharge capacity of 149 and 132 $mAh{\cdot}g^{-1}$ at 0.1 and 0.5 °C, respectively, at room temperature and showed stable cycling properties (Fig. 4.9), and the electrochemical performance of the system is displayed in Fig. 4.10. It was proposed that the performance of these membranes with the respective RTILs was attributed to the strong affinity of the PVdF-*co*-HFP co-polymer with the relatively polar electrolytes and the available surface area by the ~1 μm fiber diameters of the membranes, which resulted in high electrolyte retention throughout the separator network.

*4.4.1 Additives and Surface Modifications to PVdF-*co*-HFP Separator Membranes*

One of the most important intrinsic properties that allows for a favorable separator and thus good LIB performance is that of room temperature ionic conductivity. Ionic conductivity, which is measured in $S{\cdot}cm^{-1}$ (or $mS{\cdot}cm^{-1}$) and influences the power and energy of the battery, is mostly correlated to the degree of porosity and thus also related to electrolyte uptake of a membrane. [20, 46] The minimum value

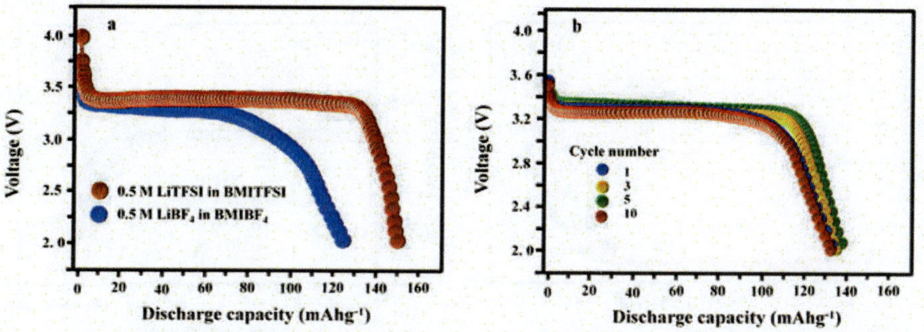

Fig. 4.10 **a** Initial discharge capacities of electrospun PVdF-*co*-HFP membranes with 0.5 M LiTFSI in BMITFSI and 0.5 M $LiBF_4$ in $BMIBF_4$ (0.1 C, RT). **b** Discharge capacity of electrospun PVdF-*co*-HFP membrane with 0.5 M LiTFSI in BMITFSI (0.5 C, RT). Adapted and reproduced from Ref. [37]. Copyright 2007 Elsevier

of ionic conductivity to ensure suitable battery performance is $>1\times10^{-4}$ S·cm^{-1}, and the maximum value of ionic conductivity is essentially limited by that of the ionic conductivity of the respective liquid electrolyte used for activating the polymer membrane [46]. The electrolyte uptake is not only determined by the membrane porosity, but also degree of crystallinity and morphology (via the filling of lattice voids) of the separator material, in order to yield good battery performance. It has been shown that separators that have a degree of crystallinity above 50% exhibit lower compatibility with liquid electrolytes [46, 47].

Although there is not a specific, defined relationship between ionic conductivity and electrolyte uptake, the low uptake values result in high resistivity since there are lack of conduction paths for Li^+ ion migration leading to poor cycling performance [39, 46]. With respect to porosity, which is characterized by the bulk fraction of void space in a porous sample, studies regarding PVdF-based separators have shown that for good battery performance >80% with pore diameters <1 μm are ideal [42, 46]. In order to improve on these properties of PVdF-*co*-HFP membranes, many studies involving the addition of ceramic fillers and surface modifications have been carried out.

Raghavan et al. [48] reported the direct addition of metal oxides such as aluminum oxide (Al_2O_3), silicon dioxide (SiO_2), and barium titanate ($BaTiO_3$) into PVdF-*co*-HFP solutions (6% by weight to total polymer solution) and electrospun into fibrous membranes having thicknesses of ~50 μm [46]. This process yielded activated fibrous membranes with ionic conductivities of 5.92×10^{-3}, 6.45×10^{-3}, and 7.21×10^{-3} S·cm^{-1} for Al_2O_3, SiO_2, and $BaTiO_3$, respectively. These enhancements of ionic conductivity in comparison to previously PVdF-*co*-HFP (such as those carried out by Kim et al. [39] can be considered a result of the Lewis acid/base-type interactions between the O/OH functional groups from the ceramic oxides interacting with the amorphous phase of the polymer fibers [39, 48, 49]. The addition of the ceramic fillers and subsequent electrospinning also assisted in the formation of a highly interconnected pore structure within the membrane that leads to electrolyte uptake

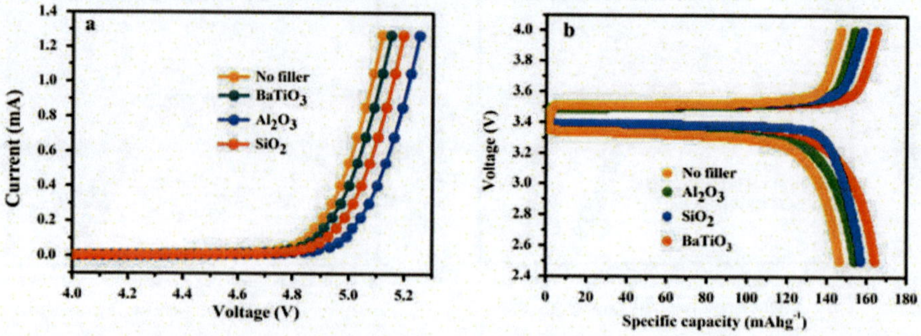

Fig. 4.11 **a** Electrolyte uptake (%) with 1 M $LiPF_6$ in EC/DMC (1/1 v/v), **b** Anodic stability using LSV of PVdF-*co*-HFP fibrous membranes with different ceramic fillers. Adapted and reproduced from Ref. [48]. Copyright 2008 Elsevier

percentages of ~460% as opposed to pristine, electrospun PVdF-*co*-HFP membrane [48]. In the same study, it was found that the addition of the ceramic fillers overall enhance mechanical properties such as tensile strength and modulus as well as slightly enhancing the anodic stability to a potential of 4.9 V, and in the case of $BaTiO_3$ anodic stability was 4.7 V (both vs. Li/Li^+), which was that of the pristine polymer fibers (Fig. 4.11a). The cell with $BaTiO_3$-based electrolyte exhibits a discharge capacity of 164 mAh·g^{-1} which is 96.5% utilization of material (Fig. 4.11b).

The $BaTiO_3$ incorporated PVdF-*co*-HFP membranes also lead to an enhancement in charge–discharge capacity (164 mAh·g^{-1}, which is 20 mAh·g^{-1} higher than pristine PVdF-*co*-HFP membrane) and stable cyclability with Li/$LiFePO_4$ cathode in lithium cells at room temperature.

Zhang et al. [50] carried out a study that consisted of incorporating different concentrations of a SiO_2 sol–gel [prepared by hydrolyzing SiO_2 nanoparticles in a tetraethoxysilane (TEOS): ethanol (1:1, v:v) solution] into a homopolymer PVdF solution (DMAC:acetone, 7:3 w:w) prior to electrospinning. Between the two concentrations (one consisting of 9:1, PVdF solution: SiO_2 sol–gel and the other, 10:1 PVdF solution: SiO_2 sol–gel, w:w), it was found that the higher concentration of SiO_2 in the PVdF membranes led to more porous fibers within the membranes, potentially yielding more tortuous paths for ion movement and thus yielding a better ionic conductivity. The higher SiO_2 concentration-based fibrous membranes also exhibited higher electrolyte uptakes, leading to improved discharge capacities and rate capabilities over various C-rates (Fig. 4.12) that could further be considered a significantly enhanced membrane in comparison to Celgard® PP membranes.

In addition, Fig. 4.13a, b show that the 9:1, PVdF solution: SiO_2 sol–gel nanocomposite fibrous membrane retains its discharge capacity of 159 mAh·g^{-1} which is nearly 100% and delivers high efficiency over 50 charge/discharge cycles. The principles in this study can most likely be used as a rationale for the incorporation of SiO_2 sol–gels in PVdF-*co*-HFP solutions prior to electrospinning.

The direct addition of SiO_2 nanoparticles into PVdF-*co*-HFP pre-cursor solutions (as opposed to SiO_2 sol–gels) has actually been more recently studied by

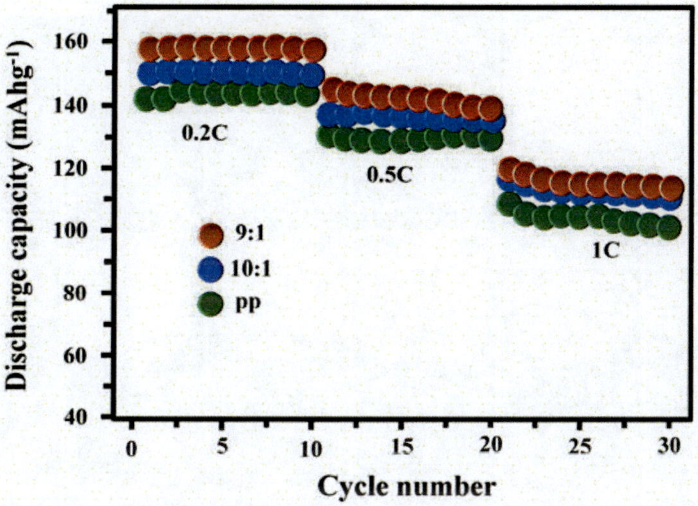

Fig. 4.12 Discharge rate capabilities of SiO_2 sol–gel: PVdF composite membranes and PP membranes. Adapted and reproduced from Ref. [50]. Copyright 2014 Elsevier

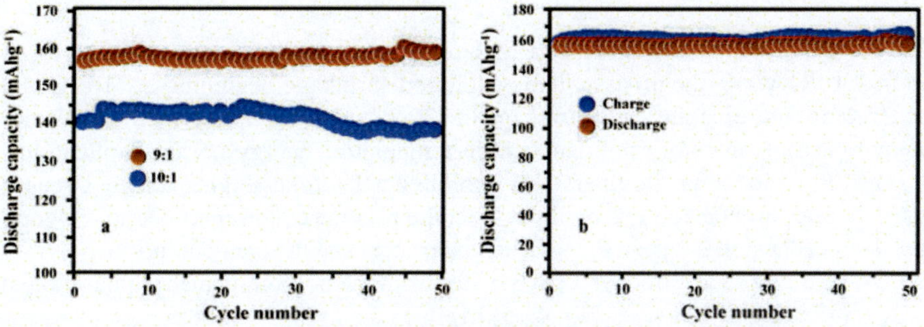

Fig. 4.13 **a** Cycle performance of a Celgard PP membrane and (9:1) PVdF/SiO_2 composite membrane separator at a 0.2C rate, **b** Charge-discharge efficiency for the (9:1) PVDF/SiO_2 composite membrane separator at a 0.2 C rate. Adapted and reproduced from Ref. [50]. Copyright 2020 Elsevier

Solarajan et al. [51] In this report, different weight percentages (0, 3, 5, 7, and 10%) of SiO_2 nanoparticles (~15 nm in size) in a 16% solution of PVdF-*co*-HFP in acetone/DMAC (7:3, w:w) were electrospun and activated with 1 M $LiPF_6$ in EC: DMC (1:1 v:v). It was found that the weight percentage of 7% showed the optimal porosity (~86%) as well as electrolyte uptake (412%), and the highest ionic conductivity (2.26×10^{-3} $S{\cdot}cm^{-1}$) in comparison to the other weight percentages of PVdF-*co*-HFP/SiO_2 nanocomposite fibers. In addition, the 7%-based nanocomposite improved the thermal properties such as thermal shrinkage as well as other mechanical properties in comparison to the pristine PVdF-*co*-HFP electrospun fibers.

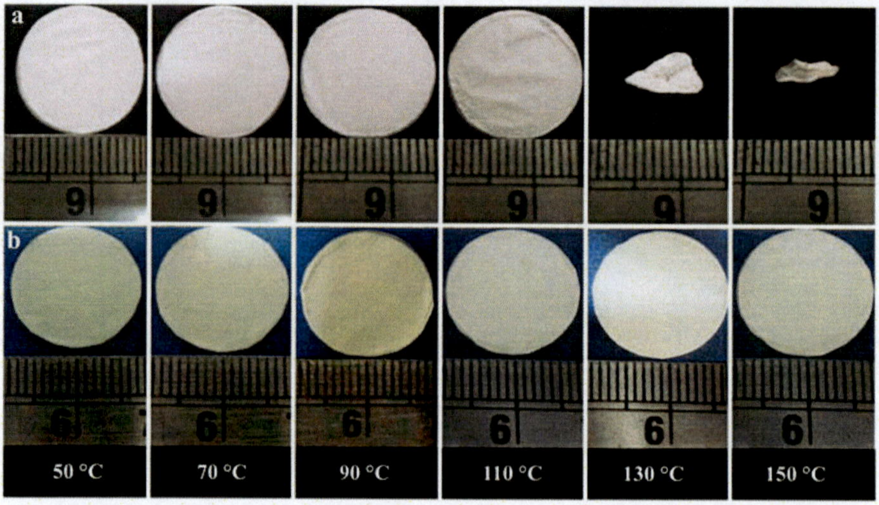

Fig. 4.14 Thermal shrinkage of **a** Pristine PVdF-*co*-HFP nanofibrous membranes **b** 7% SiO_2 incorporated into PVdF-*co*-HFP as functions of temperature. Adapted and reproduced from Ref. [51]. Copyright 2017 John Wiley and Sons

The 7%-based nanocomposites were able to maintain their dimensional integrity of up to 150 °C, while the pristine fibers exhibited shrinkage beginning at 110 °C and undergoing further significant shrinkage at 130 °C (Fig. 4.14), rendering the former membrane as a separator candidate in high-temperature energy storage applications.

The 7% nanocomposite fibers also exhibited a 1.6 times higher tensile strength and 1.3 times higher Young's modulus than the pristine counterpart due to the good interfacial adhesion between the SiO_2 nanoparticles and the nanocomposite polymer matrix, resulting in an effective load transfer from the polymeric matrix phase to the SiO_2 reinforcing filler phase (Fig. 4.15) [51].

Padmaraj et al. [52] investigated the oxide ceramic, zinc aluminate ($ZnAl_2O_4$) by incorporating it into electrospun PVdF-*co*-HFP membranes by way of addition of $ZnAl_2O_4$ (in weight-based concentrations of 1, 2, 3, 5, and 8%) into a PVdF-*co*-HFP pre-cursor solution to be electrospun. The results of this study found that per the addition of the $ZnAl_2O_4$ nanoparticles, the formation of an additional amorphous phase, which leads to a potential increase in Li^+-ions via fine pore structures, occurred. However, concentrations as high as 8% of the respective nanoparticles led to the formation of a crystalline $ZnAl_2O_4$ phase in the nanocomposite fibers, in turn, impeding the path of Li^+-ions, and thus, it was found that a concentration of 5% was optimal. It was therefore reported that the 5% concentration yielded ionic conductivities as high as ~1.6×10^{-3} $S{\cdot}cm^{-1}$ and electrolyte uptakes >400%. When utilized with 1 M $LiPF_6$ in EC:DEC (1:1, v/v) in a $Li/LiCoO_2$ battery setup, the 5% $ZnAl_2O_4$ nanofibers withstood potentials up to 4.3 V (therefore rendering their potential capability with high-voltage cathode materials such as $LiCoO_2$ and $LiMnO_2$) and exhibited discharge capacities of 110 $mAh{\cdot}g^{-1}$ for the 30th cycle (at

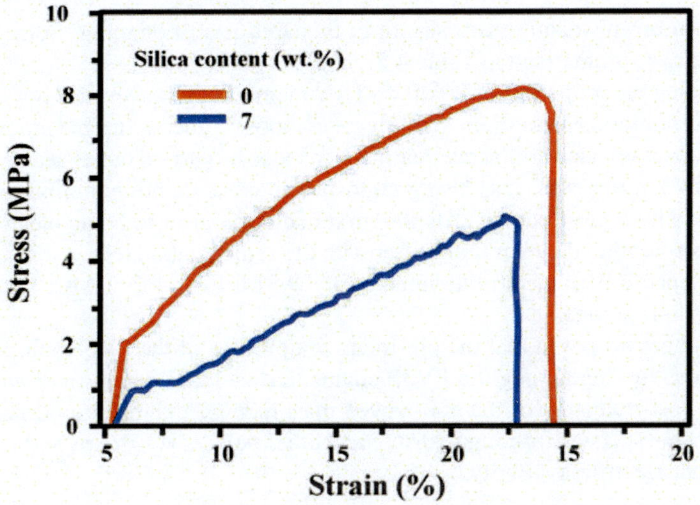

Fig. 4.15 Stress versus strain curves of electrospun polymer membrane (esPM) of PVdF-*co*-HFP and electrospun composite polymer membrane (esCPM) of 7% weight SiO_2. Adapted and reproduced from Ref. [51]. Copyright 2017 John Wiley and Sons

0.1 °C), which is in contrast to the discharge capacity of pristine PVdF-*co*-HFP (102 mAh·g^{-1}) (Fig. 4.16a). These aforementioned capacities of the 5% $ZnAl_2O_4$ nanofibers also displayed less fading in comparison to the pristine nanofibers over 30 cycles (Fig. 4.16b). In a separate report, Padmaraj et al. [53] also investigated another aluminate nanocomposite ($MgAl_2O_4$) as filler in fibrous PVdF-*co*-HFP membranes and found that the ionic conductivities, discharge capacities, and voltage stabilities were even higher than that of $ZnAl_2O_4$ nanocomposites.

Although studies specific to nanocomposite incorporation into PVdF-*co*-HFP precursor solutions to be specifically electrospun are limited, these types of membranes

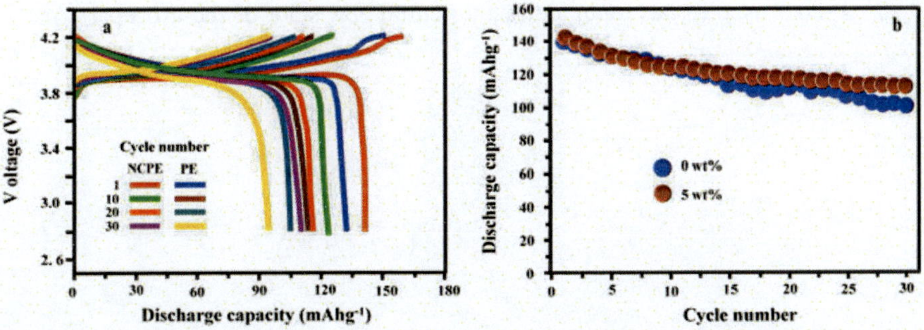

Fig. 4.16 **a** Charge/discharge capacities, **b** Cycle performance of over 30 cycles of Li/$LiCoO_2$ cells assembled with 5% $ZnAl_2O_4$/PVdF-*co*-HFP fibrous composite membrane as separator (at room temperature). Adapted and reproduced from Ref. [52]. Copyright 2014 John Wiley and Sons

have shown promise and thus research in this area is still ongoing. Some of those mentioned are summarized in Table 4.2.

Enhancing the properties of PVdF-*co*-HFP membranes, however, is not limited to utilizing ceramics as nanofillers to fabricate composite fibers. Surface modifications through processes such as plasma treatment can also be carried out as seen by a study reported by Laurita et al. [54] Using atmospheric pressure non-equilibrium plasma induction to both pre-treat a PVdF pre-cursor solution prior to being electrospun as well as post-treating the membrane after with the same method (Fig. 4.17) led to vast improvements to both electrolyte uptake (of 1 M $LiPF_6$ in EC: DMC, 1:1 v:v) and mechanical properties.

The authors proposed that the pre-treatment process of the PVdF solution led to further cross-linking among the PVdF chains that yielded 'defect-free' nanofibers, while the post-treatment of the non-woven mats induced chemical modifications of the membrane surface, forming hydroxy and other polar C=O groups and increasing the polymer electrolyte affinity.

4.5 Conclusions

Given that separators play an important role in not only the efficiency, but also safety in LIB systems, it has been seen that PVdF-*co*-HFP-based separators can serve as a much improved alternative to commercial and widely used PP/PE Celgard® separators. From this co-polymer's properties that make it ionically conducive to be utilized in energy storage applications as well as its very feasible processability via electrospinning, it remains an attractive and potential candidate as a separator. Furthermore, in order to improve on properties that are considered disadvantageous for separator applications, methods including the addition of nanoparticles to form nanocomposites as well as surface modifications can lend to facile studies to easily observe potential enhancements in room temperature ionic conductivity and electrochemical performance. There remains much room for further studies in the use of PVdF-*co*-HFP as a fibrous template as a potential new state-of-the-art separator.

Table 4.2 Properties of various nano-fillers in electrospun PVdF-*co*-HFP composite fibers

Nanofiller	Electrolyte uptake (%)	Porosity (%)	Ionic conductivity @ R.T. (mS cm^{-1})	LIB system (mixed electrolyte in 1:1 v/v)	Electrochemical activity features	Physical/mechanical properties	Ref.
No filler	425	84	4.2	$LiPF_6$ in EC/DMC(1:1, v:v) with Li/LiFePO4 Assembly	Discharge capacity of 145 mAh g^{-1} up to 30th cycle (at 0.1C-rate) Very high Coulombic efficiency (~100%). Favorable discharge capacity. Anodic Stability up to >4.5 V	Tensile strength of 6.5 MPa and Young's Modulus of 9.2 MPa,	[48]
*SiO_2 using TEOS: ethanol (1:1, v:v) 10%	646	85	7.5	$LiPF_6$ in EC/DEC with Li/$LiFePO_4$ cell	Discharge capacity of 159 mAh·g^{-1} up to 50th cycle (at 0.2C-rate). Very high Coulombic efficiency (~100%). Good cycling stability	N/A	[50]
Al_2O_3 6%	459	85	4.2	$LiPF_6$ in EC/DMC with Li/$LiFePO_4$ cell	Discharge capacity of 153 mAh·g^{-1} (1st cycle) at 0.1C-rate. Very high Coulombic efficiency (~100%). Anodic stability up to > 4.9 V.	Tensile strength of 9.2 MPa and Young's modulus of 15.8 MPa	[48]

(continued)

Table 4.2 (continued)

Nanofiller	Electrolyte uptake (%)	Porosity (%)	Ionic conductivity @ R.T. (mS cm^{-1})	LIB system (mixed electrolyte in 1:1 v/v)	Electrochemical activity features	Physical/mechanical properties	Ref.
SiO_2 6%	459	85	6.5	$LiPF_6$in EC/DMC with Li/$LiFePO_4$ cell	Discharge capacity of 156 mAh·g^{-1} (1st cycle) at 0.1 C-rate. Very high Coulombic efficiency (~100%). Anodic stability up to > 4.8 V	Tensile strength of 10.3 MPa and Young's Modulus of 16.9 MPa	[48]
$BaTiO_3$ 6%	462	87	7.2	$LiPF_6$in EC/DMC with Li/$LiFePO_4$ cell	Discharge capacity of 164 mAh·g^{-1} (1st cycle) and 156 mAh·g^{-1} (30th cycle) at 0.1C-rate. Very high coulombic efficiency (~100%). Good cycling stability Anodic stability up to 4.9 V	Tensile strength of 12.5 MPa and Young's Modulus of 17.3 MPa	[48]
$MgAl_2O_4$ 5%	415	N/A	2.8	$LiPF_6$ in EC/DEC with Li/$LiCoO_2$ cell	Discharge capacities of 158 mAh·g^{-1} (1st cycle) and 128 mAh·g^{-1} (30th cycle) at 0.1 C-rate. Very high Coulombic efficiency (~100%) Favorable cyclability	Sufficient mechanical strength with self-standing properties after saturation with liquid electrolyte	[53]

(continued)

Table 4.2 (continued)

Nanofiller	Electrolyte uptake (%)	Porosity (%)	Ionic conductivity @ R.T. ($mS\ cm^{-1}$)	LIB system (mixed electrolyte in 1:1 v/v)	Electrochemical activity features	Physical/mechanical properties	Ref.
$ZnAl_2O_4$ 5%	400	N/A	1.6	$LiPF_6$ in EC/DEC with $Li/LiCoO_2$ cell	Discharge capacities of 141 $mAh \cdot g^{-1}$ (1st cycle) and 110 $mAh \cdot g^{-1}$ (30th cycle) at 0.1 C-rate. Very high coulombic efficiency (~100%) favorable cyclability high stability window up to 4.3 V	Sufficient mechanical strength with self-standing properties after saturation with liquid electrolyte	[52]

*For electrospun homopolymer PVdF fibrous membranes

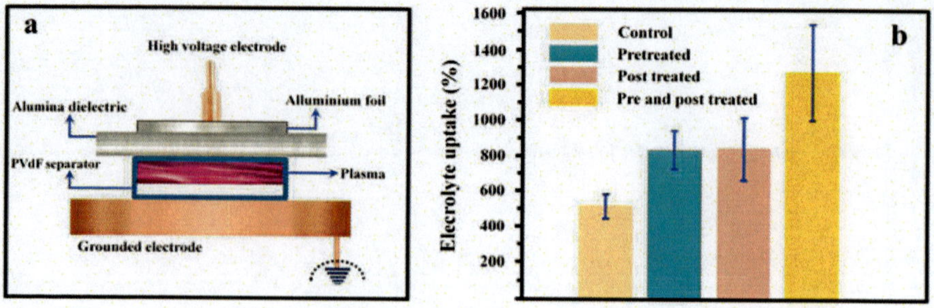

Fig. 4.17 **a** Schematic of post-treated PVdF non-woven membranes via a nano-second pulsed dielectric barrier discharge operating in ambient air. **b** Electrolyte uptake of PVdF electrospun separators. Adapted and reproduced from Ref. [54]. Copyright 2020 John Wiley and Sons

References

1. Li M, Lu J, Chen Z, Amine K (2018) 30 Years of lithium-Ion batteries. Adv Mater 30:1–24. https://doi.org/10.1002/adma.201800561
2. Alias N, Mohamad AA (2015) Advances of aqueous rechargeable lithium-Ion battery: a review. J Power Sources 274:237-251. https://doi.org/10.1016/j.jpowsour.2014.10.009
3. Wang Y, Liu B, Li Q et al (2015) Lithium and lithium Ion batteries for applications in microelectronic devices: a review. J Power Sources 286:330–345. https://doi.org/10.1016/j.jpowsour.2015.03.164
4. Lee H, Yanilmaz M, Toprakci O et al (2014) A review of recent developments in membrane separators for rechargeable lithium Ion batteries. Energy Environ Sci 7:3857–3886. https://doi.org/10.1039/c4ee01432d
5. Wang Q, Mao B, Stoliarov SI, Sun J (2019) A Review of Lithium Ion Battery Failure Mechanisms and Fire Prevention Strategies. Prog Energy Combust Sci 73:95–131. https://doi.org/10.1016/j.pecs.2019.03.002
6. Yang M, Hou J (2012) Membranes in lithium Ion batteries. membranes (Basel) 2:367–383. https://doi.org/10.3390/membranes2030367
7. Miao Y, Zhu G, Hou H et al (2013) Electrospun polyimide nano fiber-based nonwoven separators for lithium-Ion batteries. J Power Sources 226:82–86. https://doi.org/10.1016/j.jpowsour.2012.10.027
8. Love CT (2011) Thermomechanical analysis and durability of commercial micro-porous polymer li-Ion battery separators. J Power Sources 196:2905–2912. https://doi.org/10.1016/j.jpowsour.2010.10.083
9. Li Y, Pu H, Wei Y (2018) Polypropylene/polyethylene multilayer separators with enhanced thermal stability for lithium-Ion battery via multilayer coextrusion. Electrochim Acta 264:140–149. https://doi.org/10.1016/j.electacta.2018.01.114
10. Huang X (2014) Performance evaluation of a non-woven lithium Ion battery separator prepared through a paper-making process. J Power Sources 256:96–101. https://doi.org/10.1016/j.jpowsour.2014.01.080
11. Xu W, Wang Z, Shi L et al (2015) Layer-by-layer deposition of organic-inorganic hybrid multilayer on microporous polyethylene separator to enhance the electrochemical performance of lithium-ion battery. ACS Appl Mater Interfaces 7:20678–20686. https://doi.org/10.1021/acsami.5b05457
12. Cardoso VF, Correia DM, Ribeiro C et al (2018) Fluorinated polymers as smart materials for advanced biomedical applications. Polymers (Basel) 10:1–26. https://doi.org/10.3390/polym10020161

13. Ruan L, Yao X, Chang Y et al (2018) Properties and applications of the β phase poly(vinylidene fluoride). Polymer (Guildf) 10:1–27. https://doi.org/10.3390/polym10030228
14. Dutta B, Kar E, Bose N, Mukherjee S (2015) Significant enhancement of the electroactive β-phase of PVDF by incorporating hydrothermally synthesized copper oxide nanoparticles. RSC Adv 5:105422–405434. https://doi.org/10.1039/c5ra21903e
15. Wang X, Sun F, Yin G et al (2018) Tactile-sensing based on flexible PVDF nanofibers via electrospinning: a review. Sensors 18:330. https://doi.org/10.3390/s18020330
16. Barbosa JC, Dias JP, Lanceros-Méndez S, Costa CM (2018) Recent advances in poly(Vinylidene Fluoride) and its copolymers for Lithium-Ion battery separators. Membranes (Basel) 8:1–36. https://doi.org/10.3390/membranes8030045
17. Wan C, Bowen CR (2017) Multiscale-structuring of polyvinylidene fluoride for energy harvesting: the impact of molecular, micro- and macro-structure. R Soc Chem 5:3091–3128. https://doi.org/10.1039/C6TA09590A
18. Nunes-Pereira J, Costa CM, Lanceros-Méndez S (2015) Polymer composites and blends for battery separators: state of the art, challenges and future trends. J Power Sources 281:378–398. https://doi.org/10.1016/j.jpowsour.2015.02.010
19. Angulakshmi N, Stephan AM (2014) Electrospun trilayer polymeric membranes as separator for lithium-ion batteries. Electrochim Acta 127:167–172. https://doi.org/10.1016/j.electacta.2014.01.162
20. Michot T, Nishimoto A, Watanabe M (2000) Electrochemical properties of polymer gel electrolytes based on poly (Vinylidene Fluoride) copolymer and homopolymer. Electrochem Acta 45:1347–1360. https://doi.org/10.1016/S0013-4686(99)00343-6
21. Jung JW, Lee CL, Yu S, Kim ID (2016) Electrospun nanofibers as a platform for advanced secondary batteries: a comprehensive review. J Mater Chem A 4:703–750. https://doi.org/10.1039/c5ta06844d
22. Li X, Cheruvally G, Kim J et al (2007) Polymer electrolytes based on an electrospun poly (Vinylidene Fluoride- co -Hexafluoropropylene) membrane for lithium batteries. J Power Sources 167:491–498. https://doi.org/10.1016/j.jpowsour.2007.02.032
23. Zhai Y, Wang N, Mao X et al (2014) Sandwich-structured PVdF/PMIA/PVdF nanofibrous separator with robust mechanical strength and thermal stability for lithium ion batteries. J Mater Chem A 2:14511-14518. https://doi.org/10.1039/C4TA02151G
24. Orendorff CJ, Lambert TN, Chavez CA, Bencomo M, Fenton KR (2012) Polyester separators for lithium-ion cells: improving thermal stability and abuse tolerance. Adv Energy Mater 3:314–320. https://doi.org/10.1002/aenm.201200292
25. Persano L, Camposeo A, Tekmen C, Pisignano D (2013) Industrial upscaling of electrospinning and applications of polymer nanofibers: a review. Macromol Mater Eng 298: 504–520. https://doi.org/10.1002/mame.201200290
26. Garg K, Bowlin GL (2011) Electrospinning jets and nanofibrous structures. Biomicrofluidics 5:1–19. https://doi.org/10.1063/1.3567097
27. Li D, Xia Y (2004) Electrospinning of nanofibers: reinventing the wheel? Adv Mater 16:1151--1170. https://doi.org/10.1002/adma.200400719
28. Sun Y, Cheng S, Lu W et al (2019) Electrospun fibers and their application in drug controlled release, biological dressings, tissue repair, and Enzyme immobilization. RSC Adv 9:25712–25729. https://doi.org/10.1039/C9RA05012D
29. Fridrikh SV, Yu DH, Brenner MP, Rutledge GC (2003) Controlling the fiber diameter during electrospinning. Phys Rev Lett 90:144502–144505. https://doi.org/10.1103/PhysRevLett.90.144502
30. Cramariuc B, Cramariuc R, Scarlet R et al (2013) Fiber diameter in electrospinning process. J Electrostat 71:189–198. https://doi.org/10.1016/j.elstat.2012.12.018
31. Beachley V, Wen X (2009) Effect of electrospinning parameters on the Nanofiber diameter and length. Mater Sci Eng C 29:663–668. https://doi.org/10.1016/j.msec.2008.10.037
32. Motamedi AS, Mirzadeh H, Hajiesmaeilbaigi F et al (2017) Effect of electrospinning parameters on morphological properties of PVDF nanofibrous scaffolds. Prog Biomater 6:113–123. https://doi.org/10.1007/s40204-017-0071-0

33. Gao K, Hu X, Dai C, Yi T (2006) Crystal structures of electrospun PVDF membranes and its separator application for rechargeable lithium metal cells. Mater Sci Eng B 131:100–105. https://doi.org/10.1016/j.mseb.2006.03.035
34. Zulfikar MA, Afrianingsin I, Nasir M, Alni A (2018) Effect of processing parameters on the morphology of PVDF electrospun nanofiber. J Phys Conf Ser 987:1–5
35. Reneker DH, Yarin AL, Fong H, Koombhongse S (2000) Bending instability of electrically charged liquid jets of polymer solutions in electrospinning. J Appl Phys 87:4531–4547. https://doi.org/10.1063/1.373532
36. Tarascona J, Gozdz AS, Schmutz C et al (1996) Performance of bellcore's plastic rechargeable li-ion batteries. Solid State Ionics 2738:49–54. https://doi.org/10.1016/0167-2738(96)00330-X
37. Cheruvally G, Kim JK, Choi JW et al (2007) Electrospun polymer membrane activated with room temperature lonic liquid: novel polymer electrolytes for lithium batteries. J Power Sources 172:863–869. https://doi.org/10.1016/j.jpowsour.2007.07.057
38. Choi SW, Jo SM, Lee WS, Kim YR (2003) An electrospun poly(Vinylidene Fluoride) nanofibrous membrane and its battery applications. Adv Mater 15:2027–2032. https://doi.org/10.1002/adma.200304617
39. Kim JR, Choi SW, Jo SM et al (2005) Characterization and properties of P(VdF-HFP)-based fibrous polymer electrolyte membrane prepared by electrospinning. J Electrochem Soc 152:295–300. https://doi.org/10.1149/1.1839531
40. Xia L, Yu L, Hu D et al (2017) Electrolytes for electrochemical energy storage. Mater Chem Front 1:584–618. https://doi.org/10.1039/c6qm00169f
41. Younesi R, Veith GM, Johansson P et al (2015) Lithium salts for advanced lithium batteries: Li-Metal, Li-O_2 and Li-S. Energy Environ Sci 8:1905–1922. https://doi.org/10.1039/c5ee01215e
42. Costa CM, Silva MM, Lanceros-Mendez S (2013) Battery separators based on vinylidene fluoride (VDF) polymers and copolymers for lithium Ion battery applications. RSC Adv 3:11404–11417. https://doi.org/10.1039/c3ra40732b
43. Choi SW, Kim JR, Ahn YR et al (2007) Characterization of electrospun PVdF fiber-based polymer electrolytes. Chem Mater 19:104–115. https://doi.org/10.1021/cm060223
44. Song JM, Kang HR, Kim SW et al (2003) Electrochemical characteristics of phase-separated polymer electrolyte based on poly(Vinylidene Fluoride-co-Hexafluoropropane) and ethylene carbonate. Electrochim Acta 48:1339–1346. https://doi.org/10.1016/S0013-4686(02)00846-0
45. Ghandi K (2014) A review of ionic liquids, Their Limits and Applications. Green Sustain Chem 4:44–53. http://dx.doi.org/10.4236/gsc.2014.41008
46. Costa CM, Lee Y, Kim J et al (2019) Recent advances on separator membranes for lithium-ion battery applications: From Porous Membranes to Solid Electrolytes. Energy Storage Mater 22:346–375. https://doi.org/10.1016/j.ensm.2019.07.024
47. Quartarone E, Mustarelli P, Magistris A (2002) Transport properties of porous PVDF membranes. J Phys Chem B 106:10828–10833. https://doi.org/10.1021/jp0139843
48. Raghavan P, Zhao X, Kim J et al (2008) Ionic conductivity and electrochemical properties of nanocomposite polymer electrolytes based on electrospun poly (Vinylidene Fluoride-co-Hexafluoropropylene) with nano-sized ceramic fillers. Electrochem Acta 54:228–234. https://doi.org/10.1016/j.electacta.2008.08.007
49. Dissanayake MAKL, Jayathilaka PARD, Bokalawala RSP (2003) Effect of concentration and grain size of alumina filler on the ionic conductivity enhancement of the $(PEO)_9LiCF_3SO_3$: Al_2O_3 Composite polymer electrolyte. J Power Sources 121:409–414. https://doi.org/10.1016/S0378-7753(03)00262-3
50. Zhang F, Ma X, Cao C et al (2014) Poly (Vinylidene Fluoride)/ SiO_2 composite membranes prepared by electrospinning and their excellent properties for nonwoven separators for lithium-ion batteries. J Power Sources 251:423–431. https://doi.org/10.1016/j.jpowsour.2013.11.079
51. Solarajan AK, Murugadoss V, Angaiah S (2017) High performance electrospun PVdF-HFP SiO_2 nanocomposite membrane electrolyte for Li-Ion capacitors. J Appl Polym Sci 45177:1–10. https://doi.org/10.1002/app.45177

52. Padmaraj O, Venkateswarlu M, Satyanarayana N (2014) Characterization and electrochemical properties of P(VdF-co-HFP) based electrospun nanocomposite fibrous polymer electrolyte membrane for lithium battery applications. Electroanalysis 26:2373–2379. https://doi.org/10.1002/elan.201400232
53. Padmaraj O, Rao BN, Jena P et al (2014) Electrochemical studies of electrospun organic/inorganic hybrid nanocomposite fibrous polymer electrolyte for lithium battery. Polymer (Guildf) 55:1136–1142. https://doi.org/10.1016/j.polymer.2014.01.015
54. Laurita R, Zaccaria M, Gherardi M et al (2016) Plasma processing of electrospun Li-Ion battery separators to improve electrolyte uptake. Plasma Process Polym 13:124–133. https://doi.org/10.1002/ppap.201500145

Chapter 5
Electrospun Polyacrylonitrile (PAN)-Based Polymer Gel Electrolytes for Lithium-Ion Batteries

Neethu T. M. Balakrishnan, Jarin D. Joyner, N. S. Jishnu, Akhila Das, M. J. Jabeen Fatima, and Raghavan Prasanth

5.1 Introduction

The desire of long lasting, efficient and compact power sources are upturning day by day. Powering the electronic devices are became a serious concern when there arise problems regarding the extinction of non-renewable sources of energy that were constantly used for the energy production. From this scenario, as a solution of this uneasiness the exploitation of renewable sources was introduced and for the complete usage of these sources, energy storage devices were introduced that can effectively store the energy produced from the green and sustainable energy sources and can use it whenever there is a requirement of it [1–3]. Such energy storage devices include capacitors [4], supercapacitors [5], fuel cells [6] and batteries [7]. In most of the electronic devices, especially for portable electronic gadgets, battery serves as the prime powering source. Rechargeable batteries are considered to be worth in use than non-rechargeable systems because, it can be reused by recharging and thereby

N. T. M. Balakrishnan · A. Das · M. J. Jabeen Fatima (✉) · R. Prasanth (✉)
Department of Polymer Science and Rubber Technology, Cochin University of Science and Technology (CUSAT), Cochin 682022, India
e-mail: jabeen@cusat.ac.in

R. Prasanth
e-mail: prasanth@cusat.ac.in

J. D. Joyner
Department of Materials Science and NanoEngineering, Rice University, 6100 Main St, Houston, TX 77005, USA

N. S. Jishnu
Rubber Technology Centre, Indian Institute of Technology-Kharagpur (IIT-KGP), Kharagpur, West Bengal 721302, India

R. Prasanth
Department of Materials Engineering and Convergence Technology, Gyeongsang National University, 501 Jinju-Daero, Jinju 52828, Republic of Korea

N. T. M. Balakrishnan and R. Prasanth (eds.), *Electrospinning for Advanced Energy Storage Applications*, Materials Horizons: From Nature to Nanomaterials,
https://doi.org/10.1007/978-981-15-8844-0_5

reduce the waste and cost which is not possible with non-rechargeable batteries [8, 9]. Apart from this, the desire of electric vehicle further demands the research over the rechargeable battery systems that can efficiently full fill the advanced need of human beings. This headed to the discovery of lithium-ion batteries which is the most promising rechargeable system that can ensure good energy density and power density [10, 11]. Even though the main problem that faced while using the lithium-ion battery is its safety concern which limits its application, there are several thermal issues reported with the LIBs. On 2013, Boeing 787 Airlines experienced a thermal issue due to the presence of lithium-ion battery. This forced to remove the battery pack from all Nippon Airways. The main issue was related with the flammable electrolyte inside the battery. Similar thermal issues were reported with Samsung Note 7 and Tesla car. A series of fire issues were reported with Tesla Model S car in 2013 and in 2016 fire explosion was reported with the same in France. Tesla car driver by a British TV director Michael Morris burst into flames in 2018 and very recently the issue that took place with the parked Tesla car in Shanghai. These all incidents light up to the necessity of having a thermally stable system which can acquire by replacing the liquid electrolytes that used in LIBs.

In this scenario, electrolytes are the crucial part of lithium-ion batteries also known as the heart of battery. When LIBs are commercialized in 1990 by Sony the electrolyte system that introduced was a liquid electrolyte that can provide good ionic conduction as well can result a flexible system [12, 13]. But the atmospheric changes can impart an improper effect in the battery components even they are sealed inside an insulating envelop, the problem was raised at extreme conditions such as high or low temperature. During high temperature application or even during the continuous or long time charge discharge process the battery can heated up that will cause the heating up of the electrolyte which will get evaporate due to its low boiling point and finally leads to thermal runway and explosion. This scenario leads to the research over the development of electrolytes systems that could be safer and well performing [14, 15].

The use of polymer as electrolytes in LIBs, helps to reduce the safety problems that caused by the use of liquid electrolytes. Because, compared to liquid electrolytes they can withstand more temperature and thereby limit the chances for thermal issues [16, 17]. While replacing the liquid electrolytes with polymer electrolytes, the main difficulty facing is the room temperature ionic conductivity. In order to ensure better electrochemical properties a proper electrolyte with good ionic conduction is prime important. Choosing a proper materials and the right synthesis method is one of the best options that can bead opted to develop a well performing polymer electrolyte (PEs) with enhanced electrochemical properties. Different polymers such as polyethylene oxide (PEO) [18], polyacrylonitrile (PAN) [19], polymethyl methacrylate (PMMA) [20], polyvinylidene difluoride (PVdF) [21] and poly(vinylidene fluoride-*co*-hexafluoro propylene)(PVdF-*co*-HFP) [22, 23] are extensively studied as PEs in LIBs. PVdF-*co*-HFP is one of the best's polymers for electrolytes that can provide good ionic conductivity and dielectric constant ($\varepsilon \sim 8.4$) at room temperature. But the problem with the PVdF-*co*-HFP is that, a deterioration in battery performance is occurs due to the formation of a stable LiF and >

C=CF unsaturated bonds that formed by the interaction between the fluorine atom of PVdF and lithium ions of lithiated graphite. Further it causes safety problems due to the exothermic reactions taking place. Furthermore the inhibition of lithium ion transport by the crystalline part of P(VdF-*co*-HFP) results low charge-discharge capacities and poor Crates [24]. So the research is extended to other polymers and among the different systems studied. Polyacrylonitrile (PAN) is preeminent material for the preparation of polymer electrolytes in LIBs. Being a polymer material PAN possess good thermal stability along with good processability, oxidation resistance, electrochemical stability and high thermal stability [25]. The PAN based polymer electrolytes membranes are prepared by any conventional methods such as solvent casting or phase inversion normally observed to be highly brittle in nature due to the presence of adjacent cyanide group that will interact with each other and limits the chain flexibility [26]. So developing a flexible as well as porous electrolyte becomes an alarm. Among the various methods studied for the synthesis of PAN based polymer electrolytes electrospinning is considered to be the promising and simple method for the production of fibrous polymer electrolytes, especially for PAN, as they provide flexible membranes with good electrochemical properties. Electrospinning results the formation of PAN based polymer electrolytes (PEs) membranes with high porousity as well as good flexibility. It is because of the fact that the nitrile group that present in the PAN can hinder the alignment of macromolecules there by results the formation of flexible porous polymer electrolytes [27]. This chapter is discussing the various studies related with the development of PAN based PEs through electrospinning method in detail.

5.2 Preparation of Electrospun Polyacrylonitrile-Based Polymer Electrolytes

Electrospun PAN based fibrous polymer electrolytes are prepared by the electrospinning of PAN solution in organic solvents. Different organic solvents are employed for the preparation of electrospun polymer membrane. Choice of solvent is important because they can influence the morphology of the resulting membrane [28, 29] For PAN based electrospun membranes, DMF is mostly studied and dimethyl acetamide (DMAc) is seldom studied solvent [30]. The reason for the choice of DMF as the best solvent for PAN arise from its good interaction with the polymer as well as the polarity hence easy to forms a homogeneous solution [31]. As per the literature, only one study is reported with DMAc as the solvent for PAN based polymer electrolyte synthesized by electrospinning. About 12 wt% solution of PAN in DMAc forms a microporous fibrous membrane with smooth surface and high porosity of about 90% [32]. Similar to the solvents, the experimental parameters such as the collector distance and applied voltage have a great influence over the diameter of the resulting fibers [30]. The effect of electrospinning parameters over the fibrous morphology of PAN was studied by Prasanth et al. [24] by spinning the PAN solution in DMF. The effect of polymer concentration as well as applied voltage was systematically evaluated in the study. Four different concentrations, 10, 12, 14, 16 and 18 wt% and applied voltage 10, 15, 20 and 25 kV were used for the preparation of fibrous membrane. The membrane morphology (SEM images) shows the effect of solution concentration and applied voltage on the morphology and fiber diameter of the membranes. The study reveals that the concentration of polymer solution and applied voltage has significant influence over the average fiber diameter and fiber morphology. With increasing the concentration of solution and applied voltage the average fiber diameter (AFD) get increased which is the direct effect of increase in the viscosity of solution that will lead to large drops to come out from the syringe and deposited as fibers, results fibers having large diameter. The AFD of the fibers increases with increasing solution concentration, due to the higher viscosity which is directly related to the solution concentration. Higher solution viscosity leads to the ejection of larger drops of the dope solution from the tip of the spinning needle which forms fibers having larger diameter and collected on the collection drum [33]. At lower solution concentration (10, 12 and 14 wt%), the membranes displayed a broad range of fiber diameter and the morphology was not uniform. With 10 wt% solution the membrane showed poor morphology and partially beaded fibers with wide range of fiber diameter distribution. This shows that the concentration of 10 wt% solution might be the critical solution concentration to obtain good nanofibers with narrow fiber diameter distribution for electrospinning of PAN.

Similarly, the higher applied voltage ejects more fluid from the needle resulted a thicker fluid jet and leads to the formation of fibers with larger fiber [33]. In Prasanth et al. [24] study an applied voltage of 20 kV is found to be the optimum for obtaining the most uniform fiber morphology with straight fibers having smooth fiber surface and narrow distribution of fiber diameter. Application of higher voltage results in the formation of considerable amount of large fibers with narrow range of fiber diameter. At lower voltage the AFD of the fibers was lower, but it shows wide range of fiber diameter. Also the fibers are not completely bead free and as straight as those with higher applied voltages. In Prasanth et al. [24] study they were optimized 16 wt% polymer solution under 20 kV voltage as the best with uniform morphology and AFD of 350 nm (Fig. 5.1) [24]. The surface morphology of electrospun polyacrylonitrile fibrous membrane soaked in 1 M $LiPF_6$ in EC/DMC (1:1 v/v) after 1 h and 750 h is displayed in Fig. 5.2. The high porous nature of the membrane results high electrolyte absorption which is attributed by the presence of polar functional groups of PAN.

Followed by Prasanth et al. [24] similar study was reported by Carol et al. [34] and process optimization was done by changing the three important parameters: polymer concentration (6–14 wt%), flow rate (0.5–2.0 ml h^{-1}) and applied voltage (8–20 kV). Under each condition, PAN membranes were spun to about 200–250 nm thickness, which is lower than that reported by Prasanth et al. [24]. In Carol et al. [34] study observed that at low concentration (<8%) the fiber formed consists of large number of beads and are very difficult to handle due to the puffy nature. When the solution concentration increased to >12%, resulted in thicker fibers with non-uniform morphology. For an applied voltage 10–18 kV and a flow rate 1.0 and 2.0 ml

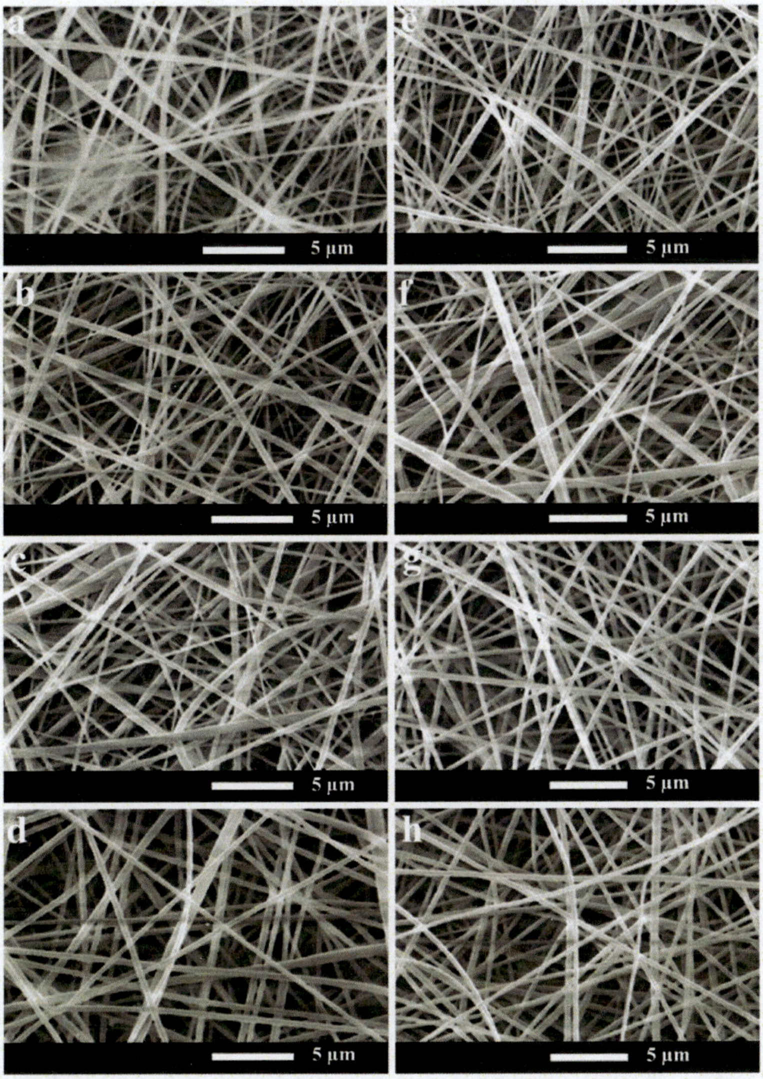

Fig. 5.1 FE-SEM images on the surface morphology of electrospun polyacrylonitrile fibrous membranes prepared with different polymer concentrations (wt%): **a** 10, **b** 12, **c** 14 and **d** 18; (applied voltage: 20 kV); and with different applied voltage (kV): **e** 10, **f** 15, **g** 20, **h** 25 (polymer concentration: 16 wt%) Adapted and reproduced from Ref. [24]. Copyright 2011 Elsevier

h^{-1} resulted smooth fiber structure and bead free uniform morphology [34]. The effect of electrospinning parameters on the characteristics of membranes in the study are displayed in Table 5.1.

Arifeen et al. [35] studied the influence of thermal pressing and pressure over the morphology of the electrospun PAN fibers produced from 10 wt% solution of PAN in

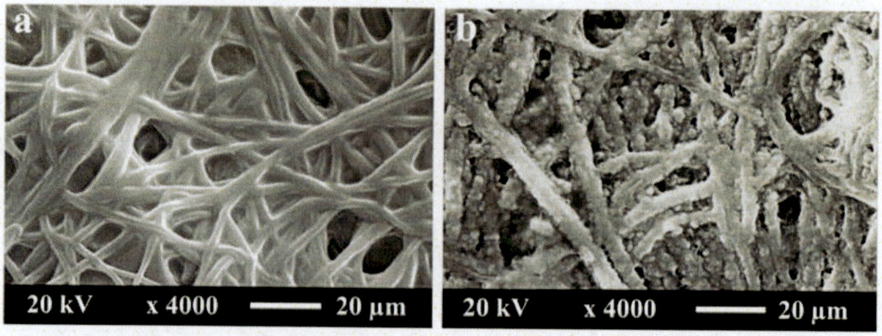

Fig. 5.2 FE-SEM images on the surface morphology of electrospun polyacrylonitrile fibrous membrane soaked in 1 M $LiPF_6$ in EC/DMC (1:1 v/v) after **a** 1 h, **b** 750 h. Adapted and reproduced from Ref. [24]. Copyright 2011 Elsevier

DMF. By varying the processing conditions three different membranes were prepared denoted as PAN 1, PAN 2 and PAN 3. The obtained fibers are thermally pressed at various pressures (20–50 MPa) and constant temperature and finally cross linked fibers were obtained. With increasing applied pressure flattened fibers are formed even though most fibers retained their shape (Fig. 5.3). The diameter exhibited by the fibers comes in between 320 and 550 nm [35].

Similar result in Prasanth et al. [24] and Carol et al. [34] studies, PAN solution having different concentration 7, 10 and 12 wt% were electrospunned and the fibers prepared at lowest concentration of 7 wt% exhibit a beaded morphology while the polymer solution concentration greater than 12 wt%, thicker fibers having a diameter close to 1 micro meters are formed, which is much higher compared to Prasanth et al. [36] (AFD was 350 nm) or Carol et al. [34] (AFD 900 nm) studies. Porous PAN based separator system developed by Chao et al. [37] using PAN in DMF exhibit an AFD of 350 nm and a better thermal stability than that of commercially using celgard separators. Other than PAN/DMF systems, a ternary system composed of PAN/DMF/Water was developed by Sabetzadeh et al. [37] appears to exhibit better electrochemical performance. The porous membrane fabricated by using this ternary system was compared with non-porous PAN membrane synthesized with similar diameter (700–800 nm). The porous PAN separator system exhibit an interconnected porous structure with a specific surface area of 52.6 m^2g^{-1} which is three times higher than that of the non-porous membrane [37].

5.3 Preparation of Electrospun Polyacrylonitrile-Based Composite Polymer Electrolytes

For a polymer electrolyte system the number of charge carriers as well as their mobility will determine the conductivity and electrochemical stability. Lithium salts

Table 5.1 Effect of electrospinning parameters on the characteristics of membranes obtained

Sample	Solution conc. (wt%)	Voltage (kV)	Flow rate (ml h^{-1})	Observations	Ref.
PAN-01	10	20	6	AFD of 0.28 μm, and wide fiber diameter distribution and partially beaded fibers having more fluffy structure	[24]
PAN-02	12	20	6	AFD of 0.3 μm, and narrow fiber diameter distribution having highest number of thin fibers, and bead free and uniform morphology	[24]
PAN-03	14	20	6	AFD of 0.32 μm, and wider fiber diameter distribution, and bead free morphology with more smooth fibers	[24]
PAN-04	16	20	6	AFD of 0.35 μm, very narrow fiber diameter distribution with significantly higher number of thin fibers and very straight fibers having very smooth fiber having very uniform morphology surface	[24]
PAN-05	18	20	6	AFD of 0.44 μm, wide fibers diameter distribution having higher fiber diameter, bead free morphology	[24]
PAN-06	16	10	6	AFD of 0.32 μm, wide fiber diameter distribution have large number of thin fibers, and bead free and straight fibers	[24]

(continued)

Table 5.1 (continued)

Sample	Solution conc. (wt%)	Voltage (kV)	Flow rate (ml h^{-1})	Observations	Ref.
PAN-07	16	15	6	AFD of 0.33 μm, wide range of fiber diameter distribution than other membranes, but large number of thin fibers than PAN-03 and PAN-05	[24]
PAN-08	16	25	6	AFD 0.40 m, narrow fiber diameter distribution but more number of larger fibers, straight and smooth fibers having uniform morphology	[24]
PAN-09	8	10	2.0	Fibers formed with beads in between, nearly uniform distribution, flimsy membrane	[34]
PAN-10	8	17	2.0	Thin fibers formed, nearly uniform distribution, better strength compared to PAN-09, difficult to handle	[34]
PAN-11	10	15	1.5	Fibers formed with uniform distribution, reasonably good handling strength	[34]
PAN-12	10	15	2.0	Fibers formed with uniform distribution and closer packing, stronger membrane with better handling strength than PAN-11	[34]
PAN-12	12	18	1.0	Thicker fibers formed with non-uniform distribution	[34]

have a remarkable impact over the determination of ionic conductivity of polymer electrolytes for LIBs. For a PAN based system PAN-$LiClO_4$ system is appeared to exhibit good ionic conductivity than the other lithium salts because of the smaller ionic radius of perchlorate ion [38]. Another approach for enhancing the performance of electrolyte is addition of fillers. Different fillers are employed in order to enhance

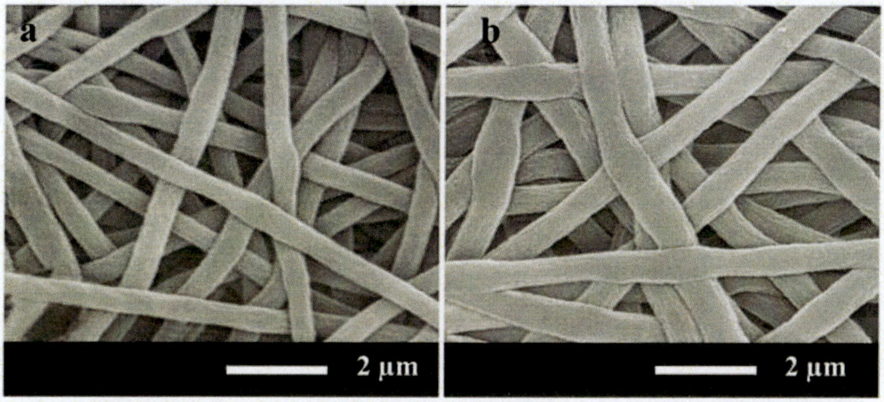

Fig. 5.3 FE-SEM images on the surface morphologies of the optimized polyacrylonitrile (PAN) membranes at different processing conditions: **a** PAN 1, **b** PAN 2. Adapted and reproduced from Ref. [35]. Copyright 2019 Elsevier

the property of fibrous PAN membrane. By the addition of fillers, the mobility of main chain will get reduced that will influence the glass transition temperature of the system as well as it can enhance the dissociation of lithium salts by Lewis acid-base interaction (Fig. 5.4) [39]. Moreover electrospinning results the formation of microporous structures rather than mesoporous structures. This leads to the polymer degradation and leakage of liquid electrolytes during cycling and finally results a poor discharge capability at high discharge rate. Addition of fillers to form composite membrane is chosen to be a best method to resolve this issue [40]. Synthesis of composite porous polymer membrane is rather simple by electrospinning method, because the filler can be simply added to the polymer solution employed [41].

Carbon allotropes such as carbon nanotubes (CNT) [42, 43] graphene and graphene derivatives [44, 45] are also employed for the property enhancement of electrospunned PAN but the application of such composites in lithium-ion batteries are not reported yet. For PAN based polymer electrolytes, incorporation of ceramic materials are extensively reported as one of the effective methods to enhance the electrochemical properties. Ceramic fillers are classified into active fillers [46] and inactive fillers [47]. The fillers that will not participate in lithium-ion conduction process are called inactive fillers such as silica and alumina while active fillers like silicon carbides are capable to participate in the lithium-ion conduction process [48]. Silica is one of the promising filler that can influence the electrochemical properties of polymer electrolytes up to a greater extend. Jung et al. [40] studied the electrochemical properties of PAN/fumed silica composite synthesized by electrospinning. The resulting composite polymer consists of homogeneously distributed silica which is the direct influence of electrospinning process. Fumed silica is hydrophilic in nature and hence provides silanol groups at the surface of the PAN fibers. The presence of silanol group enhances the zeta potential that will nullify the effect of increase in viscosity caused by silica and there by produce very fine fibers. The resulting fibers

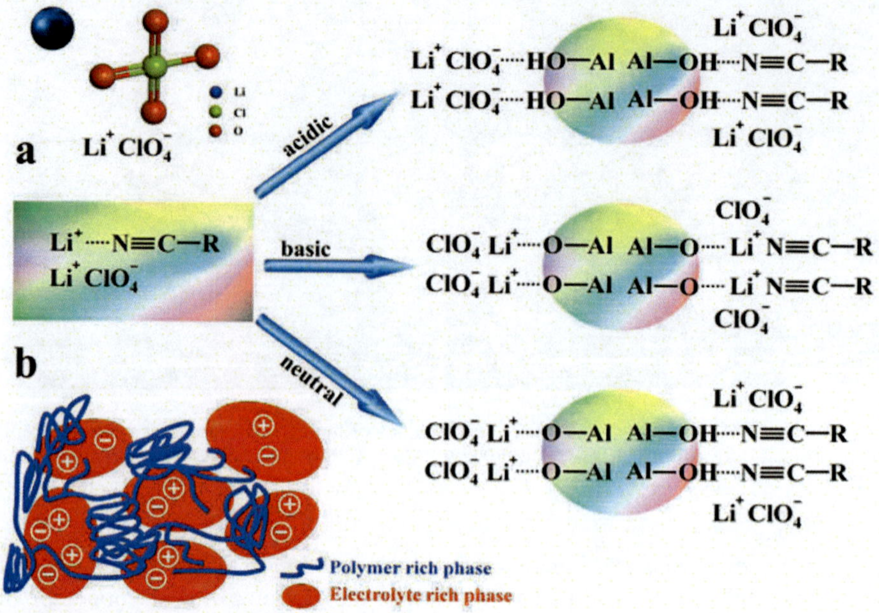

Fig. 5.4 **a** Schematic representation of the dissociation effects of Lewis acidic, basic and neutral surface groups on nano-Al_2O_3 particles. **b** Schematic representation of polymer electrolyte Adapted and reproduced from Ref. [39], Copyright 2005 Elsevier

exhibit a unique pore size distribution with the combination of both mesoporous and microporous structure [40]. Electrolytic separator membrane synthesized using silica/PAN by the combination of sol gel and electrospinning method results the formation of membrane having high thermal stability. Compared to the polypropylene separates this silica/PAN system is best that it can withstand the dimensional stability even up to 150 °C (Fig. 5.5) which is results by the presence of silica. Samples containing different silica loading were prepared by making use of TEOS along with HCl. Resulting PAN/silica solution is electrospunned to obtain the electrolytic separator. With increasing the silica content the electrolyte uptake and ionic conductivity of the resulting membrane was drastically increase due to the highly porous structure results by the silica content [49].

Aluminum oxide is well studied as nano-sized filler in PEs of LIBs [50] but there is only limited studies were reported in electrospunned PAN based PEs. PAN/alumina composite membrane prepared by electrospinning of PAN solution in DMF mixed with alumina particles exhibit excellent mechanical strength, thermal along with high porosity and electrolyte stability uptake. With increasing the alumina content the porosity is observed to be increased (Fig. 5.6) as a result of this, electrolyte uptake is gradually increased. An electrolyte uptake of greater than 400% is exhibited by all the membranes and a maximum of 561% is shown by the membrane consist 40 wt% alumina content. This is due to the large surface area of alumina that results high

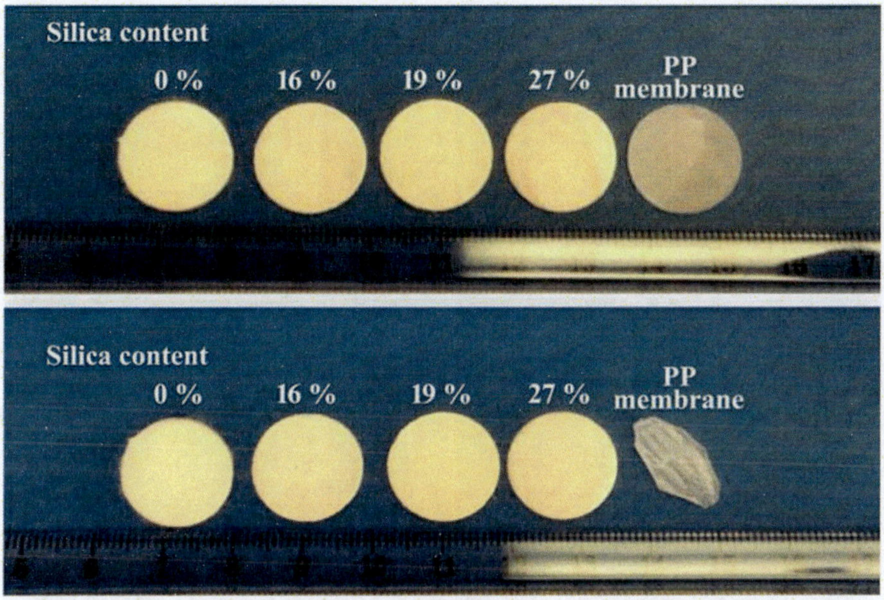

Fig. 5.5 Photographs of SiO_2/PAN hybrid nanofiber membranes with different SiO_2 contents and microporous PP membrane before and after thermal exposure at 150 °C for 30 min. Adapted and reproduced from Ref. [49]. Copyright 2016 Elsevier

hydrophilicity. Similar effect is also observed in electrolyte retention ability. The 3D network structure results by this method provide large porous space for the entrapment of triethylene glycol diacetate-2-proponoic acid ester electrolyte (TEGDA-BA) used in this study and thereby provide large ionic channel [51].

Liang et al. [52] evaluated the electrochemical characterization of lithium lanthanum titanate oxide (LLTO)/PAN submicron composite fiber membrane prepared by electrospinning. LLTO is a conducting material that exhibits a pervoskite structure [53]. The composite membrane prepared by the combination of LLTO along with PAN can effectively use as the electrolytic separator due to its unique porous structure. The LLTO particles chosen exhibit a particle size ranging from ten of nm to hundreds of nm and possess an average particle size of about 200 nm. By the addition of LLTO particles, the membrane can trap impurities from liquid electrolytes and there by provide smaller interfacial resistance that will enhance the battery performance [52]. Lithium aluminum titanium phosphate ($LiTi_2$ $(PO_4)_3$ simply LATP) is a promising filler that consist of Ti^{4+} ions that get partially replaced with Al^{3+} ions. This replacement reduces the unit cell dimension and thereby enhances the ionic conductivity. Moreover its structure consists of cavities and bottle neck at which lithium ions can occupy and move so that it can ensure the proper ionic mobility [54, 55] LATP/PAN composite polymer electrolytic separator membranes prepared by electrospinning and with increasing the LATP content the electrolyte uptake and electrochemical performance increases [56]. PAN based

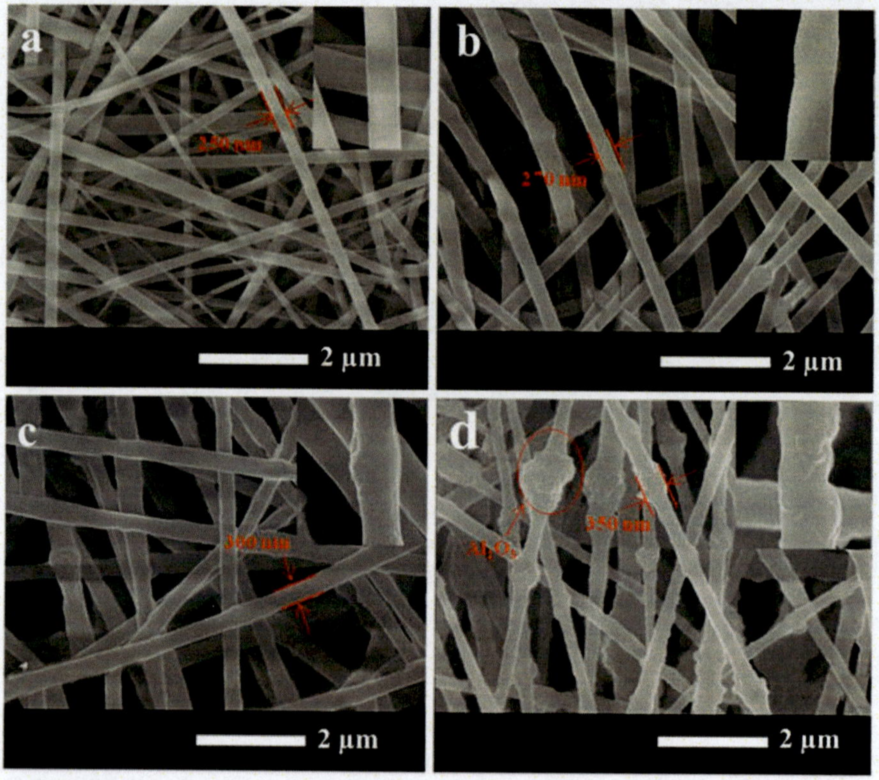

Fig. 5.6 FE-SEM images on the suface morphology of electrospun polyacrylonitrile (PAN) based composite membranes with varying PAN: Al_2O_3 contents of **a** 100:0 **b** 90:10 **c** 80:20 **d** 60:40. Adapted and reproduced from Ref. [51]. Copyright 2015 Elsevier

composites membranes synthesized by the methods other than electrospinning also explored for energy storage devices [57]. PAN/organic clay composite membrane prepared by solvent induced phase inversion show superior thermal and mechanical stability that induced by the presence of layered organic clay [58]. Even though the ceramic nanoparticles can enhance the electrochemical properties of polymer electrolytes, one of the challenge that facing is the formation of efficient network for the ionic conduction. Polymer composite synthesized by the incorporation of ceramic nanowires can efficiently form such network structures and they are extensively studied for the solid polymer electrolyte systems. $Li_{0.33}La_{0.557}TiO_3$(LLTO) nanofibers were fabricated by electrospinning of polyvinyl pyrrolidone (PVP) with relevant salts followed by calcination to produce LLTO nanowires and 15% of the same is dispersed in PAN lithium per chlorite ($LiClO_4$) (Fig. 5.7). Resulting composite exhibit a continuous ionic transport pathway and exhibit an ionic conductivity of about 2.4×10^{-4} S cm^{-1} at room temperature which is a superior conductivity reported for solid polymer electrolytes [59].

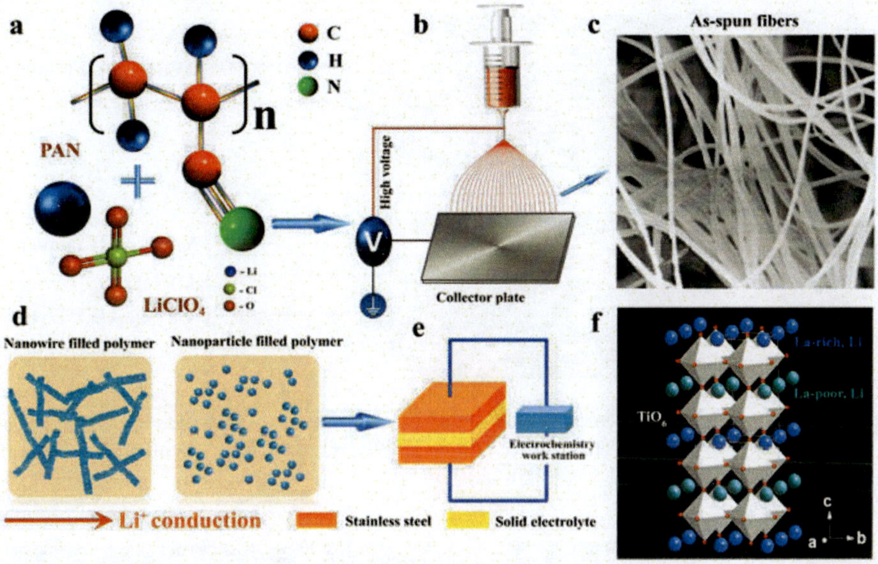

Fig. 5.7 **a–e** Schematic illustration for the synthesis of ceramic nanowire-filled polymer-based composite electrolytes **c** as spun fibers **e** illustration of the electrode configuration for the AC impedance spectroscope measurement **f** perspective view of lithium lanthanum titanate oxide (LLTO). Adapted and reproduced from Ref. [59]. Copyright 2015 American Chemical Society

A novel tailor made PAN based sandwich structured composite polymer electrolyte membrane with unique architecture is introduced to make use in LIBs by Prasanth et al. [60, 61] and the same method of preparation is adopted to prepare the electrospun polymer electrolytes by other groups by using polysulfonamide (PSA) and polyacrylonitrile [60]. The schematic and surface morphology of the tailor made electrospun polymer electrolyte prepared by Prasanth et al. [60, 61] is displayed in Figs. 5.8 and 5.9. The study elucidated the synthesis, characterization, and electrochemical investigation of a novel composite polymer electrolyte based on electrospun membranes of PVdF-*co*-HFP [60], poly(methyl methacrylate) (PMMA), poly(vinyl acetate) (PVAc), or poly(ethylene oxide) (PEO) [61] sandwiched between the layers of fibrous PAN membranes fabricated using continuous electrospinning. The polymer membranes were synthesized in a layer by layer fashion, stacked in layers to incorporate the advantages of the different polymers. The study introduced a novel method of preparation of layer-by-layer structure, instead of conventional assembly of separate layers of the membranes one over the other. The sandwiched structure of the electrolyte membranes are prepared by adopting continuous electrospinning, in which the polymeric doping solutions are continuously fed to the tip of the needle one after another. As a result, the membranes had good adhesion between the layers. In the tailor-made composite membranes, the thickness of each layer was carefully controlled during the process of electrospinning by optimizing the spinning time and other spinning parameters. In the study they demonstrated that the utilization of

Fig. 5.8 Schematic representation of electrospun poly(vinyldeneflouride-*co*-hexaflouropropylene)/polyacrylonitrile (PVdF-*co*-HFP/PAN) composite membranes. Adapted and reproduced from Ref. [60]. Copyright 2010 Elsevier

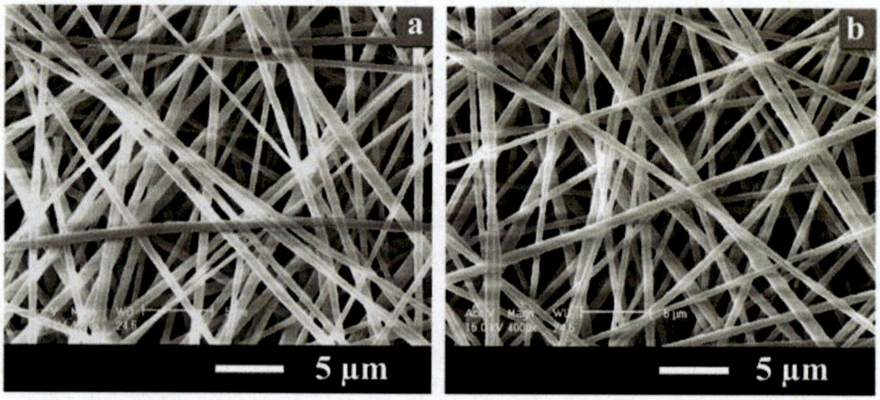

Fig. 5.9 FE-SEM images on the surface morphology of electrospun PVdF-*co*-HFP/PAN blend/composite membranes: **a** pristine PVdF-*co*-HFP and **b** 50:50 blend of PVdF-*co*-HFP and PAN. Adapted and reproduced from Ref. [60]. Copyright 2010 Elsevier

certain polymers such as PMMA or PEO which are completely soluble in common battery electrolyte solvents, or very fragile fibers like PMMA could not handle or even transfer from the collector plate or drum without breaking, could make to use as fibrous membranes gel electrolytes in LIBs. In these membranes the out pan layers act as mechanical support and the inner PEO or PMMA will offer better ionic conductivity and better electrolyte uptake as well as electrolyte retention, thereby make it good leak proof fabrication of LIBs and better safety.

Based on Prasanth et al. [60] poly(methyl methacrylate) (PMMA), poly(vinyl acetate) (PVAc),or poly(ethylene oxide) (PEO) [61] study, similar layer-by-layer structured polymer electrolyte of PAN sandwiched between PSA is recently reported

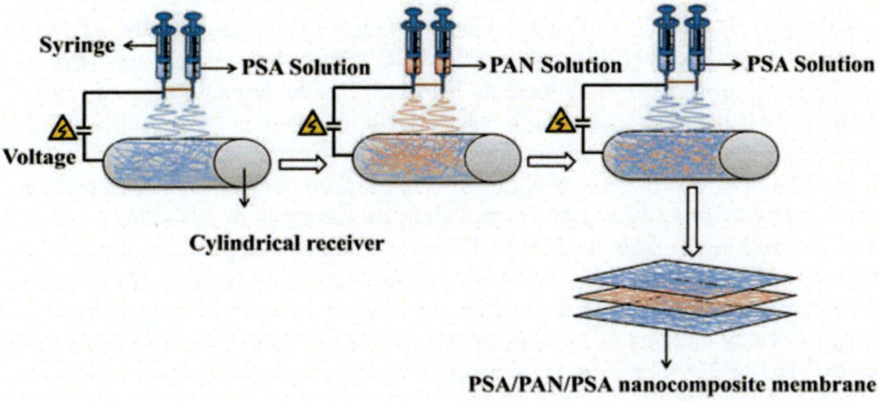

Fig. 5.10 Schematic representation of the preparation process of the polysulfonamide/polyacrylonitrile/polysulfonamide (PSA/PAN/PSA) composite nanofibrous membranes. Adapted and reproduced from Ref. [62]

[62]. In the study a uniform solution of PSA was electrospunned and a solution of PAN is spunned over the formed fiber, again the same PSA solution is spunned in order to get a sandwich structure that consist of PSA/PAN/PSA (Fig. 5.10). The sandwiched structure exhibit enhanced mechanical property due to the presence of PAN nanofibers that are capable to exhibit a maximum stress of about 41.2 MPa. Compared to commercial Celgard® separator this sandwich membrane exhibit high porosity, high thermal stability and electrolyte uptake [62].

5.4 Preparation of polyacrylonitrile-Based Blend Polymer Electrolytes

Blending is an effective method to improve the electrochemical properties of PEs. Different polymers are employed for the preparation of polymer blends based on PAN. Blends of PAN with PMMA [63], PS [64], PVdF [65] or PVdF-*co*-HFP [60] are widely studied for the preparation of PEs in LIBs. These polymers individually have certain characteristics that are important for use it as the electrolyte host in batteries. Even though each system exhibit its own characteristics that could not be exhibited by other. So blending can deliver the synergetic advantage of each of the component. For the blend polymer electrolyte systems both binary and tertiary blends were explored for lithium-ion batteries. Prasanth et al. [66] first demonstrated a multi component polymer blend based on PAN/PMMA/PS. Also he demonstrated a tailor made gel polymer electrolyte based on blends and composites of PVdF-*co*-HFP and PAN [60]. For the evaluation of the property of the system, membranes with different concentration of individual polymers were synthesized which is denoted as GPE-01, GPE-02, GPE-03 and GPE-04 which consist of PAN/PMMA/PS in the ratios 80:10:10, 90:05:05, 90:10:00 and 90:0:10, respectively. According to Prasanth et al. [66] the microporous membrane that results by the homogeneous blending of individual components results high electrolyte uptake due to the presence of PAN, PMMA which results the swelling or gelation rather than dissolution. PAN is capable to impart a mechanical stability while PS is introduced in order to provide the amorphous path for providing toughness as well as for the easier penetration of Li^+-ions. However with increasing the PMMA/PS content the resulting membrane shows an irregularity and non- uniformity, membrane with high concentration of both PS and PMMA exhibit higher AFD (950 nm) and an increment in AFD is observed with a surge in PS content. This multicomponent blend electrolyte exhibit a thermal stability lower than 300 °C which is lower than that of PAN, attributed by the reduction in crystallinity, which is further confirmed with the X-ray diffraction analysis (Fig. 5.11a). On evaluating the variation of ionic conductivity with time (Fig. 5.11b) of the blend, an increment in conductivity is observed with respect to time and from the data it is clear that the average pore size is independent of the activation energy of ionic conduction.

The amorphous nature of the membrane promotes the lithium ion mobility and thereby results in better electrochemical performance in electrospun fibrous electrolyte [66]. The advantage of both blending and composite preparation was bringing in a single system by Prasanth et al. [60] with the fabrication of blend and composite of PVdF-*co*-HFP and PAN. The resulting fibrous membrane exhibit a uniform morphology with AFD of 390–450 nm [59]. Followed by Prasanth et al. [66], Tan et al. [64] developed a tertiary blend based on PAN along with thermoplastic polyurethane (TPU) and polystyrene (PS). Electrospinning is carried out with the solution of DMF that containing PAN: TPU: PS in the ration 5:5:1. The presence of TPU can enhance the ionic conductivity of the resulting electrolyte due to the presence hard as well as

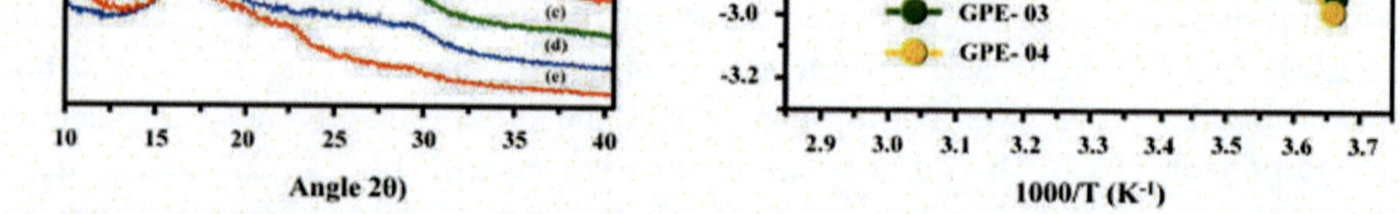

Fig. 5.11 **a** XRD spectra of polyacrylonitrile/polymethylmethacrylate/polystyrene (PAN/PMMA/PS) blend in ratio; 80:10:10 (GPE-01), 90:05:05 (GPE-02), 90:10:0 (GPE-04), pristine PAN and pristine PMMA, and inset shows the XRD of pristine polystyrene (PS). **b** Effect of temperature on ionic conductivity of polymer electrolytes based on electrospun PAN/PMMA/PS blend membranes. Adapted and reproduced from Ref. [66]. Copyright 2012 Elsevier

the soft segments that it contains. Soft segments are responsible for the dissociation of salts and thereby result in an enhancement in the ionic conductivity. The mechanical strength is provided by PS. It ensures good fatigue resistance, mechanical strength and spatial stability. Furthermore smooth and completely stretched fiber morphology (Fig. 5.12) and high thermal stability is observed (Fig. 5.13). The enhancement in thermal stability is arise because of the interaction between PAN and TPU a slight increase in thermal stability is observed in between 300 and 450 °C for this tertiary blend electrolyte compared to that of the binary system [64]. Wang et al. [63] fabricated PAN based polymer electrolytes using a copolymer of PAN, polyacrylonitrile-co-vinyl acetate (PAV) that is blended with PMMA to form a polymer host for the gel polymer electrolyte that can exhibit reduced crystallinity with improved mechanical strength and good adhesion with the electrode that is induced by the presence of PAV which is formed by the linear conjugation of PAN with PVAc (Fig. 5.14a). An improved ion transfer at the electrode-electrolyte interface is delivered by the system due to the presence of PMMA. Moreover it can help to reduce the crystallinity of

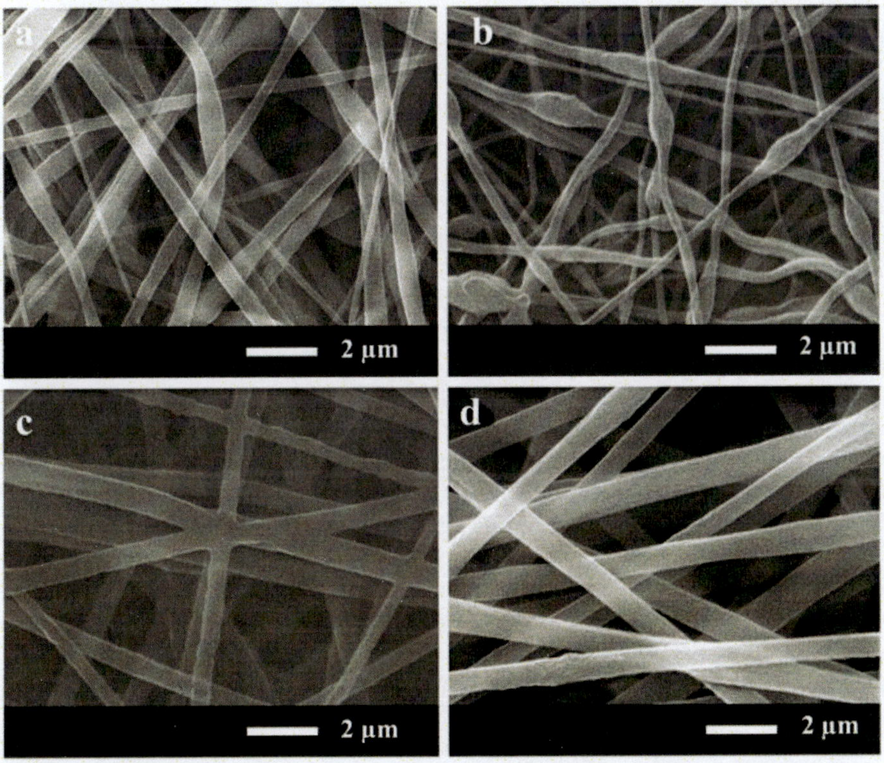

Fig. 5.12 FE-SEM images on the surface morphology of PAN based polymer blend electrospun films with different weight ratio (wt.%): **a** PAN/PS (5:1), **b** PAN/TPU(5:5), **c** TPU/PS (5:1) and **d** PAN/TPU/PS (5:5:1). Adapted and reproduced from Ref. [64]

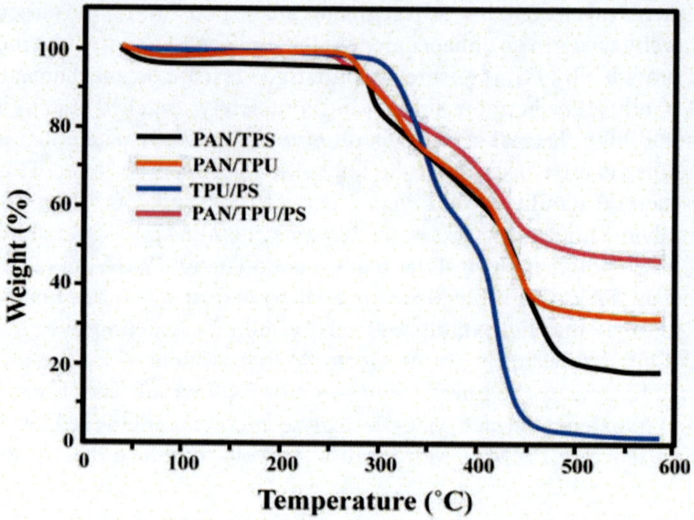

Fig. 5.13 TGA curves of the PAN based polymer blend electrospun films. Adapted and reproduced from Ref. [64]

PAN. On evaluating the temperature dependence over the ionic conductivity, the gel polymer electrolyte is more conducting than the celgard separator activated with the liquid electrolyte (1 M $LiPF_6$ EC/DMC/DEC solution) and at 30 °C GPE-AVM and GPE-AV exhibit a conductivity of about 3.5 and 2.5 mS cm^{-1}, respectively (Fig. 5.14b). This improved conductivity is attributed by the presence of interstitial space- that caused by the segregation of PAN fibers in the presence of PMMA [63].

In the binary blend system PAN/PMMA blends are extensively studied for the development of polymer electrolytes. It has been found that PAN/PMMA GPEs with acrylonitrile to methyl methacrylate ratio 4:1 exhibit high ionic conductivity and better electrochemical properties [67, 68].Another polymer that commonly blended with PAN is PVdF which exhibit a good electrochemical stability along with an excellent affinity toward the lithium ions. On synthesizing its blend along with PAN the issue comes across is the miscibility of the individual system with each other. During the mixing of PAN and PVdF a phase separated mixture is get formed. So electrospinning is used effectively that can ensure the phase mixing of both of the polymer matrixes and there by avoid the phase separation. Gopalan et al. [64] studied the electrochemical properties of electrospun PAN-PVdF electrolyte membrane synthesized with different porousity. The proper phase mixing of the components is evaluated by using the FTIR spectra [65].

Polyvinyl chloride (PVC) based polymer electrolytes are also studied for LIBs due to its dielectric properties that can induce good the ionic conductivity [69, 70]. PAN/PVC blend polymer electrolytes synthesized by solution casting followed by activation with lithium perchlorate ($LiClO_4$) in propylene carbonate (PC) exhibit ionic conductivity 7.01×10^{-5} S cm^{-1} [41] Zhong et al. [71] developed PAN/PVC

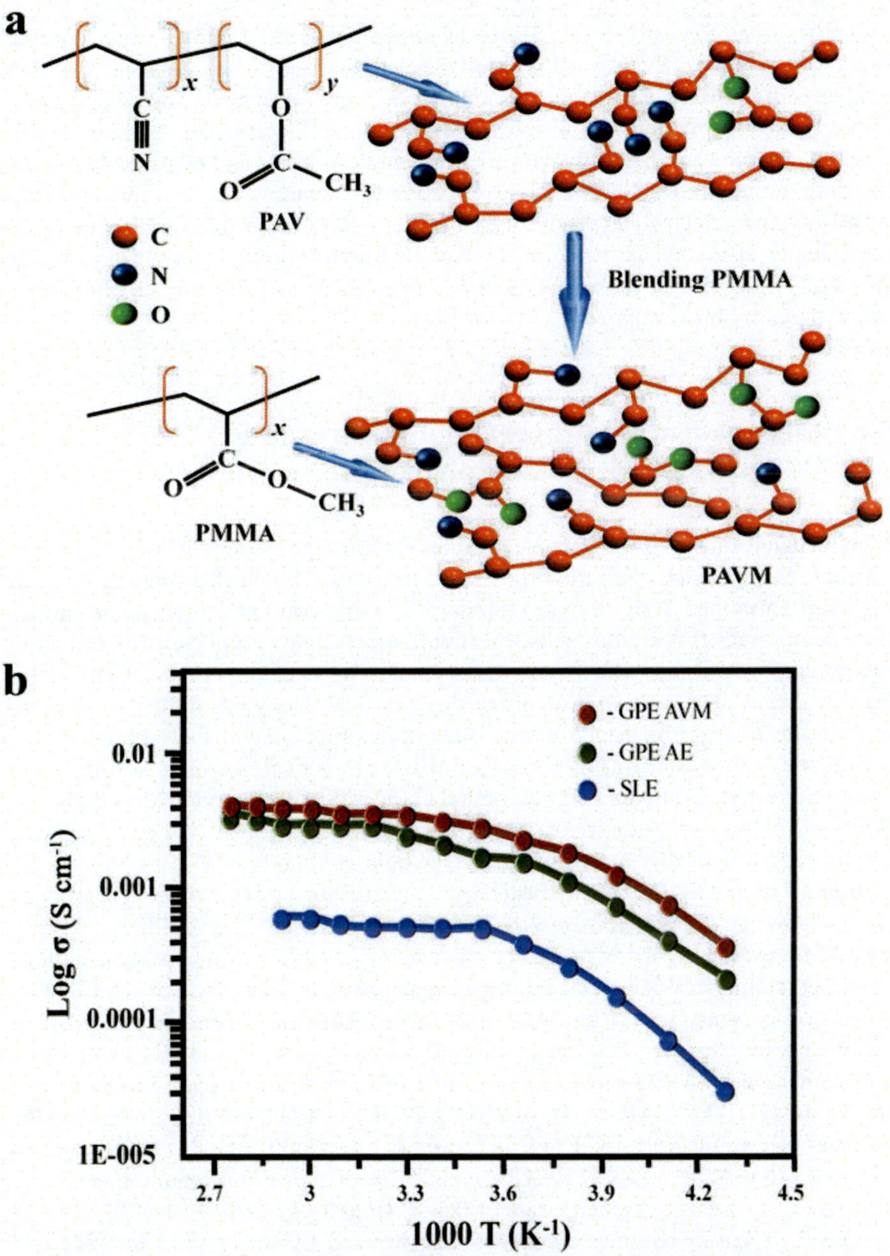

Fig. 5.14 **a** Conceptual Illustration of the polymer frameworks of PAV and PAVM **b** ionic conductivities of the GPEs and SLE at temperatures of −40 to 90 °C. Adapted and reproduced from Ref. [63]. Copyright 2014 American chemical Society

(8:2 w/w) based electrolytes for LIBs by electrospinning that exhibit an ionic conductivity of 1.05×10^{-3} S cm^{-1} at 25 °C and the electrolyte system is stable up to a 4.9 V [71]. Incorporation of ionic liquids is another method that can be adopted in order to improve the electrochemical performance of polymer electrolytes. Room temperature ionic liquids are thermally as well as chemically stable and exhibit a low vapor pressure, flammability and a wide electrochemical window [72, 73]. PAN/PMMA blend polymer electrolyte synthesized by the incorporation of room temperature ionic liquid N-methyl N-butyl pyrrolidenium bis(trifluoromethanesulfonyl imide) ($PYR_{14}TFSI$) exhibit a room temperature conductivity of 3.5×10^{-3} S cm^{-1} and a stable potential higher than 5 V (Li^+/Li) [74].

5.5 Electrochemical Properties of Electrospun Polyacrylonitrile-Based Polymer Electrolytes

Electrospun PAN based polymer electrolytes exhibit high electrochemical properties owing to the presence of microporous structure formed by electrospinning. Lithium salts that entrapped at these micro pores results improved electrochemical properties. Along with the lithium salts, the liquid electrolytes employed for activating the polymer membrane also have greater impact in its electrochemical properties. So the choice of salt as well as organic electrolytes has great influence in determining the final performance of the electrolyte. Electrospun PAN based fibrous membrane with a diameter of about 880–1260 nm activated with 1 M $LiPF_6$ in a mixture of ethyl carbonate (EC), dimethyl carbonate (DMC) and diethyl carbonate (DEC) (0.25:1:1) exhibit a ionic conductivity of about 1.7×10^{-5} S cm^{-1} at 20 °C [34]. Compared to this a peak increase in ionic conductivity is observed for the membrane with AFD 350 nm activated by 1 M $LiPF_6$ in a mixture of ethyl carbonate (EC), diethyl carbonate (DEC) ($>2\times10^{-3}$ S cm^{-1}) as well as an oxidative stability >4.7 V. On evaluating the battery performance using lithium iron phosphate cathode ($LiFePO_4$ simply LFP), the cell exhibit a capacity 150 mAh g^{-1} which is 88% of the theoretical capacity [24]. Choi et al. [32] investigated the dependence of different organic liquids over the electrochemical performance of PAN based PEs by activating the membrane with one molar solution of $LiPF_6$ in different organic electrolytes (EC/DMC (2:1), EC/DMC (1:1), EC/EMC, EC/DEC, EC/DMC/DEC). Among this polymer electrolyte with EC:DMC in the ratio 1:1 exhibit high discharge capacity of about 145 mAh g^{-1}(LCO/PE/LCO) and 94.1% of capacity is retention even after 150 cycles at a charge discharge rate of 0.5 °C (Fig. 5.15) [32].

The PAN based polymer electrolyte synthesized by Prasanth et al. [24] with 16 wt% of PAN in DMF activated with 1 M $LiFP_6$ in EC/DC (1:1) display a discharge capacity of 150 mAh g^{-1}at 0.1C (Li/PE/LFP) which is 88% of the theoretical value and even after 50 cycles, the cell retained 90% of the initial discharge capacity. The PAN based separator membrane fabricated by electrospinning of 10.5 wt% solution in DMF exhibits voltage stability up to 4.5 V versus Li and a discharge capacity of

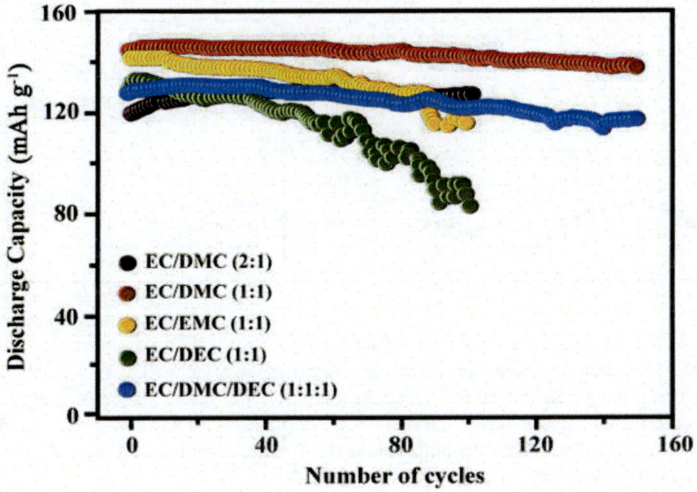

Fig. 5.15 Cycle performances of graphite/electrospun PAN-based fibrous polymer electrolytes in 1 M $LiPF_6$ in varying electrolyte compositions and $LiCoO_2$ cells at 0.5C/0.5C rate of charge/discharge condition. Adapted and reproduced from Ref. [32]. Copyright 2005 Electro Chemical Society

about 125 mAh g^{-1} which is higher than that of commercially used PP separator (120.4 mAh g^{-1}) and PE separator (120.7 mAh g^{-1}) [75]. Compared to this a greater capacity was exhibited by the separator fabricated by Carol et al. [34] which is about of 128 mAh g^{-1} (Li/PE/LCO). Moreover it exhibit 86% of capacity retention after 30 cycles. The presence of water has a greater influence over the electrochemical performance of the system. The porous PAN membrane formed by the tertiary system consist of water shows a discharge capacity of 130 mAh g^{-1} which is higher than that of non-porous PAN membrane synthesized in the same diameter. This is attributed by the interconnected porous structures formed by the spinodal phase separation that happens during the electrospinning process [37]. For the further enhancement of electrochemical performance different fillers are incorporated to the matrix, and silica is one of the extensively studied filler for the polymer electrolytes in LIBs. Composite polymer electrolyte formed by the incorporation of fumed silica to the PAN matrix and activated with the $LiPF_6$ in EC/DMC mixture exhibit excellent discharge capacity and cyclic stability. 12 wt% of SiO_2 in PAN exhibits a discharge capacity of 139 mAh g^{-1} (graphite/PE/$LiCoO_2$), which is 98% of the theoretical value. Even after 150 cycles, the system retains a capacity of 127 mAh g^{-1} which corresponds to 89% of the theoretical value, which is much better than that of electrospunned pure PAN based electrolytes for LIBs [40]. With further increase in silica content an enhancement in electrochemical performance was observed. A 7 wt% silica containing system shows discharge capacity of 163 mAh g^{-1} for the cell cycled with LFP, which is close to the theoretical capacity of cathode (170 mAh g^{-1}) and this same system exhibit very low capacity fading with increase in the C-rate as compared

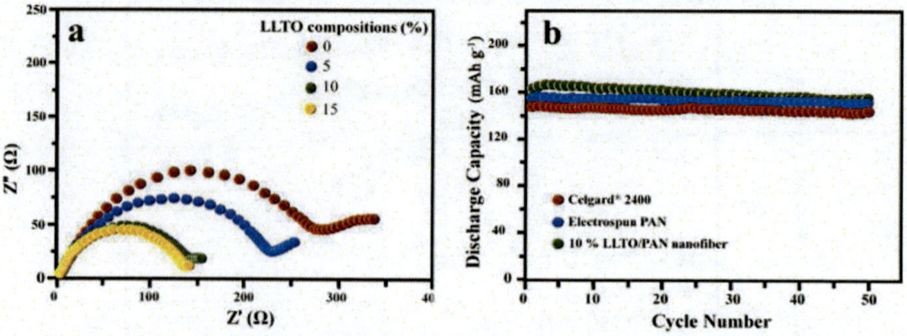

Fig. 5.16 **a** Electrochemical impedance spectra of liquid electrolyte-soaked LLTO/PAN composite fiber-based membranes with different LLTO contents, **b** cycle performance of Li/$LiFePO_4$ cells containing liquid electrolyte-soaked Celgard 2400 separator, PAN fiber-based membrane and LLTO/PAN composite fiber- based membrane at 0.2 C rate. Adapted and reproduced from Ref. [52]. Copyright 2011 Elsevier

to that of commercial PP separators. Even at 8C, 27 wt% silica containing system exhibit a discharge capacity 82 mAh g^{-1} [49]. Lithium lanthanum titanate (LLTO) is the best to use as filler due to the high bulk Li^+-ion conductivity (10^{-3} S cm^{-1}). The system containing 15 wt% of LLTO exhibit very low interfacial resistance, which arise because of their ability to trap the impurities present in the liquid electrolyte (Fig. 5.16a). The separator membrane fabricated by incorporation of LLTO exhibit a discharge capacity of about 160 mAh g^{-1} when cycled by using LFP cathode and it persist almost this capacity up to 50 cycles (Fig. 5.16b) [52].

Similar electrochemical characteristics were observed in lithium aluminum titanium phosphate (LATP) imparted PAN based polymer matrix. LATP exhibit an ionic conductivity of about 10^{-4} S cm^{-1} at room temperature. The cell (Li/PE/LFP) with 15% LATP exhibits an initial discharge capacity in between 145 and 165 mAh g^{-1}. Compared to commercial celgard separator this cell can exhibits a first cycle efficiency of 93%, which is for the celgard separators are about 91.9% [56]. Composite membrane that containing alumina (PAN: Al_2O_3: 80:20) and TEGDA-BA as liquid organic electrolyte exhibit an ionic conductivity of 2.35 mS cm^{-1} at 25 °C and an electrochemical stability greater than 4.5 V. It also shows a discharge capacity of 249 Ah g^{-1} when the cell is cycled with half cells that containing electrodes such as Li $[Li_{1/6}Ni_{1/4}Mn_{7/12}]O_{7/4}F_{1/4}$. Apart from this a capacity retention of 240.4 mAh g^{-1} is observed after 50 cycles which is excellent capacity than that of pure PAN membrane [51]. A different approach for the fabrication of composite electrolyte was made by Tian et al. [62] using PSA along with PAN by spinning PSA/PAN/PSA. This system exhibits capacity 154 mAh g^{-1} along with 85.94% of capacity retention from 0.2 to 2C. Moreover, about 96.2% of capacity retention was observed even after 70 cycles [62].

While composite polymer electrolytes enhance the electrochemical performance by the increase in the amorphous content that is induced by the presence of filler and there by the Li^+-ion movement, polymer blend based use different polymers to blend together to produce the synergetic advantage of individual ones and thereby increase the final performance. For the blend based polymer electrolytes both binary and tertiary blends were extensively studied, for the PAN based polymer electrolytes. The first multicomponent blends system proposed by Prasanth et al. [66] composed of PAN, PMMA, PS activated by soaking in liquid electrolyte containing 1 M $LiPF_6$ in EC: DEC (1:1). The charge discharge profile of resulting electrolyte membrane was evaluated using Li/$LiMn_2O_4$ coin cell. During the first discharge about 82% of theoretical capacity was achieved The gel polymer electrolyte with high concentration of PMMA (GPE-01) exhibits a discharge capacity of 122 mAh g^{-1} which is close to the theoretical value of $LiMn_2O_4$ [66]. These electrochemical performances are the effect of PVdF and PS employed that can improve the electrochemical performance of PAN membrane. At 0.1 C the cell (Li/GPE/$LiMn_2O_4$) exhibit a discharge capacity of 120mAh g^{-1} at a discharging rate of 0.1C, with increasing the cycling the discharge capacity get decreased slightly and the system containing PAN/PVdF/PS in ratio 80:10:10 exhibit a discharge capacity of 113 mAh g^{-1} which corresponds to 92.7% of initial capacity and it is higher than that of cell fabricated with the polymer ratio 90:05:05, 90:10:0 and 90:10:0 [66].

Followed by this the blend polymer membrane was fabricated by Tan et al. [64] using PAN/TPU/PS was examined using Li/GPE/LFP cell. This cell exhibits a charge discharge capacity of 161 mAh g^{-1} with a stable cycle performance and high coulombic efficiency and about 94% of capacity retention is observed even after 50 cycles [64]. It is observed that the cell exhibits an excellent electrochemical stability of 5.8 V which is attributed by the interaction between the C=O group of the carbonate, –NH group of the TPU and–CN group of the PAN [64]. Instead of using the PAN separately the copolymer of PAN with PVAc (polyacrylonitrile vinyl acetate simply PAV) was explored with PMMA. Activating with 1 M solution of $LiFP_6$ in carbonate solvent, the cell shows a capacity of 140 mAh g^{-1} at a discharge rate of 0.2C (Fig. 5.17a). Cycling at 17C, 93% of capacity retention is observed for the system and even after a 1000 cycles also 63% of capacity retention is observed

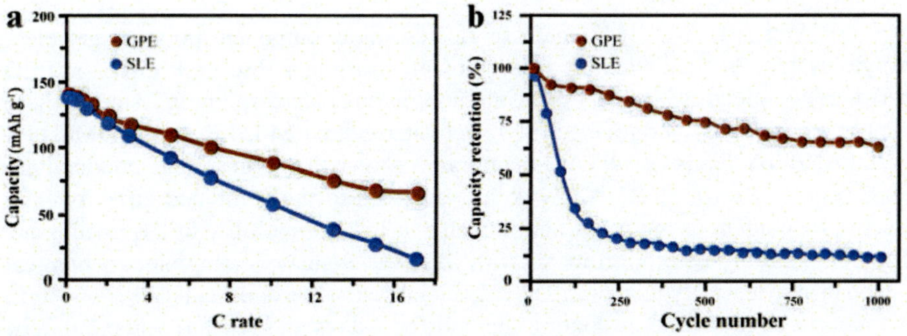

Fig. 5.17 **a** Discharge capacity and **b** cycle performance of GPE and SLE. Adapted and reproduced from Ref. [63]. Copyright 2014 American Chemical Society

(Fig. 5.17b) which is higher than that of celgard separator (SLE) activated with the carbonate solvent containing $LiFP_6$ salt [63].

A novel multicomponent electrolyte system fabricated Prasanth et al. [60] as a three layer structure, of electrospun PVdF-*co*-HFP and PAN or a the blend composite membrane of PVdF-*co*-HFP along with PAN having a three-dimensional network structure is explored owing to its good electrochemical properties [60]. In the three layer system one layer of electrospun PAN membrane is sandwiched between the layers of electrospun PVdF-*co*-HFP fibrous membrane or a layer of electrospun PVdF-*co*-HFP sandwiched between two layers of electrospun PAN membrane is fabricated by continuous electrospinning which facilitate good adhesion between the layers. The tailor made gel polymer electrolyte can entrap liquid electrolyte to form a swollen fibrous matrix with large surface area. It exhibits an anodic stability greater than 4.6 V versus Li/Li^+ (Fig. 5.18a) ascribed by the excellent affinity of the membrane to the carbonate electrolyte (EC, DMC etc.). The electrochemical stability exhibited by the membrane is in the order $4.6 < 4.7 < 4.9$ V for PAN/PVdF-*co*-HFP/PAN layer by layer-membrane (PE-LM1), (PVdF-*co*-HFP/PAN/PVdF- *co*-HFP layer by layer membrane (PE-LM2), and PAN/PVdF-*co*-HFP blend (50:50 wt./wt.) membrane (PE-BM), respectively. With increase in the storage time an urge in the interfacial resistance is observed which is reported highest for the sample PE-LM1 and the least for PE-BM (Fig. 5.18b). The similar composition of PAN/PVdF-*co*-HFP blend (50:50 wt./wt.) is observed to exhibits a discharge capacity 140 mAh g^{-1} (Fig. 5.19a) which corresponds to 85% of the utilization of the active material, but for the remaining, the capacity reported was 19–23% low than the theoretical capacity. Even after 50 cycles the blend membrane (PE-BM) could retain 94% of its initial discharge capacity that corresponds to 81% of its theoretical capacity (Fig. 5.19b). This enhanced performance is attributed by the higher ionic conductivity (6.5×10^{-3} S cm^{-1} at 25 °C) resulted from the three dimensional network structure provided by the electrospun membranes blend compared to that of layer by layer structure (3.9×10^{-3}, 4.9×10^{-3} S cm^{-1}, respectively, for PAN/PVdF-*co*-HFP/PAN layer by layer membrane (PE-LM1), (PVdF-*co*-HFP/PAN/PVdF-*co*-HFP layer by layer membrane (PE-LM2) at 25 °C) [60].

Later, Prasanth et al. [61] again fabricated multicomponent polymer gel electrolyte system by sandwiching a layer of electrospun PVAc or PMMA or PEO between two electrospun PAN by using continuous electrospinning, Among these membranes the polymer gel electrolyte activated with 1 M $LiPF_6$ in EC/DMC (1:1 v/v), PAN/PVAc/PAN (PE-PVAc) membranes shows the highest ionic conductivity (4.83×10^{-3}S cm^{-1} at 25 °C), which is ascribed to the combined effect of high electrolyte content at the pores and increase in Li^+ ion mobility in the membrane. The compatibility of lithium metal with PEs was evaluated using electrochemical impedance spectra, Nyquist plot. The interfacial resistance of the cell Li/PE-PVAc/Li is found to be increased evaluating the cell for 7 days and reaches a maximum of 200% of the initial value and then get reduced. Comparatively the resistance exhibited by the cell is low and on evaluating the variation of interfacial resistance (R_f) with temperature (between 20 and 60 °C), a reduction in R_f is observed with temperature which favors the facile charge transfer [61]. An initial discharge capacity of 137 and

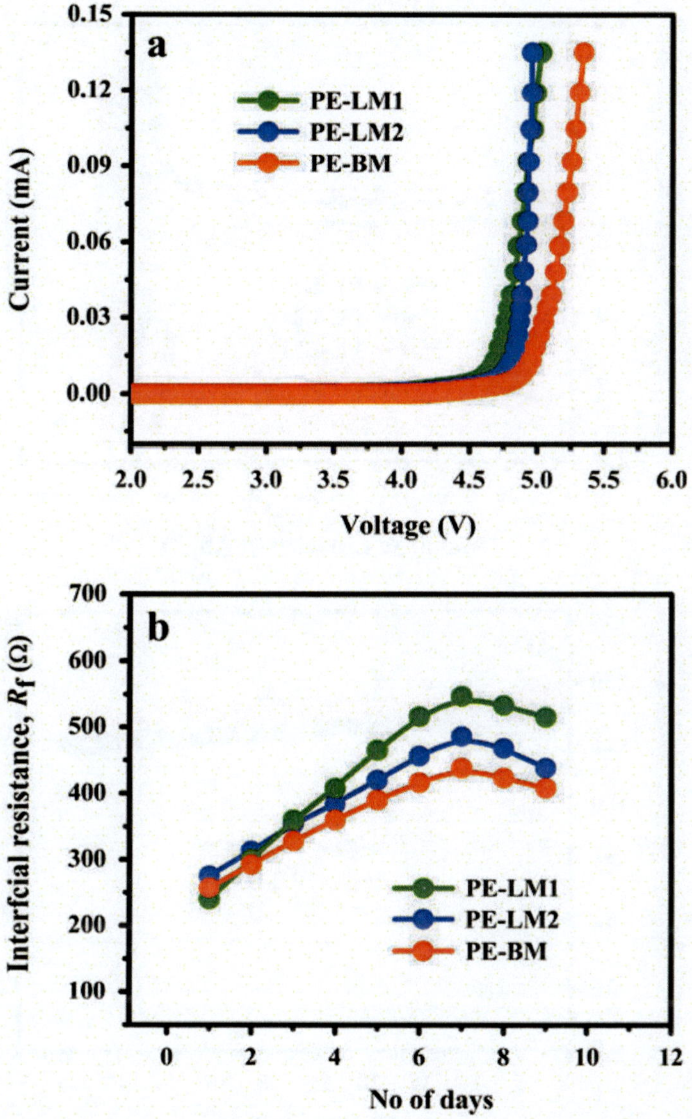

Fig. 5.18 Electrochemical properties of polymer electrolytes based on electrospun P(VdF-*co*-HFP)/PAN blend/composite membranes activated with 1 M $LiPF_6$ in EC/DMC (1:1 v/v) at room temperature: PVdF-*co*-HFP/PAN layer by layer membrane (PE-LM1), PVdF-*co*-HFP/PAN/PVdF-*co*-HFP layer by layer membrane (PE-LM2) and PAN/PVdF-*co*-HFP blend membrane (50:50 wt./wt.) (PE-BM): **a** anodic stability by LSV (Li/PE/SS cells, 1 mV s^{-1}, 2–5.5 V) and **b** variation of interfacial resistance as the function of time (Li/PE/Li cells, frequency range 100–2 MHz). Adapted and reproduced from Ref. [60]. Copyright 2010 Elsevier

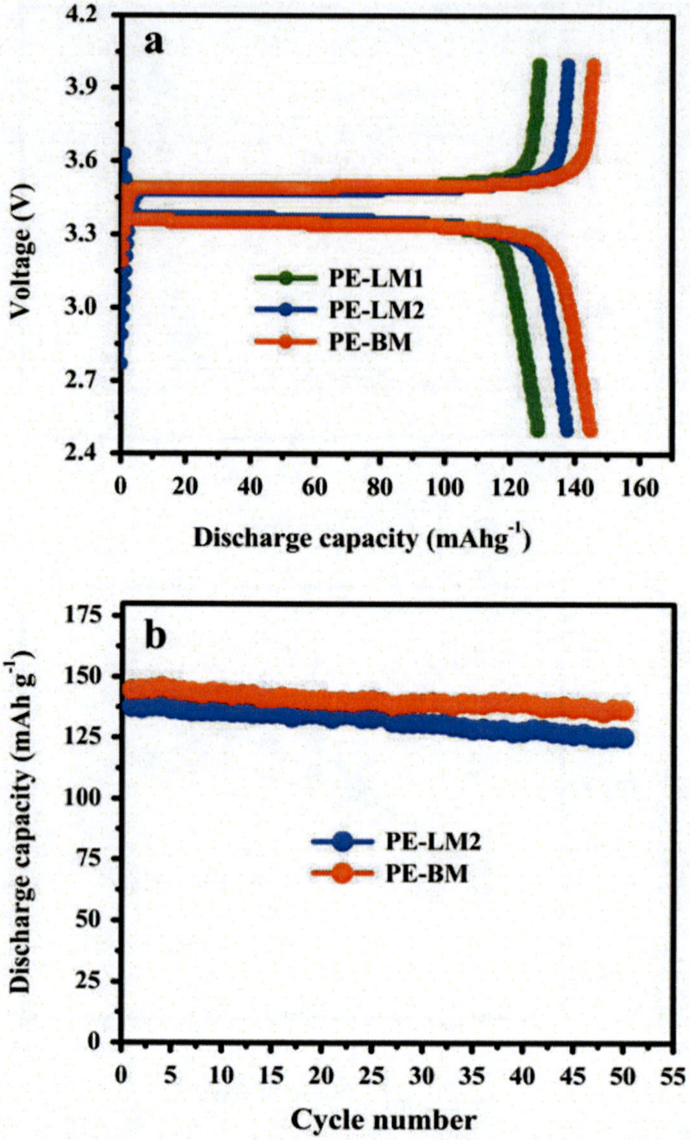

Fig. 5.19 Charge–discharge properties of Li/PE/$LiFePO_4$ cells with polymer electrolytes based on electrospun P(VdF-*co*-HFP)/PAN blend/composite membranes activated with 1 M $LiPF_6$ in EC/DMC (1:1 v/v): **a** Initial charge-discharge profiles and **b** cycling stability (at 25 °C, 0.1 C-rate, 2.5 to 4.0 V). PVdF-*co*-HFP/PAN layer by layer membrane (PE-LM1), PVdF-*co*-HFP/PAN/PVdF-*co*-HFP layer by layer membrane (PE-LM2) and PAN/PVdF-*co*-HFP blend membrane (PE-BM) (50:50 wt./wt.). Adapted and reproduced from Ref. [60]. Copyright 2010 Elsevier

145 mAh g^{-1} is shown by the cell at 0.1 C using LFP cathodes, which is 80–85% of the theoretical capacity and the highest discharge capacity, was exhibited by the PE-PVAc cell. This same cell is observed to be exhibits high retention in capacity (140 mAh g^{-1}) even after 50 cycles. The percentage capacity fade was higher for the cell with PE-PMMA which is attributed by the brittle nature of PMMA that results poor mechanical strength [61].

For the binary blend system PVC based blends of PAN are less studied for LIBs. PAN/PVC polymer electrolyte (8:2) obtained by activating the membrane in 1 M solution of $LiClO_4$ in PC solution exhibit a voltage stability above 4.5 V as well as good interfacial stability with lithium electrode [71]. PMMA binary blends formed exhibit enhanced electrochemical properties than that of PVC based blends. PAN/PMMA blend electrolyte along with ionic liquid $PYR_{14}TFSI$ a good discharge capacity 139, 134, 120 and 101 mAh g^{-1} at discharge rate of 0.1, 0.2, 0.5 and 1C, respectively (Li/GPE/LFP) [74]. Similarly PAN/PVdF blend electrolyte explored with 25% of PAN also capable to exhibit high electrochemical stability of 5.1 V and exhibit a discharge capacity of 112.38 mAh g^{-1} at 0.2C (Li/PE/LCO) [65]. Even though the capacities exhibited by these binary systems are lower than that of the tertiary systems explored, the electrochemical performance is much better than that of pure electrospun PAN based electrolytes. Hence blending with polymers is considered to be the best method for the enhancement of electrochemical performance.

5.6 Conclusion

PAN based polymer electrolytes are seems to better for the lithium-ion batteries due to the high electrochemical property that it can deliver. PAN is thermally stable polymer and can withstand very high temperature. Even though it does not have a sharp melting temperature, it is found to be thermally stable up to 300 °C. Regarding the safety concern of lithium-ion battery a thermally stable and electrochemically good performing electrolytes is prime important. Being a polymer and polymer electrolyte PAN can ensure this by its structural features, physical and chemical properties. For the development of polymer electrolytes systems highly porous membranes were prepared. Due to the brittleness of the porous PAN membranes, electrospinning is chosen to be the best method. Electrospun PAN based membranes can result sufficient porous membranes with high flexibility by hindering the alignment of cyanide groups. This membrane is capable to exhibit better electrochemical properties due than the PAN based electrolytes prepared by other conventional techniques. Even though in order to reach high electrochemical properties different structural modification such as blends and composite preparation were proposed for the PAN based polymer electrolytes. Many polymers such as PVdF, PMMA, and PVdF-*co*-HFP were blended with PAN in order to enhance its electrochemical properties. Moreover composite systems composed of PAN along with ceramic fillers such as silica and alumina can results electrolytes with improved ionic conduction. Reviewing this chapter, we can ensure that PAN based electrolytes are the best and well performing system for safer lithium-ion batteries.

Acknowledgements Authors Dr. Jabeen Fatima M. J and Dr. Raghavan Prasanth would like to acknowledge Kerala State Council for Science, Technology and Environment (KSCSTE), Government of Kerala for financial assistance.

References

1. Ren J, Xu Q, Li YG (2018) Flexible fiber-shaped energy storage devices: principles, progress, applications and challenges. Flex Print Electron 3. https://doi.org/10.1088/2058-8585/aaaa79
2. Yao L, Yang B, Cui H et al (2016) Challenges and progresses of energy storage technology and its application in power systems. J Mod Power Syst Clean Energy 4:519–528. https://doi.org/10.1007/s40565-016-0248-x
3. Liu J, Wang J, Xu C, et al (2018) Advanced Energy Storage Devices: Basic Principles, Analytical Methods, and Rational Materials Design. Adv Sci 5. https://doi.org/10.1002/advs.201700322
4. Kötz R, Carlen M (2000) Principles and applications of electrochemical capacitors. Electrochim Acta 45:2483–2498. https://doi.org/10.1016/S0013-4686(00)00354-6
5. Zhao X, Sánchez BM, Dobson PJ, Grant PS (2011) The role of nanomaterials in redox-based supercapacitors for next generation energy storage devices. Nanoscale 3:839–855. https://doi.org/10.1039/c0nr00594k
6. Smith W (2000) Role of fuel cells in energy storage. J Power Sources 86:74–83. https://doi.org/10.1016/S0378-7753(99)00485-1
7. Divya KC, Østergaard J (2009) Battery energy storage technology for power systems-an overview. Electr Power Syst Res 79:511–520. https://doi.org/10.1016/j.epsr.2008.09.017
8.. (2011) Rechargeable batteries. 2
9. Cheng F, Liang J, Tao Z, Chen J (2011) Functional materials for rechargeable batteries. Adv Mater 23:1695–1715. https://doi.org/10.1002/adma.201003587
10. Vetter J, Novák P, Wagner MR et al (2005) Ageing mechanisms in lithium-ion batteries. J Power Sources 147:269–281. https://doi.org/10.1016/j.jpowsour.2005.01.006
11. Choi JW, Aurbach D (2016) Promise and reality of post-lithium-ion batteries with high energy densities. Nat Rev Mater 1. https://doi.org/10.1038/natrevmats.2016.13
12. Scrosati B, Garche J (2010) Lithium batteries: status, prospects and future. J Power Sources 195:2419–2430. https://doi.org/10.1016/j.jpowsour.2009.11.048
13. Sloop SE, Kerr JB, Kinoshita K (2003) The role of Li-ion battery electrolyte reactivity in performance decline and self-discharge. J Power Sources 119–121:330–337. https://doi.org/10.1016/S0378-7753(03)00149-6
14. Arya A, Sharma AL (2017) Polymer electrolytes for lithium ion batteries: a critical study. Ionics (Kiel) 23:497–540. https://doi.org/10.1007/s11581-016-1908-6
15. Ketkar PM, Shen KH, Hall LM, Epps TH (2019) Charging toward improved lithium-ion polymer electrolytes: exploiting synergistic experimental and computational approaches to facilitate materials design. Mol Syst Des Eng 4:223–238. https://doi.org/10.1039/c8me00105g
16. Armand M, Tarascon JM (2008) Building better batteries. Nature 451:652–657. https://doi.org/10.1038/451652a
17. Wakihara M (2001) Recent developments in lithium ion batteries. Mater Sci Eng R Reports 33:109–134. https://doi.org/10.1016/S0927-796X(01)00030-4
18. Xue Z, He D, Xie X (2015) Poly(ethylene oxide)-based electrolytes for lithium-ion batteries. J Mater Chem A 3:19218–19253. https://doi.org/10.1039/c5ta03471j

19. Hu P, Chai J, Duan Y et al (2016) Progress in nitrile-based polymer electrolytes for high performance lithium batteries. J Mater Chem A 4:10070–10083. https://doi.org/10.1039/c6ta02907h
20. Tan CG, Siew WO, Pang WL et al (2007) The effects of ceramic fillers on the PMMA-based polymer electrolyte systems. Ionics (Kiel) 13:361–364. https://doi.org/10.1007/s11581-007-0126-7
21. Wang F, Li L, Yang X et al (2018) Influence of additives in a PVDF-based solid polymer electrolyte on conductivity and Li-ion battery performance. Sustain Energy Fuels 2:492–498. https://doi.org/10.1039/c7se00441a
22. Chen G, Zhang F, Zhou Z et al (2018) A flexible dual-ion battery based on PVDF-HFP-modified gel polymer electrolyte with excellent cycling performance and superior rate capability. Adv Energy Mater 8:1–7. https://doi.org/10.1002/aenm.201801219
23. Miao R, Liu B, Zhu Z et al (2008) PVDF-HFP-based porous polymer electrolyte membranes for lithium-ion batteries. J Power Sources 184:420–426. https://doi.org/10.1016/j.jpowsour.2008.03.045
24. Raghavan P, Manuel J, Zhao X et al (2011) Preparation and electrochemical characterization of gel polymer electrolyte based on electrospun polyacrylonitrile nonwoven membranes for lithium batteries. J Power Sources 196:6742–6749. https://doi.org/10.1016/j.jpowsour.2010.10.089
25. Gupta AK, Paliwal DK, Bajaj P (1998) Melting behavior of acrylonitrile polymers. J Appl Polym Sci 70:2703–2709. https://doi.org/10.1002/(sici)1097-4628(19981226)70:13%3c2703:aid-app15%3e3.3.co;2-u
26. Wu G, Yang HY, Chen HZ et al (2007) Novel porous polymer electrolyte based on polyacrylonitrile. Mater Chem Phys 104:284–287. https://doi.org/10.1016/j.matchemphys.2007.03.013
27. Chen R, Hu Y, Shen Z et al (2017) Facile fabrication of foldable electrospun polyacrylonitrile-based carbon nanofibers for flexible lithium-ion batteries. J Mater Chem A 5:12914–12921. https://doi.org/10.1039/c7ta02528a
28. Wannatong L, Sirivat A, Supaphol P (2004) Effects of solvents on electrospun polymeric fibers: Preliminary study on polystyrene. Polym Int 53:1851–1859. https://doi.org/10.1002/pi.1599
29. Pattamaprom C, Hongrojjanawiwat W, Koombhongse P et al (2006) The influence of solvent properties and functionality on the electrospinnability of polystyrene nanofibers. Macromol Mater Eng 291:840–847. https://doi.org/10.1002/mame.200600135
30. Yördem OS, Papila M, Menceloğlu YZ (2008) Effects of electrospinning parameters on polyacrylonitrile nanofiber diameter: an investigation by response surface methodology. Mater Des 29:34–44. https://doi.org/10.1016/j.matdes.2006.12.013
31. Sim LN, Sentanin FC, Pawlicka A et al (2017) Development of polyacrylonitrile-based polymer electrolytes incorporated with lithium bis(trifluoromethane) sulfonimide for application in electrochromic device. Electrochim Acta 229:22–30. https://doi.org/10.1016/j.electacta.2017.01.098
32. Choi SW, Kim JR, Jo SM et al (2005) Electrochemical and spectroscopic properties of electrospun PAN-based fibrous polymer electrolytes. J Electrochem Soc 152:A989–A995. https://doi.org/10.1149/1.1887166
33. Huang ZM, Zhang YZ, Kotaki M, Ramakrishna S (2003) A review on polymer nanofibers by electrospinning and their applications in nanocomposites. Compos Sci Technol 63:2223–2253. https://doi.org/10.1016/S0266-3538(03)00178-7
34. Carol P, Ramakrishnan P, John B, Cheruvally G (2011) Preparation and characterization of electrospun poly(acrylonitrile) fibrous membrane based gel polymer electrolytes for lithium-ion batteries. J Power Sources 196:10156–10162. https://doi.org/10.1016/j.jpowsour.2011.08.037
35. Arifeen WU, Kim M, Choi J et al (2019) Optimization of porosity and tensile strength of electrospun polyacrylonitrile nanofibrous membranes. Mater Chem Phys 229:310–318. https://doi.org/10.1016/j.matchemphys.2019.03.020
36. Ma X, Kolla P, Yang R et al (2017) Electrospun polyacrylonitrile nanofibrous membranes with varied fiber diameters and different membrane porosities as lithium-ion battery separators. Electrochim Acta 236:417–423. https://doi.org/10.1016/j.electacta.2017.03.205

37. Sabetzadeh N, Gharehaghaji AA, Javanbakht M (2018) Porous PAN micro/nanofiber separators for enhanced lithium-ion battery performance. Solid State Ionics 325:251–257. https://doi.org/10.1016/j.ssi.2018.08.013
38. Osman Z, Md Isa KB, Ahmad A, Othman L (2010) A comparative study of lithium and sodium salts in PAN-based ion conducting polymer electrolytes. Ionics (Kiel) 16:431–435. https://doi.org/10.1007/s11581-009-0410-9
39. Wang Z, Hu Y, Chen L (2005) Some studies on electrolytes for lithium ion batteries. J Power Sources 146:51–57. https://doi.org/10.1016/j.jpowsour.2005.03.092
40. Jung HR, Ju DH, Lee WJ et al (2009) Electrospun hydrophilic fumed silica/polyacrylonitrile nanofiber-based composite electrolyte membranes. Electrochim Acta 54:3630–3637. https://doi.org/10.1016/j.electacta.2009.01.039
41. Yang Z, Peng H, Wang W, Liu T (2010) Crystallization behavior of poly (ε-caprolactone)/layered double hydroxide nanocomposites. J Appl Polym Sci 116:2658–2667. https://doi.org/10.1002/app
42. Chae HG, Choi YH, Minus ML, Kumar S (2009) Carbon nanotube reinforced small diameter polyacrylonitrile based carbon fiber. Compos Sci Technol 69:406–413. https://doi.org/10.1016/j.compscitech.2008.11.008
43. Newcomb BA, Chae HG, Gulgunje PV et al (2014) Stress transfer in polyacrylonitrile/carbon nanotube composite fibers. Polymer (Guildf) 55:2734–2743. https://doi.org/10.1016/j.polymer.2014.04.008
44. Zhang J, Xue Q, Pan X et al (2017) Graphene oxide/polyacrylonitrile fiber hierarchical-structured membrane for ultra-fast microfiltration of oil-water emulsion. Chem Eng J 307:643–649. https://doi.org/10.1016/j.cej.2016.08.124
45. Liu Z, Xu Z, Hu X, Gao C (2013) Lyotropic liquid crystal of polyacrylonitrile-grafted graphene oxide and its assembled continuous strong nacre-mimetic fibers. Macromolecules 46:6931–6941. https://doi.org/10.1021/ma400681v
46. Walls HJ, Riley MW, Singhal RR et al (2003) Nanocomposite electrolytes with fumed silica and hectorite clay networks: passive versus active fillers. Adv Funct Mater 13:710–717. https://doi.org/10.1002/adfm.200304333
47. Jung S, Kim DW, Lee SD et al (2009) Fillers for solid-state polymer electrolytes: highlight. Bull Korean Chem Soc 30:2355–2361. https://doi.org/10.5012/bkcs.2009.30.10.2355
48. Dissanayake MAKL, Jayathilaka PARD, Bokalawala RSP et al (2003) Effect of concentration and grain size of alumina filler on the ionic conductivity enhancement of the (PEO)9LiCF_3SO_3: Al_2O_3 composite polymer electrolyte. J Power Sources 119–121:409–414. https://doi.org/10.1016/S0378-7753(03)00262-3
49. Yanilmaz M, Lu Y, Zhu J, Zhang X (2016) Silica/polyacrylonitrile hybrid nanofiber membrane separators via sol-gel and electrospinning techniques for lithium-ion batteries. J Power Sources 313:205–212. https://doi.org/10.1016/j.jpowsour.2016.02.089
50. Kim M, Lee L, Jung Y, Kim S (2013) Study on ion conductivity and crystallinity of composite polymer electrolytes based on poly(ethylene oxide)/poly(acrylonitrile) containing nano-sized Al_2O_3 fillers. J Nanosci Nanotechnol 13:7865–7869. https://doi.org/10.1166/jnn.2013.8107
51. Wang Q, Song WL, Fan LZ, Song Y (2015) Facile fabrication of polyacrylonitrile/alumina composite membranes based on triethylene glycol diacetate-2-propenoic acid butyl ester gel polymer electrolytes for high-voltage lithium-ion batteries. J Memb Sci 486:21–28. https://doi.org/10.1016/j.memsci.2015.03.022
52. Liang Y, Ji L, Guo B et al (2011) Preparation and electrochemical characterization of ionic-conducting lithium lanthanum titanate oxide/polyacrylonitrile submicron composite fiber-based lithium-ion battery separators. J Power Sources 196:436–441. https://doi.org/10.1016/j.jpowsour.2010.06.088
53. Abhilash KP, Christopher selvin P, Nalini B, et al (2016) Study of the temperature dependent transport properties in nanocrystalline lithium lanthanum titanate for lithium ion batteries. J Phys Chem Solids 91:114–121. https://doi.org/10.1016/j.jpcs.2015.12.015
54. Zhang LC, Chen P, Hu Z, Chen CH (2012) Electrical properties of NASICON-type structured Li1.3$Al_{0.3}Ti_{1.7}(PO_4)_3$ solid electrolyte prepared by 1,2-propylene glycol-assisted Sol-gel method. Chinese J Chem Phys 25:703–707. https://doi.org/10.1088/1674-0068/25/06/703-707

55. Geier S (2015) $Li_{1.4}$Al<sub>0.4<sub>$Ti_{1.6}(PO_4)_{,3}$ used as solid electrolyte for structural supercapacitors. ASME 2015 Conf Smart Mater Adapt Struct Intell Syst 6:0–7. https://doi.org/10.13140/RG.2.1.3581.4169
56. Liang Y, Lin Z, Qiu Y, Zhang X (2011) Fabrication and characterization of LATP/PAN composite fiber-based lithium-ion battery separators. Electrochim Acta 56:6474–6480. https://doi.org/10.1016/j.electacta.2011.05.007
57. Murugadoss V, Arunachalam S, Elayappan V, Angaiah S (2018) Development of electrospun PAN/CoS nanocomposite membrane electrolyte for high-performance DSSC. Ionics (Kiel) 24:4071–4080. https://doi.org/10.1007/s11581-018-2540-4
58. He C, Liu J, Cui J et al (2018) A gel polymer electrolyte based on Polyacrylonitrile/organic montmorillonite membrane exhibiting dense structure for lithium ion battery. Solid State Ionics 315:102–110. https://doi.org/10.1016/j.ssi.2017.12.014
59. Liu W, Liu N, Sun J et al (2015) Ionic conductivity enhancement of polymer electrolytes with ceramic nanowire fillers. Nano Lett 15:2740–2745. https://doi.org/10.1021/acs.nanolett.5b00600
60. Raghavan P, Zhao X, Shin C et al (2010) Preparation and electrochemical characterization of polymer electrolytes based on electrospun poly(vinylidene fluoride-co-hexafluoropropylene)/polyacrylonitrile blend/composite membranes for lithium batteries. J Power Sources 195:6088–6094. https://doi.org/10.1016/j.jpowsour.2009.11.098
61. Lim D-H, Haridas AK, Figerez SP et al (2018) Tailor-made electrospun multilayer composite polymer electrolytes for high-performance lithium polymer batteries. J Nanosci Nanotechnol 18:6499–6505. https://doi.org/10.1166/jnn.2018.15689
62. Tian X, Xin B, Lu Z et al (2019) Electrospun sandwich polysulfonamide/polyacrylonitrile/polysulfonamide composite nanofibrous membranes for lithium-ion batteries. RSC Adv 9:11220–11229. https://doi.org/10.1039/C8RA10229E
63. Wang SH, Kuo PL, Te Hsieh C, Teng H (2014) Design of poly(acrylonitrile)-based gel electrolytes for high-performance lithium ion batteries. ACS Appl Mater Interfaces 6:19360–19370. https://doi.org/10.1021/am505448a
64. Tan L, Deng Y, Cao Q et al (2019) Gel electrolytes based on polyacrylonitrile/thermoplastic polyurethane/polystyrene for lithium-ion batteries. Ionics (Kiel) 25:3673–3682. https://doi.org/10.1007/s11581-019-02940-7
65. Gopalan AI, Santhosh P, Manesh KM et al (2008) Development of electrospun PVdF-PAN membrane-based polymer electrolytes for lithium batteries. J Memb Sci 325:683–690. https://doi.org/10.1016/j.memsci.2008.08.047
66. Prasanth R, Aravindan V, Srinivasan M (2012) Novel polymer electrolyte based on cob-web electrospun multi component polymer blend of polyacrylonitrile/poly(methyl methacrylate)/polystyrene for lithium ion batteries - Preparation and electrochemical characterization. J Power Sources 202:299–307. https://doi.org/10.1016/j.jpowsour.2011.11.057
67. Rao MM, Liu JS, Li WS et al (2008) Preparation and performance analysis of PE-supported P(AN-co-MMA) gel polymer electrolyte for lithium ion battery application. J Memb Sci 322:314–319. https://doi.org/10.1016/j.memsci.2008.06.004
68. Pu W, He X, Wang L et al (2006) Preparation of P(AN-MMA) microporous membrane for Li-ion batteries by phase inversion. J Memb Sci 280:6–9. https://doi.org/10.1016/j.memsci.2006.05.028
69. Ramesh S, Yahaya AH, Arof AK (2002) Dielectric behaviour of PVC-based polymer electrolytes. Solid State Ionics 152–153:291–294. https://doi.org/10.1016/S0167-2738(02)00311-9
70. Rajendran S, Uma T (2001) FTIR and conductivity studies of PVC based polymer electrolyte systems. Ionics (Kiel) 7:122–125. https://doi.org/10.1007/BF02375478
71. Zhong Z, Cao Q, Jing B et al (2012) Novel electrospun PAN-PVC composite fibrous membranes as polymer electrolytes for polymer lithium-ion batteries. Ionics (Kiel) 18:853–859. https://doi.org/10.1007/s11581-012-0682-3
72. Cheruvally G, Kim JK, Choi JW et al (2007) Electrospun polymer membrane activated with room temperature ionic liquid: novel polymer electrolytes for lithium batteries. J Power Sources 172:863–869. https://doi.org/10.1016/j.jpowsour.2007.07.057

73. Osada I, De Vries H, Scrosati B, Passerini S (2016) Ionic-liquid-based polymer electrolytes for battery applications. Angew Chemie - Int Ed 55:500–513. https://doi.org/10.1002/anie.201504971
74. Rao M, Geng X, Liao Y et al (2012) Preparation and performance of gel polymer electrolyte based on electrospun polymer membrane and ionic liquid for lithium ion battery. J Memb Sci 399–400:37–42. https://doi.org/10.1016/j.memsci.2012.01.021
75. Cho TH, Sakai T, Tanase S et al (2007) Electrochemical performances of polyacrylonitrile nanofiber-based nonwoven separator for lithium-ion battery. Electrochem Solid-State Lett 10:A159. https://doi.org/10.1149/1.2730727

Chapter 6
Electrospun-Based Nonwoven 3D Fibrous Composite Polymer Electrolytes for High-Performance Lithium-Ion Batteries

M. A. Krishnan, Neethu T. M. Balakrishnan, Akhila Das, Leya Rose Raphael, M. J. Jabeen Fatima, and Raghavan Prasanth

6.1 Introduction

The increased energy demand and need for more reliable and safer energy storage devices have led researchers to explore more on lithium-ion batteries. Electrolytes are considered as the prime content of the battery as the performance of the energy storage device is relied on the performance and the conductivity of the electrolytic species. High-energy solid-state batteries are replacing conventional liquid electrolyte with polymer electrolytes in order to prevent leakage. Polymer electrolytes possess the advantageous properties of both liquid and solid electrolytes. The basic polymer electrolytes have several drawbacks such as low electrolyte uptake and low conductivity; hence, to enhance the performance of these systems, several additives have been incorporated into the matrix. These composite polymer electrolytes possess higher ionic conductivity at ambient conditions. They also are having improved lithium interfacial stability, better cycling capability, comparable or superior ambient-temperature ionic conductivity and good mechanical strength [1]. Hence, composite polymers are

M. A. Krishnan
Department of Mechanical Engineering, Amrita Vishwa Vidyapeetham, Amritapuri, India

Department of Electrical Engineering, Pennsylvania State University, Electrical Engineering East Building 311, University Park, Pennsylvania 16802, USA

N. T. M. Balakrishnan · A. Das · L. R. Raphael · M. J. Jabeen Fatima (✉) · R. Prasanth (✉)
Department of Polymer Science and Rubber Technology (PSRT), Cochin University of Science and Technology (CUSAT), Cochin, Kerala 682022, India
e-mail: jabeen@cusat.ac.in

R. Prasanth
e-mail: prasanth@cusat.ac.in

R. Prasanth
Department of Materials Engineering and Convergence Technology, Gyeongsang National University, Jinju-Daero 501, Jinju 52828, Republic of Korea

N. T. M. Balakrishnan and R. Prasanth (eds.), *Electrospinning for Advanced Energy Storage Applications*, Materials Horizons: From Nature to Nanomaterials,
https://doi.org/10.1007/978-981-15-8844-0_6

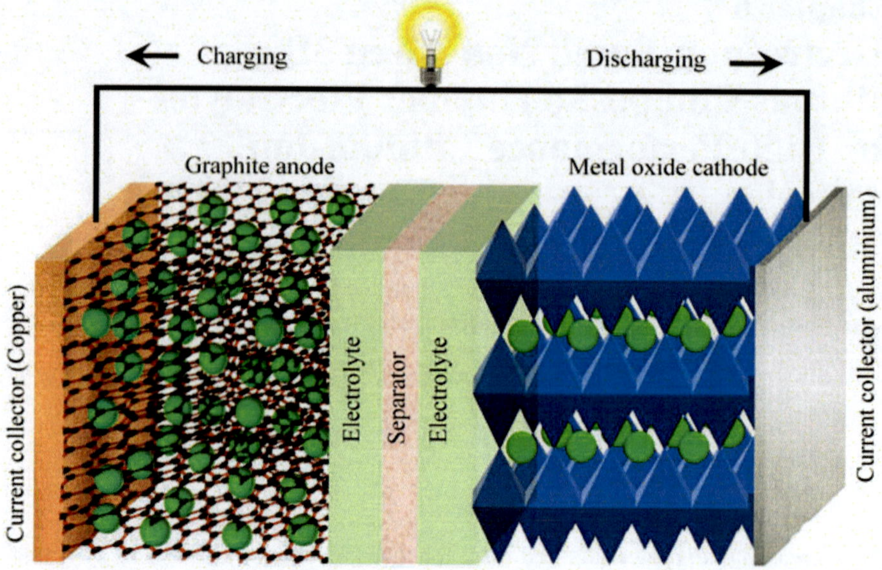

Fig. 6.1 Schematic representation of lithium-ion battery

more effective electrolytes for high-performance lithium-ion battery (LIB). A typical schematic representation of LIB is given in Fig. 6.1.

The main advantages of composite polymeric membrane as electrolyte component in lithium cell are (i) suppression of dendrite growth, (ii) enhance endurance to varying electrode volume during cycling performance, (iii) reduced reactivity with liquid electrolytes, (iv) improved safety and (v) better shape flexibility and manufacturing integrity. There are mainly two directions in the field of solid polymer electrolytes [2]. One is the highly conductive materials via the cross-linking of mobile chains to form network, which are then swollen (gel electrolytes) by lithium salt solutions. The matrix in this type of SPE is liquid-like to transport ions. The other is the development of SPE by supramolecular architectures, which intrinsically gives to much improved mechanical strength. The use of these materials as separators in LIBs has advantage of minimizing the separator thickness, and this will in turn lower the internal resistance of a lithium solid-state cell. They have strong tendency to self-organize into supramolecular architectures when cast from solution. This property will overcome the compatibility limitations of rigid and flexible polymer materials, and these materials can thus be considered as molecular composites. To obtain maximal results, both these directions should merge. The gel electrolytes exhibit excellent conductivities but still lack chemical stability under working conditions, and it is weak in mechanical strength, whereas the reinforced SPEs are mechanically strong but lack in ionic conductivities.

In polymer electrolytes, both cations and anions may contribute to the ionic conductivity of the electrolytes. Studies in the near period reveal that the composite polymer electrolytes alone can improve the electrolyte compatibilities and safety hazards. The electrolytes of lithium power resources are mainly classified into liquid polymer and solid ones, that is, crystalline compounds with chemical bonds of ionic nature [3]. The lithium power sources comprised of nanocomposite polymer electrolytes are a modification of solid polymer electrolytes added by foreign component, that is, inorganic oxide particle. The polymer electrolyte plays a crucial role in the isolation of lithium-ion batteries electrodes from short-circuiting; it helps in ion transportation by providing a medium between the electrodes.

The additives incorporated in the polymer matrix may be organic or inorganic moieties. The organic compounds such as cellulose [4–7], cellulose derivatives [8], kraft lignin [9], chitin [10], chitosan [11], carbon nanotube (CNT) [12] and graphene oxide (GO) [13] have been widely employed as fillers for polymer electrolytes in LIBs. Inorganic/ceramic fillers such as aluminium oxide (Al_2O_3) [14–18], titania (TiO_2) [19–23], silica (SiO_2) [24–28] and nickel oxide (NiO) [29] have also been tried out as electrolytes. The synthesis method of these polymer systems may vary. One of the promising and effective techniques for synthesis of polymer separators in LIBs is electrospinning method, because it was able to retain 500 wt% of liquid electrolyte. The conductivity of electrolyte was improved by many methods; the most popular one is the addition of low molecular weight solvents that make the polymer chain and lithium ions more mobile. These are called as the polymer gel electrolytes (PGE). Solid polymer electrolytes incorporating nanofillers are termed as nanocomposite polymer electrolyte (NCPE). A SEM image of honeycomb structure of polymer electrolyte membrane of PVdF-*co*-HFP is given in Fig. 6.2.

There are two types of ceramic fillers: active and passive fillers. The active fillers involve in the conduction process, e.g., Li_2N and $LiAl_2O_3$, whereas passive fillers are inactive compounds which do not involve in conduction process, e.g., Al_2O_3, SiO_2 and MgO [31]. Dispersing fillers in polymer electrolytes generating NCPE is known to have a very stable lithium electrode interface and has an increased ionic conductivity even in low temperatures. This exceptional interfacial stability is due to the addition of fillers which involves in residual removal of liquid impurities from the surface and favouring the formation of smooth surfaces [32]. The resistance offered by filler coated composite polymer electrolyte is lower than a battery using filler-free electrolyte. In addition to the increased ionic conductivity, it also widens the temperature range, which is indeed a major factor [33]. In addition, to enhance the value of ionic conductivity of the resulting composite polymer electrolyte, ceramic filler also widens its wanted temperature range. The composite polymer electrolyte operates at temperature values, which are normally excluded for electrolytes of conventional polymer batteries.

Some researchers have found that the cationic transfer number can be increased by synthesizing polymers in which anions are covalently bonded to polymer backbone as exemplified [34]. This kind of polymers is named as 'polyelectrolytes'. A typical difference between polymer electrolyte and polyelectrolyte has been depicted in Fig. 6.3. The increase in ionic conductivity is due to the cationic transport, and

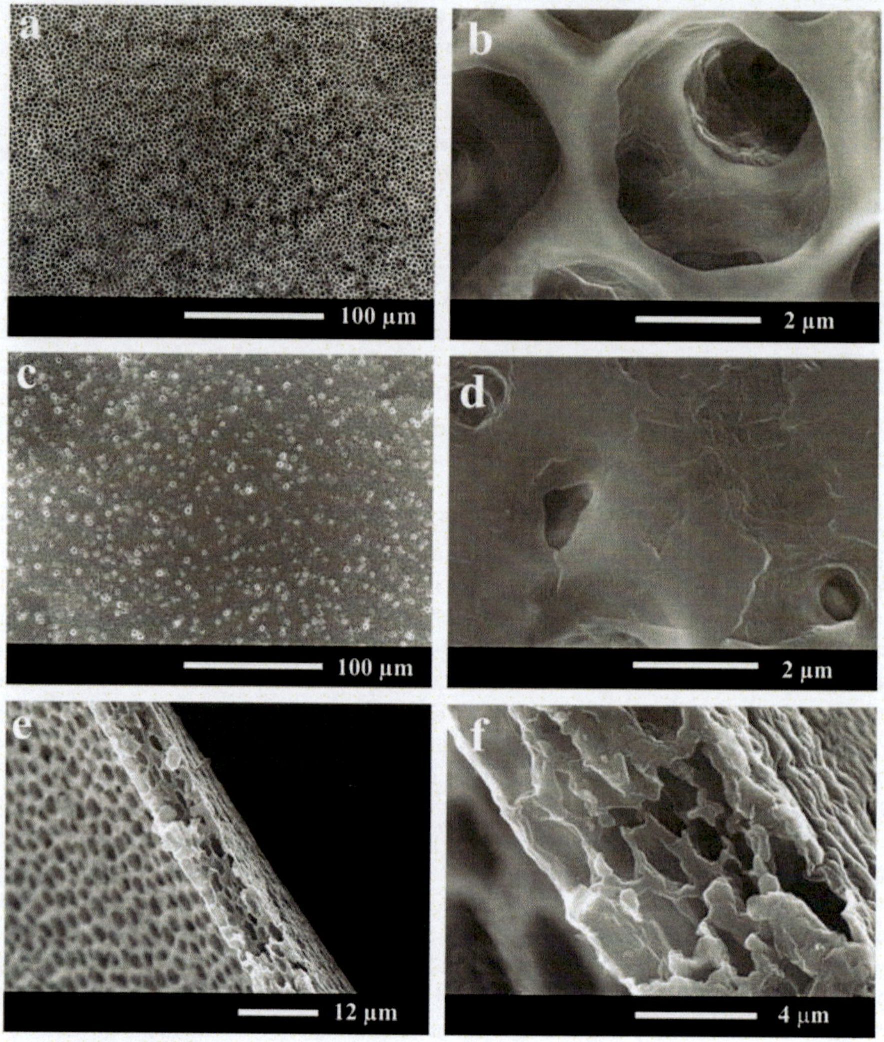

Fig. 6.2 FESEM images of **a**, **b** the front side, **c**, **d** the back side, and **e**, **f** the cross-section of the PVdF-*co*-HFP honeycomb-like porous polymer membrane. Adapted and reproduced from Ref. [30]. Copyright 2014 Springer Nature

the anions are being effectively immobilized. These materials are very much flexible and exhibit ambient-temperature conductivities. Polyelectrolytes are not susceptible to the build-up of potential resistive layers of high or low salt concentration at the electrode/electrolyte interfaces during charging and discharging unlike polymer electrolytes.

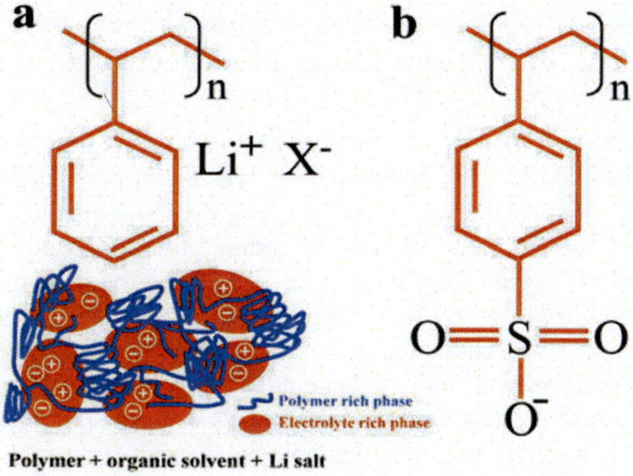

Fig. 6.3 Chemical structure of **a** polymer electrolyte, **b** polyelectrolyte

Advancement in the polymer electrolytes is achieved by the introduction of substantial amounts of plasticizers in the electrolyte. Plasticizers reduce the crystalline content and increase the polymer segmental mobility. They also exhibit greater ion dissociation, which helps greater numbers of charge carriers for ionic transport. Polar organic solvent and low molecular weight polyethers are common types of plasticizers. Another category of polymer electrolyte is the 'gelled polymer electrolytes' and is characterized by a higher ambient ionic conductivity but poorer mechanical properties when compared with pure solid polymer electrolyte. They are usually obtained by incorporating a larger quantity of liquid plasticizer. The gel is used in polymer electrolyte application because of its duality. Croche et al. [32] conducted a study on lithium polymer electrolytes formed by dissolving a lithium salt LiX in poly(ethylene oxide) PEO, useful for applications as separators in lithium rechargeable polymer batteries. In this study, a new nanocomposite of PEO-LiX polymer electrolyte is characterized by enhanced interfacial stability as well as by improved ambient-temperature transport properties. The interfacial stability is obtained due to the removal of residual traces of liquid impurities from the interface. In another study by Krawiec et al. [35], the polymer nanocomposite polymer electrolyte is developed by nanosized alumina particles with PEO and lithium tetrafluoroborate. The influence of nanoparticles of alumina increases the interfacial stability of overall electrolyte. This study concludes that the interfacial stability of a composite polymer electrolyte mainly depends on the particle size and ceramic content. The increased ionic mobility and ionic concentration at grain boundaries suggest that the ionic conductivity is a function of particle size. Several synthesis methods are known for the synthesis of polymer composite electrolytes. The most accepted and best method is electrospinning. The main advantage of electrospinning is the formation of continuous fibrous and interconnected network which provide a pathway for ionic movement. Hence, herein we are focusing on the synthesis and application of electrospun polymer composite as electrolytes for lithium-ion batteries.

6.2 Effect of Size of Ceramic Fillers on the Properties of Polymer Composite Electrolyte

The incorporation of ceramic fillers on the matrix of polymer enhances the physical and conducting properties of polymer gel composite electrolyte [36]. The presence of additive shows improvement in the ionic conductivity and transference number of Li^+ ions. The addition of ceramic fillers decreases the interfacial impedance of electrode-electrolyte interface, increasing the ionic conductivity of Li^+ ions. These additives also reinforce the polymer matrix leading to enhancement in the mechanical properties. The additives enter the polymeric matrix flexblizing the interlinking of the polymeric chains enhancing the amorphicity of the polymer. The ionic movement is fastened through the amorphous part of the polymer chain; hence, the increase in amorphicity enhances the ionic conductivity of the ions through the membrane. The cationic charges on the ceramic fillers like SiO_2, TiO_2, clay, Al_2O_3, etc. enable the Lewis acid-base interaction with Li^+ ions. The cationic charges on the filler compete with active ions like Li^+ enhancing the ionic conduction in the matrix.

6.3 Electrospinning Technique for Synthesis of Polymer Composite Electrolyte

Electrospinning of composite electrolytes enhances the properties of the electrolyte as it produces fibrous composite membranes, which are of high porosity and uniform morphology. This enhances the ionic conductivity and interfacial stability and improves the efficiency of electrolytes in high temperatures. To prepare the porous membrane, to tailor easy ion conduction path for the movement of ions, electrospinning method is adopted. The membrane properties such as membrane morphology, porosity, pore density and thickness can be monitored by changing the spinning parameters such as solution concentration, feed rate, bore size of the needle, applied spinning voltage and distance between the collection drum and tip of the needle. The pore structure of electrospun membranes is fully interconnected which helps in easy penetration, and absorption of large amounts of liquid electrolyte and cobweb structure formed by stacking of fibres helps to hold the absorbed electrolyte with minimum leakage or electrolyte retention, which makes it an improved polymer composite electrolyte. The polymeric matrix provides a good mechanical substrate and enhances the transfer rate by offering active ion for lithium ions. The surface morphology of membrane when studied exhibits three-dimensional network of fibres, with large number of fully connected pores formed by interlaying fibres with uniform bead free morphology. This shows that the evaporation of solvent is faster during the electrospinning method.

Polymers are the medium used in the electrospinning technique. A variety of polymers have been attempted for synthesis and application in wide area including energy conversion and storage devices. Polyacrylonitrile (PAN), polyvinylidene difluoride (PVdF), poly(ethylene oxide) (PEO), polypropylene oxide (PPO), polyethylene terephthalate (PET), polymethyl methacrylate (PMMA), polyvinylpyrrolidone (PVP), polyvinylidene difluoride-*co*-hexafluoropropylene (PVdF-*co*-HFP), acrylonitrile butadiene styrene (ABS), etc. are some of the polymer matrix mainly used as matrix for electrospinning. Some of the electrolytes are modified with ceramic compounds to enhance the properties for the application of the membranes as electrolytes for energy storage application.

6.4 Electrospun Polyacrylonitrile-Based Composite Polymer Electrolyte

Polyacrylonitrile (PAN) is also known as polyvinyl cyanide, which is formed by polymerization of acrylonitrile monomer. The general formula of PAN can be represented as C_3H_3N. The typical structure formation has been depicted in Fig. 6.4. The backbone of PAN is hydrocarbon chain, and it provides hard and high melting polymer. The solubility of PAN is low in aqueous medium. It is highly stable and requires very high temperature to undergo carbonization [37]. PAN exhibits good electrochemical stability and fire-resistant properties compared to other polymer electrolytes. Persisting the aforementioned qualities of the polymer, they also possess the major disadvantage of low electrolyte uptake retarding the property of material as electrolyte for LIB.

In 2000, Tsutsumi et al. [38] reported PAN-based electrolyte with enhanced ionic conductivity for the application as electrolytes in LIB. The conductivity of the samples was in the order of 10^{-5} S cm^{-1}. Later in 2011, Raghavan et al. [39] reported electrospun PAN polymer electrolyte at varying concentrations (10, 12, 14, 16 and 18 wt%) for application as electrolytes in LIBs. The maximum electrolyte (1 M lithium hexafluorophosphate ($LiPF_6$) in ethylene carbonate (EC)/dimethyl carbonate (DMC)) uptake was given by PAN with 16 wt%. They optimized the effect of applied

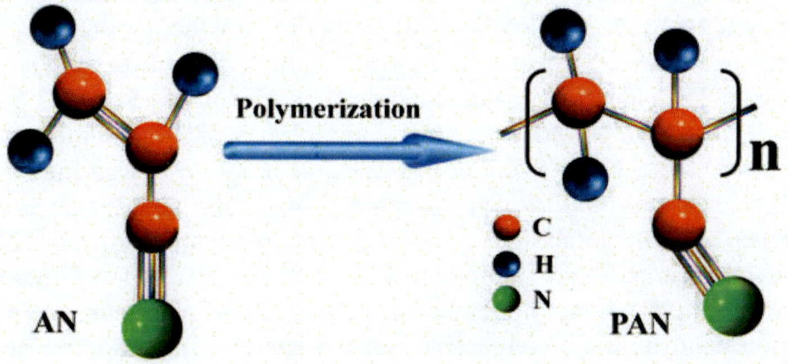

Fig. 6.4 Structure and formation of polyacrylonitrile

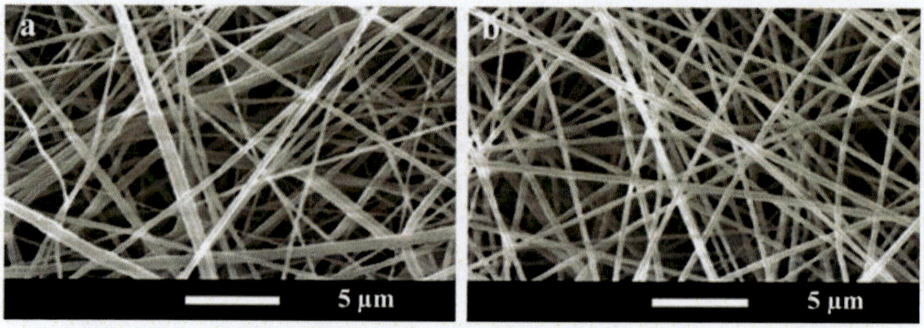

Fig. 6.5 SEM image electrospun membrane of 16 wt% PAN biased at **a** 15 kV with an average diameter of 330 nm, **b** 20 kV with an average diameter of 350 nm. Adapted and reproduced from Ref. [39], Copyright 2011 Elsevier

voltage on the size and shape of the fibre. On varying the applied voltage from 10 to 25 kV, the diameter of samples increased gradually from 80 nm to ten times the initial diameter. The electrospun PAN membrane with 16 wt% of PAN spun at 15 kV which gave porosity of 84% and 20 kV which delivered a porosity of 86% is given in Fig. 6.5, where the membrane spun at 20 kV gave uniform fibres with an ionic mobility in the range of 10^{-3} S cm^{-1} and a discharge capacity ~150 mAh g^{-1} [39]. A similar study on electrospun PAN membrane for LIB was reported by Ma et al. [40] in 2017. They varied the concentration of PAN from 7 to 12 wt%. They confirmed that the samples formed from 7 wt% of PAN showed bead-like formation which was not much identifiable in low-resolution microscopic images, but on increasing the concentration the samples formed uniform morphology fibres.

6.4.1 *Electrospun Polyacrylonitrile/Alumina Composite Polymer Electrolyte*

Alumina is one of the important ceramic compounds introduced into the polymer matrix for enhancing the electrochemical properties for the application in energy storage devices. In 2002, Yang et al. [41] reported the α-alumina (α-Al_2O_3) incorporated composite PAN electrolyte with an ionic conductivity of ~5×10^{-4} S cm^{-1}. In 2015, Wang et al. [42] reported the electrospun synthesis of PAN/alumina composite membranes for LIBs. Nano-alumina (10–20 nm) 0–40 wt% was mixed in PAN matrix. The composite mixture was electrospun at 21 kV at a distance of 25 cm. The nonwoven membranes showed increase in electrolyte uptake from 452 to 561% on increase in concentration of alumina in the matrix from 0 to 40 wt%, respectively. The porosity of the electrospun membrane also increased with increase in alumina concentration. But the ionic conductivity of the samples increased on increase in alumina concentration up to 20 wt% (2.35×10^{-3} S cm^{-1}) compared to pristine PAN membrane (1.86×10^{-3} S cm^{-1}). Further increase in concentration of alumina led

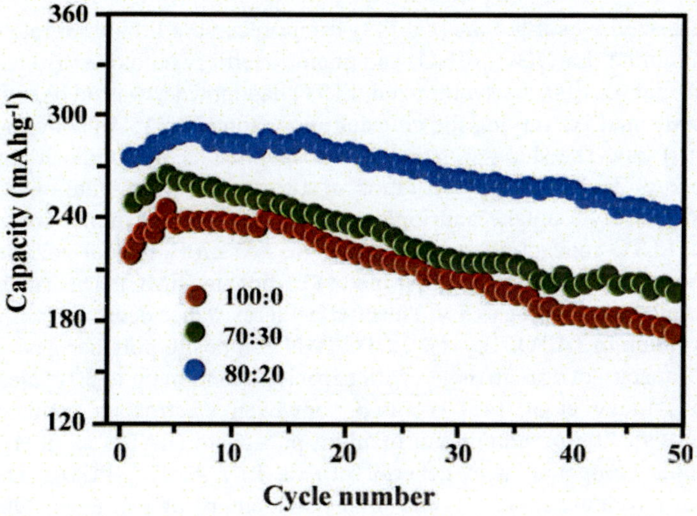

Fig. 6.6 Cycling performance of cells using $Li[Li_{1/6}Ni_{1/4}Mn_{7/12}]O_{7/4}F_{1/4}$ as electrode. Adapted and reproduced from Ref. [42], Copyright 2015 Elsevier

to a marked decline of ionic conductivity indicating the agglomeration of alumina particles at higher concentrations hindering the ionic movement. The PAN/Al_2O_3 (20 wt%)/electrospun membrane showed a maximum electrochemical stability up to 5 V. PAN/Al_2O_3 polymer composite electrospun membrane was immersed in gel electrolyte with triethylene glycol diacetate (TEGDA)–2-propenoic acid butyl ester (BA). The Li-cell studies indicate that the samples with 20 wt% of alumina gave a maximum discharge capacity of 283 mAh g^{-1} and an ionic conductivity of 2.35×10^{-3} S cm^{-1}. The cycling performance of the cells has been given in Fig. 6.6 [42]. The charge/discharge curves indicated a long plateau region above 4.5 V. The plateau regions are resultant of the extraction of lithium ions along with the release of oxygen from Li_2MnO_3 cathode. The electrospun PAN/Al_2O_3 electrolyte cell showed an initial discharge capacity ranging from 223 to 277 mAh g^{-1}. The maximum discharge capacity was obtained for 20 wt% alumina incorporated membrane and further increase in concentration of alumina decreased the performance of the cells. The cycle stability of the cell indicates retention of 86.7% whereas pristine electrospun PAN membrane showed retention of 78.2% (Fig. 6.6).

6.4.2 *Electrospun Polyacrylonitrile/Ternary Metal Oxide/Phosphate Composite Polymer Electrolyte*

Ternary metal oxides such as lithium lanthanum titanate oxide (LLTO) have been reported as composite fillers in PAN matrix. The importance of pervoskite type ceramic ternary oxides is their ability to conduct lithium ions. LLTO nanoparticles were synthesized by sol-gel method [43] from precursors lithium nitrate ($LiNO_3$), lanthanum nitrate ($La(NO_3)_3.6H_2O$) and titanium tertiary butoxide ($Ti(OC_4H_9)_4$) in the ratio 0.35:0.55:1.00. As synthesized, LLTO nanopowder was mixed with 8 wt% PAN in DMF matrix, varying the concentration from 0 to 15 wt% by mechanical stirring for 1 day. The slurry was then electrospunned to membrane at and applied bias voltage of 18 kV with a separation of 18 cm and a flow rate of 0.5 ml h^{-1} [44]. The electrolyte uptake and ionic conductivity of the samples increase with increase in LLTO concentration with electrolyte 1 M $LiPF_6$ in equi-volume mixture of ethylene carbonate (EC) and ethyl methyl carbonate (EMC). The sample gave a electrochemical stability up to 5 V. The interfacial resistance decreases with increase in concentration of LLTO. 10 wt% LLTO/PAN composite polymer gave a highest discharge capacity of 156 mAh g^{-1} with capacity retention up to 50 cycles [44].

In 2011, Liang et al. [45] reported nonwoven electrospun composite PAN membrane with lithium aluminium titanium phosphate (LATP) as filler. LATP is glass-ceramic compound with general formula $Li_{1+x}Al_xTi_{2-x}(PO_4)_3$. The importance of this ceramic moiety is the high conductivity of Li^+ ions. Synthesis of LATP is achieved by sol-gel method by mixing the precursor in the molar ratio Li(1.4):Al(0.4):Ti(1.6):P(3) followed by pyrolysis at higher temperatures to obtain LATP nanopowder. For electrospinning LATP/PAN composite membrane, 8% PAN in dimethylformamide was mixed with different wt% of LATP (0, 5, 10, 15 wt%) followed by stirring at 70 °C for 12 h. The resulting dispersion was electrospun at 14 kV with at flow rate of 0.75 ml h^{-1}. The needle to collector distance was maintained at 18 cm. The membrane obtained was vacuum dried at 60 °C for 12 h. The SEM image of the samples revealed that the fibre diameter decreases with increase in the LATP concentration from 1 to 15 wt%, and also some projection was visualized on the fibre owing to the agglomeration of LATP as lumps. The ionic conductivity of the samples increased about 10 times on addition of LATP at 15 wt% concentration due to increase in the porosity of the fibre. The membrane also showed an enhancement in the ionic conductivity on raising the temperature to 90 °C. The battery performance of the polymer composite was compared with pure PAM membrane and commercial Celgard 2400 separator. The LATP/PAN composite membrane with 15 wt% of LATP gave a discharge capacitance of 165 mAh g^{-1} at 0.2 C rate which was almost stable up to 50 cycles with a capacity retention of 91% which was in turn better than that of commercial Celgard 2400 separator (84%) (Fig. 6.7) [45].

6.5 Electrospun PVdF-Based Composite Polymer Electrolyte

Polyvinylidene difluoride (PVdF) is one of the main polymer systems widely explored as separator-cum-polymer gel electrolytes for LIBs. Many polymeric blend systems are also being investigated owing to the ability to conduct Li^+ ions through the membrane. Polyvinylidene difluoride-*co*-hexaflouropropylene

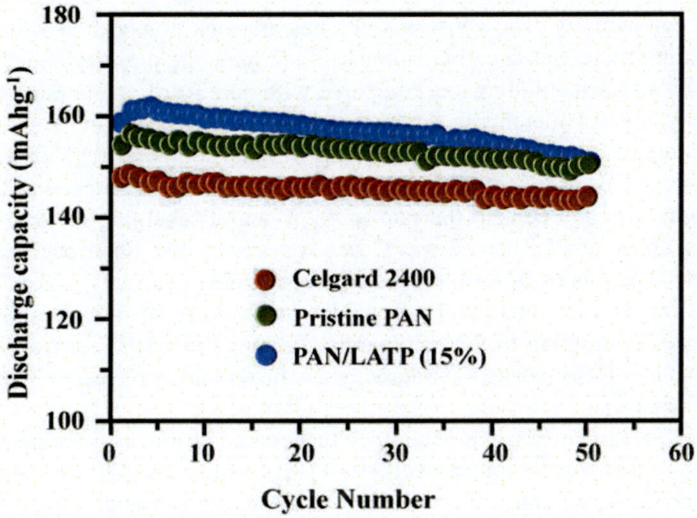

Fig. 6.7 Cycling performance of Li/LiFePO4 cells containing liquid electrolyte-soaked Celgard 2400 separator, PAN fibre-based membrane and 15 wt% LATP/PAN composite fibre-based membrane at 0.2 C rate. Adapted and reproduced from Ref. [45], Copyright 2011 Elsevier

(PVdF-*co*-HFP), polyvinylidene fluoride-*co*-chlorotrifluoroethylene (PVdF-*co*-CTFE), polyvinylidene difluoride/polyacrylonitrile (PVdF-PAN), polyvinylidene difluoride/polymethyl methacrylate (PVdF-PMMA), polyvinylidene difluoride/polyvinylchloride (PVdF-PVC), etc. are some of the polymeric compounds based on PVdF used as polymeric electrolyte for LIBs. To enhance the ionic conductivity, thermal stability and tensile strength of the polymer, certain ceramic fillers such as titania, clay, alumina and magnesium aluminate have been composited with the matrix.

6.5.1 Electrospun Polyvinylidene Fluoride/Alumina Composite Polymer Electrolyte

Electrospun hybrid composite membranes have been reported on PVdF and PVdF-based polymers with alumina (Al_2O_3) as separators-cum-electrolyte for LIBs. Recently in 2017, Wu et al. [46] reported the synthesis and application of electrospun PVdF-alumina composite as electrolytes. The PVdF membrane has been fabricated by tip-induced electrospinning process (TIE). In this process, several polymeric spikes have been mechanically generated on the PVdF stock solution which is subsequently focused on to the collection drum via potential biasing. The major advantage of the membranes synthesized by the method is the enhancement in the porosity of the membranes increasing the percentage of electrolyte uptake and ionic conductivity. The PVdF membranes are then composited with alumina by dip-coating method. On compositing with alumina, the porosity and electrolyte uptake reduced substantially when compared with pure PVdF membrane. The ionic conductivity of PVdF membrane was reported to be 3.32 and 2.23 mS cm^{-1}. The rate capacity comparison of PVdF membrane and alumina composited PVdF membrane with polypropylene is given in Fig. 6.8. The rate capability comparison was done with charging dummy cells at 0.2 rate to 4.2 V and discharging them at 0.5, 1, 2 and 4 rates up to 3.6 V. The cell was C rate was cycled five times each to confirm the stability. The rate comparison revealed that discharge capacity at 0.5 C is ~120 for both PVdF and alumina composite which has been retained even after cycling at 4.0 C rate. Comparing to the commercial PP membrane, the electrospun PVdF and alumina PVdF composite membranes show better and stable performance in the discharge capacity.

Lee et al. [47] reported a hybrid composite prepared by electrospinning PVdF-*co*-CTFE with Al_2O_3 powder on a polyethylene (PE) membrane [47]. PE membrane is having good electrochemical properties and adhesion to electrodes. Al_2O_3 ceramic filler added thermal stability of PE membrane and retention capacity of liquid electrolyte. The addition of inorganic inert ceramic filler like Al_2O_3 can prevent thermal deformations which caused the performance of the membrane at higher temperatures owing to the heat resistant property of the ceramic powder in the membrane network. The hybrid composite incorporating 5 wt% of alumina was electrospun on to PE membrane of 20 μm thickness. The composite with alumina improves efficiency by enhancing the gelation of liquid electrolyte during electrolyte uptake. The performance of the hybrid composite membrane was compared with

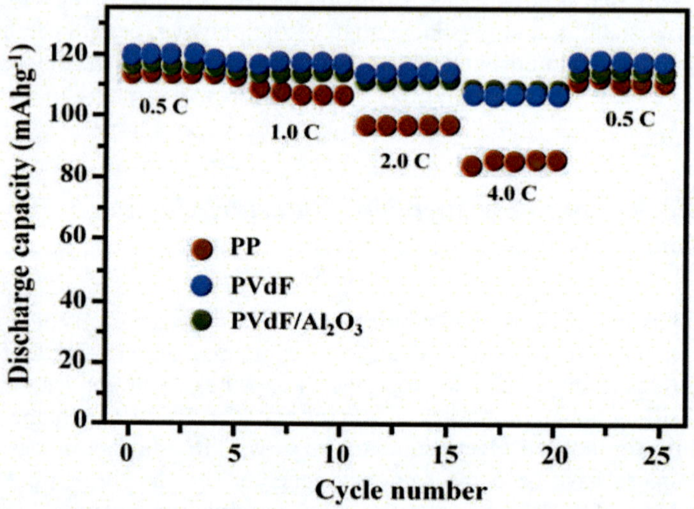

Fig. 6.8 Rate capacities of cells made by the PP, PVdF and PVdF/Al_2O_3 composite separator. Adapted and reproduced from Ref. [46]. Copyright 2017 Elsevier

PE membrane. The electrolyte uptake was enhanced from 160 to ~220% on using hybrid composite membrane. The cycling performance and discharge capacities of the hybrid membrane have improved. The electrospun fibre consisting of Al_2O_3 powder in composite membrane with high surface area and hydrophilic PVdF-*co*-CTFE can hold the solvent effectively, and the reactivity between the organic solvent and electrode is decreased. High discharge capacity of 133 mAh g^{-1} even at a 3.0 C rate was delivered by the cell, where its discharge capacity was 93% compared to that at a 0.1 C rate. The composite membrane also gave a cycling stability up to 200 cycles.

6.5.2 Electrospun Polyvinylidene Difluoride/Magnesium Aluminate Composite Polymer Electrolyte

Magnesium aluminate ($MgAl_2O_4$) is a spinel compound with strong Lewis acid character. Many researchers have explored the importance of magnesium aluminate as ceramic filler for enhancing the ionic conductivity and thermal resistive properties of the composite membranes as electrolytes for lithium-ion batteries [48]. Padmaraj et al. [49] recently reported electrospun PVdF-*co*-HFP with magnesium aluminate as filler. The filler has high dielectric constant than other fillers. Magnesium aluminate has a competing nature with the Lewis acid character of Li^+ ions in $LiPF_6$ salt in the electrolyte for the formation of complexes with the strongest Lewis base PVdF-co-HFP polymer chains. The major advantage of the usage of such Lewis acid is the creation of easy conduction pathway for Li^+ ions, which can enhance ionic transport and conductivity of the composite membrane. PVdF-*co*-HFP and $MgAl_2O_4$ fillers prepared by gel-combustion method were used as the initial materials for preparation of fibrous nanocomposite polymer membranes by electrospinning technique. The $MgAl_2O_4$ nanoparticle filler was added in variable proportions (1, 2, 3, 4, 5, 6 and 8 wt%). The optimized viscous solution was loaded in a syringe pump for the fabrication of nanocomposite fibres. Electrospinning was conducted at a potential bias of 18 kV with a rate of 1 ml h^{-1}. The distance between the tip and collector drum was maintained at 12.5 cm. The electrospun fibrous membranes with an average thickness of 80 mm were collected, dried and analysed (Fig. 6.9).

The ionic conductivity of the electrospun $MgAl_2O_4$ filler incorporated PVdF-*co*-HFP revealed an increase in ionic conductivity on increase in concentration of the sample up to 5 wt%, and on further increase in concentration, a marked decrease in ionic conductivity is observed. Similar behaviour was observed in the electrochemical impedance analysis. A decrease in resistance is observed up to 5 wt% followed by increase in the value. Hence, the performance of 5 wt% of the composite membrane was compared with standard commercial Celgard 2320® and electrospun membrane without filler. The studies revealed that the discharge capacity of composite membrane initially is obtained around 160 mAh g^{-1}, and a steady decrease

Fig. 6.9 FE-SEM images of electrospun nanocomposite fibrous membranes with various $MgAl_2O_4$ filler contents (wt%): **a** 0 (pure), **b** 2, **c** 5, **d** 8. Adapted and reproduced from Ref. [49], Copyright 2014 Elsevier

is observed up to 30 cycles. Comparing to the performance of Celgard 2320, the performance of battery is better even after 30 cycles to about 10 mAh g^{-1} [49].

6.5.3 Electrospun Polyvinylidene Fluoride/Clay Composite Polymer Electrolyte

Clay as a filler in polymer electrolytes helps in improving mechanical strength, thermal resistance and compatibility with lithium electrodes. Clay is not extensively studied as ceramic filler in polymer electrolyte in LIBs. Clay has a layered silicate structure. The clay is hydrophilic in nature. In order to convert hydrophilic system to organophilic compound, the surface modification of the clay is necessary. The surface modification of these hydrophilic surfaces can be converted to a hydrophobic surface by surfactant and organo-modified system. The surfactant system covers the

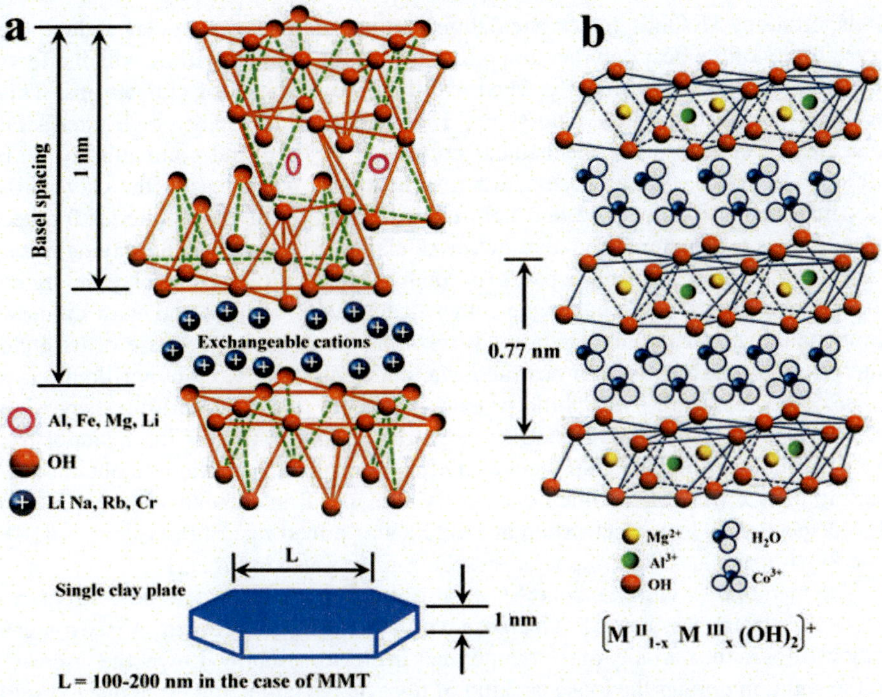

Fig. 6.10 Structure of layered silicate/nanoclay. Adapted and reproduced from Ref. [50], Copyright 2013 Elsevier

clay moiety and thereby modifies the surface with lyophilic chains of the surfactant similar to micelle formation. The structure of nanoclay is depicted in Fig. 6.10. The addition of clay as a filler in polymer electrolytes provides easy pathways for ion transportation and thus increases ionic conductivity. This also has shown results of high cyclic performances and charge-discharge properties.

Electrospinning has been used for fabricating the polymer gel electrolytes incorporated with layered nanoclay as a filler. Smectite clay/montmorillonite (hydrated sodium calcium aluminium magnesium silicate hydroxide) has been used as filler at various weight percentage with PVdF-*co*-HFP for electrospinning process [50]. The PGEs are prepared by activating the porous polymeric film with liquid electrolyte. The pores of membrane are filled with the absorbed electrolyte and help in ion conduction. The layered structure of nanoclay offers possibility of polymeric chain intercalating between the layers. This reduces the crystallinity of the composite and in turn enhances mechanical integrity, thermal resistance and compatibility with lithium electrodes. One among the commonly used inorganic layered nanoclay, montmorillonite (MMT) was studied as it has its unique properties like high aspect ratio, high cation exchange capacity, high specific surface area and appropriate interlayer charge. Completely interconnected pore structure of electrospun membrane (ESM) developed by the interlaying of ultrafine fibres that make easy pathways for ion transportation. Methods like intercalation and exfoliation of nanoclay reduce the crystallinity of the host polymer and minimize the ion-pair formation, which affects the mobility of cations directly when avoiding the mobility of counterions. This leads to increase in ionic conductivity. This also provides a very high interfacial area, which enhances the mechanical properties of PGE and compatibility with lithium electrodes. In the process, when sonication time increases, the clay particles may become pulverized and help in the formation of intercalated/exfoliated nanocomposite structure. The distribution of clay tactoids and the interlaying fibres in the membrane forms simple paths for ion transportation. Smectite clay (hydrated sodium calcium aluminium magnesium silicate hydroxide) treated with ditallow dimethylamine (quaternary amine) was taken as the nanoclay because of its good intercalation capability, lower cost and availability. Basically, montmorillonite is a polar compound with hydrophilic properties; hence, to obtain uniform dispersion of clay in the PVdF-*co*-HFP matrix, the surface modification of the material has been carried out using ionic surfactants. Different dispersions have been obtained by mixing 0, 1, 2 wt% of modified clay in PVdF-*co*-HFP in acetone/DMAc mixture. The dispersion is then electrospun at 16-20 kV at a feed rate of 0.2 mlmin^{-1} [50]. The SEM image and fibre diameter distribution are given in Fig. 6.11.

The membranes with nanoclay show uniform fibre diameter and pore structure, but surface roughness and AFD increase with increasing clay content. Viscosity is a major factor in the porosity of the membranes in electrospinning. Lower the viscosity of the solution thinner the loose packing of fibre on the collection drum due to rapid evaporation. The electrolyte uptake of PVdF-*co*-HFP increases with increase in the clay content. The thermal stability also was enhanced for the nanocomposite with the presence of clay as filler. This may be because of: (i) high thermal energy is required to initiate the chain scission due to the restricted chain motion in the nanocomposite, and (ii) the exfoliated and well-dispersed high aspect-ratio clay platelets hinder the outward diffusion of decomposed volatile products and act as an insulating surface that will slow down the pace of degradation of underlying material and loss in weight of decomposed materials. Figure 6.12 shows variation of interfacial resistance of polymer electrolytes based on PVdF-*co*-HFP with increase in the weight percentage of nanoclay as filler. The cycling performance of cell was studied and found that at the end of 50 cycles, the cell delivers a discharge capacity of 147 mAh g^{-1}, which is about 17% higher than the cell comprising polymer gel electrolyte without clay.

In another study, electrospun membrane for electrolyte based on PVdF/organo-modified clay (OC)/tripropylene glycol diacrylate (TPGDA) composite nanofibres was studied [51]. Electrospinning was used to utilize the micropores having larger surface areas. The porosity of the composite PVdF/OC/TPGDA is drastically improved due to electrospinning since electrospun nanofibres with a diameter of a few hundred nanometres are fully interconnected and provide a three-dimensional open microporous structure, which can increase the number of ionic conduction and channels to provide more surface area. A schematic representation of the cross-linking has been depicted in Fig. 6.13. The applied voltage for electrospinning was 15 kV, and the drum was rotated with speed rate of 200 rpm. The electrospun PVdF/OC/TPGDA membrane was irradiated using ultraviolet light with a wavelength of 360 nm in

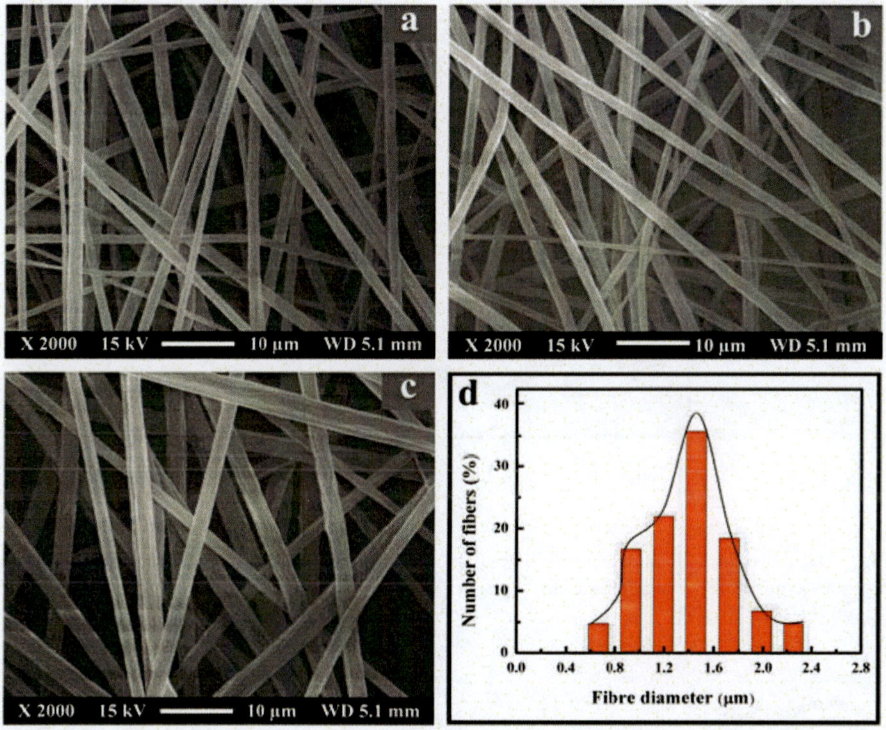

Fig. 6.11 Surface morphology and fibre diameter distribution histogram of electrospun PVdF-*co*-HFP/clay nanocomposite membranes with different clay content (wt%): **a** 0, **b** 1, **c** 2, **d** AFM of the fibre with 1% clay content. Adapted and reproduced from Ref. [50], Copyright 2013 Elsevier

order to graft on and cross-link between electrospun fibres. The current density in the composite PVdF/OC/TPGDA is 105 mAh g^{-1}. The cycling performance of PVdF electrolyte is significantly enhanced by incorporating organoclay and TPGDA. All results suggest that the composite gel polymer electrolyte obtained when PVdF doped with OC and TPGDA effectively increases the properties such as electrochemical stability, mechanical strength and cyclic performances.

6.5.4 Electrospun Polyvinylidene Difluoride/Titania Composite Polymer Electrolyte

Electrospun PVdF/TiO_2 composite and the polymeric blends based on PVdF have been composited with TiO_2 to enhance the inherent properties of the polymeric matrix. Ding et al. [52] reported the synthesis and application of PVdF/TiO_2 electrospun polymeric membrane as electrolyte for LIB. 5 wt% of tetrabutyl titanate was

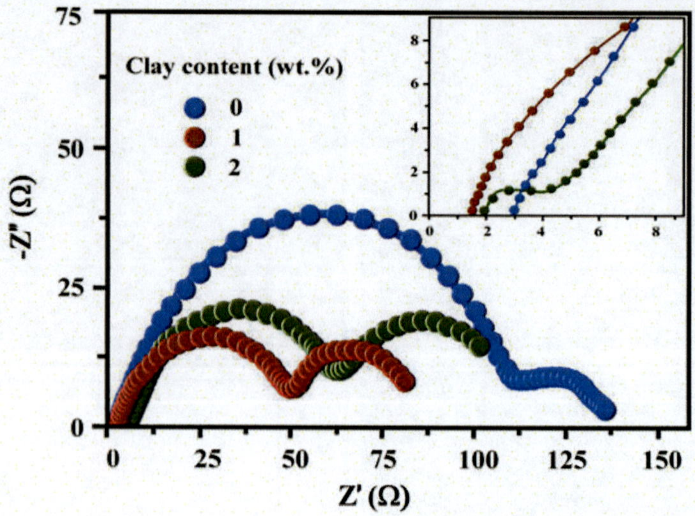

Fig. 6.12 Variation of interfacial resistance of polymer electrolytes based on PVdF-*co*-HFP-clay nanocomposite with different clay content (wt%). Adapted and reproduced from Ref. [50], Copyright 2013 Elsevier

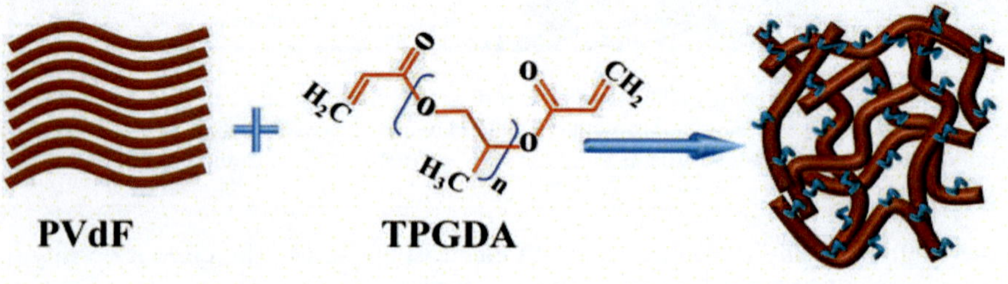

Fig. 6.13 Schematic of grafting of TPGDA onto PVdF fibre. Adapted and reproduced from Ref. [51], Copyright 2010 John Wiley and Sons

mixed with 20% PVdF solution in acetone/DMAc mixture, and the resulting slurry was fed in syringe for electrospinning process. The electrospinning was conducted at 25 kV potential bias at a separation of collector syringe distance of 15 cm. The resulting membrane was dried at 100 °C for 24 h and used as polymeric membrane for lithium-ion batteries. The AFM image of titania/PVdF polymer composite membrane has been given in Fig. 6.14. Membrane porosity, electrolyte uptake and fibre diameter of the electrospun membrane increased on incorporating titania on to the PVdF matrix. On the incorporation of titania, the ability to trap electrolyte in the membrane enhanced leading to a decrease in leakage of the electrolyte compared to the pure matrix. The composite membrane also gave an enhanced discharge capacity above 140 mAh g^{-1} compared to that of pure PVdF matrix and commercial Celgard polymeric membrane.

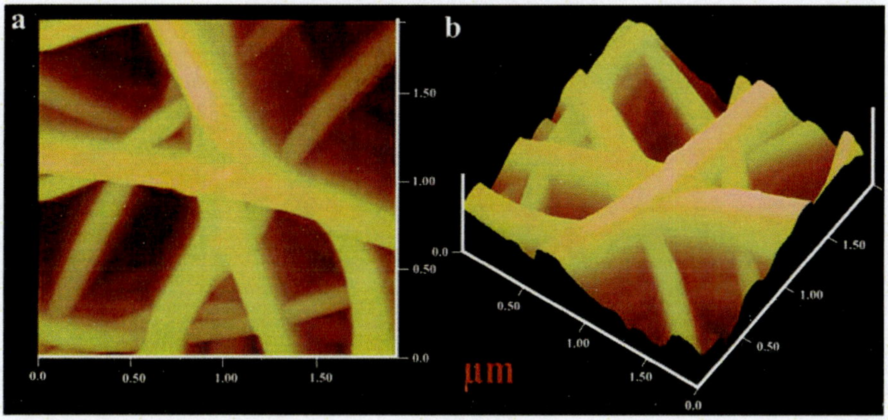

Fig. 6.14 AFM images of PVdF/TiO_2 electrospun membrane. Adapted and reproduced from Ref. [52], Copyright 2008 National Institute of Material Science

6.5.5 Electrospun Polyvinylidene Difluoride/Barium Titanate Composite Polymer Electrolyte

Barium titanate is an inorganic ceramic compound with general chemical formula of $BaTiO_3$. In 2008, Raghavan et al. [53] reported PVdF-*co*-HFP compared the performance of incorporation of fillers. They compared the effect of incorporation of $BaTiO_3$, Al_2O_3, SiO_2 nanoparticles in the polymer matrix for as electrolyte for LIBs. 6 wt% of filler was added to 14% of PVdF-*co*-HFP matrix in acetone DMAc mixture of solvents to prepare the slurry for electrospinning. The slurry was fed into the syringe kept at a distance of 16 cm from the collector and biased at a potential of 20 kV at a feed rate of 0.1 $mlmin^{-1}$. The obtained membrane was then dried at 60 °C for 12 h and used for further analysis. The average fibre diameter of $BaTiO_3$ incorporated polymer was slightly increased than that of pure electrospun polymer followed by the matrix incorporated with alumina followed by silica. The $BaTiO_3$/PVdF-*co*-HFP gave uniform fibre diameter. The electrolyte uptake percentage of the ceramic filler incorporated polymeric membrane gave almost similar results, but the relative absorption ratio of polymer composite with $BaTiO_3$ ceramic filler gave lower value. $BaTiO_3$/PVdF-*co*-HFP polymer composite membrane showed enhanced mechanical property, electrolyte uptake, porosity and ionic conductivity than other composite membranes. The electrochemical studies revealed that the sample itself showed an enhanced stability potential, and also lower interfacial resistance is offered by the sample. The polymer electrolyte with ceramic filler $BaTiO_3$ gave a discharge capacity of 164 mAh g^{-1} whereas electrolyte with Al_2O_3 and SiO_2 gave 153 and 156 mAh g^{-1}. The properties of the ceramic filler incorporated PVdF-*co*-HFP membranes have been depicted in Table 6.1.

Table 6.1 Properties of ceramic polymer composite

Properties	PVdF-*co*-HFP	Al_2O_3	SiO_2	$BaTiO_3$
Average fibre diameter (μm)	1.2	2.78	2.22	1.76
Tortuosity	14.81	12.56	12.04	11.52
Porosity (%)	84	85	85	87
Electrolyte uptake (%)	425	459	459	462
Ionic conductivity at 25°C (mS cm^{-1})	4.21	5.92	6.45	7.21
Tensile strength (MPa)	6.5	9.2	1.3	12.5
Modulus (MPa)	9.2	15.8	16.9	17.3
Elongation at break (%)	75	62	60	58
Electrochemical stability (V)	4.7	4.8	4.8	4.9
Discharge capacity (mAh g^{-1})	146	153	156	164

6.6 Other Electrospun Ceramic Composite Polymer Electrolytes

Several polymer membranes have been reported other than PAN and PVdF, among which some of the polymer have been electrospun with certain ceramic materials for application as electrolytes in LIBs. Nylon 6,6, PVA, etc. are some of the polymer matrix studied as electrospun membranes. Nylon 6,6 is a polyamide synthesized by dehydration reaction of hexamethylenediamine and adipic acid. The polymer is widely used in textile industry. In 2014, Yanilmaz et al. [54] reported the synthesis of electrospun nylon 6,6 polymer nanocomposite membrane incorporating silica nanoparticles. Commercial nylon 6,6 and silica nanoparticles of 20 nm were used for the fabrication of the membranes. 18 wt% of nylon 6,6 was mixed with various percentage by weight of silica nanoparticle (0, 3, 6, 9, 12 wt%) in formic acid medium. The slurry was then electrospun at 20 kV potential bias at a collector spinneret distance of 12 cm at a feed rate of 1 ml h^{-1}. Thus, obtained nonwoven membranes were dried at 60 °C at 24 h. The SEM image revealed that the incorporation of silica nanoparticle decreased the average fibre diameter to 56% for 12 wt% silica addition. Enhancement in the electrolyte uptake, porosity and ionic conductivity was visualized on increase in the concentration of silica nanoparticle in the polymeric matrix. The minimum interfacial resistance was obtained for 12 wt% of silica incorporated polymer composite. The silica polymer composite membranes also showed excellent thermal stability in the studies conducted at 150 °C for 30 min. LIB with polymer composite electrolyte membranes (SiO_2/nylon 6,6) containing 0, 3, 6, 9 and 12 wt% SiO_2 delivered a discharge capacity of 128, 128, 129, 129 and 130 mAh g^{-1}, respectively. The composite membranes also gave a capacity retention almost equivalent to 93%. The cycling performance of the samples at different C rate is given in Fig. 6.15. At lower C rate (0.2 C), the silica incorporated samples and pristine samples showed almost similar behaviour, but on increased C rate (8 C), 12% of SiO_2 incorporated nylon 6,6 showed better discharge capacity.

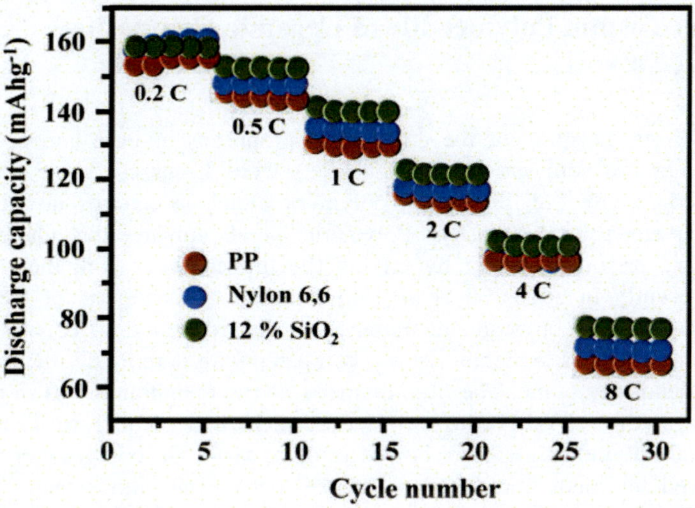

Fig. 6.15 C-rate performance of Li/$LiFePO_4$ cells containing SiO_2/nylon 6,6 nanofibre membranes with different SiO_2 contents and microporous PP membrane. Adapted and reproduced from Ref. [54], Copyright 2014 Elsevier

Chand et al. [55] nano-Al_2O_3 filled PVA-NH_4SCN electrolyte nanofibres by electrospinning reported an improvement in ionic conductivity. Commercial nanoAl_2O_3 possessing an average particle size of 40–50 nm was mixed with PVA-NH_4SCN matrix in aprotic solvent dimethyl sulfoxide (DMSO) to form the feed solution mixture. The mixture is fed in syringe for electrospinning and biased at 22 kV at a distance separation of 14 cm at ambient reaction condition. The alumina was added at various weight percentage (0, 2, 4, 6 wt%) to the polymer matrix to analyse the effect of filler in the performance of the electrolyte membrane. The analysis of SEM images revealed that the increase in the concentration of Al_2O_3 contents resulted in more uniform fibres which were obtained. The filler is dispersed homogeneously, and this promotes conduction pathways for mobile ions. There is improvement in thermal stability and ionic conductivity on the addition of Al_2O_3 as filler in the electrolyte. The increase in concentration of filler increases the conductivity, and it tends to saturate though the change remains within an order of magnitude. The enhancement is possibly due to the amorphous structure. Addition of filler also increases the conduction pathway prevailing Li^+ ion for conduction in LIBs. The electrical conductivity increased from 2.39×10^{-5} S cm^{-1} to 5.31×10^{-3} S cm^{-1} for 4 wt% of filler concentration at room temperature which is better than conductivity of corresponding gel film [55].

6.7 Electrospun Polymer Blend Ceramic Composite Electrolytes

Several reports on study of the effects and properties of pure and copolymer-based membranes with various ceramic fillers were discussed. Some researchers also studied the effect of filler on the polymeric blends as electrospun membranes as polymer electrolytes for LIBs. Polymeric blends enhance the inherent properties of the pristine polymer by mixing the properties of both the polymeric matrix. Recently in 2018, Fu et al. [56] reported the synthesis of electrospun PVdF/PMMA/SiO_2 composite membranes as electrolytes for LIB. An electrospun polymer composite membrane was compared with pure PVdF membrane and polymer blend membrane. The fibre diameter of the membranes was in the order of pristine PVdF > PVdF/PMMA > PVdF/PMMA/SiO_2. The porosity and electrolyte uptake of the samples also showed a similar trend. The composite membranes also showed minimum interfacial resistance to the ionic movement; hence, an enhanced ionic conductivity is obtained (4×10^{-3} S cm^{-1}). The electrochemical stability window exceeded 5.0 V implying the stability of the composite blend electrolyte. The discharge capacity of various polymer electrolytes was compared (PVdF/PMMA/SiO_2 −158 mAh g^{-1}, PVdF/PMMA −153 mAh g^{-1}, PVdF − 150 mAh g^{-1} and PP −140 mAh g^{-1}).

6.8 Conclusion

The polymer electrolyte is considered to be the most effective and reliable electrolyte for LIBs. The major drawbacks of the conventional electrolytes such as leakage and dendrite formation can be minimized upon the usage of polymeric electrolytes. Apart of the pros, the major con of polymer electrolyte is the decreased ionic conductivity of the electrolyte through the membrane. To enhance the conductivity of Li^+ ions, several changes have been made on the polymeric electrolyte membrane. The major changes done are (i) tuning the synthesis method to assure enhanced porosity and electrolyte uptake, (ii) polymeric blend synthesis and (iii) incorporation of fillers in the matrix. The present chapter discussed the importance and effect of several fillers in polymeric matrix enhancing the properties of the electrolyte membrane synthesized by electrospinning technique. The electrospinning is one of the most effective techniques to synthesize porous membrane. The incorporation of various fillers such as alumina, silica, titania and magnesium silicate enhances the porosity, electrolyte uptake, ionic conductivity, discharge capacity, etc. of the polymer composite membranes.

Acknowledgements Authors, Jabeen Fatima M. J and Raghavan Prasanth, would like to acknowledge Kerala State Council for Science, Technology and Environment (KSCSTE), Government of Kerala for financial assistance.

References

1. He X, Shi Q, Zhou X et al (2005) In situ composite of nano SiO_2-P(VDF-HFP) porous polymer electrolytes for Li-ion batteries. Electrochim Acta 51:1069–1075. https://doi.org/10.1016/j.electacta.2005.05.048
2. Meyer WH (1998) Polymer electrolytes for lithium-ion batteries. Adv Mater 10:439–448. https://doi.org/10.1002/(SICI)1521-4095(199804)10:6%3c439:AID-ADMA439%3e3.0.CO;2-I
3. Yarmolenko OV, Yudina AV, Khatmullina KG (2018) Nanocomposite polymer electrolytes for the lithium power sources (a review). Russ J Electrochem 54:325–343. https://doi.org/10.1134/S1023193518040092
4. Leijonmarck S, Cornell A, Lindbergh G, Wågberg L (2013) Single-paper flexible Li-ion battery cells through a paper-making process based on nano-fibrillated cellulose. J Mater Chem A 1:4671–4677. https://doi.org/10.1039/c3ta01532g
5. Sheng J, Tong S, He Z, Yang R (2017) Recent developments of cellulose materials for lithium-ion battery separators. Cellulose 24:4103–4122. https://doi.org/10.1007/s10570-017-1421-8
6. Kim JH, Gu M, Lee DH et al (2016) Functionalized nanocellulose-integrated heterolayered nanomats toward smart battery separators. Nano Lett 16:5533–5541. https://doi.org/10.1021/acs.nanolett.6b02069
7. Lavoine N, Desloges I, Dufresne A, Bras J (2012) Microfibrillated cellulose—its barrier properties and applications in cellulosic materials: a review. Carbohydr Polym 90:735–764. https://doi.org/10.1016/j.carbpol.2012.05.026
8. Ran Y, Yin Z, Ding Z et al (2013) A polymer electrolyte based on poly(vinylidene fluoride-hexafluoropylene)/hydroxypropyl methyl cellulose blending for lithium-ion battery. Ionics (Kiel) 19:757–762. https://doi.org/10.1007/s11581-012-0808-7
9. Liu B, Huang Y, Cao H et al (2018) A high-performance and environment-friendly gel polymer electrolyte for lithium ion battery based on composited lignin membrane. J Solid State Electrochem 22:807–816. https://doi.org/10.1007/s10008-017-3814-x
10. Stephan AM, Kumar TP, Kulandainathan MA, Lakshmi NA (2009) Chitin-Incorporated poly(ethylene oxide)-based nanocomposite electrolytes for lithium batteries. J Phys Chem B 113:1963–1971. https://doi.org/10.1021/jp808640j
11. Aziz NA, Majid SR, Yahya R, Arof AK (2011) Conductivity, structure, and thermal properties of chitosan-based polymer electrolytes with nanofillers. Polym Adv Technol 22:1345–1348. https://doi.org/10.1002/pat.1619
12. Tang C, Hackenberg K, Fu Q et al (2012) High ion conducting polymer nanocomposite electrolytes using hybrid nanofillers. Nano Lett 12:1152–1156. https://doi.org/10.1021/nl202692y
13. Kammoun M, Berg S, Ardebili H (2015) Flexible thin-film battery based on graphene-oxide embedded in solid polymer electrolyte. Nanoscale 7:17516–17522. https://doi.org/10.1039/c5nr04339e
14. Masoud EM, El-Bellihi A-A, Bayoumy WA, Mousa MA (2013) Organic–inorganic composite polymer electrolyte based on PEO–$LiClO_4$ and nano-Al_2O_3 filler for lithium polymer batteries: Dielectric and transport properties. J Alloys Compd 575:223–228. https://doi.org/10.1016/j.jallcom.2013.04.054
15. Liang B, Tang S, Jiang Q, et al (2015) Preparation and characterization of PEO-PMMA polymer composite electrolytes doped with nano-Al_2O_3. Electrochim Acta 169:334–341. https://doi.org/10.1016/j.electacta.2015.04.039
16. Jeong H-S, Hong SC, Lee S-Y (2010) Effect of microporous structure on thermal shrinkage and electrochemical performance of Al_2O_3/poly(vinylidene fluoride-hexafluoropropylene) composite separators for lithium-ion batteries. J Memb Sci 364:177–182. https://doi.org/10.1016/j.memsci.2010.08.012
17. Liao YH, Li XP, Fu CH, et al (2011) Polypropylene-supported and nano-Al_2O_3 doped poly(ethylene oxide)–poly(vinylidene fluoride-hexafluoropropylene)-based gel electrolyte for

lithium ion batteries. J Power Sources 196:2115–2121. https://doi.org/10.1016/j.jpowsour.2010.10.062
18. Li Z, Su G, Wang X, Gao D (2005) Micro-porous P(VDF-HFP)-based polymer electrolyte filled with Al_2O_3 nanoparticles. Solid State Ionics 176:1903–1908. https://doi.org/10.1016/j.ssi.2005.05.006
19. Lin CW, Hung CL, Venkateswarlu M, Hwang BJ (2005) Influence of TiO_2 nano-particles on the transport properties of composite polymer electrolyte for lithium-ion batteries. J Power Sources 146:397–401. https://doi.org/10.1016/j.jpowsour.2005.03.028
20. Liu Y, Lee JY, Hong L (2003) Morphology, crystallinity, and electrochemical properties of in situ formed poly(ethylene oxide)/TiO_2 nanocomposite polymer electrolytes. J Appl Polym Sci 89:2815–2822. https://doi.org/10.1002/app.12487
21. Cui W-W, Tang D-Y, Gong Z-L (2013) Electrospun poly(vinylidene fluoride)/poly(methyl methacrylate) grafted TiO_2 composite nanofibrous membrane as polymer electrolyte for lithium-ion batteries. J Power Sources 223:206–213. https://doi.org/10.1016/j.jpowsour.2012.09.049
22. Zhou L, Wu N, Cao Q, et al (2013) A novel electrospun PVDF/PMMA gel polymer electrolyte with in situ TiO_2 for Li-ion batteries. Solid State Ionics 249–250:93–97. https://doi.org/10.1016/j.ssi.2013.07.019
23. Cao J, Wang L, Shang Y, et al (2013) Dispersibility of nano-TiO_2 on performance of composite polymer electrolytes for Li-ion batteries. Electrochim Acta 111:674–679. https://doi.org/10.1016/j.electacta.2013.08.048
24. Yang CL, Li ZH, Li WJ, et al (2015) Batwing-like polymer membrane consisting of PMMA-grafted electrospun PVdF–SiO_2 nanocomposite fibers for lithium-ion batteries. J Memb Sci 495:341–350. https://doi.org/10.1016/j.memsci.2015.08.036
25. Lee Y-S, Ju SH, Kim J-H, et al (2012) Composite gel polymer electrolytes containing core-shell structured $SiO_2(Li^+)$ particles for lithium-ion polymer batteries. Electrochem Commun 17:18–21. https://doi.org/10.1016/j.elecom.2012.01.008
26. Park J-H, Cho J-H, Park W, et al (2010) Close-packed SiO_2/poly(methyl methacrylate) binary nanoparticles-coated polyethylene separators for lithium-ion batteries. J Power Sources 195:8306–8310. https://doi.org/10.1016/j.jpowsour.2010.06.112
27. Zhang P, Yang LC, Li LL, et al (2011) Enhanced electrochemical and mechanical properties of P(VDF-HFP)-based composite polymer electrolytes with SiO_2 nanowires. J Memb Sci 379:80–85. https://doi.org/10.1016/j.memsci.2011.05.043
28. Xie H, Liao Y, Sun P, et al (2014) Investigation on polyethylene-supported and nano-SiO_2 doped poly(methyl methacrylate-co-butyl acrylate) based gel polymer electrolyte for high voltage lithium ion battery. Electrochim Acta 127:327–333. https://doi.org/10.1016/j.electacta.2014.02.038
29. Rajasudha G, Nancy AP, Paramasivam T et al (2011) Synthesis and characterization of polyindole–NiO-based composite polymer electrolyte with $LiClO_4$. Int J Polym Mater Polym Biomater 60:877–892. https://doi.org/10.1080/00914037.2010.551367
30. Zhang J, Sun B, Huang X et al (2014) Honeycomb-like porous gel polymer electrolyte membrane for lithium ion batteries with enhanced safety. Sci Rep 4:1–7. https://doi.org/10.1038/srep06007
31. Manuel Stephan A, Nahm KS (2006) Review on composite polymer electrolytes for lithium batteries. Polymer (Guildf) 47:5952–5964. https://doi.org/10.1016/j.polymer.2006.05.069
32. Zhao P (2011) On the ranks of certain semigroups of orientation preserving transformations. Commun Algebr 39:4195–4205. https://doi.org/10.1080/00927872.2010.521933
33. Croce F, Sacchetti S, Scrosati B (2006) Advanced, lithium batteries based on high-performance composite polymer electrolytes. J Power Sources 162:685–689. https://doi.org/10.1016/j.jpowsour.2006.07.038
34. Song JY, Wang YY, Wan CC (1999) Review of gel-type polymer electrolytes for lithium-ion batteries. J. Power Sources 77:183–197. https://doi.org/10.1016/S0378-7753(98)00193-1
35. Krawiec W, Scanlon LG, Fellner JP et al (1995) Polymer nanocomposites: a new strategy for synthesizing solid electrolytes for rechargeable lithium batteries. J Power Sources 54:310–315. https://doi.org/10.1016/0378-7753(94)02090-P

36. Prasanth R, Shubha N, Hng HH, Srinivasan M (2013) Effect of nano-clay on ionic conductivity and electrochemical properties of poly(vinylidene fluoride) based nanocomposite porous polymer membranes and their application as polymer electrolyte in lithium ion batteries. Eur Polym J 49:307–318. https://doi.org/10.1016/j.eurpolymj.2012.10.033
37. Rahaman MSA, Ismail AF, Mustafa A (2007) A review of heat treatment on polyacrylonitrile fiber. Polym Degrad Stab 92:1421–1432. https://doi.org/10.1016/j.polymdegradstab.2007.03.023
38. Tsutsumi H, Doi S, Onimura K, Oishi T (2002) Conductivity enhancement of poly(vinylimidazoline)-based electrolytes by addition of cascade nitrile. Electrochemistry 70:94–98. https://doi.org/10.5796/electrochemistry.70.94
39. Raghavan P, Manuel J, Zhao X et al (2011) Preparation and electrochemical characterization of gel polymer electrolyte based on electrospun polyacrylonitrile nonwoven membranes for lithium batteries. J Power Sources 196:6742–6749. https://doi.org/10.1016/j.jpowsour.2010.10.089
40. Ma X, Kolla P, Yang R et al (2017) Electrospun polyacrylonitrile nanofibrous membranes with varied fiber diameters and different membrane porosities as lithium-ion battery separators. Electrochim Acta 236:417–423. https://doi.org/10.1016/j.electacta.2017.03.205
41. Chen HC, Lin FJ, Chen CC (2002) Polyacrylonitrile electrolytes 1. A novel high-conductivity composite polymer electrolyte based on PAN, $LiClO_4$ and a-Al_2O_3. Solid State Ionics 150:327–335
42. Wang Q, Song WL, Fan LZ, Song Y (2015) Facile fabrication of polyacrylonitrile/alumina composite membranes based on triethylene glycol diacetate-2-propenoic acid butyl ester gel polymer electrolytes for high-voltage lithium-ion batteries. J Memb Sci 486:21–28. https://doi.org/10.1016/j.memsci.2015.03.022
43. Vijayakumar M, Inaguma Y, Mashiko W et al (2004) Synthesis of fine powders of $Li_{3x}La_{2/3-x}TiO_3$ perovskite by a polymerizable precursor method. Chem Mater 16:2719–2724. https://doi.org/10.1021/cm049869x
44. Liang Y, Ji L, Guo B et al (2011) Preparation and electrochemical characterization of ionic-conducting lithium lanthanum titanate oxide/polyacrylonitrile submicron composite fiber-based lithium-ion battery separators. J Power Sources 196:436–441. https://doi.org/10.1016/j.jpowsour.2010.06.088
45. Liang Y, Lin Z, Qiu Y, Zhang X (2011) Fabrication and characterization of LATP/PAN composite fiber-based lithium-ion battery separators. Electrochim Acta 56:6474–6480. https://doi.org/10.1016/j.electacta.2011.05.007
46. Wu D, Deng L, Sun Y et al (2017) A high-safety PVDF/Al_2O_3 composite separator for Li-ion batteries via tip-induced electrospinning and dip-coating. RSC Adv 7:24410–24416. https://doi.org/10.1039/c7ra02681a
47. Lee YS, Jeong YB, Kim DW (2010) Cycling performance of lithium-ion batteries assembled with a hybrid composite membrane prepared by an electrospinning method. J Power Sources 195:6197–6201. https://doi.org/10.1016/j.jpowsour.2009.11.012
48. Angulakshmi N, Nahm KS, Nair JR, et al (2013) Cycling profile of $MgAl_2O_4$-incorporated composite electrolytes composed of PEO and $LiPF_6$ for lithium polymer batteries. Electrochim Acta 90:179–185. https://doi.org/10.1016/j.electacta.2012.12.003
49. Padmaraj O, Rao BN, Jena P et al (2014) Electrochemical studies of electrospun organic/ inorganic hybrid nanocomposite fi brous polymer electrolyte for lithium battery. Polymer (Guildf) 55:1136–1142. https://doi.org/10.1016/j.polymer.2014.01.015
50. Shubha N, Prasanth R, Hoon HH, Srinivasan M (2013) Dual phase polymer gel electrolyte based on non-woven poly(vinylidenefluoride-co-hexafluoropropylene)–layered clay nanocomposite fibrous membranes for lithium ion batteries. Mater Res Bull 48:526–537. https://doi.org/10.1016/j.materresbull.2012.11.002
51. Jeong KU, Chae HD, Il Lim C et al (2010) Fabrication and characterization of electrolytemembranes based on organoclay/tripropyleneglycol diacrylate/poly(vinylidene fluoride) electrospun nanofiber composites. Polym Int 59:249–255. https://doi.org/10.1002/pi.2716

52. Ding Y, Zhang P, Long Z, et al (2008) Preparation of PVdF-based electrospun membranes and their application as separators. Sci Technol Adv Mater 9. https://doi.org/10.1088/1468-6996/9/1/015005
53. Raghavan P, Zhao X, Kim J-K, et al (2008) Ionic conductivity and electrochemical properties of nanocomposite polymer electrolytes based on electrospun poly(vinylidene fluoride-co-hexafluoropropylene) with nano-sized ceramic fillers. Electrochim Acta 54:228–234. https://doi.org/10.1016/j.electacta.2008.08.007
54. Yanilmaz M, Dirican M, Zhang X (2014) Electrochimica Acta Evaluation of electrospun SiO_2/nylon 6,6 nanofiber membranes as a thermally-stable separator for lithium-ion batteries. Electrochim Acta 133:501–508. https://doi.org/10.1016/j.electacta.2014.04.109
55. Chand N, Rai N, Natarajan TS, Agrawal SL (2011) Fabrication and characterization of nano Al_2O_3 filled PVA : NH_4SCN electrolyte nanofibers by electrospinning characterization of nanofibers. Fibers Polym 12:438–443. https://doi.org/10.1007/s12221-011-0438-0
56. Qs Fu, Lin Guo, Chen Xuedan, Zuxiao Yu, Yang Ruisong, Li Mingtian, Xianguang zen JC (2018) Mechanically reinforced PVdF/PMMA/SiO_2 composite membrane and its electrochemical properties as lithium-ion batteries separator. Energy Technol 6:144–152. https://doi.org/10.1002/ente.201700347

Chapter 7
Electrospun Silica-Based Polymer Nanocomposite Electrolytes for Lithium-Ion Batteries

Akhila Das, Neethu T. M. Balakrishnan, Anjumole P. Thomas, Jou-Hyeon Ahn, M. J. Jabeen Fatima, and Raghavan Prasanth

7.1 Introduction

Current status of world fuel consumption has been raised to ~3% in the past 1 year [1]. The energy still relies mainly on coal. The environmental changes and global warming have led the officials to step ahead by replacing coal by eco-friendly systems to reduce harmful emissions. In the current context, the need for efficient and reliable energy storage devices has emerged as a necessity. The most important and widely used portable energy storage device is lithium-ion batteries. The discovery of lithium-ion batteries (LIBs) by Prof. Goodenough et al. [2] paved a way for the use of portable devices. Since the introduction of portable batteries was commercialized the era of electronic gadgets was also started to be as usual as part of our life. Mobile phones, laptops, palmtops, tablets, etc. are some of the devices working with LIBs. In the past few years, electric vehicles are being a commercialized. These e-vehicles are powered by electrical energy stored in batteries as well as hybrid e-vehicles are available which is capable of utilizing the CNG system along with stored energy. The batteries mainly employed in these systems are currently lithium-ion batteries owing to their enhanced energy density and cycling performance.

The major constituents of LIBs include an anode, cathode, and electrolyte (Fig. 7.1). Even though all the components are equally important, a major role is assigned to electrolytes and hence it is known to be the heart of the battery. Conventional batteries employed liquid phase electrolytes, which possess high ionic

A. Das · N. T. M. Balakrishnan · A. P. Thomas · M. J. Jabeen Fatima (✉) · R. Prasanth
Department of Polymer Science and Rubber Technology (PSRT), Cochin University of Science and Technology (CUSAT), Cochin 682022, India
e-mail: jabeen@cusat.ac.in

J.-H. Ahn · R. Prasanth (✉)
Department of Materials Engineering and Convergence Technology, Gyeongsang National University, Jinju 52828, Republic of Korea
e-mail: prasanth@cusat.ac.in

N. T. M. Balakrishnan and R. Prasanth (eds.), *Electrospinning for Advanced Energy Storage Applications*, Materials Horizons: From Nature to Nanomaterials,
https://doi.org/10.1007/978-981-15-8844-0_7

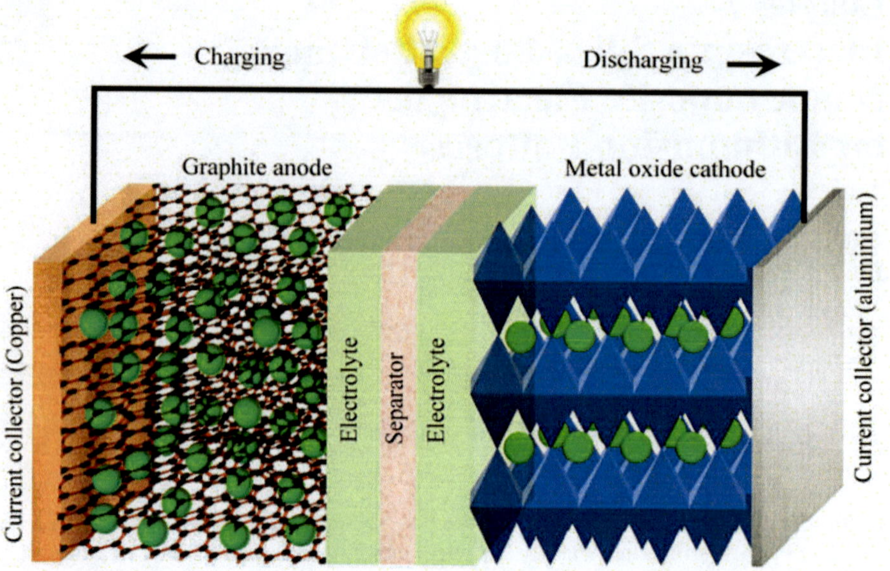

Fig. 7.1 Schematic representation of lithium-ion batteries

conductivity but have several drawbacks such as leakage and safety issues. Hence polymer electrolytes were introduced to overcome the drawbacks. Polymer electrolytes showed superior properties like high energy density, flexibility, etc. with a leakage-proof geometry.

The performance of LIBs is critically influenced by the solid electrolyte interface (SEI) formation. SEI is a passivation layer formed on the surface of the electrode in LIB preventing the chemical reactions leading to the decomposition of electrolyte and exfoliation of the active electrode [3]. An irreversible layer is grown on the electrode/electrolyte interface by the reduction reaction of the electrolyte leading to the formation of decomposition products that are capable of irreversible chemical interaction with lithium ions [3]. The formation of SEI layer is initiated during the cycling, these layer acts as a semipermeable layer allowing the transport of Li^{+}-ions while pausing the movement of electrons to the electrolyte [4]. On prolonged cycling of the battery results in the ageing of anode (graphite) leading to decomposing of the electrolyte, crack formation, solvent co-intercalation, continuous SEI growth, volume expansions leading to change in porosity, decomposition of the binder and metallic lithium plating (Fig. 7.2). Hence prolong cycling of LIBs results in an increase in the impedance and decrease in conductivity of Li^{+}-ions leading to a marked decrease in the performance. SEI growth is the major issue faced by ageing of LIBs [5]. The passivating layer tends to form a dendritic projection from the anode, piercing the separator through the electrolyte towards the cathode. These dendritic projections are formed by chemical reactions catalyzed by positive and negative ion interactions occurring in the electrolyte. The extension formed from anode to cathode is capable

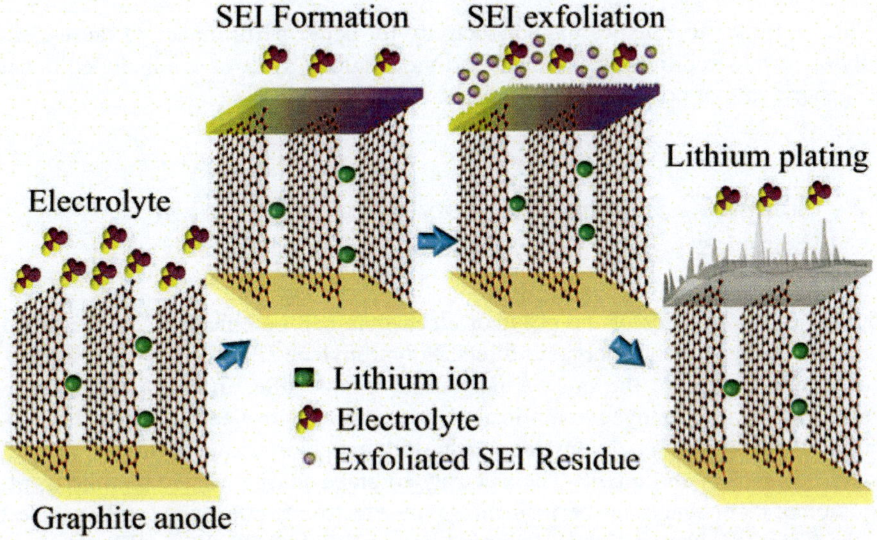

Fig. 7.2 SEI formation and changes occurring at the anode/electrolyte interface. Adapted and reproduced from Vetter et al. [5]

to cause blasting of LIBs [5]. The replacement of liquid electrolyte with polymer electrolyte is a widely accepted solution to prevent such hazards caused in LIBs.

Polymer electrolytes (PE) are ion-conducting membranes. Polyethylene oxide (PEO), poly(propylene oxide) (PPO), poly(methyl methacrylate) (PMMA), polyvinyl pyrrolidone (PVP), poly(acrylonitrile) (PAN), poly(vinylidene difluoride) (PVdF), poly(vinylidene difluoride-*co*-hexafluoropropylene) (PVdF-*co*-HFP), polyethylene terephthalate (PET), etc. are some of the polymer-based electrolytes used in LIBs. The major advantages of PEs (1) they are capable to overcome the SEI dendrite formation since they are inert towards decomposition reactions, (2) Flexibility of polymers enables the electrolyte to withstand the volume expansion occurring as a result of the continuous cycling of batteries, (3) flexibility in molding to various shapes and size depending on the application, (4) the ionic conductivity of polymer electrolyte is comparatively high ($<10^4$ S cm^{-1}). Polymer electrolytes can be characterized mainly into three based on the composition and state of matter (i) gel polymer electrolyte (ii) solid polymer electrolyte (iii) composite polymer electrolyte. Considering solid polymer electrolyte for LIBs, the low conductivity and ionic mobility are the major issues dragging them from the application.

PEO has initially developed the PE system. In 1973, Fenton [6] reported the complex formation of polyethylene oxide with alkali metal ions, and thereby increase in conductivity of the complex was been reported. The potential role of the polymer as an electrolyte for conduction of alkali metal ion was hence been suggested. PEO_{10} $LiCF_3SO_3$ polymer electrolyte has been prepared by the solvent casting method and incorporated various percentage composition of polystyrene for improving the physical properties [7]. The ionic conductivity of these polymeric systems was found to be lower. Hence to enhance the conductivity for better performance of these electrolytes certain additives were added and the resultant systems are included under the generic title of polymer composites.

7.2 Composite Polymer Electrolytes for Lithium-Ion Batteries

The ionic conductivity of the polymer electrolyte can be enhanced by the incorporation of organic and inorganic fillers. Polypropylene (PP), cellulose, cellulose-based compounds are the most widely used organic filler in LIBs. Electrospinning method was employed to fabricate polymer composite electrolytes by incorporating PP nanofibers in polyvinylidene difluoride-*co*-chlorotrifluoroethylene (PVdF-*co*-CTFE) and PVdF-*co*-HFP. The electrolyte uptake study revealed that the incorporation of PP enhances the performance of the electrolyte due to the capillary effect of the fibers [8]. Nanocellulose (NC) incorporated PVdF, PVdF-*co*-HFP, etc. have been reported as electrolytes in LIBs. In 2013, Laila et al. [9] reported nanocrystalline cellulose (2 wt%) with PVdF-*co*-HFP activated with ionic liquid electrolyte (1M of lithium bis(trifluoromethane sulfonyl)imide/1-butyl-3-methylpyrrolidnium bis(trifluoromethane sulfonyl)imide solution) in LIBs to have an ionic conductivity of 4×10^{-4} S cm^{-1} at ambient temperature. Cellulose nanofibers (CNF) was been reported as binders and electrolytes in LIBs [10]. Terpyridine functionalized CNF was coated on electrospun PVP/PAN microlayer with $LiMn_2O_4$ (LMO) was reported to have enhanced ion transport rate [11]. Synthesized terpyridine was coated on carboxy group functionalized CNF via chemical bonding owing to the reaction of the amino group and carboxyl group forming amide bond. This terpyridine cellulose nanofiber was coated on to PVP/PAN microfiber by Buchner funnel filtration method as given in Fig. 7.3. Cellulose-based compounds such as methylcellulose (MC), hydroxypropyl methylcellulose (HPMC) was also reported as organic fillers in polymer composites for LIB. PVdF coated on MC was reported to have better ionic conductivity than pure MC [12]. Natural cellulose like HPMC was blended with PVdF-*co*-HFP to obtain electrochemical stability up to 5 V Versus Li/Li^+ [13].

Inorganic fillers can be further classified into active and passive fillers. Active fillers take part in ionic conduction such as lithium nitride (Li_3N simply as LN), lithium triflate ($LiCF_3SO_3$ simply as LTF), lithium tetrafluoroborate ($LiBF_4$ simply as LTFB), lithium aluminate ($LiAlO_2$ simply as LA), etc. α and β phases of LN have good Li^+-ion conductivity of 5.8×10^{-4} S cm^{-1} and 2.0×10^{-4} S cm^{-1}, respectively [14]. Electrolyte with polymer composite of LTF, LN, and PEO was been reported to create an amorphous polymeric phase enhancing the conductivity of ions [15]. LTBF and LA in PEO matrix was been reported to have an ionic conductivity in the range of 10^{-4} S cm^{-1} and stability up to 60 °C [16]. Passive inorganic fillers such as silica (SiO_2), zirconia (ZrO_2), titania (TiO_2), alumina (Al_2O_3), zinc oxide (ZnO), and magnesium oxide (MgO) have been reported as polymer composite electrolytes

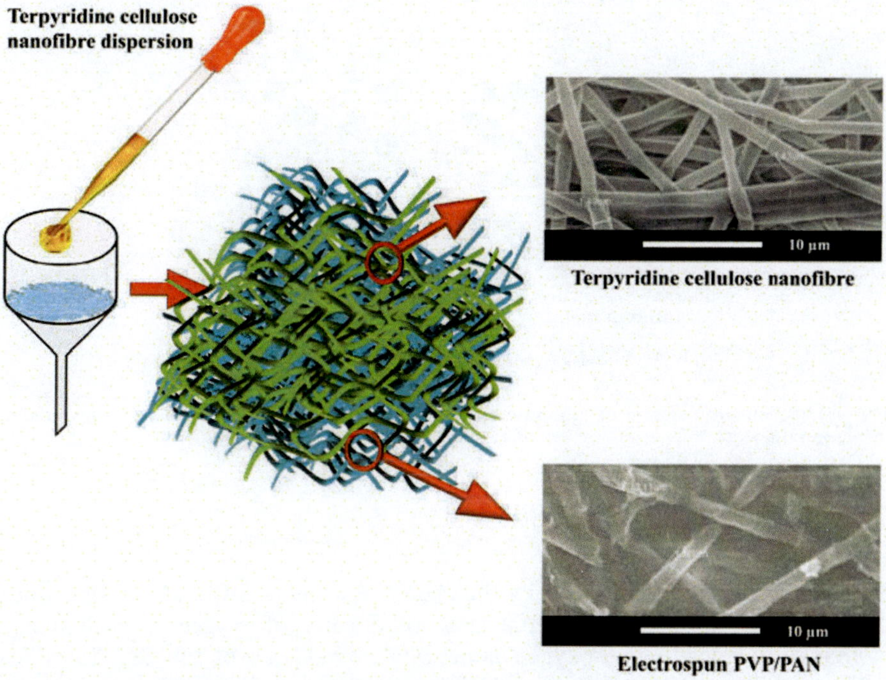

Fig. 7.3 Synthesis of terpyridine cellulose nanofiber/PVP/PAN electrolyte. Adapted and reproduced from Kim et al. [11]. Copyright 2016 American Chemical Society

in LIBs. Zirconia embedded PET showed capacity retention of 94.5% at 1 C with a discharge capacity of 143 mAh g^{-1} [17]. Nano titania synthesized via sol-gel method was incorporated as ceramic filler in the PEO matrix with LTFB as Li^{+}-ion donor with and electrochemical stability window up to 4.5 V and ionic conductivity in the range of 10^{-7} S cm^{-1} [18]. Alumina PVdF composite electrolyte was prepared via solvent casting method to obtain a specific discharge capacitance of 135 mAh g^{-1} [19]. Aravindan et al. [20] compared the performance of incorporating silica and alumina as fillers in PVdF-*co*-HFP reporting an ionic conductivity of 1.16×10^{-3} S cm^{-1} and 0.98×10^{-3} S cm^{-1} respectively. PEO–ZnO composite polymer electrolyte with lithium bis(trifluoromethanesulfonyl)imide (LTFSI) as a source of Li^{+}-ion was employed as electrolyte to obtain about 100 mAh g^{-1} [21]. Nanofillers are more effective in enhancing the conductivity in a composite electrolyte than micron-sized ceramic fillers. The electrode contact with active polymer is reduced on the addition of nanofillers owing to the decoration of ceramic moiety at the interface as given in Fig. 7.4. Herein we are focusing on the silica incorporated polymer nanocomposite, synthesis, and electronic properties for application as an electrolyte in LIBs.

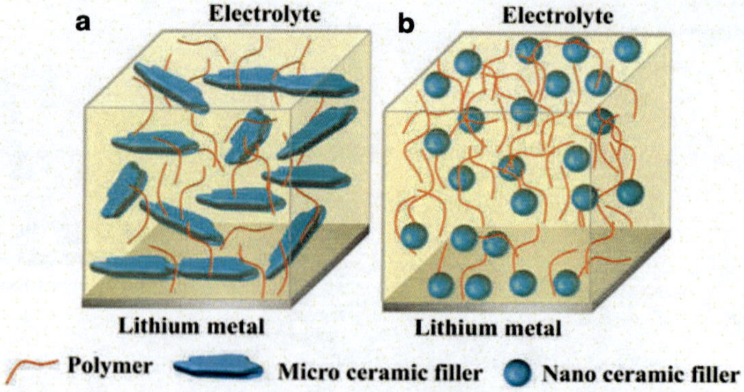

Fig. 7.4 Schematic diagram of lithium composite electrolyte **a** with micron-sized filler, **b** with nanosized filler

7.3 Silica and Forms of Silica

Silica is commonly represented by the chemical formula SiO_2. Mostly the silica compounds are formed by the tetrahedral coordination of oxygen i.e. each silicon atom is surrounded by four oxygen atoms with a bond angle of 109°28″. The silica is known to have nine polymorphs [22] (i) α-quartz (ii) β-quartz (iii) α-cristobalite (iv) β-cristobalite (v) tridymite (vi) coesite (vii) stishovite (viii) fluorite (ix) Pa3. The crystal structure of these polymorphs is depicted in Fig. 7.5. Even though several polymorphs have been reported mostly explored polymorphs are less out of which α-quartz/low quartz is of prime importance owing to its stability at ambient temperature and pressure. α-quartz is having a three-dimensional (3D) network structure crystallized with a hexagonal phase having length $a = b = 4.913$ Å $c = 5.4012$ Å and angle $\alpha = \beta = 90°$ $\gamma = 120°$ [23]. α-quartz and β-quartz posses the same crystal structure with variation only in the length of sides ($a = b = 5.010$ Å $c = 5.470$ Å). These phases can undergo a phase transition at temperature of 573 °C and pressure of 1 barr. Since this phase transition occurs as a result of the displacement of atoms in the lattice, it is also known as displacive transformation. Cristobalite and tridymite are polymorphs of silica found in the hardened rocks formed by volcanic eruption hence they form a metastable structure that is solidified and stabilized to form rocks. Cristobalite is having a 3D tetragonal structure with a side length of I $= b = 4.978$ Å $c = 6.948$ Å. Tridymite can be formed by the reconstructive transformation of β-quartz at higher temperatures. The crystallinity of this crystal is uncertain owing to the scarcity of the mineral [24]. It is assumed to be crystallized with triclinic crystal structure possessing side length of $a = 9.932$ Å $b = 17.216$ Å $c = 81.854$ Å. Coesite forms a monoclinic prismatic structure with space group B2/b with side length $a = c = 7.143$ Å $b = 12.383$ Å. On an increase in pressure, coesite crystals show compression followed by a decrease in the Si–O–Si angles as well as Si–Si bond

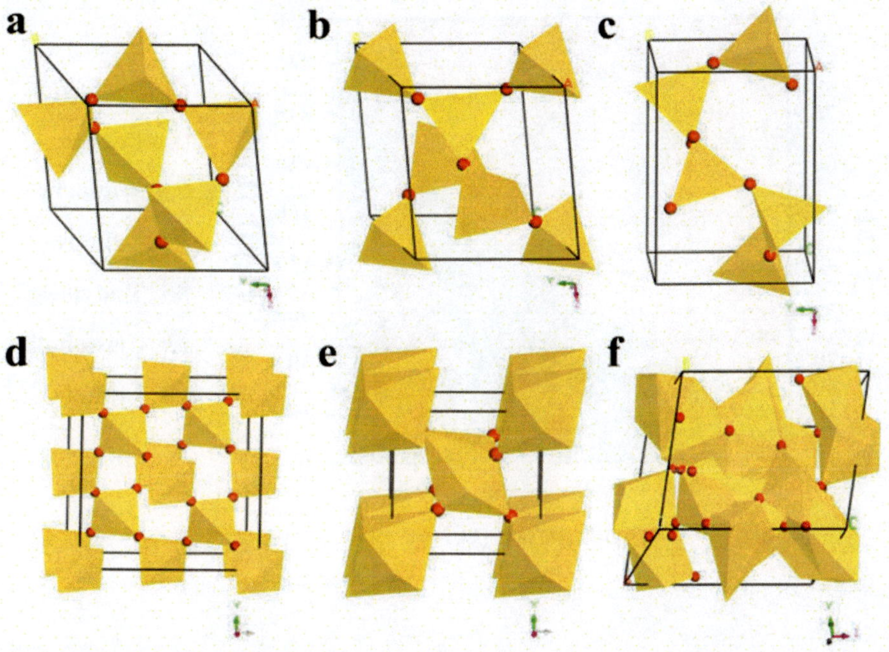

Fig. 7.5 Polyhedron crystal structure of various polymorphs of SiO_2 **a** α-quartz **b** β-quartz c α-cristobalite **d** β-cristobalite **e** stishovite **f** coesite

length, but the tetrahedra do not under any distortion and are found to be flexible almost up to 52 barr [25]. Stishovite is having a 3D cubic structure with side length of 7.160 Å.

Thermodynamic stability of the polymorphs on silica indicates that only low quartz or α-quartz is stabilized at ambient conditions and the rest of the polymorphs are stabilized at high temperature and pressure. The phase diagram of the silica polymorph is given in Fig. 7.6. confirms the thermodynamic stability of various polymorphs. Tridymite and cristobalite polymorphs are stabilized at higher temperatures above 867 and 1300 °C, respectively at 1 barr pressure. Beyond ~1700 °C silica acquires a molten phase forming liquid which is in turn stable at high-pressure conditions up to 90 kb. At a pressure range of 10–75 kb, coesite structure of silica is stabilized but on further increase in pressure stishovite, the crystal phase is formed.

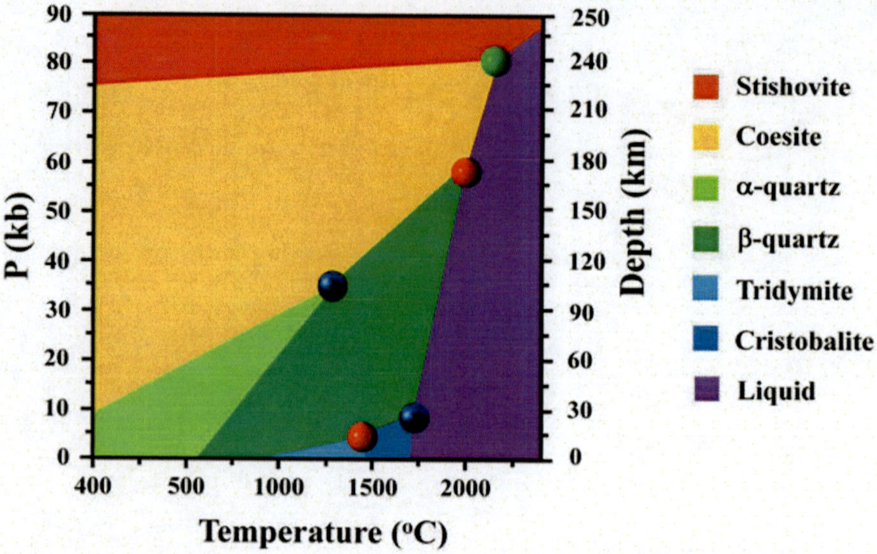

Fig. 7.6 Thermodynamic phase diagram of SiO_2

7.4 Polymer Silica Nanocomposites for Lithium-Ion Batteries

Polymer nanocomposite combines the properties of both organic and inorganic moiety enhancing the inherent properties of each other through a synergistic mechanism. Among different metal oxide systems, silica has a potential role imparted by the co-functioning of the composite membranes. Silica is a comparatively low-cost, abundant, and eco-friendly additive. Polymer nanocomposites can be synthesized by various routes in which silica can be introduced from any precursors or as silica nanoparticles/fibers. The ionic conductivity enhancement of inert inorganic additive incorporated electrolytes was reported by Nan et al. [26]. The conductivity of Li^+-ions was enhanced by the connected network structure of the nanofillers in the polymer matrix and the uniform dispersion of the silica particles in PEO. The inorganic fillers like SiO_2, increases the amorphous nature of the polymer network. The Lewis acid-base interaction also acts as an interface for the enhancement of ionic conductivity [27]. The ceramic polymer intertwined structure imparts an increase in the mechanical strength of the polymer composite. The flexibility of the polymer is maintained in the formed nanocomposite further promoting the applicability of these materials as electrolytes in LIBs. PEO-based electrolyte containing 5 wt% of silica and LTFSI gave a conductivity of $1.4 \times 10^{-4}\ \Omega^{-1}\text{cm}^{-1}$ and the lithium-ion transport number was enhanced on the addition of the filler (fumed silica (~7 nm diameter and 0.2 μm length)) from 0 to 10 wt% [28]. PEO was modified with silica via *in situ* deposition using TEOS and lithium perchlorate [29]. Composite polymer

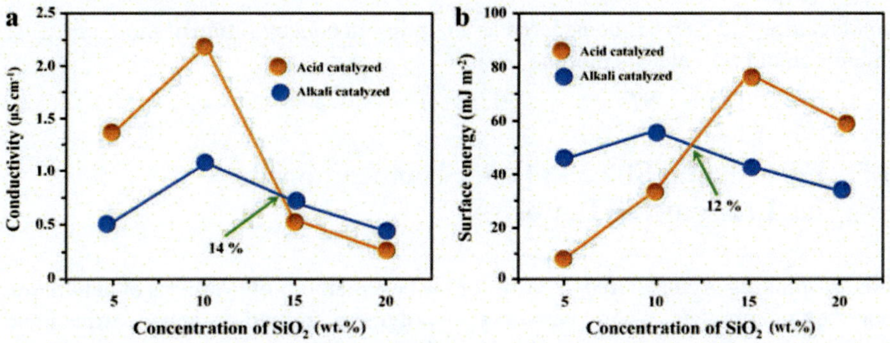

Fig. 7.7 Effect of SiO_2 content on **a** conductivity of composite polymer electrolytes at 30 °C **b** surface energy of composite polymer electrolytes. Adapted and reproduced from Pan et al. [29]. Copyright 2008 Springer Nature

matrix was prepared by mixing the precursors in suitable solvents and the membrane was prepared by solvent casting technique. Pan et al. [29] reported the comparison of acid-catalyzed silica PEO(ACSiP) electrolyte composite and alkali catalyzed silica PEO(AlCSiP) composite membranes. The ionic conductivity of AlCSiP was found to be higher than that of ACSiP indicating an increased surface energy of ACSiP than that of AlCSiP as given in Fig. 7.7. The conductivity of the samples with a different weight percentage of silica has been given in [29].

The role of nano-silica and titania in the thermal, electrochemical, and mechanical properties of PVdF-*co*-HFP was investigated with lithium triflate electrolyte [30]. Investigation on the effect of different mole concentrations of nanofiller revealed that the bulk resistance of the 8 mol% of silica incorporated polymer composite was minimum with enhanced conductivity and mechanical strength [30]. The effect of fumed silica (15 wt%) PEO nanocomposite, in the conduction of Li^+-ions was also compared TiO_2/PEO nanocomposite [31]. The activation energy of 15 wt% of silica loaded PEO was 32.58 kJ mol^{-1} whereas polymer nanocomposite loaded with the same amount of titania gave activation energy of 61.62 kJ mol^{-1} [31]. The differential scanning calorimetry (DSC) studies of the composites show that the melting temperature and glass transition temperature of silica PEO composite is slightly higher than that of the titania PEO composite membrane. The amorphicity of silica/PEO is 4-10% higher than that of titania/PEO thereby enhancing the mobility of Li^+-ions in the electrolyte [31].

PVdF-*co*-HFP/silica blend was prepared by agitating the nanoparticles of silica with polymer and stabilized with 0.2% of sodium dodecyl sulfate (SDS) followed by the addition of ionic liquid (LiTFSI-Pyr13TFSI (propyl-N-methylpyrrolidinium bis(trifluoromethane sulfonyl)-imide)) [32]. The slurry was then hot pressed using a hydraulic press at 120 °C and 10 kN. The composition of silica in the polymer ionic liquid matrix has been varied 0, 10, 20, 30 wt% [33]. The introduction of silica on to the ionic liquid polymer composite membrane enhances the ionic conductivity and Li^+-ion transport. 10 wt% of silica-added membrane gave the maximum ionic conductance and hence is considered as an optimum concentration of inorganic filler in ionic liquid polymer membrane [33].

7.5 Electrospun Silica Polymer Nanocomposite for Lithium-Ion Batteries

Electrospinning is considered to be one of the best methods of synthesis of networked polymer membranes. These networks of polymers formed to enhance the ionic conductivity of the polymers enhancing the properties of the electrolyte. These polymer chains also impart high mechanical strength. It prevents the leakage of the electrolyte by holding them in the pores of the fiber through capillary mechanism. Hence, the amount of electrolyte uptake enhances reflecting the effect in the performance of the lithium-ion batteries. Raghavan et al. [34] compared the properties of electrospun and phase inversion synthesized PVdF-*co*-HFP as an electrolyte for lithium-ion batteries. AFM images of the polymeric membranes synthesized by phase inversion method (PIM) and electrospinning method (ESM) is given in Fig. 7.8a. The AFM images was the first comparative study been reported for the surface characterization of PVdF-*co*-HFP polymeric membranes. The change in surface texture and roughness is clearly evident from the imaging. The 'lunar landscape-like' surface for electrospun fibers and enhanced roughness of the surface compared to those membranes fabricated by phase inversion method (Fig. 7.8b). The studies revealed that the ionic conductivity of the polymer membrane synthesized by the electrospinning method (ESM) showed an enhancement in the conductivity to about five times that synthesized via the phase inversion method (PIM). The discharge capacity of the battery also showed a corresponding increase (Fig. 7.9).

In 2008, Raghavan et al. [35] investigated the role of silica incorporation on the electrospun PVdF-*co*-HFP. The investigation revealed that the incorporation of silica

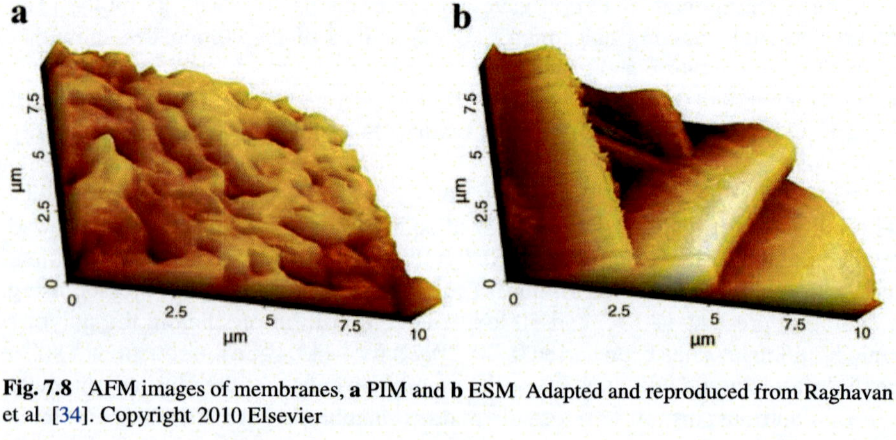

Fig. 7.8 AFM images of membranes, **a** PIM and **b** ESM Adapted and reproduced from Raghavan et al. [34]. Copyright 2010 Elsevier

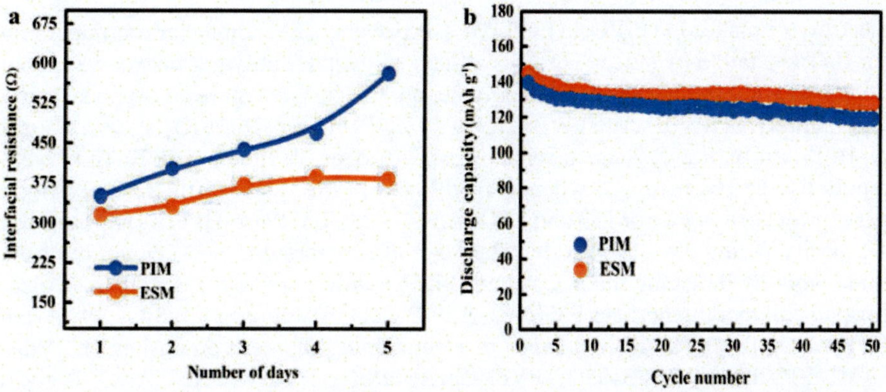

Fig. 7.9 **a** Time dependent interfacial resistance of polymer electrolytes based PVdF-*co*-HFP membranes **b** cycle performance of Li/PE/$LiFePO_4$ cells with polymer electrolytes based on PVdF-*co*-HFP membranes. Adapted and reproduced from Raghavan et al. [34] Copyright 2010 Elsevier

enhances the porosity, electrolyte uptake, ionic conductivity, and mechanic strength of the polymer membranes. The electrochemical impedance spectroscopic (EIS) analysis, clearly explains the decrease in the electrode-electrolyte interfacial resistance on the addition of silica into the PVdF-*co*-HFP polymer matrix. The silica addition into the polymer is supposed to impart amorphous nature to the polymer which can be directly related to the enhancement in the ionic conductivity and decrease in the interfacial resistance. The studies conduct on electrospun PAN/silica composite membranes [36] as well as electrospun PEC/silica membranes [37] confirmed the observations. The electrospun silica polymer nanocomposite can be synthesized by a mechanical blending of silica with the polymer carrier matrix, in situ method, and combined electrospinning electrospraying technique (EET).

7.5.1 Addition of Silica into Polymer Matrix by Mechanical Blending Method

The direct addition of silica into the polymer matrix is achieved by mixing pure silica particles in the polymer matrix at varying compositions followed by electrospinning. The parameters (applied voltage, jet to collector distance, etc.) will be variable according to the polymer and the solvent used. Several polymers such as PAN, PVdF, PEO, etc. have been used for the synthesis of electrospun membranes. Raghavan et al. [38] compared the performance of various inorganic fillers (SiO_2, $BaTiO_3$, Al_2O_3) with size ranging from 30 to 50 nm in PVdF-*co*-HFP matrix via mechanical blending method followed by electrospinning at a voltage of 20 kV at a syringe collector distance of 16 cm. The mechanical stability, ionic conductivity, electrolyte uptake (1M $LiPF_6$ in EC/DMC), and average fiber diameter was maximum for $BaTiO_3$, tailed by silica. The crystallinity of the pure polymer matrix was about 74.5% whereas the inorganic filler/polymer composite gave more amorphous nature (crystallinity of PVdF-*co*-HFP/$BaTiO_3$–47.1%, PVdF-*co*-HFP/SiO_2–47.9%, PVdF-*co*-HFP/Al_2O_3–49.2%). The discharge capacities of PVdF-*co*-HFP/$BaTiO_3$ were reported to be 164 mAh g^{-1} whereas PVdF-*co*-HFP/SiO_2 and PVdF-*co*-HFP/Al_2O_3 gave lower capacitance of 153 and 157 mAh g^{-1} [38]. PVdF-*co*-HFP/$BaTiO_3$ showed a capacity fading after 30th cycle. The team also reported a modified study of the same work by replacing the electrolyte ($LiPF_6$) with ionic liquid BMITFSI in electrospun polymer membranes PVdF-*co*-HFP/$BaTiO_3$, PVdF-*co*-HFP/SiO_2, PVdF-*co*-HFP/Al_2O_3 [39]. The SEM images of electrospun polymer composites of PVdF-*co*-HFP, PVdF-*co*-HFP/$BaTiO_3$, PVdF-*co*-HFP/SiO_2, PVdF-*co*-HFP/Al_2O_3 is given in Fig. 7.10. The initial discharge capacity of electrospun PVdF-*co*-HFP without filler was reported to be 168.5 mAh g^{-1} and PVdF-*co*-HFP/$BaTiO_3$ 165.8 mAh g^{-1}, PVdF-*co*-HFP/SiO_2 168.1 mAh g^{-1}, PVdF-*co*-HFP/Al_2O_3 165 mAh g^{-1}. The maximum electrochemical performance was shown by electrospun PVdF-*co*-HFP

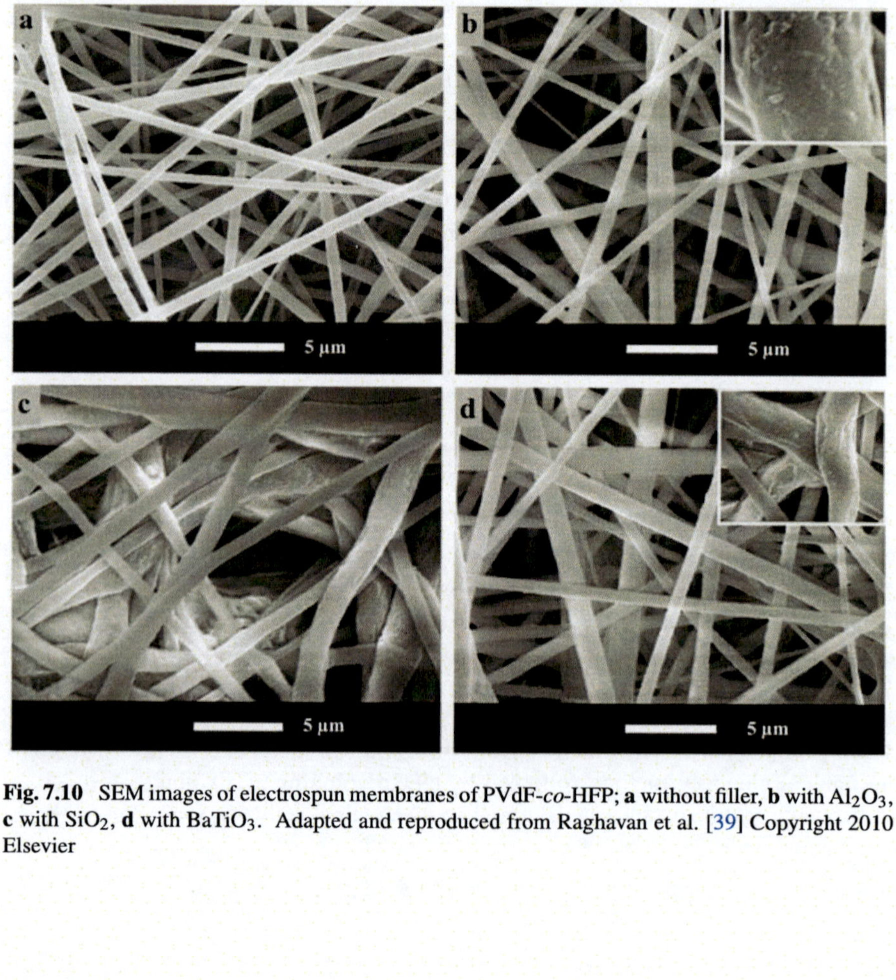

Fig. 7.10 SEM images of electrospun membranes of PVdF-*co*-HFP; **a** without filler, **b** with Al_2O_3, **c** with SiO_2, **d** with $BaTiO_3$. Adapted and reproduced from Raghavan et al. [39] Copyright 2010 Elsevier

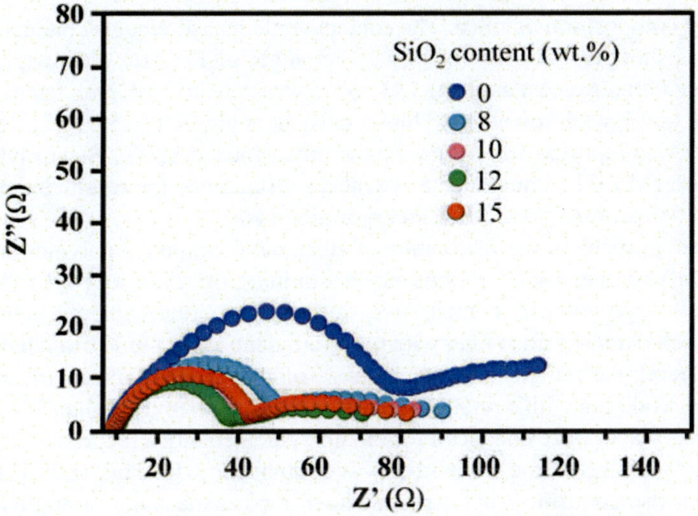

Fig. 7.11 AC impedance spectra of various electrospun SiO_2/PAN composite electrolyte membranes. Adapted and reproduced from Jung et al. [36], Copyright 2009 Elsevier

and silica incorporated nanocomposite. After 20 cycles electrospun PVdF-*co*-HFP showed capacity fading, but the capacity of PVdF-*co*-HFP/SiO_2 was retained at 164.6 mAh g^{-1} [39].

A comparative study was performed on electrospun PAN incorporated with hydrophilic fumed silica by direct addition method dispersed in N,N-dimethylformamide (DMF) solvent was electrospun with varying silica compositions (0, 8,10, 12, 15 wt%) applying a voltage of 20 kV and collector–jet distance of 17 cm at room temperature. The results revealed that the 12 wt% of fumed silica PAN composite showed the best performance with an ionic conductivity of 1.1×10^{-2} S cm^{-1}, lowest electrode-electrolyte resistance (Fig. 7.11) [36], large electrolyte uptake of 490%, high discharge capacity (139 mAh g^{-1}) almost approaching the theoretic value [36]. The increase in silica composition also enhanced anodic stability. The stability of electrospun PAN membrane was about 4.0 V whereas 12 wt% silica PAN composite polymer was obtained about 5.0 V [36]. Commercially available silica nanopowder (10 wt%) was mixed with PEO in aqueous medium and electrospun at a voltage of 13 kV and needle to collector distance of 30 cm [40]. The incorporation of silica on to PEO matrix in presence of polysulfide reduced the performance of electrospun PEO. The decrease in the performance of the solid electrolyte can be attributed to the interaction of inorganic filler and the polymer medium leading to the formation of incomplete chain networks reducing the electrolyte properties of the material [40].

Fumed silica (0, 3, 5, 7 wt%) was blended with PVdF in DMF and electrospun applying high voltage of 18 kV at 15 cm distance (jet to collector) on aluminum foil [41]. The ionic conductivity varied from 7.7 to 0.24 S cm^{-1} on an increase in the percentage composition of silica. The contact angle measurements conducted on the membranes show electrospun PVdF gave an angle of 127.10°, electrospun 3 wt% silica/PVdF blend gave an angle of 133.33°, 5 wt% silica/PVdF blend gave an angle of 134.80° and 7 wt% silica/PVdF blend gave an angle of 136.50° [41]. Maximum wettability was reported for 7 wt% fumed silica loaded PVdF. Electrolyte uptake studies with 1M $LiPF_6$ liquid electrolyte also substantiated the results indicating the maximum uptake for 7 wt% silica incorporation [41].

Another approach of the fabrication of silica nanocomposite polymer electrolyte was by synthesizing silica fibers by electrospinning of silica sol [42] followed by mixing with a polymer. A comparative study of particulate silica, calcined silica fiber, and non-calcined silica fiber with poly[(ethylene oxide)-ran-(propylene oxide)] (P(EO/PO)) relived that the ionic conductivity of non-calcined silica with an average diameter of 1000 nm with 5 wt% composition gave a maximum value. The mechanical properties also were enhanced for the non-calcined silica incorporated polymer composites [42]. The same silica fiber was composited with PEC and LiTFSI [37]. The studies also confirmed that the silica fiber imparts mechanic strength and ionic conductivity to the polymer matrix.

7.5.2 *Addition of Silica into Polymer Matrix by in Situ Method*

In situ synthesis of silica, polymer composites are carried out by dispersing the precursor of silica (tetraethoxy silane (TEOS)) in polymer matrix followed by hydrolyzing the precursor to form silicon hydroxide (Eq. 7.1). Silicon hydroxide is gradually converted to silica in the polymer matrix (Eq. 7.2). The advantage of this method is the uniform distribution of silica nanoparticle in the polymer. The polymer matrix acts as a template for the controlled growth of silica as well as prevents agglomeration of the fillers. This dispersion is subjected to electrospinning process. The parameters of the electrospinning are varied accordingly based on the polymer matrix and solvents used. The electrospun membranes prepared is used as polymer electrolyte in lithium-ion batteries.

$$Si(OC_2H_5)_2 + 4H_2O \rightarrow Si(OH)_4 + 4C_2H_5OH \tag{7.1}$$

$$Si(OH)_4 \rightarrow SiO_2 + 2H_2O \tag{7.2}$$

In 2008, Raghavan et al. [35] compared the performance silica PVdF-*co*-HFP polymer electrolyte synthesized by in situ silica synthesis (3, 6, 9 wt%), direct added silica (6 wt%), and polymer in the absence of silica (0 wt%). In situ synthesis of silica gave superior ionic conductivity and mechanical strength owing to a reduction in the nucleation of the silica forming aggregates in the polymer matrix. Uniform dispersion of silica was been reported which showed a substantial enhancement in the properties. The optimum composition of silica was obtained to be 6 wt% with an ionic conductivity of 8.06 mS cm^{-1} [35]. The cell fabricated with lithium iron) as a cathode and lithium metal as anode gave a discharge capacitance of 170 mAh g^{-1} which is equal to the theoretical capacity of LFP whereas the direct addition of silica gave a capacity of 153 mAh g^{-1} (Fig. 7.12) [35]. The structural, mechanical, and electrochemical properties of the PVdF-*co*-HFP based polymer composites have been depicted in Table 7.1.

PVdF/silica composite polymer membrane separator was synthesized by in situ method with silica sol PVdF ratio of 1:9 and 1:10, respectively, electrospun at an applied bias of 20 kV at a distance of 20 cm [43]. The synthesized membrane showed better performance than the standard commercial Celgard® PP membrane.

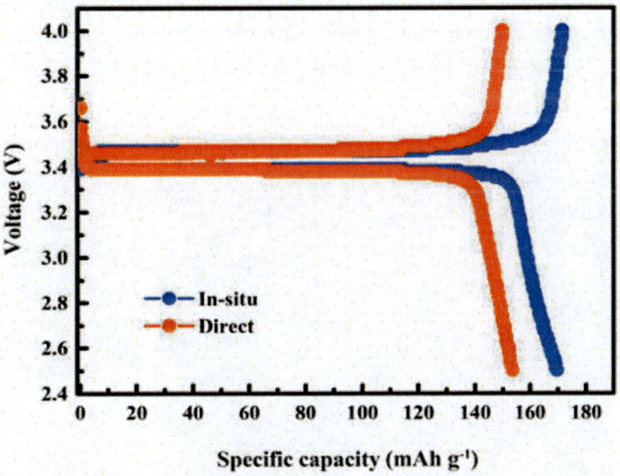

Fig. 7.12 Initial charge-discharge properties of Li/polymer electrolyte/$LiFePO_4$ cells with polymer electrolytes based on electrospun PVdF-*co*-HFP membrane containing 6% of in situ generated silica and directly added silica (25 °C, 0.1 C rate, 2.5–4.0 V). Adapted and reproduced from Raghavan et al. [35], Copyright 2008 Elsevier

Table 7.1 Properties of electrospun membranes and polymer electrolytes based on membranes activated with 1M $LiPF_6$ in EC/DMC Raghavan et al. [35]

Properties	Silica content (wt%)				
		In situ			Direct
	0	3	6	9	6
Fiber diameter range (µm)	0.4–2.0	0.4–3.3	0.7–2.0	0.7–3.3	0.9–3.5
AFD (µm)	1.1	1.2	1.2	1.6	1.9
Porosity (%)	88	91	92	93	89
Electrolyte uptake (%)	425	556	592	620	491
Ionic conductivity at 20 °C (mS cm^{-1})	4.59	5.13	8.60	6.61	5.72
Electrolyte retention ratio (R)	0.83	0.84	0.85	0.82	0.84
Tensile strength (MPa)	6.5	7.4	10.9	12.3	9.0
Modulus (MPa)	9.2	14.5	14.8	15.7	14.9
Elongation at break (%)	71	69	58	50	56

The membrane with ratio 9:1 gave the maximum discharge capacity of 159 mAh g^{-1} at 0.2 C rate [43]. The TEM images of the composited prepared with 9:1 and 10:1 composition are given in Fig. 7.13.

In 2011, Wu et al. [44] reported a comparative study on the performance of electrospun polyurethane (PU)/PVdF blend without filler and with in situ addition of silica (3 wt%) and titania (3 wt%) as the fillers. Silica addition was achieved by in situ hydrolysis of TEOS (Eqs. 7.1 and 7.2) and in situ addition of titania was succeeded using hydrolysis of titanium tertiary butoxide (TTB) (Eqs. 7.3 and 7.4). Electrospinning of the bend dispersion was conducted at high voltage of 28.5 kV [44] and the SEM images are depicted in Fig. 7.14. The polymer membrane was activated by immersing the liquid in unimolar lithium perchlorate ($LiClO_4$) solution in mixture of solvents [ethylene carbonate (EC) and propylene carbonate (PC)]. The ionic conductivity study of these gel polymer electrolytes shows that the addition of ceramic inorganic fillers enhances the performance of the electrolytes. The pure polymer blend gave an ionic conductivity of 3.2×10^{-3} cm^{-1} at ambient temperature

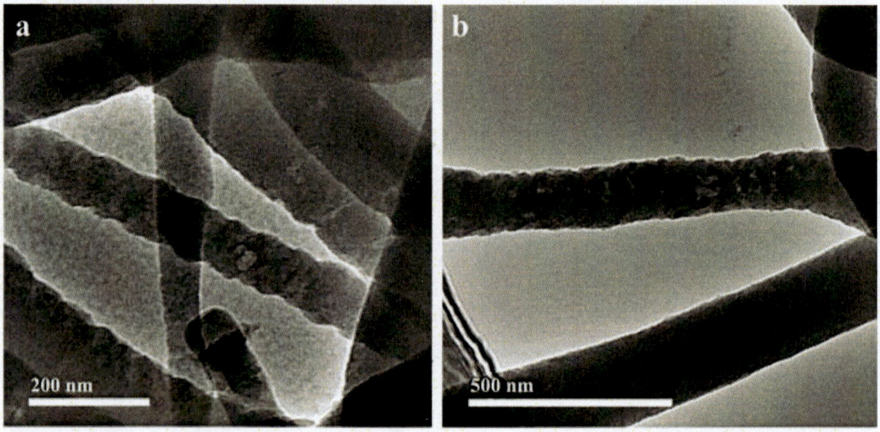

Fig. 7.13 TEM images of composite membranes **a** 1:9, **b** 1:10. Adapted and reproduced from Zhang et al. [43], Copyright 2014 Elsevier

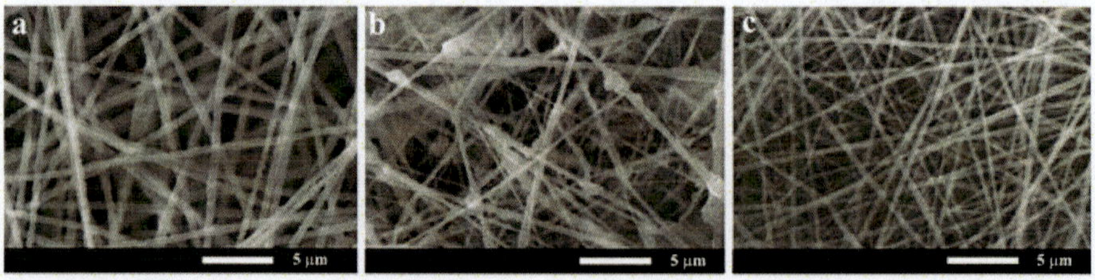

Fig. 7.14 SEM images of electrospun TPU–PVdF membranes **a** without in situ ceramic fillers, **b** with 3 wt% in situ SiO_2 and **c** with 3 wt% in situ TiO_2. Adapted and reproduced from Wu et al. [44], Copyright 2011 Elsevier

[44]. The in situ addition of silica in the polymer blend resulted in ionic conductivity of 3.8×10^{-3} S cm^{-1} and that for TiO_2 gave 4.8×10^{-3} S cm^{-1}. The bulk resistance of the samples varied from 1.4 to 0.8 Ω on the incorporation of additive [44].

$$Ti(OC_2H_5)_4 + 4H_2O \rightarrow Ti(OH)_4 + 4C_2H_5OH \tag{7.3}$$

$$Ti(OH)_4 \rightarrow TiO_2 + 2H_2O \tag{7.4}$$

7.5.3 *Electrospinning Electrospraying Technique (EET)*

Electrospinning electrospraying technique is a combined process in which polymer is electrospun on a metallic drum and simultaneously inorganic additive dispersion is sprayed using high voltage on the same drum. Thus, on the drum, the additive deposited polymer membrane is formed. The additives are agglomerated and decorated on the polymer network forming bead-like morphology.

In 2014, Yanilmaz et al. [45] reported silica decorated PVdF hybrid membrane synthesized by the EET method. Silica nanoparticles were dispersed via stirring and ultrasonication in methanol varying compositions (0, 1, 3, 5 wt%) was used for electrospraying. PVdF in DMF/acetone mixture was electrospun at 20 kV at a tip-collector distance of 17 cm. The ICP analysis of the EET membrane indicated that the composition of silica in the membranes was 0, 7, 14, 24 wt%, respectively [45]. The C rate analysis of the EET membranes was compared with standard PP membrane (Fig. 7.15) revealed that the increase in silica concentration subsequent increase in the discharge capacity is observed. The maximum discharge capacity was obtained

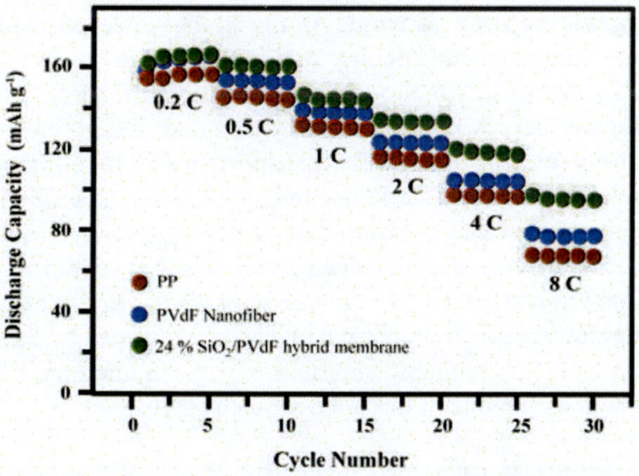

Fig. 7.15 C-rate performance of Li/$LiFePO_4$ cells containing SiO_2/PVdF nanoparticle/nanofiber hybrid membranes and microporous PP membrane. Adapted and reproduced from Yanilmaz et al. [45], Copyright 2014 Elsevier

for 24 wt% loaded EET membrane. The minimum impedance and maximum ionic conductivity were also given by the same sample [45].

7.5.4 *Addition of Silica-Based Compounds into Polymer Matrix*

Silica particle was modified to enhance the electrochemical, mechanical, and physical properties of the electrolyte membrane. An ionic complex of silica nanoparticle was prepared by chemical method forming polyacrylic acid (PAA) silica which was further neutralized using lithium hydroxide to form ionic complex named SiPAALi [45]. The complex was dispersed in PVdF (in DMF Acetone mixture) and electrospunned at high voltage of 15 kV at a distance of 18 cm. The ionic complex incorporated membrane (SiPAALi/PVdF) was compared with silica incorporated electrospun PVdF (SiO_2/PVdF) and bare electrospun PVdF. The electrolyte uptake of the SiPAALi/PVdF was enhanced. The charge/discharge cycle of Li/electrolyte membrane/$LiCoO_2$ batteries at room temperature delivered an initial capacity of 156.5 mAh g^{-1} for SiPAALi/PVdF, 152 mAh g^{-1} for SiO_2/PVdF, and 149.1 mAh g^{-1} for electrospun PVdF membranes [45]. PVdF was blended with PMMA and SiO_2 via the electrospinning technique. The PVdF/PMMA/SiO_2 blend enhanced the ionic conductivity to 2.0×10^{-3} S cm^{-1} from 1.7×10^{-3} S cm^{-1} [46].

7.6 Conclusion

Polymer nanocomposites are promising candidates for next-generation energy storage devices. The energy storage devices are of prime importance owing to the growing energy demand. As the global energy crisis is targeted to overcome by utilizing renewable energy sources such as solar energy, wind energy hydro energy, etc. The need for the efficient and reliable energy storage devices is of greatest importance. Apart from energy demand, the global policy has been developed to utilize electric energy for transportation. The e-transport also demands sustained release of energy for better mileage. Hence, a good electrolyte is required for safe and efficient energy storage devices. Polymer electrolyte overcomes the demerits of the liquid conventional electrolytes. The only deficiency of these electrolytes is their ionic conductivity. In comparison with the liquid electrolytes, polymer electrolyte has lower ionic mobility which can be improved by the addition of additives. Several inorganic and organic additives have been attempted and achieved enhancement in the performance of the polymer electrolyte.

Acknowledgements Authors Dr. Jabeen Fatima M. J and Dr. Raghavan Prasanth, would like to acknowledge Kerala State Council for Science, Technology, and Environment (KSCSTE), Kerala for financial assistance.

References

1. Dudley B (2019) BP statistical review of world energy statistical review of world. BP Stat Rev world Energy 68:1–61
2. Mizushima K, Jones PC, Wiseman PJ, Goodenough JB (1981) $LixCoO_2$ ($0 < x \leq 1$): a new cathode material for batteries of high energy density. Solid State Ionics 3–4:171–174. https://doi.org/10.1016/0167-2738(81)90077-1
3. Hamidah NL, Wang FM, Nugroho G (2019) The understanding of solid electrolyte interface (SEI) formation and mechanism as the effect of flouro-o-phenylenedimaleimaide (F-MI) additive on lithium-ion battery. Surf Interface Anal 51:345–352. https://doi.org/10.1002/sia.6586
4. Wang A, Kadam S, Li H et al (2018) Review on modeling of the anode solid electrolyte interphase (SEI) for lithium-ion batteries. npj Comput Mater 4. https://doi.org/10.1038/s41524-018-0064-0
5. Vetter J, Nov P, Wagner MR, Veit C (2005) Ageing mechanisms in lithium-ion batteries. J Power Sources 147:269–281. https://doi.org/10.1016/j.jpowsour.2005.01.006
6. Fenton DE, Parker JM, Wright PV (1973) Complexes of alkali metal ions with poly(ethylene oxide). Polym (Guildf) 14:589. https://doi.org/10.1016/0032-3861(73)90146-8
7. Fallis A (1986) Poly(ethylene oxide)—$LiCF_3SO_3$—polystyrene electrolyte systems. Solid State Ionics 18:282–286. https://doi.org/10.1017/CBO9781107415324.004
8. Lee H, Alcoutlabi M, Watson JV, Zhang X (2013) Polyvinylidene fluoride-co-chlorotrifluoroethylene and polyvinylidene fluoride-co-hexafluoropropylene nanofiber-coated polypropylene microporous battery separator membranes. J Polym Sci Part B Polym Phys 51:349–357. https://doi.org/10.1002/polb.23216
9. Jung S, Kim DW, Lee SD et al (2009) Fillers for solid-state polymer electrolytes: highlight. Bull Korean Chem Soc 30:2355–2361. https://doi.org/10.5012/bkcs.2009.30.10.2355
10. Leijonmarck S, Cornell A, Lindbergh G, Wågberg L (2013) Single-paper flexible Li-ion battery cells through a paper-making process based on nano-fibrillated cellulose. J Mater Chem A 1:4671–4677. https://doi.org/10.1039/c3ta01532g
11. Kim JH, Gu M, Lee DH et al (2016) Functionalized nanocellulose-integrated heterolayered nanomats toward smart battery separators. Nano Lett 16:5533–5541. https://doi.org/10.1021/acs.nanolett.6b02069
12. Xiao SY, Yang YQ, Li MX et al (2014) A composite membrane based on a biocompatible cellulose as a host of gel polymer electrolyte for lithium ion batteries. J Power Sources 270:53–58. https://doi.org/10.1016/j.jpowsour.2014.07.058
13. Ran Y, Yin Z, Ding Z et al (2013) A polymer electrolyte based on poly(vinylidene fluoride-hexafluoropylene)/hydroxypropyl methyl cellulose blending for lithium-ion battery. Ionics (Kiel) 19:757–762. https://doi.org/10.1007/s11581-012-0808-7
14. Li W, Wu G, Araujo C et al (2010) Li^+ ion conductivity and diffusion mechanism in α-Li_3N and β-Li_3N. Energy Environ Sci 3:1524–1530. https://doi.org/10.1039/C0EE00052C
15. Skaarup S, West K, Zachau-Christiansen B (1988) Mixed phase solid electrolytes. Solid State Ionics 30:975–978
16. Scrosati B, Croce F, Panero S (2001) Progress in lithium polymer battery R&D. J Power Sources 100:93–100. https://doi.org/10.1016/S0378-7753(01)00886-2
17. Xiao W, Gong Y, Wang H et al (2015) Preparation and electrochemical performance of ZrO_2 nanoparticle-embedded nonwoven composite separator for lithium-ion batteries. Ceram Int 41:14223–14229. https://doi.org/10.1016/j.ceramint.2015.07.048
18. Liu Y, Lee JY, Hong L (2002) Morphology, crystallinity, and electrochemical properties of in situ formed poly (ethylene oxide)/TiO_2 nanocomposite polymer electrolytes. J Appl Polym Sci 89:2815–2822
19. Wang H, Li H, Lijun Y, Yanmei Jiang KW (2013) Synthesis of porous Al_2O_3-PVDF composite separators and their application in lithium-ion batteries. J Appl Polym Sci 130:2886–2890. https://doi.org/10.1002/app

20. Aravindan V, Vickraman P (2008) Characterization of SiO_2 and Al_2 O_3 incorporated PVdF-HFP based composite polymer electrolytes with $LiPF_3$ (CF_3 CF_2)$_3$ V. J Appl Polym Sci 108:1314–1322. https://doi.org/10.1002/app
21. Xiong HM, Wang ZD, Xie DP et al (2006) Stable polymer electrolytes based on polyether-grafted ZnO nanoparticles for all-solid-state lithium batteries. J Mater Chem 16:1345–1349. https://doi.org/10.1039/b514346b
22. Keskar NR, Chelikowsky JR (1992) Structural properties of nine silica polymorphs. Phys Rev B 46:1–13. https://doi.org/10.1103/PhysRevB.46.1
23. Saha P, Bandyopadhyay T (1979) Alpha-Quartz 2. Crystal chemistry, and nature and distribution of impurities. Bull Mater Sci 1:79–93. https://doi.org/10.1007/BF02908570
24. Gibbs RE (1926) The polymorphism of silicon dioxide and the structure of tridymite. Proc R Soc A Math Phys Eng Sci 113:351–368. https://doi.org/10.1098/rspa.1926.0160
25. Levien LPCT (1981) High-pressure crystal structure and compressibility of coesite. Am Mineral 66:324–333
26. Nan CW, Fan L, Lin Y, Cai Q (2003) Enhanced ionic conductivity of polymer electrolytes containing nanocomposite SiO_2 particles. Phys Rev Lett 91:1–4. https://doi.org/10.1103/PhysRevLett.91.266104
27. Scrosati B, Croce F, Persi L (2000) Impedance spectroscopy study of PEO-based nanocomposite polymer electrolytes. J Electrochem Soc 147:1718–1721. https://doi.org/10.1149/1.1393423
28. Capiglia C, Mustarelli P, Quartarone E et al (1999) Effects of nanoscale SiO_2 on the thermal and transport properties of solvent-free, poly(ethylene oxide) (PEO)-based polymer electrolytes. Solid State Ionics 118:73–79. https://doi.org/10.1016/S0167-2738(98)00457-3
29. Pan C, Zhang Q, Feng Q, Gao J, Zhao Y (2008) Effect of catalyst on structure of $(PEO)_8LiClO_4$-SiO_2 composite polymer electrolyte films. J Cent South Univ Technol 15:438–442. https://doi.org/10.1007/s11771–008–0082–z
30. Hema M, Tamilselvi P (2016) Lithium ion conducting PVA: PVdF polymer electrolytes doped with nano SiO_2 and TiO_2 filler. J Phys Chem Solids 96–97:42–48. https://doi.org/10.1016/j.jpcs.2016.04.008
31. Wang W, Yi E, Fici AJ et al (2017) Lithium Ion conducting Poly(ethylene oxide)-based solid electrolytes containing active or passive ceramic nanoparticles. J Phys Chem C 121:2563–2573. https://doi.org/10.1021/acs.jpcc.6b11136
32. Caimi S, Wu H, Morbidelli M (2018) PVdF-HFP and Ionic-liquid-based, freestanding thin separator for lithium-ion batteries. ACS Appl Energy Mater. https://doi.org/10.1021/acsaem.8b00860
33. Caimi S, Klaue A, Wu H, Morbidelli M (2018) Effect of SiO_2 nanoparticles on the performance of PVdF-HFP/ionic liquid separator for lithium-ion batteries. Nanomaterials 8. https://doi.org/10.3390/nano8110926
34. Raghavan P, Zhao X, Manuel J et al (2010) Electrochemical studies on polymer electrolytes based on poly(vinylidene fluoride-co-hexafluoropropylene) membranes prepared by electrospinning and phase inversion—a comparative study. Mater Res Bull 45:362–366. https://doi.org/10.1016/j.materresbull.2009.12.001
35. Raghavan P, Choi JW, Ahn JH et al (2008) Novel electrospun poly(vinylidene fluoride-co-hexafluoropropylene)-in situ SiO_2 composite membrane-based polymer electrolyte for lithium batteries. J Power Sources 184:437–443. https://doi.org/10.1016/j.jpowsour.2008.03.027
36. Jung HR, Ju DH, Lee WJ et al (2009) Electrospun hydrophilic fumed silica/polyacrylonitrile nanofiber-based composite electrolyte membranes. Electrochim Acta 54:3630–3637. https://doi.org/10.1016/j.electacta.2009.01.039
37. Kimura K, Matsumoto H, Hassoun J et al (2015) A quaternary poly(ethylene carbonate)-lithium bis(trifluoromethanesulfonyl)imide-ionic liquid-silica fiber composite polymer electrolyte for lithium batteries. Electrochim Acta 175:134–140. https://doi.org/10.1016/j.electacta.2015.03.117

38. Raghavan P, Zhao X, Kim J-K et al (2008) Ionic conductivity and electrochemical properties of nanocomposite polymer electrolytes based on electrospun poly(vinylidene fluoride-co-hexafluoropropylene) with nano-sized ceramic fillers. Electrochim Acta 54:228–234. https://doi.org/10.1016/j.electacta.2008.08.007
39. Raghavan P, Zhao X, Manuel J et al (2010) Electrochemical performance of electrospun poly(vinylidene fluoride-co-hexafluoropropylene)-based nanocomposite polymer electrolytes incorporating ceramic fillers and room temperature ionic liquid. Electrochim Acta 55:1347–1354. https://doi.org/10.1016/j.electacta.2009.05.025
40. Navarra MA, Lombardo L, Bruni P et al (2018) Gel polymer electrolytes based on silica-added poly(Ethylene oxide) electrospun membranes for lithium batteries. Membranes (Basel) 8:. https://doi.org/10.3390/membranes8040126
41. Sethupathy M, Sethuraman V, Manisankar P (2013) Preparation of PVDF/SiO_2 composite nanofiber membrane using electrospinning for polymer electrolyte analysis. Soft Nanosci Lett 03:37–43. https://doi.org/10.4236/snl.2013.32007
42. Ishibe S, Anzai K, Nakamura J et al (2014) Ion-conductive and mechanical properties of polyether/silica thin fiber composite electrolytes. React Funct Polym 81:40–44. https://doi.org/10.1016/j.reactfunctpolym.2014.04.004
43. Zhang F, Ma X, Cao C et al (2014) Poly (vinylidene fluoride)/ SiO_2 composite membranes prepared by electrospinning and their excellent properties for nonwoven separators for lithium-ion batteries. J Power Sources 251:423–431. https://doi.org/10.1016/j.jpowsour.2013.11.079
44. Wu N, Cao Q, Wang X et al (2011) In situ ceramic fillers of electrospun thermoplastic polyurethane/ poly(vinylidene fluoride) based gel polymer electrolytes for Li-ion batteries. J Power Sources 196:9751–9756. https://doi.org/10.1016/j.jpowsour.2011.07.079
45. Yanilmaz M, Lu Y, Dirican M et al (2014) Nanoparticle-on-nanofiber hybrid membrane separators for lithium-ion batteries via combining electrospraying and electrospinning techniques. J Memb Sci 456:57–65. https://doi.org/10.1016/j.memsci.2014.01.022
46. Wu XL, Lin J, Wang JY, Guo H (2015) Electrospun PVDF/PMMA/SiO_2 membrane separators for rechargeable lithium-ion batteries. Key Eng Mater 645–646:1201–1206. https://doi.org/10.4028/www.scientific.net/KEM.645-646.1201

Chapter 8
Electrospun PVdF and PVdF-*co*-HFP-Based Blend Polymer Electrolytes for Lithium Ion Batteries

N. S. Jishnu, S. K. Vineeth, Akhila Das, Neethu T. M. Balakrishnan, Anjumole P. Thomas, M. J. Jabeen Fatima, Jou-Hyeon Ahn, and Raghavan Prasanth

8.1 Introduction

Advancement in technology has been in a booming rate with proliferation in population, which has adversely affected the rapid depletion of fossil fuels. Serious environmental issues related on overexploitation of these nonrenewable resources led to greenhouse effect, ozone layer depletion, global warming and climate changes. In addition to the above environmental issues, energy crisis is a serious issue. The post-fossil fuel era hence needs to find alternative source of energy. Sustainable and renewable energy source need to be explored to satisfy the exponentially increasing demand of energy. The effective utilization of energy from the renewable resources such as water (hydropower and hydrokinetic), wind, solar (power and hot water), biomass (biofuel and biopower), geothermal (power and heating) are the current possible solution. Globally, renewables made up 24% of electricity generation in 2014, in which the major contribution from hydropower (17%). Renewable energy is the fastest-growing energy source in the USA, increasing 67% from 2000 to

---

N. S. Jishnu
Rubber Technology Centre, Indian Institute of Technology-Kharagpur (IIT-KGP), Kharagpur, West Bengal 721302, India

S. K. Vineeth
Department of Polymer and Surface Engineering, Institute of Chemical Technology, Nathalal Parekh Marg, Matunga, Mumbai 400019, India

A. Das · N. T. M. Balakrishnan · A. P. Thomas · M. J. Jabeen Fatima · R. Prasanth (✉)
Department of Polymer Science and Rubber Technology, Cochin University of Science and Technology (CUSAT), Cochin 682022, India
e-mail: prasanth@cusat.ac.in

J.-H. Ahn (✉) · R. Prasanth
Department of Materials Engineering and Convergence Technology, Gyeongsang National University, 501 Jinju-Daero, Jinju 52828, Republic of Korea
e-mail: jhahn@gnu.ac.kr

N. T. M. Balakrishnan and R. Prasanth (eds.), *Electrospinning for Advanced Energy Storage Applications*, Materials Horizons: From Nature to Nanomaterials,
https://doi.org/10.1007/978-981-15-8844-0_8

2016. Renewables made up 24% of global electricity generation in 2014, and that is expected to rise to 31% by 2040. Most of the increase will likely come from wind energy and hydropower. The estimated global renewable energy share of total final energy consumption in 2015 is shown in Fig. 8.1, and renewable energy indicator in 2016 is shown in Table 8.1

In this aspect, for the efficient storage and effective utilization of renewable energy required advanced high power energy storage devices. Among the different types of energy storage devices such as supercapacitors, batteries and fuel cells, the possibilities provided by the batteries are significant due to their high energy density, power density, easy to fabricate and portability. Compared to any other battery technology such as nickel–cadmium or nickel–metal halide batteries, etc., lithium-ion batteries are promising due to their superior energy density, output voltage, long cycle life

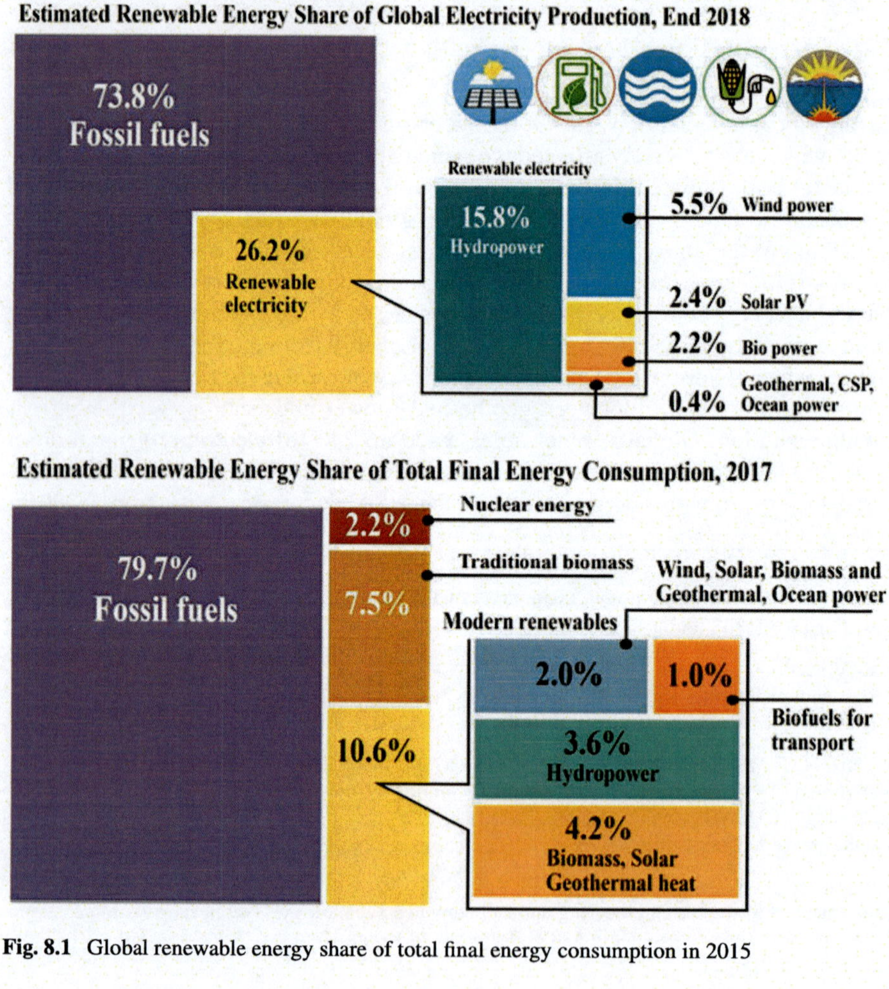

Fig. 8.1 Global renewable energy share of total final energy consumption in 2015

Table 8.1 Estimated global renewable energy share of total final energy consumption (2015)

		2015	2016
Investment			
New investment (annual) in renewable power and fuel	Billion USD	312.2	241. 6
Power			
Renewable power capacity (total, not including hydro)	GW	785	921
Renewable power capacity (total, including hydro)	GW	1856	2017
Hydropower capacity	GW	1071	1096
Biopower capacity	GW	106	112
Biopower generation (annual)	TWh	464	504
Geothermal power capacity	GW	12	13.5
Solar PV capacity	GW	228	303
Concentrating solar thermal power capacity	GW	4.7	4.8
Wind power capacity	GW	433	487
Heat			
Solar hot water capacity	GW_{th}	435	456
Transport	GW		
Ethanol production (annual)	GW	98.3	98.6
Biodiesel production (annual)	GW	30.1	30.8

and electrochemical properties. The energy density of LIBs (250 Wh kg^{-1}, 650 Wh L^{-1}) is typically twice that of the standard nickel–cadmium batteries, and there is potential for higher energy densities. The load characteristics are reasonably good and behave similarly to nickel–cadmium in terms of discharge. The high cell voltage of 3.6 V allows battery pack designs with only one cell. Most of today's mobile phones run on a single cell, while nickel-based pack would require three 1.2 V cells connected in series. Also LIBs is a low maintenance battery, an advantage that most other chemistries cannot claim. There is no memory, and no scheduled cycling is required to prolong the cycle life of the battery. In addition, the self-discharge is less than half compared to nickel–cadmium, making lithium ion well suited for modern fuel gauge applications [1]. Lithium-ion batteries (LIBs) were first commercialized by Sony Corporation in 1991, and this year, 2019, the Nobel Prize in Chemistry is awarded to John B Goodenough, M. Stanley Wittingham and Akira Yoshino "for the development of lithium-ion batteries" (Fig. 8.2).

Last two decades witnessed significant development of LIBs as its demand reflected in portable electronic gadgets, computers and related devices, and hybrid electric vehicles. Furthermore, the global market of lithium-ion batteries is currently growing, and it is expected that in 2022, the market value will reach $ 46.21 billion, with an annual growth rate of 10.8%. The major components of a battery are anode (negative electrode, where reduction occurs), cathode (positive electrode, where oxidation occurs) and electrolytes. For the development of high energy density

Fig. 8.2 Images of Nobel laureates 2019 (From right to left—John. B. Goodenough, Stanley M. Wittingham, Akira Yoshino) Adapted and reproduced from Ref. [2]. Copyright © Nobel Media 2019. Illustration: Niklas Elmehed

battery having improved safety, the major role played by the electrolyte, in which the conventional liquid electrolyte, is replaced with gel electrolyte as it have no leakage, good ionic conductivity and good dimensional stability. The gel electrolyte is a heterogeneous system in which a microporous polymeric membrane having more than 40% porosity is activated with a liquid electrolyte (typically 1 M solution) prepared by dissolving suitable lithium slat ($LiPF_6$, $LiClO_4$, LiTFSI, LIBOB) in a polar (propylene carbonate) or nonpolar (ethylene carbonate (EC), diethyl carbonate (DEC), methyl ethyl carbonate (EMC) and dimethyl carbonate (DMC) aprotic solvent or ionic liquids (1-butyl-3-methylimidazolium bis(trifluoromethanesulfonyl)imide (BMITFSI) and 1-butyl-3-methylimidazolium tetrafluoroborate ($BMIBF_4$) 1-butyl-4-methylpyridinium bis(trifluoromethanesulfonyl)imide [BMPyTFSI]). Most of the polymer electrolytes have lithium hexafluorophosphate ($LiPF_6$) in ethylene carbonate (EC), dimethyl carbonate (DMC), propylene carbonate (PC), or diethyl carbonate (DEC) or their mixtures of various compositions. Polyvinylidene fluoride (PVdF) [3–7] and its copolymer polyvinylidene difluoride-*co*-hexafluoro propylene) (PVdF-*co*-HFP) [7–17], polyethylene glycol [18], polyurethane acrylate [19], polyacrylonitrile (PAN) [20–23], polymethyl methacrylate (PMMA) [22, 24, 25] and polyethylene oxide (PEO) [7, 17, 22, 26] polyvinyl acetate (PVAc) [25], polystyrene (PS) [22], etc., have been widely used as polymer matrices for the preparation of PEs. For improving the ionic conductivity, mechanical strength, and thermal stability, polymer nanocomposite electrolyte incorporating different ceramic nanomaterials such as SiO_2 [27], Al_2O_3, [28] $BaTiO_3$ [11],TiO_2 [29, 30], ZrO [31], nanoclay [3, 9], etc., are also reported. The selection of materials and their proportions significantly influences the ionic conductivity, thermal stability and electrochemical properties of the polymer gel electrolytes [7].

In order to achieve high ionic conductivity at ambient temperature, many investigations have been reported. In recent years, much effort has been devoted to enhancing the low-temperature conductivity of PEs via various approaches such as the preparation of blends [32–34], copolymers [35–37], nanocomposites [38–40] and cross-linked networks [41, 42]. Improvements in ionic conductivity and electrochemical properties have been achieved either by reducing the crystallinity of the polymers or by lowering their T_g. Polymer blending and polymer nanocomposite preparation are found to be the most feasible among the techniques [43]. In the polymer blend electrolytes, polymers having complementary properties are used, such as one polymer have good electrochemical properties, and the other must be with good mechanical properties. The ionic conductivity and electrochemical/physical properties of polymer blend electrolytes are superior to that of the individual constituent polymers. In the blending, two or more different polymers are blend together to form a uniform mixture. Hence, the advantageous physical and electrochemical properties of different polymers can be making use in development of polymer gel electrolyte having improved properties. For example, PAN is blended with PMMA, the resulting polymer blend electrolytes of the CN groups in PAN could interact with CO groups of the liquid electrolytes such as propylene carbonate (PC), ethylene carbonate (EC), etc. as well with lithium ions [41], while the amorphous structure of PMMA is beneficial to ionic conduction and chemical cross-linking of the PMMA matrix can remarkably increase the mechanical strength and retention ability of the electrolyte solution [44]. In other words, the right materials choice to produce the gel electrolyte or battery separator for LIB is essential to achieve the appropriate and desired electrolyte characteristics [45]. The performance of a separator in a lithium-ion battery is determined by some requirements such as porosity, chemical and thermal stability, electrical insulator, wettability, dimensional stability, and resistance to degradation by chemical reagents and electrolytes [30, 46–48].

The membrane properties such as porosity, tortuosity and uniformity of pore distribution are strongly dependent on its processing method. Different methods such as solvent casting [49], plasticizer extraction [50], phase inversion [51] and electrospinning [52] have been reported for the membrane preparation. Compared to cast membranes, it is reported that the GPE based on electrospun membranes shows higher porosity, electrolyte uptake and ionic conductivity due to the presence of fully interconnected pore structure [51]. Electrospinning is a simple and versatile method which is gaining importance in recent years as membranes prepared by employing this method have controlled properties. The electrospun polymer membranes consist of thin fibers of micron/submicron diameters with high specific surface area. The interlaying of fibers generates large porosity (> 80%) with fully interconnected pore structure that can function as efficient channels for ion conduction thereby facilitates easy transport of Li^+-ions and serves as good host matrices for GPEs [22]. Earlier electrochemical studies of PEs based on electrospun membranes of PVdF [53], PVdF-*co*-HFP [51, 52], PAN [54] and PVdF-*co*-HFP/PAN blend [23] have demonstrated their potential application in lithium-ion batteries.

This chapter explains how the polymer gel electrolytes composed of two or more polymers are advantageously utilizing the beneficial electrochemical and physical properties for the individual polymers enhance the battery performance of LIBs [55–57]. The chapter covers two types of methods of preparation: (i) blending of polymers (binary blends and ternary blends) and (iii) layer-by-layer assembly of the thin layers of different polymers (trilayer assembly).

Electrospinning is one such technique which has received great attention due to its versatility in spinning of ultrafine fibers with a wide range of polymeric fibers also due to its consistency with uniform morphology in producing fibers ranging from micrometers to nanorange with porous structure [58]. Electrolyte membrane prepared by electrospinning has been reported in many literatures [59]. These membranes have large porosity with large surface area and interconnected network structure and uniform morphology. Electrospun blend electrolytes have the advantages of greater porosity with uniformity which enables easy pathways for ionic conduction, with these advantages, electrospun membranes as a host matrix for electrolyte.

8.2 Polymer Blend Electrolytes

8.2.1 PVdF or PVdF-co-HFP-Based Blend Electrolytes

Polyvinylidene difluoride or polyvinylidene difluoride (PVdF) is a highly nonreactive thermoplastic fluoropolymer produced by the polymerization of vinylidene difluoride, having the chemical formula $-(C_2H_2F_2)_n-$. PVdF is a specialty plastic used in applications requiring the highest purity, as well as resistance to solvents, acids and hydrocarbons. Currently, PVdF resins are widely used in LIBs as a binder material, both in anode and cathodes. PVdF is electrochemically stable in contact with electrolyte mixtures and has good affinity to liquid electrolyte. Also PVdF requires NMP (*N*-methyl 2-pyrrolidone) as a solvent and offers the possibility of high-voltage operation. PVdF is also electrochemically stable in contact with electrolyte mixtures. Hence, it is well-studied polymer matrix for the preparation of polymer electrolyte, and Sony is commercialized. PVdF and its copolymers (polyvinylidene difluoride-*co*-trifluoroethylene), polyvinylidene difluoride-*co*-trifluoroethylene (PVdF-*co*-TrFE), polyvinylidene difluoride-*co*-hexafluoropropylene (PVdF-*co*-HFP) and polyvinylidene fluoride-*co*-chlorotrifluoroethylene (PVdF-*co*-CTFE) show exceptional properties and characteristics for the development of battery separators, highlighting high polarity, excellent thermal and mechanical properties, wettability by organic solvents, being chemically inert and stable in the cathodic environment, and possessing tailorable porosity through binary and ternary solvent/non-solvent systems [60, 61]. Both the PVdF and its copolymers are partially fluorinated semi-crystalline polymers

Table 8.2 Major properties of these PVdF-based polymers

Polymer	Melting temp. (°C)	Degree of crystallinity (%)	Young's modulus (MPa)	Dielectric constant
PVdF	~170	40–60	1500–3000	6–12
PVdF-*co*-TrFE	~120	20–30	1600–2200	18
PVdF-*co*-HFP	130–140	15–35	500–1000	11
PVdF-*co*-CTFE	~165	15–25	155–200	13

where the amorphous phase is located between the crystalline lamellae arranged in spherulites. It can crystallize in different crystalline phase, depending on the temperature and processing conditions [62, 63]. In relation to the crystalline phases of PVdF and its copolymers, the most important phases are the β-phase, since it presents ferroelectric, piezoelectric, and pyroelectric properties, and the α-phase, which is the most stable thermodynamically, when material is obtained directly from the melt book [62]. PVdF has a strong electron withdrawing fluorine atom in its back bone and high dielectric constant ($\varepsilon = 8.4$). Thus, the polymer blend effectively dissociates lithium salts to generate a large quantity of charge carriers for conduction [64, 65]. The major physical and electrical properties of these PVdF-based polymers are presented in Table 8.2 [55, 66–68].

8.2.1.1 Blend with PMMA

Polymethyl methacrylate (PMMA), a synthetic resin produced from the polymerization of methyl methacrylate having the chemical formula –($C_5O_2H_8$)–. PMMA is a transparent and rigid thermoplastic material often used as a substitute for glass in products such as lightweight or shatterproof windows, skylights, illuminated signs and aircraft canopies. PMMA is an economical alternative to polycarbonate (PC) when tensile strength, flexural strength, transparency, polishability and UV tolerance are more important than impact strength, chemical resistance and heat resistance. PMMA is one of the host polymers previously used in plasticized PEs, which was first reported by Iijima et al. and more recently by Bohnke et al. [69] and others Zhang et al. [22, 25] Appetecchi et al. [70] studied the kinetics and stability of the lithium electrode in PMMA-based gel electrolytes. PMMA is a common thermoplastic polymer with well-known chemistry. Its amorphous structure is beneficial to ionic conduction and PMMA-based gel electrolytes have shown excellent interfacial stability toward lithium metal. The pendant –$COOCH_3$ is not likely to crystallize around Li^+ as PEO does. PMMA-based PEs exhibit high electrolyte uptake, ionic conductivity and good electrochemical stability [71]. Unlike PAN, PMMA has ability to make chemical cross-linking, which will remarkably increase the mechanical strength and the electrolyte solution retention ability of the PE [72, 73]. Also it has been reported that gel electrolytes based on cross-linked PMMA can suppress lithium dendrite formation [72]. However, they suffer from a gel-like mechanical property: They do not form a free-standing film at high plasticizer content or the electrospun PMMA fibers are very brittle. Thus, an improvement in the mechanical properties of PMMA-based PEs is required to make them suitable for commercial applications. A blend of PMMA with PVdF [74, 75] and PVdF-*co*-HFP [76]-based electrospun polymer gel electrolytes are prepared by activating the fibrous membrane with 1 M $LiClO_4$ in PC (propylene carbonate) [74] or 1 M $LiPF_6$ (lithium hexafluorophosphate) in EC:DMC (1:1 v/v) (ethylene carbonate and dimethyl carbonate) solution [75, 76] were reported.

A blend of PVdF-PMMA (80:20 wt./wt.) electrospun fiber showed the fiber diameter of 183 [75] and 325 nm [74]. The higher fiber diameter observed for the same PVdF-PMMA blend is due to the higher spinning voltage and higher feed rate. They showed an electrolyte uptake of 275 (1 M $LiPF_6$ in EC:DMC) [75] and 285% (1 M $LiClO_4$-PC) [74] even both the membranes have similar porosity of about 85%. The membrane activated with 1 M $LiClO_4$-PC is showed higher ionic conductivity (2.95 $\times 10^{-3}$ S cm^{-1}) at RT [74]. This difference in uptake may come from the difference in affinity of the membrane to the different electrolyte, thereby shows the difference in ionic conductivity. The average fiber diameter of PVdF and PVdF–PMMA (50:50) is found to be 647 and 179 nm (Fig. 8.3). PVdF–PMMA (50:50) polymer electrolyte membrane showed an electrolyte uptake 290% (porosity 87%) and ionic conductivity of 0.15 S cm^{-1} and at room temperature, while the pristine PVdF membrane showed lower porosity (78%), electrolyte uptake (260%) and ionic conductivity (0.1 S cm^{-1}) (Table 8.3). The Li/PE/$LiFePO_4$ cells with 1 M $LiPF_6$ in EC: DMC as electrolyte delivered an initial discharge capacity of 138.4, 144.7 and 150.3 mAh g^{-1} for separators of PVdF, PVdF/PMMA (80:20) and PVdF/PMMA (50:50) membranes, respectively. After 50 cycles, the cell with PVdF–PMMA (50:50) electrolyte retained a discharge capacity of 140 mAh g^{-1} which is about around 93% retention of the initial discharge capacity, while the pristine PVdF membrane retained a discharge capacity of only 130 mAh g^{-1} (~94% retention). Nanofibrous PVdF–PMMA (50:50) polymer electrolyte membrane was found to be a potential separator for lithium-ion batteries. Compared with cells having different PEs, discharged capacity for the cell with PVdF–PMMA (50:50) separator was found to be relatively high cathode utilization corresponding to 88.4% of theoretical capacity which is 3.5% higher than that of PVdF–PMMA (80:20) polymer blend electrolyte [75].

The PVdF-*co*-HFP blend with PMMA could be expected to have higher ionic conductivity than that with PVdF, due to the lower crystallinity of PVdF-*co*-HFP result from the presence of HFP in the copolymer. The PMMA blend polymer electrolyte having ~66% PVdF-*co*-HFP shows an ionic conductivity of 2 mS cm^{-1} [76], while the blend having only 50% PVdF showed an ionic conductivity of only 0.15 S cm^{-1} [75].

The fiber diameter of pristine PVdF-*co*-HFP nanofibers was found to be 100–250 nm, and the surface of the fibers was very smooth due to its homogeneous polymeric texture. The fibers are randomly oriented, and most of the fibers were not interconnected. However, by blending with PMMA (about 33%), the average diameter of the fibers increased (200–350 nm) and the nanofibers were frequently

Fig. 8.3 SEM images on the surface morphology of electrospun polymer blend nanofibrous membranes of **a**, **d** pristine PVdF, **b** PVdF–PMMA (80:20), **c** PVdF–PMMA (50:50) Adapted and reproduced from Ref. [75]. Copyright 2018 Springer

Table 8.3 Melting enthalpy (ΔH_f), crystallinity (χ_c), bulk resistance (R_b) and ionic conductivity (σ) of PVdF, PVdF/PMMA (80:20) and PVdF/PMMA (50:50) nanofibrous membranes [75]

Polymer membranes	ΔH_f (J/g)	χ_c (%)	R_b (Ω)	σ (S/cm)
PVdF	53.29	50.75	0.50	0.10
PVdF:PMMA (80:20)	43.84	41.75	0.29	0.13
PVdF:PMMA (50:50)	25.33	24.12	0.22	0.15

interconnected by fusing the attaching point of two or three individual fibers. Furthermore, in comparison with the pristine PVdF-*co*-HFP nanofibers, the surface of the composite nanofibers became rough, which will be beneficial for getting better interfacial contact with electrode and lowering the interfacial resistance between the electrode and electrolyte. The PVdF-*co*-HFP/PMMA blend polymer membrane

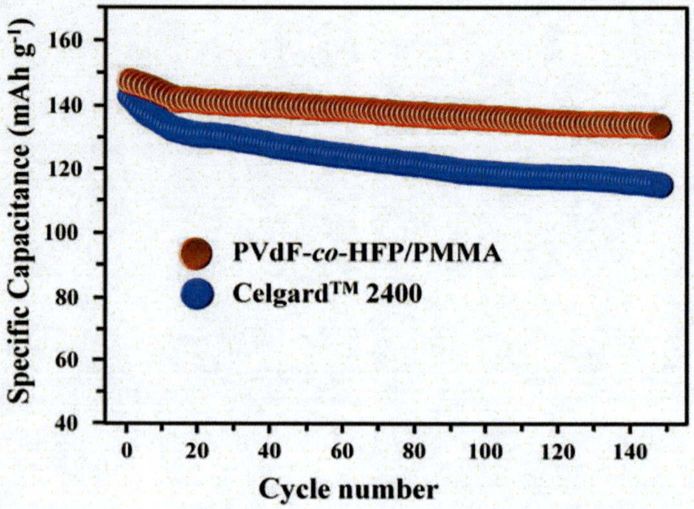

Fig. 8.4 Cycle performance of the prototype cells. Li/PE/$LiFePO_4$ prototype cells with cutoff voltages of 2.4 and 4.1 V at 0.1*C*-rate. Adapted and reproduced from Ref. [76]. Copyright 2009 Elsevier

showed an uptake and leakage for electrolyte solution of 377% and 87% which is about 75% and 11% higher than pristine PVdF-*co*-HFP membrane, respectively. The Li/PE/$LiFePO_4$ cell with PVdF-*co*-HFP/PMMA blend polymer electrolyte or commercial Celgard TM 2400® shows comparable initial discharge capacities ~145 mAh g^{-1}; however, the cells with electrospun PVdF-*co*-HFP/PMMA electrolyte showed a stable discharge behavior and little capacity loss under constant current conditions, which still retained a capacity of 133.5 mAh g^{-1} (92% retention) after 150 cycles. However, the cell with Celgard TM 2400® showed a remarkable capacity fading and the capacity decreased to 115 mAh g^{-1} (79% retention) after 150 cycles (Fig. 8.4). The improved cycle performance of the prototype cells with electrospun PVdF-*co*-HFP/PMMA could be ascribed to the lower leakage in the sandwich structure of the cell and higher porosity [76]. The crystallinity of both PVdF and PVdF-*co*-HFP has been depressed with increasing PMMA content, because the motion of PVdF or PVdF-*co*-HFP chains during the crystallization was hindered by the bulky CH_3OCO—group in PMMA chains. Moreover, the crystallinity has been depressed by the hydrogen bonds between the PMMA and PVdF or PVdF-*co*-HFP chains. Lowering the crystallinity leads to higher ionic conductivity [75, 76].

8.2.1.2 Blend with PEO

Polyethylene oxide (PEO) is also known as PEG was the first example of a single-chain single crystals having the chemical formula H–(O–CH_2–CH_2)$_n$–OH. $C_{2n}H_{4n+2}O_{n+1}$ PEO-based polymer electrolytes have received extensive attention for their potential usage as alternative materials for traditional organic liquid electrolytes (OLEs) since Fenton et al. [77] found that the complex of PEO with alkaline salts possesses good ionic conductivity. Followed by this report, (PEO)-based electrolyte has been widely studied as all solid polymer electrolytes in LIBs. PEO has low glass transition temperature (T_g), good chain flexibility, thermal properties, mechanical properties, superior electrochemical stability to lithium metal, and excellent solubility with conductive lithium salts are poised to be an enabler for solid-state lithium batteries, but their application is restricted by low room-temperature ionic conductivity and poor mechanical strength at elevated temperatures [78, 79]. It has ether linkages, with oxygen atoms present at a suitable inter-atomic separation to allow segmental motion of the polymeric chain which is beneficial for facile ionic conduction [80–82]. It is widely accepted that PEO can form complexes with a large number of lithium salts and transport Li^+ -ions in PEs. Thus, significant ion conduction occurs in the amorphous phase where the conductivity is two or three orders of magnitude higher than in thc crystalline region [83]. The high crystallinity of PEO leads to low ion conductivity (10-8 $\times$ 10^{-6} S cm^{-1}) and inferior Li^+ transference numbers (0.2–0.3) at room temperature, which seriously affects the high rate capability of LIBs. This necessitates operation at higher temperatures (generally, >70 °C) for their successful utilization in practical battery applications The Li^+-ion transport in these all solid-state polymer electrolytes has been associated with the local relaxation and segmental motion of the amorphous regions in the PEO chains. Ethylene oxide (EO) segment in the PEO is soluble in common battery electrolytes like ethylene carbonate (EC) and diethyl carbonate (DEC), which limits its use in dimensionally stable PGEs, therefore, blending of PEO with other polymers are the widely adopted method the preparation of PEO-based polymer gel electrolytes [83, 84]. Prasanth et al. reported electrospun-based polymer gel electrolyte prepared by blending of PEO with PVdF [7] and the polymer blend incorporated with ceramic filler [17]. The aim of the study was to develop polymer gel electrolyte (PGE) which utilizes the advantageous electrochemical properties of PEO to make them for the practical application in LIBs and study the effect of PEO on ionic conductivity and electrochemical properties of PVdF-based PGEs. The presence of PVdF in the polymer blend retains the mechanical properties of the membrane, while the ether linkages with oxygen atoms in PEO help to improve the segmental motion and dissociation of lithium salts. The surface morphology of PVdF/PEO blend membranes is displayed in Fig. 8.5.

The PVdF/PEO blend (5–20 wt% PEO content) electrospun membranes with [17] and without [7] nanosized ceramic filler (LAGP) were prepared first. Then, the blend polymer membrane incorporated with nanosized LAGP is immersed in the boiling water for the preferential isolated and removed removal of PEO from the blend fiber [17]. The surface morphology of electrospun membrane before and after in situ porosity generation (Fig. 8.6) and the Fig. 8.7 shows the surface morphology of the PVdF/PEO blend membranes before and after soaking in the hot water. The immersing of membrane in hot water (70 °C) for 48 h, the PEO partially leach out from the micron-sized fibers. PEO is water soluble, while PVdF is not; therefore, PEO was preferentially isolated and removed from the membranes, leaves nanopores

Fig. 8.5 FE-SEM images on the surface morphology of electrospun membrane; **a** pristine PVdF, **b** PVdF/PEO (90:10) blend; magnified image of fiber; **c** pristine PVdF, **d** PVdF/PEO (90:10) polymer blend fiber. Adapted and reproduced from Ref. [7]. Copyright 2014 Elsevier

uniformly distributed on the surface of the nanofibers. The pore density was directly related to the PEO content in the polymer blend. Porosity is one of the key factors which determine the ionic conductivity and electrochemical performance of polymer electrolytes based on electrospun membranes in lithium-ion batteries [9]. Higher porosity can enlarge the contact area between the polymer and liquid electrolyte, which ensures that the electrolyte well retained in the polymer membrane leads to

Fig. 8.6 Surface morphology of electrospun PVdF/PEO polymer belnd with membranes with lithium aluminum germanium phosphate (LAGP) glass ceramic lithium ion conducting filler (PVdF:LAGP, 94:6 wt%) having x wt% PEO on the total weight of PVdF and LAGP: before **a** ESM-01 (x=5), **c** ESM-02 (x=10), **e** ESM-03 (x=20) and after **b** IPG-01 (x=5), **d** IPG-02 (x=10) and **f** IPG-03 (x=20) soaking in hot water. (ESM-as spun membrane, IPG-in-situ porosity generated membrane). Adapted and reproduced from [17]. Copyright 2014 Elsevier

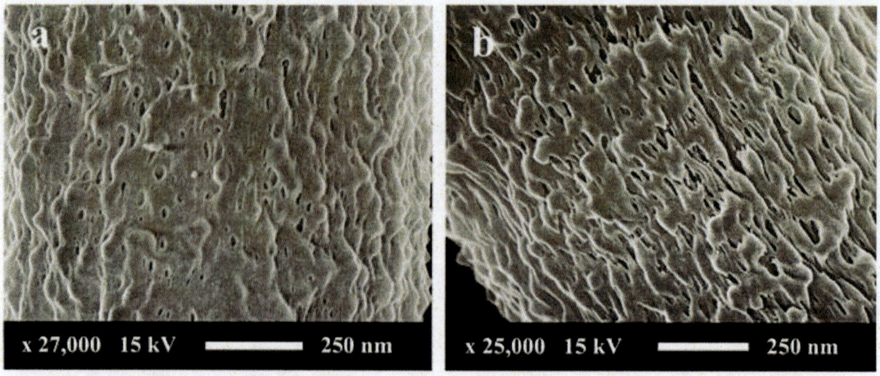

Fig. 8.7 Magnified images of surface morphology of electrospun PVdF/PEO polymer blend fibers with lithium aluminum germanium phosphate (LAGP) glass ceramic lithium ion conducting filler (PVdF:LAGP, 94:6 wt%) having x wt% PEO on the total weight of PVdF and LAGP; **a** IPG-02 (x = 10) and **b** IPG-03 (x = 20). (IPG-in-situ porosity generated membrane). Adapted and reproduced from [17]. Copyright 2014 Elsevier

the long term cycling stability of LIBs. The effect of PEO on the pore formation and electrochemical properties of the blend electrolyte is shown in Table 8.4.

The blend polymer electrolytes showed higher charge–discharge and cycling stability compared to pristine PVdF electrolyte or commercial Celgard® separator. The lithium cell assembled with Li metal/$LiFePO_4$ electrode and PVdF/PEO blend (10% PEO) PGE based on LiTFSI or $LiClO_4$ delivers an initial discharge capacity of 165–168 mAh g^{-1}, which corresponds to w 98 ± 1% utilization of the active material. While the cell with PGE based on electrospun PVdF showed a discharge capacity of 148 mAh g^{-1}, which is about 11% less than the cell with PGE based on PVdF/PEO polymer blend membrane at a current density of 0.1C. After 50 cycles, the cell comprised of PVdF membrane activated with LiTFSI and $LiClO_4$ delivers a discharge capacity of ~128 mAh g^{-1}. The capacity retention ratio of the cell with PVdF/PEO polymer blend membrane activated with LiTFSI electrolyte is 95%, and $LiClO_4$ is 89% (calculated based on initial discharge capacity), while the cell with PVdF membrane shows only w86% retention. [7]. However, the PVdF/PEO blend electrolyte, from which PEO is leached out, showed lower initial specific capacity compared to the one without leaching process. The initial capacity was found to be about 130 to 150 mAh g^{-1} with LiTFSI, where the membrane prepared with higher PEO content delivered higher capacity. After 50 cycles, the cells delivered a discharge capacity of 126, 137 and 150 mAh g^{-1} respectively, for the PVdF/PEO blend electrolyte prepared with 5, 10 and 20 wt% PEO and soaked in hot water for the preferential leaching. The rate capability studies reported that at low C-rates (0.1–0.6 C), both the cell with conventional Celgard® or PVdF/PEO blend electrolyte membranes delivered similar capacities, but at higher C-rates (1.5–3.0 C), the blend electrolyte membrane shows considerably higher capacity compared to the conventional Celgard® and retain similar capacity after the current density is changed back

Table 8.4 Properties of polymer electrolyte membranes activated with different lithium salt in electrolytes (electrolyte: 1 M solution of lithium salt in1:1 w/w EC/DEC)

Membrane	Lithium salt in EC/DEC	Porosity (%)	Electrolyteuptake (%)	Crystallinity (%)	Ionic conductivity at 30 °C (mS cm^{-1})
PVdF/PEO (90:10)	LiTFSI	85	750	42.3	4.9
PVdF/PEO (90:10)	$LiPF_6$	85	690	42.3	4.5
PVdF/PEO (90:10)	$LiClO_4$	85	675	42.3	4.2
PVdF/PEO (90:10)	$LiBF_4$	85	630	42.3	2.8
PVdF/PEO (90:10)	$LiCF_3SO_3$	85	590	42.3	1.8
PVdF/PEO (95:05) + 6 wt% LAGP	LiTFSI	78	632	39.1	6.3
PVdF/PEO (90:10) + 6 wt% LAGP	LiTFSI	80	652	40.4	7.0
PVdF/PEO (80:20) + 6 wt% LAGP	LiTFSI	83	703	42.6	7.7
PVdF/PEO (95:05) + 6 wt% LAGP (after leaching)	LiTFSI	85	723	36.8	8.0
PVdF/PEO (90:10) + 6 wt% LAGP (after leaching)	LiTFSI	89	768	38.2	9.1
PVdF/PEO (80:20) + 6 wt% LAGP (after leaching)	LiTFSI	93	831	38.8	10.9
					at 35 °C
PVdF	$LiPF_6$	80	650	40.5	3.2
PVdF/PEO	$LiPF_6$	85	690	42.3	5.3
Celgard (2320)	$LiPF_6$	41	200		2.1×10^{-2}

to 0.2 C-rate after 40 cycles, which clearly demonstrates high capacity retention of the cells [17]. Again by blending PVdF with PEO, the LSV of the blend electrolyte is >5 V, which is considerably higher than PVdF-co-HFP-based electrospun polymer gel electrolytes (4.5 V) [13, 14, 51] Also the interfacial resistance with Li/Li metal was found to be 10–22 Ω [7, 17], which is much lower than that of 750–800 Ω for PVdF-*co*-HFP-based electrospun polymer gel electrolytes [11, 14].

8.2.1.3 Blend with Polydiphenylamine (PDPA)

Polydiphenylamine (PDPA), a polymer of *N*-substituted aniline preparing by oxidative polymerization [85, 86]. It is more soluble in common organic solvents and exhibits different redox characteristics than other poly(*N*-substituted anilines) [87, 88]. The backbone units of PDPA can be grafted with other polymeric chains to have novel functional properties [89] To utilize these unique properties, Gopalan et al. [90] blended PDPA with PVdF or its copolymer PVdF-*co*-HFP for the preparation of polymer gel electrolyte. Electrospun membranes blended PDPA/PVdF or PDPA/PVdF-*co*-HFP having different thickness ranging from 20 to 60 μm were prepared by electrospinning. The PDPA content in the blend was 0.5–2 wt%. The electrospun membranes were transformed to pole electrolyte by activating with 1 M $LiClO_4$–PC solution. The electrospun polymer blend membrane exhibited morphological variations with PDPA content Fig. 8.8. Electrospun pristine PVdF fibrous membranes have a nearly straightened and tubular structure with an average fiber diameter of ~500 nm. The PDPA/PVdF has interconnected multifibrous layers with ultrafine porous structures. The presence of PDPA significantly affected average fiber diameter. The average fiber diameter decreased with increasing PDPA content and the fibers appeared to be uniform in composition without having any phase separated microstructure or beads morphology reveals the miscibility and compatibility of PVdF and PDPA. By blending PVdF with 0.5 wt% PDPA, the average fiber diameter is down to 200 nm. There is inter-fiber twisting in the PVdF/PDPA

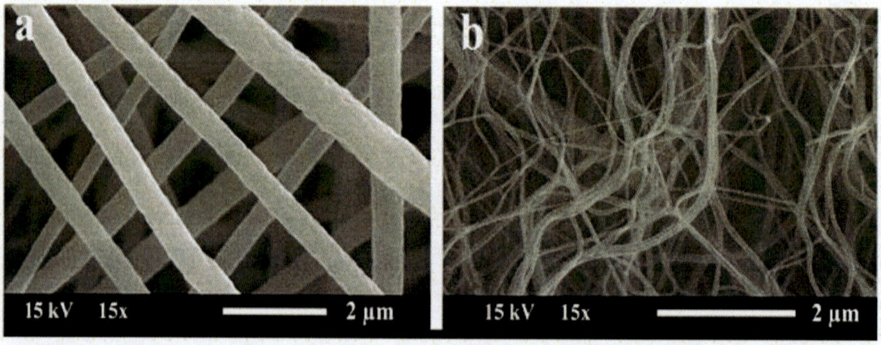

Fig. 8.8 FE-SEM images on the surafec morphology of electrospun **a** PVdF, **b** PVdF–PDPA-CFM with PDPA (wt%) **a.** 0.5 and **b.** 1 Adapted and reproduced from Ref. [90]. Copyright 2008 Elsevier

that generates microcavities and entanglements between the fibers are more with higher PDPA content. The difference morphology and fiber diameter between the pristine and PDPA/PVdF blend membranes can be possible due to the difference in the spinning parameters (distance between the nozzle and the collector, applied voltage and feed rate); however, all these parameters are kept constant in the study. Hence, the following factors may be influenced on the morphological difference in the membranes [90];

(i) Difference in viscosity of the spinning solutions (the incorporation of PDPA can increase the viscosity of the polymer solution due to the presence of aromatic ring in the polymer)
(ii) Difference in dielectric constant of the spinning solutions (the incorporation of PDPA can increase the dielectric of the medium for electrospinning)

Because of higher dielectric solution (in comparison to a lesser dielectric pristine PVdF solution) could easy to form a Taylor cone at the tip of the spinning needle of the syringe which result in the formation of smooth thin fibers without any beads [15]. The formation of interconnected and entangled fibrous morphology of the PDPA/PVdF blend membrane could be arise from the probable intermolecular interactions between the protonated amine or imine groups in PDPA [7, 8] and electronegative fluorine atoms in PVdF.

The electrolyte uptake, ionic conductivity, lithium transference number, anodic stability (>5 V) and interfacial properties are increased with PDPA content. PVdF/PDPA with 2 wt% PDPA showed an electrolyte uptake >280 wt% which is 80% higher than pristine PVdF membrane. The uptake value is considerably higher than reported values for PVdF, [20–24] but significantly less compared to the uptake for PVDF or PVdF-*co*-HFP membranes reported by Prasanth et al. [11, 14]. The polymer blend electrolyte with 2% PDPA (60-μm thickness) exhibits an ionic conductivity of 3.6 mS cm^{-1} at 25 °C, which is higher than reported value for PVdF-based electrolyte [91–95] and the lithium transference number (t_+ value) of 0.48. The reported t_+ values for SPEs range from 0.06 to 0.2 [96]. For gel polymer systems, t_+ values in the order of 0.4–0.5 have been reported. Cationic transport number ranging between 0.2 and 0.4 has been reported for PVdF-*co*-HFP-based electrolyte membrane [97–100].

The higher ionic conductivity for PVdF/PDPA fibrous electrolyte is attributed to arise from the combined influence of higher content of electrolyte (>280 wt%) incorporated into the pores of the membrane and augmented lithium-ion mobility in the membrane. Doped PDPA has positively charged nitrogen sites (protonated diphenoquinone diamine units) [87, 88]. That have molecular-level interactions with the electronegative fluorine atoms present in PVdF. This environment provides (i) new path for Li^+-ion migration in the composite, (ii) interconnected network morphology between PDPA and PVdF and (iii) compact porous structure to hold more amount of the liquid electrolyte. The recombination possibility of Li^+-ion with ClO_4^- ion is expected to be hindered in the presence of PDPA, and hence, facile Li^+-ion mobility becomes possible. A portion of the ClO_4^- ions is expected to be immobilized at the protonated amine or imine sites of PDPA by replacing the organic dopant, naphthalene sulfonate anions. The bulky naphthalene sulfonate anion has lesser mobility than ClO_4^- ion. As a result, Li^+-ions in the PVdF/PDPA membrane matrix can freely move to result higher ionic conductivity. In the pouch cell, the charge–discharge studies of PVdF blend electrolyte with 1 and 2 wt% PDPA content, in the full cell configuration using graphite anode and $LiCoO_2$ cathode under different C-rate shows similar discharge capacity in lower C-rate (0.2–0.5 C), but at the higher C-rate (at a C-rate higher than 0.5 C) the cell with blend electrolyte having higher PDPA content delivers higher discharge capacity which is due to the high ionic conductivity and highly porous morphology of PVdF/PDPA electrospun membrane. Typically, at 1 C-rate, the cell can deliver about 90% of its 0.2 C capacity, and even the C-rate is doubled (at a high rate of 2 C), and the cell can still deliver about ~70% of its 0.2 C capacity. Also, the cell delivered about 95% of its initial capacity after 50 continuous charge–discharge cycles at 1 C (25 °C).

8.2.1.4 Blend with Polyvinyl Chloride (PVC)

Polyvinyl chloride having the chemical formula $(C_2H_3Cl)_n$ is the world's third-most widely produced synthetic plastic, after polyethylene and polypropylene, which is used for the preparation of commercial battery separator Celgard®. About 40 million tones are produced per year. The presence of chlorine in the polymer backbone of PVC makes them as a self-extinguishing polymer. PVC is an inexpensive material has low dielectric constant ($\varepsilon \approx 3$), good processability, which can act as good mechanical stiffener. In the recent years, PVC-based PEs gained much popularity due to their good compatibility with many polymers and easiness to prepare a hard brittle plastic to a rubbery material by simply plasticizing with suitable plasticizer [71, 101, 102]. Many literatures [103–105] showed that PVdF and PVC can be blended as the PEs. The addition of PVC was used to suppress the crystallinity, thereby enhancing the ionic conductivity, to reduce the thermal stability [102] due to the decomposition of PVC in the membrane. The decomposition temperature of PVC is about 285 °C which is significantly lower than that of PVdF (~450 °C) [102].

PVC-based gel polymer electrolyte, blend with PVdF, PAN, poly (butyl methacrylate) [PBMA], PEMA, ionic liquids, etc., are reported either cast films or electrospun-based membranes. The PEs based on PVdF/PVC prepared by casting technology have been earlier reported by Rajendran et al. [103] and Muniyandi et al. [105]. Their results demonstrated that the PEs prepared by casting technology had a low ionic conductivity and poor cycling behavior. Thus, nanofibrous membranes based on PVdF/PVC (8:2, w/w) were prepared by electrospinning and then activated with 1 M $LiClO_4$ in PC/EC (1:1 v/v) to transform polymer gel electrolytes. The electrolyte uptake and ionic conductivity studies revealed that the blending of PVdF with PVC is beneficial to enhance the polymer gel electrolyte characteristics. Both the electrolyte uptake and ionic conductivity of the blend electrolyte increase with the PVC content. The higher ionic conductivity of the blend electrolyte is directly related to the higher electrolyte uptake and lowering of crystallinity by the PVC phase in the blend polymer gel electrolyte. By blending the PVdF with 20 wt% of PVC reduces about 22% crystallinity of PVdF (38.4%), which leads to the formation of more gelled phase in the blend polymer electrolyte. The lower fiber diameter also enhances the gelled phase in the electrolyte. Even though both the membranes identical morphology having fully interconnected porous structure composed of ultrafine straight fibers, the average fiber diameter (AFD) and fiber diameter distribution are significantly different (Fig. 8.9). The diameters of PVdF nanofibers are ranging from 257 to 1380 nm (AFD is 932 nm) (Fig. 8.9a) for the pristine PVdF membrane, while that is in the range of 385–875 nm (AFD is 624 nm) (Fig. 8.9b) for the PVdF/PVC blend membrane [106]. The considerably lower average fiber diameter could offer higher surface area which beneficial for the absorption of large quantity of liquid electrolyte and the formation of more gelled phase in the polymer electrolyte. The lower ionic conductivity of pristine PVdF-based electrospun gel electrolyte compared to PVdF/PVC blend counterpart can be attributed to the slower conduction path in the swollen polymer phase in the nanofiber due to the higher crystallinity. Hence, the

Fig. 8.9 SEM photographs on the surface morphology of **a** pristine PVdF, **b** PVdF-PVC (8:2) membrane, Adapted and reproduced from Ref. [107], Copyright 2012 Elsevier PAN–PVC polymer membrane, **c** pristine PAN, **d** PAN: PVC (8:2), Adapted and reproduced from Ref. [106], Copyright 2012 Springer

lower crystallinity of the blend polymer electrolyte promotes the mobility of the Li^+-ion, thereby lowering the activation energy for the l Li^+-ion transportation in the electrolyte [21]. The PVdF/PVC polymer blend gel electrolyte displayed high ionic conductivity up to 2.25 mS cm^{-1}, which is about 35% higher than that of while pristine PVdF membranes at 25 °C. The ionic conductivity of pristine PVdF and PVC electrospun gel electrolyte was found to be 1.47 and 0.66 mS cm^{-1} at 25 °C. The Li/PE/$LiFePO_4$ charge–discharge studies showed about 85.4% cathode unitization (145 mAh g^{-1}) in the first cycle at a current density of 0.1 C with a Columbic efficiency of (ratio between charge and discharge capacity) is 98.6%. After 50 cycles, the discharge capacity of the cell retained a discharge capacity of 130.8 mAh g^{-1} which is still 90.1% of the initial discharge capacity, which is an indication of good cycling stability and good interfacial stability between the electrode and the PVdF/PVC blend polymer membranes [90]. These results suggest that the nanofibrous polymer gel electrolytes based on PVdF/PVC have great potential in polymer lithium-ion batteries [107]. Blending of PAN with PVC results the three-dimensional network structure (Fig. 8.9c, d) that able to absorb large amounts of liquid electrolyte. The average fiber diameter (AFD) of the electrospun fibrous membranes is 890, 1,100, respectively, that contains PAN and PVC in the ratio 10:0 and 8:2. This variation in AFD is attributed by the difference in viscosity of the polymer solution and its rheological properties. Even though the ionic conductivity of this blend is lower than that of pure PAN, the mechanical strength is observed to be increased, that make it for the practical use. However, the blend film with PAN: PVC (8:2) exhibits an ionic conductivity of 1.05×10^{-3} S cm^{-1} which is apt to use in lithium-ion batteries [106].

8.2.1.5 Blend with Polypropylene Carbonate (PPC)

Polypropylene carbonate (PPC) having a chemical formula $[CH_9CH_3)CH_2OCO_2]_n$ has recently received considerable attention. Since this unique polymer is synthesizing by copolymerizing carbon dioxide (CO_2) and propylene oxide (PO), it can effectively reduce CO_2 emissions and mitigate the greenhouse effect [108–112]. Aside from this, it is a completely biodegradable and a good substrate material for the preparation of environmentally friendly polymer electrolytes [111, 112]. PPC-based gel electrolytes are widely tested in LIBs. PPC has similar structure to carbonate-based solvents, which is commonly applied in conventional gel electrolytes as a plasticizer, and carbonate solvent for dissolving the lithium salt [35] suggests that it could offer good compatibility as well as interfacial contact with commonly using battery electrodes. In addition, the polar ester group in the polymer backbone can effectively trap and store liquid electrolyte, thereby showing high electrolyte absorption, ionic conductivity and lithium transference number. The higher lithium transference number results from the presence of amorphous aliphatic polycarbonate with low glass transition temperature (T_g) that leads to the local relaxation and segmental motion of PPC chain, which is favorable for the Li^+-ions transportation at lower activation energy. In addition, the polar ester groups facilitate PPC chains to a solvent gelator owing to their strong interactions with the common organic liquid electrolyte. Even though PPC shows high electrolyte uptake, the low mechanical properties and poor thermal stability of pristine PPC-based polymer gel electrolytes restrict its successful application in LIBs. To overcome this problem and make their practical use in LIBs, the most straightforward approach is to modify the PPC to achieve enough mechanical strength and thermal stability while retaining a high volume fraction of the conductive amorphous phase and the outstanding electrolyte uptake. Among different methods such as structural modification [30], incorporation of reinforcing nanoceramic fillers [11], blending with mechanically strong and dimensionally stable polymers having good thermal stability, is the simple and effective method to polymer gel electrolyte which have superior electrochemical properties and charge–discharge cycling stability even at high C-rate [32, 113]. PPC blended with different polymers such as poly(lactic acid) (PLLA) [114, 115]. poly (3-hydroxybutrate) (PHB), triethyl citrate (TEC) [115] thermoplastic polyurethane (TPU) [113] and PVdF [113, 116, 117] for different applications such as mechanical applications[114] water purification, [115] battery electrolyte, [113, 116, 118] binder for electrodes in LIBs [117], etc. Electrospun PPC/PVdF binary blend [116] and PPC/PVdF/TPU tri-polymer blend [113] fibrous membrane activated with 1 M $LiPF_6$ in EC/DMC (1:1, v/v) are reported as polymer gel electrolyte for LIBs. As the PVdF has very stable polar fluorine atoms (-CF groups) in their backbone and high dielectric constant ($\varepsilon = 8.4$) [7], blending of PPC with PVdF is beneficial in dissociating more lithium salt and conducting lithium ions in the polymer electrolytes, thereby improving the mobility and transference number of Li^+-ions. In addition, the presence of PVdF in the polymer blend imparts the mechanical integrity, dimensional and thermal stability of the electrospun membranes which again make the membrane free-standing and easy to handle. Given the high crystallinity of PVdF, blending with the amorphous PPC is an effective way to reduce the crystallinity and facilitate Li^+-ion transportation in the 3D polymeric framework. The effect of the ratio of PPC to PVdF (0 to 40 wt% PPC) [116] or PVdF/TPU blend (50:50 blend of PPC/PVdF with 12.5 wt% TPU) [113] on the properties of the highly porous membrane including morphology, porosity, liquid uptake capability, thermal stability, ionic conductivity and electrochemical behavior was systematically studied and reported. Table 8.5 summarizes the physical, thermal and electrochemical properties of these binary and ternary polymer blend electrolytes. It can be found that the incorporation of TPU in the PVdF/PPC blend has pronounced effect on ionic conductivity thereby Li^+-ion transportation. The ternary blend shows higher ionic conductivity than PVdF/PPC, and it shows higher ionic conductivity at 25 °C [113] than the ionic conductivity of PVdF/PPC blend at higher temperature (30 °C) [116]. The evaluation of blend polymer gel electrolytes in Li/$LiFePO_4$ comprising 1 M $LiPF_6$ in EC/DMC (1:1, v/v) the PVdF/PPC blend electrolyte showed higher charge–discharge cycling stability and rate capability. The PVdF/PPC binary blend is having 20 wt% PPC delivered an initial discharge capacity of 156 mAh g^{-1} (92% of theoretical capacity of $LiFePO_4$), which is about 4.5% higher than the pristine PVdF fibrous gel electrolyte. In comparison with Celgard®, the PVdF or PVdF/PPC blend electrolyte delivered 2.5 to 7.6 higher capacities at a current density of 0.1 A g^{-1} at room temperature. After 100 cycles, the capacity of pristine PVdF polymer gel electrolyte and Celgard® is faded

Table 8.5 Physical, thermal and electrochemical properties of these binary and ternary blend electrolytes

Sample	ΔH_m (J/g)	Crystallinity, X_c (%)	Porosity(%)	Electrolyte uptake(wt%)	Ionic conductivity (mS cm^{-1})	
Celgard			42	147	0.66	at 30 °C
PVdF/PPC (100:0)	21.4	20.4	77	422	2.11	
PVdF/PPC (90:10)	20	19.1	80	492	3.87	
PVdF/PPC (80:20)	16.3	15.6	85	501	4.05	
PVdF/PPC (70:30)	15.4	14.7	85.4	471	3.15	
PVdF/PPC(60:40)	11.7	11.2	87	450	2.43	
PVdF/TPU/PPC (9 wt%)	18.22	17.4	78.2	335	2.17	at 25 °C
PVdF/TPU/PPC (10 wt%)	17.67	16.9	87	379	3.27	
PVdF/TPU/PPC (11 wt%)	16.92	16.2	88	422	4.27	
PVdF/TPU/PPC (12 wt%)	14.55	13.9	85.5	449	5.32	
PVdF/TPU/PPC (13 wt%)	19.30	18.4	76	319	2.13	

to 93 and 90% of its initial capacity, but there was no capacity fade observed for PBDF/PPC binary blend (100th cycle capacity is 157 mAh g^{-1}). The rate capability studies at a current density of 0.1–2C showed the capacity is decreasing with current rate and when the current rate was reduced back to 0.2 C from 2C, after 25 cycles, a reversible capacity of 149 mAh g^{-1} was obtained, showing the good rate capability of PVdF, PPC binary blend electrolyte [32, 116].

TPU is a polymer with good mechanical properties. It is commonly used to improve the elongation at break of other materials. It is a typical block polymer that contains two phases in the molecule [119]. The hard segment provides support function, so that it has good mechanical properties, such as tensile strength, wear resistance, high elasticity, etc. [120]. The soft segment can dissolve the cations in the lithium salts and facilitate the transportation of ions. When it is mixed with PVdF, the amino group (–NH) in its molecular chain can form hydrogen bonds with fluorine atoms, which makes the two polymers better compatible. The ternary blend consist of PVdF/PPC/TPU exhibits a three-dimensional network with randomly oriented nanofibers which enhance the porous structure (Fig. 8.10). When the concentration of ternary PVdF/PPC/TPU polymer blend increases from 9 to 13%, the initial charge–discharge capacity of the battery is increased to peak value from ~158.3 to

Fig. 8.10 Surface morphology of differnt concentration (wt%) of electrospun PVdF/TPU/PPC blend membranes **a** 9, **b** 10, **c** 11, **d** 13. Adapted and reproduced from Ref [113]. Copyright 2019 Springer

~165.8 mAh g^{-1} with concentration, however concentrate on of the blend reached to 13% the peak value decayed to 157.5 mAh g^{-1} at a current density of 0.1 C. The charge–discharge capacity values are concurrent with ionic conductivity of the ternary blends electrolyte, and the entire tri-polymer blend delivered a capacity corresponding to more than 90% of the theoretical capacity of $LiFePO_4$. After 50 continuous charge–discharges cycling, all the cell retained about 95–97% of its initial capacity suggesting the good cycling stability. Again, the rate capability studies on the PVdF/PPC/TPU-based blend polymer electrolyte PE (12 wt%) at different rates display good charge–discharge specific capacity and Coulombic efficiency at the current densities sweep between 0.1 and 1.0 C. At a current density of 1 C, the cell delivered a specific capacity of about 142 mAh g^{-1}, which corresponds to the 83.5% utilization of active material, after 200 cycle (Fig. 8.11) [113]. The capacity retention of these PVdF/PPC/TPU tri-polymer blends' polymer electrolyte is significantly higher than that of PVdF/PPC binary blends, which delivered a maximum capacity of about 105 mAh g^{-1} at 1 C-rate after 25 cycles [32]. The discharge capacities are the capacity retention percentages are 95.47%, 92.4%, 90.4% and 83.3%, respectively.

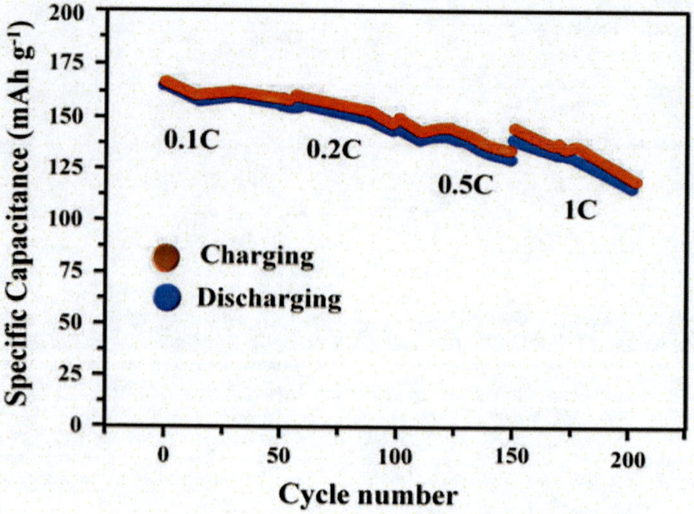

Fig. 8.11 Cycle performance and Coulombic efficiency diagram of the button battery with PVDF/TPU/PPC (12 wt%) based GPE at different rates. Adapted and reproduced from Ref [113]. Copyright 2019 Springer

The surface morphology of the electrospun PVdF and PVdF/PPC membranes with different weight ratios is shown in Fig. 8.12. The electrospun membranes result a three-dimensional network which formed by the random orientated nanofibers, which results in a porous structure. These porous structures result an increase the specific surface area that will results an enhanced electrolyte absorption and gelatinization efficiently as it soaked in electrolyte solution. The porous structure enables numerous ionic transport channels and that will provide a path way for fast Li^+-ions transport. The AFD measured for the membranes is between 300 and 850 nm, and the AFD appears to decrease with increase in pPC to PVdF ratio.

Figure 8.13 shows the SEM images of the surface for the pristine PVdF and PVdF/PPC binary blend electrolyte having 20 wt% PPC membranes after the 100th charge/discharge cycle. The appearance of pristine PVdF and PVdF/PPC binary was well retained without appreciable signs of mechanical stress, maintaining the original three-dimensional network and hierarchically mesoporous structure. Specifically, the electrospun membranes with a porous structure readily relax the strain originated from mechanical stress during cycling. The electrospun membranes that offer ionic transport channels and provide a pathway for fast Li^+ ions transportation enable one to firmly maintain the shape of network and nanofibrous structure. The SEM images of the PVdF/PPC nanofibers show the AFD of about 700 nm, similar to the pristine electrospun membrane before cycling, as shown in Fig. 8.13b. By contrast, the VGE-100 nanofibers show volume expansion after cycling, which can be attributed to the presence of the swollen polymer chains [116].

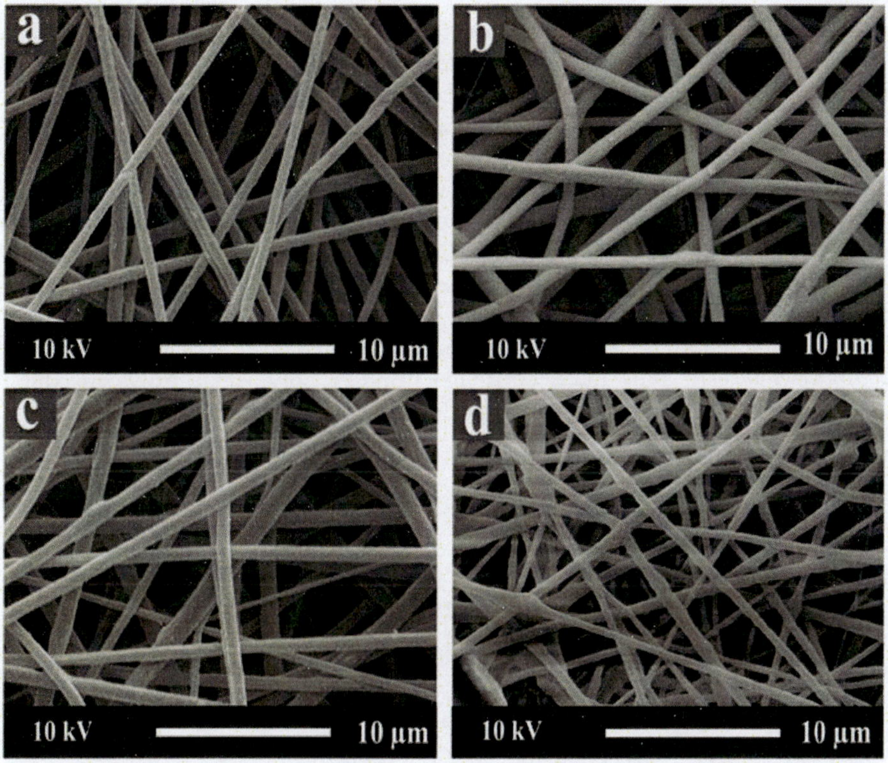

Fig. 8.12 SEM images on the surface morphology of the electrospun **a**. pristine PVdF, PVDF/PPC polymer blend membranes **b** 90:10; **c** 80:20; **d** 70:30. Adapted and reproduced from Ref. [116]. Copyright 2015 American Chemical Society

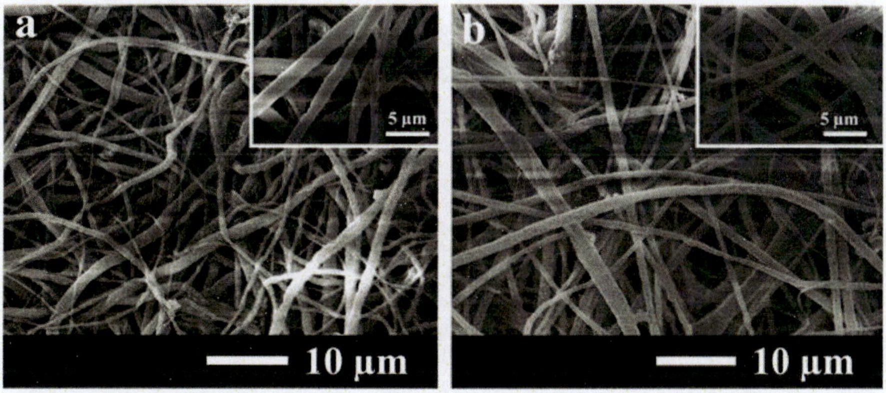

Fig. 8.13 Typical SEM images on the surface morphology of polymer membrane after activating with liquid electyte **a** VGE-100 (pristine PVdF) and **b** CGE-20 (PVdF/PPC (80:20)) after 100 cycles. Adapted and reproduced from Ref. [116]. Copyright 2015 American Chemical Society

The results show that our developed route has several distinct advantages: (1) The electrospun membranes present a three-dimensional network porous structure, which can greatly enlarge the specific surface area, promote the adsorption electrolyte and gelatinize efficiently; (2) the segregation of PVdF chains by polymeric chains of PPC can reduce the crystallization of PVdF-based polymers and increase the segmental mobility of the polymer, which benefit for the transport of lithium ions; (3) the formation of $Li^{+}\cdots(\delta-)$ F-C $(\delta+)$ and $PF_6\cdots(\delta+)$ C = O $(\delta-)$ complexes can separate the Li^+ and PF_6^- anions, which prevent the reconnection between Li^+ and PF_6^- anions simultaneously and create more free Li^+ (Fig. 8.14); thus, the ability of lithium-ion transference is improved. Therefore, the large surface area, well-developed microporous structure, sufficient electrolyte uptake, low crystallinity and appropriate porosity allow the electrospun PVdF/PPC polymer electrolyte to exhibit a significantly higher ionic conductivity and excellent electrochemical performances [116].

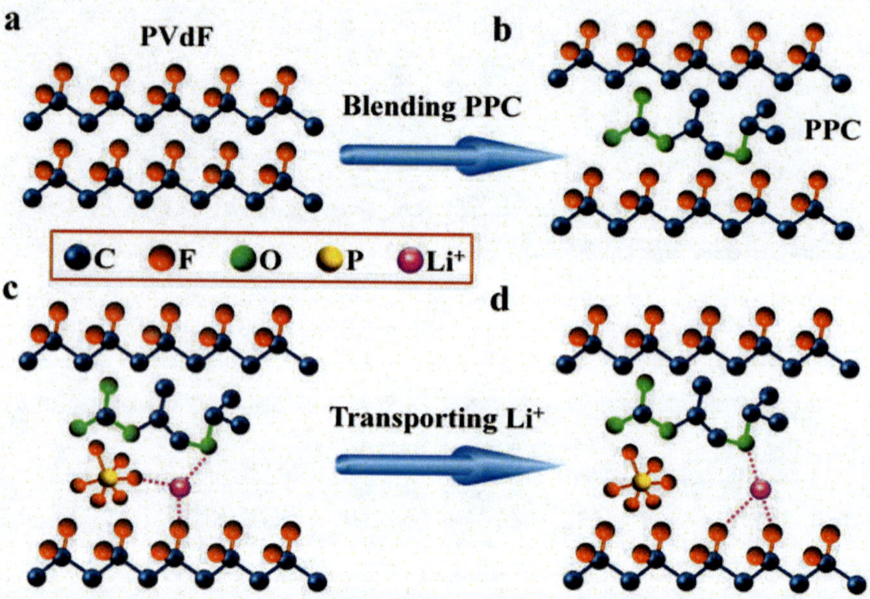

Fig. 8.14 Conceptual illustration of the polymer frameworks of **a** PVdF and **b** PVdF/PPC. Schematic illustration of main interaction forms between the ions and the polymer in the GPE. **c** Interaction of Li^+ between PF_6^- and the polymer chains; **d** transport of Li^+ -ions was associated with the ether linkages in the PPC and the fluorine atom in the PVdF. Adapted and reproduced from Ref. [116]. Copyright 2015 American Chemical Society

8.3 Conclusion

Electrospinning technique was widely used in energy storage devices such as batteries, supercapacitors. Electrospinning is one of the effective and attractive techniques for the preparation of fibrous polymer membranes having high surface area and porosity. Interconnected fibrous polymer electrolytes can be obtained by these techniques which exhibit high liquid electrolyte uptake leading to the high ionic conductivity. Different polymer electrolytes were used for the synthesis of electrolytes in which PVdF grabbed the attention as a favorable polymer matrix because of its high dielectric constant, high electrochemical stability, chemical compatibility, etc. The major drawback of this matrix is the hindrance of the migration of the doping salt since PVdF has a crystalline part offering low ionic conductivity. Blending other polymers along with PVdF can hinder the crystalline nature and pave an easy migration of lithium doping salt. PVdF blend polymer electrolytes prepared by this electrospinning technique show higher ionic conductivity.

References

1. Manthiram A (2017) An outlook on lithium ion battery technology. ACS Cent Sci 3:1063–1069. https://doi.org/10.1021/acscentsci.7b00288
2. Prize N History. In: Nobel media. https://www.nobelpeaceprize.org/History
3. Prasanth R, Shubha N, Hng HH, Srinivasan M (2013) Effect of nano-clay on ionic conductivity and electrochemical properties of poly(vinylidene fluoride) based nanocomposite porous polymer membranes and their application as polymer electrolyte in lithium ion batteries. Eur Polym J 49:307–318. https://doi.org/10.1016/j.eurpolymj.2012.10.033
4. Chiang C-Y, Jaipal Reddy M, Chu PP (2004) Nano-tube TiO_2 composite PVdF/$LiPF_6$ solid membranes. Solid State Ionics 175:631–635. https://doi.org/10.1016/j.ssi.2003.12.039
5. Wang Y-J, Kim D (2007) Crystallinity, morphology, mechanical properties and conductivity study of in situ formed PVdF/$LiClO_4$/TiO_2 nanocomposite polymer electrolytes. Electrochim Acta 52:3181–3189. https://doi.org/10.1016/j.electacta.2006.09.070
6. Gentili V, Panero S, Reale P, Scrosati B (2007) Composite gel-type polymer electrolytes for advanced, rechargeable lithium batteries. J Power Sources 170:185–190. https://doi.org/10.1016/j.jpowsour.2007.04.008
7. Prasanth R, Shubha N, Hng HH, Srinivasan M (2014) Effect of poly(ethylene oxide) on ionic conductivity and electrochemical properties of poly(vinylidenefluoride) based polymer gel electrolytes prepared by electrospinning for lithium ion batteries. J Power Sources 245:283–291. https://doi.org/10.1016/j.jpowsour.2013.05.178
8. Kim JR, Choi SW, Jo SM et al (2006) Characterization of electrospun PVdF fiber-based polymer electrolytes. Chem Mater 19:104–115. https://doi.org/10.1021/cm060223+
9. Shubha N, Prasanth R, Hoon HH, Srinivasan M (2013) Dual phase polymer gel electrolyte based on non-woven poly(vinylidenefluoride-co-hexafluoropropylene)-layered clay nanocomposite fibrous membranes for lithium ion batteries. Mater Res Bull 48:526–537. https://doi.org/10.1016/j.materresbull.2012.11.002
10. Raghavan P, Zhao X, Manuel J et al (2010) Electrochemical performance of electrospun poly(vinylidene fluoride-co-hexafluoropropylene)-based nanocomposite polymer electrolytes incorporating ceramic fillers and room temperature ionic liquid. Electrochim Acta 55:1347–1354. https://doi.org/10.1016/j.electacta.2009.05.025

11. Raghavan P, Zhao X, Kim JK et al (2008) Ionic conductivity and electrochemical properties of nanocomposite polymer electrolytes based on electrospun poly(vinylidene fluoride-co-hexafluoropropylene) with nano-sized ceramic fillers. Electrochim Acta 54:228–234. https://doi.org/10.1016/j.electacta.2008.08.007
12. Raghavan P, Zhao X, Choi H et al (2014) Electrochemical characterization of poly(vinylidene fluoride-co-hexafluoro propylene) based electrospun gel polymer electrolytes incorporating room temperature ionic liquids as green electrolytes for lithium batteries. Solid State Ionics 262:77–82. https://doi.org/10.1016/j.ssi.2013.10.044
13. Shubha N, Prasanth R, Hoon HH, Srinivasan M (2014) Plastic crystalline-semi crystalline polymer composite electrolyte based on non-woven poly(vinylidenefluoride-co-hexafluoropropylene) porous membranes for lithium ion batteries. Electrochim Acta 125:362–370. https://doi.org/10.1016/j.electacta.2014.01.024
14. Raghavan P, Choi J-W, Ahn J-H et al (2008) Novel electrospun poly(vinylidene fluoride-co-hexafluoropropylene)–in situ SiO_2 composite membrane-based polymer electrolyte for lithium batteries. J Power Sources 184:437–443. https://doi.org/10.1016/j.jpowsour.2008.03.027
15. Cheruvally G, Kim J-K, Choi J-W et al (2007) Electrospun polymer membrane activated with room temperature ionic liquid: novel polymer electrolytes for lithium batteries. J Power Sources 172:863–869. https://doi.org/10.1016/j.jpowsour.2007.07.057
16. Kim J-K, Cheruvally G, Li X et al (2008) Preparation and electrochemical characterization of electrospun, microporous membrane-based composite polymer electrolytes for lithium batteries. J Power Sources 178:815–820. https://doi.org/10.1016/j.jpowsour.2007.08.063
17. Shubha N, Prasanth R, Hng HH, Srinivasan M (2014) Study on effect of poly (ethylene oxide) addition and in-situ porosity generation on poly (vinylidene fluoride)-glass ceramic composite membranes for lithium polymer batteries. J Power Sources 267:48–57. https://doi.org/10.1016/j.jpowsour.2014.05.074
18. Kalyana Sundaram NT, Subramania A (2007) Nano-size $LiAlO_2$ ceramic filler incorporated porous PVDF-co-HFP electrolyte for lithium-ion battery applications. Electrochim Acta 52:4987–4993. https://doi.org/10.1016/j.electacta.2007.01.066
19. Jiang G, Maeda S, Yang H et al (2005) All solid-state lithium-polymer battery using poly(urethane acrylate)/nano-SiO_2 composite electrolytes. J Power Sources 141:143–148. https://doi.org/10.1016/j.jpowsour.2004.09.004
20. Raghavan P, Manuel J, Zhao X et al (2011) Preparation and electrochemical characterization of gel polymer electrolyte based on electrospun polyacrylonitrile nonwoven membranes for lithium batteries. J Power Sources 196:6742–6749. https://doi.org/10.1016/j.jpowsour.2010.10.089
21. Gopalan AI, Santhosh P, Manesh KM et al (2008) Development of electrospun PVdF-PAN membrane-based polymer electrolytes for lithium batteries. J Memb Sci 325:683–690. https://doi.org/10.1016/j.memsci.2008.08.047
22. Prasanth R, Aravindan V, Srinivasan M (2012) Novel polymer electrolyte based on cob-web electrospun multi component polymer blend of polyacrylonitrile/poly(methyl methacrylate)/polystyrene for lithium ion batteries—preparation and electrochemical characterization. J Power Sources 202:299–307. https://doi.org/10.1016/j.jpowsour.2011.11.057
23. Raghavan P, Zhao X, Shin C et al (2010) Preparation and electrochemical characterization of polymer electrolytes based on electrospun poly(vinylidene fluoride-co-hexafluoropropylene)/polyacrylonitrile blend/composite membranes for lithium batteries. J Power Sources 195:6088–6094. https://doi.org/10.1016/j.jpowsour.2009.11.098
24. Liao YH, Zhou DY, Rao MM et al (2009) Self-supported poly(methyl methacrylate–acrylonitrile–vinyl acetate)-based gel electrolyte for lithium ion battery. J Power Sources 189:139–144. https://doi.org/10.1016/j.jpowsour.2008.10.027
25. Lim D-H, Haridas AK, Figerez SP et al (2018) Tailor-made electrospun multilayer composite polymer electrolytes for high-performance lithium polymer batteries. J Nanosci Nanotechnol 18:6499–6505. https://doi.org/10.1166/jnn.2018.15689

26. Choi J-W, Cheruvally G, Kim Y-H et al (2007) Poly(ethylene oxide)-based polymer electrolyte incorporating room-temperature ionic liquid for lithium batteries. Solid State Ionics 178:1235–1241. https://doi.org/10.1016/j.ssi.2007.06.006
27. Caimi S, Klaue A, Wu H, Morbidelli M (2018) Effect of SiO_2 nanoparticles on the performance of PVdF-HFP/ionic liquid separator for lithium-ion batteries. Nanomaterials 8:926. https://doi.org/10.3390/nano8110926
28. Polu AR, Kumar R (2013) Effect of Al_2O_3 ceramic filler on peg-based composite polymer electrolytes for magnesium batteries. Adv Mater Lett 4:543–547. https://doi.org/10.5185/amlett.2012.9417
29. Wu N, Cao Q, Wang X et al (2011) In situ ceramic fillers of electrospun thermoplastic polyurethane/poly(vinylidene fluoride) based gel polymer electrolytes for Li-ion batteries. J Power Sources 196:9751–9756. https://doi.org/10.1016/j.jpowsour.2011.07.079
30. Arya A, Sharma AL (2017) Polymer electrolytes for lithium ion batteries: a critical study. Ionics (Kiel) 23:497–540. https://doi.org/10.1007/s11581-016-1908-6
31. Stephan AM (2006) Review on gel polymer electrolytes for lithium batteries. Eur Polym J 42:21–42. https://doi.org/10.1016/j.eurpolymj.2005.09.017
32. Liu Y, Peng X, Cao Q et al (2017) Gel polymer electrolyte based on poly(vinylidene fluoride)/thermoplastic polyurethane/polyacrylonitrile by the electrospinning technique. J Phys Chem C 121:19140–19146. https://doi.org/10.1021/acs.jpcc.7b03411
33. Liang YZ, Cheng SC, Zhao JM et al (2013) Preparation and characterization of electrospun PVDF/PMMA composite fibrous membranes-based separator for lithium-ion batteries. Adv Mater Res 750–752:1914–1918. https://doi.org/10.4028/www.scientific.net/amr.750-752.1914
34. Manjuladevi R, Thamilselvan M, Selvasekarapandian S et al (2017) Mg-ion conducting blend polymer electrolyte based on poly(vinyl alcohol)-poly (acrylonitrile) with magnesium perchlorate. Solid State Ionics 308:90–100. https://doi.org/10.1016/j.ssi.2017.06.002
35. Zhou D, Zhou R, Chen C et al (2013) Non-volatile polymer electrolyte based on poly(propylene carbonate), Ionic liquid, and lithium perchlorate for electrochromic devices. J Phys Chem B 117:7783–7789. https://doi.org/10.1021/jp4021678
36. Ruan L, Yao X, Chang Y et al (2018) Properties and applications of the β phase poly(vinylidene fluoride). Polymer (Guildf) 10:1–27. https://doi.org/10.3390/polym10030228
37. Abbrent S, Plestil J, Hlavata D et al (2001) Crystallinity and morphology of PVdF—HFP-based gel electrolytes. Polymer 42:1407–1416
38. Lewandowski A, Swiderska M (2013) Ionic liquids as electrolytes for Li-ion batteries—an overview of electrochemical studies. Aviat Psychol Appl Hum Factors 3:1–8. https://doi.org/10.1016/j.jpowsour.2009.06.089
39. Yu R, Bao JJ, Chen TT et al (2017) Solid polymer electrolyte based on thermoplastic polyurethane and its application in all-solid-state lithium ion batteries. Solid State Ionics 309:15–21. https://doi.org/10.1016/j.ssi.2017.06.013
40. In situ composite of nano SiO_2–P(VDF-HFP) porous polymer.pdf
41. Abraham KM, Alamgir M (1990) Li^+-conductive solid polymer electrolytes with liquid-like conductivity. J Electrochem Soc 137:1657–1658. https://doi.org/10.1149/1.2086749
42. Wang F, Li L, Yang X et al (2018) Influence of additives in a PVDF-based solid polymer electrolyte on conductivity and Li-ion battery performance. Sustain Energy Fuels 2:492–498. https://doi.org/10.1039/c7se00441a
43. Rajendran S, Mahendran O, Kannan R (2002) Characterisation of [(1 2 x) PMMA ± x PVdF] polymer blend electrolyte with Li 1 ion q. Fuel 81:1077–1081
44. Kim CS, Oh SM (2002) Spectroscopic and electrochemical studies of PMMA-based gel polymer electrolytes modified with interpenetrating networks. J Power Sources 109:98–104. https://doi.org/10.1016/S0378-7753(02)00055-1
45. Lu H, Sun S (2018) Polyimide electrode materials for Li-ion batteries via dispersion-corrected density functional theory. Comput Mater Sci 146:119–125. https://doi.org/10.1016/j.commatsci.2018.01.029

46. Song JY, Wang YY, Wan CC (1999) Review of gel-type polymer electrolytes for lithium-ion batteries. J Power Sources 77:183–197. https://doi.org/10.1016/S0378-7753(98)00193-1
47. Lee H, Yanilmaz M, Toprakci O et al (2014) A review of recent developments in membrane separators for rechargeable lithium-ion batteries. Energy Environ Sci 7:3857–3886. https://doi.org/10.1039/c4ee01432d
48. Costa CM, Rodrigues HM, Gören A et al (2017) Preparation of poly(vinylidene fluoride) lithium-ion battery separators and their compatibilization with ionic liquid—a green solvent approach. ChemistrySelect 2:5394–5402. https://doi.org/10.1002/slct.201701028
49. Wang Z, Tang Z (2003) Characterization of the polymer electrolyte based on the blend of poly(vinylidene fluoride-co-hexafluoropropylene) and poly(vinyl pyrrolidone) for lithium ion battery. Mater Chem Phys 82:16–20. https://doi.org/10.1016/S0254-0584(03)00241-4
50. Choi J, Kim J, Cheruvally G et al (2006) Microporous poly (vinylidene fluoride-co -hexafluoropropylene) polymer electrolytes for lithium/sulfur cells. J Ind Eng Chem 12:939–949
51. Raghavan P, Zhao X, Manuel J et al (2010) Electrochemical studies on polymer electrolytes based on poly(vinylidene fluoride-co-hexafluoropropylene) membranes prepared by electrospinning and phase inversion—a comparative study. Mater Res Bull 45:362–366. https://doi.org/10.1016/j.materresbull.2009.12.001
52. Li X, Cheruvally G, Kim J-K et al (2007) Polymer electrolytes based on an electrospun poly(vinylidene fluoride-co-hexafluoropropylene) membrane for lithium batteries. J Power Sources 167:491–498. https://doi.org/10.1016/j.jpowsour.2007.02.032
53. Kim JR, Choi SW, Jo SM et al (2004) Electrospun PVdF-based fibrous polymer electrolytes for lithium ion polymer batteries. Electrochim Acta 50:69–75. https://doi.org/10.1016/j.electacta.2004.07.014
54. Choi SW, Kim JR, Jo SM et al (2005) Electrochemical and spectroscopic properties of electrospun PAN-based fibrous polymer electrolytes. J Electrochem Soc 152:A989. https://doi.org/10.1149/1.1887166
55. Sousa RE, Nunes-Pereira J, Ferreira JCC et al (2014) Microstructural variations of poly(vinylidene fluoride co-hexafluoropropylene) and their influence on the thermal, dielectric and piezoelectric properties. Polym Test 40:245–255. https://doi.org/10.1016/j.polymertesting.2014.09.012
56. Ahmad Z, Al-Awadi NA, Al-Sagheer F (2007) Morphology, thermal stability and visco-elastic properties of polystyrene-poly(vinyl chloride) blends. Polym Degrad Stab 92:1025–1033. https://doi.org/10.1016/j.polymdegradstab.2007.02.016
57. Nunes-Pereira J, Costa CM, Lanceros-Méndez S (2015) Polymer composites and blends for battery separators: state of the art, challenges and future trends. J Power Sources 281:378–398
58. Yee WA, Nguyen AC, Lee PS et al (2008) Stress-induced structural changes in electrospun polyvinylidene difluoride nanofibers collected using a modified rotating disk. Polymer (Guildf) 49:4196–4203. https://doi.org/10.1016/j.polymer.2008.07.032
59. Bansal D, Meyer B, Salomon M (2008) Gelled membranes for Li and Li-ion batteries prepared by electrospinning. J Power Sources 178:848–851. https://doi.org/10.1016/j.jpowsour.2007.07.070
60. Kim JF, Jung JT, Wang HH et al (2016) Microporous PVDF membranes via thermally induced phase separation (TIPS) and stretching methods. J Memb Sci 509:94–104. https://doi.org/10.1016/j.memsci.2016.02.050
61. Ribeiro C, Costa CM, Correia DM et al (2018) Electroactive poly(vinylidene fluoride)-based structures for advanced applications. Nat Protoc 13:681–704. https://doi.org/10.1038/nprot.2017.157
62. Martins P, Lopes AC, Lanceros-Mendez S (2014) Electroactive phases of poly(vinylidene fluoride): determination, processing and applications. Prog Polym Sci 39:683–706. https://doi.org/10.1016/j.progpolymsci.2013.07.006
63. Edmonds EA, Acharya UR, Bonjour E, et al (2007) Founding editor, 193. https://doi.org/10.1016/S0950-7051(18)30220-X

64. Du Pasquier A, Warren PC, Culver D et al (2000) Plastic PVDF-HFP electrolyte laminates prepared by a phase-inversion process. Solid State Ionics 135:249–257. https://doi.org/10.1016/S0167-2738(00)00371-4
65. Li ZH, Xiao QZ, Zhang P et al (2008) Porous nanocomposite polymer electrolyte prepared by a non-solvent induced phase separation process. Funct Mater Lett 1:139–143. https://doi.org/10.1142/S1793604708000253
66. Costa CM, Silva MM, Lanceros-Méndez S (2013) Battery separators based on vinylidene fluoride (VDF) polymers and copolymers for lithium ion battery applications. RSC Adv 3:11404–11417. https://doi.org/10.1039/c3ra40732b
67. Sousa RE, Ferreira JCC, Costa CM et al (2015) Tailoring poly(vinylidene fluoride-co-chlorotrifluoroethylene) microstructure and physicochemical properties by exploring its binary phase diagram with dimethylformamide. J Polym Sci, Part B: Polym Phys 53:761–773. https://doi.org/10.1002/polb.23692
68. Costa CM, Rodrigues LC, Sencadas V et al (2012) Effect of degree of porosity on the properties of poly(vinylidene fluoride–trifluorethylene) for Li-ion battery separators. J Memb Sci 407–408:193–201. https://doi.org/10.1016/j.memsci.2012.03.044
69. Bohnke O, Frand G, Rezrazi M et al (1993) Fast ion transport in new lithium electrolytes gelled with PMMA. 1. Influence of polymer concentration. Solid State Ionics 66:97–104. https://doi.org/10.1016/0167-2738(93)90032-X
70. Appetecchi GB, Croce F, Scrosati B (1995) Kinetics and stability of the lithium electrode in poly(methylmethacrylate)-based gel electrolytes. Electrochim Acta 40:991–997. https://doi.org/10.1016/0013-4686(94)00345-2
71. Rhoo H-J, Kim H-T, Park J-K, Hwang T-S (1997) Ionic conduction in plasticized PVCPMMA blend polymer electrolytes. Electrochim Acta 42:1571–1579. https://doi.org/10.1016/S0013-4686(96)00318-0
72. Tatsuma T, Taguchi M, Oyama N (2001) Inhibition effect of covalently cross-linked gel electrolytes on lithium dendrite formation. Electrochim Acta 46:1201–1205. https://doi.org/10.1016/S0013-4686(00)00706-4
73. Gray FM, MacCallum JR, Vincent CA (1986) Poly(ethylene oxide)—$LiCF_3SO_3$—polystyrene electrolyte systems. Solid State Ionics 18–19:282–286. https://doi.org/10.1016/0167-2738(86)90127-X
74. Mahant YP, Kondawar SB, Bhute M, Nandanwar DV (2015) Electrospun poly (vinylidene fluoride)/poly (methyl methacrylate) composite nanofibers polymer electrolyte for batteries. Procedia Mater Sci 10:595–602. https://doi.org/10.1016/j.mspro.2015.06.011
75. Mahant YP, Kondawar SB, Nandanwar DV, Koinkar P (2018) Poly(methyl methacrylate) reinforced poly(vinylidene fluoride) composites electrospun nanofibrous polymer electrolytes as potential separator for lithium ion batteries. Mater Renew Sustain Energy 7:1–9. https://doi.org/10.1007/s40243-018-0115-y
76. Ding Y, Zhang P, Long Z et al (2009) The ionic conductivity and mechanical property of electrospun P(VdF-HFP)/PMMA membranes for lithium ion batteries. J Memb Sci 329:56–59. https://doi.org/10.1016/j.memsci.2008.12.024
77. Tominaga Y, Nanthana V, Tohyama D (2012) Ionic conduction in poly(ethylene carbonate)-based rubbery electrolytes including lithium salts. Polym J 44:1155–1158. https://doi.org/10.1038/pj.2012.97
78. Deng F, Wang X, He D et al (2015) Microporous polymer electrolyte based on PVDF/PEO star polymer blends for lithium ion batteries. J Memb Sci 491:82–89. https://doi.org/10.1016/j.memsci.2015.05.021
79. Polu AR, Rhee H-W (2015) Nanocomposite solid polymer electrolytes based on poly(ethylene oxide)/POSS-PEG ($n = 13.3$) hybrid nanoparticles for lithium ion batteries. J Ind Eng Chem 31:323–329. https://doi.org/10.1016/j.jiec.2015.07.005
80. Konnik OV, Shul'gin VF, Bekirova ZZ et al (2014) Coordination compounds of dysprosium(III) with 3-methyl-1-phenyl-4-formylpyrazol-5-one diacyldihydrazones. Russ J Inorg Chem 59:1237–1243. https://doi.org/10.1134/S0036023614110126

81. Appetecchi GB, Croce F, Hassoun J et al (2003) Hot-pressed, dry, composite, PEO-based electrolyte membranes: I. Ionic conductivity characterization. J Power Sources 114:105–112. https://doi.org/10.1016/S0378-7753(02)00543-8
82. Shin JH, Lim YT, Kim KW et al (2002) Effect of ball milling on structural and electrochemical properties of (PEO)nLiX (LiX = $LiCF_3SO_3$ and $LiBF_4$) polymer electrolytes. J Power Sources 107:103–109. https://doi.org/10.1016/S0378-7753(01)00990-9
83. Ramrakhiani M, Nogriya V (2012) Synthesis and characterization of zinc sulfide nanocrystals and zinc sulfide/polyvinyl alcohol nanocomposites for luminescence applications. Polym Process Charact 394:109–138. https://doi.org/10.1201/b13105
84. Shi J, Yang Y, Shao H (2018) Co-polymerization and blending based PEO/PMMA/P(VDF-HFP) gel polymer electrolyte for rechargeable lithium metal batteries. J Memb Sci 547:1–10. https://doi.org/10.1016/j.memsci.2017.10.033
85. Nagarajan S, Santhosh P, Sankarasubramanian M et al (2005) UV–vis spectroscopy for following the kinetics of homogeneous polymerization of diphenylamine in p-toluene sulphonic acid. Spectrochim Acta Part A Mol Biomol Spectrosc 62:420–430. https://doi.org/10.1016/j.saa.2005.01.010
86. Wen T-C, Chen J-B, Gopalan A (2002) Soluble and methane sulfonic acid doped poly(diphenylamine)—synthesis and characterization. Mater Lett 57:280–290. https://doi.org/10.1016/S0167-577X(02)00779-6
87. Orlov AV, Ozkan SZ, Karpacheva GP (2006) Oxidative polymerization of diphenylamine: a mechanistic study. Polym Sci Ser B 48:11–17. https://doi.org/10.1134/S1560090406010039
88. Santhosh P, Sankarasubramanian M, Thanneermalai M et al (2004) Electrochemical, spectroelectrochemical and spectroscopic evidences for copolymer formation between diphenylamine and m-toluidine. Mater Chem Phys 85:316–328. https://doi.org/10.1016/j.matchemphys.2004.01.021
89. Hua F, Ruckenstein E (2003) Water-soluble conducting poly (ethylene oxide)—grafted polydiphenylamine synthesis through a "Graft Onto" process. Macromolecules 36:9971–9978. https://doi.org/10.1021/ma030431c
90. Gopalan AI, Lee K-P, Manesh KM, Santhosh P (2008) Poly(vinylidene fluoride)–polydiphenylamine composite electrospun membrane as high-performance polymer electrolyte for lithium batteries. J Memb Sci 318:422–428. https://doi.org/10.1016/j.memsci.2008.03.007
91. Choi SW, Jo SM, Lee WS, Kim YR (2003) An electrospun poly(vinylidene fluoride) nanofibrous membrane and its battery applications. Adv Mater 15:2027–2032. https://doi.org/10.1002/adma.200304617
92. Chiang C-Y, Shen YJ, Reddy MJ, Chu PP (2003) Complexation of poly(vinylidene fluoride): LiPF6 solid polymer electrolyte with enhanced ion conduction in 'wet' form. J Power Sources 123:222–229. https://doi.org/10.1016/S0378-7753(03)00514-7
93. Panero S, Ciuffa F, D'Epifano A, Scrosati B (2003) New concepts for the development of lithium and proton conducting membranes. Electrochim Acta 48:2009–2014. https://doi.org/10.1016/S0013-4686(03)00179-8
94. Choi S-S, Lee YS, Joo CW et al (2004) Electrospun PVDF nanofiber web as polymer electrolyte or separator. Electrochim Acta 50:339–343. https://doi.org/10.1016/j.electacta.2004.03.057
95. Choe HS, Giaccai J, Alamgir M, Abraham KM (1995) Preparation and characterization of poly(vinyl sulfone)- and poly(vinylidene fluoride)-based electrolytes. Electrochim Acta 40:2289–2293. https://doi.org/10.1016/0013-4686(95)00180-M
96. Gozdz AS, Schmutz CN, Tarascon J-M, Warren PC (1995) Method of making an electrolyte activatable lithium-ion rechargeable battery cell
97. Quartarone E, Brusa M, Mustarelli P et al (1998) Preparation and characterization of fluorinated hybrid electrolytes. Electrochim Acta 44:677–681. https://doi.org/10.1016/S0013-4686(98)00182-0
98. Stolarska M, Niedzicki L, Borkowska R et al (2007) Structure, transport properties and interfacial stability of PVdF/HFP electrolytes containing modified inorganic filler. Electrochim Acta 53:1512–1517. https://doi.org/10.1016/j.electacta.2007.05.079

99. Dai H, Zawodzinski TA (1998) The dependence of lithium transference numbers on temperature, salt concentration and anion type in poly (vinylidene fluoride)–hexafluoropropylene copolymer-based gel electrolytes (Presented at the Electrochemical Society Symposium: Processes in Polymers and Polymers). J Electroanal Chem 459:111–119. https://doi.org/10.1016/S0022-0728(98)00373-8
100. Kataoka H, Saito Y, Miyazaki Y, Deki S (2002) Ionic mobilities of PVDF-based polymer gel electrolytes as studied by direct current NMR. Solid State Ionics 152–153:175–179. https://doi.org/10.1016/S0167-2738(02)00297-7
101. Ramesh S, Winie T, Arof AK (2007) Investigation of mechanical properties of polyvinyl chloride–polyethylene oxide (PVC–PEO) based polymer electrolytes for lithium polymer cells. Eur Polym J 43:1963–1968. https://doi.org/10.1016/j.eurpolymj.2007.02.006
102. Rajendran S, Shanker Babu R, Sivakumar P (2008) Investigations on PVC/PAN composite polymer electrolytes. J Memb Sci 315:67–73. https://doi.org/10.1016/j.memsci.2008.02.007
103. Rajendran S, Sivakumar P, Shanker R (2007) Studies on the salt concentration of a PVdF—PVC based polymer blend electrolyte. J Power Sources 164:815–821. https://doi.org/10.1016/j.jpowsour.2006.09.011
104. Vickraman P, Ramamurthy S (2006) A study on the blending effect of PVDF in the ionic transport mechanism of plasticized PVC–$LiBF_4$ polymer electrolyte. Mater Lett 60:3431–3436. https://doi.org/10.1016/j.matlet.2006.03.028
105. Muniyandi N, Kalaiselvi N, Periyasamy P et al (2001) Optimisation of PVdF-based polymer electrolytes. J Power Sources 96:14–19. https://doi.org/10.1016/S0378-7753(01)00562-6
106. Zhong Z, Cao Q, Jing B et al (2012) Novel electrospun PAN – PVC composite fibrous membranes as polymer electrolytes for polymer lithium-ion batteries. Ionics 18:853–859. https://doi.org/10.1007/s11581-012-0682-3
107. Zhong Z, Cao Q, Jing B et al (2012) Electrospun PVdF—PVC nanofibrous polymer electrolytes for polymer lithium-ion batteries. Mater Sci Eng B 177:86–91. https://doi.org/10.1016/j.mseb.2011.09.008
108. Zeng S, Wang S, Xiao M et al (2011) Preparation and properties of biodegradable blend containing poly (propylene carbonate) and starch acetate with different degrees of substitution. Carbohydr Polym 86:1260–1265. https://doi.org/10.1016/j.carbpol.2011.06.023
109. Du L, Qu B, Meng Y, Zhu Q (2006) Structural characterization and thermal and mechanical properties of poly(propylene carbonate)/MgAl-LDH exfoliation nanocomposite via solution intercalation. Compos Sci Technol 66:913–918. https://doi.org/10.1016/j.compscitech.2005.08.012
110. Sugimoto H, Inoue S (2004) Copolymerization of carbon dioxide and epoxide. J Polym Sci Part A Polym Chem 42:5561–5573. https://doi.org/10.1002/pola.20319
111. Li XH, Meng YZ, Wang SJ et al (2004) Completely biodegradable composites of poly(propylene carbonate) and short, lignocellulose fiber Hildegardia populifolia. J Polym Sci Part B Polym Phys 42:666–675. https://doi.org/10.1002/polb.10761
112. Shi X, Gan Z (2007) Preparation and characterization of poly(propylene carbonate)/montmorillonite nanocomposites by solution intercalation. Eur Polym J 43:4852–4858. https://doi.org/10.1016/j.eurpolymj.2007.09.024
113. Xu J, Liu Y, Cao Q et al (2019) A high-performance gel polymer electrolyte based on poly(vinylidene fluoride)/thermoplastic polyurethane/poly(propylene carbonate) for lithium-ion batteries. J Chem Sci 131:1–10. https://doi.org/10.1007/s12039-019-1627-4
114. Gao J, Bai H, Zhang Q et al (2012) Effect of homopolymer poly(vinyl acetate) on compatibility and mechanical properties of poly(propylene carbonate)/poly(lactic acid) blends. Express Polym Lett 6:860–870. https://doi.org/10.3144/expresspolymlett.2012.92
115. El-Hadi AM, Alamri HR (2018) The new generation from biomembrane with green technologies for wastewater treatment. Polymers (Basel) 10:1174. https://doi.org/10.3390/polym10101174
116. Huang X, Zeng S, Liu J et al (2015) High-performance electrospun poly(vinylidene fluoride)/poly(propylene carbonate) gel polymer electrolyte for lithium-ion batteries. J Phys Chem C 119:27882–27891. https://doi.org/10.1021/acs.jpcc.5b09130

117. Zheng M, Fu X, Wang Y et al (2018) Poly(vinylidene fluoride)-based blends as new binders for lithium-ion batteries. ChemElectroChem 5:2288–2294. https://doi.org/10.1002/celc.201800553
118. Li H, Zhang H, Liang Z-Y et al (2014) Preparation and properties of poly (vinylidene fluoride)/poly(dimethylsiloxane) graft (poly(propylene oxide)-block-poly(ethylene oxide)) blend porous separators and corresponding electrolytes. Electrochim Acta 116:413–420. https://doi.org/10.1016/j.electacta.2013.11.076
119. van Heumen JD, Stevens JR (1995) The role of lithium salts in the conductivity and phase morphology of a thermoplastic polyurethane. Macromolecules 28:4268–4277. https://doi.org/10.1021/ma00116a030
120. Wu N, Jing B, Cao Q, Wang X, Kuang H, Wang Q (2010) A novel electrospun TPU/PVdF porous fibrous polymer electrolyte for lithium ion batteries. J Appl Polym Sci 116:2658–2667. https://doi.org/10.1002/app

Chapter 9
Electrospun Nanostructured Iron Oxide Carbon Composites for High-Performance Lithium Ion Batteries

Neethu T. M. Balakrishnan, Akhila Das, N. S. Jishnu, M. A. Krishnan, Sabu Thomas, M. J. Jabeen Fatima, Jou-Hyeon Ahn, and Raghavan Prasanth

9.1 Introduction

In the current global scenario, energy and environment play key role for the socio-economic development and long-term healthy survival of human beings. The constantly increasing energy demand associated with the growth of the global population and the rapidly changing living standards of human beings, coupled to the need for decreasing greenhouse gas emissions to limit the environmental pollution, global

N. T. M. Balakrishnan · A. Das · M. J. Jabeen Fatima (✉) · R. Prasanth
Department of Polymer Science and Rubber Technology, Cochin University of Science and Technology (CUSAT), Cochin 682022, India
e-mail: jabeen@cusat.ac.in

N. S. Jishnu
Rubber Technology Centre, Indian Institute of Technology-Kharagpur (IIT-KGP), Kharagpur, West Bengal 721302, India

M. A. Krishnan
Department of Mechanical Engineering, Amrita Vishwa Vidyapeetham, Amritapuri, India

Department of Electrical Engineering, Pennsylvania State University, Electrical Engineering East Building, 311, City College, PA 16802, USA

S. Thomas
School of Chemical Sciences, Mahatma Gandhi University, Athirampuzha, Kerala 686560, India

J.-H. Ahn (✉) · R. Prasanth (✉)
Department of Materials Engineering and Convergence Technology, Gyeongsang National University, 501, Jinju-daero, Jinju 52828, Republic of Korea
e-mail: jhahn@gnu.ac.kr

R. Prasanth
e-mail: prasanth@cusat.ac.in

N. T. M. Balakrishnan and R. Prasanth (eds.), *Electrospinning for Advanced Energy Storage Applications*, Materials Horizons: From Nature to Nanomaterials,
https://doi.org/10.1007/978-981-15-8844-0_9

warming, climate changes and rapid depletion of fossil fuels, has strongly encouraged the utilization of clean and renewable energy sources such as solar, wind or tidal energy. However, stochastic and discontinuous nature of these sources demands the use of efficient energy storage systems for its effective utilization. Progressively increasing serious environmental pollution and carbon emission, battery technologists have seen expanding their research activity to develop an ultimate solution for clean energy sources and efficient energy conversion and storage techniques. The development of cost-effective, efficient and large-scale energy storage devices for the expected upcoming growth of renewable energy production is one of the key challenges of the current millennium. Among different energy storage devices based on electrochemical methods such as supercapacitors, batteries or fuel cells, rechargeable batteries gained much attention due to its higher energy density moderately good power density and compact size, suitable to meet the different needs of power supply. There are different types of secondary batteries such a nickel-cadmium, nickel-metal hydride, lead-acid or lithium ion batteries which are in use, but among them, lithium ion batteries (LIBs), which deliver 1.7 GW [1], are gained much attention by the battery technologist and electronic gadgets industry and became the most commonly used potential energy conversion and storage devices. Lead-acid batteries have a relatively stable charge/discharge state; however, due to the heavyweight, large volume and higher maintenance, they are far from the portable and light electric devices [2]. Ni-Cd battery is one of the most rugged and enduring batteries, earlier used in mobile phones, cameras and other portable electronic devices which had a high-specific capacity, long service life, high discharge current, safely operate even in extreme temperatures and the only chemistry that allows ultra-fast charging with minimal stress, but it suffers from severe memory effects leading to shortened battery life coupled with significant toxicity issues [3, 4]. Hence, due to environmental concerns, Ni-Cd is being replaced with other chemistries, but it retains its status in aircraft due to its good safety record. Another battery type, Ni-MH, serves as a replacement for Ni-Cd to find applications in medical instruments, portable devices, hybrid cars, musical instruments, etc. Ni-MH, as it has only mild toxic metals and provides higher-specific energy, longer lifetime but has serious leakage problems [5]. Therefore, the lithium ion battery was pioneered in the field with great potential.

In the last decade, LIBs have been extensively investigated for a wide range of applications including information/telecommunication technology, portable electronic gadgets, electric/hybrid vehicles, aeronautical, space, etc. Endowed with unique properties such as high energy density, long cycle life, small size, lightweight, minimal memory effects, good shelf life, negligible self-discharge and low pollution, LIBs have been recognized as the most attractive electrochemical device to store electrical power in the future heavy duty applications. However, the relatively low power density, short battery life and poor rate capability still LIBs impede the further development for indispensable electronic gadgets and electric/hybrid vehicles. The ever-demanding popularity of EVs and HEVs studies about improving the properties of LIBs is necessary and critical. Among the different components of the battery, anode plays a crucial role in the lithium ion battery as the physical and electrochemical characteristics of the anode determine battery behavior and are directly influence the battery performance.

Some unprecedented characteristics of anodic materials admittedly provide convenience for their application as active anode materials, but inherent defects in materials hinder their optimized utilization. Hence, in addition to the choice of material and method of preparation, appropriate architectural modification and/or design, morphology and size is essential for better battery performance. Currently in commercial battery, the most popularly using anode materials are one-dimensional (1D), two-dimensional (2D), three-dimensional (3D) carbon-based materials as well as porous and core–shell structures. The breakthrough occurred in 1983, when Rachid Yazami demonstrated the Li metal anode could be replaced by a graphite-based carbon material in which lithium was reversibly intercalated and deintercalated [6]. Carbon-based materials are preferred as anode material for long years due to their abundance, low cost, easy processing and good electrochemical performance. The carbon materials have outstanding stability and protect the lithium from dendrite formation, while recharging the battery, a serious issue, hence usually, in commercial Li-ion batteries graphite is using as the anode material. Again the carbon materials are available in various forms, such as graphite, hard and soft carbons, carbon fibers, carbon nanotubes and graphene. However, their theoretical maximum capacity is limited to 372 mAh g^{-1}, corresponding to the formation of LiC_6 [7]. It also suffer from poor cycling behavior, a consequence of the easy movement of the graphene planes along the a-axis direction during the intercalation and de-intercalation of lithium. In addition to the problem of solvated lithium intercalation this issue ruled this material out of practical lithium ion battery applications. On the other hand, silicon anodes with higher theoretical capacity, which is about ten times the capacity of carbon anodes, have attracted considerable research attention; however, the severe volumetric change is a great issue, which is a common drawback in alloy anodes. Transition metal oxide anodes are the choice; however, the low Coulombic efficiency limits their application in LIBs. To overcome the disadvantages of different materials, the researchers combined different anodic materials such as alloys, transition metal oxides or silicon with carbon complexes to prepare composite anodes, which is found to be an effective and promising way to improve electrochemical behavior.

9.2 Anode Materials in Lithium Ion Batteries

The primary functional components of a lithium ion battery are the positive electrode (cathode), negative electrode (anode) and a polymeric microporous separator film which protect these electrodes from short-circuiting and an ion-conductive electrolyte. The schematic of structure of LIBs is displayed in Fig. 9.1. The positive electrode (where reduction happens during discharge) is a metal oxide such as $LiCoO_2$ and $LiFePO_4$, and the electrolyte is a lithium salt such $LiPF_6$ and $LiClO_4$ in an organic solvent such as ethylene carbonate (EC), dimethyl carbonate (DMC) and propylene

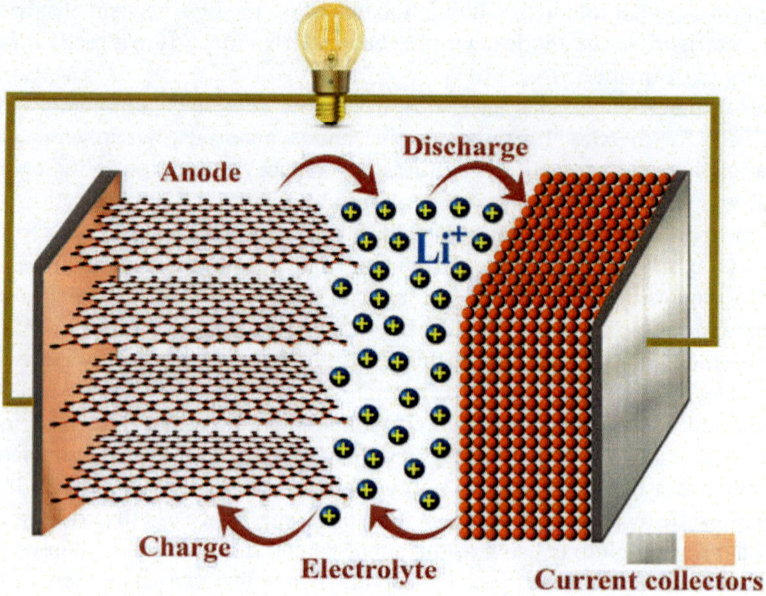

Fig. 9.1 Schematic representation and operating principle of lithium ion batteries

carbonate (PC). Generally, the negative electrode (where oxidation happens during discharge) of a conventional lithium ion cell is made from carbon without any safety issues [5, 6]. Numerous alternative materials including silicon [7, 8], transition metal oxides such as Fe_2O_3 [9, 10], Fe_3O_4 [11, 12] and TiO_2 [13, 14, 15] and post-transition metal oxide SnO_2 have been widely used for improving anode performance. As a crucial component of the lithium secondary battery, the anode significantly influences the performance of the battery. Currently, in most of the commercially available LIBs, the anodes are made up of graphite due to its special hierarchical structure. When the Li^+ ions insert into the anode material (lithiation), the relatively wide interspace between two adjacent carbon layers facilitates enough Li^+-ion insertion locations, thus avoiding the size, structure and shape variations of the anode during the charge–discharge cycling process [16–19]. Apart from the conventional lithiation mode, there are also other novel electrochemical mechanisms for lithium interactions such as the displacement reaction in alloy anodes and the redox reaction in transition metal oxide anodes [20, 21]. Based on the electrochemical reaction mechanism, the anode materials are broadly classified as carbon-based anodes, alloy anodes, transition metal oxide anodes and silicon anodes. The major properties and the pros and cause are displayed in Table 9.1. The battery performance not only hinges on the inherent properties of the anode material including the physical and/or chemical properties and energy storage capacity, but also depends on the crystallinity or amorphous structure of the anode material as well as the shape, size and component state.

Table 9.1 Most common anode materials used for lithium ion batteries

Active anode material	Theoretical capacity (mAh g^{-1})	Advantages	Common issues	Reference
Insertion/de-insertion materials				
A. Carbonaceous		Good working potential Low cost good safety	Low Coulombic efficiency High voltage hysteresis High irreversible capacity	
(a) Hard carbons	200–600			[22–24]
(b) CNTS	1116			[25–28]
(c) Graphene	780/1116			[29]
B. Titanium oxides		Extreme safety Good cycle life low cost High power capability	Very low capacity Low energy density	
(a) $LiTi_4O_5$	175			[30]
(b) TiO_2	330			[30]
Alloy/de-alloy materials		Higher-specific capacities High energy density good safety	Large irreversible capacity Huge capacity fading poor cycling	
(a) Silicon	4212	High-specific capacity Good cycle life high-rate capability	Poor capacity retention High volume change during cycling Thicker SEI formation	[31]
(b) Germanium	1624	High power density High electrical conductivity Fast lithium diffusion	Low specific capacity Expensive high volume Change during cycling	[32, 33]
(c) Tin	993	High theoretical capacity	Suffer from electrode degradation	[34]

(continued)

Table 9.1 (continued)

Active anode material	Theoretical capacity (mAh g^{-1})	Advantages	Common issues	Reference
(d) Antimony	660	High theoretical capacity Low reaction potential Small volume change during cycling	Low cycle stability poor rate performance	[35]
(e) Tin oxide	790	High volumetric capacity high cycle life prevent lithium dendrite formation	Suffer from sever gassing	[36]
(f) SiO_2	1600	Low cost easy to synthesis High theoretical capacity	Low intrinsic electrical conductivity Drastic volume fluctuation During lithiation/delithiation Low initial Coulombic efficiency (ICE)	[37]
Conversion materials				
(a) Metal oxides (Fe_2O_3, Fe_3O_4, CoO, Co_3O_4, MnxOy, Cu_2O/CuO, NiO, Cr_2O_3, RuO_2, MoO_2/MoO_3, etc.)	500–1200	High capacity High energy Low cost Environmentally compatibility	Low Coulombic efficiency Unstable SEI formation Large potential hysteresis Poor cycle life	[38] [36] [39]
(b) Metal phosphides/sulfides/nitrides (MXy; M ¼ Fe, Mn, Ni, Cu, Co, etc. and X ¼ P, S, N)	50–1800	High-specific capacity Low operation potential and low polarization than counter oxides	Poor capacity retention Short cycle life High cost of production	[40] [41] [42]

Iron-based oxides are most popular from the group of conversion or transition metal oxide (TMO) anode materials and have been extensively studied for rechargeable lithium batteries due to their tremendously high theoretical capacity, environmentally friendly properties, enhanced safety, low cost, non-toxicity, biocompatibility and high natural abundance. Iron oxides of diverse types, haematite (α-Fe_2O_3)

and magnetite (Fe_3O_4) are capable of participating in reversible conversion reactions with lithium, providing a theoretical capacity of 1007 and 926 mAh g^{-1}, respectively [43]. However, iron oxides suffer from poor cycling stability due to low electrical conductivity, low Li^+-ion diffusion coefficient, high volume expansion and iron aggregation during charging and discharging. Therefore, to overcome the above-identified limits, many recent investigations have been focused on developing new methods for the preparation of iron oxide nanomaterials as well as for modifying their structure, size, shape (nanoparticles, nanowires, nanospheres, etc.) and porosity [44–48] or novel methods to stabilize their structure and to improve the electrochemical kinetics and power capability, mainly by carbon coating or by carbon-based composites of α-Fe_2O_3 and Fe_3O_4 [49–51]. In 2019, Kim et al. [52] prepared hematite (α-Fe_2O_3) nanocapsules with a length of 65 nm, diameter of 15 nm and thickness of 5 nm via wrapping–baking–peeling method, in which the three steps are involved as silicon wrapping, annealing 500 °C for 5 h under an air atmosphere to produce the silica shell/hollow hematite nanostructures and removal of the silicon shell by immersing iron oxide/silica nanostructures in 0.1M of NaOH solution with sonication for 5 h. The nanocapsule structure demonstrated an initial capacity of 888 mAh g^{-1} and maintaining 83% of initial capacity after 30 cycles at a constant current density of 0.1 C. This good cycling stability is achieved as a result of the thin shell and large surface area with interior space to allow for volume variation.

Liu et al. [53] and Muraliganth et al. [54] prepared α-Fe_2O_3 nanowires via a facile hydrothermal method [53] or carbon-decorated crystalline Fe_3O_4 nanowires with diameter ranging from 20 to 50 nm and length of several micrometers through microwave-assisted hydrothermal approach, using PEG-400 as soft template and polysaccharide as carbon source [54]. α-Fe_2O_3 [53] shows initial charge/discharge capacities are 947/1303 mAh g^{-1} at the rate of 0.1 C, and the reversible capacity was maintained at 456 mAh g^{-1} after 100 cycles, while carbon-decorated crystalline Fe_3O_4 nanowires showed excellent electrochemical stability with a good reversible capacity of 830 mAh g^{-1} without any capacity damage over 50 cycles at the current density of 0.1 C.

The high capacity and good capacity retention may be attributed to the positive role played by the short Li^+-ion diffusion length, large contact area and/or lower contact resistance between electrodes and the electrolyte, enhanced lithium permeability and electrochemical stability. The carbon coating provides good electronic conductivity and facilitates the fast and highly reversible conversion reaction of iron oxides with lithium [52, 53]. The uniformly coated carbon buffer layer around the Fe_3O_4 nanowires acts as an electronically conducting barrier to prevent the direct contact among the adjacent Fe_3O_4 nanowires, thereby minimizing aggregation of the nanowires during continuous electrochemical charge–discharge cycling. The uniform nanometer thick carbon layer over the nanowire also provides an elastic inactive matrix that can absorb the massive volume expansion and contraction occurring during charge–discharge cycling. More importantly, the carbon-decorated Fe_3O_4 nanowires exhibit improved rate performance compared to the as-synthesized Fe_3O_4 nanowires [54].

Liu et al. [45] prepared haematite α-Fe_2O_3 nanotubes with outer diameter ranging from 200 to 300 nm by using ZnO as sacrificial template, and Kang et al. [55] synthesized nanoparticulate Fe_2O_3 porous tubes from microporous organic nanotubes (MONTs) used as template. Both the Fe_2O_3 nanotubes showed good lithium insertion and de-insertion reversibility with a capacity close to 750 mAh g^{-1} (α-Fe_2O_3) over 150 charge–discharge cycles at a current rate of 0.2 C, [45] while the Fe_2O_3 porous tubes exhibited excellent electrochemical performances with large capacities such as 918 and 882 mAh g^{-1} at current densities of 500 and 1000 mA g^{-1}, respectively [46].

Similarly, α-Fe_2O_3 hollow spheres synthesized by a facile quasi-emulsion soft template method [56] and spindle-like porous α-Fe_2O_3 nanoparticles with dimensions around 20 nm from an iron-based metal organic structure [57]. These novel α-Fe_2O_3 structures showed great enhancement of lithium storage capacity with a reversible capacity of 911 mAh g^{-1} at 0.2 C current rate for more than 50 charge–discharge cycles and reversible capacity of 424 mAh g^{-1} was achieved at the higher current rate of 10 C [57]. It was found that α-F_2O_3 hollow spheres exhibited lower lithium storage capacity compared to spindle-like porous α-Fe_2O_3 nanoparticles, which is higher than 700 mAh g^{-1} for about one hundred of charge–discharge cycles at a current density of 200 mA g^{-1} [56]. Wu et al. [58] have studied the size and morphology effect of α-Fe_2O_3 nanorods in the range from 300 to 500 nm length with and without porosity, and Sohn et al. synthesized nanocomposites of Fe_3O_4 cores and porous carbon-silicate layers through an aerosol-assisted process, followed by vapor coating, which shows enhanced specific capacity, cycling stability and rate capability compared to the micro-sized or bare counterpart (without carbon coating) of α-Fe_2O_3 or Fe_3O_4 [59]. All these investigations on α-Fe_2O_3 or Fe_3O_4 revealed that the preparation method, morphology, composition, carbon coating or size of the Fe_2O_3/Fe_3O_4 play a significant role in the reactivity with lithium [56] and strongly affect both rate and lithium storage capability of iron oxides. Hence, recently electrospinning is widely used to prepare iron oxide-based electrode having unique architectures and fibrous morphology for enhancing the electrochemical performance of LIBs. Figure 9.2 illustrates the self-standing active material/carbon composite electrodes having nanofibrous morphology and conventional active material/carbon composite electrode for lithium ion battery applications. Electrospinning is a simple, efficient and versatile method to prepare continuous nanofibers and flexible electrode having very high porosity by the aid of a high potential electric filed. By this method, nanofibers with diameters ranging between a few nanometers and a few hundred nanometers can be produced [60]. Majorly, there are two types of electrospinning technique: (i) nozzle electrospinning and (ii) nozzle-less electrospinning. This technique can cost-effectively and easily employed in the laboratory as well as can scale up to an industrial process. Electrospun nanofibers have many advantages over the regular fibers such as higher surface areas, modulus and strength. The electrospun nonwoven mats have small pore size, high porosity, fully interconnected pore structure and high surface area. There is also the advantage to control the nanofiber composition to achieve the desired property or functionality, offering more flexibility in surface functionalities. The electrospinning setup, process of electrospinning and

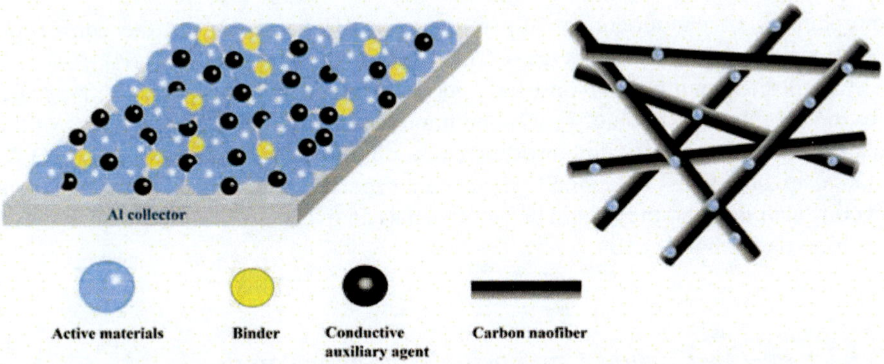

Fig. 9.2 Schematic illustration of self-standing active material composite carbon nanofibrous electrodes for lithium- ion batteries. Adapted and reproduced from Ref. [66], Copyright 2017 Royal Society of Chemistry

the factors affecting the properties of electrospun fibers are detailedly explained in the previous chapters; hence, this chapter is straightaway discussing the electrospinning of iron oxide-based electrodes and their electrochemical properties.

9.3 Electrospun Fe_2O_3/Fe_3O_4 Anode

Yet carbon nanofibers have not been well investigated for this application. On the other hand, nanosized transition metal oxides represent another type of promising alternative as the LIB anode material thanks to their very high capacity and interesting cycling performance. Therefore, it is worthwhile to investigate a nanoscale carbon/metal oxide composite as the electrode material, which may combine the merits of both components. Electrospinning is an inexpensive, versatile and simple method to synthesis flexible film electrode, in which transition metal oxides can be conveniently introduced into 1D carbon matrix [16–22]. The 1D carbon matrix can effectively improve the electron conductivity and ion transmission of the active material and restrict the agglomeration of transition metal oxides during lithium insertion. Chen et al. used electrospinning technique to the synthesis of C/Fe_3O_4 composite nanofiber, which exhibits a high reversible capacity, excellent rate capability and good cycling performance [2]. Hence, it is necessary to design and fabricate new style composite nanofibers, which can provide sufficient buffer space for the volume expansion of transition metal oxides during cycling, thereby maintaining structural stability and flexibility of composite nanofibers [24–27].

Carbon-based nanofibers can be used as anode materials for lithium ion batteries. Both pure carbon nanofiber and C/Fe_3O_4 composite nanofibers were prepared by electrospinning and subsequent carbonization processes. The composite fiber is designed for two purposes: On the one hand, the carbon component and fibrous morphology of nanofibers can digest the volume change during the conversion reaction composite nanofibers (Fe_3O_4/CNFs) with internal voids [61–64].

Composite nanofibers not only have good flexibility, but also possess buffer space for the volume expansion of Fe_3O_4 and maintain the morphological stability of the nanofibers during the Li^+-ion insertion/extraction process. As a result, the Fe_3O_4/C composite nanofibers exhibited high-specific capacity, rate capability and excellent cycling stability, and they could be employed as an excellent flexible anode material for high-performance LIBs.

9.3.1 Fe_2O_3–Carbon Composite Nanofiber Anode

Combination of metal oxides and electronically conducting carbon has been a favorable practice for their applications in high-rate energy storage mesoscopic electrodes having high-rate capability and cycling stability. C/Fe_2O_3 composite [65, 66] $CoFe_2O_4$/C composite fibers [67] C/Fe_3O_4 composite nanofibers [67, 68] Fe_2O_3@CCNFs [70] Fe_3O_4 nanoparticles encapsulated in porous carbon fibers (Fe_3O_4@PCFs) [71] porous Fe_3O_4/C microbelts [72] prepared by electrospinning are reported as anode in LIBs. Zhang et al. [65] reported a quasi-1D Fe_2O_3–carbon composite nanofibers obtained by the electrospinning using different iron source as $Fe(AcAc)_3$-PAN composite and $Fe(NO_3)_3$ $9H_2O$-PAN by Abe et al. [66]. The as-spun membranes are then stabilized at 280 °C for 2–3 h under air and then carbonized in an atmosphere of argon (Ar) or nitrogen (N_2) as shown in Fig. 9.3.

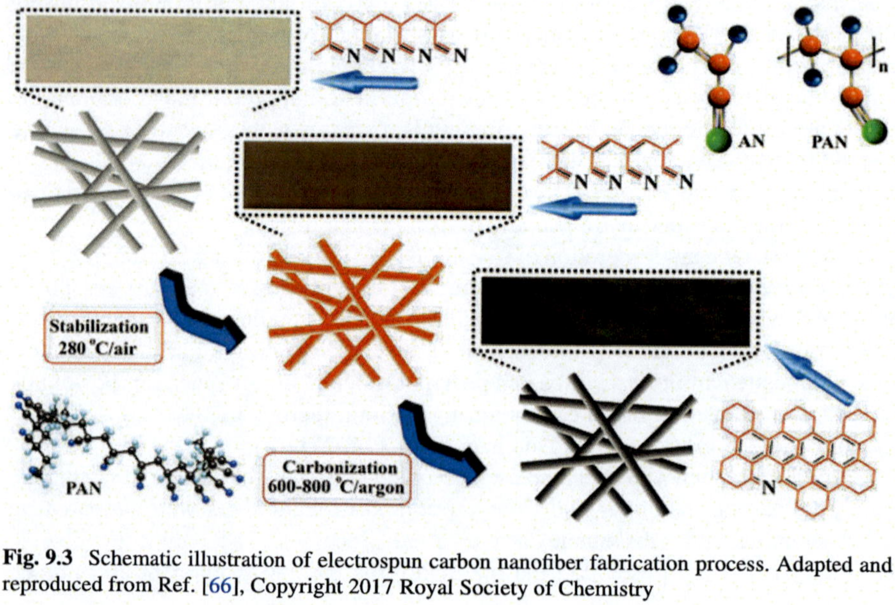

Fig. 9.3 Schematic illustration of electrospun carbon nanofiber fabrication process. Adapted and reproduced from Ref. [66], Copyright 2017 Royal Society of Chemistry

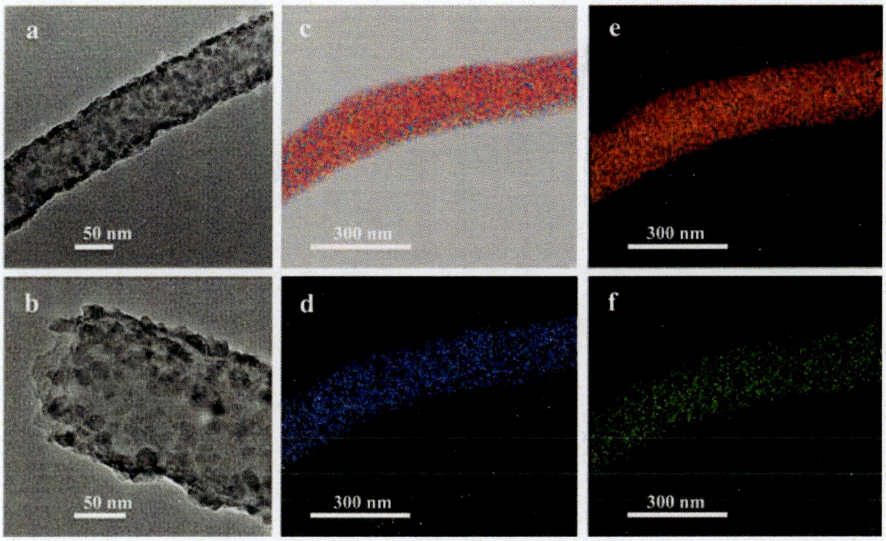

Fig. 9.4 TEM characterization of Fe_2O_3–C composite nanofibers: **a** typical structure of composite nanofiber, **b** the tip of one composite nanofiber, **c** TEM image of single Fe_2O_3–C composite nanofibers with element mapping, **d–f** element mapping of oxygen (blue), carbon (red), iron (green) respectively. Adapted and reproduced from Ref. [65], Copyright 2014 Royal Society of Chemistry

During the stabilization process, the PAN counterpart was converted to an aromatic cyclized ladder type structure by cyclization, dehydrogenation, aromatization and cross-linking reactions, which convert the CH_2 and C≡N groups to infusible C=N and C–H groups [73]. The as-stabilized composite nanofiber mat further transformed to Fe_2O_3–C composite nanofibers after a further calcination process. The TEM images and elemental mapping of the resulted composite fiber are shown in Fig. 9.4

Galvanostatic cycling profiles of Fe_2O_3–C composite nanofiber anode in half-cell configuration were performed to understand the electrochemical performance and cyclability. The electrochemical studies were conducted in two-electrode coin cell CR 2016 [65] or CR 2032 Fig. 9.5 from [66] configuration using iron oxide/carbon composite nanofibers as anode against lithium metal counter electrode and a glass microporous fiber filter (Whatman, cat. no. 1825-047) or microporous monolayer membrane (Celgard 2500®) as separator. The electrolyte employed was 1M $LiPF_6$ in ethylene carbonate (EC)–diethyl carbonate (DEC)/dimethyl carbonate (1:1 by volume). In the half-cell configuration, the Fe_2O_3–C composite nanofiber anode exhibits a reversible capacity of 820 mAh g^{-1} at a current rate of 0.2 C up to 100 cycles. At a higher current density of 5 C, the cells comprising Fe_2O_3–C composite nanofiber anode delivers a specific capacity of 262 mAh g^{-1}. To the detailed analysis electrochemical reactions, the first discharge curve of the Li/Fe_2O_3–C cells is divided into three regions labeled as I, II and III. It is clear from the discharge profile; first, the cell was discharged from an open-circuit voltage (OCV) of 2.8 V to intercalate lithium ions into the Fe_2O_3 matrix as per Eq. (9.1) [7475]. In the region I, a plateau with a

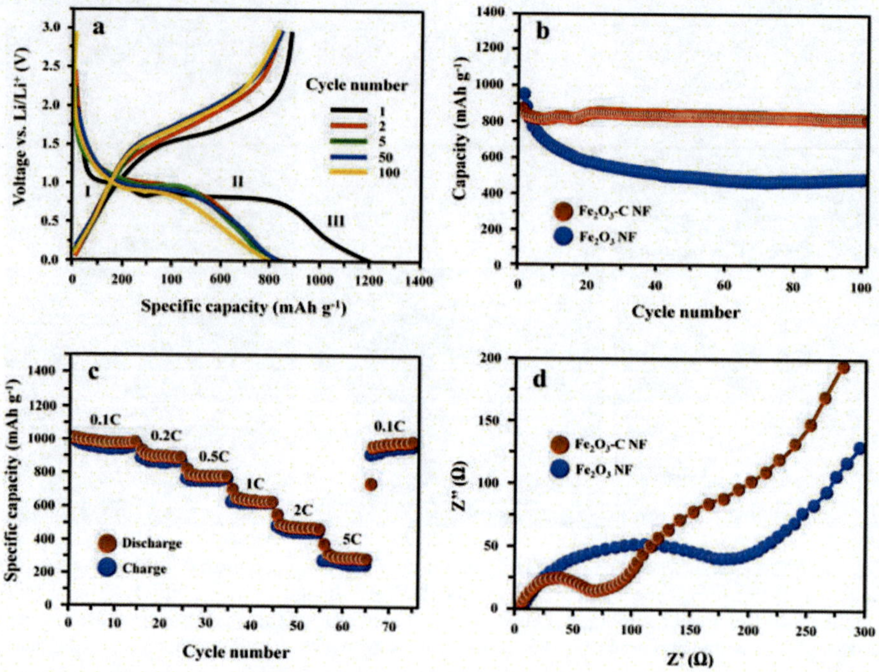

Fig. 9.5 **a** Galvanostatic charge–discharge curves of the Fe_2O_3–C composite nanofibers; **b** cyclic performance (charge capacity) of the Fe_2O_3–C composite nanofiber and pure Fe_2O_3 nanofiber electrodes at 0.2 C rate; **c** rate capability of Fe_2O_3–C composite nanofiber electrodes at different rates; **d** Nyquist plots of the Fe_2O_3–C composite nanofiber and pure Fe_2O_3 nanofiber electrodes. Adapted and reproduced from Ref. [65]. Copyright 2014 Royal Society of Chemistry

wide slope can be observed at 1.1 V, which is ascribed to the phase transformation from hexagonal $Li_xFe_2O_3$ to cubic $Li_2Fe_2O_3$ as per Eq. (9.2) [74, 75]. In the region II, a distinct plateau can be observed at 0.85 V due to the complete reduction of Fe(III) to Fe(0) as per Eq. (9.3) [74, 75], and the Fe nanocrystals were dispersed into the Li_2O matrix. In the discharge region III, the electrolyte gets reduced below 0.8 V and forms the SEI film, which led to the further lithium storage via an interfacial reduction at the metal–Li_2O boundary. The initial specific capacity of Fe_2O_3–C composite nanofiber anode is found to be 1214 mAh g^{-1}, which exceeds the theoretical capacity of Fe_2O_3, 1007 mAh g^{-1} (6 mol of Li per 1 mol of a-Fe_2O_3) and corresponds to an uptake of 7.2 mol Li per mol of the a-Fe_2O_3–C composite.

The initial discharge capacity of the cell having freestanding iron oxide/carbon composite electrospun nanofibers carbonized at different temperature from 600 to 800 °C is observed during the first discharge for all the samples owing to the initial SEI formation at the surface of the electrode and the presence of mesopores in the carbon nanofibers [76]. It was observed that the subsequent charge/discharge curves of the cells overlap, which suggesting that the SEI is formed during initial discharge and becomes stable after the first charge–discharge cycle. The capacity after stabilization of the SEI is the sum of the electrode capacities of the carbon nanofibers and iron oxide, and the corresponding electrochemical reaction can be represented by the following equation:

$$Fe_2O_3 + 6Li^+ + 6e^- \rightarrow 2Fe + 3Li_2O$$

or

$$Fe_3O_4 + 8Li^+ + 8e^- \rightarrow 3Fe + 4Li_2O$$

The capacity per unit weight and volume of the self-standing iron oxide/carbon composite electrospun nanofibers under all the preparation conditions showed that carbonization and press processing increase the capacity per unit volume of the nanofibers. The carbonization at high temperature (600–800 °C) leads to the shrinking of the nanofibers, whereas pressing of the as-spun samples in between the heated plates of hot press set at zero pressure until the temperature of the plates reached 110 °C increases the nanofiber density due to the close packing of fibers, which increases the electrode density. The study also showed that the sample having low electrode density bears low energy density and also found that the carbonization temperature has a pronounced effect on the charge–discharge capacity of the electrode. Higher graphitization temperature produces the fibers having higher electronic conductivity, thereby improving the charge–discharge capacity. The membrane carbonized at 600 °C showed the lowest charge–discharge capacity which is about 50 and 375 mAh g^{-1}, respectively, for the electrode with and without pressing processing, while the electrode carbonized at 800 °C delivered a capacity of about 575 and 675 mAh g^{-1}, respectively, for the electrode with and without pressing processing, suggesting that a certain level of graphitization is necessary to achieve a certain capacity.

The electrospun electrode prepared by carbonization at 800 °C without press processing exhibits the highest capacity retention of 61% at a current density of 2500 mA g^{-1}. Increasing the carbonization temperature increases both charge–discharge capacity and capacity retention, whereas press processing decreases the capacity and capacity retention, which is attributed to the differences in electrode resistance caused by the difference degree of graphitization and pore structure. The graphitization at higher temperature causes more ordered graphitic crystallite growth on the fiber during the carbonization [77]. The good capacity and cycling stability are due to the suppressing of electrode destruction by the nano-iron oxide and carbon nanofiber composites in combination with the unique fibrous structure, thereby suppressing the increase in cell resistance due to cycling. Compared to pure electrospun Fe_2O_3 nanofibers, the capacity retention of Fe_2O_3–C composite nanofiber electrodes is drastically improved. The good electrochemical performance is associated with the homogenous dispersed Fe_2O_3 nanocrystals on the carbon nanofiber support. Such a structure prevents the aggregation of active materials, maintains the structural integrity and thus enhances the electronic conductivity during lithium insertion and extraction. In short, the enhanced electrochemical performance of iron oxide/carbon composite electrospun nanofibers is ascribed to the unique structure of the fiber mat with a stable structural integrity and improved electrical conductivity rendered by the carbon fiber network.

9.3.2 Fe_3O_4–Carbon Composite Nanofibers

Fe_3O_4/C nanofibers with PVdF binder (as a coating on copper foil) [68] and without PVdF binder (as freestanding flexible) [69] are reported electrode for LIBs [100]. Compared to conventional electrode fabricated on current collector, the freestanding binder-free electrodes can effectively improve the energy density of LIBs. The ferri-ferrous oxide (Fe_3O_4) has high theoretical capacity (924 mAh g^{-1}), which is economically viable and environmental friendliness material. In both studies, polyacrylonitrile (PAN) was used as the polymer precursor and N, N-dimethylformamide (DMF) as solvent. For the preparation of flexible electrode, the rod-shaped a-FeOOH was prepared by a simple hydrothermal [69] wherein the other study ferric acetylacetonate designated as $Fe(acac)_2$ was used as source for Fe_3O_4 [68]. The electrospinning was performed at an applied voltage of 12-15 kV. After electrospinning process, the as-prepared PAN/α-FeOOH or PAN/$Fe(acac)_2$ nanofibers were pre-oxidized at about 250 °C h in air and further graphitized at 500, 600 to 700 °C for 8–10 h under inert atmosphere. The electrochemical properties of the Fe_3O_4/C nanofibers film were systematically investigated by assembling them into coin-type 2032 cell with lithium metal foil as the counter electrode and Celgard as separator. The electrolyte was used as 1M $LiPF_6$ in a binary or ternary mixture (1:1:1 v/v) of carbonate solvents such as ethylene carbonate, dimethyl carbonate or methyl ethyl carbonate.

Wang et al. [68] found that after an annealing temperature of 500–700 °C, the carbon has disordered structure while $Fe(acac)_2$ is converted as Fe_3O_4 has nanocrystalline structure with a particle size from 8.5 to 52 nm. An in-depth electrochemical study of the electrode fabricated on copper foil current collector using PVdF as the binder reveals that the C/Fe_3O_4 composite nanofiber obtained at a graphitization temperature of 600 °C exhibits a high reversible capacity, good cycling performance and excellent rate capability. The incorporation of Fe_3O_4 nanoparticles in the composite plays an important role in increasing the effective surface area, electronic conductivity and wettability of the electrode with electrolyte. The charge–discharge studies of the cell having C/Fe_3O_4 composite nanofibers at a current density of around 200 mA g^{-1} show different voltage profiles with a long plateau at about 0.75 V in the discharge (lithiation) process and a short plateau at about 1.5 V in the charge (delithiation) process, due to the presence of Fe_3O_4 and the heterogeneous reaction between lithium and metal oxide $8Li + Fe_3O_4$ $3Fe + 4Li_2O$ [78, 79]. The authors fragmentized the discharge profiles into two segments above 0.7 V, as a sloping part from about 2.0–0.75 V, and a plateau at 0.75 V. The former can be attributed to the reaction $Fe_3O_4 + xLi \rightarrow LixFe_3O_4$; the latter corresponds to the conversion reaction $Li_xFe_3O_4 + (8 - x)Li \rightarrow 3Fe + 4Li_2O$. However, the plateaus are not pronounced for the C/Fe_3O_4 nanofibers graphitized at a temperature lower than 600 °C, due to the exceptionally lower particle size (8.5 nm) of Fe_3O_4 in these electrodes. The particle size of Fe_3O_4 is found to be 17.1 and 52.0 nm for the composite fibers graphitized at 600 and 700 °C. Also, they found that the capacity associated with the high voltage sloping part (or x value in the above reaction) decreases with increasing the graphitization temperature.

The charge–discharge cycling study at a current density of around 200 mAg^{-1}, the C/Fe_3O_4 composite nanofiber graphitized at 600 °C, exhibits the largest discharge capacity of 1551 mAh g^{-1} in the first cycle, while the composite fibers carbonized at 500, 550, 700 °C deliver a capacity of 1318, 1151 and 957 mAh g^{-1}, respectively, while pristine carbon nanofiber graphitized at 600 °C delivers a first-cycle discharge capacity of 1041 mAh g^{-1}, which is higher than the C/Fe_3O_4 composite electrode graphitized at 700 °C. It is well known that a disordered carbon obtained at a relatively low temperature (<1000 °C) has a large capacity but they suffer from low electronic conductivity [80, 81]. In other words, at lower the annealing temperature, more defects are formed on the surface of the carbon, which in turn allow more Li^+ ions to store. Because of the optimum defects and electronic conductivity, the 600 °C-carbonized C/Fe_3O_4 nanofibers gives rise to the highest capacity compared to those graphitized at 500, 550 or 700 °C. It was also reported that reversible capacity of the 600 °C-carbonized C/Fe_3O_4 composite nanofibers reaches about 1000 mAh g^{-1} at the 80th cycle, which is significantly higher than that of pristine carbon nanofiber graphitized at 600 C or other C/Fe_3O_4 composite nanofiber or the state-of-the-art graphite anode material in the commercial LIBs (around 340 mAh g^{-1}).

Another pristine carbon nanofiber graphitized at 600 °C gradual decrease in the capacity about 40 cycles and thereafter keeps virtually stable, while all of the C/Fe_3O_4 composite nanofibers experience a gradual capacity rise from 10th cycle to the end of the measurement (80th cycle). For example, the specific capacity of C/Fe_3O_4 composite nanofiber graphitized at 600 °C reaches a minimum of 763 mAh g^{-1} at the 7th cycle and rises to 1007 mAh g^{-1} after 80 cycles. In both cases (for pristine carbon nanofiber and C/Fe_3O_4 composite nanofiber electrodes), initial capacity decrease could be explained by the structure re-organization of the carbon [82]. While it has been concluded that increase in capacity after a certain number of charge–discharge cycle may due to the progressive wetting of the electrode with electrolyte until the cell gets stabilized. The morphological change also affects the performance of electrodes to be delivered. The SEM photographs of the nanofibers after 80 cycles are displays in Fig. 9.6, which shows that the pristine carbon nanofiber retains its original morphology (Fig. 9.6a) while the C/Fe_3O_4 composite nanofiber showed a marked morphology change (Fig. 9.6b). In short, the fibers in the pristine carbon nanofiber still retain the smooth and long fiber structure, while the C/Fe_3O_4 composite nanofiber becomes powderized and Fe_3O_4 or fiber particles are loosely agglomerated and arranged as a line on the electrode surface. This indicates during the lithium insertion and extraction in the C/Fe_3O_4 composite nanofibers which undergoes a large volume change. Such a continuous volume expansion and contraction lead to the breaking of the fibers into small particles, and it was calculated that when Fe_3O_4 is completely reduced into Fe by lithium, the volume increases theoretically by 80.8%. Another factor that induces the powderization of the fiber is related to

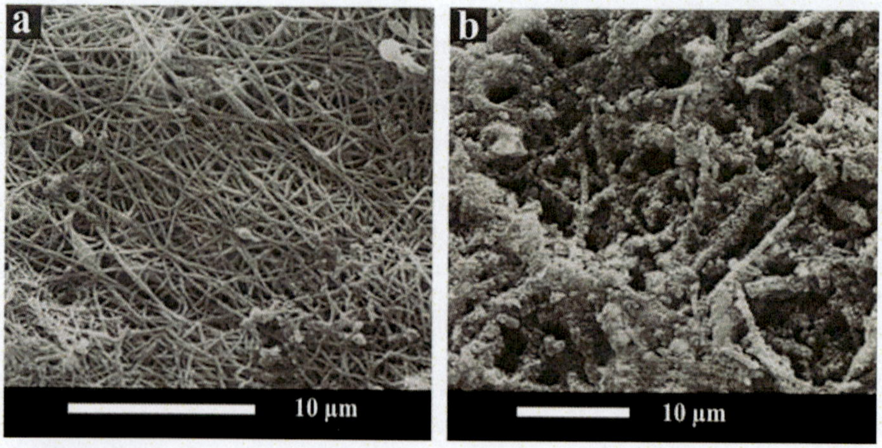

Fig. 9.6 **a** FE-SEM image on the surface morphology of a 600 °C-carbonized pure carbon nanofibers, **b** C/Fe_2O_3 nanofibers after 80 cycles. Adapted and reproduced from Ref. [68], Copyright 2008 Elsevier

the presence of highly conducting iron produced in situ after deeply discharging the cell down to a low voltage below 0.8 V. The iron inclusion should enhance the electronic conduction along with the C/Fe interface and thus further activate the carbon component for lithium insertion. This is also likely one of the reasons why the 600 °C-carbonized composite nanofiber has a higher capacity than the 600 °C-carbonized pure carbon nanofiber. The morphological studies reported that the powderization is comparatively much less on the electrode side which is not come in contact with the current collector, than the electrode side in contact with current collector, i.e., the copper foil. This difference in degree of powderization on the two sides of the fiber electrode is explained by the difference in the strain release during the charge–discharge cycling process. Because of the restriction of the current collector, the strain caused by the volume change cannot be released easily on the fiber electrode side that is in contact with the current collector leads higher degree of powderization, while the other side of the fibrous electrode can easily release the strain makes the fibrous structure be retained during the charge–discharge cycling. Another possible reason for the difference in powderization between the current collector side and the other side of the fibrous electrode might be related to the transient potential difference between these two locations during cycling process. The potentials are lower in discharging and higher in charging in the vicinity of current collector side compared to the other side of the electrode. Such a potential difference may cause different powderization behavior of the electrode. However, the powderization effect in the cell does not cause negative impact on the electrochemical performance of the C/Fe_3O_4 composite nanofiber electrodes. The studies show the powderization of nanofiber actually improves the cycling property by increasing the effective surface area of the electrodes, which increase the available active site for electrochemical reaction and thereby accommodating the inevitable periodical volume change during lithium insertion and extraction. Such an improvement effect on the electrochemical performance caused by the powderization has been also reported for Cu_2O–Li_2O composite system [83, 84].

The charge–discharge cycling at different current density ranges from 200 to 1600 mA g^{-1}, and C/Fe_3O_4 composite nanofiber graphitized at 600 °C delivers a capacity of 1096, 1022 and 623 mAh g^{-1} at a current density of 200, 400 and 1600 mA g^{-1}, respectively. It is worthy to note that the C/Fe_3O_4 composite nanofiber delivers this specific capacity at higher C rate, after cycled for 100 continuous charge–discharge cycles at 200 mA g^{-1}. Surprisingly, when the current density is sweeped from 200 to 400 mA g^{-1}, only a little capacity loss of 74 mAh g^{-1} is observed.

Wu et al. [69] study reports the electrochemical properties of a Fe_3O_4/C composite nanofibers which have the unique internal voids between Fe_3O_4 nanoparticles and carbon matrix. Such Fe_3O_4/C composite nanofibers films with good flexibility and excellent electrical conductivity facilitate the direct fabrication of lithium cells without any current collector, binder and additional conductive agent. The electrochemical charge–discharge studies of the half-cells using lithium metal counter electrode and the freestanding Fe_3O_4/C composite nanofiber graphitized at 600 °C as the working electrode showed first charge/discharge capacities of 830 and 1161 mAh g^{-1}, which is higher than that of Fe_3O_4/C composite nanofiber graphitized at 500 °C (charge capacity is 683 mAh g^{-1} and discharge capacity is 1003 mAh g^{-1}) at a current density of 0.1 A g^{-1} with a poor Coulombic efficiency of 68.1% and 71.5%, respectively. However, the Coulombic efficiency of the subsequent charge–discharge cycles is high. The formation of an irreversible solid SEI film at the electrode–electrolyte interface results in the low Coulombic efficiency in the first cycle. In concurrent with Wang et al. [68] study, Fe/C3-700 shows the lowest specific capacity (383.6 and 619.1 mAh g^{-1}) in the first charge–discharge curve, indicating Fe/C3-700 is not ideal electrode material. The charge and discharge capacities of Fe_3O_4/C3 composite fiber graphitized at 600 °C are found to be 780 mAh g^{-1} after 150 cycles at 0.5 A g^{-1} which is 84.2% of the theoretical specific capacity and the Coulombic efficiency up to 99.5%, the highest specific capacity and Coulombic efficiency compared to all other samples graphitized at different temperatures. The long-term cycling profiles of the Fe_3O_4/C3-500 and Fe_3O_4/C3-600 flexible electrodes at 0.5 A g^{-1} are showed that with the increase of cycle number, the capacity of Fe_3O_4/C3-500 flexible electrode is decreased and the capacity of Fe_3O_4/C3-600 is almost unchanged. Also the reversible capacities at a current density of 0.5 A g^{-1} and 1 A g^{-1} after 300 cycles (Fig. 9.7a) are still remained as 761 and 611 mAh g^{-1} without significant attenuation, and the Coulombic efficiencies are close to 100% which indicate an excellent cycling performance of Fe_3O_4/C3-600 flexible electrode. This confirms the Fe_3O_4/C3-600 electrode has better cycling stability than the Fe_3O_4/C3-500 electrode owing to larger internal void space of Fe_3O_4/C3-600.

The rate capability of Fe_3O_4/C3-600 NFs is tested at different current densities ranging from 0.1 to 1 Ag^{-1}, and the result is shown in Fig. 9.7b. The specific capacities of the sample graphitized at 600 °C are lies between 982 and 636 mAh g^{-1} which is 105–133 mAh g^{-1} higher than Fe_3O_4/C3-500 NFs as shown in Fig. 9.7a. Moreover, after continuous charge–discharge cycling at current densities 0.1–1 A g^{-1}, when

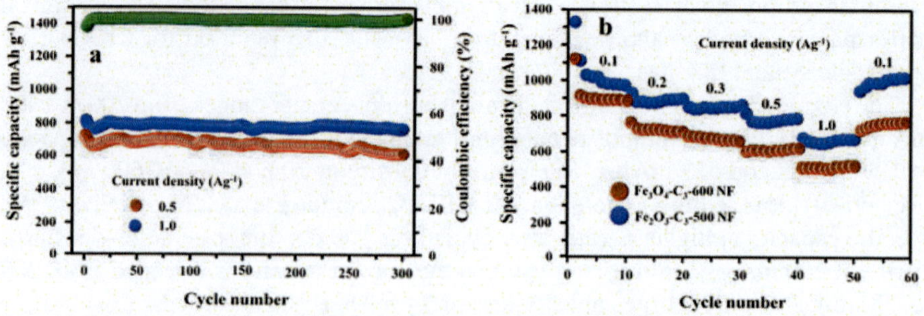

Fig. 9.7 **a** Long-cycling profiles of the Fe_3O_4/C3-600 flexible electrode at current densities of 0.5 A g^{-1} and 1 A g^{-1}. **b** The rate capability of Fe_3O_4/C3-500 and Fe_3O_4/C3-600 at different current density. Adapted and reproduced from Ref. [69], Copyright 2017 Elsevier

the current rate returns back to 0.1 A g^{-1} after 80 cycles, the specific capacity of Fe_3O_4/C3-600 NFs recovers to 1000 mAh g^{-1} (slightly higher than the initial capacity 982 mAh g^{-1}).

Electrochemical impedance spectroscopy studies measurements confirm that the samples graphitized at 600 °C exhibit the lowest charge-transfer resistances (R_{ct}). Charge-transfer resistances of Fe_3O_4/C composite nanofiber graphitized at 600 °C are found to be 53.8 Ω, while the samples graphitized at 500 °C are found to be 88.9 Ω and pristine Fe_3O_4 is found to be 271.2 Ω. The lower charge-transfer resistances of Fe_3O_4/C composite nanofiber graphitized at 600 °C is attributed to the higher degree of graphitization and low solid-state interface layer resistance which significantly improves the rate capability. The morphological changes of the Fe_3O_4/C graphitized at 600 °C after cycling test were recorded by SEM and TEM. As seen in Fig. 9.8a, b, the morphology of Fe_3O_4/C3-composite electrode still exhibit homogeneous and continuous fibrous structure which is consistent in the overall electrode area after 300 continuous charge–discharge cycling. The TEM images (Fig. 9.8c, d) show that the Fe_3O_4 nanoparticles suffered volume expansion and broke up into small nanoparticles after repeated charge–discharge cycling. However, because of the buffer of internal space and the protective of carbon matrix, the Fe_3O_4 nanoparticles are still well constrained in the carbon nanofibers and the overall structure of composite nanofibers maintains intact. This further proves the Fe_3O_4/C3-600 which has a good cycling stability.

To evaluate the flexibility and electrochemical performance of the Fe_3O_4/C nanofibers film under bending states, the Fe_3O_4/C nanofibers were used to fabricate flexible symmetrical supercapacitors and the device level performance is studied and displayed that the supercapacitors exhibit high flexibility and maintain the structural integrity under the bending test. The CV studies of the flexible device under different bending modes at a scanning rate of 100 mV s^{-1} was investigated and found that there is no change in the CV curves at different bending angles suggesting that the Fe_3O_4/C composite nanofibrous electrode can work properly at different bending state.

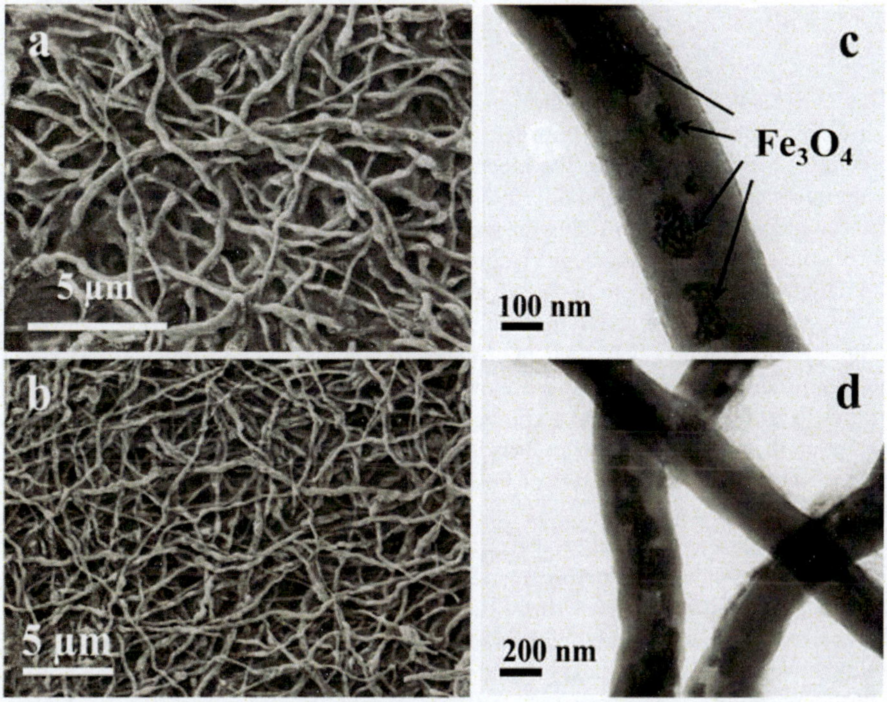

Fig. 9.8 **a**, **b** SEM images on the morphology and **c**, **d** TEM image of Fe_3O_4/C3 600 after 150 cycles. Adapted and reproduced from Ref. [69], Copyright 2017 Elsevier

It is clear that our Fe_3O_4/C3-600 anode is a competitive candidate in terms of the capacity output and the cycle stability. We can conclude that the Fe_3O_4/C3-600 NFs exhibit excellent rate performance. These superior electrochemical properties of the composite nanofiber indicate that this unique composite may find promising applications in high-performance Li-ion batteries. On the other hand, a disordered carbon has a low electronic conductivity compared with that of a graphitic carbon. For a carbon-based composite material, three factors can be considered to determine its capability of lithium storage: electronic conductivity, concentration of surface defects and the interfacial area between the two components. This result suggests that this C/Fe_3O_4 composite nanofiber represents a promising candidate of anode material for high-performance LIBs which can be used in (hybrid) electric vehicles and electric tools.

9.4 Hollow α-Fe_2O_3 Electrospun Nanofibers

The electrospun nanofibers have high active surface area; flexibility and the nonwoven fibrous structure improve the electrode integrity. As discussed in the previous section (Sect. 3.4), the electrospun fibrous electrode provides relatively wide interspace between two adjacent fiber layers which facilitate enough Li^+-ion insertion locations, thus avoiding the size, structure and shape variations of the anode during the charge–discharge cycling process. In addition, the long fibrous structure, the nanofibers show excellent electrical conductivity. Owing to the unique internal void space between the active material and carbon matrix, the electrode exhibits a better cycling stability than their micro-sized counterpart. The active surface could be again increased by preparing hollow fibers-based anodes, which increase electrode–electrolyte contact area and thus would enhance electrochemical properties. However, there are only a few reports in the literature, dealing with the electrospinning of interconnected α-Fe_2O_3 or Fe_3O_4 hollow fibers for anode material used in lithium ion batteries.

Various methods including self-assembly [85], template-engaged etching process [45] and electrochemical anodization [86] are employed for the synthesis of hollow iron oxide nanofibers. However, these methods typically involve sacrificial templates, expensive equipment and complex experimental procedures. The electrospinning is a simple, versatile and practical method for low cost, facile fabrication of iron hollow nanofibers (hNFs). The important strategies adopted for the preparation of hollow nanofiber via electrospinning are (i) tubes by fiber template process, (ii) coaxial electrospinning and (iii) single-nozzle co-electrospinning [87]. In the coaxial electrospinning, fiber templates were first fabricated by electrospinning and then coated with various precursors. Hollow fibers are subsequently obtained by removing fiber templates via extraction with an appropriate solvent or calcination at high temperature. The coaxial electrospinning uses two coaxial capillaries in a spinneret containing different precursor solutions, while single-nozzle co-electrospinning uses single capillary in a spinneret containing a solution of two immiscible polymers dissolved in solvent to generate core–shell composite fibers that results in hollow fibers via removal of core by either extraction or calcination at high temperature. In the single-nozzle co-electrospinning, when the fiber jet is evolved and stretching out from the needle tip or Taylor cone, the solution of the immiscible polymers results in phase separation owing to the intrinsic polymer properties yielding core–shell composite fibers.

Hollow fibers of α-Fe_2O_3 [88] and carbon-coated Fe_3O_4 [89] are prepared by a single spinneret electrospinning process. The iron source ferric acetylacetonate ($Fe(acac)_3$) [88] or iron(III) nitrate nonahydrate ($Fe(NO_3)_3 \cdot 9H_2O$ [89] was dissolved in a solution of polyvinylpyrrolidone (PVP) in a suitable solvent and electron spun using a 25-gauge steel injection needle with an applied voltage of $\geq$10 kV. The resulting core–shell structured nanofiber is then calcined at 500 C. In contrast, the Fe_3O_4 hNFs were prepared by annealing the Fe_2O_3 hNFs at 350 °C for 2 h under Ar/10% H_2 atmosphere. To prepare the carbon-coated hollow nanofiber, a uniform layer of carbon was coated over the as-spun core-shell fiber by hydrothermal method using glucose as the carbon source. The schematic on the fabrication process of carbon-coated iron oxide hollow fiber is shown in Fig. 9.9. The surface morphology of the Fe_2O_3 and Fe_3O_4 hollow nanofibers with and without carbon coating is shown in Fig. 9.10 [89]. In both studies, the as-spun as well as the annealed Fe_2O_3 and Fe_3O_4 hollow nanofibers are randomly oriented, interconnected and have a large aspect ratio with the length of individual fibers ranging up to tens of micrometers. Also the fibers are continuous and exhibited uniform diameter [90]. All the fibers have very smooth surfaces owing to their amorphous nature and have an average diameter of ~360 nm. Unique internal voids and inter-fiber space are formed between the fibers. Even after

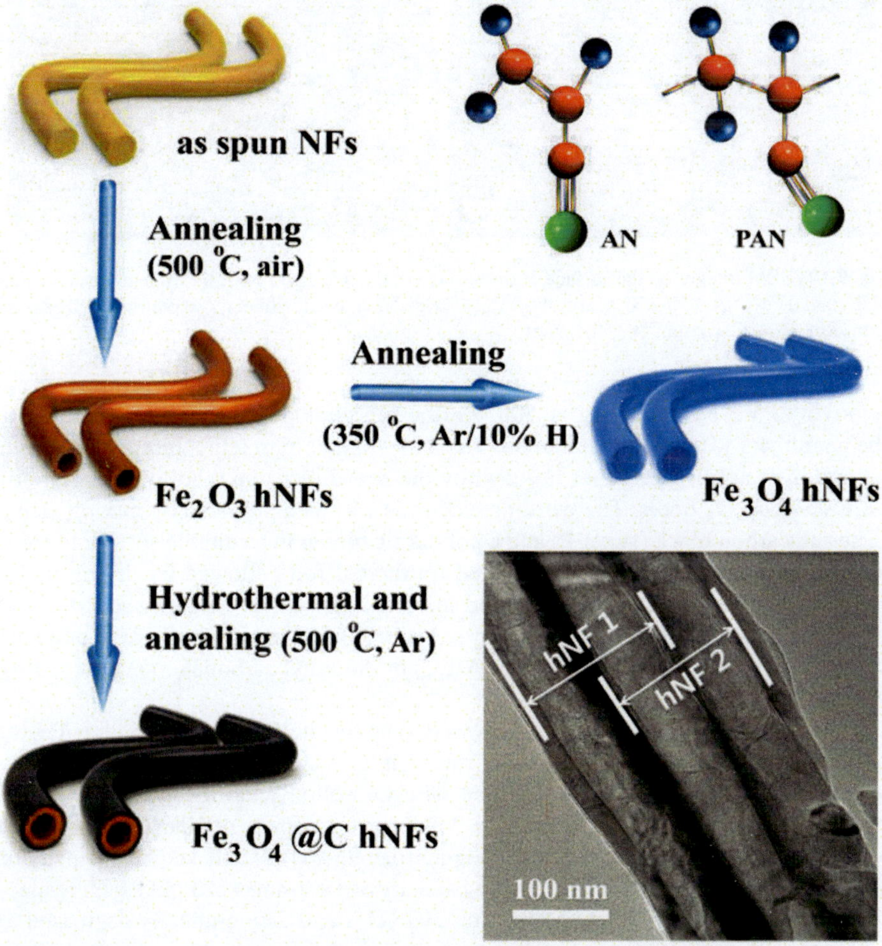

Fig. 9.9 Schematic of the fabrication process of the as-spun iron oxide polymeric nanofibers (as-spun NFs), Fe_2O_3 hNFs (Fe_2O_3 hNFs), Fe_3O_4 hNFs (Fe_3O_4 hNFs) and carbon-coated Fe_3O_4 hNFs (Fe_3O_4/C hNFs). Adapted and reproduced from Ref. [89], Copyright 2017 Elsevier

Fig. 9.10 SEM images on the surface of the surface of the **a** as-spun NFs, **b** Fe_2O_3 hNFs, **c** Fe_3O_4 hNFs and **d** Fe_3O_4/C hNFs. The inset shows an enlarged image of the fibers. Adapted and reproduced from Ref. [89], Copyright 2015 Elsevier

removing PVP by annealing at 500 °C, the morphology and hollow structure of the nanofibers were well maintained. The annealing at high temperature under both air and inert atmosphere does not destroy the one-dimensional hollow structure of Fe_2O_3 and Fe_3O_4 fibers. The working electrode was then prepared by mixing active material, carbon black (super P) and a polymeric binder in a suitable solvent to form uniform slurry and subsequently coated on the Cu foil followed by drying under vacuum. The test cells were assembled in coin cells with a Celgard separator and lithium foil as a counter electrode using 1M $LiPF_6$ dissolved in a mixture of ethylene carbonate (EC) and diethyl carbonate (DEC) or dimethyl carbonate (DMC) (1: 1 by volume) is used as electrolyte.

Electrochemical measurements showed that the hollow structure of α-Fe_2O_3 hollow nanofibers played an important role in improving the specific capacity, cycle stability and rate capability in LIBs. The α-Fe_2O_3 hollow nanofiber anodes exhibit a high reversible capacity of 1293 mAh g^{-1} at a current density of 60 mA g^{-1} (0.06 C) after 40 cycles (Coulombic efficiency higher than 95%) [88] and a current density of 1000 mA g^{-1} first discharge and charge capacity of 1876 and 1223 mAh g^{-1}, respectively, which faded to 470 mAh g^{-1} after 30 cycles and then gradually decreased to 230 mAh g^{-1} at the 100th cycle. The rapid decay in the capacity of the Fe_2O_3 hollow nanofibrous electrodes are observed due to the large volume change during lithiation and delithiation process and the formation of the stable solid electrolyte interface (SEI) layer on at the electrode–electrolyte interface during charge/discharge cycling [89]. This result showed that the discharge capacity of α-Fe_2O_3 hollow fibers observed after 40 cycles at a current density of 60 mA g^{-1} is almost four times higher than the reversible capacity of commercial graphite and exhibited Coulombic efficiency higher than 95%. At a current density of 60 mA g^{-1}, the α-Fe_2O_3 nanofibers exhibited an initial discharge capacity of 938 mAh g^{-1} above 0.8 V (5.6 mol of Li per mole of α-Fe_2O_3), which is very close to its theoretical capacity of 1007 mAh g^{-1} (6 mol of Li per mol of α-Fe_2O_3). In the first discharge cycle, the α-Fe_2O_3 hollow fibers electrode shows a total specific capacity of 1795 mAh g^{-1}, corresponding to 10.7 mol of Li per mol of α-Fe_2O_3, which is much higher than the theoretical capacity. The phenomenon that the first discharge capacity of α-Fe_2O_3 or other transition metal oxide electrode considerably exceeds the theoretical capacity has been widely reported [91–93]. As explained in the previous section (Sect. 9.3.1 Fe_2O_3–carbon composite nanofiber anode), the extra capacity delivered by the Fe_2O_3 hollow nanofiber electrode has been explained as the electrolyte being reduced at low voltage (below 0.8 V vs. Li^+/Li) 25 to form an irreversible solid-electrolyte interface (SEI) layer and which led to the further lithium storage via an interfacial reduction due to charge separation at the metal–Li_2O phase boundary [94]. However, the formation of this irreversible solid electrolyte interface film at the electrode–electrolyte interface results in the low Coulombic efficiency in the initial charge–discharge cycles. The cycle stability of α-Fe_2O_3 hollow fibers electrode was investigated at the current rate of 60 mA g^{-1} (0.06 C) (Fig. 9.11a). Even after 40 cycles, α-Fe_2O_3 hollow fibers exhibit a stable capacity of 1293 mAh g^{-1} which is much better. The continuous charge–discharge cycling stability of α-Fe_2O_3 hollow fibrous electrode at different current densities ranging from 0.2 to 08 A g^{-1} (Fig. 9.11b) displays that there is always a sudden drop on capacity after switching from a lower current rate to a higher current rate. This phenomenon can be explained by the concentration polarization of Li^+ ions in the α-Fe_2O_3 hollow fibers resulting from a diffusion-limited process. Under the studied conditions, the α-Fe_2O_3 hollow fibers displayed an excellent rate capability and the displayed capacity was as high as 1001 mAh g^{-1} even at a current density of 0.8 C. After the continuous cycling at different current density,

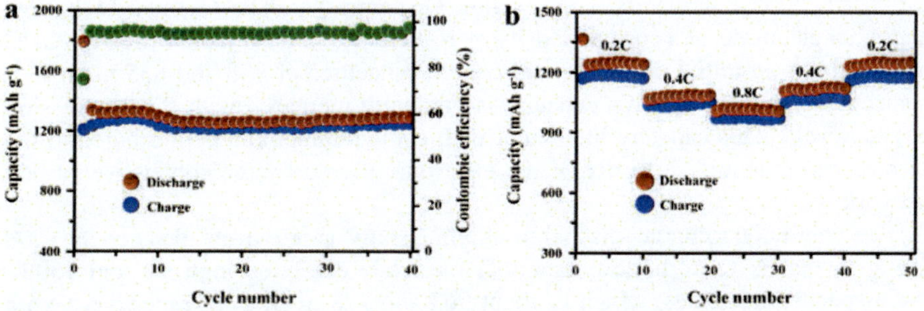

Fig. 9.11 **a** Cycling performance of the electrode made from a-Fe_2O3 hollow fibers at a current density of 60 mA g^{-1} (0.06 C), **b** rate capability of α-Fe_2O_3 hollow fibers at different current densities. Adapted and reproduced from Ref. [88]. Copyright 2012 Royal Society of Chemistry

α-Fe_2O_3 hollow fibrous electrode resumes nearly the full capacity when returning to the initial current density of 0.2 C. The capability of the electrode resumes the original capacity after a long rating, confirming the exceptional capacity and cycling stability of α-Fe_2O_3 hollow fibers, which father indicate the ability of the electrode to keep its integrity not only for a long number of charge–discharge cycles, but also at high rates [88].

Mi et al. [89] made a comparison study on the electrochemical properties of Fe_3O_4 hollow nanofiber with and without carbon coating against Fe_2O_3 hollow nanofibers. The carbon coating on the Fe_3O_4 prepared by the electrospinning is first confirmed by XRD studies and Raman spectroscopy. The XRD and Raman spectroscopic study showed that Fe_3O_4 hNFs crystalline phase was completely transformed from the Fe_2O_3 hollow nanofibers phase by annealing under the reducing atmosphere. However, the peak positions in the XRD remained same even after the Fe_3O_4 hollow nanofiber was hydrothermally coated with carbon and annealed in argon atmosphere to produce Fe_3O_4/C hNFs. Both the Fe_2O_3 and Fe_3O_4 hollow nanofibers possess highly crystalline structure, and the layer of carbon coated over the hollow nanofiber is not well crystallized at the annealing temperature of 500 °C. The ID/IG value measured from Raman spectroscopy is reported as 1.02 for the Fe_3O_4/C hollow nanofibers, which suggest that the electrospun hollow nanofibers contain a relatively high amount of disordered defects. However, for the sample Fe_3O_4 hollow nanofiber, the D and G bands were not observed which implies the removal of carbon by complete combustion of the polymer during annealing.

The electrochemical studies reported that the Fe_3O_4 hNF electrode had a first discharge and charge capacity of 1374 and 1069 mAh g^{-1}, respectively, at a current of 1000 mA g^{-1}. Similar to the rapid decay in the charge–discharge capacity of Fe_2O_3 hollow nanofiber electrode, Fe_3O_4 hNF electrode is suffering from poor cycling stability. After 150 cycles, the specific discharge capacity faded rapidly to ~150 mAh g^{-1}, which is only 11% of the initial capacity of Fe_3O_4 hNF electrode, indicates the poor capacity retention of Fe_3O_4 hNF electrode. The major reason for the rapid capacity decay of the Fe_2O_3 hNF and Fe_3O_4 hNF electrodes may be due to the large volume change and the formation of the irreversible solid electrolyte interface (SEI) layer during charge/discharge cycling. On contrast with Fe_2O_3 or Fe_3O_4 hollow nanofiber electrode, at a current density of 1000 mA g^{-1}, the template-free Fe_3O_4/C hNFs exhibit an initial discharge and charge capacities of 948 and 835 mAh g^{-1}, respectively, and the specific capacity is gradually increased with the progression of each cycle. This capacity increment with cycle number can be attributed to the surface and structural integrity of the electrode offered by the conductive carbon network.

The continuous charge–discharge cycling study at a current density of 1000 mA g^{-1}, the Fe_3O_4/C hollow nanofiber electrode exhibited high 1st and 150th-cycle-specific capacities, which is about 963 and 978 mAh g^{-1}, respectively, with a Coulombic efficiency of more than 99%. Also the continuous charge–discharge studies showed that Fe_3O_4/C hNFs have excellent and stable rate capability, compared to that of the Fe_3O_4 hNFs counterpart. At a current density of 2000 mA g^{-1}, the carbon-coated Fe_3O_4/C hNFs delivered a capacity of 704 mAh g^{-1}.

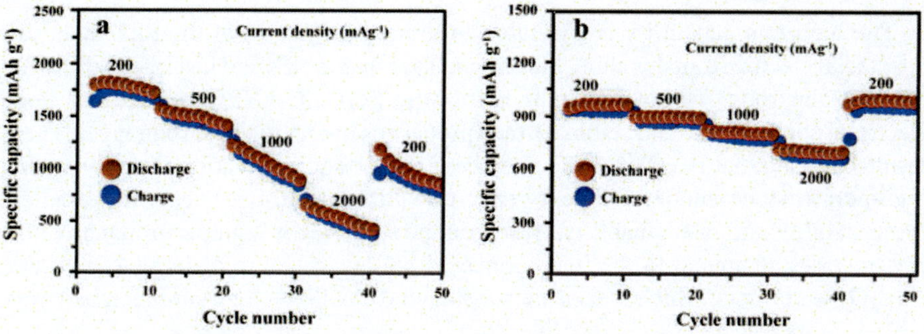

Fig. 9.12 Rate capability of **a** Fe_3O_4 hNFs, **b** Fe_3O_4/C hNFs, respectively. Adapted and reproduced from Ref. [89], Copyright 2015 Elsevier

Rate capability of Fe_3O_4/C hNFs is conducted in continuous 10 number of cycles for current densities ranging from 200 to 2000 mA g^{-1} (200, 500, 1000 and 2000 mA g^{-1}) and compared with that of Fe_3O_4 hNFs, as shown in Fig. 9.12 For Fe_3O_4 hollow nanofiber, the continuous capacity fade is observed in each cycle at any current density. The capacity fade from cycle to cycle is more pronounced in the higher current density. Also, the capacity of the Fe_3O_4 hNF electrode decreases precipitously to 47% of the initial value when the current restores to 200 mA g^{-1} (Fig. 9.12a) while the Fe_3O_4/C hNF electrode exhibits an excellent and stable rate capability (Fig. 9.12b). The Fe_3O_4/C hNF electrode exhibits a specific capacity of 774 mAh g^{-1} even at a high current density of 2000 mA g^{-1}, and the capacity was fully recovered when switch back the current density from higher value to the initial current density of 200 mA g^{-1}.

The stable capacity retention and high-rate capability of the Fe_3O_4/C hNF electrode are attributed to its unique hollow structure, limited formation of the irreversible SEI layer at the electrolyte electrode interface, fast interfacial electron transfer along the carbon layer on the NFs and fast Li-ion diffusion into the electrode [95–97]. The hollow structure of the Fe_3O_4/C hNF alleviates volume expansion/contraction of the interior upon Li-ion insertion/extraction and can also prevent physical detachment of active materials falling from the current collector owing to the interconnectivity as well as entanglement of the fiber [98, 99] while the carbon coating acts both as an efficient conducting and clamping layer that increases the electronic conductivity of the electrode and maintains the structural integrity during the charge/discharge process cycling process [55, 90, 93, 94]. The significantly higher-specific capacity, charge–discharge cycling stability and rate capability of Fe_3O_4/C hNFs are also due to the higher Li-ion diffusion coefficient than Fe_3O_4 hNFs or Fe_2O_3 hNFs. The Li-ion diffusion coefficient of the Fe_3O_4/C hNFs is reported to be 8.1×10^{-14} cm^2 s^{-1}, which is 60 times higher than that (1.33×10^{-14} cm^2 s^{-1}) of the Fe_3O_4 hNFs. The difference in Li-ion diffusion coefficient can be attributed to the presence of uniformly coated nanosized carbon layer over the Fe_2O_3 hollow nanofiber.

The high-rate capability is attributed to limited formation of the SEI layer, fast interfacial electron transfer along the carbon layer on the NFs and fast Li-ion diffusion into the electrode. This result implies that the Fe_3O_4/C hNF electrode has good electrical conductivity and stable surface properties owing to the carbon layer. These results indicate that Fe_3O_4/C hNFs may have important implications for developing high-performance anodes for next-generation lithium ion batteries with excellent cycle stability and rate capability. Based on the studies, the high performance of the electrodes is attributed to the interconnected hollow structure of large aspect ratio a-Fe_2O_3 nanofibers, which makes them a potential candidate for lithium ion batteries.

9.5 Graphene-Doped Carbon/Fe_3O_4 Porous Nanofibers

Graphene is a one atom thick (3.4 Å thick) planar sheet consisting of single layer of carbon atoms bound into a two-dimensional (2D) sp^2-sp^3 hybridized honeycomb lattice. It has received great excitement from the scientific and technological community, especially after uncovering some of its astonishing properties by many researchers over time and notably by A. K. Geim and K. S. Novoselov [100–102]. Graphene is known as the "mother of all graphite forms", as this base structure when introduced with defects can be molded into any of the 0–3-dimensional carbon structures [103, 104]. Particular interest by the scientific community on this 2D material comes from the fact that it possesses a package of astounding properties [102]. Theoretically, it has a large gravimetric specific surface area (2630 $m^2\ g^{-1}$), high intrinsic mobility (200,000 $cm^2v^{-1}\ s^{-1}$) [105, 106], high Young's modulus (~1.0 TPa) [107], thermal conductivity (~5000 $Wm^{-1}K^{-1}$) [108] and optical transmittance (~97.7%) [109]. Generally, several techniques, such as micromechanical cleavage [110], epitaxial growth on silicon carbide or metal [111] chemical vapor [112], thermal exfoliation, [113] mechanical exfoliation, [114] electrochemical exfoliation and chemical reduction of graphene oxide (GO) [115, 116], have been developed to produce graphene.

Recently, graphene has attracted considerable interest for utilization in encapsulating metal oxides with controlled microstructures to prepare graphene-based nanocomposite anodes and cathodes in rechargeable LIBs [117]. Graphene-based composite anodes such as graphene/Si [118], graphene/Sn [119], graphene/Co_3O [112, 114–116], graphene/Mn_3O [120], graphene/Fe_3O [120, 122], graphene/SnO [123, 124], graphene/TiO_2 [123, 125, 126] and graphene/CuO [127] have enhanced electrochemical properties, large reversible capacity, long-term cycling stability and good rate capability which were reported. These significantly improved properties of the composite electrodes mainly due to the spectacular electrochemical and physical properties of the graphene along with the combinative virtues between the heterogeneous components, which opens a new avenue for the exploration and exploitation of new electrochemical behaviors within a confined two-dimensional nanostructure interspace [128–130]. A considerable number of hierarchical structured anodes based on Fe_2O_3/graphene composite [51, 131–133] and Fe_3O_4/graphene composite having good electrochemical properties have been reported as anode for LIBs [134–137]. In all these studies, the composite is prepared by simply mixing with the active electrode material such as SnO_2, Fe_2O_3 or Fe_3O_4., or in situ synthesis the active electrode material in a graphene dispersed solution by hydrothermal, sol-gel, CVD method, etc. and resulted materials have the morphology of nanoparticle [138], nanosphere [139], nanorod [140], nanotube [135], nanoflake, nanoribbon, nanofiber [68], etc. Recently, Jianxin et al. [141] reported graphene-doped carbon/Fe_3O_4 porous nanofibers having relatively smooth surfaces with average fiber diameter of about 300 nm by in situ electrospinning and thermal treatment. GN@C/Fe_3O_4 porous nanofibers were prepared from a uniform solution of PAN/PMMA/GO/Fe(acac)$_3$. The electrospinning was carried out at an applied voltage of 20 kV, solution flow rate of 0.5 mL h^{-1}, and the distance from the needle to the collector was kept constant at 15 cm. The fibers were first oxidized at 250 °C followed by annealing at 650 °C under argon blanket. The electrochemical performance of the GN@C/Fe_3O_4 porous nanofibers was compared with that of C/Fe_3O_4 and porous C/Fe_3O_4 and nanocomposite nanofibers. PAN/Fe(acac)$_3$ and PAN/PMMA/Fe(acac)$_3$ were used as precursor for the preparation of C/Fe_3O_4 and porous C/Fe_3O_4 nanocomposite nanofibers, respectively, and they are prepared via the same procedures used to prepare the porous GN@C/Fe_3O_4 nanofibers. The methods of preparation of nanocomposite nanofibers are displayed in Fig. 9.13 [141].

The morphological studies on as-prepared C/Fe_3O_4, porous C/Fe_3O_4 and porous GN@C/Fe_3O_4 samples have an average fiber diameter of 300 nm and are uniformly decorated with Fe_3O_4 nanoparticles. After the graphitization, the fiber diameter is reduced to 250 nm due to the thermal decomposition of PMMA and PAN with different char yields and forms pores in the carbon matrix. The fiber surface constitute of Fe_3O_4 nanoparticle embedded in the porous carbon matrix ranging from 10 to 20 nm along with a few agglomerated nanoparticles of particle size ranges from 20 to 30 nm. Compared to other nanocomposite fiber, the C/Fe_3O_4 nanocomposite nanofibers have more smooth surface morphology. Because of the improved electronic conductivity of the spinning solution imparted by doping with graphene, the porous GN@C/Fe_3O_4 nanofibers exhibited uniform fiber topography and reduced average fiber diameter of 150 nm after graphitization. In addition to the embedded Fe_3O_4 nanoparticles in the carbon matrix, some Fe_3O_4 nanoparticles of size range from 20 to 30 nm are extruded from the fiber surface. Also the nanoparticles having the particle size of about 10 nm are arranged along the fiber axis to form bands on as-prepared porous GN@C/Fe_3O_4 nanocomposite nanofibers. The formation of this distinct band structure made wide channels between the bands and micro/mesoporous within the bands. In porous GN@C/Fe_3O_4 nanocomposite nanofibers, the Fe_3O_4 nanocrystals anchored onto bent graphene nanosheets embedded in the amorphous carbon matrix and Fe_3O_4 particles protrude from the fiber surface are wrapped in 2D graphene nanosheets. The porous GN@C/Fe_3O_4 nanocomposite nanofiber has high-specific surface area and pore volume of 323.0 $m^2 g^{-1}$ and 0.337 $cm^3 g^{-1}$, respectively, which is about 220 and 280% higher than C/Fe_3O_4 nanocomposite

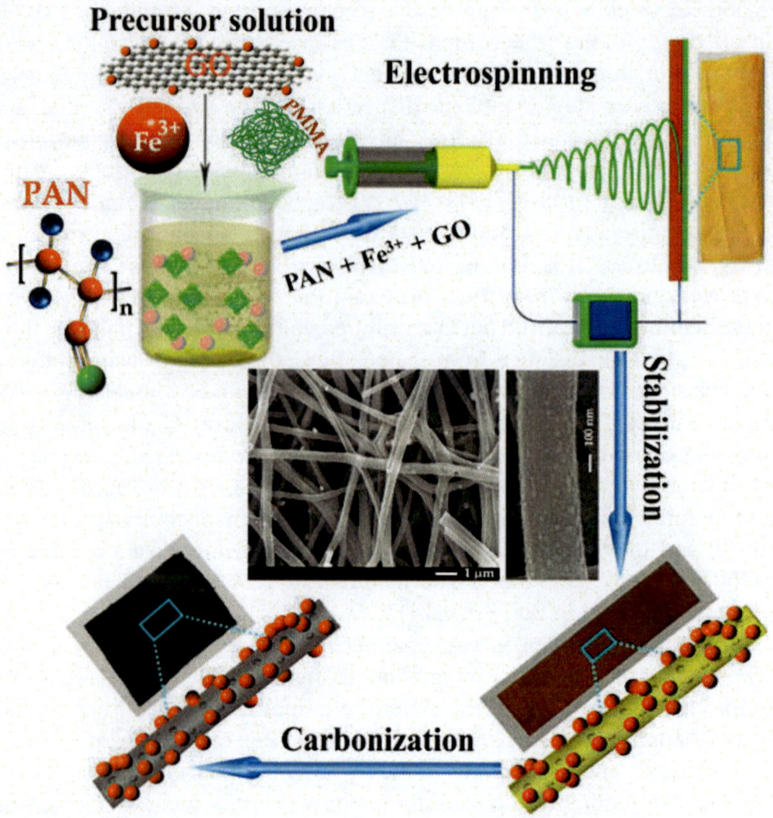

Fig. 9.13 Schematic illustration of the detailed synthesis procedure and structural diagram for the porous GN@C/Fe_3O_4 composites. Adapted and reproduced from Ref. [141]. Copyright 2017 Elsevier

nanofiber. The C/Fe_3O_4 sample only contains mesopores with size ranges of 3–15 nm while the porous C/Fe_3O_4 and porous GN@C/Fe_3O_4 have not only meso-porous (4 nm in size), but also micropores. Again, porous GN@C/Fe_3O_4 has a higher proportion of micropores than porous C/Fe_3O_4. The higher-specific surface area and pore volume of porous C/Fe_3O_4 and GN@C/Fe_3O_4 nanocomposite nanofibers are attributed to the decomposition of PMMA. The hierarchical pore distribution and the porosity in porous GN@C/Fe_3O_4 are suitable for accommodating the large volume changes in Fe_3O_4 active material and rapid Li-ion diffusion that occur during charge/discharge. An illustration of the possible formation process of porous GN@C/Fe_3O_4 nanofibers is displayed in Fig. 9.14. The Raman spectroscopic studies showed that both C/Fe_3O_4 and nanocomposite nanofiber have highly disordered carbon structure, which is ascribed to the relatively low carbonization temperature. The graphitization of GN@C/Fe_3O_4 nanocomposite nanofibers at 650 °C under argon blanket thermally reduces the GO to rGO.

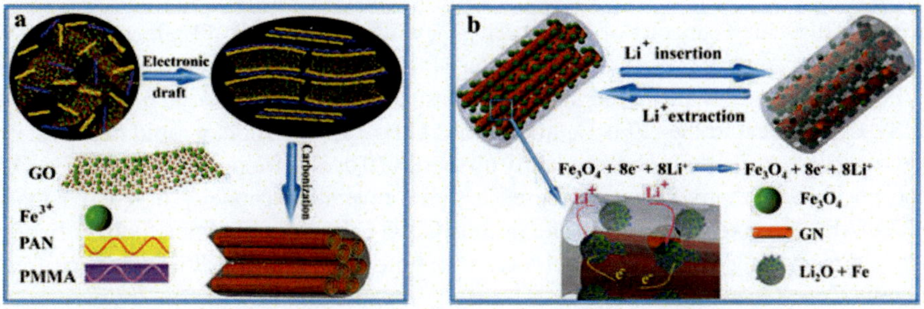

Fig. 9.14 **a** Possible formation mechanism of porous GN@C/Fe_3O_4 composite, **b** schematic illustration of electron transmission and the mechanism of Li^+ -ion insertion/extraction in porous GN@C/Fe_3O_4. Adapted and reproduced from Ref. [141], Copyright 2017 Elsevier

The electrochemical studies in CR2032 coin half-cells using lithium metal foil as counter electrode and Celgard 2400 films wet with 1 mol $LiPF_6$ in a mixture of ethylene carbonate/dimethyl carbonate/ethyl methyl carbonate (EC/DEC/EMC, 1:1:1, v/v) as separators. Working electrodes were prepared by directly attaching the composite nanofibers onto a copper foil with a PVdF binder kept in the weight ratio of the active materials to binder at 8:1. The first specific discharge capacities of the C/Fe_3O_4, porous C/Fe_3O_4 and porous GN@C/Fe_3O_4 nanocomposites are 1252, 1389 and 1412 mAh g^{-1}, respectively. The higher initial specific capacity than theoretical capacity of Fe_3O_4 (924 mAh g^{-1}) or theoretical predictions of the specific capacity for the GN@C/Fe_3O_4 (586 mAh g^{-1}), porous C/Fe_3O_4 (637 mAh g^{-1}) and C/Fe_3O_4 (657.3 mAh g^{-1}) is responsible for formation of the SEI film and possibly by interfacial Li-ion storage during the first discharge process [142]. Compared with C/Fe_3O_4, the porous C/Fe_3O_4 and porous GN@C/Fe_3O_4 composites show initially high charge/discharge capacities, which are associated with the porous structure of the composites and the extra lithium storage capacity caused by their high-specific surface areas [143, 144] improved specific capacity of porous GN@C/Fe_3O_4 nanocomposites obviously due to the higher-specific surface area, pore volume, the hierarchical pore distribution and the porosity, the combination of mesoporous and microporous structure in combination with higher electronic conductivity offered by the graphene nanosheets. The porous C/Fe_3O_4 nanofibers [145] showed a higher reversible specific capacity than most of C/Fe_3O_4 nanofibers, graphene@Fe_3O_4 and C/Fe_3O_4 nanoparticles [141, 142] electrodes reported. A few of the C/Fe_3O_4 samples show a higher reversible specific capacity than the GN@C/Fe_3O_4 sample we prepared, which may be due to the much higher content of Fe_3O_4 and graphene in their samples. The initial Coulombic efficiency (CE) of the porous GN@C/Fe_3O_4 electrode (74.4%) is also higher other electrode, that may ascribed from the protection of the flexible graphene nanosheets, higher electronic conductivity and the unique banded structure of the material.

Under continuous charge–discharge cycling at a current density of 100 mA g^{-1}, the specific capacity of the C/Fe_3O_4 electrode rapidly decreases to 480 mAh g^{-1} from an initial discharge capacity of 1320 mAh g^{-1}, while porous C/Fe_3O_4 electrode shows a reversible capacity of 759 mAh g^{-1} after 30. The better cycling stability of porous C/Fe_3O_4 electrode is attributed to the porous structure of the electrode material [135, 148]. On contrast to the C/Fe_3O_4 and porous C/Fe_3O_4 electrode, a rapid decrease in the reversible capacity is observed only up to first 10 cycles for porous GN@C/Fe_3O_4 composite, and thereafter, it delivered a stable reversible capacity of greater than 872 mAh g^{-1} up to 100 cycles. The porous GN@C/Fe_3O_4 nanofibers deliver better long-term cycling stability and higher reversible capacity than the graphene-free porous C/Fe_3O_4 nanofibers.

The rate performance of C/Fe_3O_4, porous C/Fe_3O_4 and porous GN@C/Fe_3O_4 electrodes from 0.1 to 5 C shows (Fig. 9.15a) the porous GN@C/Fe_3O_4 good rate capability of these electrodes. The porous C/Fe_3O_4 composite shows better rate performance than the Fe_3O_4/C composite. At current density of 5C, the porous C/Fe_3O_4 and C/Fe_3O_4 nanocomposite fibrous electrode delivers a reversible capacity of 330 and 248 mAh g^{-1}, respectively, while the porous GN@C/Fe_3O_4 electrode delivers a higher capacity with excellent cycling stability at each current density. Even at a high C rate of 5C, the discharge capacity of the porous GN@C/Fe_3O_4 electrode is maintained in 455 mAh g^{-1}. When the current rate is switched back to 0.1 C, after 70 cycles at higher C rate, the porous C/Fe_3O_4 electrode recovers the reversible capacity to 700 mAh g^{-1}, whereas the recovered value of the C/Fe_3O_4 electrode is 512 mAh g^{-1}, which corresponds to the 75–80% of its initial capacity. But in the case of porous GN@C/Fe_3O_4 electrode, when the current density is switched back to 0.1 C from 5C, the porous GN@C/Fe_3O_4 electrode is completely recovered (100%) its initial capacity. The porous GN@C/Fe_3O_4 electrode delivers a higher capacity with excellent cycling stability than porous C/Fe_3O_4 electrode, at each current density (Fig. 9.15b) which is attributed by advantageous structural features.

The enhanced cycling and rate performance of the porous C/Fe_3O_4 electrode tan C/Fe_3O_4 electrode are attributed to its micro/mesoporous structure. It is widely recognized that the porous structure can improve the specific surface area, leading

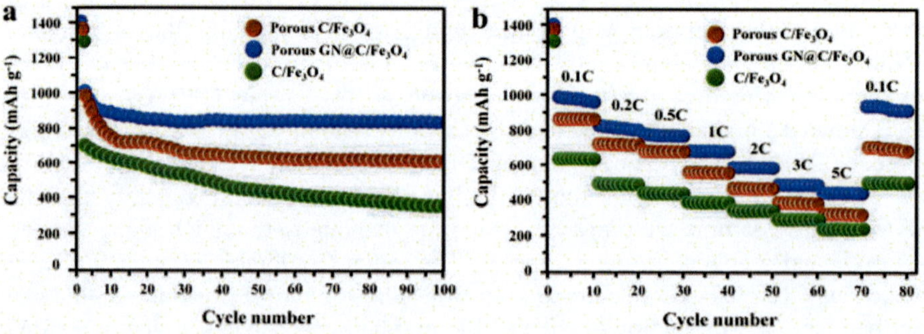

Fig. 9.15 **a** Cycling performance at a current density of 100 mAh g^{-1}, **b** rate performance at different current densities of C/Fe_3O_4, porous C/Fe_3O_4 and porous GN@C/Fe_3O_4 composite electrodes. Adapted and reproduced from Ref. [141]. Copyright 2017 Elsevier

to larger reversible pseudo capacitive surface for Li-ion storage. The porous structure can effectively accommodate the volume changes resulting from Li-ion insertion/extraction during the continuous charge–discharge cycling and buffers larger structural changes at higher current densities. The excellent long-term cycling stability and rate capability of porous GN@C/Fe_3O_4 nanocomposite electrode are associated with the novel banded structure of the porous GN@C/Fe_3O_4 composite electrode. The band morphologies with hierarchical porous structure of the porous GN@C/Fe_3O_4 composite electrode not only increase the number of active sites for lithium intercalation, but also make sufficient electrode/electrolyte interface contact, providing more short pathways for lithium ion insertion/extraction. Again the graphene-doped carbon nanofiber webs can serve as efficiency channels for electron transportation. The graphene nanosheets wrapped around the active material could prevent the aggregation of the Fe_3O_4 nanoparticles and provide better mechanical integrity to the electrode. The unique banded structure, hierarchical porous structure and presence of flexible graphene enable to buffer the volume expansion/contraction which occurs during the charge/discharge process. Therefore, the continual rupturing and growth of the SEI film caused by mechanical strain and continuous consumption of the electrolyte are substantially less than that on the C/Fe_3O_4 or porous C/Fe_3O_4 nanocomposite electrode. Hence, the porous GN@C/Fe_3O_4 nanocomposite electrode exhibited higher Coulombic efficiency, rate capability and cycling stability than the C/Fe_3O_4 or porous C/Fe_3O_4 nanocomposite electrode, because the cyclical rupture and growth of SEI films can lead to reduced Coulombic efficiency [149]. In addition, the zero-valent iron and the graphene nanosheets serve as conductive agents to significantly improve the conductivity of the electrode. The formation of continuous electronically conducting channel formed by the graphene nanosheets on the nanofiber facilitates an efficient transport pathway for ions and electrons diffusion during the charge/discharge process and effective transfer of electron to the current collector. Table 9.2 represents the electrochemical performance of the α-Fe_2O_3 hollow nanofibers electrode with those of α-Fe_2O_3 with different structures synthesized by different methods reported in the literature.

9.6 Conclusion

For the advanced energy storage application and to meet the high performance of LIBs, developing battery components that can deliver the best electrochemical performance is prime important. For the best performance, the electrode material chose should exhibit a fast lithium ion insertion/release kinetics, high lithium ion storage and good electronic as well as ionic conductivity and should exhibit superior mechanical strength for the stabilization of lithium ion transfer. Being an anode material in LIBs, the application of iron oxide was limited by the low electrical conductivity, poor capacity retention and unstable SEI formation. On the effort of developing improved electrochemical performance in iron oxide-based anodes, carbon-based composites

Table 9.2 Comparison of the electrochemical performance of the α-Fe_2O_3 hollow nanofibers electrode with those of α-Fe_2O_3 with different structures synthesized by different methods reported in the literature

Samples	Synthesis method	Potential (V)	Current rate (C)	Initial capacity (mAh g^{-1})	Capacity retention (mAh g^{-1})/(cycles)	Ref.
α-Fe_2O_3	Electrospinning	3.0–0.005	C/16	1795	1293 (40)	[88]
			1C	1276	900 (30)	
			2 C	948	527 (30)	
			10 C	538	370 (30)	
α-Fe_2O_3 mesoporous hollow spheres	Solvothermal	3.0–0.5	C/16	81,279	830 (15)	[150]
α -Fe_2O_3 nanospheres	Hydrothermal	3.0–0.01	C/50.3	1248	586 (30)	[151]
α -Fe_2O_3 nanowires				1199	549 (30)	
α -Fe_2O_3 nanosheets				1190	442 (30)	
α- Fe_2O_3 nanocuboids				1067	360 (30)	
α-Fe_2O_3 nanorods	Hydrothermal	3.0–0.01	C/10	1332	763 (30)	[152]
α-Fe_2O_3 submicro-flower	Hydrothermal	3.0–0.01	C/50.3	1248	650 (50)	[153]
			C/25.1	1226	575 (50)	
			C/16.8	1171	520 (50)	
			C/12.6	1021	421 (50)	
			C/10.1	906	325 (50)	
α-Fe_2O_3 CMC	Ball milling	3.0–0.005	C/20 then C/5	1280	795 (90)	[93]
α-Fe_2O_3 nanotubes	Thermal decomposition	3.0–0.01	C/10.1	1415	510 (100)	[92]
α-Fe_2O_3	Molten salts	3.0–0.01	C/2	1688	903 (100)	[154]

(continued)

Table 9.2 (continued)

Samples	Synthesis method	Potential (V)	Current rate (C)	Initial capacity (mAh g^{-1})	Capacity retention (mAh g^{-1})/(cycles)	Ref.
C/Fe_3O_4 composite nanofibers	Electrospinning		200	1551	1007 (80)	[68]
Porous Fe_3O_4/C Microbelts	Electrospinning		100	1751	710 (50)	[72]
Fe_3O_4/carbon composites	Electrospinning		100	876.6	508.2 (100)	[155]
Fe_3O_4@PCF composite	Electrospinning		200	1409	857 (80)	[156]
Fe_3O_4@C hNFs	Electrospinning		1000	963	978 (150)	[89]
Carbon@Fe_3O_4 core–shell nanofiber			200	1520	847 (100)	[157]
Fe_3O_4–TiO_2–carbon composite nanofiber			100	1340	525 (100)	[158]
Ferroferric oxide/porous carbon nanofiber composites	Electrospinning		500	1422.1	717.2 (100)	[71]
GN@C/Fe_3O_4 nanofibers	Electrospinning		100	1412	872 (100)	[141]
Flexible Fe_3O_4/C nanofibers	Electrospinning		500	1161	762 (300)	[69]

and one-dimensional hollow structure were developed using electrospinning. Electrospinning is considered as the best method that can result in 1D structure with high lithium ion kinetics which will directly influence the performance of battery. For most of the electrode materials, presence of carbon is considered to be best for delivering an enhanced electronic conductivity. In iron oxide-based anodes as well, carbon-based composites are capable to exhibit high electronic conductivity, specific capacity and superior cycling stability. This is attributed by the presence of conducting carbon. Similarly, 1D hollow-structured iron oxide synthesized using electrospinning results in an increase in the electrode–electrolyte contact area that will enhance the electrochemical properties. Compare to other techniques, the electrospunned iron oxide-based hollow spheres are superior to exhibit excellent initial capacity and capacity retention. From this chapter it is evident that the future LIBs are at its best was possible only with the exploration of transition metal oxide electrodes and iron oxide based electrodes that can be a proising anode since it can ensure good electrochemical performance by a reversible conversion reaction.

References

1. Santangelo S (2019) Electrospun nanomaterials for energy applications: recent advances. Appl Sci 9:1049. https://doi.org/10.3390/app9061049
2. Manwell JF, McGowan JG (1993) Lead acid battery storage model for hybrid energy systems. Sol Energy 50:399–405. https://doi.org/10.1016/0038-092X(93)90060-2
3. Sebastián R, Alzola RP (2010) Effective active power control of a high penetration wind diesel system with a Ni–Cd battery energy storage. Renew Energy 35:952–965. https://doi.org/10.1016/j.renene.2009.11.029
4. Hammouche A, Karden E, De Doncker RW (2004) Monitoring state-of-charge of Ni-MH and Ni-Cd batteries using impedance spectroscopy. J Power Sources 127:105–111. https://doi.org/10.1016/j.jpowsour.2003.09.012
5. Feng F (2001) Electrochemical behaviour of intermetallic-based metal hydrides used in Ni/metal hydride (MH) batteries: a review. Int J Hydrogen Energy 26:725–734. https://doi.org/10.1016/S0360-3199(00)00127-0
6. Yazami R, Touzain P (1983) A reversible graphite-lithium negative electrode for electrochemical generators. J Power Sources 9:365–371. https://doi.org/10.1016/0378-7753(83)87040-2
7. Brandt K (1994) Historical development of secondary lithium batteries. Solid State Ionics 69:173–183. https://doi.org/10.1016/0167-2738(94)90408-1
8. Wen Z, Lu G, Mao S et al (2013) Silicon nanotube anode for lithium-ion batteries. Electrochem Commun 29:67–70. https://doi.org/10.1016/j.elecom.2013.01.015
9. Ma Z, Li T, Huang YL et al (2013) Critical silicon-anode size for averting lithiation-induced mechanical failure of lithium-ion batteries. RSC Adv 3:7398–7402. https://doi.org/10.1039/c3ra41052h
10. Song W, Nie Y, Sun S (2016) Author' s accepted manuscript porous Fe_2O_3 nanotubes as advanced anode for high performance lithium ion batteries. Ceram Int. https://doi.org/10.1016/j.ceramint.2016.09.166
11. Liu N, Shen J, Liu D (2013) A Fe_2O_3 nanoparticle carbon aerogel compsite for use as an electrode material for lithium ion batteries. Electrochim Acta 97:271–277. https://doi.org/10.1016/j.electacta.2013.02.111

12. Xiong QQ, Lu Y, Wang XL, et al (2012) Improved electrochemical performance of porous Fe_3O_4/ carbon core/shell nanorods as an anode for lithium-ion batteries. 536:219–225. https://doi.org/10.1016/j.jallcom.2012.05.034
13. Zhou G, Wang D, Li F, et al (2010) Graphene-Wrapped Fe_3O_4 anode material with improved reversible capacity and cyclic stability for lithium ion batteries. 5306–5313. https://doi.org/10.1021/cm101532x
14. Park SJ, Kim YJ, Lee H (2011) Synthesis of carbon-coated TiO_2 nanotubes for high-power lithium-ion batteries. J Power Sources 196(11):5133–5137
15. Zhang Z, Zhou Z, Nie S, Wang H, Peng H, Li G, Chen K (2014) Flower-like hydrogenated TiO_2 (B) nanostructures as anode materials for high-performance lithium ion batteries. J Power Sources 267:388–393
16. Dos Santos MC, Kesler O, Reddy ALM (2012) Nanomaterials for energy conversion and storage. J. Nanomater. 2012:2–4
17. Sun MH, Huang SZ, Chen LH et al (2016) Applications of hierarchically structured porous materials from energy storage and conversion, catalysis, photocatalysis, adsorption, separation, and sensing to biomedicine. Chem Soc Rev 45:3479–3563. https://doi.org/10.1039/c6cs00135a
18. Osiak M, Geaney H, Armstrong E, O'Dwyer C (2014) Structuring materials for lithium-ion batteries: advancements in nanomaterial structure, composition, and defined assembly on cell performance. J Mater Chem A 2:9433–9460. https://doi.org/10.1039/c4ta00534a
19. Wang Y, Li H, He P et al (2010) Nano active materials for lithium-ion batteries. Nanoscale 2:1294–1305. https://doi.org/10.1039/c0nr00068j
20. Zhang WJ (2011) A review of the electrochemical performance of alloy anodes for lithium-ion batteries. J Power Sources 196:13–24. https://doi.org/10.1016/j.jpowsour.2010.07.020
21. Kim Y, Song W, Lee SY et al (2011) Low-temperature synthesis of graphene on nickel foil by microwave plasma chemical vapor deposition. Appl Phys Lett 98:1–4. https://doi.org/10.1063/1.3605560
22. Fujimoto H, Tokumitsu K, Mabuchi A et al (2010) The anode performance of the hard carbon for the lithium ion battery derived from the oxygen-containing aromatic precursors. J Power Sources 195:7452–7456. https://doi.org/10.1016/j.jpowsour.2010.05.041
23. Yang J, Zhou XY, Li J et al (2012) Study of nano-porous hard carbons as anode materials for lithium ion batteries. Mater Chem Phys 135:445–450. https://doi.org/10.1016/j.matchemphys.2012.05.006
24. Bridges CA, Sun XG, Zhao J et al (2012) In situ observation of solid electrolyte interphase formation in ordered mesoporous hard carbon by small-angle neutron scattering. J Phys Chem C 116:7701–7711. https://doi.org/10.1021/jp3012393
25. Meunier V, Kephart J, Roland C, Bernholc J (2002) Ab initio investigations of lithium diffusion in carbon nanotube systems. Phys Rev Lett 88:4. https://doi.org/10.1103/PhysRevLett.88.075506
26. Schauerman CM, Ganter MJ, Gaustad G et al (2012) Recycling single-wall carbon nanotube anodes from lithium ion batteries. J Mater Chem 22:12008–12015. https://doi.org/10.1039/c2jm31971c
27. Nishidate K, Hasegawa M (2005) Energetics of lithium ion adsorption on defective carbon nanotubes. Phys Rev B Condens Matter Mater Phys 71:1–6. https://doi.org/10.1103/PhysRevB.71.245418
28. Zhao J, Buldum A, Han J, Lu JP (1999) First principles study of Li intercalated carbon nanotube ropes. Phys Rev Lett 85:1–12. https://doi.org/10.1103/PhysRevLett.85.1706
29. Hou J, Shao Y, Ellis MW et al (2011) Graphene-based electrochemical energy conversion and storage: fuel cells, supercapacitors and lithium ion batteries. Phys Chem Chem Phys 13:15384–15402. https://doi.org/10.1039/c1cp21915d
30. Chen Z, Belharouak I, Sun YK, Amine K (2013) Titanium-based anode materials for safe lithium-ion batteries. Adv Funct Mater 23:959–969. https://doi.org/10.1002/adfm.201200698
31. Szczech JR, Jin S (2011) Nanostructured silicon for high capacity lithium battery anodes. Energy Environ Sci 4:56–72. https://doi.org/10.1039/C0EE00281J

32. Rudawski NG, Yates BR, Holzworth MR et al (2013) Ion beam-mixed Ge electrodes for high capacity Li rechargeable batteries. J Power Sources 223:336–340. https://doi.org/10.1016/j.jpowsour.2012.09.056
33. Chockla AM, Klavetter KC, Mullins CB, Korgel BA (2012) Solution-grown germanium nanowire anodes for lithium-ion batteries. ACS Appl Mater Interfaces 4:4658–4664. https://doi.org/10.1021/am3010253
34. Bruce PG, Scrosati B, Tarascon J-M, Bruce PG (2008) Lithium batteries nanomaterials for rechargeable lithium batteries. Reviews 47:2930–2946. https://doi.org/10.1002/anie.200702505
35. Park CM, Kim JH, Kim H, Sohn HJ (2010) Li-alloy based anode materials for Li secondary batteries. Chem Soc Rev 39:3115–3141. https://doi.org/10.1039/b919877f
36. Wang Z, Zhou L, Lou XW (2012) Metal oxide hollow nanostructures for lithium-ion batteries. Adv Mater 24:1903–1911. https://doi.org/10.1002/adma.201200469
37. Yang J, Takeda Y, Imanishi N et al (2002) SiO_x -based anodes for secondary lithium batteries. Solid State Ionics 152–153:125–129. https://doi.org/10.1016/S0167-2738(02)00362-4
38. Jiang J, Li Y, Liu J et al (2012) Recent advances in metal oxide-based electrode architecture design for electrochemical energy storage. Adv Mater 24:5166–5180. https://doi.org/10.1002/adma.201202146
39. Prosini PP, Carewska M, Loreti S et al (2000) Lithium iron oxide as alternative anode for Li-ion batteries. Int J Inorg Mater 2:365–370. https://doi.org/10.1016/S1466-6049(00)00028-3
40. Lai CH, Lu MY, Chen LJ (2012) Metal sulfide nanostructures: synthesis, properties and applications in energy conversion and storage. J Mater Chem 22:19–30. https://doi.org/10.1039/c1jm13879k
41. Boyanov S, Annou K, Villevieille C et al (2008) Nanostructured transition metal phosphide as negative electrode for lithium-ion batteries. Ionics (Kiel) 14:183–190. https://doi.org/10.1007/s11581-007-0170-3
42. Ji L, Lin Z, Alcoutlabi M, Zhang X (2011) Recent developments in nanostructured anode materials for rechargeable lithium-ion batteries. Energy Environ Sci 4:2682–2689. https://doi.org/10.1039/c0ee00699h
43. Xu JS, Zhu YJ (2012) Monodisperse Fe_3O_4 and γ-Fe_2O_3 magnetic mesoporous microspheres as anode materials for lithium-ion batteries. ACS Appl Mater Interfaces 4:4752–4757. https://doi.org/10.1021/am301123f
44. Koo B, Xiong H, Slater MD et al (2012) Hollow iron oxide nanoparticles for application in lithium ion batteries. Nano Lett 12:2429–2435. https://doi.org/10.1021/nl3004286
45. Liu J, Li Y, Fan H et al (2010) Iron oxide-based nanotube arrays derived from sacrificial template-accelerated hydrolysis: large-area design and reversible lithium storage. Chem Mater 22:212–217. https://doi.org/10.1021/cm903099w
46. Kang N, Park JH, Choi J et al (2012) Nanoparticulate iron oxide tubes from microporous organic nanotubes as stable anode materials for lithium ion batteries. Angew Chemie - Int Ed 51:6626–6630. https://doi.org/10.1002/anie.201202244
47. Ma XH, Feng XY, Song C et al (2013) Facile synthesis of flower-like and yarn-like α-Fe_2O_3 spherical clusters as anode materials for lithium-ion batteries. Electrochim Acta 93:131–136. https://doi.org/10.1016/j.electacta.2013.01.096
48. Mitra S, Poizot P, Finke A, Tarascon JM (2006) Growth and electrochemical characterization versus lithium of Fe_3O_4 electrodes made via electrodeposition. Adv Funct Mater 16:2281–2287. https://doi.org/10.1002/adfm.200500753
49. Zhu X, Wu W, Liu Z et al (2013) A reduced graphene oxide-nanoporous magnetic oxide iron hybrid as an improved anode material for lithium ion batteries. Electrochim Acta 95:24–28. https://doi.org/10.1016/j.electacta.2013.02.029
50. Hwang JK, Lim HS, Sun YK, Do SuhK (2013) Monodispersed hollow carbon/Fe_3O_4 composite microspheres for high performance anode materials in lithium-ion batteries. J Power Sources 244:538–543. https://doi.org/10.1016/j.jpowsour.2013.02.017

51. Kim IT, Magasinski A, Jacob K et al (2013) Synthesis and electrochemical performance of reduced graphene oxide/maghemite composite anode for lithium ion batteries. Carbon N Y 52:56–64. https://doi.org/10.1016/j.carbon.2012.09.004
52. Kim HS, Piao Y, Kang SH et al (2010) Uniform hematite nanocapsules based on an anode material for lithium ion batteries. Electrochem Commun 12:382–385. https://doi.org/10.1016/j.elecom.2009.12.040
53. Liu H, Wexler D, Wang G (2009) One-pot facile synthesis of iron oxide nanowires as high capacity anode materials for lithium ion batteries. J Alloys Compd 487:24–27. https://doi.org/10.1016/j.jallcom.2009.08.043
54. Muraliganth T, Vadivel Murugan A, Manthiram A (2009) Facile synthesis of carbon-decorated single-crystalline Fe_3O_4 nanowires and their application as high performance anode in lithium ion batteries. Chem Commun 7360–7362. https://doi.org/10.1039/b916376j
55. Kang N, Park JH, Jin M et al (2013) Microporous organic network hollow spheres: useful templates for nanoparticulate Co_3O_4 hollow oxidation catalysts. J Am Chem Soc 135:19115–19118. https://doi.org/10.1021/ja411263h
56. Wang B, Chen JS, Bin WuH et al (2011) Quasiemulsion-templated formation of α-Fe_2O_3 hollow spheres with enhanced lithium storage properties. J Am Chem Soc 133:17146–17148. https://doi.org/10.1021/ja208346s
57. Xu X, Cao R, Jeong S, Cho J (2012) Spindle-like mesoporous α-Fe_2O_3 anode material prepared from MOF template for high-rate lithium batteries. Nano Lett 12:4988–4991. https://doi.org/10.1021/nl302618s
58. Wu C, Yin P, Zhu X et al (2006) Synthesis of hematite (α-Fe_2O_3) nanorods: diameter-size and shape effects on their applications in magnetism, lithium ion battery, and gas sensors. J Phys Chem B 110:17806–17812. https://doi.org/10.1021/jp0633906
59. Sohn H, Chen Z, Jung YS et al (2013) Robust lithium-ion anodes based on nanocomposites of iron oxide-carbon-silicate. J Mater Chem A 1:4539–4545. https://doi.org/10.1039/c2ta00443g
60. Raghavan P, Manuel J, Zhao X et al (2011) Preparation and electrochemical characterization of gel polymer electrolyte based on electrospun polyacrylonitrile nonwoven membranes for lithium batteries. J Power Sources 196:6742–6749. https://doi.org/10.1016/j.jpowsour.2010.10.089
61. Lindsay MJ, Wang GX, Liu HK (2003) Al-based anode materials for Li-ion batteries. J Power Sources 119–121:84–87. https://doi.org/10.1016/S0378-7753(03)00130-7
62. Fan J, Wang T, Yu C et al (2004) Ordered, nanostructured tin-based oxides/carbon composite as the negative-electrode material for lithium-ion batteries. Adv Mater 16:1432–1436. https://doi.org/10.1002/adma.200400106
63. Wang Y, Lee JY, Zeng HC (2005) Polycrystalline SnO_2 nanotubes prepared via infiltration casting of nanocrystallites and their electrochemical application. Chem Mater 17:3899–3903. https://doi.org/10.1021/cm050724f
64. Wang L, Yu Y, Chen PC, Chen CH (2008) Electrospun carbon-cobalt composite nanofiber as an anode material for lithium ion batteries. Scr Mater 58:405–408. https://doi.org/10.1016/j.scriptamat.2007.10.024
65. Zhang X, Liu H, Petnikota S et al (2014) Electrospun Fe_2O_3-carbon composite nanofibers as durable anode materials for lithium ion batteries. J Mater Chem A 2:10835–10841. https://doi.org/10.1039/c3ta15123a
66. Abe J, Kawase K, Tachikawa N et al (2017) Influence of carbonization temperature and press processing on the electrochemical characteristics of self-standing iron oxide/carbon composite electrospun nanofibers. RSC Adv 7:32812–32818. https://doi.org/10.1039/c7ra05301k
67. Wu L, Xiao Q, Li Z et al (2012) $CoFe_2O_4$/C composite fibers as anode materials for lithium-ion batteries with stable and high electrochemical performance. Solid State Ionics 215:24–28. https://doi.org/10.1016/j.ssi.2012.03.044
68. Wang L, Yu Y, Chen PC et al (2008) Electrospinning synthesis of C/Fe_3O_4 composite nanofibers and their application for high performance lithium-ion batteries. J Power Sources 183:717–723. https://doi.org/10.1016/j.jpowsour.2008.05.079

69. Wu Q, Zhao R, Zhang X et al (2017) Synthesis of flexible Fe_3O_4/C nanofibers with buffering volume expansion performance and their application in lithium-ion batteries. J Power Sources 359:7–16. https://doi.org/10.1016/j.jpowsour.2017.05.020
70. Xu X, Wan Y, Liu J et al (2017) Encapsulating iron oxide@carbon in carbon nanofibers as stable electric conductive network for lithium-ion batteries. Electrochim Acta 246:766–775. https://doi.org/10.1016/j.electacta.2017.06.078
71. Zhu S, Chen M, Sun J et al (2016) Novel highly conductive ferroferric oxide/porous carbon nanofiber composites prepared by electrospinning as anode materials for high performance Li-ion batteries. RSC Adv 6:58529–58540. https://doi.org/10.1039/c6ra04090j
72. Lang L, Xu Z (2013) In situ synthesis of porous Fe_3O_4/C microbelts and their enhanced electrochemical performance for lithium-ion batteries. ACS Appl Mater Interfaces 5:1698–1703. https://doi.org/10.1021/am302753p
73. Hamideh Mortazavi S, Pilehvar S, Ghoranneviss M et al (2013) Plasma oxidation and stabilization of electrospun polyacrylonitrile nanofiber for carbon nanofiber formation. Appl Phys A Mater Sci Process 113:703–712. https://doi.org/10.1007/s00339-013-7707-2
74. Larcher D, Bonnin D, Cortes R et al (2003) Combined XRD, EXAFS, and Mössbauer studies of the reduction by lithium of α-Fe_2O_3 with various particle sizes. J Electrochem Soc 150:1643–1650. https://doi.org/10.1149/1.1622959
75. Verrelli R, Brescia R, Scarpellini A, Manna L, Scrosati B, Hassoun J (2014) A lithium ion battery exploiting a composite Fe_2O_3 anode and a high voltage Li 1.35 Ni 0.48 Fe 0.1 Mn 1.72 O 4 cathode. RSC Adv 4(106):61855–61862
76. Cabana J, Monconduit L, Larcher D, Palacín MR (2010) Beyond intercalation-based Li-ion batteries: the state of the art and challenges of electrode materials reacting through conversion reactions. Adv Mater 22:170–192. https://doi.org/10.1002/adma.201000717
77. Lee JK, An KW, Ju JB et al (2001) Electrochemical properties of PAN-based carbon fibers as anodes for rechargeable lithium ion batteries. Carbon N Y 39:1299–1305. https://doi.org/10.1016/S0008-6223(00)00237-2
78. Poizot P, Laruelle S, Grugeon S, Tarascon JM (2002) Rationalization of the low-potential reactivity of 3d-metal-based inorganic compounds toward Li. J Electrochem Soc 149. https://doi.org/10.1149/1.1497981
79. Courtney IA, Dahn JR (1997) Electrochemical and in situ X-ray diffraction studies of the reaction of lithium with tin oxide composites. J Electrochem Soc 144:2045–2052. https://doi.org/10.1149/1.1837740
80. Xue JS, Dahn JR (1995) Dramatic effect of oxidation on lithium insertion in carbons made from epoxy resins. J Electrochem Soc 142:3668–3677. https://doi.org/10.1149/1.2048397
81. Kim C, Yang KS, Kim YJ, Endo M (2003) Heat treatment temperature effects on structural and electrochemical properties of PVDC-based disordered carbons. J Mater Sci 38:2987–2991. https://doi.org/10.1023/A:1024429828872
82. Gnanaraj JS, Levi MD, Levi E, et al (2001) Comparison between the electrochemical behavior of disordered carbons and graphite electrodes in connection with their structure. J Electrochem Soc 148. https://doi.org/10.1149/1.1368096
83. Yu Y, Shi Y, Chen CH (2007) Nanoporous cuprous oxide/lithia composite anode with capacity increasing characteristic and high rate capability. Nanotechnology 18. https://doi.org/10.1088/0957-4484/18/5/055706
84. Yu Y, Shi Y, Chen CH, Wang C (2008) Facile electrochemical synthesis of single-crystalline copper nanospheres, pyramids, and truncated pyramidal nanoparticles. from lithia/cuprous oxide composite thin films. J Phys Chem C 112:4176–4179. https://doi.org/10.1021/jp800071h
85. Zhang WD, Xiao HM, Zhu LP et al (2010) Facile one-step synthesis of electromagnetic functionalized polypyrrole/Fe_3O_4 nanotubes via a self-assembly process. J Polym Sci, Part A: Polym Chem 48:320–326. https://doi.org/10.1002/pola.23787
86. Xie K, Lu Z, Huang H et al (2012) Iron supported C@Fe_3O_4 nanotube array: a new type of 3D anode with low-cost for high performance lithium-ion batteries. J Mater Chem 22:5560–5567. https://doi.org/10.1039/c2jm15955d

87. Cheng Y, Zou B, Wang C et al (2011) Formation mechanism of Fe_2O_3 hollow fibers by direct annealing of the electrospun composite fibers and their magnetic, electrochemical properties. CrystEngComm 13:2863–2870. https://doi.org/10.1039/c0ce00379d
88. Chaudhari S, Srinivasan M (2012) 1D hollow α-Fe_2O_3 electrospun nanofibers as high performance anode material for lithium ion batteries. J Mater Chem 22:23049–23056. https://doi.org/10.1039/c2jm32989a
89. Im ME, Pham-Cong D, Kim JY et al (2015) Enhanced electrochemical performance of template-free carbon-coated iron(II, III) oxide hollow nanofibers as anode material for lithium-ion batteries. J Power Sources 284:392–399. https://doi.org/10.1016/j.jpowsour.2015.03.024
90. Zhang Z, Li X, Wang C et al (2009) ZnO hollow nanofibers: fabrication from facile single capillary electrospinning and applications in gas sensors. J Phys Chem C 113:19397–19403. https://doi.org/10.1021/jp9070373
91. Morimoto H, Tobishima SI, Iizuka Y (2005) Lithium intercalation into α-Fe_2O_3 obtained by mechanical milling of α-FeOOH. J Power Sources 146:315–318. https://doi.org/10.1016/j.jpowsour.2005.03.036
92. Hang BT, Doi T, Okada S, Yamaki J (2007) Effect of carbonaceous materials on electrochemical properties of nano-sized Fe_2O_3-loaded carbon as a lithium battery negative electrode. J Power Sources 174:493–500. https://doi.org/10.1016/j.jpowsour.2007.06.031
93. Liu S, Zhang L, Zhou J et al (2008) Fiberlike Fe_2O_3 macroporous nanomaterials fabricated by calcinating regenerate cellulose composite fibers. Chem Mater 20:3623–3628. https://doi.org/10.1021/cm703623v
94. Jamnik J, Maier J (2003) Nanocrystallinity effects in lithium battery materials: Aspects of nano-ionics. Part IV. Phys Chem Chem Phys 5:5215–5220. https://doi.org/10.1039/b309130a
95. Song K, Lee Y, Jo MR, et al (2012) Comprehensive design of carbon-encapsulated Fe_3O_4 nanocrystals and their lithium storage properties. Nanotechnology 23. https://doi.org/10.1088/0957-4484/23/50/505401
96. He C, Wu S, Zhao N et al (2013) Carbon-encapsulated Fe_3O_4 nanoparticles as a high-rate lithium ion battery anode material. ACS Nano 7:4459–4469. https://doi.org/10.1021/nn401059h
97. Ma Y, Huang Y, Wang X, et al (2014) One-pot synthesis of Fe_3O_4/C nanocomposites by PEG-assisted co-precipitation as anode materials for high-rate lithium-ion batteries. J Nanopart Res 16. https://doi.org/10.1007/s11051-014-2614-9
98. Maier J (2010) Nanoionics: Ion transport and electrochemical storage in confined systems. Mater Sustain Energy A Collect Peer Rev Res Rev Artic from Nat Publ Group 4:160–170. https://doi.org/10.1142/9789814317665_0023
99. Hu YS, Kienle L, Guo YG, Maier J (2006) High lithium electroactivity of nanometer-sized rutile TiO_2. Adv Mater 18:1421–1426. https://doi.org/10.1002/adma.200502723
100. Geim AK, Novoselov KS (2010) The rise of graphene. In nanoscience and technology: a collection of reviews from nature journals, pp 11–19
101. Geim AK (2009) Graphene: status and prospects. Prospects 324:1–8. https://doi.org/10.1126/science.1158877
102. Zhu BY, Murali S, Cai W et al (2010) Graphene and graphene oxide: synthesis, properties, and applications. Adv Mater 22:3906–3924. https://doi.org/10.1002/adma.201001068
103. Blake P, Hill EW, Castro Neto AH, Novoselov KS, JiangD, Yang R, Geim AK (2007) Making graphene visible. Appl Phys Lett 91(6):063124
104. Novoselov KS, Morozov SV, Mohinddin TMG, Ponomarenko LA, Elias DC, Yang R, Giesbers J (2007) Electronic properties of graphene. Phys Status Solidi (B) 244(11):4106–4111
105. Bolotin KI, Sikes KJ, Jiang Z, Klima M, Fudenberg G, Hone JE, Stormer HL (2008) Ultrahigh electron mobility in suspended graphene. Solid State Commun 146(9–10):351–355
106. Morozov SV, Novoselov KS, Katsnelson MI et al (2008) Giant intrinsic carrier mobilities in graphene and its bilayer. Phys Rev Lett 100:016602. https://doi.org/10.1103/PhysRevLett.100.016602

107. Lee C, Wei X, Kysar JW, Hone J (2008) Measurement of the elastic properties and intrinsic strength of monolayer graphene. Science (80-) 321:385–388. https://doi.org/10.1126/science.1157996
108. Balandin AA, Ghosh S, Bao W et al (2008) Superior thermal conductivity of single-layer graphene. Nano Lett 8:902–907. https://doi.org/10.1021/nl0731872
109. Cai W, Zhu Y, Li X et al (2009) Large area few-layer graphene/graphite films as transparent thin conducting electrodes. Appl Phys Lett 95:2007–2010. https://doi.org/10.1063/1.3220807
110. Ganguly S, Ghosh JJ (1979) Steroid hormone induced alterations in endometrium: I. Changes in lipid content, swelling pattern & lipid peroxidation of mitochondria. Indian J Biochem Biophys 16:61–65
111. Berger C, Wu X, Brown N, et al (2006) Electronic confinement and coherence in patterned epitaxial graphene. Science 312(80):1191–1196. https://doi.org/10.1126/science.1125925
112. Mattevi C, Kim H, Chhowalla M (2011) A review of chemical vapour deposition of graphene on copper. J Mater Chem 21:3324–3334. https://doi.org/10.1039/c0jm02126a
113. Lv W, Tang DM, He YB et al (2009) Low-temperature exfoliated graphenes: vacuum-promoted exfoliation and electrochemical energy storage. ACS Nano 3:3730–3736. https://doi.org/10.1021/nn900933u
114. Yuan L, Ge J, Peng X et al (2016) A reliable way of mechanical exfoliation of large scale two dimensional materials with high quality. AIP Adv 6:125201. https://doi.org/10.1063/1.4967967
115. Chua CK, Pumera M (2014) Chemical reduction of graphene oxide: a synthetic chemistry viewpoint. Chem Soc Rev 43:291–312. https://doi.org/10.1039/c3cs60303b
116. Azizighannad S, Mitra S (2018) Stepwise reduction of Graphene Oxide (GO) and its effects on chemical and colloidal properties. Sci Rep 8:1–8. https://doi.org/10.1038/s41598-018-28353-6
117. Zhu J, Yang D, Yin Z, Yan Q, Zhang H (2014) Graphene and graphene-based materials for energy storage applications. Small 10(17):3480–3498
118. Lee JK, Smith KB, Hayner CM, Kung HH (2010) Silicon nanoparticles-graphene paper composites for Li ion battery anodes. Chem Commun 46:2025–2027. https://doi.org/10.1039/b919738a
119. Wang G, Wang B, Wang X et al (2009) Sn/graphene nanocomposite with 3D architecture for enhanced reversible lithium storage in lithium ion batteries. J Mater Chem 19:8378–8384. https://doi.org/10.1039/b914650d
120. Kim MG, Cho J (2009) Reversible and high-capacity nanostructured electrode materials for Li-ion batteries. Adv Funct Mater 19:1497–1514. https://doi.org/10.1002/adfm.200801095
121. Yang S, Feng X, Wang L et al (2010) Graphene-based nanosheets with a sandwich structure. Angew Chemie - Int Ed 49:4795–4799. https://doi.org/10.1002/anie.201001634
122. Zhou G, Wang DW, Li F et al (2010) Graphene-wrapped Fe_3O_4 anode material with improved reversible capacity and cyclic stability for lithium ion batteries. Chem Mater 22:5306–5313. https://doi.org/10.1021/cm101532x
123. Baxter J, Bian Z, Chen G et al (2009) Nanoscale design to enable the revolution in renewable energy. Energy Environ Sci 2:559–588. https://doi.org/10.1039/b821698c
124. Paek SM, Yoo EJ, Honma I (2009) Enhanced cyclic performance and lithium storage capacity of SnO_2/graphene nanoporous electrodes with three-dimensionally delaminated flexible structure. Nano Lett 9:72–75. https://doi.org/10.1021/nl802484w
125. Wang D, Choi D, Li J et al (2009) Self-assembled TiO_2—graphene hybrid insertion. ACS Nano 3:907–914
126. Kim HK, Bak SM, Kim KB (2010) $Li_4Ti_5O_{12}$/reduced graphite oxide nano-hybrid material for high rate lithium-ion batteries. Electrochem Commun 12:1768–1771. https://doi.org/10.1016/j.elecom.2010.10.018
127. Wang B, Wu XL, Shu CY et al (2010) Synthesis of CuO/graphene nanocomposite as a high-performance anode material for lithium-ion batteries. J Mater Chem 20:10661–10664. https://doi.org/10.1039/c0jm01941k

128. Chen D, Tang L, Li J (2010) Graphene-based materials in electrochemistry. Chem Soc Rev 39:3157–3180. https://doi.org/10.1039/b923596e
129. Pumera M (2010) Graphene-based nanomaterials and their electrochemistry. Chem Soc Rev 39:4146–4157. https://doi.org/10.1039/c002690p
130. Brownson DAC, Banks CE (2010) Graphene electrochemistry: an overview of potential applications. Analyst 135:2768–2778. https://doi.org/10.1039/c0an00590h
131. Zhu X, Zhu Y, Murali S et al (2011) Nanostructured reduced graphene oxide/Fe_2O_3 composite as a high-performance anode material for lithium ion batteries. ACS Nano 5:3333–3338. https://doi.org/10.1021/nn200493r
132. Du M, Xu C, Sun J, Gao L (2012) One step synthesis of Fe_2O_3/nitrogen-doped graphene composite as anode materials for lithium ion batteries. Electrochim Acta 80:302–307. https://doi.org/10.1016/j.electacta.2012.07.029
133. Jang B, Chae OB, Park SK et al (2013) Solventless synthesis of an iron-oxide/graphene nanocomposite and its application as an anode in high-rate Li-ion batteries. J Mater Chem A 1:15442–15446. https://doi.org/10.1039/c3ta13717a
134. Kim DW, Bach LG, Hong SS et al (2014) A facile route towards the synthesis of Fe_3O_4/graphene oxide nanocomposites for environmental applications. Mol Cryst Liq Cryst 599:43–50. https://doi.org/10.1080/15421406.2014.935919
135. Yoon T, Kim J, Kim J, Lee JK (2013) Electrostatic self-assembly of Fe_3O_4 nanoparticles on graphene oxides for high capacity lithium-ion battery anodes. Energies 6:4830–4840. https://doi.org/10.3390/en6094830
136. Su J, Cao M, Ren L, Hu C (2011) Fe3O4-graphene nanocomposites with improved lithium storage and magnetism properties. J Phys Chem C 115:14469–14477. https://doi.org/10.1021/jp201666s
137. Ji L, Tan Z, Kuykendall TR et al (2011) Fe_3O_4 nanoparticle-integrated graphene sheets for high-performance half and full lithium ion cells. Phys Chem Chem Phys 13:7170–7177. https://doi.org/10.1039/c1cp20455f
138. Liu J, Zhou Y, Liu F et al (2012) One-pot synthesis of mesoporous interconnected carbon-encapsulated Fe_3O_4 nanospheres as superior anodes for Li-ion batteries. RSC Adv 2:2262–2265. https://doi.org/10.1039/c2ra01241c
139. Chen D, Ji G, Ma Y et al (2011) Graphene-encapsulated hollow Fe_3O_4 nanoparticle aggregates as a high-performance anode material for lithium ion batteries. ACS Appl Mater Interfaces 3:3078–3083. https://doi.org/10.1021/am200592r
140. Zhou Q, Zhao Z, Wang Z et al (2014) Low temperature plasma synthesis of mesoporous Fe_3O_4 nanorods grafted on reduced graphene oxide for high performance lithium storage. Nanoscale 6:2286–2291. https://doi.org/10.1039/c3nr05423c
141. He J, Zhao S, Lian Y et al (2017) Graphene-doped carbon/Fe_3O_4 porous nanofibers with hierarchical band construction as high-performance anodes for lithium-ion batteries. Electrochim Acta 229:306–315. https://doi.org/10.1016/j.electacta.2017.01.092
142. Poizot P, Laruelle S, Grugeon S et al (2000) Nano-sized transition-metal oxides as negative-electrode materials for lithium-ion batteries. Nature 407:496–499. https://doi.org/10.1038/35035045
143. Chen Y, Li X, Park K et al (2013) Hollow carbon-nanotube/carbon-nanofiber hybrid anodes for Li-ion batteries. J Am Chem Soc 135:16280–16283. https://doi.org/10.1021/ja408421n
144. Zhang B, Xu ZL, He YB et al (2014) Exceptional rate performance of functionalized carbon nanofiber anodes containing nanopores created by (Fe) sacrificial catalyst. Nano Energy 4:88–96. https://doi.org/10.1016/j.nanoen.2013.12.011
145. Wang B, Zhang S, Wang G et al (2019) The morphology and electrochemical properties of porous Fe_2O_3@C and FeS@C nanofibers as stable and high-capacity anodes for lithium and sodium storage. J Colloid Interface Sci 557:216–226. https://doi.org/10.1016/j.jcis.2019.08.071
146. Yang S, Cao C, Li G, Sun Y, Huang P, Wei F, Song W (2015) Improving the electrochemical performance of Fe_3O_4 nanoparticles via a double protection strategy through carbon nanotube decoration and graphene networks. Nano Res 8(4):1339–1347

147. Hsieh C, Lin J, Mo C (2011) Electrochimica Acta Improved storage capacity and rate capability of Fe_3O_4—graphene anodes for lithium-ion batteries. Electrochim Acta 58:119–124. https://doi.org/10.1016/j.electacta.2011.09.008
148. Peng YT, Lo CT (2015) Electrospun porous carbon nanofibers as lithium ion battery anodes. J Solid State Electrochem 19:3401–3410. https://doi.org/10.1007/s10008-015-2976-7
149. Zhao N, Wu S, He C et al (2013) One-pot synthesis of uniform Fe_3O_4 nanocrystals encapsulated in interconnected carbon nanospheres for superior lithium storage capability. Carbon N Y 57:130–138. https://doi.org/10.1016/j.carbon.2013.01.056
150. Chen J, Xu L, Li W, Gou X (2005) α-Fe_2O_3 nanotubes in gas sensor and lithium-ion battery applications. Adv Mater 17:582–586. https://doi.org/10.1002/adma.200401101
151. Jannik J, Maier J (2004) Nanocrystallinity effects in lithium battery materials. Aspects of nano-ionics. J Phys Chem Solids 5:70569
152. Sharma Y, Sharma N, Subba Rao GV, Chowdari BVR (2007) Nanophase $ZnCo_2O_4$ as a high performance anode material for Li-ion batteries. Adv Funct Mater 17:2855–2861. https://doi.org/10.1002/adfm.200600997
153. Grugeon S, Laruelle S, Dupont L, Tarascon JM (2003) An update on the reactivity of nanoparticles Co-based compounds towards Li. In: Solid state sciences, pp 895–904
154. Laruelle S, Grugeon S, Poizot P et al (2002) On the origin of the extra electrochemical capacity displayed by MO/Li cells at low potential. J Electrochem Soc 149:627–634. https://doi.org/10.1149/1.1467947
155. Gu S, Liu Y, Zhang G et al (2014) Fe_3O_4/carbon composites obtained by electrospinning as an anode material with high rate capability for lithium ion batteries. RSC Adv 4:41179–41184. https://doi.org/10.1039/c4ra06888b
156. Qin X, Zhang H, Wu J et al (2016) Fe_3O_4 nanoparticles encapsulated in electrospun porous carbon fibers with a compact shell as high-performance anode for lithium ion batteries. Carbon N Y 87:347–356. https://doi.org/10.1016/j.carbon.2015.02.044
157. Chen Z, Zhou J, Wang X et al (2016) Natural collagen fiber-enabled facile synthesis of carbon@Fe_3O_4 core-shell nanofiber bundles and their application as ultrahigh-rate anode materials for Li-ion batteries. RSC Adv 6:10824–10830. https://doi.org/10.1039/c5ra22481k
158. Li S, Wang M, Luo Y, Huang J (2016) Bio-inspired hierarchical nanofibrous Fe_3O_4-TiO_2-Carbon composite as a high-performance anode material for Lithium-Ion batteries. ACS Appl Mater Interfaces 8:17343–17351. https://doi.org/10.1021/acsami.6b05206

Chapter 10
Electrospun Nanostructured Iron Oxides for High-Performance Lithium Ion Batteries

Neethu T. M. Balakrishnan, Akhila Das, N. S. Jishnu, M. A. Krishnan, Sabu Thomas, M. J. Jabeen Fatima, Jou-Hyeon Ahn, and Raghavan Prasanth

10.1 Introduction

The rechargeable Li-ion battery (LIB) has attracted intensive research interest due to their large spectrum of applications as energy storage devices for electric, electric/hybrid electric vehicles, and intermittent renewable energy sources [1–5]. The LIB is referred to as a rocking-chair battery, because Li^+-ions "rock" back and forth between the anode and the cathode during cycling and they possess high energy and power densities, no memory effect and long cycle life. The current generation commercial LIBs are utilizing the electrode materials, which could store Li^+-ions

N. T. M. Balakrishnan · A. Das · M. J. Jabeen Fatima · R. Prasanth
Department of Polymer Science and Rubber Technology, Cochin University of Science and Technology (CUSAT), Cochin 682022, India
e-mail: jabeen@cusat.ac.in

N. S. Jishnu
Rubber Technology Centre, Indian Institute of Technology-Kharagpur (IIT-KGP), Kharagpur, West Bengal 721302, India

M. A. Krishnan
Department of Mechanical Engineering, Amrita Vishwa Vidyapeetham, Amritapuri, India

Department of Electrical Engineering, Pennsylvania State University, 311, Electrical Engineering East Building, University Park, PA 16802, USA

S. Thomas
School of Chemical Sciences, Mahatma Gandhi University, Athirampuzha, Kerala 686560, India

J.-H. Ahn (✉) · R. Prasanth (✉)
Department of Materials Engineering and Convergence Technology, Gyeongsang National University, 501 Jinju-daero, Jinju 52828, Republic of Korea
e-mail: jhahn@gnu.ac.kr

R. Prasanth
e-mail: prasanth@cusat.ac.in

N. T. M. Balakrishnan and R. Prasanth (eds.), *Electrospinning for Advanced Energy Storage Applications*, Materials Horizons: From Nature to Nanomaterials,
https://doi.org/10.1007/978-981-15-8844-0_10

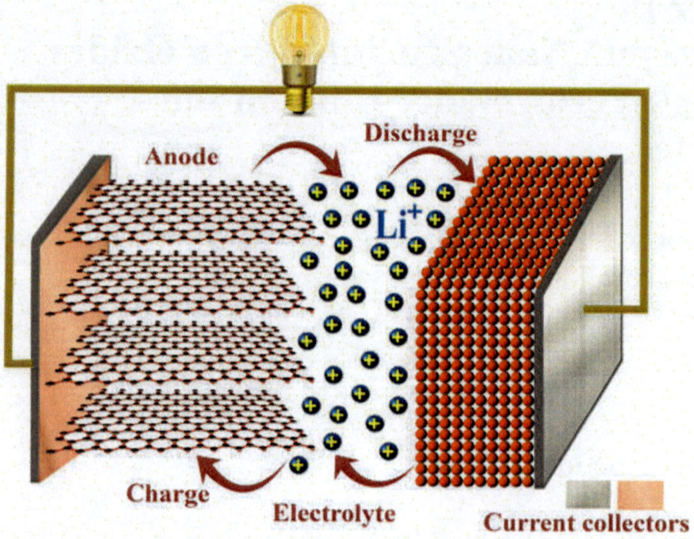

Fig. 10.1 Schematic representation and operating principle of lithium ion batteries

by insertion between their structural layers during charging and extracted out from the layers during discharging (Fig. 10.1) without any significant structural change leading to excellent cycling performance. In recent years, globally great efforts have been paid by the researchers and battery technologists to develop and design high-performance electrode materials in terms of energy density, cycling stability and rate capability. Among different classes of anode materials such as intercalation, conversion reaction, alloying/de-alloying reaction-type materials, transition metal oxides, Co_3O [6–10], FeOx [11–17], TiO_2 [18, 19], MnO_2 [20–22], and SnO_2 [23–25], iron oxide micro-/nanomaterials, such as hematite (α-Fe_2O_3) [26–28] and magnetite (Fe_3O_4) [29–31], have been extensively studied as potential electrode materials in LIBs. Iron oxide is popular in their higher theoretical capacities (1004 mAh g^{-1} for α-Fe_2O_3 and 924 mAh g^{-1} for Fe_3O_4), has low toxicity, and is economically viable.

In spite of their lower cost and better safety, the capacity retention of Fe_2O_3/Fe_3O_4 remains a major drawback, due to the huge volumetric expansion/contraction during the lithiation/delithiation process which ultimately leads to pulverization of the electrode from the current collector, resulting in loss of electrical contact and loss of morphological structure of the active material [32]. The electrochemical performance is highly dependent on their diverse morphologies and micro-/nanostructure of an anode material [33–37] The confining dimension effect and high surface area of the nanostructured materials lead to the short lithium diffusion lengths and increased active sites for Li^+-ion insertion/extraction reactions [38]. Also, the cyclability of the nanostructured electrodes significantly improved due to the sufficient free spaces to relax the large volume changes during the continuous charge–discharge process

[29]. Hence, various types of nanostructures have been employed as anode materials for LIBs. In order to mitigate the large-volume variation problem of transition metal oxides and to increase the electronic conductivity, carbon coatings have been extensively explored [39–50]. However, the compact carbon coatings on nanoparticles cannot allow residual buffer space to accommodate the large volume change of Fe_3O_4 nanoparticles during Li^+ insertion/extraction. Thus, it remains necessary to exploit an approach for the fabrication of suitable carbon matrix to accommodate volume expansion upon Li^+-ion insertion as well as to increase the electronic conductivity [51–53]. One-dimensional (1D) structure can efficiently improve the performance of Fe_2O_3 as anode material in LIBs due to its excellent electron transport along the lengthways direction and large surface–volume ratio [54–58]. Electrospinning is now a convenient, inexpensive, simple, and versatile method to manufacture the 1D structure including polymer, metal oxide, and organic–inorganic composites [59–62], especially the robust electrode for LIBs. The material obtained by electrospinning can take full advantage of 1D architectures as well as the material can form metal oxide nanoparticles/carbon nanofibers (CNFs) after being calcined under an inert atmosphere. Uniformly dispersing the nano-sized metal oxide into CNFs matrix can significantly enhance the electronic conductivity, buffer the large volume change and pulverization of the electrode, and prevent the agglomerates of nanoparticles [63–71]. The electrospinning technique facilitates to develop 1D metal oxide nano-/microstructure with various morphologies including porous nanowires, nanotubes, nanorods, and tube-in-tube by using non-coaxial electrospinning. This chapter is presenting a detailed overview on the facile fabrication and electrochemical performance of hierarchal Fe_2O_3/Fe_3O_4 nanostructured anode for LIBs.

10.2 Principle of Lithium Ion Batteries

The global lithium-ion battery market is forecasted to grow from USD 36.20 billion in 2018 to USD 109.72 billion by 2026, at a compound annual growth rate (CAGR) of 13.4%, during the forecast period. In recent years, lithium-ion batteries are increasingly being used as the power source for hybrid (HEV) and full-battery electric vehicle (BEV). Over the past couple of years, maximum sales of electric vehicles have accounted by China, the USA, and the European region, which are primarily high-end electric vehicles. Roughly, 1.6 million electric cars were on the roads in China in 2018, followed by 810,000 in the USA. By March 2018, BEV production and sales in China reached 27,673 and 24,127 units, rising 88.35 and 69.21% year-on-year; such figures for plug-in hybrid electric vehicle (PHEV) were 11,210 and 11,171 units, rising 291.21 and 201.47% year-on-year [72]. The industry produced about 660 million units of cylindrical lithium-ion cells in 2012; the 18650 size is by far the most popular for cylindrical cells. The Tesla's Model S electric cars SUVs under 40,000 USD with 85 kWh battery uses 7,104 of lithium ion cells. A 2014 study projected that the Model S alone would use almost 40 percent of estimated global cylindrical battery production during 2014. Production of the cell was gradually shifted to higher-capacity 3000+ mAh cells.

Based on the use type, LIBs are categorized as primary LIBs and secondary or rechargeable LIBs. A primary battery is one-direction galvanic device designed to be used once and discarded when it is fully discharged, and not recharged with electricity and reused like a secondary or rechargeable battery; i.e., the electrochemical reaction occurring in the cell is not reversible, or it has only discharging process. Lithium primary battery has metallic lithium as anode. Hence, these types of batteries are also referred to as lithium-metal batteries. Presently represent the primary EES systems, with a production higher than 100 million cells/month and about 1500 tons/month of electrode materials. Lithium–manganese dioxide, lithium iron disulfide, lithium thionyl chloride, and lithium iodine batteries are the common lithium primary batteries. Among different lithium primary batteries, lithium thionyl chloride battery has the highest energy density of all lithium-type cells and has a service life of 15–20 years, while lithium iodine batteries provide excellent safety and long service life. In batteries, during discharging, reduction happens on the cathode gaining electrons and oxidation happens on the anode, which is losing electrons, as per the electrochemical reaction shown below [73].

$$\begin{aligned}
&\text{Cathode: } MS_2 + Li^+ + e^- \xrightarrow{\text{discharge}} LiMS_2 \\
&\text{Anode: } Li \xrightarrow{\text{discharge}} Li^+ + e^- \\
&\text{Full cell: } Li + MS_2 \xrightarrow{\text{discharge}} LiMS_2 \\
&(M = \text{Ti or Mo})
\end{aligned}$$

In contrast to lithium primary batteries, lithium secondary batteries, referred as lithium-ion batteries, are rechargeable batteries in which lithium ions move from the negative electrode to the positive electrode during discharge and opposite action happens during charging. Research on LIBs started in the early 1980s, and the principle of the current LIB was completed in 1985 and then first commercialized in 1991 by Sony. Most of the technological developments to date have been directed toward the needs of portable electronics, but now the focus tends to be on the performance demands of medium- and large-scale applications. As shown in Fig. 10.1, typically, LIB consists of three layers: (i) cathode or positive electrode which commonly consists of $LiCoO_2$ [74, 75], $LiNiO_2$ [76], $LiMn_2O_4$ [77], etc., (ii) anode or negative electrode consists of graphitic carbon [78], TiO_2 [79], Fe_2O_3/Fe_3O_4 [80], etc., and (iii) a separating cum electrolyte called gel polymer electrolytes (GPEs) which is permeable to the ions and the electrolyte (e.g., $LiPF_6$ in an organic solvent). GPEs are prepared by immobilization of organic liquid electrolytes, e.g., a 1M solution of $LiPF_6$, $LiClO_4$, or LiTFSI into polymer structures [81–83]. Polymers such as polyethylene oxide (PEO) [84, 85], polyacrylonitrile (PAN) [86, 87], polyvinylidene difluoride (PVdF) [88, 89] and its copolymer polyvinylidene difluoride-*co*-hexafluoropropylene (PVdF-*co*-HFP) [90, 91], and polymethyl methacrylate (PMMA) [92, 93] are among the well-studied materials. As the name implies, the working of a lithium-ion battery mainly relies on repeated transfer of lithium ions between the anode and the cathode. The electrochemical properties of the electrodes are strongly influenced by the physical and chemical properties of the electrode active material, such as particle size, homogeneity, morphology, and surface area. Lithium-ion polymer batteries (LiPo batteries) are by far the most common commercialized secondary cell polymer battery, with leading technology among other types of metal-ion polymer batteries. A LiPo battery is a rechargeable battery of lithium-ion technology using a polymer electrolyte instead of a liquid electrolyte. LIBs are able to supply continuous energy due to the spontaneous oxidation–reduction reactions occurring at the electrodes. During the charging process (delithiation), Li^+ ions are extracted from the cathodic material by supplying energy by an external source. The extracted Li + ions diffuse in the electrolyte and enter the anodic material (according to the reaction $C_x + LiMO_2 \rightarrow Li_{(1-y)}MO_2 + C_xLi_y$, in the case of a traditional LIB [94, 95]), while electrons are simultaneously the electrons transferred to the positive electrode through the external circuit. In the discharge process (lithiation), the opposite process takes place (i.e., $Li_{(1-y)}MO_2 + C_xLi_y \rightarrow C_x + LiMO_2$, in the considered example): i.e., Li^+-ions, extracted from the anodic material, are re-inserted into the cathodic material, and the cell provides energy. The oxidation and reduction process occurred at two electrodes in the lithium rechargeable batteries as shown below [96, 97].

$$\text{Cathode: } LiMn_2O_4 \rightarrow Li_{1+x}Mn_2O_4 + xLi^+ + xe^-$$
$$\text{Anode: } xLi^+ + xe^- + C_6 \rightarrow Li_xC_6$$
$$\text{Full Cell: } LiMn_2O_4 + C_6LiC_6 + LiMn_2O_4$$

Each combination of the aforementioned materials and compound will slightly influence cost, voltage, cycle durability, and other characteristics of the LiPo batteries. The secondary lithium-ion batteries, in general, operate 3.7 V and demonstrate a capacity of 150 mAh g^{-1} [98].

10.3 Electrode Materials for Lithium Ion Batteries

LIB primarily contains four essential components, namely the anode, the electrolyte, the separator, and the cathode. Typically, LIBs use an intercalated lithium compound as the positive electrode and graphite as the negative electrode. However in the earliest configuration of LIBs metallic lithium or Li–Al alloys was used as the negative electrode, with a variety of chalcogenides (TiS_2, MoS_2, etc.), [73] as the positive electrode in several prototypes and commercial products. Due to safety concerns, lithium metal as an anode material in rechargeable batteries was ultimately rejected. The unavoidable dendrite growth on the lithium metal surface during the repeated cycling cause lithium plating that leads to internal short circuits.

10.3.1 Positive Electrode (Cathode) for Lithium Ion Batteries

The cathode material requires a stable crystalline structure over wide ranges of composition because during the process of lithiation, the oxidation reaction leads to large compositional changes and therefore to unfavorable phase changes [99]. Also, the cathode performance directly depends not only on the electrode microstructure and morphology, but also on the inherent electrochemical properties of the cathode material due to the fact that Li^+-ion exchange with the electrolyte only happens at the electrode–electrolyte interface [100, 101]. The development of electrochemically stable $LiCoO_2$ as a positive electrode leads to the commercialization of the lithium ion battery by Sony, Japan, paved a path to hunt novel electrode materials which provided a step change in the approach to the development of advanced energy storage based on lithium technology [102, 103]. Combining $LiCoO_2$ with graphitic carbon which provided a host for Li^+-ions at low potential thus successfully removed metallic lithium from the LIBs. The domination of LIBs as the power source in the portable electronic and automobile market leads to the subsequent improvements in the LIBs, which forced the scientific community to focus on developing new cathode materials; thus, LMO, NMC, LFP, etc., were introduced. But recently, the introduction of nanocomposites comprised of Sn (theoretical capacity 992 mAh g^{-1}) or Si (theoretical capacity 4200 mAh g^{-1}) led to major developments in anode materials, which require higher-capacity cathode materials to provide optimum utilization of the storage properties.

In LIBs, cathode materials can store energy through two different electrochemical reaction mechanisms, (i) intercalation and (ii) conversion reaction [100]. Conversion-type cathodes undergo a solid-state redox reaction during lithiation/delithiation process, in which there is a change in the crystalline structure, accompanied by the breaking and recombining chemical bonds, while the intercalation cathode materials act as a host for Li^+-ions, so that the ions can insert in or extracted out from the cathode material reversibly. Metal halides such as FeF_2, CoFe, and NiF_2 are examples of conversion-based cathode materials. Due to the high volume expansion, poor electronic conductivity, and hysteresis issues, development of conversion-based cathode materials has faced a lot of challenges [100].

Intercalation-based cathode materials are mainly divided into three categories: chalcogenides, transition metal oxides, and polyanion compounds. Due to the higher operating voltage and higher specific capacity, most of research on intercalation cathode materials is focused on transition metal oxides [100]. $LiCoO_2$, $LiNiO_2$ (LNO) [104], $LiNi_xMn_yCo_zO_2$ (with $x + y + z = 1$ or NMC) [105, 106] are some of the examples for the transition metal oxide-based cathodes in LIBs. Even though the layered crystal structure of LNO is similar to LCO, LNO delivers 20–30% more reversible capacity than LCO, but due to its inherent electrochemical properties [104].

Over-lithiated oxides (OLOs) are relatively new replacement material for the cathode in high-capacity LIBs due to their very high capacity over 250 mAh g^{-1} at high-voltage charge over 4.5 V along with many other enhanced properties. OLOs have been developed with a variety of stoichiometric variations of the general composition: Li_2MnO_3-$(1-x)LiMO_2$ [107]. The major limitation of OLOs is their initial irreversibility caused by Li_2O formation [105]. Another class of important cathode materials is spinel oxides with a general formula of AB_2O_4. The most famous cathode in this group is $LiMn_2O_4$ (LMO), which is a low-cost, reliable, non-toxic, and a high electrochemical potential material, and delivers a practical capacity of 100–120 mAh g^{-1} (theoretical specific capacity is 148 mAh g^{-1}); however, its reversible capacity is less than that of LCO or LNO. The spinel structure of LMO creates a 3D framework which promotes the easy movement of Li^+-ions [105, 108] and undergoes less damage during continuous charge–discharge cycles in comparison with the anode materials having layered structure, because the continuous two-way transportation of Li^+-ions in the spinel structure does not make large volume change at room temperature. However, at high-temperature LMO shows poor cycle as well as calendar life [105]. The manganese dissolution, structural fatigue, and microcracks are other issues associated with LMO, which lead to poor cycling stability and capacity fading. Because of its low specific capacity and low practical capacity, LMO is not recommended to be used as a single cathode material. It also suffers severely from the Jahn–Teller effect, which is a geometric distortion of a nonlinear molecular system that reduces its symmetry and energy. Hence, the cathode material in the new generation of commercially produced batteries consists of a complementary blend of spinal LMO and layered NMC [109].

$LiFePO_4$ (LFP) olivines are polyanionic compounds developed by John B. Goodenough, who received Nobel Prize, in 2019, at the University of Texas in 1996, and that have attracted a lot of attentions due to their thermal stability, environmental friendliness, very flat potential during charge–discharge processes, and high-power capabilities [105]. Again even in harsh environments, the release of oxygen from the active cathode material is inhibited due to the strong P–O bond in phosphate and this structurally stable material guarantees their safety [110, 111].

10.3.2 Anode Materials for Lithium Ion Batteries

Traditionally in commercial LIB anode, the negative electrode from which electrons flow out toward the external part of the circuit is constructed from graphite and other carbon materials coated on a thin copper foil current collector. However, they suffer from serious safety problems, which have hampered their further development [98, 112]. To replace conventional carbon-based anodes, several studies have focused on exploiting novel anode materials. As a result of extensive research, the anode in LIBs can be fabricated from three distinguished groups of materials having very different electrochemical energy storage mechanisms broadly categorized as (i) intercalation-based materials, (ii) conversion-reaction-based materials, and (iii) alloying-reaction-based materials. The properties of most common anode materials used for lithium-ion batteries are summarized in Table 10.1.

10.3.2.1 Intercalation-Based Materials as Anodes in Lithium Ion Batteries

Graphite in its natural or artificial form is the best representative material in the intercalation-based anode materials first introduced by Rachid Yazami, a Moroccan scientist and engineer in 1980. He established the reversible intercalation of lithium into graphite in an electrochemical cell using a polymer electrolyte. Eventually, his discovery led to the lithium–graphite anode now used in commercial lithium-ion batteries, over US$20B value market. In the intercalation-based anode materials, the lithium ions are electrochemically intercalated into the space between the layers of the active materials. Lithium-ion intercalation in graphite can be described by the

Table 10.1 Summarization of most common anode materials used for lithium-ion batteries

Anode material	Theoretical capacity (mAh g^{-1})	Energy storage mechanism	Properties	
			Advantages	Disadvantages
Graphite	372	Intercalation	High electronic conductivity Nice hierarchical structure Abundant and low-cost resources	Low specific capacity Low rate capacity Safety issues
Nanostructured carbonaceous materials (e.g., carbon nanotube/graphene/carbon nanofibers/porous carbons)	Up to 1750	Intercalation		
Metal oxides (Cu_2O, Fe_3O_4, Co_3O_4, MoO_3, etc.)	375–1170	Conversion reaction	High specific capacity Nice stability	Low Coulombic efficiency Large potential hysteresis
Metal nitrides (M_xN_y, M: Fe, CO, Ni, Cu, Cr, V, Ti, etc.)	400–1300	Conversion reaction		
Metal sulfides (M_xS_y) (Ni_3S_2, FeS_2, MoS_2, SnS, SnS_2, etc.)	447–1230	Conversion reaction		
Metal phosphides $Li_xM_yP_4$ (M: V, Ti, Cu, Fe, Mn) (CoP_3, NiP_3, MnP_4, etc.)	700–1800	Conversion reaction		

(continued)

Table 10.1 (continued)

Anode material	Theoretical capacity ($mAh\ g^{-1}$)	Energy storage mechanism	Properties	
			Advantages	Disadvantages
Si	4200	Alloying/de-alloying	Highest specific capacity Rich, low-cost, clean resources	Low electronic conductivity Large volume change (100%)
Germanium	1384	Alloying/de-alloying	High specific capacity Good security	Low electronic conductivity Large volume change (100%)
Tin	960	Alloying/de-alloying	Highest specific capacity Rich, low-cost, clean resources High specific capacity Good security	
Phosphorus	2596	Alloying/de-alloying		
Antimony	660	Alloying/de-alloying		
Indium	1012	Alloying/de-alloying		

equation

$$x\text{Li}^+ + \text{C}_6(\text{in graphite}) + xe^- \longleftrightarrow \text{Li}_x\text{C}_6$$

The reversible lithiation/delithiation (intercalation/deintercalation) reaction proceeds less than 0.25 V versus Li/Li^+, with a practical reversible capacity greater than 360 mAh g^{-1} (theoretically at 372 mAh g^{-1} or 975 mAh cm^{-3}) with high Coulombic efficiency approaching 100% [113, 114]. However, one of the drawbacks with graphitic anode is that some irreversible reactions happen during the first charge (lithiation) process causing a cathodic decomposition of some constituents of the electrolyte. Another major downside of graphite anodes is their low specific capacity, which is addressed in great extent by increasing the surface area of the carbonaceous materials; therefore, the active material can provide more space for the intercalation of Li^+-ions between the graphitic layers leading to the higher specific capacity. Different carbon allotropes such as carbon nanotubes (CNTs), buckminsterfullerene (buck balls), and graphene nanosheets (GNSs) [115] or carbon nanofibers (CNFs) [116] are vastly studied as an alternative to graphite due to their larger surface area as well as higher electronic conductivity which makes them suitable for high rate charging/discharging [117]. Single-wall CNTs are expected to exhibit reversible capacities somewhere around 300–600 mAh g^{-1} [118], and for graphene the theoretical capacity is about 744 mAh g^{-1}. A reversible specific capacity is as high as 1264 mAh g^{-1} at a current density of 100 mA g^{-1}, and a capacity retention of 718 mAh g^{-1} is reported even at a high current density of 500 mA g^{-1} [115, 116, 119].

10.3.2.2 Conversion Reaction-Based Materials as Anodes in Lithium Ion Batteries

The simple conversion electrochemistry of transition metal oxides (TMOs), sulfides, phosphides, and similar compounds of p-block metalloids shares some interesting and useful electrochemical features with other anode materials [120]. It was already well understood, and different types of anode electrochemistries that qualify the required norms set for the battery application, namely the operating potential of <2 versus Li/Li^+, are categorized as intercalation, alloying, and conversion types. The conversion reaction-based materials are based on the Faradaic reaction represented as follows.

$$M_aX_b + (b \cdot n)Li^+ + ae^- \longleftrightarrow aM + bLi_nX$$

where M is the transition metal such as Ti, V, Cr, Mn, Fe, Co, Ni, Cu, Mo, W, and Ru, X is the anion such as O, N, F, S, and P, and n is the number of negative charges of X [121]. A variety of anode materials are possible in this group by a simple combination of candidates of M and X even without considering multiple oxidation states of M. The theoretical capacity of the conversion-reaction-based anode materials ranges from 350 mAh g^{-1} for Cu_2S to 1800 mAh g^{-1} for MnP_4 [121]. The relatively high theoretical capacity of conversion reaction-based compounds as compared to graphite (372 mAh g^{-1}) makes these materials as ideal anode materials. However, compared to graphitic carbon, these materials have some major down steps including lower Coulombic efficiency, electronic conductivity, inferior cycling stability, and rate capability, which must be dealt with before being used as anode material. In addition, the conversion-reaction-based anode materials undergo large volume changes during lithiation and the following delithiation, which could lead to pulverization or electric isolation leads to the fast capacity fade under cycling. This means that the conversion reactions in conversion-reaction-based anodes would have intrinsically limited reversibility [122, 123]. Again the reaction potentials of conversion-reaction-based anode materials (E_{conv}), at which the potential profiles reach a plateau, are relatively higher than that of graphite. Higher E_{conv} leads to lower cell potential which results in a lower energy density than expected only from capacities [121]. Although intercalation and alloying materials have received the preferred focus by the battery technologist, conversion materials have thus far been left out for any type of practical industrial applications owing to the aforementioned specific shortcomings [124].

10.3.2.3 Alloying Reaction-Based Materials

The third category of anode materials after intercalation and conversion reaction-based materials is the alloying reaction-based materials. This group consists of metals that can be alloyed with lithium such as silicon (Si), germanium (Ge), tin (Sn), and their alloys [114, 125]. In this category, the Li ions are inserted into the structure of anode material during the charge cycle, making an alloy with the anode. The reversible alloying reaction is shown in Eq. 4, where M is the anode material [125].

$$M + xLi^{+} + xe^{-} \longleftrightarrow Li_{x}M$$

Alloying reaction-based materials are most famous for their high theoretical capacity: 4200 mAh g^{-1} for Si in $Li_{4.4}Si$, 1600 mAh g^{-1} for Ge in $Li_{4.4}Ge$, 993 mAh g^{-1} for Al in LiAl, 992 mAh g^{-1} for Sn in $Li_{4.4}Sn$, and 660 mAh g^{-1} for Sb in Li_3Sb [126]. However, the major disadvantage of these materials is their extremely large volume change during charge and discharge [118]. They experience serious pulverization resulting in electrical isolation of the active materials from electric contact with the conducting agent (carbon black) and the current collector. Among the alloying elements which can be used in this group of anode materials, the vast majority of research and development has been focused on silicon because of its highest capacity and its most serious detrimental volumetric change [126–129].

10.4 Electrospun-Based Iron Oxide Anodes for Lithium Ion Batteries

Iron oxide-based nanocomposites including Fe_3O_4/carbon nanocomposite [43, 49, 130, 131] and Fe_2O_3/carbon nanocomposites [132–134] have been fabricated and studied extensively as electrode materials for lithium-ion batteries. Iron oxide is a paramagnetic material which has only two unpaired electrons. Because the lower number of unpaired electrons irons oxide is less magnetic than iron, which has four unpaired electrons. Iron (III) oxide or ferric oxide is the inorganic compound with the formula Fe_2O_3. It is one of the three main oxides of iron, the other two being iron (II) oxide (FeO), which is rare, and iron (II,III) oxide (Fe_3O_4), which also occurs naturally as the mineral magnetite. Fe_2O_3 can be obtained in various polymorphs. In the main ones, α and γ, iron adopts octahedral coordination geometry; i.e., each Fe center is bound to six oxygen ligands. α-Fe_2O_3 has the rhombohedral, corundum (α-Al_2O_3) structure and is the most common form, which occurs naturally as the mineral hematite while γ-Fe_2O_3 has a cubic structure and occurs naturally as the mineral maghemite. γ-Fe_2O_3 is metastable and converted from the alpha phase at high temperatures. γ-Fe_2O_3 is ferromagnetic, and however ultrafine particles smaller than 10 nm are superparamagnetic. α-Fe_2O_3, γ-Fe_2O_3, and Fe_3O_4 are explored as electrode materials in LIBs. The crystal structure of Fe_2O_3 is displayed in Fig. 10.2

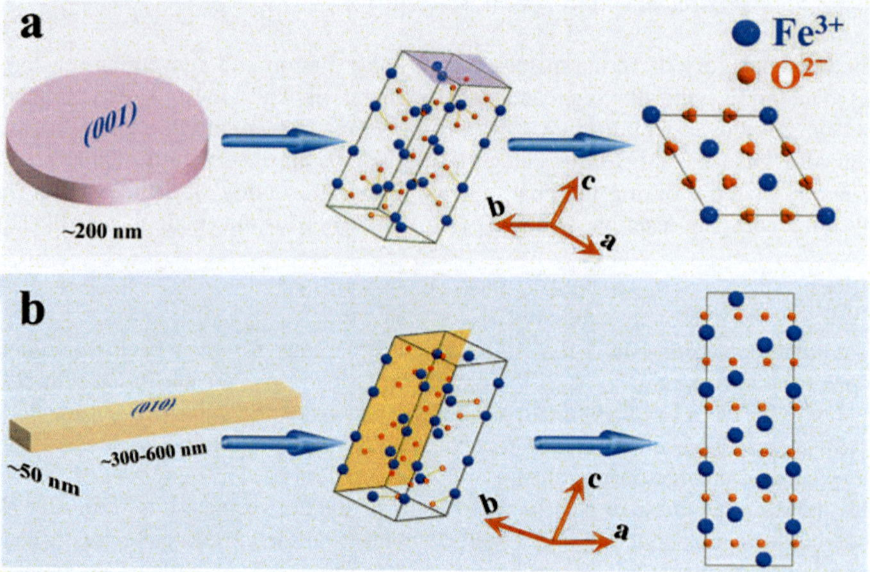

Fig. 10.2 The surface atomic configurations in **a** the (001) plane and schematic hematite structure project along {001} and **b** the (010) plane and schematic hematite structure projected along {010}. Adapted and reproduced from Ref. [162]. Copyright 2016 Springer

[162]. In principle, lithium storage capacity of iron oxides is mainly delivered through the reversible conversion reaction between lithium ions and metal oxide forming metal nanocrystals dispersed in a Li_2O matrix. Even though this material is abundant, inexpensive, and environmentally friendly and has high theoretical capacity (Fe_3O_4: 924 mAh g^{-1} [135]; Fe_2O_3: 1005 mAh g^{-1} [136]), they are fraught with several problems such as poor electronic conductivity, higher volume expansion, and poor cycling stability. Among the major disadvantages, the cycle deterioration is the most important one which is caused by several factors such as the decomposition of the electrolyte solution and loss of the conductive path owing to electrode collapse [137]. In addition, large volume change of the anode material during the continuous charge–discharge cycling causes the cracking of the electrode, loose the contact with current collector, and destruction of the solid electrolyte interface (SEI), which result in the breaking of the continuous conductive path. Reducing the extent of volume expansion and contraction would suppress these problems in a great extent. Nanoscale processing of active material particles has been studied for this purpose [138–140]. Carbon coating [141], mixing with electronically conducting materials such as conducting carbon [142], carbon nanotubes [143], and graphene [144], is the commonly adopted technique to control the extent of volume change and for improving the electronic conductivity of iron oxides. Although the volume change ratio depends on material-specific quantities, it is possible to control the extent of volume change by controlling the particle size of the active material. Owing to the

formation of short Li^+-ion diffusion length by the size reduction of the active material to nanoscale, the chemical reaction resistance also get reduced. However, there are various processing difficulties associated in the use of nano-sized active material particles and nanoscale processing such as poor dispersibility due to an increase in the van der Waals force, low initial Coulombic efficiency due to high specific surface area, and low safety due to high chemical activity. To solve out these issues, electrodes prepared using composites of nanoscale active materials and carbon materials have been reported extensively [140], owing to the reduction in interfacial area between the electrolyte and active material by the composite effect, which suppresses the SEI growth, thereby improving the Coulombic efficiency.

Carbon coating on iron oxide nanoparticles can form 0D nanospheres [145, 146] 1D nanowires [36, 147], 2D nanoflakes [148, 149], and 3D structures of porous carbon foam loading iron oxide [43, 45]. Different methods such as solid-state reaction [150], hydrothermal process [151], solgel method [152], spray-drying technique [153], vacuum sintering [154], and vacuum decomposition are some of the commonly used methods for doing the carbon coating. Electrospinning is one of the unique carbon coating methods and has been used to fabricate 1D hybrid carbon coating iron oxide nanofibrous composites [79, 155–158], 1D iron oxide nanofibers [12, 159], and 1D carbon nanofibers for Li-ion batteries [160, 161]. The method can embed the iron oxides into a conductive carbon by subsequent heat treatment which enhances the electrochemical properties effectively due to the increased electronic conductivity. The uniformly coated nanometer-thick carbon layer on the active material (iron oxide) acts as the mechanical buffer, which prevents or minimizes the large volume expansion and cracking of electrode during the continuous charge–discharge cycles.

10.4.1 Electrospun Fe_2O_3/Fe_3O_4 Nanostructures

Different coating methods for the preparation of nanostructures such as chemical vapor deposition (CVD) [163], atomic layer deposition (ALD) [164, 165], electrochemical deposition (ECD) [166], and chemical bath deposition (CBD) [167–169] or the conventional synthetic strategies such as hydrothermal method [170] and scaffold-assisted synthesis method [170] are reported for the fabrication of nanostructures of Fe_2O_3/Fe_3O_4. Unfortunately, these methods need to involve the multistep growth of designed shell materials on various removable or sacrificial templates, which suffers from severe drawbacks. These methods are time consuming; tedious, high-temperature, and complicated processes need expensive equipment. Compared to the other synthesis methods, electrospinning is a facile, cost-effective, and flexible platform for one-dimensional (1D) robust Fe_2O_3/Fe_3O_4 nanostructures. This versatile method offers several advantages, such as high yield, mass production, easy control over the morphology, and high degree of reproducibility of the obtained materials [171–177]. More importantly, the electrospun nanofibers possess high surface-to-volume ratios due to the formation of super long scale in length of thin fibers, the completely interconnected pore structure, the porous substructure formed on the fiber during annealing, nano- to submicron size fiber diameters, and the great control over the morphology by simply changing the electrospinning process, solution, or ambient parameters [172]. When used these electrospun Fe_2O_3/Fe_3O_4 nanostructured materials as electrode in lithium-ion batteries, the large specific surface area and sufficient void spaces not only tolerate the volume change during the Li^+-ion intercalation, but also endow with more open channels for ions and electrons to migrate rapidly, resulting in the improved electrochemical performances. There are significant number of studies reported on the preparation of F_2O_3-/Fe_3O_4-based nanostructured anodes having different morphologies such as nanotubes [159, 178], nanorods [162, 179–182], porous structure [158, 183], microbelt [9], or hollow fibers [12, 184].

10.4.1.1 Porous Fe_2O_3/Hollow Fe_3O_4 Nanotube

A nanotube is a nanoscale material that has a seamless tubelike structure. Among the various nanostructures of Fe_2O_3 such as nanoparticles [185], nanorods [26, 186] and nano fibers [187, 188] etc., (1D) Fe_2O_3/Fe_3O_4 nanotubes [159, 170, 189] gained munch attraction due to their advantages properties, including the increase contact surface area between electrolyte and active materials, shorten migration path for Li+ and electron, and accommodate the volume variations via additional void space during cycling.

Sun et al. [170] prepared 1D porous Fe_2O_3 nanotubes with 2-μm length, 220-nm outer diameter, and 65-nm wall thickness via a low-temperature hydrothermal method followed by thermal treatment. The resulted porous Fe_2O_3 nanotubes exhibited enhanced electrochemical properties in terms of lithium storage capacity (1050 mAh g^{-1} at 100 mA g^{-1} rate), initial Coulombic efficiency (78.4%), cycle performances (90.6% capacity retention at 50th cycle), and rate capability (613.7 mAh g^{-1} at 1000 mA g^{-1} rate). 1D porous Fe2O3 nanostructures have also been synthesized via a SiO2 scaffold method, exhibiting the initial discharge and charge capacities of 1304.3 and 950.9 mAh g^{-1} at a current density of 100 mA g^{-1}, respectively [170]. Also, the porous Fe_2O_3 nanorods deliver a capacity of 671 and 541 mAh g^{-1} at current densities of 1000 and 2000 mA g^{-1}, respectively, showing good rate capability. Although the aforementioned porous Fe2O3 nanotubes showed the enhancement of lithium storage capacities, to fabricate porous Fe2O3 nanotubes with satisfied properties via a facile technology is still an appealing challenge. Hence, the robust electrospun Fe_2O_3 nanotubes are prepared and studied its electrochemical properties in LIBs.

Porous Fe_2O_3 [159] and Fe_3O_4/C [178] nanotubes were prepared by electrospinning a solution of iron (III) acetylacetonate and PVP (for porous Fe_2O_3 nanotubes) or iron (III) acetylacetonate and polyacrylonitrile (PAN) (10:8 wt./wt.) along with a 40% mineral oil on weight of iron (III) acetylacetonate (for Fe_3O_4/C nanotubes). First, the Fe_2O_3 or Fe_3O_4/C precursor fiber was prepared and then the fibers are heat-treated to transform to the porous hollow nanotubes. Fe_2O_3 precursor nanofibers (Fig. 10.3a, b) having an average fiber diameter of 520 nm possess continuous fibrous geometry

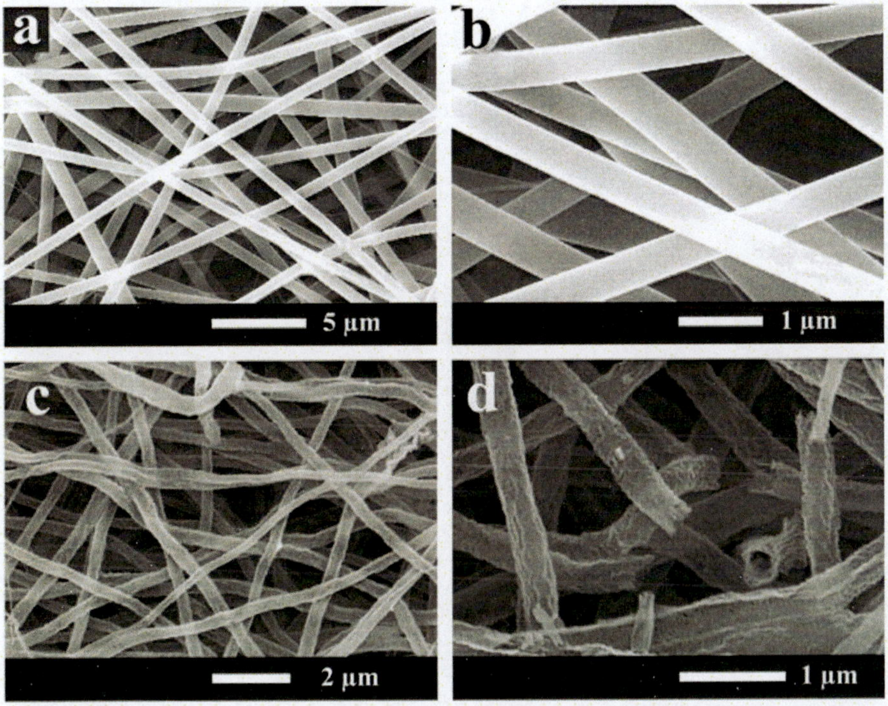

Fig. 10.3 **a, b** Low- and high-resolution images of precursor fibers, **c, d** FE-SEM images of porous Fe_2O_3 nanotubes. Adapted and reproduced from Ref. [159]. Copyright 2015 Elsevier

with a relatively smooth surface and without any pores or hierarchical nanostructure were first prepared by electrospinning and then annealed at 500 °C. After annealing, the as-prepared sample inherits the continuous 1D nanostructures from the precursor fibers and shows a large quantity of tubelike structures having average diameter of 400 nm (Fig. 10.3c, d); i.e., the diameter of the precursor fiber gets reduced by the shrinkage at higher temperature. For the production of Fe_3O_4/C nanotubes, the as-spun nanofibers were pre-oxidized in air at 250 °C for 2 h to follow by the carbonization at 600 °C for 2 h under high-purity argon atmosphere. The resultant nanotubes have outer diameter range from 200 to 400 nm and length of several millimeters.

During the electrospinning process, when the fibers are spun out from the spinning needle, the solvent will evaporate rapidly from the surface of fibers. This leads to the formation of a concentration gradient of solvent along the radial direction of the fibers, as illustrated in Fig. 10.4a; hence, the concentration at the center of the fibers is usually high than that of the surface. Due to the rapid evaporation of the solvent, the PAN will solidify on the surface of fiber. During the solidification process, the AAI and PAN will be extracted and reside at the shell of the fiber due to the rapid phase separation result from their poor solubility in mineral oil. Hence, the rapid evaporation of solvent produces a region near the fiber surface enriched in

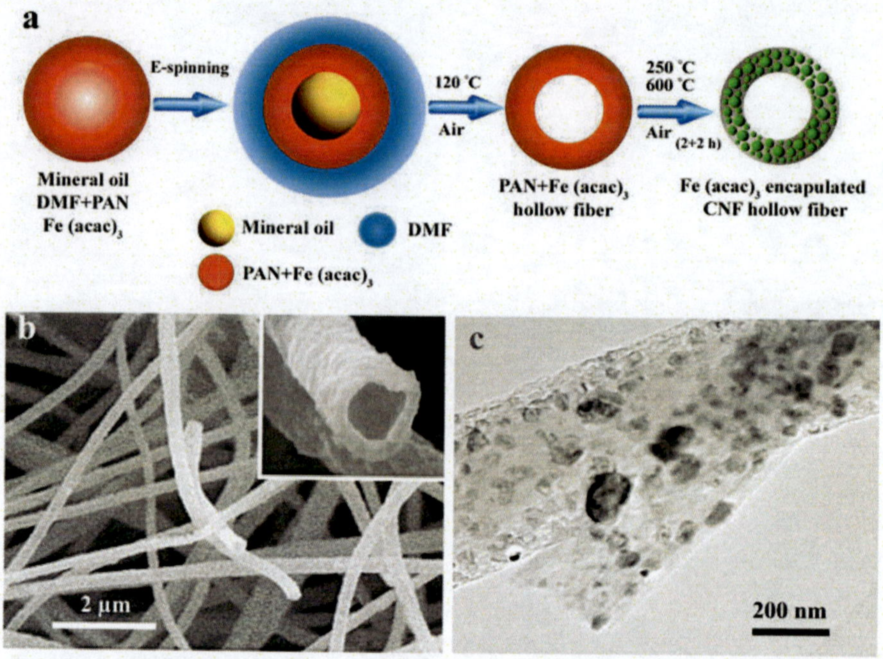

Fig. 10.4 **a** Schematic illustration of one step method to fabricate 1D Fe_3O_4/C nanotubes by electrospinning, **b** SEM image on the surface morphology of the final Fe_3O_4/C nanofibers, and the lower-left corner inset is an enlarged view of a single nanotube, and the scale bar is 200 nm, **c** TEM images of the Fe_3O_4/C nanotube. Adapted and reproduced from Ref. [178]. Copyright 2014 Elsevier

PAN/AAI, and the mineral oil would diffuse from the surface to the core of the fiber. With the continuing evaporation of the solvent, the concentration of PAN/AAI decreases continuously from the surface to the center of fiber, and the mineral oil tends to congregate at the center of fibers, as displayed in Fig. 10.4a. Further, the mineral oil at the center of fiber might evaporate through the wall of the nanotubes and PAN/AAI nanotubes were collected on the target. During the carbonization at 600 °C in Ar atmosphere, PAN would be decomposed and carbonized completely leading to the formation of Fe_3O_4/C nanotubes obtained. The tubes have a wall thickness of 40 and 150 nm hollow cores as shown in Fig. 10.4b, and it shows a roughness surface that can be distinctly observed from Fig. 10.4c.

On contrary, during the production of porous Fe_2O_3, the electrospun precursor nanofibers showed a smooth surface without porous or hollow structures. When the precursor fiber is sintering in air atmosphere, the degradation of the side chain of PVP formed through the intermolecular cross-linking reaction and the decomposition of iron acetylacetonate occur simultaneously. Upon increasing the temperature, both PVP and iron acetylacetonate would be oxidized. The oxidation of PVP results in the volatilization and evolution of CO_2, while the oxidation of iron acetylacetonate resulted in the formation of Fe_2O_3 nanoparticles. The outward diffusion of CO_2

generates a pressure to compress the Fe_2O_3 nanoparticles on the surface region of the fiber and forms a porous shell. Then, the porous shell allows CO_2 effusion from the regions below the shell in the fiber and the iron precursor in the inner part of the fibers would move toward the surface which is presumably accelerated by gaseous species that are produced by the oxidation of PVP, and crystallized into Fe_2O_3 nanoparticles. Finally, these Fe_2O_3 nanoparticles get connected together to generate porous Fe_2O_3 nanotubes. However, the formation of porous annotate greatly depends on the concentration iron precursor (iron acetylacetonate) in the electrospunned precursor fiber. To get more insight into the actual evolution process of the Fe_2O_3 nanostructures, we carry out a series of concentration-dependent experiments which samples are prepared by adjusting the concentration of iron acetylacetonate in the electrospun precursor solution. The precursor fiber having iron acetylacetonate higher than about 45% in the precursor electrospun fibers resulted in the formation of Fe_2O_3 nanotubes. If the concentration of iron precursor is less than 45% obviously, it results in the formation of a few Fe_2O_3 nanoparticles, which is difficult to keep the robust frame of the nanotube shell. As a result, the collapse of the tube structure results in the formation of the nanobelts (Fig. 10.5a). When the iron precursor concentration is lower than 37%, only nanobelts are formed (Fig. 10.5b), while the increase in concentration to 47% forms both nanobelts and nanotubes (Fig. 10.5c). When the iron precursor concentration is further extended to about 50%, the increase progressively encourages the formation of a large number of Fe_2O_3 nanoparticles; as a result, the Fe_2O_3 porous nanotubes appear (Fig. 10.5d). Hence, the evolution of Fe_2O_3 nanostructures could be controllably synthesized by adjusting the concentration of the electrospun precursor solution (Fig. 10.5e).

The cyclic voltammetry studies showed three cathodic peaks correspond to the potentials 1.55, 0.89, and 0.55 V, indicating the different lithiation steps [12, 156, 190, 191]. The peaks at 1.55 and 0.89 V correspond to the intercalation of Li + ions into the crystal structure of porous Fe_2O_3 nanotubes and the transformation from hexagonal α-$Li_xFe_2O_3$ to cubic $Li_2Fe_2O_3$ without any crystal structure destruction. The high intensity peak observed at 0.55 V corresponds to the crystal structure destruction accompanied by the complete reduction of iron from Fe(III) to Fe(0) and the decomposition of electrolyte. The anodic polarization peaks observed at 1.85 V correspond to the oxidation of Fe(0) to Fe(II) and Fe(III) to re-form Fe_2O_3. Compared to the first cycles, the subsequent cycles are significantly different, which is due to irreversible phase transformation during lithium insertion and extraction in the initial cycle. During the second cathodic process, the peaks at 1.55 and 0.89 V disappear, which indicates lithium insertion and irreversible phase transformation of hexagonal α-$Li_xFe_2O_3$ to cubic $Li_2Fe_2O_3$. Also, a decrease in the peak intensity with the number of cycles is observed, which indicates that the capacity is decreased during cycling. However, the CV curves starts to overlap after the third cycle cathodic scan, which indicate the reversibility and capacity stability the continuous charge-discharge processes. The galvanostatic charge–discharge profiles of the porous Fe_2O_3 are consistent with CV curves [159].

The porous Fe_2O_3/Li metal half cells assembled by using Celgard 2400 membrane as the separator and 1M $LiPF_6$ in ethylene carbonate/dimethyl carbonate

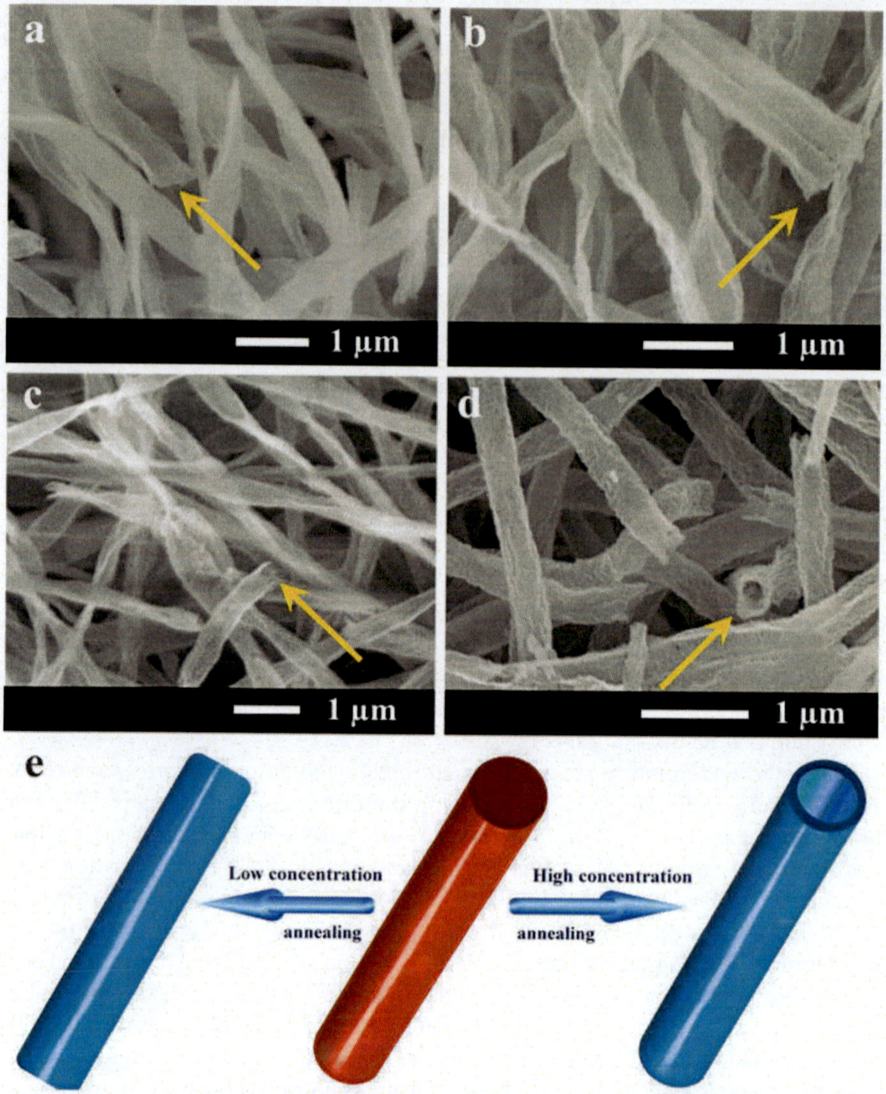

Fig. 10.5 SEM images on the surface morphology of Fe_2O_3 nanostructures obtained by adjusting the quantity of iron acetylacetonate (g): **a** 0.2, **b** 0.3 **c** 0.4, and **d** 0.5, respectively in the electrospun precursor solution. **e** Possible formation mechanism of Fe_2O_3 nanostructures. Adapted and reproduced from Ref. [159]. Copyright 2015 Elsevier

(EC/DMC,1:1 v/v) as the electrolyte delivered an initial charge and discharge capacities of around 1045 mAh g^{-1}, at a current density of 100 mA g^{-1}. Particularly, the porous Fe_2O_3 nanotubes still exhibit an excellent cyclic performance at a much higher current density of 200 mA g^{-1}, and the capacity reaches 988 mAh g^{-1} after 250 discharge and charge cycles. The Coulombic efficiency rises rapidly in the subsequent cycles, reaching up to 95% at the 5th cycle, and remains above 98% after 50 cycles, suggesting an excellent reversible Li^+-ion intercalation/extraction performance. The specific reversible charge and discharge capacity decrease slightly up to the initial 50 cycles and reach 513 and 524 mAh g^{-1}, respectively, and then increase significantly and reach over 995 and 988 mAh g^{-1} by 250th cycle. The porous Fe_2O_3 also showed good rate capability [159]. Similarly, the Fe_3O_4/C nanotubes also showed good charge–discharge cycling stability and rate capability. In the half cell studies using M $LiPF_6$ in ethylene carbonate, diethyl carbonate and ethyl methyl carbonate (EC/DMC/EMC, 1:1:1 vol) as the electrolyte and Celgard 2400 polypropylene as separator showed an initial discharge and charge capacity of 1102 and 727 mAh g^{-1}, with a Coulombic efficiency of only 66% at a current density of 0.15 °C. The significantly higher capacity loss (34% loss) during the first cycle [178] is corresponding to the formation of SEI layer and the incomplete conversion reaction [192, 193]. When the cell is cycled at 1600 mA g^{-1}, the cell delivers a very stable capacity of 350 mAh g^{-1}, and when the current density is switched back to 0.15 °C, a capacity about 600 mAh g^{-1} is delivered, which is about 100 mAh g^{-1} less compared to the first cycles at 0.15 °C [178].

In both porous Fe_2O_3 and Fe_3O_4/C electrodes, an initial capacity loss (25–35%) is observed. Similar irreversible capacity loss was noted with other metal oxides or metal oxide combinations reported in previous literature [12, 193, 194].

Compared to Fe_3O_4/C nanofibers prepared by a controlled fabrication process, sample fabricated according to Wang's reports [157] good electrochemical performance. The charge–discharge studies at a current density of 0.15 °C displayed a quick drop-down in discharge capacity to 300 mAh g^{-1} after 100 cycles, which is about 50% discharge capacity of the F_3O_4/C. More clearly, F_3O_4/C nanotube shows negligible capacity decreases from the 2nd cycle onward and displayed a discharge capacity as high as 600 mAh g^{-1} (85% of the second cycle capacity) after 100 cycles at a current density of 0.15 °C [178]. The lackluster cycling performance of Fe_3O_4/C nanofibers is due to the solid structure of the nanofiber; hence, there is no enough space to accommodate the mechanical stress of Fe_3O_4 nanoparticles along the fiber axis of the nanofibers. The substantially subdued capacity fade in Fe_3O_4/C nanotubes is due to the hollow nanotube structure having larger surface-to-volume ratio than that of 1D nanofibers, which could effectively accommodate large volume changes associated with Li^+-ions insertion/extraction. In addition, the tubular structure can increase the surface area accessible to the electrolyte facilitating the diffusion of Li^+ ions at the interior and exterior of the nanotube. The charge–discharge capacity fade under continuous cycling is mainly due to the pulverization of original aggregation of Fe_2O_3 particles by the huge volume expansion and contraction during the Li^+-ion intercalation/extraction process and resulting in the loss of electrical connectivity between the particles and current collector.

A novel process for the preparation of aggregate-free metal oxide nanopowders with spherical (0D) and non-spherical (1D) hollow nanostructures was introduced. Carbon nanofibers embedded with iron selenide (FeSe) nanopowders with various nanostructures are prepared via the selenization of electrospun nanofibers. Ostwald ripening occurs during the selenization process, resulting in the formation of a FeSe-C composite nanofiber exhibiting a hierarchical structure. These nanofibers transform into aggregate-free hollow Fe_2O_3 powders via the complete oxidation of FeSe and combustion of carbon. Indeed, the zero-dimensional (0D) and one-dimensional (1D) FeSe nanocrystals transform into the hollow-structured Fe_2O_3 nanopowders via a nanoscale Kirkendall diffusion process, thus conserving their overall morphology. The discharge capacities for the 1000th cycle of the hollow-structured Fe_2O_3 nanopowders obtained from the FeSe–C composite nanofibers prepared at selenization temperatures of 500, 800, and 1000 °C at a current density of 1 A g^{-1} are 932, 767, and 544 mAh g^{-1}, respectively; their capacity retentions from the second cycle are 88, 92, and 78%, respectively. The high structural stabilities of these hollow Fe_2O_3 nanopowders during repeated lithium insertion/desertion processes result in superior lithium-ion storage performances.

Figure 10.6 outlines the mechanism of the formation of Fe_2O_3 nanopowders exhibiting hollow nanostructures of different dimensions via the nanoscale Kirkendall diffusion process. Following the selenization processes at different temperatures (i.e., 500, 800, or 1000 °C), the electrospun nanofibers (Fig. 10.6a) were transformed into the hierarchical nanostructures. Selenization of the iron components located close to the nanofiber surface resulted in the formation of FeSe nanocrystals during the early stages of the process. Ostwald ripening then occurred during further selenization to yield the hierarchical FeSe–C composite nanofiber. In this process, the ultrafine FeSe nanocrystals formed inside the carbon nanofiber diffused to the surface to produce FeSe crystals via Ostwald ripening. Complete selenization at 500 °C resulted in the carbon nanofiber being uniformly embedded with ultrafine FeSe nanocrystals (Fig. 10.6b). However, at higher selenization temperatures (Fig. 10.6c), crystal growth occurred via the segregation of nanocrystals and spheroidization due to melting. Finally, the hierarchical FeSe–C nanofibers transformed into hollow aggregate-free Fe_2O_3 nanopowders (Fig. 10.6d, e) via the complete combustion of carbon and oxidation of FeSe. Furthermore, as shown in Fig. 10.7, FeSe nanocrystals with 0D and 1D structures transformed into the hollow-structured Fe_2O_3 nanopowders via a nanoscale Kirkendall diffusion process, thus conserving their overall morphology. For simplicity, the hollow Fe_2O_3 nanopowders obtained from the FeSe-C composite nanofibers prepared at 500, 800, and 1000 °C are referred to as "Sel.500-Oxi.600," "Sel.800-Oxi.600," and "Sel.1000-Oxi.600," respectively [1].

10.4.1.2 α-Fe_2O_3 Nanorods

Nanorods are one of the nanostructured entities, reported as electrodes in LIBs. Different synthesis techniques such as forced hydrolysis, solgel synthesis, template

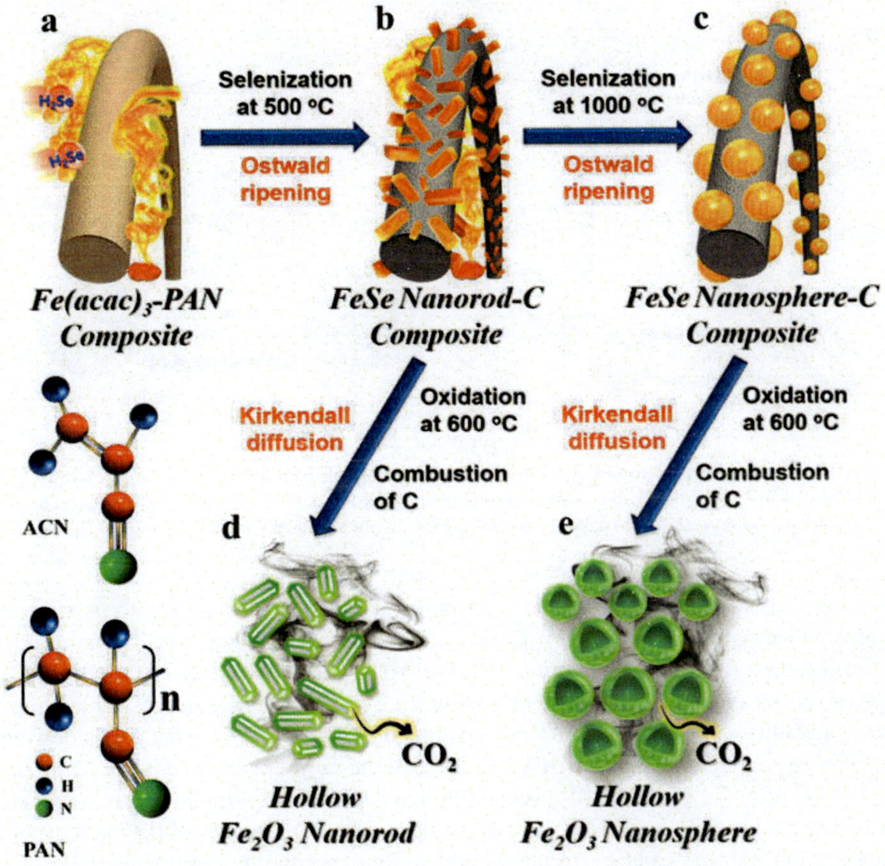

Fig. 10.6 Formation mechanism of the hollow-structured Fe_2O_3 nanopowders with 0D and 1D structure. Adapted and reproduced from Ref. [182]. Copyright 2016 Springer

methods, molten salt process, spray pyrolysis, hot plate method, hydrothermal method, and co-precipitation technique have been adopted to prepare various nanostructures of hematite [34, 134, 191, 195–200]. Nanoscale Kirkendall diffusion and Ostwald ripening processes, in which filled structures are transformed into hollow structures during heat treatment, have been applied recently for the preparation of hollow nanospheres (0D) in the absence of templates [182, 201–205] α-Fe_2O_3nanorods [156] and bubble-nanorod-structured Fe_2O_3-carbon nanofibers [179] prepared by electrospinning are reported as high performance anode in LIBs.

The α-Fe_2O_3 nanorods are synthesized by electrospinning of polyvinylpyrrolidone (PVP)/ferric acetylacetonate (Fe(acac) $_3$) composite precursors and subsequent annealing at 500 °C for 5 h. The phase separation of Fe(acac)$_3$ and PVP process involving a polymer/precursor forms "islands" that account for the generation of α-Fe_2O_3 nanorods having average diameter of the nanorods is found to be 150 nm

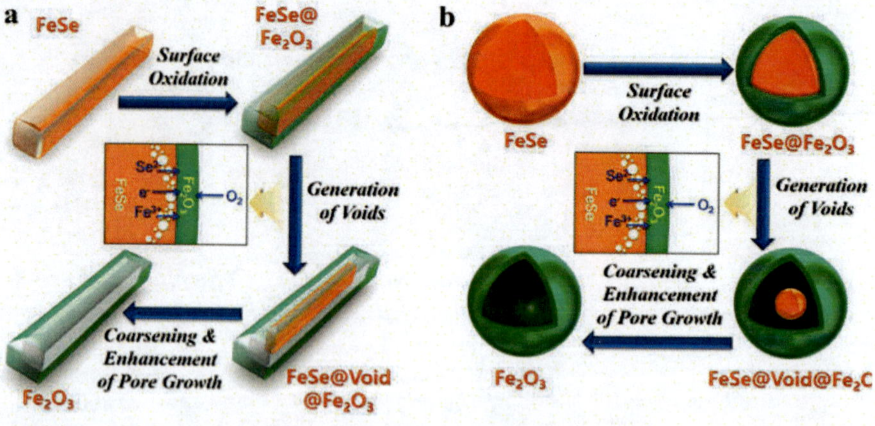

Fig. 10.7 Conversion reaction of the FeSe filled structure into Fe_2O_3 hollow structure by nanoscale Kirkendall diffusion effect, **a** hollow-structured Fe_2O_3 nanopowder with 1D and **b** hollow-structured Fe_2O_3 nanopowder with 0D. Adapted and reproduced from Ref. [182]. Copyright 2016 Springer

upon heat treatment. The annealed electrospun α-Fe_2O_3 nanorods are composed of agglomerates of nano-sized α-Fe_2O_3 particles. The electrospun α-Fe_2O_3 nanorods exhibit a high reversible capacity of 1095 mAh g^{-1} at 0.05 °C, are stable up to 50 cycles (with capacity retention of 93% between 2 and 50 cycles), and also show high rate capability, up to 2.5 °C. At a current rate of 2.5 °C, α-Fe_2O_3 nanorods deliver a discharge capacity of 765 mAh g^{-1} and when the current rate is reduced from 2.5 to 0.1 °C after 70 cycles, still a reversible capacity of 1090 mAh g^{-1} is obtained showing the good rate capability of the material. The high rate capability and excellent cycling stability can be attributed to the unique morphology of the macroporous nanorods comprised of interconnected nano-sized particles [156].

The structure denoted as "bubble-nanorod composite" is synthesized by introducing the Kirkendall effect into the electrospinning method. Bubble-nanorod-structured Fe_2O_3/C composite nanofibers, which are composed of nano-sized hollow Fe_2O_3 spheres uniformly dispersed in an amorphous carbon matrix, are synthesized as the target material using Fe(acac)$_3$-PAN composite solution as the precursor. Post-treatment of the electrospun precursor nanofibers at 500 °C under 10% H_2/Ar mixture gas atmosphere produces amorphous FeO_x/carbon composite nanofibers, and the further post-treatment at 300 °C under air atmosphere produces the bubble-nanorod-structured Fe_2O_3/C composite nanofibers. The solid Fe nanocrystals formed by the reduction of FeO_x are converted into hollow Fe_2O_3 nanospheres during the further heating process by the well-known Kirkendall diffusion process [179].

The formation mechanism of bubble-nanorod-structured Fe_2O_3/C composite nanofibers is schematically displayed in Figs. 10.8 and 10.9. During the carbonization at 500 °C, the PAN gets decomposed to form the carbon matrix and the decomposition of iron acetylacetonate produced FeO_x, which is uniformly dispersed in carbon matrix resulting in the carbon composite nanofibers. The significantly large amount

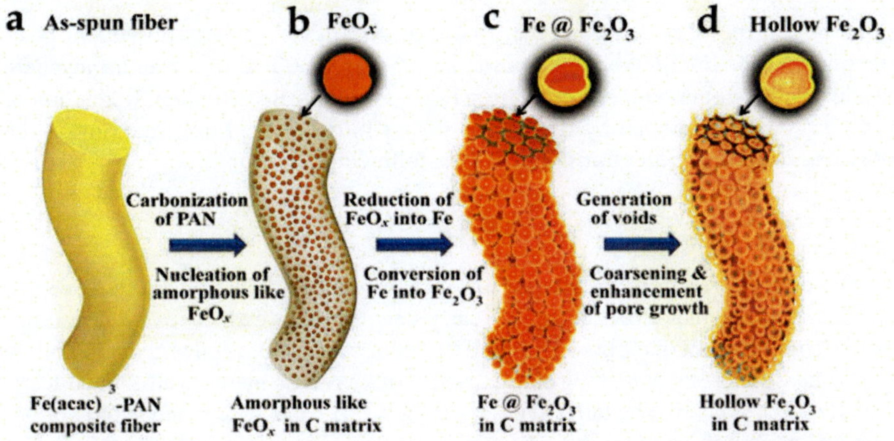

Fig. 10.8 Formation mechanism of bubble-nanorod-structured Fe_2O_3/C composite nanofiber by Kirkendall-type diffusion. Adapted and reproduced from Ref. [179]. Copyright 2015 American Chemical Society

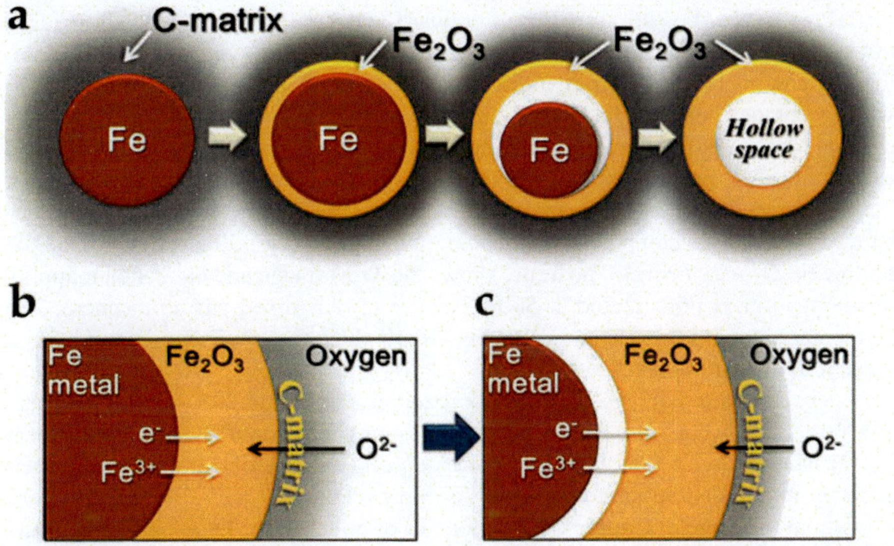

Fig. 10.9 **a** Formation mechanism of hollow Fe_2O_3 nanosphere in the bubble-nanorod-structured Fe_2O_3 carbon composite nanofiber by Kirkendall effect, **b**, **c** its chemical conversion process in the surface region of a sphere. Adapted and reproduced from Ref. [179]. Copyright 2015 American Chemical Society

of carbon in the fiber disturbed the crystal growth of FeO_x. The subsequent post-treatment of the FeO_x/carbon composite nanofibers at 300 °C under air atmosphere produced the bubble-nanorod-structured Fe_2O_3C composite nanofiber. Reduction of FeO_x crystals surrounded by the carbon matrix into Fe metal occurred during the post-treatment under air atmosphere by the following equation:

$$FeO_x(s) + xC(s) \rightarrow Fe(s) + xCO(g)$$

The crystal growth of Fe formed ultrafine Fe nanocrystals uniformly dispersed within the carbon nanofibers during the early stage of post-treatment by the consumption of some amount of carbon. The solid Fe nanocrystals were converted into hollow Fe_2O_3 nanospheres during the subsequent heating process by the well-known Kirkendall effect. The Kirkendall effect, a vacancy flux, and subsequent void formation process resulting from diffusivity differences at inorganic interfaces were first reported by Aldinger [206]. The Kirkendall effect results in the formation of a thin Fe_2O_3 layer on the Fe metal surface (Fig. 10.8c), followed by simultaneous outward diffusion of Fe cations through the oxide layer and inward diffusion of oxygen into the nanospheres, creating an intermediate $Fe@Fe_2O_3$ core–shell structure (Fig. 10.9b). Fe cations diffused outward more quickly than oxygen diffused inward, which is consistent with the larger ionic radius of oxygen anions (140 pm) than Fe cations (Fe^{2+} is 76 pm, and Fe^{3+} is 65 pm). Accordingly, Kirkendall voids were generated near the Fe/Fe_2O_3 interface during vacancy-assisted exchange of the material via bulk interdiffusion (Fig. 10.9c), which gave rise to coarsening and enhancement of pore growth in the spheres (Fig. 10.8d). Complete conversion of Fe metal into Fe_2O_3 by Kirkendall-type diffusion resulted in the bubble-nanorod-structured Fe_2O_3C composite nanofibers (the highly crystalline structure of the hollow Fe_2O_3 nanofibers). The figure shows the TEM image and elemental mapping of the nanofibers post-treated at 500 °C under H_2/Ar mixed gas atmosphere. The elemental mapping images shown in Fig. 10.10 show the trace amounts of carbon present in the nanofibers.

The cell studies displayed discharge capacities of 812 and 285 mAh g^{-1}, respectively, for bubble-nanorod-structured Fe_2O_3/C composite nanofibers and hollow bare Fe_2O_3 nanofibers for the 300th cycles at a current density of 1.0 A g^{-1}, and their capacity retentions measured from the second cycle are 84 and 24%, respectively. The initial (first cycle) Coulombic efficiencies of hollow bare Fe_2O_3 nanofibers and bubble-nanorod-structured Fe_2O_3/C composite nanofibers were 81 and 69%, respectively. The initial irreversible capacity loss of the hollow bare Fe_2O_3 nanofibers is ascribed to the formation of an SEI layer on the surface of the nanofibers and the incomplete restoration of metallic Fe into the original oxide during the initial charging process [156, 207]. The high amount of amorphous carbon, which has a low lithium storage capacity and a large initial irreversible capacity loss, decreased the initial discharge capacity and Coulombic efficiency of the bubble-nanorod-structured Fe_2O_3/C composite fibers [208, 209]. The rate capability studies showed that the stable reversible discharge capacities of the bubble-nanorod-structured Fe_2O_3C nanofibers decreased from 913 to 491 mAh g^{-1} as the current density increased

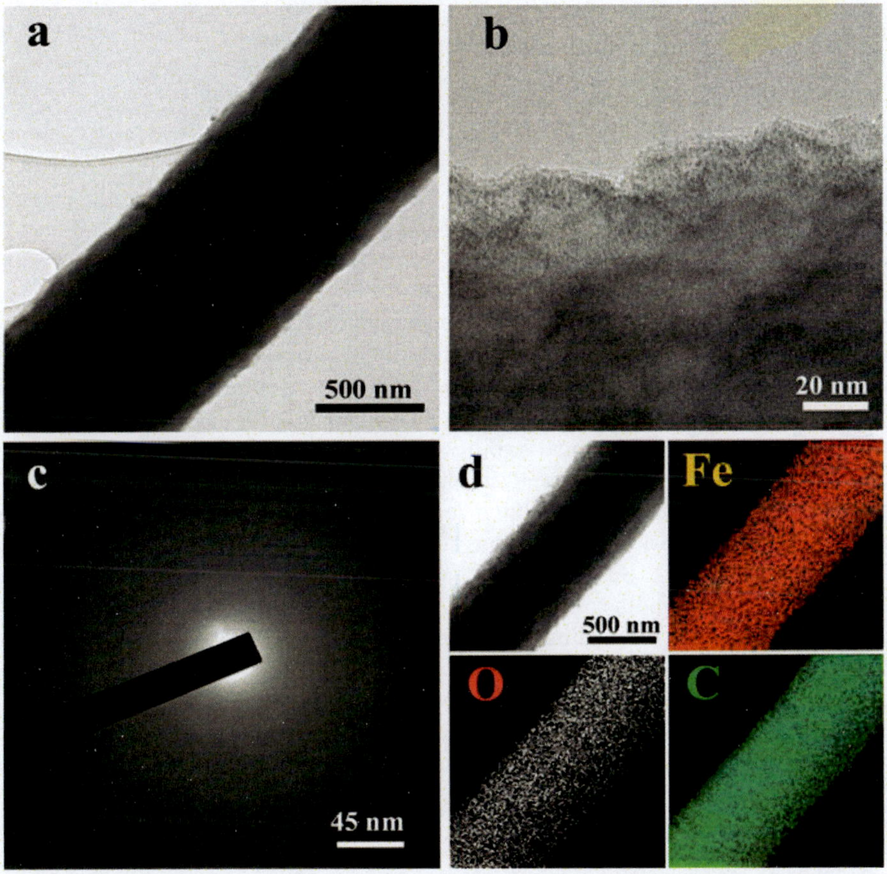

Fig. 10.10 Morphologies, SAED, and elemental mapping images of the nanofibers post-treated at 500 °C under H_2/Ar gas atmosphere: **a**, **b** TEM images, **c** SAED pattern, and **d** elemental mapping. Adapted and reproduced from Ref. [179], Copyright 2015 American Chemical Society

from 0.5 to 5.0 A g^{-1}. Furthermore, the discharge capacity recovered to 852 mAh g^{-1} as the current density was restored to 0.5 A g^{-1}. The electrochemical studies showed that the bubble-nanorod-structured Fe_2O_3C nanofibers showed superior electrochemical properties as an anode material for LIBs as compared with the hollow bare Fe_2O_3 nanofibers. The synergetic effect of hollow nanospheres and a carbon matrix resulted in the superior cycling and rate performance of the bubble-nanorod-structured Fe_2O_3/C nanofibers. The hollow nanospheres accommodate the volume change that occurs during the continuous charge–discharge cycling which leads to the long-term cycling stability. The unique structure of the bubble-nanorod-structured Fe_2O_3/C composite nanofibers results in their superior electrochemical properties by improving the structural stability during long-term cycling.

10.4.1.3 Metal/Fe_2O_4 or Fe_3O_4 Composite Nanofibers

In spite of their lower cost and better safety, the capacity retention of transition metal oxides especially Fe_2O_3/Fe_3O_4 and turning of their energy/power density remain as major drawbacks. However, transition metal oxide spinels (AB_2O_4) with two transition elements (both A- and B-sites) provide the feasibility to tune the energy density and working voltages by varying the transition metal content [210, 211]. Owing to it, zinc [212–215], Co, Cu, etc., have been substituted into iron oxide to construct a ternary metal ferrite ($ZnFe_2O_4$/$CuFe_2O_4$/$CoFe_2O_4$, etc.) and possess a lower working voltage to effectively enhance the total output voltage of the LIBs. $CuFe_2O_4$ nanoparticles, nanorods, and hollow spheres have been fabricated through wet chemical routes, and their electrochemical properties have been investigated [216–218]. Cobalt ferrite ($CoFe_2O_4$) has been well regarded as a distinguished anode due to its low cost, high chemical stability, and good environmental benignity [219–222]. Especially, $CoFe_2O_4$ can deliver a high theoretical capacity of 916 mAh g^{-1} based on an eight-electron conversion reaction. Similarly, zinc ferrite ($ZnFe_2O_4$) generates high capacity as lithium ions form an alloy with Zn and dealloy, while Fe and Zn react with Li_2O to absorb/release Li during lithiation/delithiation. Thus, $ZnFe_2O_4$ implements both conversion [6, 223] and alloy/dealloy [224, 225] reaction, simultaneously. Also, fabrication of nanostructured binary metal oxides could buffer the mechanical strain during the cycling process. In particular, 1D hollow nanostructure could not only provide sufficient void spaces to tolerate the volume change during cycling process, but also allow for efficient electron transport along the longitudinal direction, resulting in the improved electrochemical performance. Hence, $CoFe_2O_4$ nanotubes [226], $CuFe_2O_4$ nanotubes and nanorods [227], $CoFe_2O_4$/C composite fibers [228], interwoven $ZnFe_2O_4$ nanofibers [229], and *N*-doped amorphous carbon-coated Fe_3O_4/SnO_2 coaxial nanofibers [230] were prepared by electrospinning and reported as anode in LIBs having enhanced electrochemical properties. When evaluated as anode materials for LIBs, the $CoFe_2O_4$ nanotubes exhibited good electrochemical performance with high specific capacity of 1228 and 693.9 mAh g^{-1} at a current density of 50 and 200 mA g^{-1}, respectively, long cycling stability over 160 cycles, and good rate capability (214.7 mAh g^{-1} at 2 A g^{-1}) [226], while the $CuFe_2O_4$ nanotubes delivered a high reversible capacity of 1399.4 mAh g^{-1} at a current density of 200 mA·g^{-1} and a capacity retention of ~816 mAh g^{-1} after 50 cycles with a good rate capability (450 mAh g^{-1} at 2.5 A g^{-1}) [227] which is much higher compared to $CoFe_2O_4$ nanotubes (wall thickness of ~50 nm, presented diameters of ~150 nm, and lengths up to several millimeters) [226]. The $CuFe_2O_4$ nanorods showed a discharge capacity of only ~489 mAh g^{-1} after 50 cycles for the same current density of 2.5 A g^{-1} which clearly suggest that the nanotubes have superior electrochemical properties than their counterpart nanotubes [227] which can be attributed to the continuous one-dimensional (1D) hollow nanostructure and their higher surface area. However, compared to $CoFe_2O_4$ nanotubes [226], $CoFe_2O_4$/C composite fibers consist of $CoFe_2O_4$ nanoparticles with a diameter of about 42 nm well dispersed in the carbon matrix as anode material prepared by electrospinning

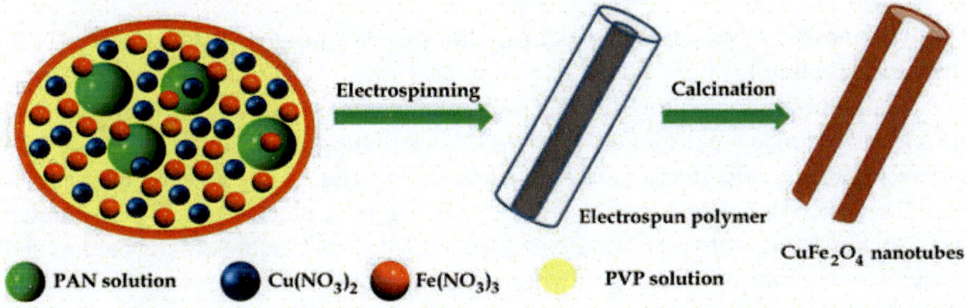

Fig. 10.11 Scheme of the procedure for producing $CuFe_2O_4$ nanotubes. Adapted and reproduced from Ref. [227], Copyright 2014 Elsevier

and thermal technique displayed a stable and reversible capacity of over 490 mAh g^{-1} after 700 cycles at a rate of 2.0 C and good rate capability [228].

The (1D) $CuFe_2O_4$ nanotubes and nanorods were fabricated by a single spinneret electrospinning method followed by thermal decomposition for removal of polymers from the precursor fibers. It was found that phase separation between the electrospun composite materials occurred during the electrospinning process, while the as-spun precursor nanofibers composed of polyacrylonitrile (PAN), polyvinylpyrrolidone (PVP), and metal salts might possess a core–shell structure (PAN as the core and PVP/metal salt composite as the shell) and then transformed to a hollow structure after calcinations as shown in Fig. 10.11 [227]. Based on the literature and the above analysis, the electrochemical reactions involved in the cycling process are believed to proceed as follows:

$$
\begin{aligned}
&\mathrm{CuFe_2O_4} + \mathrm{nLi} + {+}\mathrm{ne}^- \rightarrow \mathrm{Li}_n\mathrm{CuFe_2O_4}\\
&\mathrm{Li}_n\mathrm{CuFe_2O_4} + (8-n)\mathrm{Li}^+ + (8-n) \rightarrow x\mathrm{Cu} + (2-x)\mathrm{Fe}\\
&\quad + \mathrm{Fe}_x\mathrm{Cu}(1-x) + 4\mathrm{Li_2O}(0 \le 1 \le 4)\\
&\mathrm{Cu} + \mathrm{Li_2O} \rightarrow \mathrm{CuO} + 2\mathrm{Li}^+ + 2\mathrm{e}^-\\
&\mathrm{Fe} + 3\mathrm{Li_2O} \rightarrow \mathrm{Fe_2O_3} + \mathrm{Li}^+ + \mathrm{e}^-
\end{aligned}
$$

As a heavy metal, Co, is a toxic material, hence nanowebs consisting of interwoven $ZnFe_2O_4$ nanofibers are synthesized by a simple electrospinning technique, to be employed as an environmentally friendly anode in lithium-ion batteries. The morphological studies showed self-assembly of electrospun $ZnFe_2O_4$ nanofibers into intertwined porous nanowebs with a continuous framework. Benefitting from the one-dimensional functional nanostructured architecture, the application of electrospun nanowebs with $ZnFe_2O_4$ nanofiber (ZFO-NF) anodes in LIBs delivers the first charge capacity of 925 mAh g^{-1}, exhibits excellent cyclability, and retains a reversible capacity of 733 mAh g^{-1} up to 30 cycles at 60 mA g^{-1} as compared to $ZnFe_2O_4$ nanorods (ZFO-NR) with a capacity of 200 mAh g^{-1}. The $ZnFe_2O_4$ nanowebs also displayed a high capacity of 400 mAh g^{-1} at 800 mA g^{-1} (1C). The enhanced capacity releases at higher current densities have shown the importance of having a well-connected electronic wiring during lithium insertion/extraction especially in prolonged cycling [229].

N-doped amorphous carbon-coated Fe_3O_4/SnO_2 coaxial nanofibers were prepared via a facile approach and studied as binder-free self-supported anode for lithium-ion batteries and their electrochemical performance in LIBs. In the process of preparation, the core composite nanofibers were first made by electrospinning technique, and then the shells were conformally coated using the chemical bath deposition and subsequent carbonization using polydopamine as a carbon source (Fig. 10.12). The coaxial nanofibers displayed an enhanced electrochemical storage capacity of 1223,

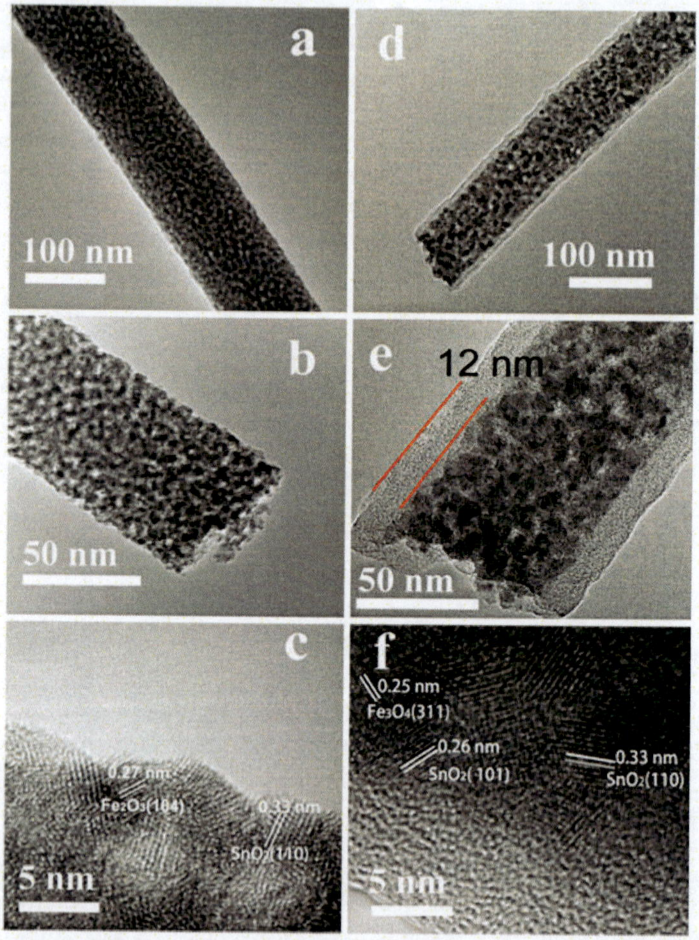

Fig. 10.12 **a**, **b** TEM, **c** high-resolution TEM images of the prepared single Fe_2O_3/SnO_2 composite nanofiber, **d**, **e** TEM, **f** high-resolution TEM images of the carbonized polydopamine-coated single Fe_3O_4/SnO_2 coaxial nanofiber. Adapted and reproduced from Ref. [230]. Copyright 2014 American Chemical Society

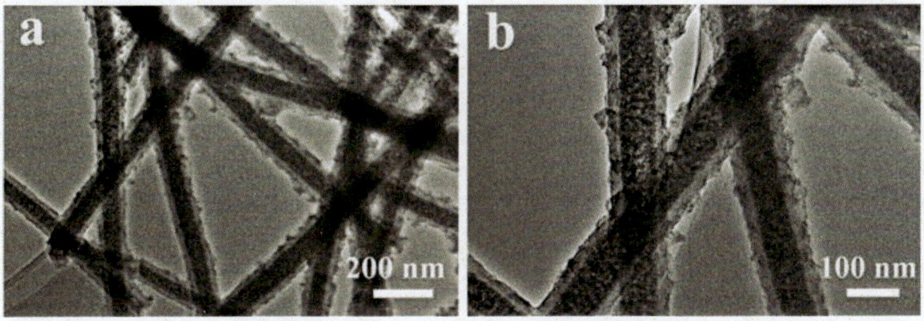

Fig. 10.13 **a**, **b** TEM images of the carbonized polydopamine-coated Fe_3O_4/SnO_2 coaxial nanofibers after rate performance test. Adapted and reproduced from Ref. [230]. Copyright 2014 American Chemical Society

1030, 862, and 640 mAh g^{-1} at 100, 200, 400, and 800 mAg^{-1}, respectively. Even after the current density is increased to 1600 mA g^{-1}, it still maintains a high charge capacity of 402 mAh g^{-1}. When the current density is restored to the initial setting of 100 mA g^{-1}, the carbonized polydopamine-coated Fe_3O_4/SnO_2 coaxial nanofiber electrode (Fig. 10.13) leads to a reversible capacity of 1070 mAh g^{-1} displaying not only its superior capacity retention but also excellent capacity recovery performance. Also, it was found that the morphology of the interwoven nanofibers was maintained even after the rate cycle test. The superior electrochemical performance originates from the structural stability of the N-doped amorphous carbon shells formed by carbonizing polydopamine [230]. On the basis of the literature, the electrochemical reactions of the carbonized polydopamine-coated Fe_3O_4/SnO_2 coaxial nanofibers can be described as follows

$$SnO_2 + 4Li^+ + 4e^- \rightarrow Sn + 2LiO$$
$$Sn + xLi^+ + xe^- \rightarrow Li_xSn(0 \leq x \leq 4.4)$$
$$Fe_3O_4 + 8Li^+ + 8e^- \rightarrow 3Fe + 4Li_2O$$

10.5 Conclusion

For the achievement of best performing lithium-ion batteries, different materials were greatly explored that can deliver a best performing system. Metal oxides possess great significance in lithium-ion batteries since they are capable of exhibiting better electrochemical properties. Similar to carbon base materials, currently, metal oxides are also widely explored as anodes in LIBs. Iron oxide-based anode materials such as Fe_2O_3 and Fe_3O_4 are promising due to its high capacity and lithium storage properties. Even though, despite of its high capacity, the poor cycling stability and high polarization during lithiation and delithiation process limit its practical use in LIBs.

Modifications of structural and surface characteristics are considered to be the best method for the enhancement of electrochemical properties of these anodes. Electrospinning is considered to be the best method for this. Electrospinning is considered to be the most versatile and simple method that provides a flexible platform for the fabrication of nanostructures that can deliver better battery performance. Porous and hollow iron oxide structure results by the electrospinning technique facilitate the fast lithium-ion transfer as well as it accommodates the volume change that results in unique electrochemical properties. Similarly, the hollow nanotubes and nanorods that results by this technique can deliver high performance in battery owing to its high surface area, controlled shape, and a low density. Further performance can be enhanced by the structural modification by incorporating different metals to form the composite structures which can result an enhancement in conductivity which is significant for a better performing electrode material. These potential features make iron oxide-based materials as the best performing anode material in LIBs.

References

1. Nam KT, Kim DW, Yoo PJ et al (2006) Virus-enabled synthesis and assembly of nanowires for lithium ion battery electrodes. Science 312(80):885–888. https://doi.org/10.1126/science.1122716
2. Zhang H, Yu X, Braun PV (2011) Three-dimensional bicontinuous ultrafast-charge and-discharge bulk battery electrodes. Nat Nanotechnol 6:277–281. https://doi.org/10.1038/nnano.2011.38
3. Wang H, Yang Y, Liang Y et al (2011) $LiMn_1$-$xFexPO_4$ nanorods grown on graphene sheets for ultrahigh-rate-performance lithium ion batteries. Angew Chemie Int Ed 50:7364–7368. https://doi.org/10.1002/anie.201103163
4. Hwang TH, Lee YM, Kong BS et al (2012) Electrospun core-shell fibers for robust silicon nanoparticle-based lithium ion battery anodes. Nano Lett 12:802–807. https://doi.org/10.1021/nl203817r
5. O'Heir J (2017) Building better batteries. Mech Eng 139:10–11. https://doi.org/10.1038/451652a
6. Poizot P, Laruelle S, Grugeon S et al (2000) Nano-sized transition-metal oxides as negative-electrode materials for lithium-ion batteries. Nature 407:496–499. https://doi.org/10.1038/35035045
7. Lou XW, Deng D, Lee JY et al (2008) Self-supported formation of needlelike Co_3O_4 nanotubes and their application as lithium-ion battery electrodes. Adv Mater 20:258–262. https://doi.org/10.1002/adma.200702412
8. Xiong S, Chen JS, Lou XW, Zeng HC (2012) Mesoporous Co_3O_4 and CoO@C topotactically transformed from chrysanthemum-like $Co(CO_3)0.5(OH)\cdot 0.11H_2O$ and their lithium-storage properties. Adv Funct Mater 22:861–871. https://doi.org/10.1002/adfm.201102192
9. Lang L, Xu Z (2013) In situ synthesis of porous Fe_3O_4/C microbelts and their enhanced electrochemical performance for lithium-ion batteries. ACS Appl Mater Interfaces 5:1698–1703. https://doi.org/10.1021/am302753p
10. Li Y, Tan B, Wu Y (2006) Freestanding mesoporous quasi-single-crystalline Co_3O_4 nanowire arrays. J Am Chem Soc 128:14258–14259. https://doi.org/10.1021/ja065308q
11. Xiong QQ, Tu JP, Lu Y et al (2012) Three-dimensional porous nano-Ni/Fe_3O_4 composite film: Enhanced electrochemical performance for lithium-ion batteries. J Mater Chem 22:18639–18645. https://doi.org/10.1039/c2jm33770c

12. Chaudhari S, Srinivasan M (2012) 1D hollow α-Fe_2O_3 electrospun nanofibers as high performance anode material for lithium ion batteries. J Mater Chem 22:23049–23056. https://doi.org/10.1039/c2jm32989a
13. Wang B, Chen JS, Bin WuH et al (2011) Quasiemulsion-templated formation of α-Fe_2O_3 hollow spheres with enhanced lithium storage properties. J Am Chem Soc 133:17146–17148. https://doi.org/10.1021/ja208346s
14. Zhong J, Cao C, Liu Y et al (2010) Hollow core-shell η-Fe_2O_3 microspheres with excellent lithium-storage and gas-sensing properties. Chem Commun 46:3869–3871. https://doi.org/10.1039/c0cc00204f
15. Xiong QQ, Tu JP, Lu Y et al (2012) Synthesis of hierarchical hollow-structured single-crystalline magnetite (Fe_3O_4) microspheres: the highly powerful storage versus lithium as an anode for lithium ion batteries. J Phys Chem C 116:6495–6502. https://doi.org/10.1021/jp3002178
16. Chen J, Xu L, Li W, Gou X (2005) α-Fe_2O_3 nanotubes in gas sensor and lithium-ion battery applications. Adv Mater 17:582–586. https://doi.org/10.1002/adma.200401101
17. Xu JS, Zhu YJ (2012) Monodisperse Fe 3O 4 and γ-Fe_2O_3 magnetic mesoporous microspheres as anode materials for lithium-ion batteries. ACS Appl Mater Interfaces 4:4752–4757. https://doi.org/10.1021/am301123f
18. Li J, Wan W, Zhou H et al (2011) Hydrothermal synthesis of TiO_2(B) nanowires with ultrahigh surface area and their fast charging and discharging properties in Li-ion batteries. Chem Commun 47:3439–3441. https://doi.org/10.1039/c0cc04634e
19. Wang K, Wei M, Morris MA et al (2007) Mesoporous titania nanotubes: their preparation and application as electrode materials for rechargeable lithium batteries. Adv Mater 19:3016–3020. https://doi.org/10.1002/adma.200602189
20. Sayle TXT, Maphanga RR, Ngoepe PE, Sayle DC (2009) Predicting the electrochemical properties of MnO_2 nanomaterials used in rechargeable Li batteries: simulating nanostructure at the atomistic level. J Am Chem Soc 131:6161–6173. https://doi.org/10.1021/ja8082335
21. Jiao F, Bruce PG (2007) Mesoporous crystalline β-MnO_2—a reversible positive electrode for rechargeable lithium batteries. Adv Mater 19:657–660. https://doi.org/10.1002/adma.200602499
22. Zhao J, Tao Z, Liang J, Chen J (2008) Facile synthesis of nanoporous γ-MnO_2 structures and their application in rechargeable Li-ion batteries. Cryst Growth Des 8:2799–2805. https://doi.org/10.1021/cg701044b
23. Meduri P, Pendyala C, Kumar V et al (2009) Hybrid tin oxide nanowires as stable and high capacity anodes for li-ion batteries. Nano Lett 9:612–616. https://doi.org/10.1021/n1802864a
24. Deng D, Lee JY (2008) Hollow core-shell mesospheres of crystalline SnO_2 nanoparticle aggregates for high capacity Li+ ion storage. Chem Mater 20:1841–1846. https://doi.org/10.1021/cm7030575
25. Ye J, Zhang H, Yang R et al (2010) Morphology-controlled synthesis of SnO2 nanotubes by using 1D silica mesostructures as sacrificial templates and their applications in lithium-ion batteries. Small 6:296–306. https://doi.org/10.1002/smll.200901815
26. Balogun MS, Wu Z, Luo Y et al (2016) High power density nitridated hematite (α-Fe2O3) nanorods as anode for high-performance flexible lithium ion batteries. J Power Sources 308:7–17. https://doi.org/10.1016/j.jpowsour.2016.01.043
27. Song Y, Qin S, Zhang Y et al (2010) Large-scale porous hematite nanorod arrays: Direct growth on titanium foil and reversible lithium storage. J Phys Chem C 114:21158–21164. https://doi.org/10.1021/jp1091009
28. Brandt A, Balducci A (2013) A study about the use of carbon coated iron oxide-based electrodes in lithium-ion capacitors. Electrochim Acta 108:219–225. https://doi.org/10.1016/j.electacta.2013.06.076
29. Taberna PL, Mitra S, Poizot P et al (2006) High rate capabilities Fe3O4-based Cu nano-architectured electrodes for lithium-ion battery applications. Nat Mater 5:567–573. https://doi.org/10.1038/nmat1672

30. Behera SK (2011) Facile synthesis and electrochemical properties of Fe3O 4 nanoparticles for Li ion battery anode. J Power Sources 196:8669–8674. https://doi.org/10.1016/j.jpowsour.2011.06.067
31. Bruck AM, Cama CA, Gannett CN et al (2016) Nanocrystalline iron oxide based electroactive materials in lithium ion batteries: the critical role of crystallite size, morphology, and electrode heterostructure on battery relevant electrochemistry. Inorg Chem Front 3:26–40. https://doi.org/10.1039/c5qi00247h
32. Wu LL, Zhao DL, Cheng XW et al (2017) Nanorod Mn3O4 anchored on graphene nanosheet as anode of lithium ion batteries with enhanced reversible capacity and cyclic performance. J Alloys Compd 728:383–390. https://doi.org/10.1016/j.jallcom.2017.09.005
33. Zeng S, Tang K, Li T et al (2008) Facile route for the fabrication of porous hematite nanoflowers: Its synthesis, growth mechanism, application in the lithium ion battery, and magnetic and photocatalytic properties. J Phys Chem C 112:4836–4843. https://doi.org/10.1021/jp0768773
34. Reddy MV, Yu T, Sow CH et al (2007) α-Fe2O3 nanoflakes as an anode material for li-ion batteries. Adv Funct Mater 17:2792–2799. https://doi.org/10.1002/adfm.200601186
35. Morales J, Sánchez L, Martín F et al (2005) Synthesis and characterization of nanometric iron and iron-titanium oxides by mechanical milling: Electrochemical properties as anodic materials in lithium cells. J Electrochem Soc 152:1748–1754. https://doi.org/10.1149/1.1972812
36. Muraliganth T, Vadivel Murugan A, Manthiram A (2009) Facile synthesis of carbon-decorated single-crystalline Fe3 O4 nanowires and their application as high performance anode in lithium ion batteries. Chem Commun 7360–7362. https://doi.org/10.1039/b916376j
37. Magasinski A, Dixon P, Hertzberg B et al (2010) Erratum: high-performance lithium-ion anodes using a hierarchical bottom-up approach. Nat Mater 9:353–358. https://doi.org/10.1038/nmat2749
38. Guo B, Fang X, Li B et al (2012) Synthesis and lithium storage mechanism of ultrafine MoO 2 nanorods. Chem Mater 24:457–463. https://doi.org/10.1021/cm202459r
39. Zhou G, Wang DW, Li F et al (2010) Graphene-wrapped Fe_3O_4 anode material with improved reversible capacity and cyclic stability for lithium ion batteries. Chem Mater 22:5306–5313. https://doi.org/10.1021/cm101532x
40. Gu M, Li Y, Li X et al (2012) In situ TEM study of lithiation behavior of silicon nanoparticles attached to and embedded in a carbon matrix. ACS Nano 6:8439–8447. https://doi.org/10.1021/nn303312m
41. Wang H, Cui LF, Yang Y et al (2010) Mn_3O_4-graphene hybrid as a high-capacity anode material for lithium ion batteries. J Am Chem Soc 132:13978–13980. https://doi.org/10.1021/ja105296a
42. Ban C, Wu Z, Gillaspie DT et al (2010) Nanostructured Fe_3O_4/SWNT electrode: Binder-free and high-rate Li-ion anode. Adv Mater 22:145–149. https://doi.org/10.1002/adma.200904285
43. Yoon T, Chae C, Sun YK et al (2011) Bottom-up in situ formation of Fe_3O_4 nanocrystals in a porous carbon foam for lithium-ion battery anodes. J Mater Chem 21:17325–17330. https://doi.org/10.1039/c1jm13450g
44. Li Y, Zhu S, Liu Q et al (2012) Carbon-coated SnO_2@C with hierarchically porous structures and graphite layers inside for a high-performance lithium-ion battery. J Mater Chem 22:2766–2773. https://doi.org/10.1039/c1jm14290a
45. Kang E, Jung YS, Cavanagh AS et al (2011) Fe_3O_4 nanoparticles confined in mesocellular carbon foam for high performance anode materials for lithium-ion batteries. Adv Funct Mater 21:2430–2438. https://doi.org/10.1002/adfm.201002576
46. Li B, Cao H, Shao J, Qu M (2011) Enhanced anode performances of the Fe_3O_4-Carbon-rGO three dimensional composite in lithium ion batteries. Chem Commun 47:10374–10376. https://doi.org/10.1039/c1cc13462k
47. Piao Y, Kim HS, Sung YE, Hyeon T (2010) Facile scalable synthesis of magnetite nanocrystals embedded in carbon matrix as superior anode materials for lithium-ion batteries. Chem Commun 46:118–120. https://doi.org/10.1039/b920037a

48. Li H, Zhou H (2012) Enhancing the performances of Li-ion batteries by carbon-coating: present and future. Chem Commun 48:1201–1217. https://doi.org/10.1039/c1cc14764a
49. Zhu T, Chen JS, Lou XW (2011) Glucose-assisted one-pot synthesis of FeOOH nanorods and their transformation to Fe_3O_4@carbon nanorods for application in lithium ion batteries. J Phys Chem C 115:9814–9820. https://doi.org/10.1021/jp2013754
50. Zhang XF, Wang KX, Wei X, Chen JS (2011) Carbon-coated V_2O_5 nanocrystals as high performance cathode material for lithium ion batteries. Chem Mater 23:5290–5292. https://doi.org/10.1021/cm202812z
51. Chen D, Ji G, Ma Y et al (2011) Graphene-encapsulated hollow Fe_3O_4 nanoparticle aggregates as a high-performance anode material for lithium ion batteries. ACS Appl Mater Interfaces 3:3078–3083. https://doi.org/10.1021/am200592r
52. Chen JS, Zhang Y, Lou XW (2011) One-pot synthesis of uniform Fe_3O_4 nanospheres with carbon matrix support for improved lithium storage capabilities. ACS Appl Mater Interfaces 3:3276–3279. https://doi.org/10.1021/am201079z
53. Zhao X, Xia D, Zheng K (2012) Fe_3O_4/Fe/carbon composite and its application as anode material for lithium-ion batteries. ACS Appl Mater Interfaces 4:1350–1356. https://doi.org/10.1021/am201617j
54. Wu J, Qin X, Miao C et al (2016) A honeycomb-cobweb inspired hierarchical core-shell structure design for electrospun silicon/carbon fibers as lithium-ion battery anodes. Carbon N Y 98:582–591. https://doi.org/10.1016/j.carbon.2015.11.048
55. Bonino CA, Ji L, Lin Z et al (2011) Electrospun carbon-tin oxide composite nanofibers for use as lithium ion battery anodes. ACS Appl Mater Interfaces 3:2534–2542. https://doi.org/10.1021/am2004015
56. Kong J, Tan HR, Tan SY et al (2010) A generic approach for preparing core-shell carbon-metal oxide nanofibers: morphological evolution and its mechanism. Chem Commun 46:8773–8775. https://doi.org/10.1039/c0cc03006f
57. Tran T, McCormac K, Li J et al (2014) Electrospun SnO_2 and TiO_2 composite nanofibers for lithium ion batteries. Electrochim Acta 117:68–75. https://doi.org/10.1016/j.electacta.2013.11.101
58. Zhou D, Song WL, Fan LZ (2015) Hollow core-shell SnO_2/C fibers as highly stable anodes for lithium-ion batteries. ACS Appl Mater Interfaces 7:21472–21478. https://doi.org/10.1021/acsami.5b06512
59. Li X, Zhang H, Feng C et al (2014) Novel cage-like α-Fe_2O_3/SnO_2 composite nanofibers by electrospinning for rapid gas sensing properties. RSC Adv 4:27552–27555. https://doi.org/10.1039/c4ra03307h
60. Wang X, Zhang K, Zhu M et al (2008) Continuous polymer nanofiber yarns prepared by self-bundling electrospinning method. Polymer (Guildf) 49:2755–2761. https://doi.org/10.1016/j.polymer.2008.04.015
61. Fei L, Williams BP, Yoo SH et al (2016) Graphene folding in si rich carbon nanofibers for highly stable, high capacity Li-ion battery anodes. ACS Appl Mater Interfaces 8:5243–5250. https://doi.org/10.1021/acsami.5b10548
62. Wang HG, Yuan S, Ma DL et al (2015) Electrospun materials for lithium and sodium rechargeable batteries: from structure evolution to electrochemical performance. Energy Environ Sci 8:1660–1681. https://doi.org/10.1039/c4ee03912b
63. Xu Y, Zhu Y, Han F et al (2015) 3D Si/C fiber paper electrodes fabricated using a combined electrospray/electrospinning technique for Li-ion batteries. Adv Energy Mater 5:1–7. https://doi.org/10.1002/aenm.201400753
64. Zhang M, Yan F, Tang X et al (2014) Flexible CoO-graphene-carbon nanofiber mats as binder-free anodes for lithium-ion batteries with superior rate capacity and cyclic stability. J Mater Chem A 2:5890–5897. https://doi.org/10.1039/c4ta00311j
65. Zhao B, Cai R, Jiang S et al (2012) Highly flexible self-standing film electrode composed of mesoporous rutile TiO_2/C nanofibers for lithium-ion batteries. Electrochim Acta 85:636–643. https://doi.org/10.1016/j.electacta.2012.08.126

66. Miao YE, Wang R, Chen D et al (2012) Electrospun self-standing membrane of hierarchical SiO 2atγ-AlOOH (Boehmite) core/sheath fibers for water remediation. ACS Appl Mater Interfaces 4:5353–5359. https://doi.org/10.1021/am3012998
67. McCann JT, Marquez M, Xia Y (2006) Highly porous fibers by electrospinning into a cryogenic liquid. J Am Chem Soc 128:1436–1437. https://doi.org/10.1021/ja056810y
68. Mou F, Guan JG, Shi W et al (2010) Oriented contraction: a facile nonequilibrium heat-treatment approach for fabrication of maghemite fiber-in-tube and tube-in-tube nanostructures. Langmuir 26:15580–15585. https://doi.org/10.1021/la102830p
69. Chen H, Wang N, Di J et al (2010) Nanowire-in-microtube structured core/shell fibers via multifluidic coaxial electrospinning. Langmuir 26:11291–11296. https://doi.org/10.1021/la100611f
70. Zhu C, Yu Y, Gu L et al (2011) Electrospinning of highly electroactive carbon-coated single-crystalline $LiFePO_4$ nanowires. Angew Chemie Int Ed 50:6278–6282. https://doi.org/10.1002/anie.201005428
71. McCann JT, Lim B, Ostermann R et al (2007) Carbon nanotubes by electrospinning with a polyelectrolyte and vapor deposition polymerization. Nano Lett 7:2470–2474. https://doi.org/10.1021/nl071234k
72. Reports and data, lithium ion battery market to reach USD 109.72 Billion By 2026 (2019)
73. Bhutani A, Schiller JA, Zuo JL et al (2017) Combined computational and in situ experimental search for phases in an open ternary system, Ba-Ru-S. Chem Mater 29:5841–5849. https://doi.org/10.1021/acs.chemmater.7b00809
74. Whittingham BMS (2012) WhittinghamIEEE2012energystorage.pdf. 100
75. Mehul Oswal JP, RZ A comparative study of Lithium-Ion Batteries
76. Haruyama J, Sodeyama K, Han L et al (2014) Space-charge layer effect at interface between oxide cathode and sulfide electrolyte in all-solid-state lithium-ion battery. Chem Mater 26:4248–4255. https://doi.org/10.1021/cm5016959
77. Kim Y (2012) Lithium nickel cobalt manganese oxide synthesized using alkali chloride flux: Morphology and performance as a cathode material for lithium ion batteries. ACS Appl Mater Interfaces 4:2329–2333. https://doi.org/10.1021/am300386j
78. Chen Y, Lu Z, Zhou L et al (2012) Triple-coaxial electrospun amorphous carbon nanotubes with hollow graphitic carbon nanospheres for high-performance Li ion batteries. Energy Environ Sci 5:7898–7902. https://doi.org/10.1039/c2ee22085g
79. Date I (2009) This document is downloaded from DR-NTU, Nanyang Technological. Security 299:1719–1722. https://doi.org/10.1063/1.2978249
80. Wang J, Li L, Wong CL et al (2013) Controlled synthesis of α-FeOOH nanorods and their transformation to mesoporous α-Fe_2O_3, Fe_3O_4@C nanorods as anodes for lithium ion batteries. RSC Adv 3:15316–15326. https://doi.org/10.1039/c3ra41886c
81. Chagnes A, Swiatowsk J (2012) Electrolyte and solid-electrolyte interphase layer in lithium-ion batteries. Lithium Ion Batter New Dev. https://doi.org/10.5772/31112
82. Zhang HP, Zhang P, Li ZH, Sun M, Wu YP, Wu HQ (2007) A novel sandwiched membrane as polymer electrolyte for lithium ion battery. Electrochem Commun 9(7):1700–1703
83. Chapman N (2016) Spectroscopic measurements of ionic associaition in common lithium salts and carbonate electrolytes. https://doi.org/10.23860/thesis-chapman-navid-2016
84. Scrosati B, Croce F, Persi L (2000) Impedance spectroscopy study of PEO-based nanocomposite polymer electrolytes. J Electrochem Soc 147:1718–1721. https://doi.org/10.1149/1.1393423
85. Shin JH, Henderson WA, Passerini S (2005) PEO-based polymer electrolytes with ionic liquids and their use in lithium metal-polymer electrolyte batteries. J Electrochem Soc 152:978–983. https://doi.org/10.1149/1.1890701
86. Gopalan AI, Santhosh P, Manesh KM et al (2008) Development of electrospun PVdF-PAN membrane-based polymer electrolytes for lithium batteries. J Memb Sci 325:683–690. https://doi.org/10.1016/j.memsci.2008.08.047
87. Newcomb BA, Chae HG, Gulgunje PV et al (2014) Stress transfer in polyacrylonitrile/carbon nanotube composite fibers. Polymer (Guildf) 55:2734–2743. https://doi.org/10.1016/j.polymer.2014.04.008

88. Kim JR, Choi SW, Jo SM et al (2004) Electrospun PVdF-based fibrous polymer electrolytes for lithium ion polymer batteries. Electrochim Acta 50:69–75. https://doi.org/10.1016/j.electacta.2004.07.014
89. Mohamed NS, Arof AK (2004) Investigation of electrical and electrochemical properties of PVDF-based polymer electrolytes. J Power Sources 132:229–234. https://doi.org/10.1016/j.jpowsour.2003.12.031
90. Angulakshmi N, Thomas S, Nahm KS et al (2011) Electrochemical and mechanical properties of nanochitin-incorporated PVDF-HFP-based polymer electrolytes for lithium batteries. Ionics (Kiel) 17:407–414. https://doi.org/10.1007/s11581-010-0517-z
91. Stephan AM, Nahm KS, Anbu Kulandainathan M et al (2006) Poly(vinylidene fluoride-hexafluoropropylene) (PVdF-HFP) based composite electrolytes for lithium batteries. Eur Polym J 42:1728–1734. https://doi.org/10.1016/j.eurpolymj.2006.02.006
92. Chiu CY, Yen YJ, Kuo SW et al (2007) Complicated phase behavior and ionic conductivities of PVP-co-PMMA-based polymer electrolytes. Polymer (Guildf) 48:1329–1342. https://doi.org/10.1016/j.polymer.2006.12.059
93. Rajendran S, Kannan R, Mahendran O (2001) An electrochemical investigation on PMMA/PVdF blend-based polymer electrolytes. Mater Lett 49:172–179. https://doi.org/10.1016/S0167-577X(00)00363-3
94. Doughty DH, Butler PC, Akhil AA, Clark NH, Boyes JD (2010) Batteries for large-scale stationary electrical energy storage. Electrochem Soc Interface 19(3):49
95. Corporation LB, Beach N (1999) Laub BioChem. Computer (Long Beach Calif) 18:461–472. https://doi.org/10.1016/0025-5408(83)90138-1
96. Chitra S, Kalyani P, Mohan T et al (1999) Characterization and electrochemical studies of LiMn2O4 cathode materials prepared by combustion method. J Electroceramics 3:433–441. https://doi.org/10.1023/A:1009982301437
97. Li X, Cheng F, Guo B, Chen J (2005) Template-synthesized $LiCoO_2$, $LiMn_2O_4$, and LiNi0.8Co0.2O2 nanotubes as the cathode materials of lithium ion batteries. J Phys Chem B 109:14017–14024. https://doi.org/10.1021/jp051900a
98. Tarascon JM, Armand M (2011) Issues and challenges facing rechargeable lithium batteries. In: Materials for sustainable energy: a collection of peer-reviewed research and review articles from Nature Publishing Group, pp. 171–179. https://doi.org/10.1038/35104644
99. Goodenough JB, Park KS (2013) The Li-ion rechargeable battery: a perspective. J Am Chem Soc 135:1167–1176. https://doi.org/10.1021/ja3091438
100. Nitta N, Wu F, Lee JT, Yushin G (2015) Li-ion battery materials: present and future. Mater Today 18:252–264. https://doi.org/10.1016/j.mattod.2014.10.040
101. Etacheri V, Marom R, Elazari R et al (2011) Challenges in the development of advanced Li-ion batteries: a review. Energy Environ Sci 4:3243–3262. https://doi.org/10.1039/c1ee01598b
102. Mizushima K, Jones PC, Wiseman PJ, Goodenough JB (1981) $LixCoO_2$ ($0 < x\sim l$): A new cathode material for batteries of high energy density K. Solid State Ionics 3–4:171–174. https://doi.org/10.1016/0167-2738(81)90077-1
103. Endo M, Nakamura JI, Sasabe Y, Takahashi T, Inagaki M (1995) Lithium secondary battery using vapor grown carbon fibers as a negative electrode and analysis of the electrode mechanism by TEM observation. IEEJ Trans Fundam Mater 115(4):349–356
104. Liu H, Yang Y, Zhang J (2007) Reaction mechanism and kinetics of lithium ion battery cathode material $LiNiO_2$ with CO_2. J Power Sources 173:556–561. https://doi.org/10.1016/j.jpowsour.2007.04.083
105. Kim TH, Park JS, Chang SK et al (2012) The current move of lithium ion batteries towards the next phase. Adv Energy Mater 2:860–872. https://doi.org/10.1002/aenm.201200028
106. Shaju KM, Subba Rao GV, Chowdari BVR (2002) Performance of layered $Li(Ni1/3Co1/3Mn1/3)O_2$ as cathode for Li-ion batteries. Electrochim Acta 48:145–151. https://doi.org/10.1016/S0013-4686(02)00593-5
107. Koyama Y, Makimura Y, Tanaka I et al (2004) Systematic research on insertion materials based on superlattice models in a phase triangle of $LiCoO_2$-$LiNiO_2$-$LiMnO_2$ I. First-principles

calculation on electronic and crystal structures, phase stability and new LiNi1/2Mn1/2O_2 material. J Electrochem Soc 151:1499–1506. https://doi.org/10.1149/1.1783908
108. Park OK, Cho Y, Lee S et al (2011) Who will drive electric vehicles, olivine or spinel? Energy Environ Sci 4:1621–1633. https://doi.org/10.1039/c0ee00559b
109. Myung ST, Maglia F, Park KJ et al (2017) Nickel-rich layered cathode materials for automotive lithium-ion batteries: achievements and perspectives. ACS Energy Lett 2:196–223. https://doi.org/10.1021/acsenergylett.6b00594
110. Yamada A, Chung SC, Hinokuma K (2001) Optimized $LiFePO_4$ for lithium battery cathodes. J Electrochem Soc 148:224–229. https://doi.org/10.1149/1.1348257
111. Huang ZD, Oh SW, He YB et al (2012) Porous C-$LiFePO_4$-C composite microspheres with a hierarchical conductive architecture as a high performance cathode for lithium ion batteries. J Mater Chem 22:19643–19645. https://doi.org/10.1039/c2jm33960a
112. Shen L, Uchaker E, Zhang X, Cao G (2012) Hydrogenated $Li_4Ti_5O_{12}$ nanowire arrays for high rate lithium ion batteries. Adv Mater 24:6502–6506. https://doi.org/10.1002/adma.201203151
113. Liu Y, Wang YM, Yakobson BI, Wood BC (2014) Assessing carbon-based anodes for lithium-ion batteries: a universal description of charge-transfer binding. Phys Rev Lett 113. https://doi.org/10.1103/PhysRevLett.113.028304
114. Goriparti S, Miele E, De Angelis F et al (2014) Review on recent progress of nanostructured anode materials for Li-ion batteries. J Power Sources 257:421–443. https://doi.org/10.1016/j.jpowsour.2013.11.103
115. Lian P, Zhu X, Liang S et al (2010) Large reversible capacity of high quality graphene sheets as an anode material for lithium-ion batteries. Electrochim Acta 55:3909–3914. https://doi.org/10.1016/j.electacta.2010.02.025
116. Zhang B, Xu ZL, He YB et al (2014) Exceptional rate performance of functionalized carbon nanofiber anodes containing nanopores created by (Fe) sacrificial catalyst. Nano Energy 4:88–96. https://doi.org/10.1016/j.nanoen.2013.12.011
117. Raccichini R, Varzi A, Passerini S, Scrosati B (2015) The role of graphene for electrochemical energy storage. Nat Mater 14:271–279. https://doi.org/10.1038/nmat4170
118. Liu C, Li F, Lai-Peng M, Cheng HM (2010) Advanced materials for energy storage. Adv Mater 22:28–62. https://doi.org/10.1002/adma.200903328
119. Kim J, Zhou H, Hosono E et al (2008) Large reversible Li storage of graphene nanosheet families for use in rechargeable lithium ion batteries. Nano Lett 8:2277–2282. https://doi.org/10.1021/nl800957b
120. Klein F, Jache B, Bhide A, Adelhelm P (2013) Conversion reactions for sodium-ion batteries. Phys Chem Chem Phys 15:15876–15887. https://doi.org/10.1039/c3cp52125g
121. Cabana J, Monconduit L, Larcher D, Palacín MR (2010) Beyond intercalation-based Li-ion batteries: the state of the art and challenges of electrode materials reacting through conversion reactions. Adv Mater 22:170–192. https://doi.org/10.1002/adma.201000717
122. Online VA, Afrooz M (2015) RSC Adv. https://doi.org/10.1039/b000000x
123. Zhou W, Zhu J, Cheng C et al (2011) A general strategy toward graphene@metal oxide core-shell nanostructures for high-performance lithium storage. Energy Environ Sci 4:4954–4961. https://doi.org/10.1039/c1ee02168k
124. Aravindan V, Lee YS, Madhavi S (2017) Best practices for mitigating irreversible capacity loss of negative electrodes in Li-ion batteries. Adv Energy Mater 7:1–17. https://doi.org/10.1002/aenm.201602607
125. Lu J, Chen Z, Pan F, Cui Y, Amine K (2018) High-performance anode materials for rechargeable lithium-ion batteries. Electrochemi Energy Rev 1(1):35–53
126. Yoo H, Lee JI, Kim H et al (2011) Helical silicon/silicon oxide core-shell anodes grown onto the surface of bulk silicon. Nano Lett 11:4324–4328. https://doi.org/10.1021/nl202417c
127. Yin YX, Wan LJ, Guo YG (2012) Silicon-based nanomaterials for lithium-ion batteries. Chinese Sci Bull 57:4104–4110. https://doi.org/10.1007/s11434-012-5017-2
128. Wang B, Luo B, Li X, Zhi L (2012) The dimensionality of Sn anodes in Li-ion batteries. Mater Today 15:544–552. https://doi.org/10.1016/S1369-7021(13)70012-9

129. Obrovac MN, Chevrier VL (2014) Alloy negative electrodes for Li-ion batteries. Chem Rev 114:11444–11502. https://doi.org/10.1021/cr500207g
130. Jin S, Deng H, Long D et al (2011) Facile synthesis of hierarchically structured Fe_3O 4/carbon micro-flowers and their application to lithium-ion battery anodes. J Power Sources 196:3887–3893. https://doi.org/10.1016/j.jpowsour.2010.12.078
131. Lee JE, Yu SH, Lee DJ et al (2012) Facile and economical synthesis of hierarchical carbon-coated magnetite nanocomposite particles and their applications in lithium ion battery anodes. Energy Environ Sci 5:9528–9533. https://doi.org/10.1039/c2ee22792d
132. Zhou G, Wang DW, Hou PX et al (2012) A nanosized Fe 2 O 3 decorated single-walled carbon nanotube membrane as a high-performance flexible anode for lithium ion batteries. J Mater Chem 22:17942–17946. https://doi.org/10.1039/c2jm32893c
133. Li Y, Zhu C, Lu T et al (2013) Simple fabrication of a Fe2O3/carbon composite for use in a high-performance lithium ion battery. Carbon N Y 52:565–573. https://doi.org/10.1016/j.carbon.2012.10.015
134. Chou SL, Wang JZ, Wexler D et al (2010) High-surface-area α-Fe_2O_3/carbon nanocomposite: one-step synthesis and its highly reversible and enhanced high-rate lithium storage properties. J Mater Chem 20:2092–2098. https://doi.org/10.1039/b922237e
135. Yoon T, Kim J, Kim J, Lee JK (2013) Electrostatic self-assembly of Fe_3O_4 nanoparticles on graphene oxides for high capacity lithium-ion battery anodes. Energies 6:4830–4840. https://doi.org/10.3390/en6094830
136. Xu L, Tian Y, Liu T et al (2018) α-Fe_2O_3 nanoplates with superior electrochemical performance for lithium-ion batteries. Green Energy Environ 3:156–162. https://doi.org/10.1016/j.gee.2018.01.005
137. Vetter J, Novák P, Wagner MR et al (2005) Ageing mechanisms in lithium-ion batteries. J Power Sources 147:269–281. https://doi.org/10.1016/j.jpowsour.2005.01.006
138. Chan CK, Peng H, Liu G et al (2008) High-performance lithium battery anodes using silicon nanowires. Nat Nanotechnol 3:31–35. https://doi.org/10.1038/nnano.2007.411
139. Martinet S (2016) Nanomaterials for rechargeable lithium batteries. Nanosci Technol 471–512. https://doi.org/10.1007/978-3-319-32023-6_13
140. Ji L, Lin Z, Alcoutlabi M, Zhang X (2011) Recent developments in nanostructured anode materials for rechargeable lithium-ion batteries. Energy Environ Sci 4:2682–2689. https://doi.org/10.1039/c0ee00699h
141. Fei H, Peng Z, Li L et al (2014) Preparation of carbon-coated iron oxide nanoparticles dispersed on graphene sheets and applications as advanced anode materials for lithium-ion batteries. Nano Res 7:1–9. https://doi.org/10.1007/s12274-014-0416-0
142. Ma Y, Ji G, Lee JY (2011) Synthesis of mixed-conducting carbon coated porous γ-Fe_2O_3 microparticles and their properties for reversible lithium ion storage. J Mater Chem 21:13009–13014. https://doi.org/10.1039/c1jm12070k
143. Zhang L, Bin WuH, Lou XW (2014) Iron-oxide-based advanced anode materials for lithium-ion batteries. Adv Energy Mater 4:1–11. https://doi.org/10.1002/aenm.201300958
144. Lin J, Raji ARO, Nan K et al (2014) Iron oxide nanoparticle and graphene nanoribbon composite as an anode material for high-performance li-ion batteries. Adv Funct Mater 24:2044–2048. https://doi.org/10.1002/adfm.201303023
145. Wu X, Hu J, Zhang BW et al (2015) Carbon coated Fe_3O_4 nanospindles as a superior anode material for lithium—ion batteries carbon coated Fe_3O_4 nanospindles as a superior anode material for lithium-ion batteries. J Power Sources 259:92–97. https://doi.org/10.1002/adfm.200801386
146. Zeng Z, Zhao H, Wang J et al (2014) Nanostructured Fe_3O_4@C as anode material for lithium-ion batteries. J Power Sources 248:15–21. https://doi.org/10.1016/j.jpowsour.2013.09.063
147. Wu Y, Wei Y, Wang J et al (2013) Conformal Fe_3O_4 sheath on aligned carbon nanotube scaffolds as high-performance anodes for lithium ion batteries. Nano Lett 13:818–823. https://doi.org/10.1021/nl3046409
148. He C, Wu S, Zhao N et al (2013) Carbon-encapsulated Fe_3O_4 nanoparticles as a high-rate lithium ion battery anode material. ACS Nano 7:4459–4469. https://doi.org/10.1021/nn401059h

149. Su Y, Li S, Wu D et al (2012) Two-dimensional carbon-coated graphene/metal oxide hybrids for enhanced lithium storage. ACS Nano 6:8349–8356. https://doi.org/10.1021/nn303091t
150. Hong J, Wang Y, He G, He M (2012) A new approach to $LiFePO_4/C$ synthesis: the use of complex carbon source without ball milling. Mater Chem Phys 133:573–577. https://doi.org/10.1016/j.matchemphys.2012.01.090
151. Huang X, Du Y, Qu P et al (2017) Effect of carbon coating on the properties and electrochemical performance of $LiFePO_4/C$ composites by vacuum decomposition method. Int J Electrochem Sci 12:7183–7196. https://doi.org/10.20964/2017.08.77
152. Chen H, Chen Y, Gong W et al (2011) Preparation and electrochemical performance of LiFePO4/C composite with network connections of nano-carbon wires. Mater Lett 65:559–561. https://doi.org/10.1016/j.matlet.2010.11.012
153. Huang B, Zheng X, Fan X et al (2011) Enhanced rate performance of nano-micro structured $LiFePO_4/C$ by improved process for high-power Li-ion batteries. Electrochim Acta 56:4865–4868. https://doi.org/10.1016/j.electacta.2011.02.118
154. Yao Y, Qu P, Gan X et al (2016) Preparation of porous-structured $LiFePO_4/C$ composite by vacuum sintering for lithium-ion battery. Ceram Int 42:18303–18311. https://doi.org/10.1016/j.ceramint.2016.08.158
155. Ji L, Toprakci O, Alcoutlabi M et al (2012) α-Fe_2O_3 nanoparticle-loaded carbon nanofibers as stable and high-capacity anodes for rechargeable lithium-ion batteries. ACS Appl Mater Interfaces 4:2672–2679. https://doi.org/10.1021/am300333s
156. Cherian CT, Sundaramurthy J, Kalaivani M et al (2012) Electrospun α-Fe_2O_3 nanorods as a stable, high capacity anode material for Li-ion batteries. J Mater Chem 22:12198–12204. https://doi.org/10.1039/c2jm31053h
157. Wang L, Yu Y, Chen PC et al (2008) Electrospinning synthesis of C/Fe_3O_4 composite nanofibers and their application for high performance lithium-ion batteries. J Power Sources 183:717–723. https://doi.org/10.1016/j.jpowsour.2008.05.079
158. Qin X, Zhang H, Wu J et al (2016) Fe_3O_4 nanoparticles encapsulated in electrospun porous carbon fibers with a compact shell as high-performance anode for lithium ion batteries. Carbon N Y 87:347–356. https://doi.org/10.1016/j.carbon.2015.02.044
159. Wang HG, Zhou Y, Shen Y et al (2015) Fabrication, formation mechanism and the application in lithium-ion battery of porous Fe_2O_3 nanotubes via single-spinneret electrospinning. Electrochim Acta 158:105–112. https://doi.org/10.1016/j.electacta.2015.01.149
160. Kim C, Yang KS, Kojima M et al (2006) Fabrication of electrospinning-derived carbon nanofiber webs for the anode material of lithium-ion secondary batteries. Adv Funct Mater 16:2393–2397. https://doi.org/10.1002/adfm.200500911
161. Te Peng Y, Lo CT (2015) Effect of microstructure and morphology of electrospun ultra-small carbon nanofibers on anode performances for lithium ion batteries. J Electrochem Soc 162:A1085–A1093. https://doi.org/10.1149/2.0061507jes
162. Chen M, Zhao E, Yan Q et al (2016) The effect of crystal face of Fe_2O_3 on the electrochemical performance for lithium-ion batteries. Sci Rep 6:1–9. https://doi.org/10.1038/srep29381
163. Choy KL (2003) Chemical vapour deposition of coatings. Prog Mater Sci 48:57–170. https://doi.org/10.1016/S0079-6425(01)00009-3
164. George SM (2010) Atomic layer deposition: an overview. Chem Rev 110:111–131. https://doi.org/10.1021/cr900056b
165. Leskelä M, Ritala M (2002) Atomic layer deposition (ALD): from precursors to thin film structures. Thin Solid Films 409:138–146. https://doi.org/10.1016/S0040-6090(02)00117-7
166. Anderson J (2013) Fundamentals of fundamentals 5:447–469
167. Thambidurai M, Muthukumarasamy N, Velauthapillai D, Lee C (2013) Synthesis of garland like ZnO nanorods and their application in dye sensitized solar cells. Mater Lett 92:104–107. https://doi.org/10.1016/j.matlet.2012.10.036
168. Baviskar PK, Zhang JB, Gupta V et al (2011) Nanobeads of zinc oxide with rhodamine B dye as a sensitizer for dye sensitized solar cell application. J Alloys Compd 510:33–37. https://doi.org/10.1016/j.jallcom.2011.08.034

169. Wang Y, Cui X, Zhang Y et al (2013) Preparation of Cauliflower-like ZnO Films by chemical bath deposition: photovoltaic performance and equivalent circuit of dye-sensitized solar cells. J Mater Sci Technol 29:123–127. https://doi.org/10.1016/j.jmst.2012.12.019
170. Wang Y, Sun Y, Zhang X et al (2016) Synthesis of 1D porous Fe_2O_3 nanostructures using SiO_2 scaffold towards good lithium storages. Mater Lett 171:125–128. https://doi.org/10.1016/j.matlet.2016.02.012
171. He L, Wang C, Yao X et al (2014) Synthesis of carbon nanotube/mesoporous TiO_2 coaxial nanocables with enhanced lithium ion battery performance. Carbon N Y 75:345–352. https://doi.org/10.1016/j.carbon.2014.04.013
172. Chen WS, Huang DA, Chen HC et al (2009) Fabrication of polycrystalline ZnO nanotubes from the electrospinning of Zn2+/poly(acrylic acid). Cryst Growth Des 9:4070–4077. https://doi.org/10.1021/cg900297q
173. Luo W, Hu X, Sun Y, Huang Y (2011) Electrospinning of carbon-coated MoO_2 nanofibers with enhanced lithium-storage properties. Phys Chem Chem Phys 13:16735–16740. https://doi.org/10.1039/c1cp22184a
174. Mai L, Xu L, Han C et al (2010) Electrospun ultralong hierarchical vanadium oxide nanowires with high performance for lithium ion batteries. Nano Lett 10:4750–4755. https://doi.org/10.1021/nl103343w
175. Li D, Xia Y (2003) Fabrication of titania nanofibers by electrospinning. Nano Lett 3:555–560. https://doi.org/10.1021/nl034039o
176. Xu X, Tan H, Xi K et al (2015) Bamboo-like amorphous carbon nanotubes clad in ultrathin nickel oxide nanosheets for lithium-ion battery electrodes with long cycle life. Carbon N Y 84:491–499. https://doi.org/10.1016/j.carbon.2014.12.040
177. Shang M, Wang W, Yin W et al (2010) General strategy for a large-scale fabric with branched nanofiber-nanorod hierarchical heterostructure: controllable synthesis and applications. Chem A Eur J 16:11412–11419. https://doi.org/10.1002/chem.201000639
178. Luo H, Huang K, Sun B, Zhong J (2014) Strategy to synthesize Fe_3O_4/C nanotubes as anode material for advanced lithium-ion batteries. Electrochim Acta 149:11–17. https://doi.org/10.1016/j.electacta.2014.10.086
179. Cho JS, Hong YJ, Kang YC (2015) Design and synthesis of bubble-nanorod-structured Fe_2O_3 -Carbon nanofibers as advanced anode material for li-ion batteries. ACS Nano 9:4026–4035. https://doi.org/10.1021/acsnano.5b00088
180. Courtel FM, Duncan H, Abu-Lebdeh Y, Davidson IJ (2011) High capacity anode materials for Li-ion batteries based on spinel metal oxides AMn_2O_4 (A = Co, Ni, and Zn). J Mater Chem 21:10206–10218. https://doi.org/10.1039/c0jm04465b
181. Liu Z, Tay SW, Li X (2011) Rechargeable battery using a novel iron oxide nanorods anode and a nickel hydroxide cathode in an aqueous electrolyte. Chem Commun 47:12473–12475. https://doi.org/10.1039/c1cc15022g
182. Cho JS, Park JS, Kang YC (2016) Preparation of hollow Fe_2O_3 nanorods and nanospheres by nanoscale kirkendall diffusion, and their electrochemical properties for use in lithium-ion batteries. Sci Rep 6:1–13. https://doi.org/10.1038/srep38933
183. Zhu S, Chen M, Sun J et al (2016) Novel highly conductive ferroferric oxide/porous carbon nanofiber composites prepared by electrospinning as anode materials for high performance Li-ion batteries. RSC Adv 6:58529–58540. https://doi.org/10.1039/c6ra04090j
184. Im ME, Pham-Cong D, Kim JY et al (2015) Enhanced electrochemical performance of template-free carbon-coated iron(II, III) oxide hollow nanofibers as anode material for lithium-ion batteries. J Power Sources 284:392–399. https://doi.org/10.1016/j.jpowsour.2015.03.024
185. Chen M, Liu J, Chao D et al (2014) Porous α-Fe_2O_3 nanorods supported on carbon nanotubes-graphene foam as superior anode for lithium ion batteries. Nano Energy 9:364–372. https://doi.org/10.1016/j.nanoen.2014.08.011
186. Wu C, Yin P, Zhu X et al (2006) Synthesis of hematite (α-Fe_2O_3) nanorods: diameter-size and shape effects on their applications in magnetism, lithium ion battery, and gas sensors. J Phys Chem B 110:17806–17812. https://doi.org/10.1021/jp0633906

187. Zhu Y, Zhang JC, Zhai J, Jiang L (2006) Preparation of superhydrophilic α-Fe_2O_3 nanofibers with tunable magnetic properties. Thin Solid Films 510:271–274. https://doi.org/10.1016/j.tsf.2005.09.004
188. Zou M, Li J, Wen W et al (2014) Silver-incorporated composites of Fe_2O_3 carbon nanofibers as anodes for high-performance lithium batteries. J Power Sources 270:468–474. https://doi.org/10.1016/j.jpowsour.2014.07.119
189. Manuscript A (2015) Hierarchical $NiFe_2O_4/Fe_2O_3$ nanotubes derived from metal organic frameworks for superior lithium ion battery anode. https://doi.org/10.1039/C5TA00355E
190. Larcher D, Bonnin D, Cortes R et al (2003) Combined XRD, EXAFS, and Mössbauer studies of the reduction by lithium of α-Fe_2O3 with various particle sizes. J Electrochem Soc 150:1643–1650. https://doi.org/10.1149/1.1622959
191. Larcher D, Masquelier C, Bonnin D et al (2003) Effect of particle size on lithium intercalation into α-Fe_2O_3. J Electrochem Soc 150:133–139. https://doi.org/10.1149/1.1528941
192. Liu Y, Huang K, Luo H et al (2014) Nitrogen-doped graphene-Fe_3O_4 architecture as anode material for improved Li-ion storage. RSC Adv 4:17653–17659. https://doi.org/10.1039/c4ra01080a
193. Reddy MV, Subba Rao GV, Chowdari BVR (2013) Metal oxides and oxysalts as anode materials for Li ion batteries. Chem Rev 113:5364–5457. https://doi.org/10.1021/cr3001884
194. Jiang J, Li Y, Liu J et al (2012) Recent advances in metal oxide-based electrode architecture design for electrochemical energy storage. Adv Mater 24:5166–5180. https://doi.org/10.1002/adma.201202146
195. Liu H, Wexler D, Wang G (2009) One-pot facile synthesis of iron oxide nanowires as high capacity anode materials for lithium ion batteries. J Alloys Compd 487:24–27. https://doi.org/10.1016/j.jallcom.2009.08.043
196. Wu MS, Ou YH, Lin YP (2011) Iron oxide nanosheets and nanoparticles synthesized by a facile single-step coprecipitation method for lithium-ion batteries. J Electrochem Soc 158:231–236. https://doi.org/10.1149/1.3527982
197. Liu J, Li Y, Fan H et al (2010) Iron oxide-based nanotube arrays derived from sacrificial template-accelerated hydrolysis: large-area design and reversible lithium storage. Chem Mater 22:212–217. https://doi.org/10.1021/cm903099w
198. Hassan MF, Rahman MM, Guo ZP et al (2010) Solvent-assisted molten salt process: a new route to synthesise α-Fe_2O_3/C nanocomposite and its electrochemical performance in lithium-ion batteries. Electrochim Acta 55:5006–5013. https://doi.org/10.1016/j.electacta.2010.04.006
199. NuLi Y, Zeng R, Zhang P et al (2008) Controlled synthesis of α-Fe_2O_3 nanostructures and their size-dependent electrochemical properties for lithium-ion batteries. J Power Sources 184:456–461. https://doi.org/10.1016/j.jpowsour.2008.03.004
200. Wu XL, Guo YG, Wan LJ, Hu CW (2008) α-Fe_2O_3 nanostructures: inorganic salt-controlled synthesis and their electrochemical performance toward lithium storage. J Phys Chem C 112:16824–16829. https://doi.org/10.1021/jp8058307
201. Railsback JG, Johnston-Peck AC, Wang J, Tracy JB (2010) Size-dependent nanoscale kirkendall effect during the oxidation of nickel nanoparticles. ACS Nano 4:1913–1920. https://doi.org/10.1021/nn901736y
202. Cho JS, Won JM, Lee JH, Kang YC (2015) Synthesis and electrochemical properties of spherical and hollow-structured NiO aggregates created by combining the Kirkendall effect and Ostwald ripening. Nanoscale 7:19620–19626. https://doi.org/10.1039/c5nr05930e
203. Pan JH, Zhang X, Du AJ et al (2008) Self-etching reconstruction of hierarchically mesoporous F-TiO_2 hollow microspherical photocatalyst for concurrent membrane water purifications. J Am Chem Soc 130:11256–11257. https://doi.org/10.1021/ja803582m
204. Chun Zeng H (2007) Ostwald ripening: a synthetic approach for hollow nanomaterials. Curr Nanosci 3:177–181. https://doi.org/10.2174/157341307780619279
205. Lou XW, Wang Y, Yuan C et al (2006) Template-free synthesis of SnO_2 hollow nanostructures with high lithium storage capacity. Adv Mater 18:2325–2329. https://doi.org/10.1002/adma.200600733

206. Takeuchi T, Chimura H, Hamada M, Umezawa H (1975) Controlled porosity by an extreme kirkendall effect. J Antibiot 28:1–21
207. Jeong JM, Choi BG, Lee SC et al (2013) Hierarchical hollow spheres of Fe_2O_3@polyaniline for lithium ion battery anodes. Adv Mater 25:6250–6255. https://doi.org/10.1002/adma.201302710
208. Rahman MM, Wang JZ, Hassan MF et al (2011) Amorphous carbon coated high grain boundary density dual phase $Li_4Ti_5O_{12}$-TiO_2: A nanocomposite anode material for li-ion batteries. Adv Energy Mater 1:212–220. https://doi.org/10.1002/aenm.201000051
209. Wu YP, Rahm E, Holze R (2003) Carbon anode materials for lithium ion batteries. J Power Sources 114:228–236. https://doi.org/10.1016/S0378-7753(02)00596-7
210. Lavela P, Tirado JL (2007) $CoFe_2O_4$ and $NiFe_2O_4$ synthesized by sol-gel procedures for their use as anode materials for Li ion batteries. J Power Sources 172:379–387. https://doi.org/10.1016/j.jpowsour.2007.07.055
211. Bomio M, Lavela P, Tirado JL (2008) Electrochemical evaluation of $CuFe_2O_4$ samples obtained by sol-gel methods used as anodes in lithium batteries. J Solid State Electrochem 12:729–737. https://doi.org/10.1007/s10008-007-0420-3
212. Sharma Y, Sharma N, Rao GVS, Chowdari BVR (2008) Li-storage and cyclability of urea combustion derived $ZnFe_2O_4$ as anode for Li-ion batteries. Electrochim Acta 53:2380–2385. https://doi.org/10.1016/j.electacta.2007.09.059
213. Guo X, Lu X, Fang X et al (2010) Lithium storage in hollow spherical $ZnFe_2O_4$ as anode materials for lithium ion batteries. Electrochem Commun 12:847–850. https://doi.org/10.1016/j.elecom.2010.04.003
214. Kalai Selvan R, Kalaiselvi N, Augustin CO et al (2006) $CuFe_2O_4$/SnO_2 nanocomposites as anodes for Li-ion batteries. J Power Sources 157:522–527. https://doi.org/10.1016/j.jpowsour.2005.07.030
215. NuLi YN, Chu YQ, Qin QZ (2004) Nanocrystalline $ZnFe_2O_4$ and Ag-doped $ZnFe_2O_4$ films used as new anode materials for Li-ion batteries. J Electrochem Soc 151:1–8. https://doi.org/10.1149/1.1760576
216. Fu Y, Chen Q, He M et al (2012) Copper ferrite-graphene hybrid: a multifunctional heteroarchitecture for photocatalysis and energy storage. Ind Eng Chem Res 51:11700–11709. https://doi.org/10.1021/ie301347j
217. Mao J, Hou X, Wang X et al (2015) The cubic aggregated $CoFe_2O_4$ nanoparticle anode material for lithium ion battery with good performance. Mater Lett 161:652–655. https://doi.org/10.1016/j.matlet.2015.08.102
218. Dong Y, Chui YS, Ma R et al (2014) One-pot scalable synthesis of Cu-$CuFe_2O_4$/graphene composites as anode materials for lithium-ion batteries with enhanced lithium storage properties. J Mater Chem A 2:13892–13897. https://doi.org/10.1039/c4ta02203c
219. Xia H, Zhu D, Fu Y, Wang X (2012) $CoFe_2O_4$-graphene nanocomposite as a high-capacity anode material for lithium-ion batteries. Electrochim Acta 83:166–174. https://doi.org/10.1016/j.electacta.2012.08.027
220. Zhang M, Yang X, Kan X et al (2013) Carbon-encapsulated $CoFe_2O_4$/graphene nanocomposite as high performance anode for lithium ion batteries. Electrochim Acta 112:727–734. https://doi.org/10.1016/j.electacta.2013.09.034
221. Li S, Wang B, Liu J, Yu M (2014) In situ one-step synthesis of $CoFe_2O_4$/graphene nanocomposites as high-performance anode for lithium-ion batteries. Electrochim Acta 129:33–39. https://doi.org/10.1016/j.electacta.2014.02.039
222. Zhang Z, Li W, Zou R et al (2015) Layer-stacked cobalt ferrite ($CoFe_2O_4$) mesoporous platelets for high-performance lithium ion battery anodes. J Mater Chem A 3:6990–6997. https://doi.org/10.1039/c5ta00073d
223. Bashir S, Hanumandla P, Huang HY, Liu JL (2018) Nanostructured materials for advanced energy conversion and storage devices: safety implications at end-of-life disposal. Nanostruct Mater Next-Gener Energy Storage Convers Fuel Cells 4:517–542. https://doi.org/10.1007/978-3-662-56364-9_18

224. Tarascon JM, Armand M (2001) Issues and challenges facing rechargeable batteries. Nature 414:359–367. https://doi.org/10.1038/35104644
225. Wang J, King P, Huggins RA (1986) Investigations of binary lithium-zinc, lithium-cadmium and lithium-lead alloys as negative electrodes in organic solvent-based electrolyte. Solid State Ionics 20:185–189. https://doi.org/10.1016/0167-2738(86)90212-2
226. Wang HG, Liu D, Li Y, Duan Q (2016) Single-spinneret electrospinning fabrication of $CoFe_2O_4$ nanotubes as high-performance anode materials for lithium-ion batteries. Mater Lett 172:64–67. https://doi.org/10.1016/j.matlet.2016.02.133
227. Peng S, Li L, Srinivasan M (2014) Electrospun $CuFe_2O_4$ nanotubes as anodes for high-performance lithium-ion batteries. J Energy Chem 23:301–307. https://doi.org/10.1016/S2095-4956(14)60151-0
228. Wu L, Xiao Q, Li Z et al (2012) $CoFe_2O_4$/C composite fibers as anode materials for lithium-ion batteries with stable and high electrochemical performance. Solid State Ionics 215:24–28. https://doi.org/10.1016/j.ssi.2012.03.044
229. Teh PF, Sharma Y, Pramana SS, Srinivasan M (2011) Nanoweb anodes composed of one-dimensional, high aspect ratio, size tunable electrospun $ZnFe_2O_4$ nanofibers for lithium ion batteries. J Mater Chem 21:14999–15008. https://doi.org/10.1039/c1jm12088c
230. Xie W, Li S, Wang S et al (2014) N-doped amorphous carbon coated Fe_3O_4/SnO_2 coaxial nanofibers as a binder-free self-supported electrode for lithium ion batteries. ACS Appl Mater Interfaces 6:20334–20339. https://doi.org/10.1021/am505829v

Chapter 11
Electrospun Cobalt Based Composites as Anodes for Lithium-Ion Batteries

R. Krishnapriya, Devika Laishram, Bhagirath Saini, M. J. Jabeen Fatima, and Rakesh K. Sharma

11.1 Introduction

Energy crisis and environmental pollution are the most challenging issues of this century. The growing global energy demand and the requisite of a sustainable environment triggered a renewed interest in developing very efficient methods for harvesting and storing renewable energy [1–5]. Effective implementation of these alternative renewable energy technologies (such as solar, wind, geothermal) necessitates the support of prospective energy storage systems. These include batteries, supercapacitors, and fuel cells and amongst the different possible choices, electrochemical batteries are proved to be the most accepted candidate [6]. A battery is an electrochemical device that stores electrical energy as chemical energy in its electrodes through the charging process and delivers the same as electrical energy through the discharge process when required without any gaseous emission. Among various types of batteries explored, lead-acid batteries find exceptional interest since they exhibit a relatively stable charge/discharge state. However, the large weight and volume of these batteries make them an undesirable candidate for portable, lightweight electric devices. Nickel–cadmium (Ni–Cd) or nickel–hydride (Ni–MH) based batteries, which developed later find use in mobile phones and portable electric devices, had suffered from toxicity issues and short battery life and leakage problem [7]. In this scenario, lithium-ion batteries (LIBs) benefiting from chemistry with the relatively

R. Krishnapriya (✉) · D. Laishram · B. Saini · R. K. Sharma (✉)
Sustainable Materials and Catalysis Research Laboratory (SMCRL), Department of Chemistry, Indian Institute of Technology-Jodhpur, Jodhpur, Rajasthan 342037, India
e-mail: krishnapriya@iitj.ac.in

R. K. Sharma
e-mail: rks@iitj.ac.in

M. J. Jabeen Fatima
Department of Polymer Science and Rubber Technology, Cochin University of Science and Technology, Cochin 682022, India

N. T. M. Balakrishnan and R. Prasanth (eds.), *Electrospinning for Advanced Energy Storage Applications*, Materials Horizons: From Nature to Nanomaterials,
https://doi.org/10.1007/978-981-15-8844-0_11

high energy density, power density, operating voltage, lightweight, without memory effect, outstanding rate capability, low rates of self-discharge, and low environmental pollution have accomplished remarkable developments in various advanced portable electric devices (mobile phones and laptops) as well as transportation systems like electric vehicles (EVs) and hybrid electric vehicles (HEVs) [8]. The research on LIBs began early in the 1980s and has revolutionized the portable electronics market since the first commercialization in 1990 by Sony Corp. for the Kyocera cellular phones. The key components of LIBs are anode, cathode, separators, and electrolytes. The charging and discharging of an archetypal LIB follow by a redox process in which lithium ions (Li^+-ion) shuttle among anode (negative electrode) and the cathode (positive electrode). The anode releases lithium ions into the electrolyte leading to their passage to the cathode which absorbs ions during the discharge process. The charge capacity of LIBs mainly depends on the amount of Li that can be incorporated into the electrode materials. Figure 11.1 is the schematic diagram of a rechargeable lithium-ion battery containing solid electrodes and a liquid electrolyte. In LIBs, the electrode material plays a central role in deciding battery energy/power density and cycle-life output. The most successful cathode material for LIBs is considered to be $LiCoO_2$ (LCO), which was initially reported by Prof. Goodenough et al. [9] in 1980. There are three important types of cathode materials that are widely applied in LIBs, which include olivine materials (e.g., $LiFePO_4$), spinel materials (e.g., $LiMn_2O_4$), and layered materials (e.g., $LiNi_xCo_yMn_zO_2$) [10].

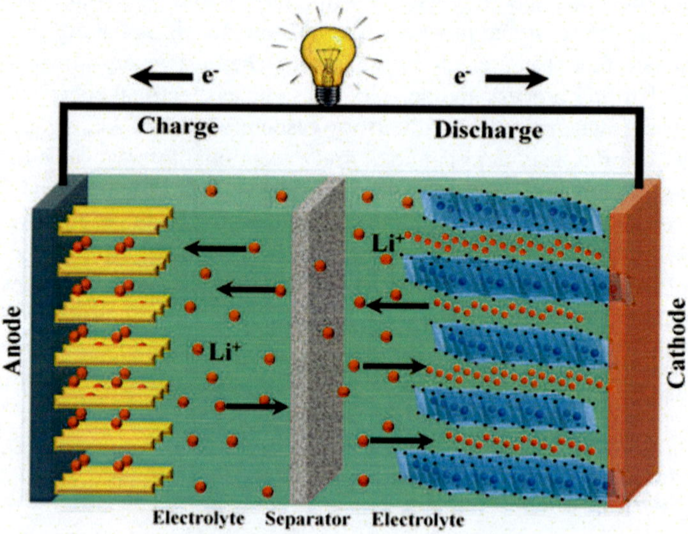

Fig. 11.1 Schematic illustration and working of lithium ion batteries

11.2 Anode Materials for Lithium-Ion Batteries

The anode plays a critical role in the entire battery performance, and it greatly relays on the physical properties (shape, size), component state, crystalline or amorphous structure, chemical properties, and energy storage capacity, etc. Owing to the low atomic weight, redox energy near to that of metallic lithium, the good theoretical capacity of 372 mAh g^{-1}, high conductivity, reversibility, and reasonably low-cost graphite has become a smart anode material in the present generation of LIBs [11]. One-dimensional (1D), two-dimensional (2D) and three-dimensional (3D) carbon-based materials, as well as porous and core-shell structures, metal alloys (Si, Sn, and Ge), Ti-based oxides ($Li_4Ti_5O_{12}$ and TiO_2) metal oxides as well as their nanocomposites, were thoroughly studied to substitute the conventional graphite which has a limited theoretical capacity of 372 mAh g^{-1} [7, 12, 13]. In order to find the wide applicability of LIBs, it is neccssary to develop potential electrode materials that possess high charge/discharge rate capability and energy/power density along with reduced production/operation costs, stability, and safety [14, 15]. Such advanced functional materials fabrication requires immense scientific innovation. The development of ideal nanostructured materials undoubtedly leads to enhance capacities in terms of gravimetric or volumetric energy densities. Recently, nanotechnology is providing novel solutions and opportunities to develop many sustainable materials with predefined properties for prospective technological applications [16–18]. Nanoscale materials have absolutely substituted the traditional anode materials in most of the technological applications. Researchers found that when the size of particles inside materials reduces to match with the wavelength of electrons, phonons, and magnons, some slight size effects occur, comprising the enhancement of electrical conductivity magnetic coercivity, mechanical, optical, superparamagnetic properties owing to the large surface area. As of now, different techniques including post-templating, sol-gel, pray-drying, co-precipitation, and hydrothermal methods have been successfully demonstrated for the fabrication of functional nanomaterials for LIBs. Among the applied nano-electrode materials, 1D nanomaterials like nanowires, nanotubes, and nanofibers are getting special attention because of the considerable enhancement in charge transportation due to the quantum confinement effects [19]. The main shortcoming with the carbon anode materials is the permanent capacity loss during the first discharge-charge cycle because of irreversible side reactions with the electrolyte that leads to the formation of solid-electrolyte interfacial (SEI) layer. To avoid these issues, nanocomposites of carbon with other active materials were demonstrated, and these nano-electrode materials should be mixed with suitable binders when applied to LIBs. However, this added binder often hampers the electrical conductivity of anodes and thereby deteriorating the battery performance.

11.3 Electrospinning Process

In order to avoid severe issues of binders used for the electrode fabrication for LIBs, binder-free carbon nanofibers should be employed, and that can be fabricated via a facile electrospinning method. This technique can achieve electrode materials with solid as well as hollow interiors possessing exceptionally long length and uniform diameter with tunable composition. Such nanofiber (NF) structures are free-standing and are prepared by merely applying an electric field among the nozzle and collector while introducing the polymeric solution, as shown in Fig. 11.2 in the schematic representation [20–22]. The process of electrospinning is based on the principle of electrostatics in which the electrostatic repulsion forces in a comparatively large electrical field are used for the fabrication of nanofibers. The material to be electrospun is taken in the form of a polymeric solution and is hosted in a syringe. Later,

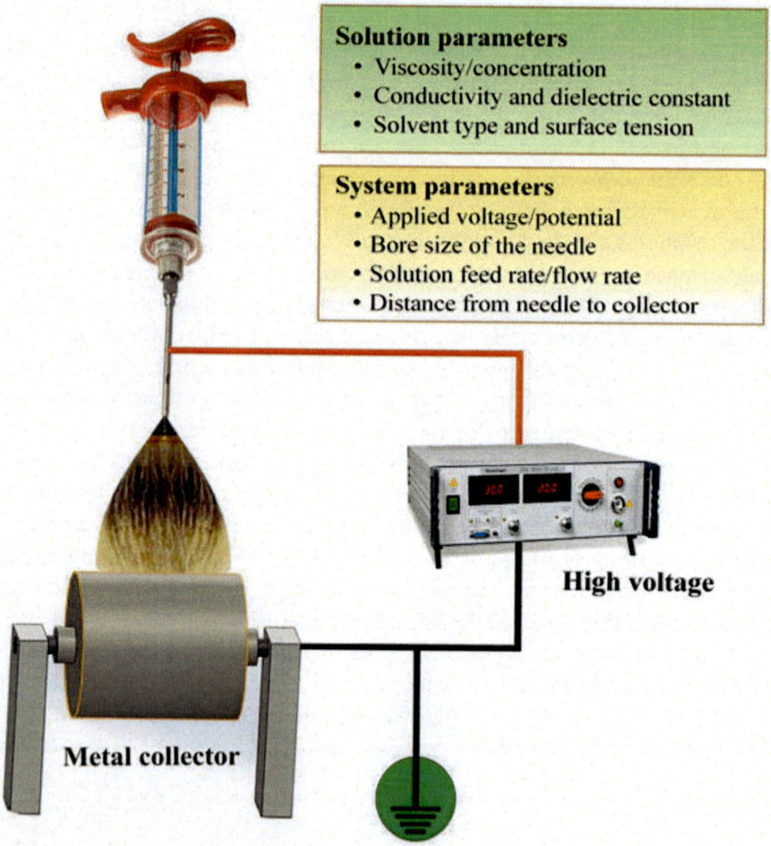

Fig. 11.2 Schematic representation of the Electrospinning process and the controllable parameters to fabricate nanofibers

a large electrical field (typically kV) is generated between the syringe nozzle and the counter electrode. When the solution is ejected through the syringe, a conical fluid-structure called the Taylor cone is created at the tip of the needle owing to the developed potential difference between the nozzle and the target [23].

At a particular voltage, the repulsive force of the charged polymer surpasses the surface tension of the solution, and a charged jet ejects from the tip of the Taylor cone. However, if the applied voltage is not large enough, the jet will break up into droplets (Rayleigh instability) [24]. Beyond the stable region, the jet erupts susceptible to bending which results in the deposition of polymer over the collector via a whipping motion. During this, the solvent evaporates which forms solid continuous nanofibers in the target [22]. As the charged solution moves towards the regions of lower potential, the evaporation of the solvent takes place that results in the high electrostatic repulsion of the charged polymer and its elongation.

The morphology and the diameter of the NFs can be tuned by careful changes of several operating parameters, including the molecular weight of the polymer used, applied voltage, solution pumping rate, spinning distance, temperature, humidity, air velocity, conductivity, viscosity, and surface tension of the solution, etc. Thus, NFs of diameters from tens to several hundred with moderately narrow size distribution can be easily fabricated. Also, this method is proved to be industry-viable, cost-effective technology, and revealed the benefits of versatility, simplicity, excellent efficiency, and yield. Moreover, the low-cost, one-step fabrication process devoid of using any hazardous chemical is found to be sustainable and can successfully be scaled up for massive production of ready to use binder-free anodes for many technological applications. The primary favorable characteristics of the electrospinning method to be used for LIBs include the tunable fiber diameter, high porosity, and specific surface area along with interconnected pore structure, which favors high electronic and ionic conductivity and is capable of enhancing the cyclability and rate capability of electrode materials [25]. By carefully choosing different polymeric blends and operational parameters, several NFs of natural polymers, polymer blends, metals, metal oxides, etc. have been spun into various NF morphologies [26–30].

11.4 Electrospun Nanofibers for Lithium-Ion Batteries

As discussed above, the electrospun NFs can apply as advanced electrode materials for LIBs with outstanding electrochemical performance, flexibility, and stretching ability. These enhanced performances are achieved through excellent conductivity due to the short ion diffusion path and a considerable length of individual NFs that can inhibit self-aggregation during the charging/discharging process. Different electrochemically active electrode materials made of carbon, metals, transition metal oxides (SnO_2, TiO_2, Fe_3O_4, MoO_2, $LiCoO_2$, CoO, $NiCo_2O_4$, and Co_3O_4), and sulphides (CuS, Cu_2S, Li_2S, MoS_2, CoS, Co_3S_4, FeS_2, Ni_3S_4) have been successfully designed and fabricated for more excellent lithium storage in LIBs [12, 31–39]. Transition metal oxides are favorable alternative anode materials owing to their high specific capacities and lithium storage capacities in comparison to commercial graphite. When multiple components are integrated into electrospun NFs, the electrochemical performance of the nanofiber-based electrode materials can be upgraded, when compared with the single counterparts. Recently, cobalt-based nanofibers and their composite electrode materials gain special attention as anode materials. The application of cobalt-based anodes for LIBs has been significantly enhanced, and it is expected that these materials will have the capability to become commercialized shortly. Each cobalt-based compounds exhibit dissimilar electrochemical performance. The redox mechanism of the cobalt-based electrodes follows the conversion mechanism instead of the intercalation mechanism of conventional carbon materials and is generally named as conversion reaction-based anode materials for LIBs [40]. They exhibit high theoretical capacity, rich redox reaction, and favorable cyclability. So this chapter gives a brief overview of advancement in electrospun cobalt-based nanofiber composites as anode materials for LIBs application.

11.5 Structure and Properties of Cobalt Oxides

Cobalt can exhibit variable oxidation states (Co^{+2}, Co^{+3}, and Co^{+4}) as well as different coordination properties (tetrahedral, pyramidal, and octahedral) in his compounds. Subsequently, cobalt oxides offer a wide range of stoichiometric and nonstoichiometric oxides which comprising a mixed-valence state and also the oxygen vacancies. The crystal chemistry of cobalt oxides is very complex. However, these oxides are of particular interest from the other 3d metal oxides because of its ability to exist in different spin states such as low spin (LS), high spin (HS), and intermediate spin (IS), respectively.

11.5.1 Structure of Cobalt Monoxide

CoO (cobalt monoxide) generally possesses rocksalt structure at room temperature (NaCl structure), which contains Co^{2+} and O^{2-} interpenetrating fcc sublattices. It exhibits paramagnetic properties. When the temperature is below 289 K, a crystal distortion to tetragonal symmetry occurs, and it becomes antiferromagnetic. The two interpenetrating sublattices Co^{2+} and O^{2-} shifted along the body diagonal by half of its length. As a result, an individual ion is surrounded by six of the other ions. The corresponding crystal structure is given in Fig. 11.3. Each cubic unit cell consists of four Co^{2+} and four O^{2-} ions and the lattice constant of CoO is found to be 4.260 Å [41].

This material exhibits distinct electrochemical properties and has shown remarkable capacitance and excellent cyclic properties owing to the distinctive charge-discharge reactions via the formation of the Li_2O phase and consequent decomposition into Li and O species. As the charging occurs, CoO reacts with Li^+ to form

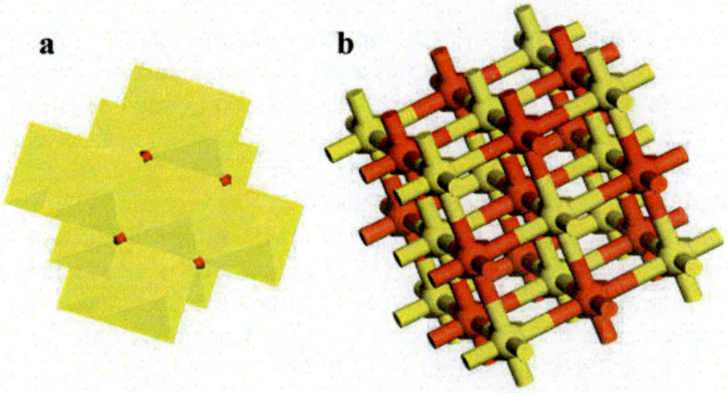

Fig. 11.3 Crystal structure of CoO consists of Co^{2+} and O^{2-} interpenetrating fcc sublattices generated using Material Studio 7.0® **a** polyhydra structure, **b** ball and stick model

metal nano-domains (M^0) dispersed in the Li_2O matrix. Whereas, during the charging steps, the M^0 and Li_2O components are transformed back to metal oxide (MO_x). This well-proved battery material is found to exhibit theoretical capacity as high as 718 mAh g^{-1} and can react reversibly with Li according to the following conversion reaction [42]:

$$CoO + 2Li^+ + 2e \rightarrow 4\,Li_2O + Co$$

Accordingly, each CoO molecule can hold two Li^+ ions per one Co ion over the reaction, which leads to greater capacity compared to the commercial graphite electrode (C_6Li) [43]. However, these anode materials often hamper the performance due to low conductivity and extreme volume change throughout the lithiation/delithiation cycles. These limitations can be overcome either by using high specific surface area nanosized electrode materials that are capable of increasing the contact area between the electrolyte and electrode materials and reducing the path length for Li-ion, which can result in a greater improvement of discharge capacities. Another approach is to fabricate the composite of CoO with highly conductive electrochemically active materials like carbon.

11.5.2 Electrospun Cobalt Monoxide for Lithium-Ion Batteries

Several composite materials of CoO with carbon, graphene, as well as NiO-CoO-carbon and CoO–graphene–carbon NFs were successfully fabricated by the electrospinning method. The electrospun composite of carbon–cobalt (C/Co) nanofibers of diameters ranging from 100 to 300 nm was studied by Wang et al. [44] when these C/Co CNFs applied as lithium storage material, a remarkable high reversible capacity >750 mAh g^{-1}, and an excellent rate capability of 578 mAh g^{-1} at 1 C rate was obtained. The improvement in the electronic conductivity and the interfacial surface area found to have resulted from the incorporated cobalt in the fiber. The presence of cobalt in the NF increased the interfacial surface area between the carbon and liquid electrolyte, and thereby improves the electronic conductivity. The first study on highly conductive porous CoO NPs embedded in porous CNFs composite material was reported by Ryu et al. [43] The highly conductive and porous CNFs that excellently wrapped distinct CoO nanoparticles of sizes ranging from 20 to 70 nm were prepared via one-step electrospinning and successive calcination in an Ar atmosphere. The homogenous distribution of CoO all over the surface of NF can be seen in the FESEM image in Fig. 11.4.

The HRTEM images in Fig. 11.4d, f clearly exhibited CoO nanoparticles of size ranging from 20 to 70 nm and is found to be discretely embedded in the CNF. The CoO embedded CNFs exhibited a high capacity of 853.5 mAh g^{-1}, stable cycling during 100 cycles with a high coulombic efficiency of over 99%, and high rate capability even at a rate of 20 C by a reversible conversion reaction of CoO nanoparticles and facile electron transport through highly conductive CNFs. It is found that the greater LIB performance is due to the combined effect of nanosize CoO particles, which is responsible for providing reversible and facile Li insertion and extraction along with conductive CNF backbones, capable of providing 1-D electron transport path and further available capacity. Zhang et al. [45] fabricated CoO–C nanofibers with the homogeneous distribution of CoO NPs with network structure by electrospinning of cobalt(II) acetate and polyacrylonitrile. The morphology of the NFs was networks structured, and after these electrode heat-treated at 650 °C, it showed a remarkable discharge capacity as high as 633 mAh g^{-1} even after 52 cycles at a current density of 0.1 A g^{-1}. The propitious properties are found to be owing to the superior structural stability of CoO nanoparticles in CNF and by the better conductance arising from the CNF and the resultant network-based structure. The flexible mats like CoO–graphene–carbon NFs were also synthesized via electrospinning followed by careful heat treatment [46]. The graphene precursor was carefully treated to get good dispersion in the N, N-dimethylformamide (DMF) solvent. The resulted amount of graphene and CoO in the fabricated nanofibers was comparatively less. Still, the material exhibited a reversible capacity of 690 mAh g^{-1} after 352 cycles at a current density of 500 mAg^{-1} (Fig. 11.5). The resultant improvement was found to be attributed to many factors such as the microstructure with large pores that provided the effective diffusion of Li^+ ion, high conductivity of mats which allows easy transfer of electrons as well as decrease the polarization. Remarkably, the graphene in the composite is found to control the particle size of CoO NPs and thereby provided excellent structural stability against fracture. The binder-free anodes reduced the internal resistance and are capable to provide a high output voltage. Moreover, the CoO and graphene introduced active defect sites into the CNF mats that improved the capacity of Li^+ storage. Another significant work that explains the direct growth of NiO–CoO nanoneedles on carbon fibers by a simple solvothermal strategy to fabricate nanobrushes [47].

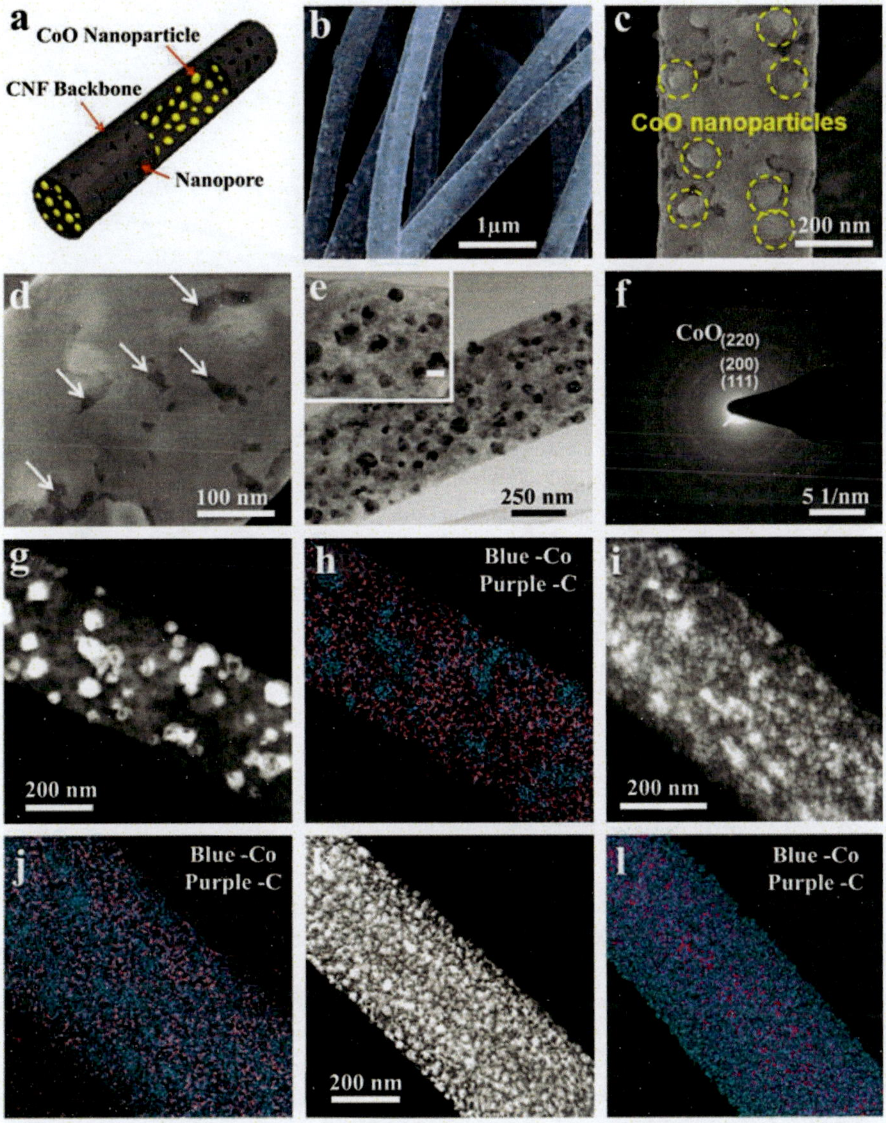

Fig. 11.4 **a** Schematic illustration of a CoO embedded CNF; **b–d** (SEM) images of CoO embedded CNFs calcined with a Co precursor/PAN ratio of 0.4 (Co precursor: 0.2 g); **e** HRTEM images and **f** SAED pattern of the CoO embedded CNF; **g** STEM image of the CoO embedded CNF; and **h** atomic distribution of cobalt and carbon analyzed from the STEM image in (**g**). STEM images of Co/CoO nanoparticles embedded in CNFs calcined with different ratios of Co precursor/PAN: **i** 1 (Co precursor: 0.5 g) and **k** 2 (Co precursor: 1 g); **j, l** atomic distribution of Co (blue color) and C (purple color) analyzed from the STEM images in (**i**) and (**k**), respectively; Co precursor/PAN ratios of (**j**) 1 and (**l**) 2. Reproduced from Ref. [43] with the permission from Royal Society of Chemistry

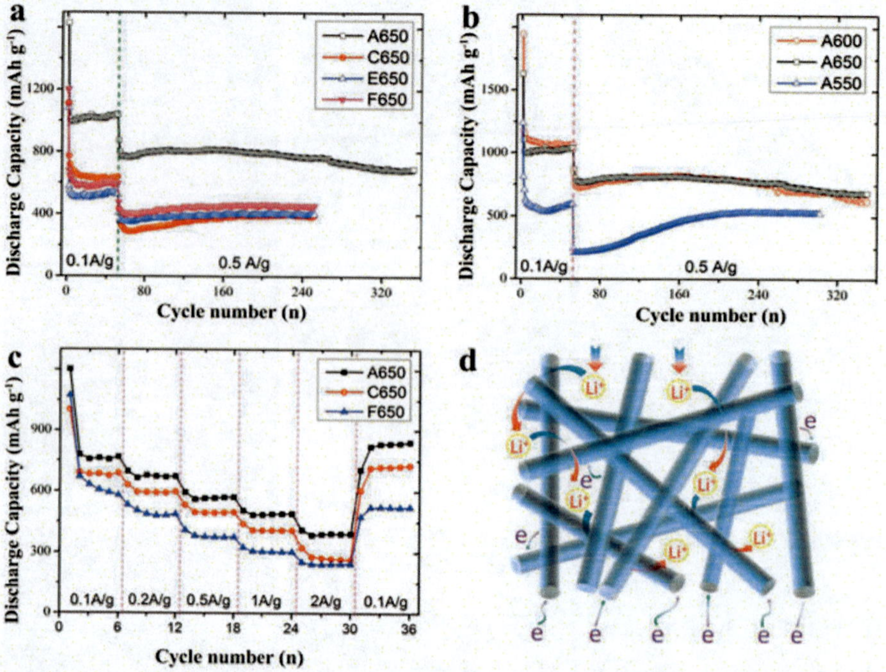

Fig. 11.5 **a** The cyclic properties of samples A650 (CoO-G-C), C650 (CoO-C), E650 (pure carbon), and F650 (graphene-carbon). **b** The discharge capacity vs. cyclic number curves of the samples A650, A600, and A550. **c** The rate capacity of the samples A650, C650, and F650. **d** A schematic diagram to show that the A650 nanofiber flexible mats are of benefit for the storage of Li^+. Reproduced from Ref. [46] with the permission from Royal Society of Chemistry

By changing the solvents from water to absolute ethanol, and finally to isopropanol, the density of nanobrushes was found to change drastically. This can be clearly observed in the SEM images of fabricated fibers in Fig. 11.6. The elctrospinned carbon fibers (Fig. 11.6a) show an interconnected continuous framework with 300–400 nm diameters. After the fibers were allowed to undergo the hydrothermal process, the surface of the fibers was effectively covered with disordered NiO–CoO nanoneedles by utilizing water as a solvent, as revealed in Fig. 11.6b. Particularly, the mixed water and absolute ethanol solvent resulted in the formation of brush-like NiO–CoO/carbon fiber composites with strong adherence to the carbon fiber substrates, as evident in Fig. 11.6c. In this study, the solvent ethanol and isopropanol were responsible for altering the morphology of NFs by reducing the surface tension and to modify the polarity and coordination potentiality of the solvent used. Thus, the nucleation and growth of nanoneedles were effected resulted in distinct morphology, as shown in the SEM image.

The particular study also reported detailed time-dependent electrospinning experiments to illustrate the mechanism of NiO–CoO/carbon fiber nanobrushes formation. The NiO–CoO/carbon fiber nanobrushes synthesized using ethanol as a solvent

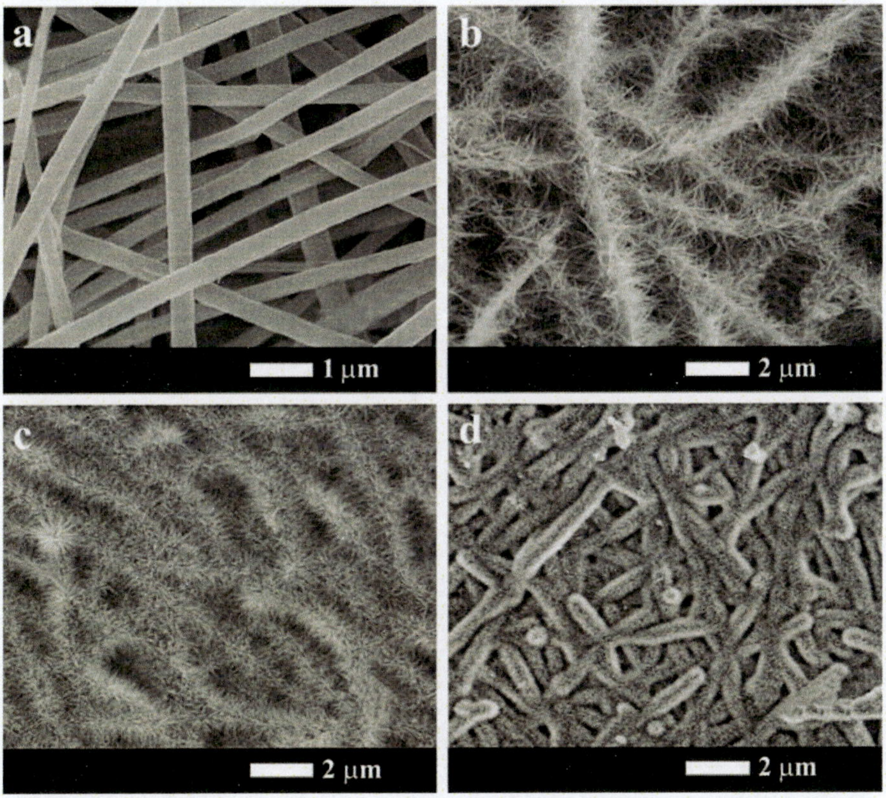

Fig. 11.6 SEM images of **a** pure carbon fibers, **b** sample synthesized water, **c** sample synthesized absolute ethanol, **d** sample synthesized isopropanol. Reproduced from Ref. [47] with permission from American Chemical Society

showed the discharge capacity as large as 801 mAh g^{-1} after 200 cycles at a current density of 200 mAg^{-1}. The electrode showed an initial discharge capacity of 961.53 mAh g^{-1} with a charge capacity of 692 mAh g^{-1}, demonstrating an irreversible capacity loss of about 38% which can be understood from Fig. 11.7. A long stable voltage stage at 1.0 V in the first discharge is exhibited by the electrode, which is found to be due to the complex phase alteration of CoO–NiO to Co–Ni and also due to the solid–electrolyte layer (SEI) formation. The obtained enhanced Li^+ storage ability and high cyclic performance for the NiO–CoO/carbon fiber is due to the high surface area, large porosity, and excellent electric conductivity of carbon fibers as well as the superior properties of NiO–CoO nanoneedles. Particularly in the presence of NiO along with unique electroactive material CoO, the better electrochemical performance was exhibited for LIBs than the single metal oxide. The interconnected NiO–CoO structures that could act as the highway for electrons and the interfacial space formed between NiO and CoO nanoneedles that is capable of accommodating

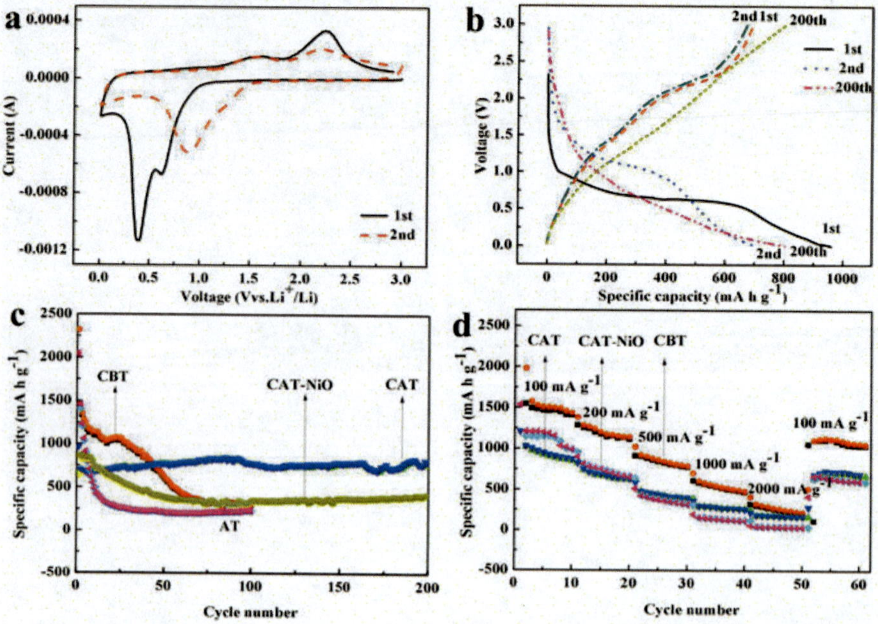

Fig. 11.7 **a** CV curves of NiO–CoO/carbon fiber nanobrushes between 0.01 and 3.0 V at the sweep speed of 0.1 mv/s, **b** Charge–discharge curves of NiO–CoO/carbon fiber nanobrushes at a current density of 200 mAg^{-1}, **c** Cycling performances of NiO–CoO/carbon fiber nanobrushes before and after heat treatment (CBT and CAT), NiO–CoO nanosphere electrode (AT), and NiO/carbon fiber nanobrushes (CAT-NiO) at 200 mAg^{-1}, respectively. **d** Rate-capability properties of NiO–CoO/carbon fiber nanobrushes before and after heat treatment (CBT and CAT) and NiO/carbon fiber nanobrushes (CAT-NiO at different current densities). Reproduced from Ref. [47] with permission from the American Chemical Society

the substantial volume change and also provided high structural stability towards the fracture.

Thus, CoO NFs has proved to be a prospective candidate for binder-free, self-standing NFs for LIBs applications, and Table 11.1 provides comprehensive electrochemical progress of CoO NF anodes for LIBs.

Table 11.1 Performance of electrospun CoO NFs for LIBs

Material	Capacity ($mAh\ g^{-1}$)	Cycles	Current density (mAg^{-1})	References
CoO–CNF	853.5	100	140	[43]
C/Co–CNF	800	50	100	[44]
CoO–C–NF	633	52	100	[45]
CoO–graphene	800	100	500	[46]
NiO–CoO–C	801	200	200	[47]

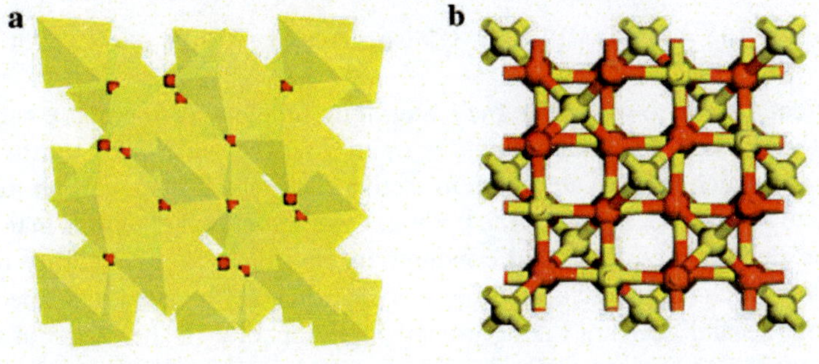

Fig. 11.8 Crystal structure of Co_3O_4 generated using Material Studio 7.0® **a** polyhydra structure, **b** ball and stick model

11.5.3 Structure of Cobalt (II, III) Oxide

Co_3O_4 is the thermodynamically favored form of cobalt oxide, which belongs to a normal cubic spinel with the Fd3m space group. Compared to CoO, the structure of Co_3O_4 is complicated with two types of Co ions tetrahedrally coordinated Co^{2+}(II) and octahedrally coordinated Co^{3+}(III), which is represented in Fig. 11.8. The distance of Co-O is 1.929 Å for the tetrahedrally coordinated species and 1.916 Å for the octahedrally coordinated species. The unit cell of Co_3O_4 consists of 8 Co^{2+}, 16 Co^{3+}, and 32 O^{2-} ions, respectively which provides a very large unit cell with 56 atoms. [40] The structure can also be defined as the stacking of two kinds of polyhedral layers along the < 111 >$_c$ direction of the cubic cell. Furthermore, when Li^+ ions are inserted into Co_3O_4, the natural black color of this material is changes to light yellow. Hence this material is applied as the most active material of cobalt.

11.5.4 Electrospun Cobalt (II, III) Oxide as Anodes for Lithium-Ion Batteries

Co_3O_4 is regarded as a competent anode material for LIBs owing to its high theoretical specific capacity of 890 mAh g^{-1} compared to commercial graphite (372 mAh g^{-1}) [48]. The material possesses a spinel structure with a Co^{2+} located in a tetrahedral coordination site and Co^{3+} located in an octahedral site. From the DFT calculations, the calculated bandgap is 1.96 eV. Owing to its wide bandgap, the electrical conductivity of this material is found to be very low, and thus material suffers from large volume change upon cycling resulting in low coulombic efficiency, low rate capability, and finally poor cycling stability. These material demerits are mainly due to the pulverization and deterioration of the electrode materials while lithiation and delithiation steps. The mechanism of operation follows the reversible conversion reaction lead to the storage of 8 mol of Li per mole of Co_3O_4 according to the following reaction:

$$Co_3O_4 + 8Li^+ + 8e \rightarrow 4Li_2O + 3CoCo_3O_4 + 8Li^+ + 8e \rightarrow 4Li_2O + 3Co$$

The effective way to overcome the material inefficiency is to prepare nanometer-sized materials with designed structures or to disperse the Co_3O_4 uniformly into carbon matrix such as activated carbon, active carbon coatings, carbon nanotubes (CNTs), and also graphene. Since the homogenous dispersion of Co_3O_4 in the carbon matrix can deliver high electrical conductivity and high cyclic stability benefited from carbon. Such composites are effectively synthesized by a one-step electrospinning technique and delivered a significant high capacity in the range of 800-952 mAh g^{-1} after 100 cycles with provided current density of 100 mAg^{-1}. An et al. [48] successfully synthesized CNF/Co_3O_4 nanopyramid core-shell nanowires (NWs) by following electrospinning, reduction, and hydrothermal methods. The FESEM images confirmed the uniform formation of the Co_3O_4 composites with CNF and CNF/Co_3O_4 nanopyramid core-shell NWs as shown in Fig. 11.9.

The FESEM image of the material exhibited smooth surfaces of CNFs with diameters in the range of 208–224 nm and revealed a uniform morphology. Figure 11.9d clearly shows the electrode material is composed of octahedral Co_3O_4 and CNFs in which the CNFs penetrated the octahedral Co_3O_4. The resultant interface formed

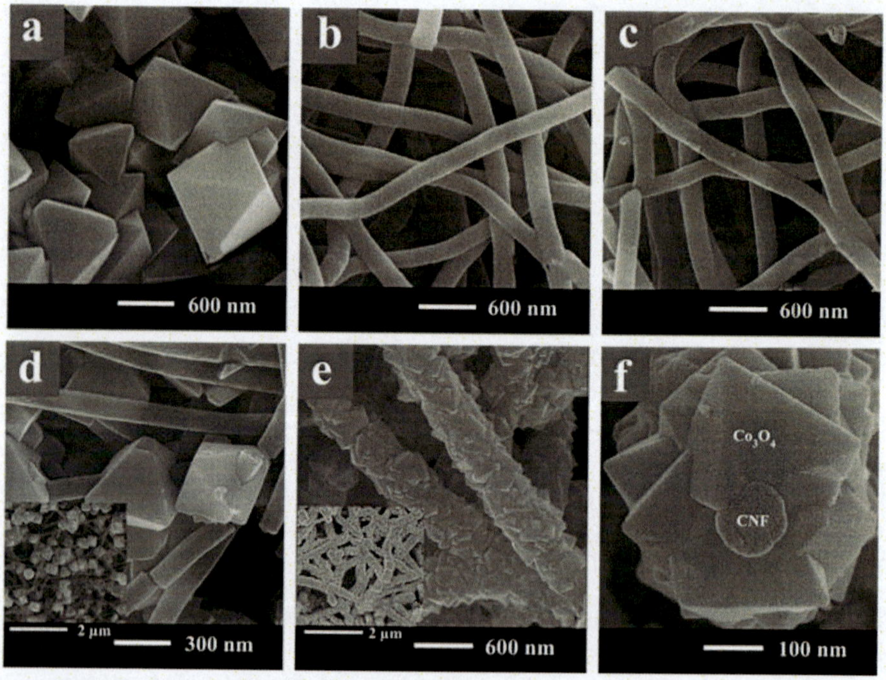

Fig. 11.9 SEM images of **a** octahedral Co_3O_4, **b** conventional CNFs, **c** CoO seeds/CNFs, **d** Co_3O_4/CNF composites, and **e** CNF/Co_3O_4 nanopyramid core shell NWs. **f** A cross sectional SEM image obtained from CNF/Co_3O_4 nanopyramid core shell NWs. Reproduced from Ref. [48] with permission from the Elsevier

between the CNFs and Co_3O_4 nanopyramids is well connected, and that is capable enough to contribute the enhanced electron transfer. The LIBs fabricated using CNF/Co_3O_4 nanopyramid core-shell NWs showed an exceptional discharge capacity of 1173 mAh g^{-1} after the first cycle and 795 mAh g^{-1} after 50 cycles. The obtained initial coulombic efficiency of 84.8% and rate capability of 570 mAh g^{-1} at 700 mAg^{-1} made this composite much suitable for LIB application. The excellent electrochemical performance exhibited by this material is due to the synergistic effect of 1-D CNFs that offer effectual electron transport and the geometric network structures which facilitate easy diffusion of Li-ion. Abouali et al. did another significant work. [49], reported a facile one-pot synthesis approach to fabricate porous carbon nanofibers (PCNF) integrated with Co and Co_3O_4 NPs embedded CoO_x/PCNF. The adopted synthesis method favored the bounding of spherical Co_3O_4 NPs with graphitic carbon shells in the CNF matrix by careful control of reaction parameters. The composite electrodes exhibited a remarkable capacity of 952 mAh g^{-1} after 100 cycles at a current density of 100 mAg^{-1}. The layers of graphene around each Co_3O_4 NPs provided an uninterrupted conductive network inside the conductive CNFs, resulted in an effective decrease of the ion/electron inter transfer resistance. Such a structure also helped to reduce the detachment of NPs from the CNF during the charge-discharge process and facilitated to retain the stable structure features. Another remarkable work was the fabrication of the hierarchically mesoporous flower-like Co_3O_4/carbon nanofiber (Co_3O_4/CNF) composites with a shell–core structure and its application as the anode material for LIBs [50]. The synthesis method involved the preparation of CNF by electrospinning followed by the electrophoretic deposition of Co_3O_4 and finally annealing in air. The morphological analysis revealed the formation of flower-like Co_3O_4 shell around the CNF core that delivers a porous, high surface area with copious inner spaces through the mesopores that enabled the easy diffusion of Li^+ and avoided the volume expansion to result in high specific capacities. The particular 1D pathway increased the rate capability and also offered good electrical conductivity as well as structural stability. The fabricated LIBs provided initial capacity as high as 1446 mAh g^{-1} and the specific capacity 911 mAh g^{-1} without capacity fading after many cycles. Later, Hu et al. [51] fabricated composite of porous Co_3O_4 NF coated with reduced graphene oxide (GNS) as a surface-modification layer to facilitate better interconnection, ability to sustain volume change, to avoid the aggregation of metal NPs and to facilitate rapid transport of Li^+ ion during charge/discharge cycling. The fibers with pore structure-function as the paths for fast electrolyte transport to the electrode. The ultrafine Co_3O_4 NCs in the fibers with grain boundaries/interfaces offered a large material/electrolyte contact area, short diffusion distance for Li^+ ions, and enabled the rapid lithiation/delithiation. Moreover, the unique porous architecture was interconnected by Co_3O_4 NCs and the rGO sheets deliver the volume changes and avoiding the agglomeration of fibers during the electrochemical reactions. Thus, the fabricated electrodes of Co_3O_4@rGO material displayed high Coulombic efficiency, high cyclic stability, and high rate capability of 900 mAh g^{-1} at 1 A g^{-1}, and 600 mAh g^{-1} at 5 A g^{-1}. Recent modification on anode material is the use of a highly porous zeolite type metal-organic framework (MOF) named as ZIF-67 to synthesize cobalt-based composites through electrospinning. Zhang et al. [52] fabricated carbon-based hierarchical porous fiber composite with uniformly distributed hollow Co_3O_4 NPs via electrospinning of ZIF-67 NPs. When the fabricated electrodes were applied for LIBs, it is found that the micropores and Co_3O_4 yielded high-density active sites. Simultaneously, the macropores in the electrode surface offered a superior mass/charge transports through the electrodes, as shown in Fig. 11.10.

Another work explored a controllable electrostatic spinning to synthesize1D Co_3O_4 nanotubes at large-scale that yielded Co_3O_4 nanotubes with tightly packed NPs as primary building blocks. The fabricated LIBs showed a high specific capacity and cycle stability of 856.4 mAh g^{-1} at 0.25 C and 677.2 mAh g^{-1} at 1 C for up to 60 cycles [53]. The Co, N-doped carbon materials can be formed by thermal treatment and the CNFs derived by this approach that exhibited high surface area, hierarchical pores, and high nitrogen content. Moreover, the presence of Co and N can catalyze the graphitization of carbon, which is highly desirable for enhancement in conductivity. The more defects created around the pores during the heat treatment process can effectively increase the layer space by the presence of cobalt. In a recent study, Liu et al. [54] modified the electrospinning method with sol-gel synthesis to prepare

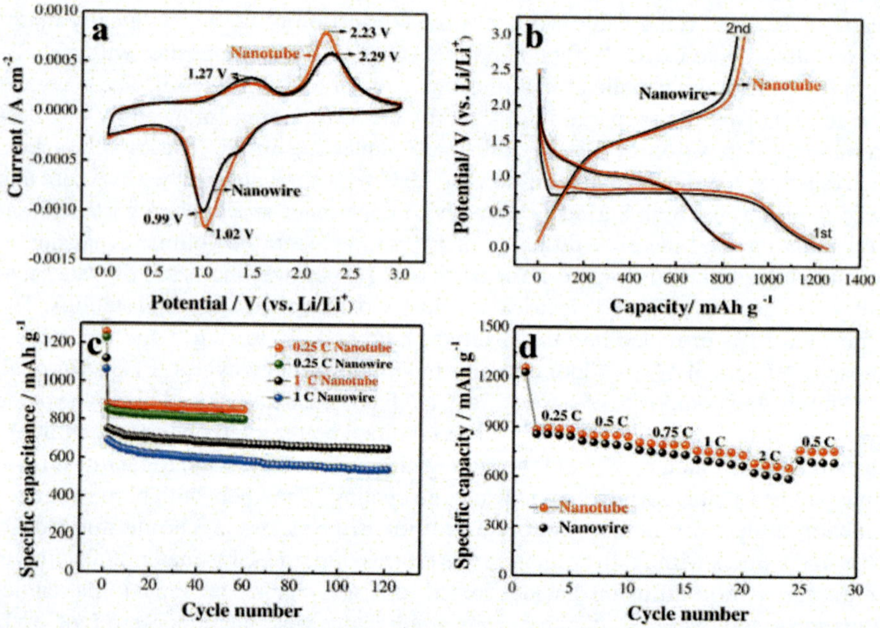

Fig. 11.10 **a** CV curves of Co_3O_4 NTs and NWs at the second cycle at a scan rate of 0.1 mV s^{-1} in the potential range 0–3.0 V (vs. Li/Li^+); **b** Charge/discharge profiles of Co_3O_4 NTs and NWs electrodes between 0.01 and 3.0 V at 0.25 C; **c** Nyquist plots of the Co_3O_4 NTs and NWs electrodes after 10 cycles in the frequency range from 100 kHz to 10 MHz; **d** Cycling performance of Co_3O_4 NTs and NWs at 0.25 C and 1 C; **e** Rate performance of Co_3O_4 NTs and NWs. Reproduced from Ref. [53]. with permission from the Elsevier

Table 11.2 Performance of electrospun Co_3O_4 NFs for LIBs

Material	Capacity (mAh g^{-1})	Cycles	Current density (mAh g^{-1})	References
CNF/Co_3O_4 nanopyramid core-shell NWs	795	50	100	[48]
Co_3O_4–PCNF	952	100	100	[49]
Co_3O_4/CNF	911	50	200	[50]
Co_3O_4@rGO	967	100	100	[51]
Co_3O_4@CPNF	1100	200	100	[52]
Co_3O_4 NT-NP	856.4	60	0.25C	[53]
Co_3O_4@NFs	330	170	0.5C	[54]
Co_3O_4@CNFs	400	200	0.5C	[54]

Co_3O_4@nanofibers (NFs) and Co_3O_4@carbon NFs (CNFs). The hollow 3D network nanostructure obtained with tiny well-dispersed Co_3O_4 NCs on the surface that could effectively avoid the aggregation. Stable cyclic performance with Li storage capacity of 330 and 400 mAh g^{-1} after 170 and 200 cycles obtained for Co_3O_4@NFs and Co_3O_4@CNFs, respectively. The study revealed that through electrospun carbon coating layer, the internal resistance was decreased by increasing electrical conductivity which resulted in an enhanced energy storage performance of LIBs. Table 11.2 summarize the progress of Co_3O_4 NF anodes for LIBs.

11.5.5 Ternary Oxides of Cobalt as Anodes for Lithium-Ion Batteries

Owing to the high lithium redox potential while using Co_3O_4, many efforts have been made by scientists to substitute the Co ion in Co_3O_4 with low-cost and eco-friendlier metal ions such as Zn, Cu, Ni, Mn, and Fe to form potential anode materials for LIBs which include $ZnCo_2O_4$, $CuCo_2O_4$, $NiCo_2O_4$, and $MnCo_2O_4$. These mixed transition metals with stoichiometric or non-stoichiometric compositions have gained copious consideration due to their fascinating electrochemical properties such as high theoretical specific capacitance, and variable oxidation states. These are generally denoted by $A_xB_yO_z$ with two metal elements of A, B = Co, Ni, Zn, Mn, Cu, Fe, Mo, etc. and both A and B play substantial roles in electrochemical energy storage devices. Depending on the ratio of A/B/O, transition metal oxides (TMOs) can be mainly classified into three groups: AB_2O_4, $ABO_{2/3/4}$, and $A_3B_2O_8$. [55] Therefore, the crystal structures of each type of TMOs are found to vary which results in different physical and chemical properties, and accordingly, their battery performances. Generally, the single-phase ternary metal oxides possessing a spinel structure consist of two different cations with variable oxidation states within their octahedral and tetrahedral sites. This specific structure facilitates the perfect tuning of its physical and chemical features. Consequently, the material shows higher electrical conductivity than simple transition metal oxides due to comparatively low activation energy for charge transfer among the two different cations [56]. Furthermore, such chemical compositions result in high electrochemical activities thereby could contribute to a great specific capacity.

Spinel AB_2O_4 electrode materials for LIBs can be again classified as spinel metal-cobalt oxides (ACo_2O_4, A = Ni, Zn, Cu, etc.), spinel metal-ferrum oxides (AFe_2O_4, A = Ni, Co, Mn, etc.), and spinel metal-manganese oxides (AMn_2O_4, A = Ni, Co, Zn, etc.) Amongst these spinel metal-cobalt oxides have been studied extensively in which specifically $NiCo_2O_4$ gains particular attention for its high theoretical capacity, benign environmental nature, cost-effectiveness, and also the earth-abundance. $NiCo_2O_4$ is an inverse spinel oxide with the nickel ions occupying the octahedral sites, and half of the cobalt ions occupying the tetrahedral sites and the other half ions hold the remaining octahedral sites [57]. When applied to LIBs, $NiCo_2O_4$ material has shown high specific capacity as high as 891 mAh g^{-1}, good conductivity, and low diffusion resistance between Li^+ ions and the electrolyte. Zhou et al. [58] reported the tufted $NiCo_2O_4$ nanoneedles hydrothermally grown on electrospun carbon nanofibers as anode material for LIBs. The unique three dimensional (3D) hybrid morphology of synthesized composite looks as if clustered acicular leaves on branches. A reversible discharge specific capacity of 1033.6 mAh g^{-1} at a current density of 200 mAg^{-1} was achieved even after 250 cycles with the coulombic efficiency >98%. The obtained high specific capacity is mainly attributed to the unique 3D morphology and the synergetic effects of hybrid CNFs. Zhu et al. [59] modified the $NiCo_2O_4$ electrode material with Au NPs. The $NiCo_2O_4$@Au nanotubes (NTs) having a mesoporous structure with hollow interiors, and these nanotubes were synthesized by the electrospinning followed by calcination in air. The composite electrode showed high reversible capacity of 732.5 mAh g^{-1} even after 200th cycles at a current density of 100 mAg^{-1} with excellent cycling stability. The embedded Au NPs penetrated the whole electrode to form 3D percolation networks and also acts as adhesion centers/mechanic anchoring points to firmly hold $NiCo_2O_4$ particles, which is capable of giving the required tolerance ability against electrode to volume expansion as shown in Fig. 11.11. An electrospun binder-free carbon fiber hybrid with Ni–Co also synthesized and applied to LIBs that have demonstrated the highest capacity of 734 mAh g^{-1} at a current density of 150 mAg^{-1} [60]. This binder-free electrode provided better interconnection by carbon fibers, which also gave a highly conductive path for effective charge transfer. Furthermore, the highly dispersed Ni–Co compounds are proved to be capable of resisting the volume change during charging and discharging and successfully maintained the electrode reliability.

Other notable work reported was the preparation of uniform $NiCo_2O_4$ nanotubes (NCO-NTs) through electrospinning followed by calcination in the air by Li et al. [35], and these NTs were successfully used as anode materials for LIBs as well as Sulphur ion batteries (SIBs). The exceptional long-term electrochemical stability and reversibility obtained with retention of 93.6 % of specific capacitance even after 3000 cycles with the advantage of the distinct porous structure with hollow 1D morphology of NCO-NTs. The unique morphology of NCO-NTs provided high surface area as

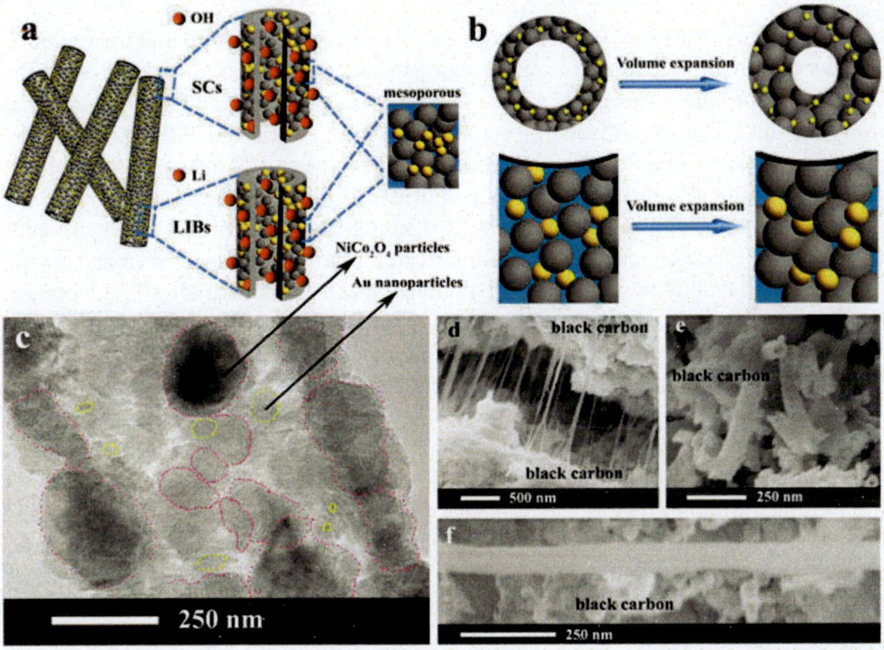

Fig. 11.11 **a** Schematic representation of the electrolyte/electrode interface for LIBs and SCs; **b** illustration of volume change of $NiCo_2O_4$@Au NTs before and after cycle; **c** TEM images of a single $NiCo_2O_4$@Au NT; **d**, **e** SEM image for the SCs of the $NiCo_2O_4$@Au NT electrodes after 10,000 cycles; and **f** SEM image for the lithium ion batteries of the $NiCo_2O_4$@Au NT electrodes after 200 cycles. Reproduced from Ref. [59] with the permission from Elsevier

well as efficient transport pathways for rapid electron transport thereby capable of enhancing the electrochemical kinetics. Mainly, the hollow structures ascertained to serve as ion-buffering reservoirs for OH^- ions and also capable of adjusting the volume change occurs by insertion and extraction of the OH^- ions. Thus, offer minimum diffusion distance to the interior surfaces, increased contact area between electrolyte to the electrode, more active sites for rapid faradaic redox reactions. A porous $ZnCo_2O_4$ NTs as anode materials for LIBs are reported for the first time by Luo et al. [61] by the electrospinning method followed by thermal annealing in air. The fabricated NTs were polycrystalline and consisted of both interconnected NCs and many nanopores in the tube walls. Such NTs when applied as anode materials for LIBs, offered a high reversible capacity of 1454 mAh g^{-1} at a current density of 100 mAg^{-1}, and even keeps 794 mAh g^{-1} at 2000 mAg^{-1} after 30 cycles. Such a unique structure facilitated fast Li-ion transport with the volume expansion/contraction during Li insertion/extraction processes. Chen et al. [62] described a simple, cost-effective, and scalable approach for the fabrication of CNFs by electrospinning and thereby anchoring these NFs with $Zn_xCo_{3-x}O_4$ (ZCO) nanocubes, which involved a hydrothermal process followed by thermal treatment. Fabricated anodes delivered high reversible capacity around 600 mAh g^{-1} at a current density of 0.5 C following 300 charge-discharge cycles. The authors found that the obtained superior electrochemical performance was accredited to the controllable size of the ZCO nanocubes and synergistic effects between ZCO and the CNFs. It is found that even after the 150 cycles, discharge capacities as high as 656 and 390 mAh g^{-1} were obtained which was much higher than the reported value of 372 mAh g^{-1} for graphite electrode. Another significant ternary cobalt-based metal oxide applied to LIBs is $CoMn_2O_4$. Yang et al. [63] fabricated hollow nanofibers of $CoMn_2O_4$ by electrospinning method. These materials also exhibited good lithium storage performances with high specific capacity and excellent reversible capacity of 526 mAh g^{-1} at a current density of 400 mAg^{-1} after 50 cycles. The unique 1D shape with the porous structure not only offered smooth electron transport pathways but also could provide enough void spaces to endure the volume change during the charge-discharge process. Hwang et al. [64] reported the synthesis of both $MnCo_2O_4$ (MCO) and $CoMn_2O_4$ (CMO) mesoporous hollow nanofibres by electrospinning and highly exothermic combustion reactions in the air. They have found the mechanism of formation of such fiber differs from the earlier reported heterogeneous contraction mechanism [65]. From the thermal analyses and structural investigation, the authors found the formation pathways of single- and double-walled hollow fibers via combustion reactions, which principally depends on the ramping rate during heat-treatment. The SEM and TEM images of Fig. 11.12 clearly shows the formation of single and double walled fibers with hollow interiors.

The high discharge capacity of 997 and 798 mAh g^{-1} were obtained for the electrodes even after the 50th cycle with a coloumbic efficiency as high as 98 and 99%. Moreover, superior charge capacity retention at various current densities and the highest current density of 5 C, the charge capacity of 696 and 575 mAh g^{-1} were achieved. The excellent lithium storage capacity and good cycling performance and rate capability for LIBs were ascribed to the large surface area, the void space that is originating from the hollow morphology which enabled to provide more active sites for Li^+ storage and easy Li^+ transportation during charge-discharge cycles. Table 11.3 summarize the electrode performance of ternary oxide NFs used for LIBs application.

11.6 Conclusion

The electrospinning approach is recognized as a very versatile methodology to fabricate various composite anode materials with tuneable morphologies, hybrid, and interconnected 3D network structures that are capable of enhancing the electrochemical performance through facile electron transfer. The achieved remarkable advancement of LIB technology would not have been conceivably deprived of the progress of many nanocomposites and nanometer-thick coatings, which is accountable to optimize the ionic and electronic conduction pathways and even to obstruct unsought and irreversible side reactions. Advances in electrospinning of cobalt-based composites

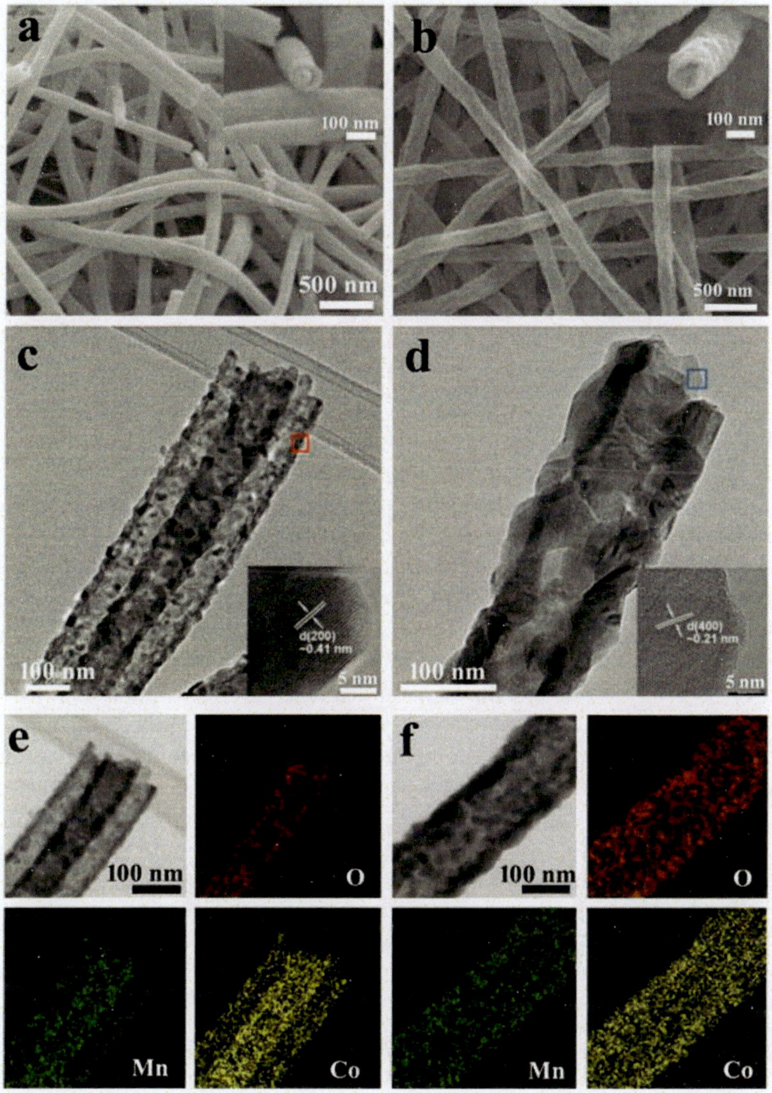

Fig. 11.12 **a**, **b** SEM, **c**, **d** TEM, and **e**, **f** EDS mapping images of MCO fibres annealed at different heating rates of 0.5 °C min^{-1} (**a**, **c**, **e**) and 5 °C min^{-1} (**b**, **d**, **f**). The insets in **a**, **b** show cross-sectional images of representative hollow fibres, and those in **c**, **d** are HRTEM images of the regions marked by squares; **e**, **f** present the elemental distributions of O, Mn, and Co. Reproduced from Ref. [64] with permission from the Royal Society of Chemistry

Table 11.3 Performance of electrospun ternary oxide NFs for LIBs

Material	Capacity (mAh g^{-1})	Cycles	Current density (mAg^{-1}) or C	References
Tufted $NiCo_2O_4$ on CNF	1033.6	250	200	[58]
$NiCo_2O_4$@Au–CNF	732.5	200	100	[59]
$NiCo_2O_4$	595	100	100	[60]
$NiCo_2O_4$ NTs	1454	30	100	[61]
$Zn_xCo_{3-x}O_4$	600	300	0.5C	[62]
$CoMn_2O_4$ hollow nanofibers	526	50	400	[63]
$MnCo_2O_4$	696	50	5C	[64]

have demonstrated that the technique affords us a remarkably simple and versatile approach for the preparation of composite nanoparticles with fibers that offers diverse functionalities for LIB applications. These composite fibers can be fabricated either by a post-treatment of the already synthesized electrospun fibers or by adding the cobalt oxide directly into the electrospun solution. The morphology of these nanofibers can be easily tunable by the control of various electrospinning parameters and also by the heat treatment steps after the fiber formation. The possibility of making 1D NFs with three dimensional (3D) networked structure provides high surface-to-volume ratio to the anode materials and the macro-pores that available in these structure facilitates easy transport of Li^+ ions with reduced diffusion path length, high surface area, easy accommodation of increasing volume changes during cycling and reaction sites. This technique can be combined with other methods such as electrophoretic deposition, hydrothermal and sputtering methods to fabricate dense or porous NFs, core-shell NFs, and hollow structured NFs. Thus, these high-performance battery materials can be combined together so as to get the best advantages of individual materials.

Acknowledgements Authors Dr. Jabeen Fatima M. J and Dr. Raghavan Prasanth, would like to acknowledge the Kerala State Council for Science, Technology, and Environment (KSCSTE), Government of Kerala for financial assistance. Dr. R. Krishnaprya and Dr. R. K. Sharma thank DBT-PAN IIT Centre for Bioenergy (BT/EB/PANIIT/2012) for financial support.

References

1. Chu S, Majumdar A (2012) Opportunities and challenges for a sustainable energy future. Nature 488:294–303. https://doi.org/10.1038/nature11475
2. Goodenough JB, Park KS (2013) The Li-ion rechargeable battery: a perspective. J Am Chem Soc 135:1167–1176. https://doi.org/10.1021/ja3091438
3. Wang W, Tade MO, Shao Z (2015) Research progress of perovskite materials in photocatalysis- and photovoltaics-related energy conversion and environmental treatment. Chem Soc Rev 44:5371–5408. https://doi.org/10.1039/c5cs00113g

4. Balzani V, Credi A, Venturi M (2008) Photochemical conversion of solar energy. Chemsuschem 1:26–58. https://doi.org/10.1002/cssc.200700087
5. Liu C, Li F, Ma LP, Cheng HM (2010) Advanced materials for energy storage. Adv Mater 22:E28–62. https://doi.org/10.1002/adma.200903328
6. Zubi G, Dufo-López R, Carvalho M, Pasaoglu G (2018) The lithium-ion battery: state of the art and future perspectives. Renew Sustain Energy Rev 89:292–308. https://doi.org/10.1016/j.rser.2018.03.002
7. Qi W, Shapter JG, Wu Q, Yin T, Gao G, Cui D (2017) Nanostructured anode materials for lithium-ion batteries: principle, recent progress and future perspectives. J Mater Chem A 5:19521–19540. https://doi.org/10.1039/c7ta05283a
8. Hall PJ, Bain EJ (2008) Energy-storage technologies and electricity generation. Energy Policy 36:4352–4355. https://doi.org/10.1016/j.enpol.2008.09.037
9. Mizushima K, Jones PC, Wiseman PJ, Goodenough JB (1980) Li_xCoO_2 (0 < x~l): a new cathode material for batteries of high energy density. Mat Res Bull 15:783–789
10. Zhang H, Zhao H, Khan MA, Zou W, Xu J, Zhang L, Zhang J (2018) Recent progress in advanced electrode materials, separators and electrolytes for lithium batteries. J Mater Chem A 6:20564–20620. https://doi.org/10.1039/c8ta05336g
11. Winter M, Besenhard JO, Spahr ME, Novak P (1998) Insertion electrode materials for lithium batteries. Adv Mater 10:725–763
12. Ji L, Lin Z, Alcoutlabi M, Zhang X (2011) Recent developments in nanostructured anode materials for rechargeable lithium-ion batteries. Energy Environ Sci 4:5053–5059. https://doi.org/10.1039/c0ee00699h
13. Roy P, Srivastava SK (2015) Nanostructured anode materials for lithium ion batteries. J Mater Chem A 3:2454–2484. https://doi.org/10.1039/c4ta04980b
14. Nitta N, Wu F, Lee JT, Yushin G (2015) Li-ion battery materials: present and future. Mater Today 18:252–264. https://doi.org/10.1016/j.mattod.2014.10.040
15. Tarascon JM, Armand M (2001) Issues and challenges facing rechargeable lithium batteries. Nature 414:359–367. https://doi.org/10.1038/35104644
16. Zhang Y, Liu N (2017) Nanostructured electrode materials for high-energy rechargeable Li, Na and Zn batteries. Chem Mater 29:9589–9604. https://doi.org/10.1021/acs.chemmater.7b03839
17. Canepa P, Sai Gautam G, Hannah DC, Malik R, Liu M, Gallagher KG, Persson KA, Ceder G (2017) Odyssey of multivalent cathode materials: open questions and future challenges. Chem Rev 117:4287–4341. https://doi.org/10.1021/acs.chemrev.6b00614
18. Mai L, Tian X, Xu X, Chang L, Xu L (2014) Nanowire electrodes for electrochemical energy storage devices. Chem Rev 114:11828–11862. https://doi.org/10.1021/cr500177a
19. Dosch H (2001) Some general aspects of confinement in nanomaterials. Appl Surf Sci 182:192–195. https://doi.org/10.1016/S0169-4332(01)00426-3
20. Li D, Xia Y (2004) Electrospinning of nanofibers: reinventing the wheel? Adv Mater 16:1151–1170. https://doi.org/10.1002/adma.200400719
21. Ramakrishna S, Fujihara K, Teo W-E, Yong T, Ma Z, Ramaseshan R (2006) Electrospun nanofibers: solving global issues. Mater Today 9:40–50. https://doi.org/10.1016/s1369-7021(06)71389-x
22. Kenry, Lim CT (2017) Nanofiber technology: current status and emerging developments. Prog Polym Sci 70:1–17. https://doi.org/10.1016/j.progpolymsci.2017.03.002
23. Yarin AL, Koombhongse S, Reneker DH (2001) Taylor cone and jetting from liquid droplets in electrospinning of nanofibers. J Appl Phys 90:4836–4846. https://doi.org/10.1063/1.1408260
24. Lee T-H, Chiu Y-J, Lai Y-C, Fan P-W, Kuo T-Y, Liau I, Chen J-T (2014) Rayleigh-instability-driven morphology transformation of electrospun polymer fibers imaged by in situ optical microscopy and stimulated Raman scattering microscopy. RSC Adv 4:51884–51892. https://doi.org/10.1039/c4ra06228k
25. Dong Z, Kennedy SJ, Wu Y (2011) Electrospinning materials for energy-related applications and devices. J Power Sources 196:4886–4904. https://doi.org/10.1016/j.jpowsour.2011.01.090

26. Fong H, Chun I, Reneker DH (1999) Beaded nanofibers formed during electrospinning. Polymer 40:4585–4592. https://doi.org/10.1016/S0032-3861(99)00068-3
27. Gates DP, Svejda SA, Onate E, Killian CM, Johnson LK, White PS, Brookhart M (2000) Synthesis of branched polyethylene using (R-Diimine)nickel(II) catalysts: influence of temperature, ethylene pressure, and ligand structure on polymer properties. Macromolecules 33:2320–2334
28. Casper CL, Stephens JS, Tassi NG, Chase DB, Rabolt JF (2004) Controlling surface morphology of electrospun polystyrene fibers: Effect of humidity and molecular weight in the electrospinning process. Macromolecules 37:573–578. https://doi.org/10.1021/ma0351975
29. McCann JT, Li D, Xia Y (2005) Electrospinning of nanofibers with core-sheath, hollow, or porous structures. J Mater Chem 15:735–738. https://doi.org/10.1039/b415094e
30. Li D, Wang YL, Xia YN (2003) Electrospinning of polymeric and ceramic nanofibers as uniaxially aligned arrays. Nano Lett 3:1167–1171. https://doi.org/10.1021/nl0344256
31. Fu Z, Li X, Xu G (2014) Novel electrospun SnO_2@carbon nanofibers as high performance anodes for lithium-ion batteries. Cryst Res Technol 49:441–445. https://doi.org/10.1002/crat.201300211
32. Sundaramurthy J, Aravindan V, Suresh Kumar P, Madhavi S, Ramakrishna S (2014) Electrospun TiO_2–δ nanofibers as insertion anode for Li-Ion battery applications. J Phys Chem C 118:16776–16781. https://doi.org/10.1021/jp412787z
33. Wu Q, Zhao R, Zhang X, Li W, Xu R, Diao G, Chen M (2017) Synthesis of flexible Fe_3O_4/C nanofibers with buffering volume expansion performance and their application in lithium-ion batteries. J Power Sources 359:7–16. https://doi.org/10.1016/j.jpowsour.2017.05.020
34. Ou Y, Wen J, Xu H, Xie S, Li J (2013) Ultrafine $LiCoO_2$ powders derived from electrospun nanofibers for Li-ion batteries. J Phys Chem Solids 74:322–327. https://doi.org/10.1016/j.jpcs.2012.10.007
35. Li L, Peng S, Cheah Y, Teh P, Wang J, Wee G, Ko Y, Wong C, Srinivasan M (2013) Electrospun porous $NiCo_2O_4$ nanotubes as advanced electrodes for electrochemical capacitors. Chemistry 19:5892–5898. https://doi.org/10.1002/chem.201204153
36. Li L, Zhu P, Peng S, Srinivasan M, Yan Q, Nair AS, Liu B, Samakrishna S (2014) Controlled growth of cus on electrospun carbon nanofibers as an efficient counter electrode for quantum dot-sensitized solar cells. J Phys Chem C 118:16526–16535. https://doi.org/10.1021/jp4117529
37. Kumuthini R, Ramachandran R, Therese HA, Wang F (2017) Electrochemical properties of electrospun MoS_2@C nanofiber as electrode material for high-performance supercapacitor application. J Alloy Compd 705:624–630. https://doi.org/10.1016/j.jallcom.2017.02.163
38. Ye F, Hou Y, Liu M, Li W, Yang X, Qiu Y, Zhou L, Li H, Xu Y, Zhang Y (2015) Fabrication of mesoporous Li_2S-C nanofibers for high performance Li/Li_2S cell cathodes. Nanoscale 7:9472–9476. https://doi.org/10.1039/c5nr00480b
39. Wu J, Lau W-M, Geng D-S (2017) Recent progress in cobalt-based compounds as high-performance anode materials for lithium ion batteries. Rare Met 36:307–320. https://doi.org/10.1007/s12598-017-0904-y
40. Cao K, Jin T, Yang L, Jiao L (2017) Recent progress in conversion reaction metal oxide anodes for Li-ion batteries. Mater Chem Front 1:2213–2242. https://doi.org/10.1039/c7qm00175d
41. Chen C, Huang Y, Zhang H, Wang X, Wang Y, Jiao L, Yuan H (2016) Controllable synthesis of Cu-doped CoO hierarchical structure for high performance lithium-ion battery. J Power Sources 314:66–75. https://doi.org/10.1016/j.jpowsour.2016.02.085
42. Cao K, Jiao L, Liu Y, Liu H, Wang Y, Yuan H (2015) Ultra-high capacity lithium-ion batteries with hierarchical CoO nanowire clusters as binder free electrodes. Adv Func Mater 25:1082–1089. https://doi.org/10.1002/adfm.201403111
43. Ryu W-H, Shin J, Jung J-W, Kim I-D (2013) Cobalt(ii) monoxide nanoparticles embedded in porous carbon nanofibers as a highly reversible conversion reaction anode for Li-ion batteries. J Mater Chem A 1:3239–3243. https://doi.org/10.1039/c3ta01443f
44. Wang L, Yu Y, Chen P-C, Chen C-H (2008) Electrospun carbon–cobalt composite nanofiber as an anode material for lithium ion batteries. Scripta Mater 58:405–408. https://doi.org/10.1016/j.scriptamat.2007.10.024

45. Zhang M, Uchaker E, Hu S, Zhang Q, Wang T, Cao G, Li J (2013) CoO–carbon nanofiber networks prepared by electrospinning as binder-free anode materials for lithium-ion batteries with enhanced properties. Nanoscale 5:12342–12349. https://doi.org/10.1039/c3nr03931e
46. Zhang M, Yan F, Tang X, Li Q, Wang T, Cao G (2014) Flexible CoO–graphene–carbon nanofiber mats as binder-free anodes for lithium-ion batteries with superior rate capacity and cyclic stability. J Mater Chem A 2:5890–5897. https://doi.org/10.1039/C4TA00311J
47. Wei Y, Yan F, Tang X, Luo Y, Zhang M, Wei W, Chen L (2015) Solvent-controlled synthesis of NiO-CoO/carbon fiber nanobrushes with different densities and their excellent properties for lithium ion storage. ACS Appl Mater Interfaces 7:21703–21711. https://doi.org/10.1021/acsami.5b07233
48. An G-H, Ahn H-J (2014) Carbon nanofiber/cobalt oxide nanopyramid core–shell nanowires for high-performance lithium-ion batteries. J Power Sources 272:828–836. https://doi.org/10.1016/j.jpowsour.2014.09.032
49. Abouali S, Akbari Garakani M, Zhang B, Luo H, Xu Z-L, Huang J-Q, Huang J, Kim J-K (2014) Co3O4/porous electrospun carbon nanofibers as anodes for high performance Li-ion batteries. J Mater Chem A 2:16939–16944. https://doi.org/10.1039/c4ta03206c
50. Park S-H, Lee W-J (2015) Hierarchically mesoporous flower-like cobalt oxide/carbon nanofiber composites with shell–core structure as anodes for lithium ion batteries. Carbon 89:197–207. https://doi.org/10.1016/j.carbon.2015.03.039
51. Hu R, Zhang H, Bu Y, Zhang H, Zhao B, Yang C (2017) Porous Co_3O_4 nanofibers surface-modified by reduced graphene oxide as a durable, high-rate anode for lithium ion battery. Electrochim Acta 228:241–250. https://doi.org/10.1016/j.electacta.2017.01.067
52. Zhang C-L, Lu B-R, Cao F-H, Yu Z-L, Cong H-P, Yu S-H (2018) Hierarchically structured Co3O4@carbon porous fibers derived from electrospun ZIF-67/PAN nanofibers as anodes for lithium ion batteries. J Mater Chem A 6:12962–12968. https://doi.org/10.1039/c8ta03397h
53. Chen M, Xia X, Yin J, Chen Q (2015) Construction of Co_3O_4 nanotubes as high-performance anode material for lithium ion batteries. Electrochim Acta 160:15–21. https://doi.org/10.1016/j.electacta.2015.02.055
54. Liu J, Fu B, Dai J, Zhu X, Wang Q (2019) Study the mechanism of enhanced Li storage capacity through decreasing internal resistance by high electronical conductivity via sol-gel electrospinning of Co_3O_4 carbon nanofibers. ChemistrySelect 4:3542–3546. https://doi.org/10.1002/slct.201804041
55. Wang R, Wu J (2017) Structure and basic properties of ternary metal oxides and their prospects for application in supercapacitors. In: Metal oxides in supercapacitors, pp 99–132
56. Yuan C, Wu HB, Xie Y, Lou XW (2014) Mixed transition-metal oxides: design, synthesis, and energy-related applications. Angew Chem Int Ed Engl 53:1488–1504. https://doi.org/10.1002/anie.201303971
57. Marco JF, Gancedo JR, Gracia M, Gautier JL, Ríos E, Berry FJ (2000) Characterization of the nickel cobaltite, $NiCo_2O_4$, prepared by several methods: an XRD, XANES, EXAFS, and XPS study. J Solid State Chem 153:74–81. https://doi.org/10.1006/jssc.2000.8749
58. Zhou G, Wu C, Wei Y, Li C, Lian Q, Cui C, Wei W, Chen L (2016) Tufted $NiCo_2O_4$ nanoneedles grown on carbon nanofibers with advanced electrochemical property for lithium ion batteries. Electrochim Acta 222:1878–1886. https://doi.org/10.1016/j.electacta.2016.12.001
59. Zhu J, Xu Z, Lu B (2014) Ultrafine Au nanoparticles decorated $NiCo_2O_4$ nanotubes as anode material for high-performance supercapacitor and lithium-ion battery applications. Nano Energy 7:114–123. https://doi.org/10.1016/j.nanoen.2014.04.010
60. Syu J-M, Hsiao M-L, Lo C-T (2017) Electrospun carbon fiber/Ni–Co composites as binder-free anodes for lithium-ion batteries. J Electrochem Soc 164:A3903–A3913. https://doi.org/10.1149/2.0021802jes
61. Luo W, Hu X, Sun Y, Huang Y (2012) Electrospun porous $ZnCo_2O_4$ nanotubes as a high-performance anode material for lithium-ion batteries. J Mater Chem 22:8916–8921. https://doi.org/10.1039/c2jm00094f
62. Chen R, Hu Y, Shen Z, Chen Y, He X, Zhang X, Zhang Y (2016) Controlled synthesis of carbon nanofibers anchored with Zn(x)Co(3-x)O_4 nanocubes as binder-free anode materials

for lithium-ion batteries. ACS Appl Mater Interfaces 8:2591–2599. https://doi.org/10.1021/acsami.5b10340
63. Yang G, Xu X, Yan W, Yang H, Ding S (2014) Single-spinneret electrospinning fabrication of $CoMn_2O_4$ hollow nanofibers with excellent performance in lithium-ion batteries. Electrochim Acta 137:462–469. https://doi.org/10.1016/j.electacta.2014.05.167
64. Hwang SM, Kim SY, Kim JG, Kim KJ, Lee JW, Park MS, Kim YJ, Shahabuddin M, Yamauchi Y, Kim JH (2015) Electrospun manganese-cobalt oxide hollow nanofibres synthesized via combustion reactions and their lithium storage performance. Nanoscale 7:8351–8355. https://doi.org/10.1039/c5nr01145k
65. Zhang G, Lou XW (2013) General solution growth of mesoporous $NiCo_2O_4$ nanosheets on various conductive substrates as high-performance electrodes for supercapacitors. Adv Mater 25:976–979. https://doi.org/10.1002/adma.201204128

Chapter 12
Electrospun Manganese Oxide-Based Composites as Anodes for Lithium-Ion Batteries

Jayesh Cherusseri and Sreekanth J. Varma

12.1 Introduction

The depletion of fossil fuels and increased energy consumption necessitates the need for a sustainable energy solution to mankind. Renewable energy sources including solar photovoltaics, wind, biomass, sea current, tidal, and geothermal have been explored extensively in the recent past [1–3]. But their intermittent nature of electrical energy generation has created trouble in utilizing the energy round the clock. As a matter of fact, the time constraint of these renewable energy systems cannot be solved in any other way [4, 5]. A plausible solution is to collect the as-produced energy in storage devices and use it whenever required [6–8]. The ceaseless availability of energy demands that the storage devices like supercapacitors and batteries have high energy storage capacities and power delivering capabilities. Supercapacitors are superior to batteries in delivering bursts of energy in fraction of seconds, but their low energy density is a major demerit [9–12]. On the other hand, metal-ion batteries such as lithium-ion batteries (LIBs) and sodium-ion batteries have excellent energy storage capabilities, the drawback being their poor power densities [13, 14]. Among the metal-ion batteries, LIBs have received great attention due to their superior electrochemical characteristics compared to other metal-ion counterparts [14, 15]. This aspect of LIBs has drawn a lot of attention owing to their excellent charge storage capabilities [16, 17]. Research on rechargeable LIBs has changed the face of new generation electronic devices [18]. The LIBs exhibit high energy density and are suitable choices for the next-generation hybrid electric vehicles, smart grids, load-levelling, etc. [19–21].

J. Cherusseri
NanoScience Technology Centre, University of Central Florida, Orlando 32826, USA

S. J. Varma (✉)
Department of Physics, Sanatana Dharma College, Alappuzha 688003, Kerala, India
e-mail: sjv@sdcollege.in

N. T. M. Balakrishnan and R. Prasanth (eds.), *Electrospinning for Advanced Energy Storage Applications*, Materials Horizons: From Nature to Nanomaterials,
https://doi.org/10.1007/978-981-15-8844-0_12

The main parts of a LIB are the anode, cathode, electrolyte, and solid electrolyte interface (SEI) [22, 23]. Each component has an important role in determining the performance of the LIB [24]. Our main focus in this chapter is on the synthesis, characterization, and electrochemical performance evaluation of electrospun MnO_x-based LIB anodes. The anode plays a critical role in determining the overall performance of the LIB [25]. The properties such as intercalation/deintercalation capability, electrical conductivity, reversibility in charging/discharging, etc. play crucial roles in achieving high energy density and good coulombic efficiency. Graphite is the most commonly used anode material in commercial LIBs due to its good electrical conductivity but the low theoretical capacity of 372 mAh g^{-1} necessitates the development of other materials [26]. The long diffusion pathways for the Li-ions in graphite anode are detrimental in obtaining high energy and power densities. The recent development in nanoscience and technology has led to the development of a variety of anode materials with nanostructured morphologies such as carbon nanomaterials [22], nanostructured transition metal oxides [20], silicon nanostructures [25], etc. for application in LIBs. The cost of active material also has an important role in the commercialization of LIBs as the total cost is associated with its components. This makes most of the electrode-active materials unsuitable for fabricating commercial LIBs. Due to these reasons, transition metal oxides have become potential alternatives for application in secondary batteries.

Transition metal oxides exhibiting appreciable electrical conductivity, significant electrochemical performance, and substantial chemical/electrochemical stabilities are economical, environment-friendly, and safe to use [27]. Nanostructured MnO_x offers large surface area to volume ratio, high capacity, and low diffusion lengths to the Li-ions to intercalate/de-intercalate from the electrode [28]. MnO_x-based anodes have advantages such as good working potential, low-cost, and provide excellent safety to the LIBs. However, they suffer from low coulombic efficiency, high voltage hysteresis, and high irreversible capacity. To overcome these shortcomings, MnO_x-based composite anodes have been developed [28–35]. In this chapter, we discuss the synthesis and characterization of electrospun MnO_x-based composite anodes for LIB applications. The fabrication of LIBs using the electrospun MnO_x-based composite anodes and their performance evaluation are also included. An emphasis is given to the recent developments rendered in the electrospun MnO_x-based anodes for the next generation LIBs.

12.2 Synthesis of Pristine Manganese Oxides by Electrospinning

Electrospinning is a simple, non-mechanical technique to synthesize a variety of nanofibers of materials from their solutions and dispersions [36]. This method has attracted much attention due to the reason that any material can be spun as nanofibers in large quantities [37]. Scalability is an important factor while dealing with the synthesis of nanomaterials as commercialization is possible only if the required materials can be synthesized in large quantities. Electrospinning is a versatile technique used in the preparation of electrode materials for use in secondary batteries [38]. A schematic diagram of electrospinning set-up is depicted in Fig. 12.1. In a typical electrospinning process, a polymer solution is filled in a syringe connected to a syringe pump. The syringe pump, in most cases, is an automated one where the pumping speed can be set initially and moves with the set speed. A metallic collector is used to gather the nanofibers and electrical connections are established with the needle of the syringe and the collector [39]. A D.C. power supply is used to generate a high voltage (in kilovolts), and the collector is always grounded. The high-voltage applied between the syringe needle and the collector charges the polymer solution to form a "Taylor cone" at the tip of the needle [40]. The spinning takes place only when the electrostatic force overcomes the surface tension of the polymer solution and once this voltage is attained, the jet starts to elongate towards the metallic collector. The jet undergoes stretching and whipping processes followed by the evaporation of solvents leading to the formation of fibers. The dried (almost) fibers are then collected on the metallic collector. Nanometer-sized fiber (in diameters) mats can be synthesized homogeneously using the electrospinning procedure by varying the processing parameters [38]. Uniform fibers for specific applications can be obtained on a metallic drum collector which rotates with a pre-set speed so that the fibers wind-up as a non-woven mat. The woven mats can be tailor-made using this technique with the help of multiple spinnerets and syringes. The main advantage of electrospinning is that the technique can be used to synthesize porous nanofibers which are in high demand for electrode applications in supercapacitors and LIBs. Various porogens are used to obtain the desired pore diameter and pore volume. Various parameters that affect the spinning process can be categorized as (i) polymer properties such as molecular weight, molecular weight distribution, architecture of

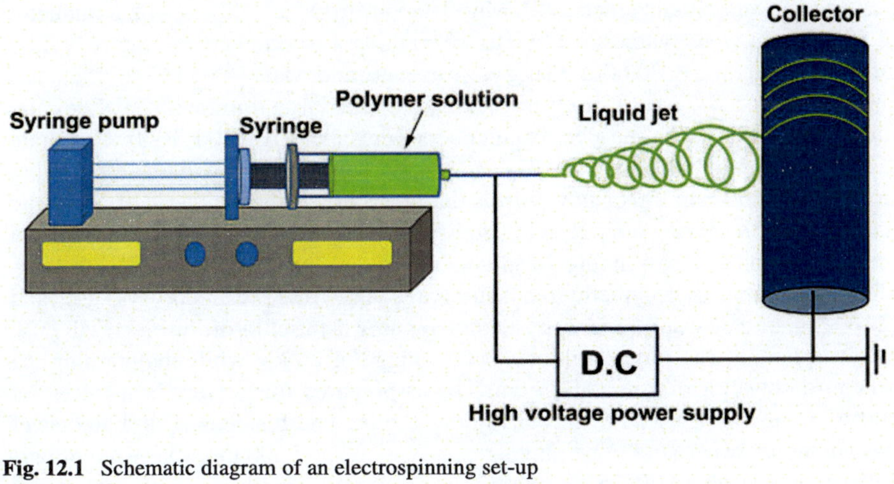

Fig. 12.1 Schematic diagram of an electrospinning set-up

the polymer (branched, linear, etc.), (ii) solution properties such as viscosity, conductivity, surface tension, etc., (iii) process parameters such as electric potential, flow rate, distance between the spinneret and the collector, motion of the drum collector, etc., (iv) ambient parameters such as temperature, humidity, air velocity, etc. Electrospinning is a comparatively low-cost technique to synthesize nanofibers and hence is widely used for manufacturing various commercial products.

Hollow MnO_x nanofibers can be synthesized by a three-step process; sol preparation, electrospinning, and calcination [41]. In this method, the first step was the preparation of the sol using polyvinylpyrrolidone (PVP), manganese acetate, and citric acid in a methanol-ethanol solvent mixture. In the second step, the as-prepared sol was electrospun on to a 5″ diameter circular mesh collector. The needle to collector distance was 8 cm and the sol pump rate was 0.01 mL/min and a voltage of 13 kV was applied. The third step was the calcination at an elevated temperature (700 °C) for 10 h. The as-synthesized fibers were found to have diameters in the range of 90–150 nm.

Mn_3O_4 in the form of nanoparticles, nanorods, and nanofibers can also be prepared by electrospinning [42]. All these nanostructures were obtained by changing the ratio of metal precursor to polymer from 0.33:1 to 2:1 in the electrospinning solution. The PVP (M_w = 360,000) and manganese acetate tetrahydrate were used as the precursors. The solution was electrospun at a voltage of 15 kV and the distance from spinneret to collector was 10 cm. When the metal precursor to polymer ratio was 0.33:1, Mn_3O_4 nanoparticles were obtained due to the destruction of fibrous morphology. When the ratio was 0.5:1, the morphology transformed to nanorods due to the breakage of nanofibers into shorter segments. A unidirectional growth of Mn_3O_4 nanofibers was obtained when the ratio was set to 1:1 and on further increasing the ratio to 2:1, dense fibers were obtained.

In addition to MnO and MnO_2 nanofibers, Mn_2O_3 and Mn_3O_4 nanofibers are also potential candidates for electrode materials [43, 44], catalysts [45, 46], nanocomposites [47], and nanomagnets [48]. Mn_2O_3 and Mn_3O_4 nanofibers with a diameter of 50–200 nm were synthesized by electrospinning a solution containing polyvinyl alcohol (PVA) (M_w = 80,000), manganese acetate, and water [49]. The solution was electrospun at a voltage of 20 kV in which a grounded iron drum with an aluminum foil collector was used. The electrospun fibers were calcined at 300–1000 °C at a rate of 240 °C/h for 10 h to obtain Mn_2O_3 and Mn_3O_4 nanofibers with diameters ranging from ~50 to 200 nm. This study reveals the tunability in the synthesis of inorganic nanofibers with varying diameters by modifying the parameters related to the sol-gel preparation and electrospinning. In another report, instead of PVA as the electrospinning solution matrix, polymethylmethacrylate (PMMA) (M_w = 35,000) was used and Mn_3O_4 nanofibers were prepared by electrospinning using chloroform and DMF [50]. Electrospinning was carried out at a D.C. voltage of 20 kV while maintaining the needle to collector distance of 25 cm. The as-prepared fibers were heat-treated at 300 °C in air for a period of 3 h to remove the organic residues. The diameter of the nanofibers was found to be ranging between 200 and 300 nm. These nanofibers with grain sizes about 20–25 nm exhibited excellent electrochemical characteristics when used as an anode in LIBs.

12.3 Synthesis of MnO_x-Based Composites by Electrospinning

One of the main bottlenecks in using pristine MnO_x as anodes in LIBs is its low intrinsic conductivity. The electrochemical performances of pristine MnO_x-based LIB anodes are affected by low intrinsic conductivity and poor cycle stability. Due to these reasons, the anode performance is not promising for a LIB with long cycle life. These critical issues can be overcome by preparing composites with highly electrically conducting materials. A nanocomposite consists of a filler and matrix in which the filler is usually conductive with excellent thermal and electrochemical stabilities and the matrix holds the fillers intact within it [51–54]. The incorporation of electronically conducting filler materials with MnO_x has proven to improve the performances of LIBs. Carbon nanomaterials such as carbon nanotube (CNT), graphene, etc. are found to be the best materials known for their electrical conductivities and are widely used for preparing composites for various applications [55–58]. They possess unique properties such as large surface area, high electronic conductivity, appreciable chemical and electrochemical resistance, and exceptional thermal stabilities. But one disadvantage of these materials is their high production cost. Although they are expensive, their unmatched electrochemical performances have made them potential candidates in energy storage devices such as supercapacitors and batteries [11, 59].

Mondal et al. have reported the synthesis of MnO_x/SnO_2 composite nanofibers by electrospinning using precursors such as PVP, tin acetate, manganese acetate (Mn:Sn = 80), and citric acid in a methanol-ethanol solvent mixture [41]. The as-prepared sol was electrospun on to a 5″ diameter circular mesh collector. The needle to collector distance was 8 cm and the sol pump rate was 0.01 mL/min and a voltage of 13 kV was applied. Later, the electrospun mats were calcined at an elevated temperature (700 °C) for 10 h. The composite fibers had diameters in the range of 60–120 nm. The authors adopted the same procedure to prepare the hybrid nanofibers of MnO_x/CNT. In this method, 0.05 wt% of CNT fibers were added into the sol mixture and ultrasonicated for an hour to get a uniform dispersion. The MnO_x/CNT hybrid nanofibers of diameters in the range of 60–90 nm were obtained. They were able to observe a higher conductivity due to the incorporation of CNTs in the hybrid nanofibers.

Nanostructured manganese dioxide (MnO_2)/carbon nanofiber (CNF) composites were synthesized by Wang et al. by combining electrospinning and hydrothermal treatment [60]. In a typical procedure, 10 wt% polyacrylonitrile (PAN) (M_w = 150,000) was dissolved in N, N-dimethylformamide (DMF) at 80 °C and stirred for 12 h. PAN nanofibers were prepared by electrospinning at an applied voltage of 25 kV, the distance between needle and collector being 25 cm. The as-prepared electrospun PAN nanofibers were then carbonized at 1000 °C for two hours under nitrogen atmosphere. These substrates were placed in 0.1 M $KMnO_4$ acidic solution for 30 and 60 min thereafter. After rinsing, these substrates were dipped in 1 M KCl solution and placed inside a Teflon-lined stainless-steel autoclave for 3 h at a temperature of 120 °C. The SEM image of pristine CNFs shown in Fig. 12.2a reveals the interconnected fibers having smooth surfaces with average diameter ranging from 50 nm to 250 nm. In the samples dipped for 30 min (MC30HT), γ-MnO_2 nanoparticles were embedded on the surface of CNFs with diameters ranging 5–20 nm as shown in Fig. 12.2b. But for the samples dipped for 60 min (MC60HT), the composite consisted of nanoneedles of γ-MnO_2/CNFs with diameters of 20–30 nm and lengths

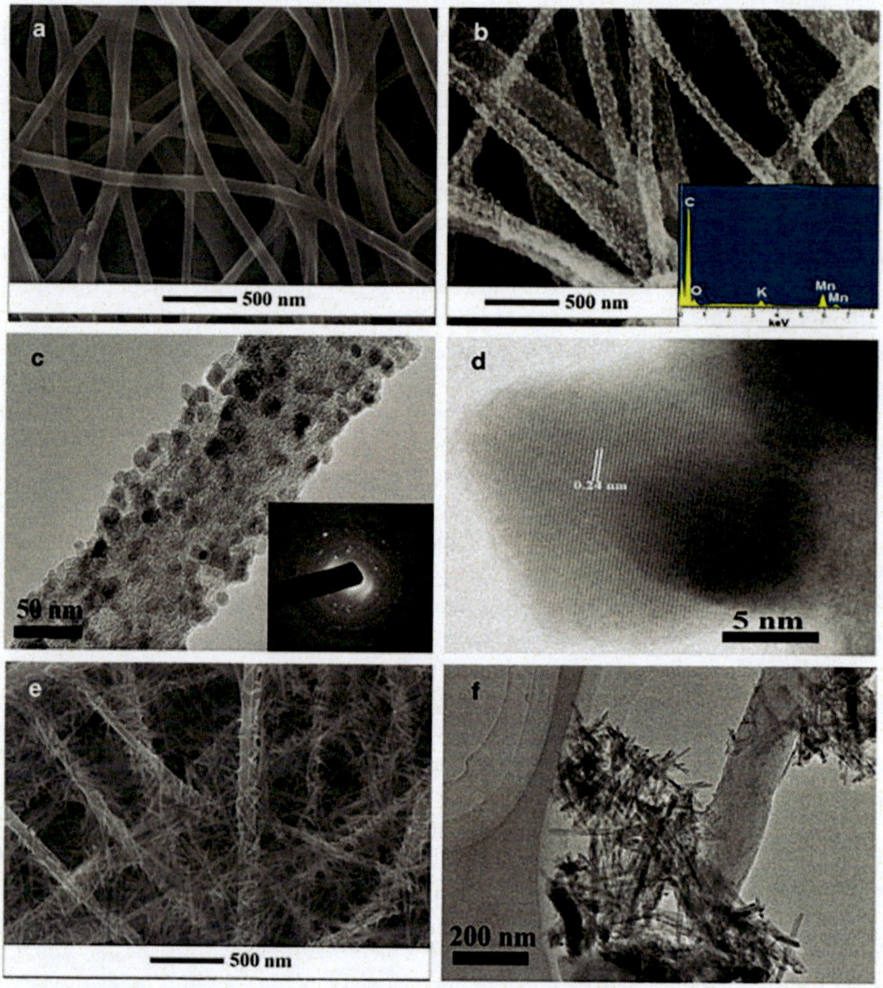

Fig. 12.2 SEM images of **a** pure CNFs, **b** MC30HT, and **e** MC60HT. TEM images of **c** MC30HT and **f** MC60HT. The insets in **b**, **c** are the corresponding EDS and SAED. **d** HRTEM of γ-MnO_2 nanoparticle. Adapted and reproduced from [60] Copyright 2012 Elsevier

of 0.3–1 μm instead of nanoparticles, as shown in Fig. 12.2e. The transmission electron microscope (TEM) images of MC30HT and MC60HT are shown in Fig. 12.2c, f, respectively. The high-resolution (HR) TEM image of the γ-MnO_2 nanoparticle in Fig. 12.2d shows a d-spacing of 0.24 nm that corresponds to the (131) plane of γ-MnO_2. The inset figures in Fig. 12.2b, c are their corresponding energy dispersive spectroscopy (EDS) and selected area diffraction analyses (SAED).

A method for preparing MnO_2-based composites by combining electrospinning and electrospraying techniques have received great attention due to the realization of 3D free-standing composite films [61]. In a typical procedure to prepare the composites films consisting of MnO, CNFs, and reduced graphene oxide (rGO), a solution was prepared by mixing MnO_2 powder in PAN/DMF and stirred for 12 h at 60 °C to obtain a homogeneous mixture. Another one was prepared by dispersing GO synthesised by Hummer's method, in ethanol. These two solutions were loaded into a plastic syringe with a stainless-steel needle for electrospinning. Electrospinning was carried out at a D.C. potential of 16 kV and at a pump rate of 1 mL h^{-1} with a needle to collector distance of 15 cm. Thus MnO_2/PAN@GO composite film was first obtained by performing electrospinning and electrospraying simultaneously. The MnO_2/PAN@GO composite film was then stabilized at 280 °C for 2 h in air atmosphere and sintered at different temperatures such as 600, 700, and 800 °C and thereafter. The as-synthesized MnO_2/CNFs@rGO (MCG) contained MnO_2 nanoparticles embedded on the surface of the nanofibers.

A new type of hybrid nanofibers based on calcium manganese oxide ($CaMnO_3$), with diameters ranging from 460 to 520 nm were prepared by electrospinning [62]. In a typical procedure, PAN ($M_w = 150{,}000$) was added to a solution of calcium nitrate and manganese (II) nitrate in DMF. The solution was electrospun at a voltage of 15 kV where the distance from needle to collector was set to 15 cm. The diameter of the $CaMnO_3$ hybrid fibers thus obtained was 816 ± 150 nm and displayed a smooth and uniform morphology as evident from Fig. 12.3a. But the $CaMnO_3$ hybrid fibers calcined at 700 °C showed a reduction in diameter as shown in Fig. 12.3b. The average diameter was reduced to 522 ± 112 nm due to the removal of PAN and other organic components. The SEM image of the $CaMnO_3$ hybrid fibers calcined at 800 °C, as shown in Fig. 12.3c, had a rough surface and a porous architecture with an average fiber diameter of 493 ± 39 nm. It was also observed that the average diameter of the fibers was further reduced to 468 ± 103 nm, when the calcination temperature was increased to 900 °C, as shown in Fig. 12.3d. From this study, it is inferred that the fiber diameter can be tuned by optimizing the calcination temperature. The bright-field TEM image and corresponding SAED pattern of $CaMnO_3$ nanofibers calcined at 900 °C are given in Fig. 12.3e, f, respectively.

Composites consisting of free-standing MnO_2 nanoflakes/CNFs were prepared by electrospinning followed by a redox reaction [63]. The porous CNFs in this study exhibited a high surface area of 1841 m^2/g. Precursors such as phenol-formaldehyde resin (as the carbon precursor) and tetraethyl orthosilicate were used in the method. A triblock copolymer Pluronic P123 was used as the porous and soft template. Porous CNFs were synthesized by electrospinning at a D.C. voltage of 18 kV with a needle to collector distance of 15 cm. The as-prepared porous CNFs were incubated at a

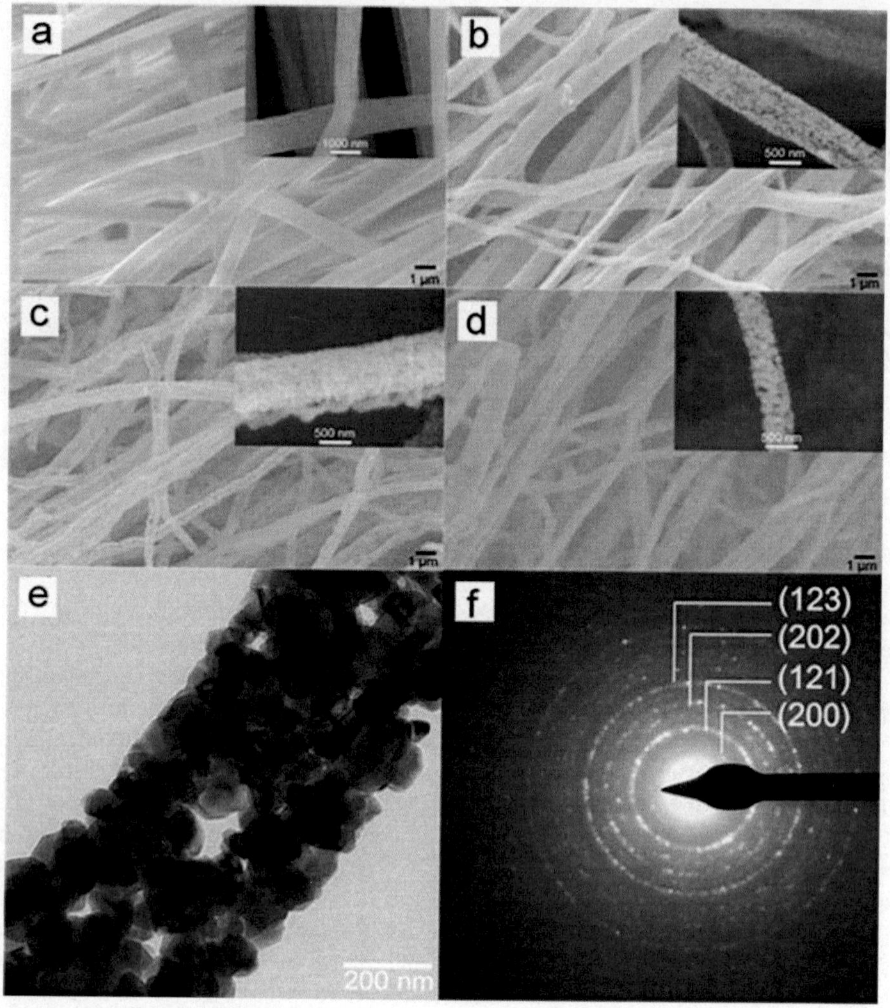

Fig. 12.3 SEM images of $CaMnO_3$ nanofibers: **a** the as-synthesized fibers, **b** fibers calcined at 700 °C, **c** fibers calcined at 800 °C, **d** fibers calcined at 900 °C. Bright-field TEM image (**e**) and corresponding SAED pattern (**f**) of $CaMnO_3$ nanofibers calcined at 900 °C. Adapted and reproduced from [62] Copyright 2016 Elsevier

temperature of 350 °C for 3 h (stabilization step) and further carbonized at 800 °C for 4 h under nitrogen atmosphere. The silica template was then removed using hydrofluoric acid (HF). The as-prepared CNFs exhibited a one-dimensional nanostructure with a uniform diameter of ~1 μm. The presence of micropores in the CNF was due to the removal of the SiO_2 after the HF treatment. Further, MnO_2 was incorporated with the porous CNFs by immersing it into a neutral $KMnO_4$ solution for 24 h while maintaining the temperature of the bath at 65 °C. Here, the porous CNFs serves as both reducing agent and porous scaffold for the growth of MnO_2 nanoflakes. Initially, MnO_2 with nanoparticle morphology was formed and it underwent a transformation to nanoflake morphology after an extended reaction on adding an excess amount of $KMnO_4$. The weight ratio of $KMnO_4$ to porous CNFs was found to influence the uniform deposition and the surface morphologies of the deposited MnO_2 as well as the specific surface area of the MnO_2/porous CNFs composite. When the ratios were 1:5 and 1:9, the flake-like MnO_2 was found to aggregate and partially coated the surface of porous CNFs. But when the ratio was set to 1:6 (termed as PCNFs-6), the surface of as-obtained PCNFs-6 was smooth and exhibited a 1D nanostructure with a uniform diameter of 1 μm. The SEM and TEM images of the PCNFs-6 are shown in Fig. 12.4a, b, respectively. The nanocomposites prepared using PCNFs-6 had fibers totally covered with large amounts of MnO_2 nanoflakes with a specific surface area of 1814 m^2/g. The coating of MnO_2 on porous CNFs possessed a thickness of about 12 nm. The SEM images of the MnO_2/PCNFs-6 nanocomposite fibers are shown in Fig. 12.4c, d. It can be noted from these images that the MnO_2/PCNFs-6 nanocomposite fibers exhibit a porous structure like in the PCNFs-6. The thickness of the MnO_2 shell coated on the PCNFs was found to be ~12 nm. The TEM images of MnO_2/PCNFs-6 nanocomposite fibers shown in Fig. 12.4e, f reveals an interplanar spacing of the lattice fringes of ~0.7 nm.

In another report, free-standing coaxial CNFs/MnO_2 nanocomposites were prepared by redox deposition and electrospinning [64]. A schematic diagram of the preparation procedure is given in Fig. 12.5a. An interconnected network was obtained by depositing birnessite-type MnO_2 on electrospun porous CNFs with a large surface area. The digital photographs of the as-prepared composite sheet before and after carbonization are shown in Fig. 12.5b and the inset shows the cross-sectional thickness of the compact film. This method has envisaged the maximum utilization of the surface area to deposit MnO_2 on the surfaces of CNFs. In this method, PAN (M_w = 150,000) was used as the carbon precursor to synthesize porous CNFs. Porous CNFs were synthesized by electrospinning at a D.C. voltage of 18 kV with a needle to collector distance of 15 cm. The as-prepared porous CNFs were incubated at a temperature of 280 °C for 2 h (stabilization step) and carbonized at 1000 °C for 2 h under nitrogen atmosphere. The SEM image of as-synthesized CNFs is shown in Fig. 12.5c. Further, the MnO_2 deposited on the porous CNFs via an in situ redox reaction between permanganate-ions and CNFs using dip-coating. The morphology of the as-deposited MnO_2 was examined by SEM imaging and Fig. 12.5d shows the CNFs/MnO_2 with a nanowhisker-type morphology. The MnO_2 nanowhisker-shell coated on the CNFs had a thickness of 40 nm at a dipping time of 60 min.

A similar strategy was adopted to synthesize MnO_2/CNTs-embedded CNF (MnO_2/CNTs-CNFs) nanocomposites [65]. The SEM images of pristine CNFs, as shown in Fig. 12.6a show fibers with an average diameter of 200 nm and lengths of a few hundreds of microns. PAN powders dissolved in the CNTs-dispersed DMF solution was used to prepare the MnO_2/CNTs composites by electrospinning. A curled morphology with a rough surface can be observed in the SEM image of MnO_2/CNFs depicted in Fig. 12.6b. In the case of CNFs/CNTs composites, the PAN to CNTs ratio was set as 10:1 (in wt%). A D.C. voltage of 25 kV was applied and the

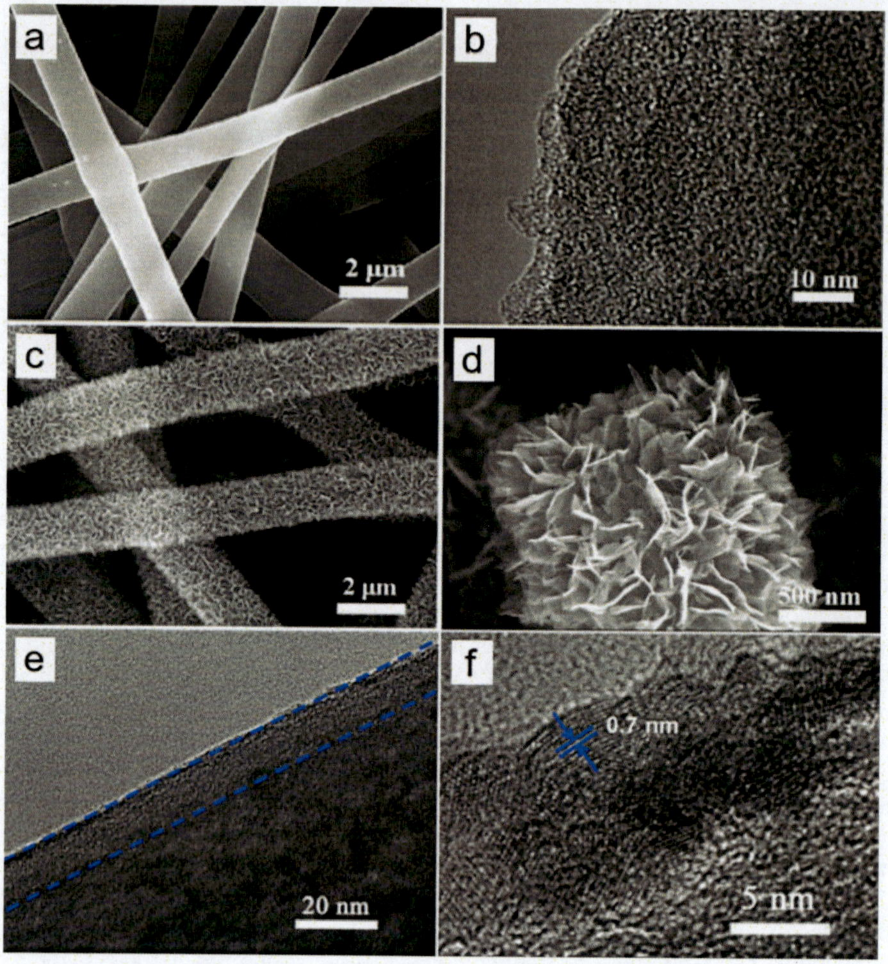

Fig. 12.4 **a** SEM and **b** TEM images of representative PCNFs. **c**, **d** SEM and **e**, **f** TEM images of MnO_2/PCNFs-6. Adapted and reproduced from [63] Copyright 2015 Elsevier

distance between the spinneret and the collector was set to 25 cm for the electrospinning process. By selecting the CNTs-CNFs instead of pure CNFs helped to increase both the conductivity and the specific surface area. The SEM image of CNFs/CNTs composite is illustrated in Fig. 12.6c. A higher conductivity value of 24.1 S/cm was obtained for the CNTs-CNFs whereas it was only 11.6 S/cm for the pure CNFs. Hierarchically porous MnO_2 was deposited on the porous network of CNTs-CNFs with nanorod morphology as shown in Fig. 12.6d. The MnO_2 nanorods were 2–3 nm in diameter and individually coated on to the CNTs-CNFs without agglomeration with an average coating thickness of ~15–30 nm.

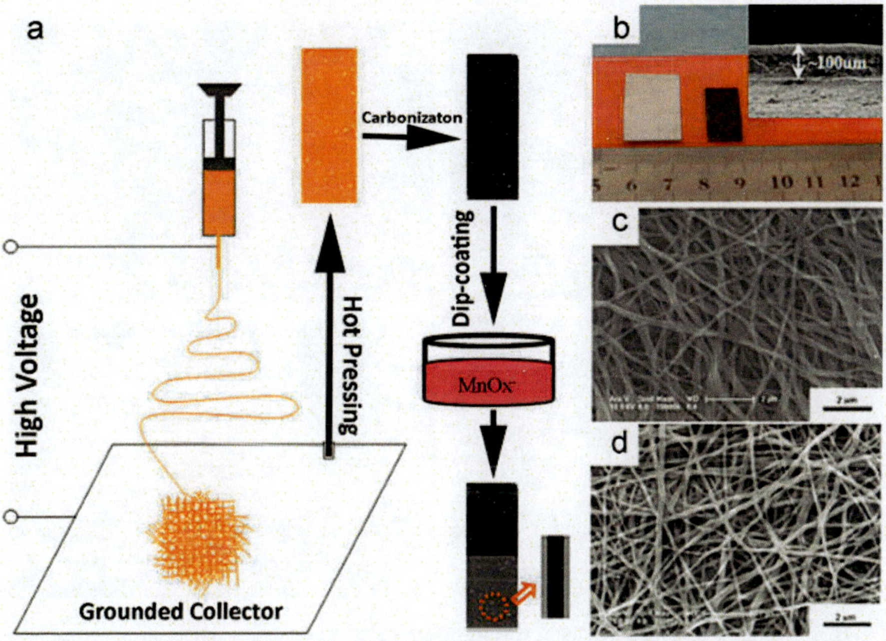

Fig. 12.5 **a** Schematic illustration of the fabricating procedures of composites. **b** A photograph of the compact sheet before and after carbonization (inset is a SEM image of cross-sectional thickness). Low magnification SEM images of the **c** CNFs (**d**) and CNFs/MnO_2 composites. Adapted and reproduced from [64] Copyright 2011 Elsevier

Hierarchically porous MnO_2/CNF composites with hollow cores can be easily prepared by electrospinning using PMMA as a stabilizer in PAN solution [66]. The concentration of the PMMA in the electrospinning solution helps in dispersing MnO_2 particles uniformly in the solution and aids the formation of numerous hollow cores. In this process, the electrospinning solution was prepared by dispersing 3 wt% of $MnCl_2$ in PAN/PMMA with two different ratios (7:3 and 8:2) in DMF solvent. Electrospinning was performed at a D.C. voltage of 20 kV with a 13 cm distance between the syringe needle tip and the metal collector. PMMA as a stabilizer was found to inhibit the possible aggregation of amorphous MnO_2 nanoparticles in the electrospinning solution. Average diameters of 700 nm and 490 nm were exhibited by the samples prepared from PAN:PMMA ratios of 10:0, and 8:2, respectively, and found to be smooth throughout the length.

Hollow CNFs with diameters about ~300–400 nm and an inner diameter of 100 nm can be prepared from electrospun MnO_2/CNF composites by hydrothermal treatment [67]. The growth of MnO_2 nanoparticles occurs via. the reduction reaction between permanganate ions and a possible low-level reduction reaction between the manganese and permanganate ions.

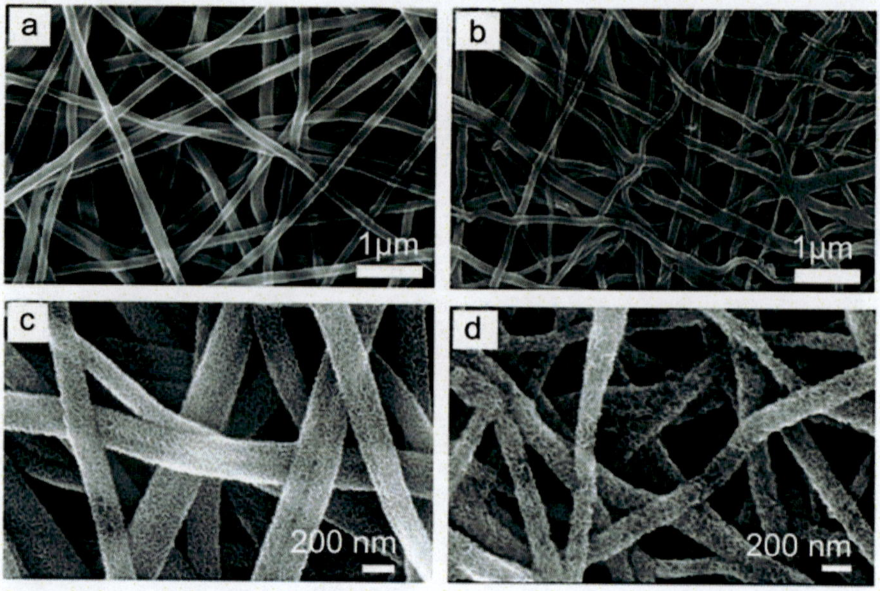

Fig. 12.6 SEM images of **a** pure CNFs, **b** MnO_2/CNFs, **c** CNTs–CNFs, and **d** MnO_2/CNTs–CNFs. Adapted and reproduced from [65] Copyright 2012 Elsevier

The synthesis of hierarchical one-dimensional composite consisting of carbon, MO_x (M = Mn, Cu, Co), and MnO_2 by a two-step process involving electrospinning and thermal treatment was reported by Nie et al. [68] CNF/MO_x@MnO_2 nanofibers were prepared by electrospinning using precursors such as PAN (M_w = 150,000), $Mn(Ac)_2.4H_2O$ in DMF solvent. To prepare CNF/CoO_x@MnO_2 composite nanofibers, $Co(Ac)_2.4H_2O$ and for CNF/CoO_x@MnO_2 composite nanofibers, $Cu(Ac)_2.H_2O$ were used along with $Mn(Ac)_2.4H_2O$. In the case of C/CuO_x/MnO_2 composite nanofibers, no CuO_x nanoparticles were found inside the CNFs due to the interaction between the Cu^{2+} and –CN groups on PAN chains. Instead, all nanoparticles were deposited on the surface of the CNFs resulting in a core-shell architecture. But in the case of C/CoO_x/MnO_2 composite nanofibers, CoO_x nanoparticles were encapsulated inside the CNFs which also exhibits a core/shell architecture as seen from Fig. 12.7a. The HRTEM image indicates a lattice spacing of ~0.66 nm (Fig. 12.7b). The SAED pattern (Fig. 12.7c) illustrates the (006) and (119) planes of the hexagonal polycrystalline MnO_2. The EDX spectrum (Fig. 12.7d) and the HAADF-STEM EDX mapping (Fig. 12.7e) indicate the composition and distribution of various elements in the C/CoO_x/MnO_2 composite nanofibers, respectively. The high conductivity of the as-prepared composite fibers is beneficial for using them as potential electrodes in energy storage applications.

Composite fibers with hollow nanostructures comprising of manganese–cobalt oxide (MCO) and cobalt-manganese oxide (CMO) were synthesized via electrospinning followed by combustion reactions during calcination [69]. Initially, the

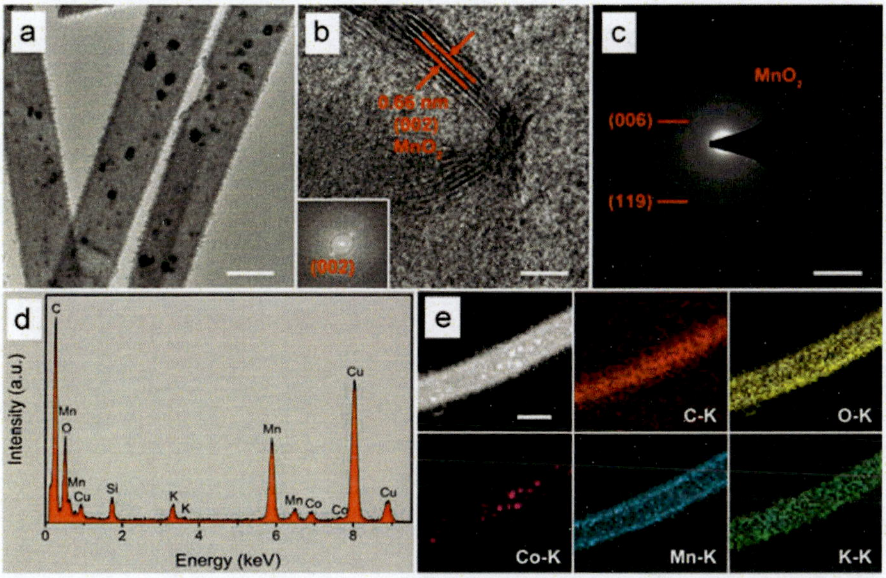

Fig. 12.7 TEM images (scale bar: **a** 500 nm and **b** 5 nm) and corresponding FFT (inset), **c** SAED pattern (scale bar: 5 nm^{-1}), **d** EDX spectrum, and **e** HAADF-STEM analysis and mapping (scale bar: 500 nm) of C/CoO_x/MnO_2 composite nanofibers. Adapted and reproduced from [68] Copyright 2018 Elsevier

precursor solutions were prepared by adding cobalt acetate and manganese acetate with a molar ratio of, 2:1 for preparing the MCO fibers and, 1:2 for the CMO fibers in DMF solvent along with PVP under constant magnetic stirring. Electrospinning was carried out using these solutions and dried at 150 °C thereafter to obtain the hybrid nanofibers.

A novel composite structure consisting of nitrogen-containing spindle-like CNFs anchored with cubic MnO phase was prepared by electrospinning followed by pre-oxidation and carbonization procedures [70]. The as-prepared composites exhibited a spindle-like morphology. These entangled beaded nanofibers help in developing a robust multilayer network and were used directly as a binder-free anode for fabricating high-performance LIB. A similar strategy was adopted to synthesize free-standing MnO nanowires/CNF composite membranes via electrospinning followed by carbonization procedures [71]. Initially, MnO_2 nanowires were synthesized hydrothermally using manganese sulfate monohydrate and $KMnO_4$. These MnO nanowires were then added into the electrospinning solution containing 8 wt% PAN/DMF and electrospun at a D.C. voltage of 20 kV with a needle to collector distance of 20 cm. Samples were prepared by changing the amount of MnO_2 nanowires with respect to PAN such as 10 wt%, 20 wt%, and 30 wt%, and were named MnP-1, MnP-2, and MnP-3, respectively. The as-obtained electrospun MnO nanowires/CNF composite membranes were stabilized at 280 °C for 2 h in air atmosphere and then carbonized at 1000 °C for 1 h under nitrogen atmosphere.

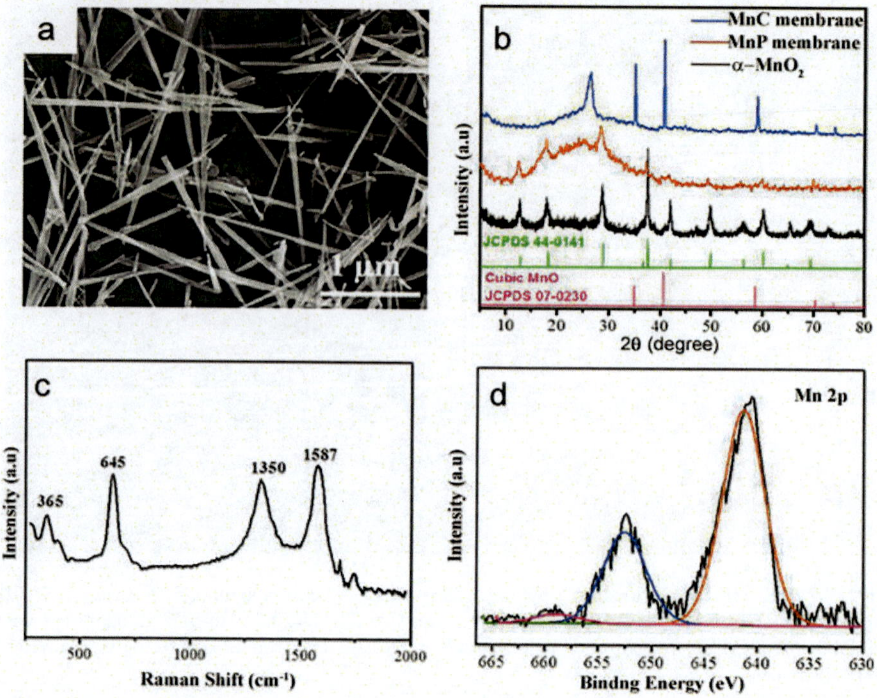

Fig. 12.8 **a** SEM image of MnO_2 nanowires; **b** XRD spectra of MnO_2 nanowires and MnP, MnC membranes prepared via. electrospinning; **c** Raman spectrum and **d** XPS spectrum of MnC membrane. Adapted and reproduced from [71] Copyright 2015 Springer Nature

The carbonized samples were named as MnC-1, MnC-2, and MnC-3, respectively. The average length of MnO_2 nanowires was ~1 to 2 μm with an average diameter of 20–30 nm, as shown in Fig. 12.8a. The XRD spectra of the samples shown in Fig. 12.8b confirm the formation of α-MnO_2 phase (JCPDS file no. 44-0141). The Raman (Fig. 12.8c) and XPS spectra (Fig. 12.8d) of MnO nanowires/CNF composite membrane illustrate the bonding features between MnO nanowires and the CNFs within the composite membrane.

Two dimensional rGO with high electronic conductivity is a potential additive to prepare LIB anodes. This additive helps not only in increasing the conductivity but also serves a strong backbone to give the structural integrity and restricts the possible pulverization process during charge-discharge. Simultaneous electrospinning electrospraying technique was adopted by Wang et al. to prepare CNFs/MnO/rGO composite thin films to perform the role of a high-performance anode for Li-ion batteries [18]. In a typical procedure, PAN/DMF was prepared by mixing 0.8 g PAN in 10 ml DMF and 1 g of $MnAc_2.4H_2O$ was dissolved thereafter. Another precursor solution was also prepared by dispersing graphene oxide in ethanol separately. These two solutions were mixed thoroughly, loaded to the plastic syringe, and electrospun at a D.C. voltage of 16 kV at a current collector-needle distance of 15 cm maintaining

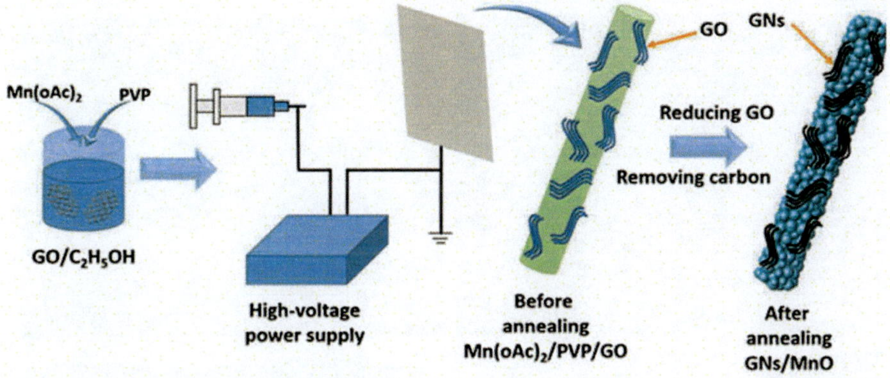

Fig. 12.9 Schematic showing the preparation of GNs/MnO nanowires composite. Adapted and reproduced from [72] Copyright 2016 American Chemical Society

a pump rate of 1 mL h^{-1}. The $MnAc_2$/PAN/GO composite fibers were stabilized via heat-treatment at 260 °C in air for a period of 2 h and sintered at 700 °C for 3 h under argon atmosphere. The as-obtained samples were CNFs/MnO/rGO composites and control samples were prepared using MnO/CNFs and pure MnO samples in a similar way.

Sun et al. have prepared graphene nanosheets (GNs) reinforced-MnO (GNs/MnO) nanowires composite by electrospinning [72]. A schematic showing the preparation of GNs/MnO nanowires composite is given in Fig. 12.9. In this procedure, 1 g of PVP was dissolved in a solution containing 10 mg of GO flakes in 10 mL ethanol. This solution was vigorously stirred and then mixed with 0.5 g of $Mn(oAc)_2 \cdot 4H_2O$. The mixture was heated at 55 °C for 15 min. This solution was then taken in a syringe with stainless steel needle and electrospun at a D.C. voltage of 16 kV while maintaining a constant distance of 16 cm between the needle and the collector. The as-obtained $Mn(oAc)_2$/PVP/GO nanowires composite was stabilized at 80 °C for 12 h in air and underwent heat-treatment at 700°C under argon/hydrogen atmosphere for 1 h. Control sample of pure MnO nanowires was also prepared in a similar manner. The FESEM images of pure MnO nanowires and GNs/MnO nanowires composite are shown in Fig. 12.10a, b, respectively, and the inset images show the FESEM images at high-magnification. The surface of pristine MnO nanowires is found to have a rough surface whereas a smooth surface can be visualized for the GNs/MnO nanowires composite. The GNs were embedded inside the MnO nanowires other than wrapping around the nanowires as evident from the TEM images (Fig. 12.10c, d).

A new strategy of synthesizing Mn-based hybrid fibers by gradient electrospinning followed by controlled pyrolysis has achieved much attention recently [73]. Different Mn-based hybrid fibers like $Na_{0.7}Fe_{0.7}Mn_{0.3}O_2$ mesoporous nanotubes, $LiNi_{1/3}Co_{1/3}Mn_{1/3}O_2$ mesoporous nanotubes, $LiMn_2O_4$ mesoporous nanotubes, and

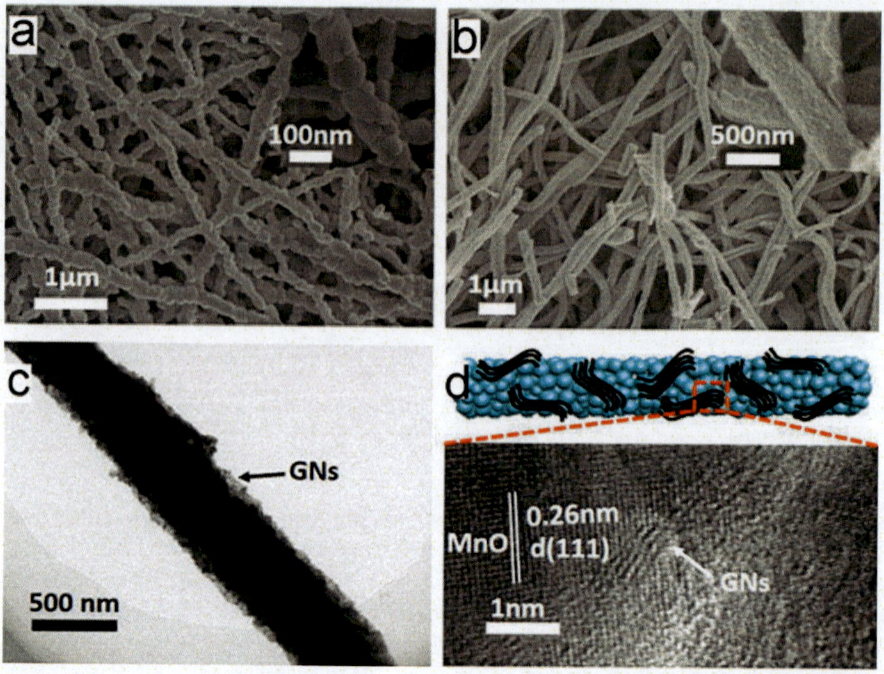

Fig. 12.10 FESEM images of the **a** pure MnO nanowires (inset shows the FESEM image at high magnification), **b** GNs/MnO nanowires composite after high-temperature treatment, **c** Low-magnification and **d** high-resolution TEM images of the GNs/MnO nanowires composite. Reprinted with permission from [72] Copyright 2016 American Chemical Society

pristine MnO_2 mesoporous nanotubes were prepared using this approach. Nanostructured morphologies such as mesoporous nanotubes and pea-like nanotubes were synthesized by this method. The method employs a precursor solution of PVA of different weights such as low (98–99% hydrolyzed), medium (86–89% hydrolyzed), and high (98–99% hydrolyzed) with a ratio of 3:2:1 (in wt%), in water. The precursors of each metal oxide were added to this solution subsequently. In the synthesis of $Na_{0.7}Fe_{0.7}Mn_{0.3}O_2$ mesoporous nanotubes, precursors such as $NaNO_3$, $Fe(NO_3)_3.9H_2O$, and $Mn(CH_3COO)2.4H_2O$ in the molar ratio of 7:7:3 was electrospun at a D.C. voltage of 20 kV. The electrospun fibers were sintered at 300 °C for 5 h under argon atmosphere.

To synthesize $LiNi_{1/3}Co_{1/3}Mn_{1/3}O_2$ mesoporous nanotubes, precursors such as $CH_3COOLi.2H_2O$, $Ni(CH_3COO)_2.4H_2O$, $Mn(CH_3COO)_2.4H_2O$, and $Co(CH_3COO)_2.4H_2O$ were used in the molar ratio of 3:1:1:1 and then electrospun at a D.C. voltage of 20 kV. These fibers were annealed at 700 °C for 5 h in the air to obtain the resultant product. The synthesis of $LiMn_2O_4$ mesoporous nanotubes involved the use of the precursors $CH_3COOLi.2H_2O$ and $Mn(CH_3COO)_2.4H_2O$ in the molar ratio 1:2. The electrospinning was carried out at a D.C. voltage of 18 kV and the so obtained fibers were annealed at 700 °C for 5 h in air. In another electrospinning experiment, pea-like nanotubes of $Na_{0.7}Fe_{0.7}Mn_{0.3}O_2$, were synthesized from precursors such as $NaNO_3$, $Fe(NO_3)_3.9H_2O$, and $Mn(CH_3COO)_2.4H_2O$ in the molar ratio of 7:7:3 spun at a D.C. voltage of 20 kV. The electrospun fibers were first stabilized at 300 °C for 1.5 h in air and further annealed at 700 °C for 8 h under argon atmosphere.

12.4 Electrochemical Performances of Electrospun Mno_x-Based Composites in LIB Anodes

To meet the ever-growing demands in hybrid electric vehicles, the energy and power densities of LIBs should be improved significantly. Specific capacity is not the only pertinent parameter in the case of LIBs. The unstable SEI layer formed after the first charging cycle deteriorates the overall performance of the LIBs. Hence, a control in the formation of SEI layer is mandatory for the LIBs to exhibit high performances. The Li-intercalation and de-intercalation processes lead to a volume change and subsequent damage of the SEI layer. The breakage of the SEI layer not only diminishes the capacity of the LIB but also degrades its cycling stability. Tremendous efforts have been made to attain the desired energies and power densities by adopting various strategies such as preparing composite electrodes consisting of transition metal oxides and carbon nanomaterials. LIB anodes using composite nanostructures have the advantage of improving the performance of the battery as well as in preventing the dramatic volume change that happens during the charge/discharge cycles. MnO_x-based LIB anodes due to their high theoretical capacity of 755 mAh g^{-1}, low voltage hysteresis, and low conversion potential [74, 75] while carbon nanomaterials are appropriate to prepare the composite anodes because of their ability to form controlled SEI layers during the initial cycles. There are many reports on the preparation of such composite fibers which, when incorporated into the LIBs, improve the overall performances. The electrochemical performances of the pristine MnO_x-based nanofibers synthesized by electrospinning method are not very attractive when compared with their composite counterparts. For example, the Mn_3O_4 nanofibers synthesized via electrospinning exhibited a discharge capacity of ~600 mAh g^{-1} for the preliminary cycles [50]. The specific capacity retention was also found very poor in which only 450–500 mAh g^{-1} was retained after 50 cycles.

A simple and facile method of synthesizing MnO_x/CNF nanofiber composite is to prepare CNF first via electrospinning, followed by carbonizing the electrospun mat and then depositing MnO_x on to the CNFs mat via electrochemical deposition route. Lin et al. have prepared a composite by adopting the same strategy by electrospinning 8 wt% PAN/DMF solution followed by carbonization at 600 °C for 8 h in argon atmosphere [76]. These electrospun CNFs mats were then used as conducting substrates for the electrochemical deposition of MnO_x. The deposition was carried out in a three-electrode cell set-up where CNFs mat, platinum wire, and Ag/AgCl (in 4 M KCl) were used as a working electrode, counter electrode, and reference electrode, respectively. Prior to the MnO_x deposition, the CNFs mat was electrochemically cycled in 1 M H_2SO_4 for surface functionalization. This step helps in the surface oxidation and linkage of various surface functional groups, such as quinoid (O), hydroxy (–OH), and carboxyl (–COOH) on the surface of CNFs mat. Electrochemical cycling in H_2SO_4 is a versatile technique to produce defect sites on the substrate, which is proven to enhance the deposition of MnO_x nanoparticles [77]. MnO_x nanoparticles were deposited on the CNFs mat by applying a potential of –0.2 V (vs. Ag/AgCl) in an electrochemical bath containing 0.1 M $Mn(CH_3COO)_2$ and 0.1 M Na_2SO_4 solution at different deposition durations such as 2.5, 5, 10, and 20 h. After 2.5 h of deposition, the size of the MnO_x reached ~21 nm and started to grow as the deposition duration increases. The size increased to ~27–37 nm after 10 h of deposition and was found to agglomerate after 20 h.

Electrospun spindle-like CNFs anchored with cubic MnO is an ideal candidate to be used as an anode in LIB to exhibit a high specific capacity [70]. The significance of these spindle-like CNFs anchored with cubic MnO is not only that it can be used directly as a binder-free anode for LIBs, but also boosts the overall capacity, rate capability, and cycling stability. A CR 2032-type coin cell was fabricated using nitrogen-containing spindle-like CNFs anchored with cubic MnO as the working electrode and Li-foil as the counter electrode. The CV curves depicted in Fig. 12.11a indicate the electrochemical performance of the cell during its first, second, and third cycles. In the first cycle, the peak positioned at about 0.34 V indicates the reduction of Mn^{2+} to Mn° and in the second and third cycles, this peak shifted to 0.37–0.39 V which may be due to the change in its microstructure after the first lithiation process. The galvanostatic charge/discharge (GCD) curves shown in Fig. 12.11b shows typical curves for a LIB. From the GCD curves, a discharge capacity of 1188.6 mAh g^{-1} was calculated, which is much higher than that of pristine MnO (755 mAh g^{-1}) and pristine carbon (372 mAh g^{-1}) as shown in Fig. 12.11c. The capacity was found to decrease after the first cycle due to the electrolyte degradation and the formation of SEI films on the electrode surface. After 200 cycles, the LIB delivered a high reversible capacity of 875.5 mAh g^{-1} at a current density of 0.2 A g^{-1} as shown in Fig. 12.11d. The LIB also exhibited a reversible specific capacity of 591 mAh g^{-1} even after 200 cycles with a coulombic efficiency of >98.9% (Fig. 12.11e).

Free-standing MnO_2 nanowires/CNF composite membranes-based LIB anode prepared by electrospinning followed by calcination displayed a reversible capacity of 987.3 mAh g^{-1} at a current density of 0.1 A g^{-1} (Fig. 12.12a) [71]. The Nyquist plots depicted in Fig. 12.12b indicate that the MnO_2 nanowires/CNF composite membrane prepared using 30wt% of MnO_2 nanowires (MnC-3) exhibits a higher charge transfer resistance when compared to that prepared with 10 wt% (MnC-1) and 20wt% (MnC-2). This is a sign of the continuous formation of SEI layer which can eventually result in a slower charge transfer rate. The rate capability of the anode was found to be 406.1 mAh g^{-1} at a current density of 3 A g^{-1} and retained an excellent cycling performance for more than 280 cycles with a capacity retention of 655 mAh g^{-1} when cycled at a current density of 0.5 A g^{-1}, as shown in Fig. 12.12c. A comparison of the discharge capacity shown by the various MnO_x-based LIB

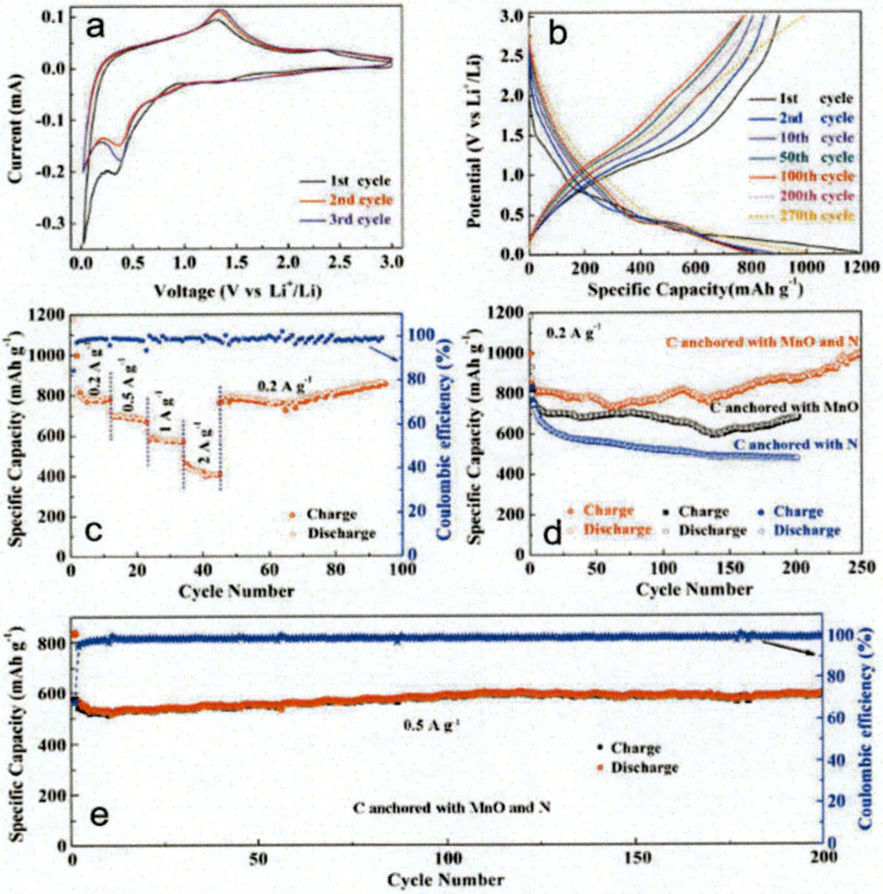

Fig. 12.11 **a** CV curves, **b** GCD curves, **c** capacity at different current densities, and **d**, **e** cycling study of the nitrogen-containing spindle-like CNFs/MnO composite anode. Adapted and reproduced from [70] Copyright 2016 Liu et al.; licensee Beilstein-Institut

anodes is depicted in Fig. 12.12d. A PET encapsulated flexible LIB cell was fabricated using the free-standing MnO_2 nanowires/CNF composite membrane as anode (Fig. 12.12e). The composite anode demonstrated an appreciable electrochemical performance even after a severe bent, as seen in Fig. 12.12f. This study proclaims the development of high-performance free-standing MnO_2 nanowires/CNF composite membrane-based anode for flexible LIB applications.

A composite anode of MnO_x/CNFs nanofiber prepared by electrochemically depositing MnO_x nanoparticles on the CNFs mat showed a discharge capacity >600 mAh g^{-1} at a deposition duration of 5 h [76]. The capacity was found to deteriorate when the deposition time was increased by more than 5 h due to the aggregation

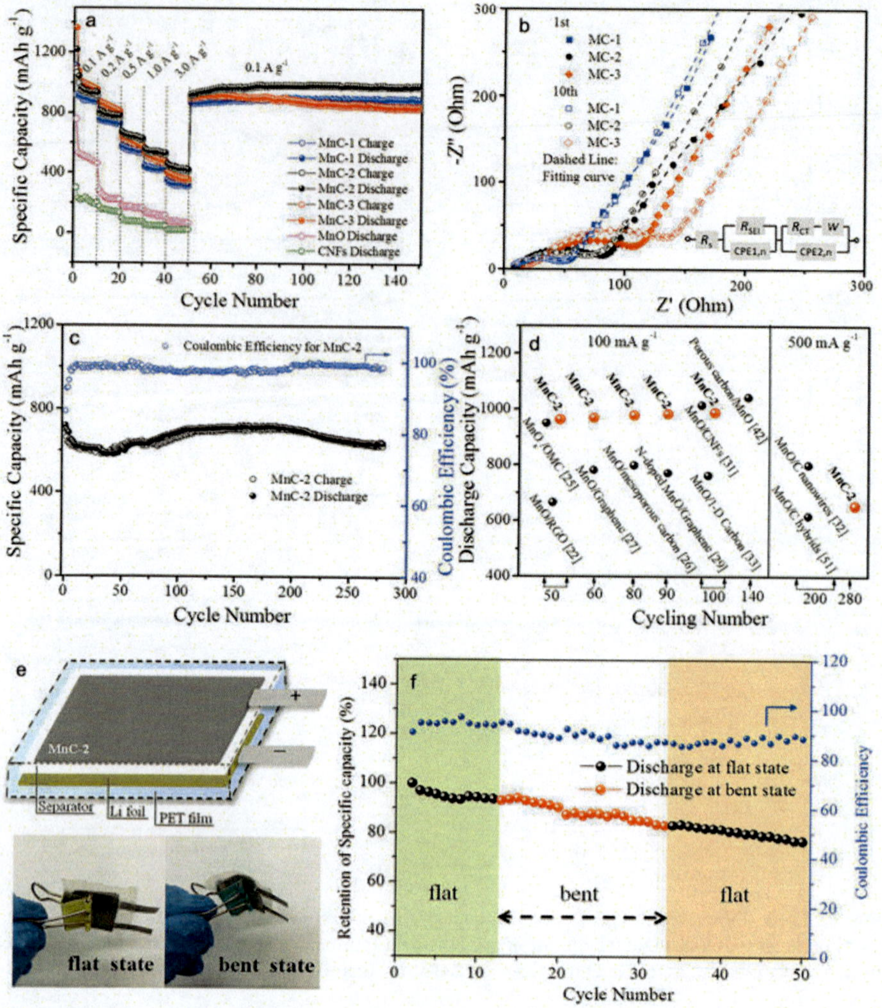

Fig. 12.12 **a** Rate capacities at different current densities and capacity at different cycles for MnC and MnO; **b** Nyquist curves of MnC samples at 1st and 10th cycle and their respective fittings with an appropriate electric equivalent circuit; **c** cycling performance of MnC-2 with columbic efficiency; **d** a comparison of the discharge capacities of MnC-2 with reported MnO-based materials at various cycles; **e** Scheme of the assembled flexible cell encapsulated by PET film and photograph of the battery tested at flat and bent states; **f** cyclic performance of the battery under flat and bent states. Reprinted with permission from [71] Copyright 2015 Springer Nature Publishing

of MnO_x nanoparticles on the CNFs surface. This study reveals that electrospinning followed by electrodeposition of MnO_x nanoparticles is a versatile method of preparing anodes for LIBs with high reversible capacity, good capacity retention, and excellent structural integrity during cycling.

An enhanced electrochemical performance was observed for the free-standing MnO_2/CNFs@rGO (MCG) composite films prepared by combining electrospinning and electrospraying techniques [61]. The MCG composites were obtained by sintering at different temperatures. The loading mass of MnO was fixed as ~19.6, 21.9, and 43.9 wt% for the samples sintered at 600, 700, and 800 °C, respectively, and were named MCG-600, MCG-700, MCG-800, respectively. Electrochemical performances of the half cells were performed by fabricating CR 2025 coin cells in which Li metal acts as the counter electrode and Celgard 2400 as the separator. The discharge-charge curves of the coin cell at different current densities (Fig. 12.13a) show that the MCG-800 sample exhibits excellent rate capability and obtained a

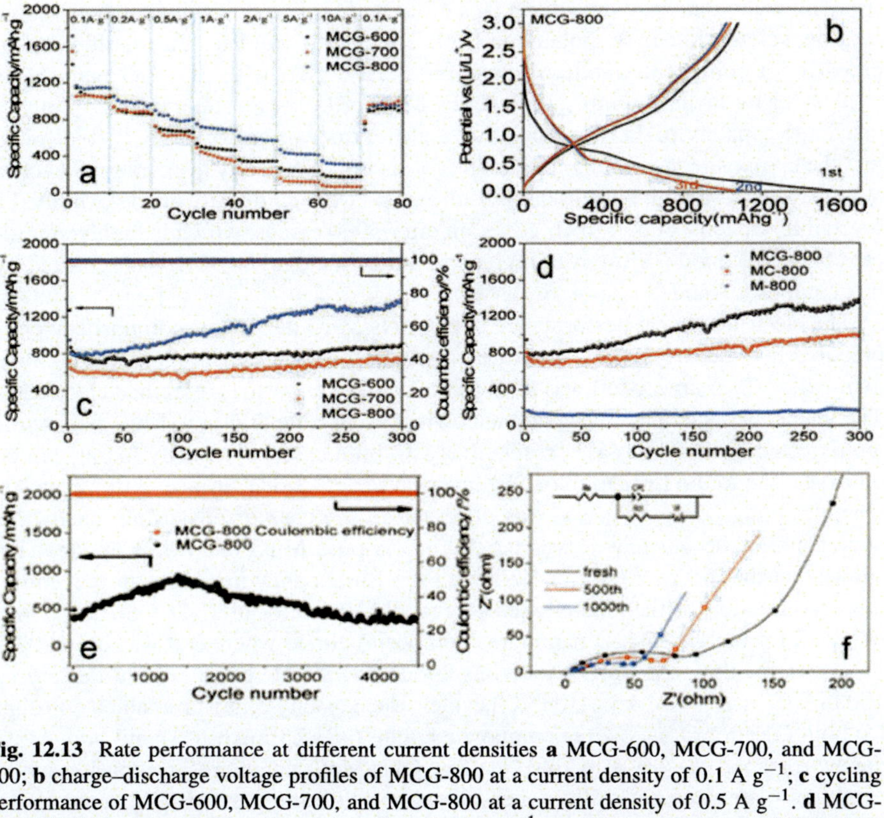

Fig. 12.13 Rate performance at different current densities **a** MCG-600, MCG-700, and MCG-800; **b** charge–discharge voltage profiles of MCG-800 at a current density of 0.1 A g^{-1}; **c** cycling performance of MCG-600, MCG-700, and MCG-800 at a current density of 0.5 A g^{-1}. **d** MCG-800, MC-800, and M-800 at a current density of 0.5 A g^{-1}; **e** cycling performance of MCG-800 at a current density of 5 A g^{-1}. **f** Nyquist plot of MCG-800. Reprinted with permission from [61] Copyright 2017 The Royal Society of Chemistry

specific capacity of 1148 mAh g^{-1} at a current density of 0.1 A g^{-1}. The MCG-800 sample exhibits an initial discharge and charge capacities of 1719 and 1159 mAh g^{-1}, respectively (Fig. 12.13b). A dominant plateau observed at ~0.6 V corresponds to the reaction between Li^+ and MnO and a minor plateau observed at 0.3–0.02 V corresponds to the formation of SEI film on the MCG-800 composite electrode surface. Figure 12.13c, d shows the discharge capacities obtained at different cycle numbers for the coin cells prepared using MCG-600, MCG-700, and MCG-800 composites. The MCG-800 cell exhibited a high reversible capacity of 1176 mAh g^{-1} with a coulombic efficiency ~100%, higher than the other two composite samples, and also their pristine counterparts. After 4000 cycles, the MCG-800 composite cell showed capacity retention of 332 mAh g^{-1} when cycled at a current density of 5 A g^{-1} (Fig. 12.13e). The Nyquist plots obtained at different cycles of MCG-800 cells are illustrated in Fig. 12.13f. The resistance is found to increase after 500 cycles and reduces on further cycling up to 1000 cycles.

The composite anodes prepared by simultaneous electrospinning and electrospraying technique have the advantages of being highly electrically conductive and flexible, suitable for LIB application [18]. This report validates the significance of preparing a composite anode by combining 1D CNFs with 2D rGO nanosheets in a layer-by-layer assembly. The CNFs/MnO/rGO composite anode exhibited a discharge capacity of 1118 mAh g^{-1} at a current density of 0.1 A g^{-1}. It is found that the composite was able to retain 98% of its initial capacity even after 80 cycles. When cycled at a high current density of 5 A g^{-1}, the composite anode exhibited a reversible capacity of 574 mAh g^{-1} even after 3000 cycles, which is highly significant for a LIB anode. Moreover, high coulombic efficiency of 99% was achieved by this composite anode.

The electrochemical performance of the GNs/MnO nanowires composite anodes prepared by electrospinning technique was monitored by assembling CR2032 type coin-cells [72]. Pure Li-foil was used as the counter as well as reference electrodes. The GCD curves of the GNs/MnO nanowires composite and pure MnO nanowires are depicted in Fig. 12.14a, b, respectively. In both cases, a capacity fading can be observed due to the formation of SEI layer on the electrode surface. After 2 cycles, a small voltage drop of 0.1 V was observed in the case of GNs/MnO nanowires composite anode whereas it was 0.5 V for the pure MnO nanowires anode. This shows that the incorporation of GNs with the MnO nanowires helps in preventing the pulverization of the LIB anode. A reversible capacity of 1347.5 mAh g^{-1} has obtained for the GNs/MnO nanowires composite anode whereas it was only 999.7 mAh g^{-1} for the pure MnO nanowires anode when tested at a current density of 100 mA g^{-1} (Fig. 12.14c). After 200 cycles, the capacity almost remained constant for both GNs/MnO nanowires composite anode (~815.3 mAh g^{-1}) and pure MnO nanowires anode (~489.4 mAh g^{-1}). The GNs/MnO nanowires composite anode exhibited better rate capability when compared with the pure MnO nanowires, as can be seen from Fig. 12.14d. A specific capacity of 285.1 mAh g^{-1} was obtained at a very high current density of 2500 mA g^{-1}. After continuous cycling for further 10 cycles at a current density of 50 mA g^{-1}, an average specific capacity of 1000.19 mAh g^{-1} was retained, which shows the excellent rate capability of the LIB anode.

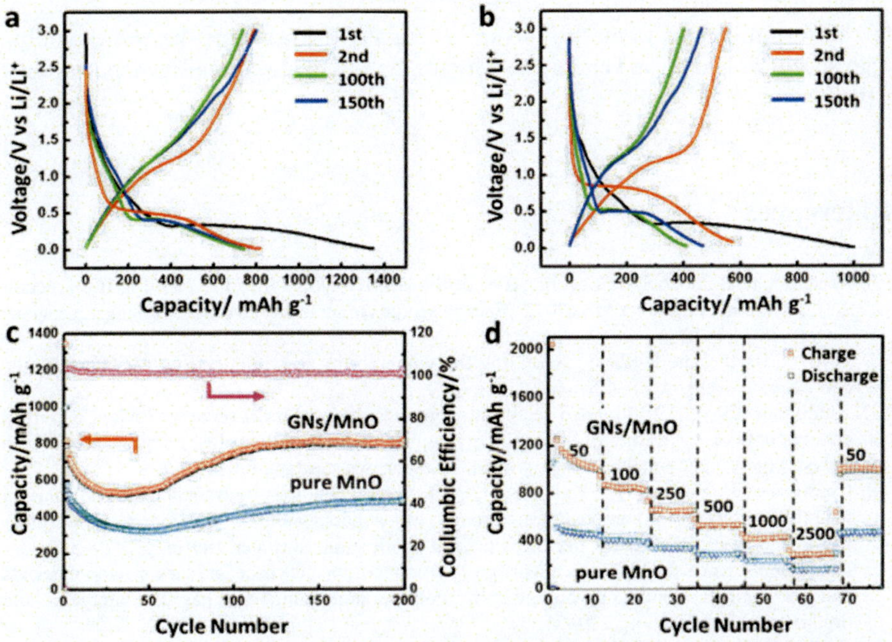

Fig. 12.14 GCD curves of **a** GNs/MnO nanowires composite and **b** pure MnO nanowires obtained at 100 mA g^{-1} within a voltage range of 0.005–3.0 V versus Li/Li^+; **c** Capacity retention of GNs/MnO nanowires composite and pure MnO nanowires at different cycle numbers; **d** rate capabilities of GNs/MnO nanowires composite and pure MnO nanowires at different cycle numbers. Reprinted with permission from [72] Copyright 2016 American Chemical Society

But in the case of pure MnO anode, only a specific capacity of 161 mAh g^{-1} at 2500 mA g^{-1} was obtained after 70 cycles.

12.5 Conclusion

In this chapter, we have discussed the recent developments in the field of electrospun Mn oxides-based LIB anodes. Electrospinning is proven to be a potential method for the preparation of Mn oxides-based composite anodes for LIBs as it provides feasibility in designing the electrode nanostructure to obtain the best performance. The preparation of the MnO_x-based composite anodes includes simple steps like electrospinning the desired precursor solution followed by the carbonization of the electrospun composite mats. The synthesis and electrochemical characterizations of various MnO_x-based composite nanofibrous anodes such as MnO_x/CNFs, MnO_x/CNTs, MnO_x/SnO_2, $CaMnO_3$, graphene/Mn_3O_4, etc. are discussed in this chapter with an emphasis on its performance parameters. Tunability in the electrode pore structure and large surface area coupled with high electrical conductivity enable low diffusion lengths to the Li-ions to intercalate/de-intercalate thereby providing high specific capacity and energy density for the Mn oxide composite anodes-based LIBs.

References

1. Schaeffer R, Szklo AS, de Lucena AFP, Borba BSMC, Nogueira LPP, Fleming FP, Troccoli A, Harrison M, Boulahya MS (2012) Energy sector vulnerability to climate change: a review. Energy 38(1):1–12
2. Pryor S, Barthelmie R (2013) Assessing the vulnerability of wind energy to climate change and extreme events. Clim Change 121(1):79–91
3. Khalilpour KR (2019) Stranded renewable energies, beyond local security, toward export: a concept note on the design of future energy and chemical supply chains. Polygeneration with polystorage for chemical and energy hubs. Elsevier, Amsterdam, pp 157–173
4. Cherusseri J, Sharma R, Kar KK (2015) Nanotechnology advancements on carbon nanotube/polypyrrole composite electrodes for supercapacitors. Handbook of polymer nanocomposites. Processing, performance and application. Springer, Berlin, pp 479–510
5. Cherusseri J, Kar KK (2015) Recent progress in nanocomposites based on carbon nanomaterials and electronically conducting polymers. In: Polymer nanocomposites based on inorganic and organic nanomaterials, pp 229–256
6. Rohit AK, Rangnekar S (2017) An overview of energy storage and its importance in Indian renewable energy sector: part II–energy storage applications, benefits and market potential. J Energy Storage 13:447–456
7. Amirante R, Cassone E, Distaso E, Tamburrano P (2017) Overview on recent developments in energy storage: mechanical, electrochemical and hydrogen technologies. Energy Convers Manag 132:372–387
8. Liu C, Han X, Cao Y, Zhang S, Zhang Y, Sun J (2018) Topological construction of phosphorus and carbon composite and its application in energy storage. Energy Storage Mater 20:343–372
9. Cherusseri J, Kumar KS, Choudhary N, Nagaiah N, Jung Y, Roy T, Thomas J (2019) Novel mesoporous electrode materials for symmetric, asymmetric and hybrid supercapacitors. Nanotechnology 30(20):202001
10. Cherusseri J, Choudhary N, Kumar KS, Jung Y, Thomas J (2019) Recent trends in transition metal dichalcogenide based supercapacitor electrodes. Nanoscale Horiz 4(4):840–858
11. Cherusseri J, Kar KK (2015) Hierarchically mesoporous carbon nanopetal based electrodes for flexible supercapacitors with super-long cyclic stability. J Mater Chem A 3(43):21586–21598
12. Cherusseri J, Kar KK (2015) Self-standing carbon nanotube forest electrodes for flexible supercapacitors. Rsc Advances 5(43):34335–34341
13. Makimura Y, Sasaki T, Nonaka T, Nishimura YF, Uyama T, Okuda C, Itou Y, Takeuchi Y (2016) Factors affecting cycling life of LiNi 0.8 Co 0.15 Al 0.05 O 2. J Mater Chem A 4(21):8350–8358
14. Nayak PK, Yang L, Brehm W, Adelhelm P (2018) From lithium-ion to sodium-ion batteries: advantages, challenges, and surprises. Angew Chem Int Ed 57(1):102–120
15. Dahbi M, Komaba S (2015) Fluorine chemistry for negative electrode in sodium and lithium ion batteries. Advanced fluoride-based materials for energy conversion. Elsevier, Amsterdam, pp 387–414
16. Kim D, Lee D, Kim J, Moon J (2012) Electrospun Ni-added SnO_2–carbon nanofiber composite anode for high-performance lithium-ion batteries. ACS Appl Mater Interfaces 4(10):5408–5415
17. Santangelo S (2019) Electrospun nanomaterials for energy applications: recent advances. Appl Sci 9(6):1049
18. Wang F, Cai J, Yu J, Li C, Yang Z (2018) Simultaneous electrospinning and electrospraying: fabrication of a carbon nanofibre/mno/reduced graphene oxide thin film as a high-performance anode for lithium-ion batteries. ChemElectroChem 5(1):51–61

19. Wu J, Pan Z, Zhang Y, Wang B, Peng H (2018) The recent progress of nitrogen-doped carbon nanomaterials for electrochemical batteries. J Mater Chem A 6(27):12932–12944
20. Kim T, Song W, Son D-Y, Ono LK, Qi Y (2019) Lithium-ion batteries: outlook on present, future, and hybridized technologies. J Mater Chem A 7(7):2942–2964
21. Andre D, Hain H, Lamp P, Maglia F, Stiaszny B (2017) Future high-energy density anode materials from an automotive application perspective. J Mater Chem A 5(33):17174–17198
22. Zhang H, Zhao H, Khan MA, Zou W, Xu J, Zhang L, Zhang J (2018) Recent progress in advanced electrode materials, separators and electrolytes for lithium batteries. J Mater Chem A 6(42):20564–20620
23. Chen S, Wen K, Fan J, Bando Y, Golberg D (2018) Progress and future prospects of high-voltage and high-safety electrolytes in advanced lithium batteries: from liquid to solid electrolytes. J Mater Chem A 6(25):11631–11663
24. Varma SJ, Sambath Kumar K, Seal S, Rajaraman S, Thomas J (2018) Fiber-type solar cells, nanogenerators, batteries, and supercapacitors for wearable applications. Adv Sci 5(9):1800340
25. Qi W, Shapter JG, Wu Q, Yin T, Gao G, Cui D (2017) Nanostructured anode materials for lithium-ion batteries: principle, recent progress and future perspectives. J Mater Chem A 5(37):19521–19540
26. Zhang W-J (2011) A review of the electrochemical performance of alloy anodes for lithium-ion batteries. J Power Sources 196(1):13–24
27. Sambath Kumar K, Cherusseri J, Thomas J (2019) Two-dimensional Mn_3O_4 nanowalls grown on carbon fibers as electrodes for flexible supercapacitors. ACS Omega 4(2):4472–4480
28. Zhang C, Wang J-G, Jin D, Xie K, Wei B (2015) Facile fabrication of MnO/C core–shell nanowires as an advanced anode material for lithium-ion batteries. Electrochim Acta 180:990–997
29. Gu X, Yue J, Li L, Xue H, Yang J, Zhao X (2015) General synthesis of MnO_x (MnO_2, Mn_2O_3, Mn_3O_4, MnO) hierarchical microspheres as lithium-ion battery anodes. Electrochim Acta 184:250–256
30. Zhong K, Xia X, Zhang B, Li H, Wang Z, Chen L (2010) MnO powder as anode active materials for lithium ion batteries. J Power Sources 195(10):3300–3308
31. Sun B, Chen Z, Kim H-S, Ahn H, Wang G (2011) MnO/C core–shell nanorods as high capacity anode materials for lithium-ion batteries. J Power Sources 196(6):3346–3349
32. Mai Y, Zhang D, Qiao Y, Gu C, Wang X, Tu J (2012) MnO/reduced graphene oxide sheet hybrid as an anode for Li-ion batteries with enhanced lithium storage performance. J Power Sources 216:201–207
33. Qin J, Zhang Q, Cao Z, Li X, Hu C, Wei B (2013) MnO_x/SWCNT macro-films as flexible binder-free anodes for high-performance Li-ion batteries. Nano Energy 2(5):733–741
34. Luo W, Hu X, Sun Y, Huang Y (2013) Controlled synthesis of mesoporous MnO/C networks by microwave irradiation and their enhanced lithium-storage properties. ACS Appl Mater Interfaces 5(6):1997–2003
35. Sun Y, Hu X, Luo W, Xia F, Huang Y (2013) Reconstruction of conformal nanoscale MnO on graphene as a high-capacity and long-life anode material for lithium ion batteries. Adv Func Mater 23(19):2436–2444
36. Wu J, Wang N, Zhao Y, Jiang L (2013) Electrospinning of multilevel structured functional micro-/nanofibers and their applications. J Mater Chem A 1(25):7290–7305
37. Li W, Li M, Adair KR, Sun X, Yu Y (2017) Carbon nanofiber-based nanostructures for lithium-ion and sodium-ion batteries. J Mater Chem A 5(27):13882–13906
38. Jung J-W, Lee C-L, Yu S, Kim I-D (2016) Electrospun nanofibers as a platform for advanced secondary batteries: a comprehensive review. J Mater Chem A 4(3):703–750
39. Qu H, Wei S, Guo Z (2013) Coaxial electrospun nanostructures and their applications. J Mater Chem A 1(38):11513–11528
40. Taylor GI (1969) Electrically driven jets. Proc R Soc London Math Phys Sci 313(1515):453–475
41. Mondal K, Tsai C-Y, Stout S, Talapatra S (2015) Manganese oxide based hybrid nanofibers for supercapacitors. Mater Lett 148:142–146

42. Bhagwan J, Sahoo A, Yadav KL, Sharma Y (2015) Porous, one dimensional and high aspect ratio Mn_3O_4 nanofibers: fabrication and optimization for enhanced supercapacitive properties. Electrochim Acta 174:992–1001
43. Nardi JC (1985) Characterization of the Li/MnO_2 multistep discharge. J Electrochem Soc 132(8):1787–1791
44. Sánchez L, Farcy J, Pereira-Ramos J-P, Hernán L, Morales J, Tirado JL (1996) Low-temperature mixed spinel oxides as lithium insertion compounds. J Mater Chem 6(1):37–39
45. Yamashita T, Vannice A (1996) NO decomposition over Mn_2O_3 and Mn_3O_4. J Catal 163(1):158–168
46. Maltha A, Kist H, Brunet B, Ziolkowski J, Onishi H, Iwasawa Y, Ponec V (1994) The active sites of manganese-and cobalt-containing catalysts in the selective gas phase reduction of nitrobenzene. J Catal 149(2):356–363
47. Silva GC, Almeida FS, Ferreira AM, Ciminelli VST (2012) Preparation and application of a magnetic composite (Mn_3O_4/Fe_3O_4) for removal of As (III) from aqueous solutions. Mater Res 15(3):403–408
48. Jiang M, Peng X (2017) Anisotropic Fe_3O_4/Mn_3O_4 hybrid nanocrystals with unique magnetic properties. Nano Lett 17(6):3570–3575
49. Shao C, Guan H, Liu Y, Li X, Yang X (2004) Preparation of Mn_2O_3 and Mn_3O_4 nanofibers via an electrospinning technique. J Solid State Chem 177(7):2628–2631
50. Fan Q, Whittingham MS (2007) Electrospun manganese oxide nanofibers as anodes for lithium-ion batteries. Electrochem Solid-State Lett 10(3):A48–A51
51. Cherusseri J, Pramanik S, Sowntharya L, Pandey D, Kar KK, Sharma S (2017) Polymer-based composite materials: characterizations. composite materials. Springer, Berlin, pp 37–77
52. Akhtar SN, Cherusseri J, Ramkumar J, Kar KK (2017) Ionic polymer metal composites. Composite materials. Springer, Berlin, pp 223–249
53. Pramanik S, Cherusseri J, Baban NS, Sowntharya L, Kar KK (2017) Metal matrix composites: theory, techniques, and applications. Composite materials. Springer, Berlin, pp 369–411
54. Mishra R, Cherusseri J, Joseph K (2017) Thermal and crystallization behavior of micro and nano fibrillar in-situ composites. Micro and nano fibrillar composites (MFCs and NFCs) from polymer blends. Elsevier, Amsterdam, pp 213–231
55. Cherusseri J, Sharma R, Kar KK (2016) Helically coiled carbon nanotube electrodes for flexible supercapacitors. Carbon 105:113–125
56. Cherusseri J, Kar KK (2016) Polypyrrole-decorated 2D carbon nanosheet electrodes for supercapacitors with high areal capacitance. RSC Adv 6(65):60454–60466
57. Cherusseri J, Kar KK (2016) Hierarchical carbon nanopetal/polypyrrole nanocomposite electrodes with brush-like architecture for supercapacitors. Phys Chem Chem Phys 18(12):8587–8597
58. Cherusseri J, Kar KK (2016) Ultra-flexible fibrous supercapacitors with carbon nanotube/polypyrrole brush-like electrodes. J Mater Chem A 4(25):9910–9922
59. Mahmood N, Zhang C, Yin H, Hou Y (2014) Graphene-based nanocomposites for energy storage and conversion in lithium batteries, supercapacitors and fuel cells. J Mater Chem A 2(1):15–32
60. Wang J-G, Yang Y, Huang Z-H, Kang F (2012) Incorporation of nanostructured manganese dioxide into carbon nanofibers and its electrochemical performance. Mater Lett 72:18–21
61. Wang F, Zhu P, Pan J, Li C, Yang Z (2017) A facile electrospinning and electrospraying synchronization technique for preparation of high performance MnO/C@ rGO composite anodes for lithium storage. RSC Adv 7(76):48294–48302
62. Nakhowong R (2016) Preparation and characterization of calcium manganese oxide ($CaMnO_3$) nanofibers by electrospinning. Mater Lett 163:222–225
63. Zhou D, Lin H, Zhang F, Niu H, Cui L, Wang Q, Qu F (2015) Freestanding MnO2 nanoflakes/porous carbon nanofibers for high-performance flexible supercapacitor electrodes. Electrochim Acta 161:427–435
64. Wang J-G, Yang Y, Huang Z-H, Kang F (2011) Coaxial carbon nanofibers/MnO_2 nanocomposites as freestanding electrodes for high-performance electrochemical capacitors. Electrochim Acta 56(25):9240–9247

65. Wang J-G, Yang Y, Huang Z-H, Kang F (2012) Synthesis and electrochemical performance of MnO_2/CNTs–embedded carbon nanofibers nanocomposites for supercapacitors. Electrochim Acta 75:213–219
66. Kim JH, Kim B-H (2016) Hierarchical porous MnO_2/carbon nanofiber composites with hollow cores for high-performance supercapacitor electrodes: Effect of poly (methyl methacrylate) concentration. Electrochim Acta 200:174–181
67. Ning P, Duan X, Ju X, Lin X, Tong X, Pan X, Wang T, Li Q (2016) Facile synthesis of carbon nanofibers/MnO_2 nanosheets as high-performance electrodes for asymmetric supercapacitors. Electrochim Acta 210:754–761
68. Nie G, Lu X, Chi M, Gao M, Wang C (2018) General synthesis of hierarchical C/MO_x@ MnO_2 (M = Mn, Cu, Co) composite nanofibers for high-performance supercapacitor electrodes. J Colloid Interface Sci 509:235–244
69. Hwang SM, Kim SY, Kim J-G, Kim KJ, Lee J-W, Park M-S, Kim Y-J, Shahabuddin M, Yamauchi Y, Kim JH (2015) Electrospun manganese–cobalt oxide hollow nanofibres synthesized via combustion reactions and their lithium storage performance. Nanoscale 7(18):8351–8355
70. Liu M, Xie W, Gu L, Qin T, Hou X, He D (2016) Improved lithium-ion battery anode capacity with a network of easily fabricated spindle-like carbon nanofibers. Beilstein J Nanotechnol 7(1):1289–1295
71. Zhao X, Du Y, Jin L, Yang Y, Wu S, Li W, Yu Y, Zhu Y, Zhang Q (2015) Membranes of MnO beading in carbon nanofibers as flexible anodes for high-performance lithium-ion batteries. Sci Rep 5:14146
72. Sun Q, Wang Z, Zhang Z, Yu Q, Qu Y, Zhang J, Yu Y, Xiang B (2016) Rational design of graphene-reinforced MnO nanowires with enhanced electrochemical performance for Li-ion batteries. ACS Appl Mater Interfaces 8(10):6303–6308
73. Niu C, Meng J, Wang X, Han C, Yan M, Zhao K, Xu X, Ren W, Zhao Y, Xu L (2015) General synthesis of complex nanotubes by gradient electrospinning and controlled pyrolysis. Nat Commun 6:7402
74. Cui X, Wang Y, Chen Z, Zhou H, Xu Q, Sun P, Zhou J, Xia L, Sun Y, Lu Y (2015) Preparation of pompon-like MnO/carbon nanotube composite microspheres as anodes for lithium ion batteries. Electrochim Acta 180:858–865
75. Xiao Y, Wang X, Wang W, Zhao D, Cao M (2014) Engineering hybrid between MnO and N-doped carbon to achieve exceptionally high capacity for lithium-ion battery anode. ACS Appl Mater Interfaces 6(3):2051–2058
76. Lin Z, Ji L, Woodroof MD, Zhang X (2010) Electrodeposited MnO_x/carbon nanofiber composites for use as anode materials in rechargeable lithium-ion batteries. J Power Sources 195(15):5025–5031
77. Rasheed A, Howe JY, Dadmun MD, Britt PF (2007) The efficiency of the oxidation of carbon nanofibers with various oxidizing agents. Carbon 45(5):1072–1080

Chapter 13
Electrospun Tin Based Composites as Anodes for Lithium-Ion Batteries

Nizao Kong, Yunhua Yu, and Xiaoping Yang

13.1 Introduction

The research on tin-based negative electrode materials first originated from Japan Seiko Electronics Industry Company, and then Sanyo Electric Company, Panasonic Electric Company, and Fuji Film Company successively carried out the research. The early research manifested that lithium and tin could form alloy and alloying/dealloying process was reversible [1]. In 1997, Fuji Film developed a SnO_x-based anode as an alternative of graphite [2]. Compared with carbon-based negative electrode materials, this amorphous tin oxide composite possessed a higher gravimetric specific capacity and volumetric specific capacity. However, it was not successfully commercial owing to poor cycling performance. Sony corporation launched a new-generation LIB, namely "Nexelion," which firstly employed a microcrystalline Sn–Co/C composite as the anode, and brought about a thirty percent of volumetric capacity output promotion in comparison with conventional LIBs in 2005 [3–5]. In 2011, Sony Corporation made the Nexelion LIB moving forward a single step, meaning that it could supply power for notebook computers. Via adjusting the structure of the microcrystalline Sn-Co/C anode, the shape change derived from the contraction/expansion of Sn during Li^+ deintercalation/intercalation process was partly prevented [6, 7]. These inspiring breakthroughs have further fired up enthusiasm from researchers towards Sn-based anode materials for LIB applications.

N. Kong · Y. Yu (✉) · X. Yang
College of Materials Science and Engineering, Beijing University of Chemical Technology, North 3rd Ring East Road, No. 15, Chaoyang District, Beijing 100029, People's Republic of China
e-mail: yuyh@mail.buct.edu.cn

N. Kong
e-mail: 15358130048@163.com

X. Yang
e-mail: yangxp@mail.buct.edu.cn

N. T. M. Balakrishnan and R. Prasanth (eds.), *Electrospinning for Advanced Energy Storage Applications*, Materials Horizons: From Nature to Nanomaterials,
https://doi.org/10.1007/978-981-15-8844-0_13

Up to now, a good deal of electrospinning-based research articles has been published on LIB applications of metallic Sn [8–10] Sn-based nonmetallic compounds (SnO_2, SnS, SnSe) [11–15] and Sn-based intermetallic compounds (SnSb, CuSn, CoSn) [16]. However, one systematic summary of this field is not made yet. So it is necessary to conclude the current development to indicate the gravity center of future research in this field. In this chapter, we center on summarizing research breakthroughs about electrospinning Sn-based anode materials for LIB applications, mainly underlying nanostructure design concept as well as the relation between nanoarchitectures and electrochemical performances associated with the anode material. In addition, the newly-presented challenges and some opinions on the development of electrospinning Sn-based anode materials for LIB applications are minutely discussed.

13.2 Metallic Sn Anode Materials Based on Electrospinning Technique

The Li^+ insertion/extraction of metallic Sn can be depicted by the following chemical equation:

$$Sn + xLi^+ \leftrightarrow Li_xSn(0 \leq x \leq 4.4) \quad (13.1)$$

With Li^+ constantly being inserted into metallic Sn, different crystal states of Li_xSn_y are formed, such as Li_2Sn_5, LiSn, Li_7Sn_3, Li_5Sn_2, $Li_{13}Sn_5$, Li_7Sn_2 and $Li_{22}Sn_5$. $Li_{22}Sn_5$ possesses a capacity theoretical of 993 mAh g^{-1}, which is almost three times of graphitic negative electrode. From characteristic discharge spline of galvanized Sn in a test environment of 1M $LiClO_4$/PC, we can see that the voltage range of Li^+-insertion is between 0.0 and 0.7 V [17].

The volumetric expansion ratio of alloy $Li_{22}Sn_5$ can reach up to 359%, which easily leads to pulverization and agglomeration after several cycles and thus capacity fades rapidly. For the sake of addressing the problems of dramatic volumetric expansion/contraction as well as consequent pulverization and aggregation, many researches based on electrospinning technique have been conducted to design different Sn/carbon composite materials possessing smaller particle size or open-framework structure. These well-designed structures could adjust the volumetric change of Sn during Li^+ insertion/deinsertion process. Besides, additional advantages, such as shortened Li^+ transfer pathway, enlarged electrolyte/electrode contact area, and increased active sites for Li^+ storage, can be acquired at the same time. Last but not least, the stable and flexible carbon nanofibers (CNFs) can prevent the immediate contact between electrolyte and Sn and thereby notably reduce the insignificant associated responses, thus enhancing the ICE [3]. The carbonaceous substrate can also function as a well-performing conductor and improve the rate property of anode materials.

In the primary stage, the majority of the research efforts have been put toward optimizing particle size of Sn (nanocrystallization) in CNFs matrix, for that nanoparticle could abbreviate the movement pathway for electrons and ions as well as offer affluent electrochemical activated sites. [19–21] For instance, Yu et al. [22] successfully developed an intricate Sn/CNF network by using precursor $SnCl_4$ as well as investigated the carbonized temperature influence onto particle size of Sn. The result indicated that Sn/CNFs web carbonized at 850 °C has a narrow size range of Sn nanoparticles (30–40 nm). And Sn grain size is influenced by carbonization temperature and then influence the electrochemical properties of the anode. An invertible capacity output of 450 mAh g^{-1} was acquired by the sample obtained at 850 °C displayed after a short-term cycle of 30 at 0.025 A g^{-1}. Nonetheless, the tie between the corresponding electrochemical properties and mass constitution of Sn and CNFs was not expounded clearly. The research finding of Wang et al. [23] showed the morphology and electrochemical properties of Sn/C samples were markedly influenced by the precursor weight percentage of $SnCl_2$ versus polyacrylonitrile ($SnCl_2$/PAN). The samples with beginning mass ratio of $SnCl_2$/PAN (3:2, 1:1, 2:3) were denoted as Sn_3Pan_2, Sn_1Pan_1 and Sn_2Pan_3, severally. And average crystal Sn particle size of the above samples calculated by Scherrer equation is 25.9, 67.2, and 36.1 nm, respectively. The initial discharge capacities of the above samples are 977.8, 1329.8, and 1137.0 mAh g^{-1}, severally. Based on Wang's research outcome, $SnCl_2$ has two functions. The first function is to adjust the countenance of CNFs as conducting ionic salt possessing different consistence and the second function is to become metallic Sn particles as active substance after carbon thermal reduction. The diameter of CNFs significantly reduced followed by the Sn2+ content in the spinning solution of $SnCl_2$/PAN increasing. And Wang et al. [24] used SnO_2 nanoparticles synthesized by a modified hydrothermal method as precursor of Sn and fabricated an Sn nanoparticle (NP)/carbon nanofiber (CNF) as binder-free anode material. Their research finding suggested that higher temperatures could be in favor of metallic Sn melting for further separating out on the CNFs surface.

Although nanoscale Sn can promote the effective Li^+ insertion/deinsertion of an alone particle and efficiently adjust significant bulk changes, the mutual affinity of Sn NPs would result in aggregation inevitably occurring behind multiple cycles (Fig. 13.1a, b) and free from the constraint of carbon matrix, which brings about the rapid capacity attenuation (Fig. 13.1c, d). The graphical representation of the anodes surface microstructural transmutation during cycling process is illustrated in Fig. 13.2.

The above-mentioned literature reports also just simply made pony-size Sn nanoparticles attached to CNFs and not took into consideration aggregation problems of Sn after cycling, so that poor short-term cyclic performances were displayed. For the sake of getting the utmost out of becoming pony dimensionally and avert above adverse impacts of nanocrystallization, investigators tried to utilize various modification methods, such as creating void space between Sn nanoparticles and control the Sn size at the same time. Yu and co-workers [26] successfully manufactured an Sn/C composite construction, meaning that encapsulating in situ formed Sn nanoparticles within bamboo-shaped hollow CNFs with internal void space via

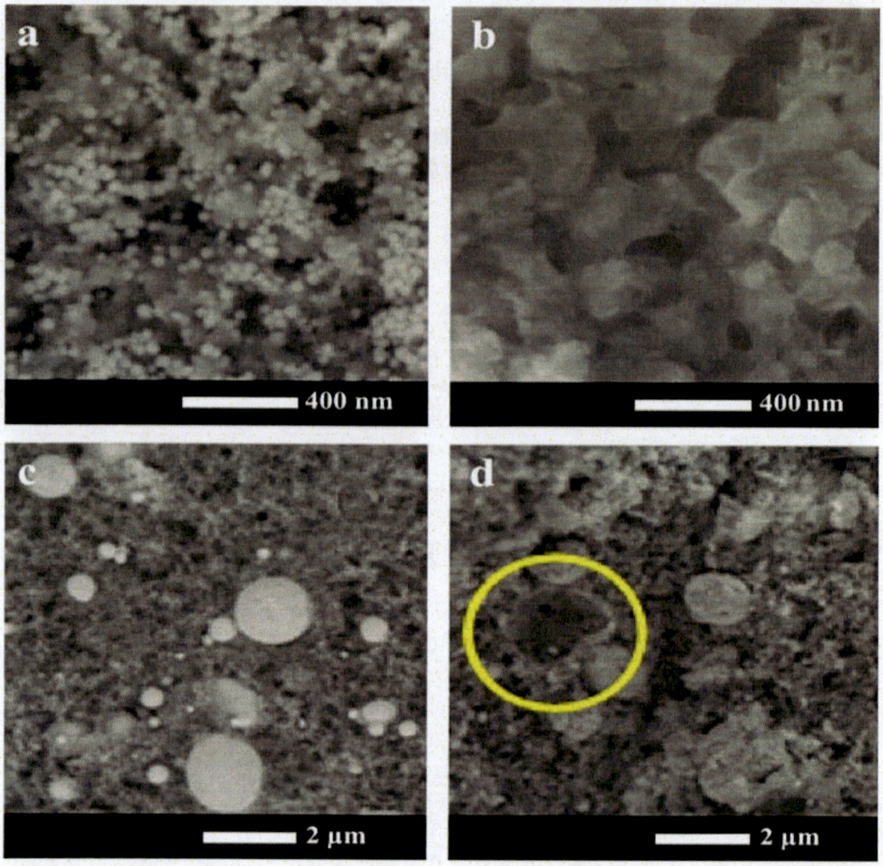

Fig. 13.1 SEM micrographs of electrode surfaces and relevant schematic illustration: 45 nm Sn nanoballs: **a** before cycling, **b** after cycling; pure Sn particles **c** ahead of cycling and **d** behind cycling. Adapted and reproduced by Wang et al. [25]. Copyright 2010 American Chemical Society

thermally decomposing coaxially electrospinning tributyltin (TBT, core)/PAN (shell) precursor fibers. The as-spun TBT/PAN precursor fibers were firstly pre-oxidized in a muffle furnace, generating SnO_2 NPs limited within multihole fistulous intermediate fibers on account of the PAN stabilizing and tin octoate decomposing. Then Sn NPs (~200 nm) were acquired and capsulated in multihole fistulous CNFs (diameter ~2 μm) behind carbonization treatment. The adequate internal void space could digest the volumetric change and subsequent aggregation, holding back the electrical isolation after the oft-repeated discharge/charge process. Profit from the abovementioned preponderance, the anode could display an invertible capacity output of more than 800 mAh g^{-1} during the initial ten discharge and acquire an invertible capacity of approximately 737 mAh g^{-1} behind 200 charge/discharge processes, which is approximately equal to 91% of the theoretical capacity calculated by mass

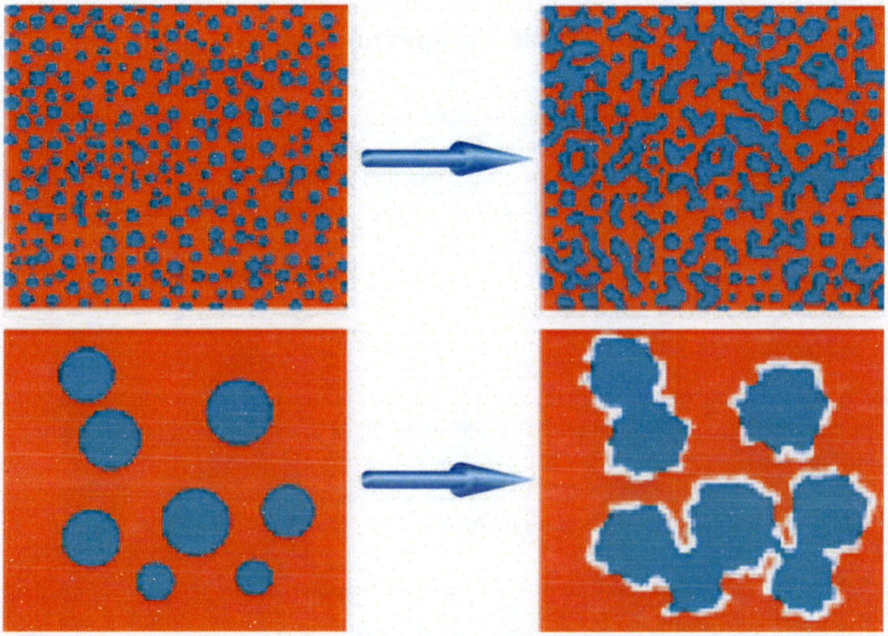

Fig. 13.2 Graphical representation of the anodes surface microstructural transmutation during cycling process. Adapted and reproduced by Wang et al. [25]. Copyright 2010 American Chemical Society

percentage of Sn and carbon. The enhanced electrochemical properties of Sn/C anode should be derived from its special morphology.

Some other modification methods also were applied to electrospun Sn/C anodes. Shen et al. [27] fabricated a Sn-PCNF composite where there was an even pores distribution on the surface of CNFs. Profiting from its unique porous structure, an invertible capacity output of ~774 mAh g^{-1} could be achieved under 0.8 A g^{-1} behind 300 times of discharge. Yang et al. [28] also synthesized a three-dimensional (3D) Sn/C core/shell hierarchical construction where component CNFs possess nanopores, mesopores, and large surface-to-volume ratio. Zhang et al. [29] synthesized a Sn NPs@CNFs composite where Sn ultrafine NPs encapsulated in N-doping CNFs. N-doping could enhance the conductivity of the carbon substrate [30]. Furthermore, compared with no heteroatom doped carbon materials, the N-doped carbon had more advantages in accommodating large changes in volume caused by Li–Sn alloying/dealloying process [31]. Wang et al. [32] fabricated a range of nanoscale constructions where Si/Sn NPs were encapsulated within multihole CNFs (SiSnC). The most optimized sample offered an invertible capacity output of 1347 mAh g^{-1} at Second cycle, as well as retained 1073 mAh g^{-1} behind 50 times of discharge. The synergistic action between Si and Sn NPs within CNFs has pivotal role in promoting the electrochemical properties of the optimized sample.

13.3 Sn-Based Nonmetallic Compound Anodes

Some kind of Sn-based nonmetallic compounds, like tin dioxide (SnO_2), stannous sulfide (SnS), and stannic selenide (SnSe) has been also proved to be as promising anode materials for LIBs. Here, we have made a summary of the electrospinning technology evolvements of Sn-based nonmetallic compounds in recent years.

13.3.1 SnO_2

SnO_2 is emerging in the aspects of its profusion, environmentally friendship, and superior theoretical capacity output (~790 mAh g^{-1}) than merchant graphite anode materials (~370 mAh g^{-1}). SnO_2 has a rutile structure (Fig. 13.3).

In a half-cell test of SnO_2//Li, Li^+ insertion/deinsertion process can be depicted by chemical transformation equations below:

$$SnO_2 + 4Li^+ + 4e^- \rightarrow 2Li_2O + Sn \quad (13.2)$$

$$Sn + xLi^+ + xe^- \leftrightarrow Li_xSn(0 \leq x \leq 4.4) \quad (13.3)$$

Equation (13.2) is on behalf of the irreversible conversion process where SnO_2 transforms into a simple substance metallic Sn. [33, 34]. The large irreversible loss of initial discharge capacity output which can be usually observed in SnO_2-based

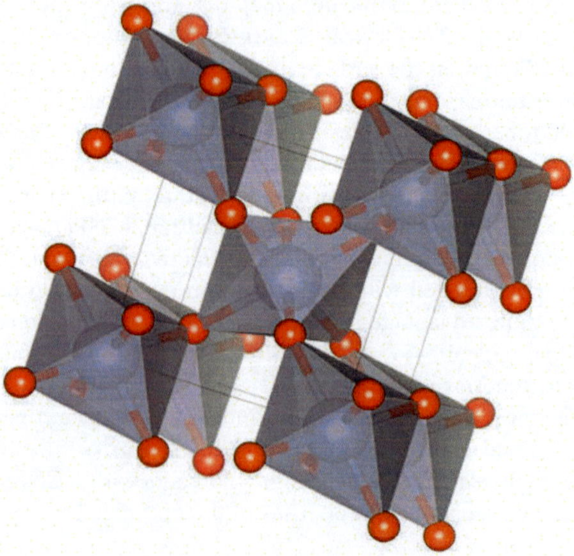

Fig. 13.3 Crystallographic of SnO_2 [18]

electrodes is partly due to above irreversible chemical conversion. Equation (13.3) is on behalf of the reversible alloying/dealloying process, which occupies the dominant percentage of the capacity output of LIBs. Its reaction voltage platform falls on ~0.6 V versus Li/Li^+ which can address potential security threats connected with metallic Li deposition. However, such a reversible alloying/dealloying process also induces over 200% change in volume [35, 36]. This will cause large internal strain/stress in the anode, which brings about collapse and efficacy loss of electronic/ionic touch, and ultimately leads to rapid capacity attenuation after multiple cycles. The two main defects of SnO_2–based materials limit their employment in industrial.

Like electrospinning metallic Sn anode materials, the majority of researches focus on designing and synthesizing smaller particle size of SnO_2. Fu et al. [37] employed polyvinylpyrrolidone (PVP) as carbon source and prepared electrospinning SnO_2@CNFs where SnO_2 nanoparticles (NPs) were attached within CNFs (Fig. 13.4a). As well, SnO_2@CNFs displayed an invertible capacity of 410.5 mAh g^{-1} behind 100 times of discharge, which is superior to contrast samples including SnO_2 nanofibers (NFs), commercial SnO_2, and pure CNFs. The hyperfine SnO_2 NPs protected by CNFs shell could display higher electrochemical activity than contrast samples like pure CNFs and pure SnO_2 NFs. Due to this synergy deriving from the hyperfine SnO_2 NPs within CNFs shell, the SnO_2@CNFs is able to demonstrate high capacity output. Bonino et al. [38] also fabricated Carbon-Tin Oxide Nanofibers (C-SnO_2) possessing individual SnO_2 NPs (Φ: ~15 nm) and SnO_2 nanoparticle clusters (~200 nm) by in situ electrospinning (Fig. 13.4b). The preponderance of C-SnO_2 composite relative to pure CNFs is evident at rate performance when current densities are more than 50 mAg^{-1}. The invertible capacity of C-SnO_2

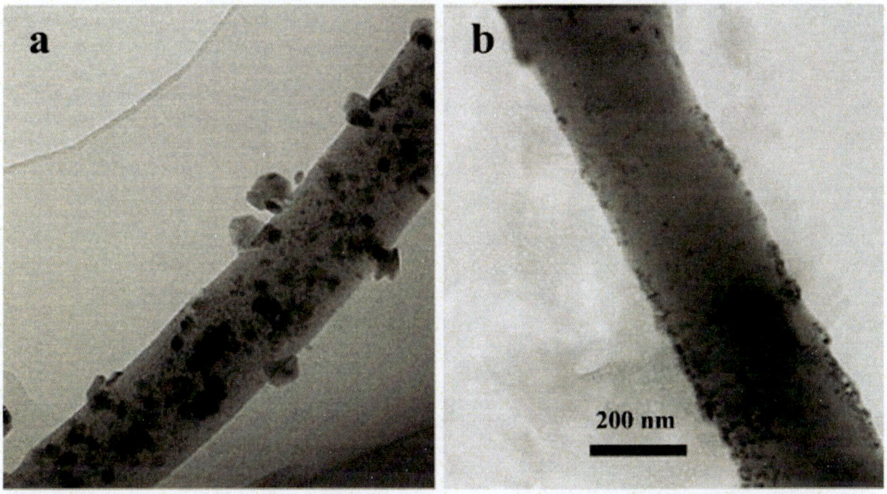

Fig. 13.4 **a** TEM image of SnO_2@CNFs. Adapted and reproduced by Fu et al.[37]. Copyright 2020 John Wiley and Sons. **b** TEM image of C-SnO_2. Adapted and reproduced by Bonino et al. [38]. Copyright 2011 American Chemical Society

composite is greater than the pure CNFs by up to 150 mAh g^{-1}. C-SnO_2 can store more Li^+ than pure CNFs for the presence of SnO_2 during rapid rates of charging.

Encapsulating SnO_2 NPs into CNFs with sufficient internal buffer space was fabricated to improve the invertible capacity output and cyclic steadiness of SnO_2/CNFs composite. Liu et al. [39] fabricated a fresh 'fiber-in-tube' hierarchical nanoscale construction of SnO_2@multihole carbon encapsulated within carbon tubes (namely SnO_2@PC/CTs). By adjusting pre-oxidation temperature and time of pilot process, they also obtained a 'particle-in-tube' nanostructure by sealing merely SnO_2 nanoparticles in CTs (namely SnO_2/CTs). The relevant graphical representation of fabrication process was illustrated in Fig. 13.5. The STEM images of as-synthesized SnO_2@PC/CTs exhibits elemental mapping images of carbon, nitrogen, oxygen, and tin, thereby confirming the homogeneous distribution of $SonO_2$ particles are within the SnO_2@PC/CTs (Fig. 13.6a–d). The HRTEM images show that within a wall thickness of about 18 nm, a large number of SnO_2 nanoparticles are well confined (Fig. 13.6e, f). Fortunately, the SnO_2@PC/CTs showed greatly enhanced cyclic performance via the uniform dispersion of SnO_2 NPs within PC skeleton for lessening nanoparticle aggregation. It showed an outstanding invertible capacity output (~1045 m Ah g^{-1}) behind 300 times of discharge (Fig. 13.6g). Xie et al. [40] firstly fabricated SnO_2/Fe_2O_3@C@CNFs composite and put sample into 3 M HCl solution to dissolve away ironoxides, then obtaining final product SnO_2/void@C nanofibers. The SnO_2 NPs were slickly decentralized as well as capsulated by internal multihole carbon and later wrapped by CNFs. And SnO_2/voids@C nanofibers anode showed an outstanding invertible capacity output (986 mAh g^{-1}) at 0.2 A g^{-1} behind 200 times of cycle.

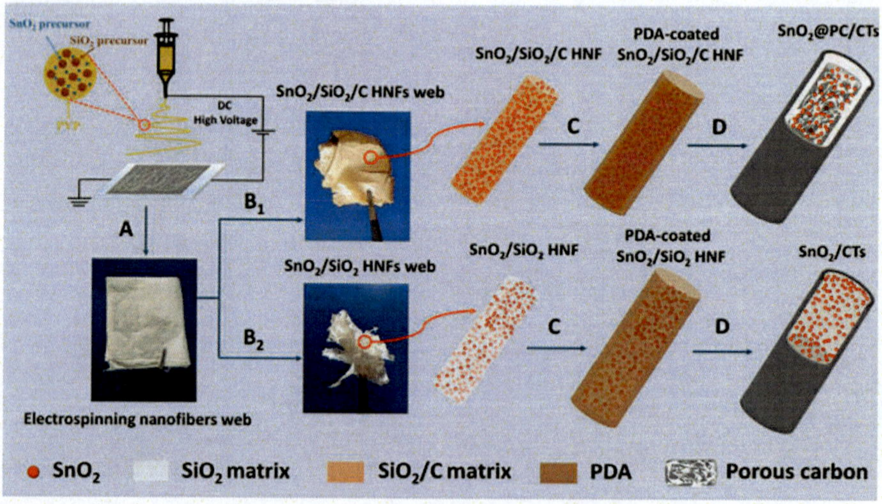

Fig. 13.5 Graphical representation preparation of SnO_2@PC/CTs and SnO_2/CTs. Adapted and reproduced by Liu et al. [39]. Copyright 2015 John Wiley and Sons

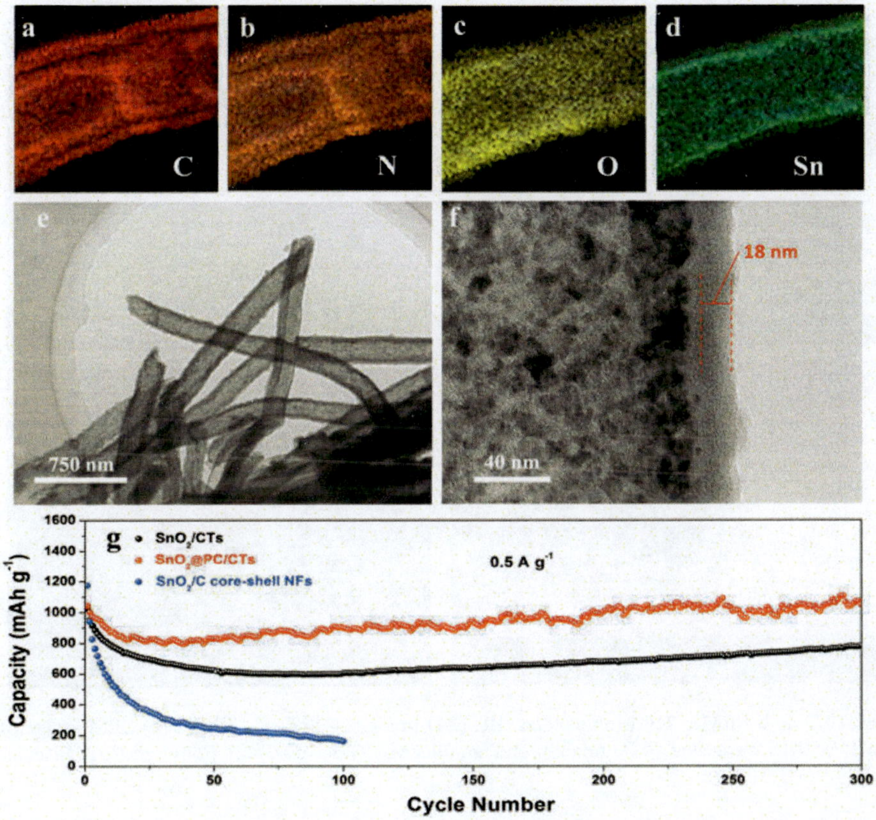

Fig. 13.6 STEM image and the corresponding element mapping images of **a** carbon, **b** nitrogen, **c** oxygen, and **d** tin, **e**, **f** HR-TEM images of SnO_2@PC/CTs, **g** Cyclic performance of SnO_2@PC/CTs and SnO_2/CTs anodes at 500 mAg^{-1}. Adapted and reproduced by Liu et al. [39]. Copyright 2015 John Wiley and Sons

Some researches combined the electrospinning method with other preparation methods to make SnO_2/CNFs composites. Ji et al. [41] firstly fabricated pure CNFs by electrospinning and subsequent carbonization and made SnO_2 NPs electrodeposited onto CNFs, namely carbon–SnO_2 core–shell composite nanofibers (CSCSCNs). By changing parameters of electrodeposition, the SnO_2 shell thicknesses can be adjusted. Wang et al. [42] work are that carbon-cladding SnO_2 NPs attached to CNFs (SnO_2@C/CNFs) were fabricated by an isochronous electrospinning-electrospraying method and subsequent carbonization. And we can see from Fig. 13.7a, b that the SnO_2@C congeries were snugly bound within the writhen CNFs network either anchoring on the cover of CNFs that prevented a harmful conductive contact. TEM image showed that the carbon layers wrapped SnO_2 NPs, confirming its core/shell structure (Fig. 13.7c). The CNFs network could provide a buffer function for a huge change in volume by SnO_2 NPs in the time of long-term Li^+ insertion/deinsertion

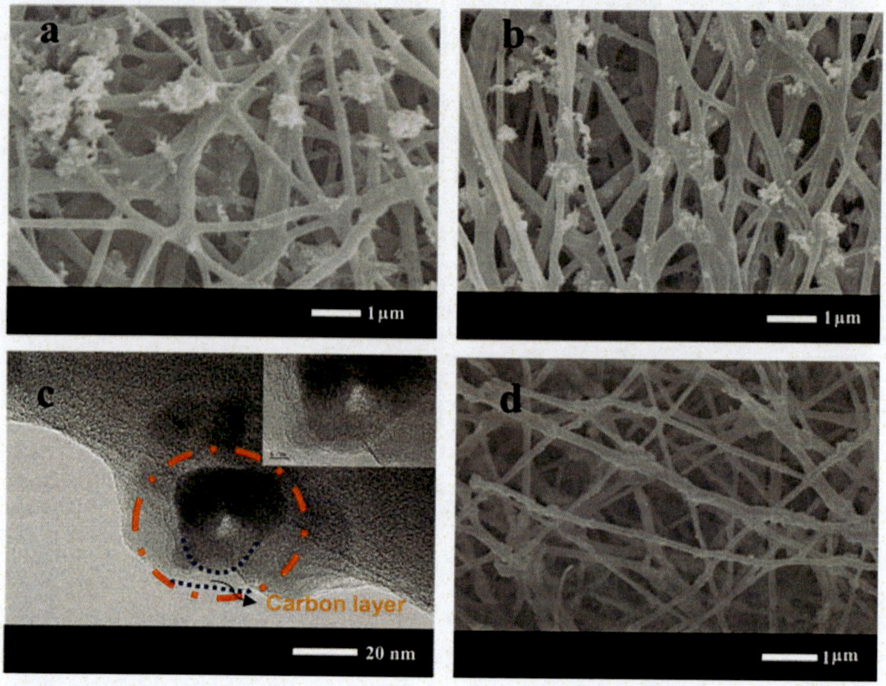

Fig. 13.7 **a**, **b** SEM **c** TEM (embedded HRTEM) images of SnO_2@C/CNFs (d) SEM image of SnO_2@CNFs Wang et al. [42]. Adapted and reproduce by Wang et al. [42]. Copyright 2019 Elsevier

process, thus redounding to an improved lifetime. In contrast to that, the sample SnO_2@CNFs, which was just fabricated by electrospinning method, possesses rough fiber surface caused by the large SnO_2 clusters (Fig. 13.7d). The big SnO_2 congeries could destroy effective touch between CNFs and SnO_2, and it could scarcely adjust large change in volume of SnO_2 NPs in the time of charge/discharge process. So the SnO_2@C/CNFs composite displayed better electrochemical property than contrast sample SnO_2@CNFs. The SnO_2@C/CNFs composite revealed an invertible capacity output (~492 mAh g^{-1}) behind 50 times of cycle, which was superior to contrast sample SnO_2@CNFs (~322 mAh g^{-1}). And for rate performance, SnO_2@C/CNFs are also better than contrast sample SnO_2@CNFs.

There are also some researches trying to introduce some heteroatom (P, N, etc) or metallic oxides (TiO_2, ZnO, Fe_3O_4, etc) into SnO_2 composites. Xia and co-workers [43] capsulated SnO_2 NPs into a N-doping CNFs film (namely SnO_2@N-CNF) by choosing polyamic acid (PAA) as carbon source (Fig. 13.8a). The self-supported SnO_2@N-CNF film anode exhibited excellent flexibility. TEM image of SnO_2@N-CNF revealed that the SnO_2 NPs with diameter size around 7.5 nm were well encapsulated inside N-CNF (Fig. 13.8b). And SnO_2@N-CNF showed excellent rate performance and cyclic performance. Kim et al. [44] made finely disseminated SnO_2 NPs loaded on CNFs developed by the existence of metallic Ni. The participation

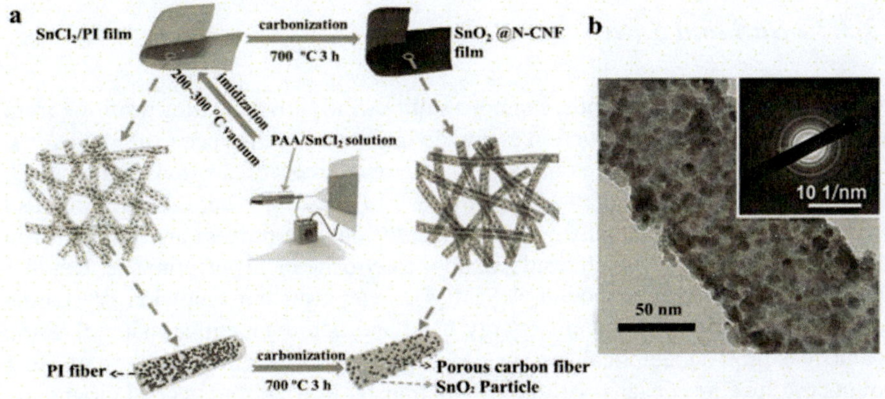

Fig. 13.8 **a** Graphical representation preparation of SnO_2@N-CNF, **b** TEM (embedded SAED pattern) image of SnO_2@N-CNF. Adapted and reproduced by Xie et al. [47]. Copyright 2014 John Wiley and Sons

of Ni could improve the steadiness of SnO_2 such that the morphologies of SnO_2 phase were optimized by the Ni composition. Tara et al. [45] fabricated SnO_2/TiO_2 composite nanofibers by burning to remove PVP after electrospinning. The incorporation of steady TiO_2 component into structurally unstable SnO_2 can significantly improve both the specific capacity output and cyclic performance. Zhao et al. [46] successfully synthesized SnO_2–ZnO nanofibers where SnO_2 NFs possessing mesoporous structures were made up of superfine SnO_2 NPs. With the addition of ZnO NPs, the beneficial effects like buffering large change in volume of SnO_2 and synergistic function between SnO_2 and ZnO could be achieved. Xie et al. [47] successfully synthesized N-doping amorphic carbon-coated Fe_3O_4/SnO_2 concentric nanofibers (Fe_3O_4/SnO_2 NFs) with a large areal mass density and BET specific surface area. At discharge stage, Li^+ inserts into Fe_3O_4 NPs while Sn could function as a conductive agent, then the formed Li_2O and Fe could function as a conductive agent while Li^+ inserts into Sn, and the reverse is also true in the charge stage. This reaction machine-made could facilitate Li^+ insertion/deinsertion and adjust the volumetric change. In addition, Sn and Fe NPs could promote electrolyte decomposing on account of their catalytic properties. [48, 49] Last but not least, N-doping dopamine carbon shell could improve Li^+-storage performance.

It should also be worth mentioning that nonquantitative compound SnO_x could be obtained via incomplete carbon thermal reduction in many electrospinning related works. [50–58] Like Sn and SnO_2, SnO_x also suffers from the severe accumulation and prominent volume expansion/constriction in the time of multiple cycles. So the same strategies were also used for SnO_x-based electrospinning. anodes.

13.3.2 SnS and SnSe

As one of Sn-based compounds, stannous sulfide (SnS) is outstanding profiting from its especial sandwich construction and wide interlamellar distance (SnS:4.04 Å vs. Li^+:0.76 Å), which can offer a rapid pathway for movement of electrons and ions [59, 60]. SnS can deliver a theoretical capacity of 1137 mAh g^{-1} but the bare SnS also suffers from large volume change of alloying/dealloying process and low intrinsic electronic conductivity which results in poor electrochemical performance. Double-atom semiconductive compound SnSe has a representative sandwich crystalline construction and incommodious energy band and is a competitive material. SnSe is able to exhibit a theoretical specific capacity output of 847 mAh g^{-1} and exciting volumetric specific capacity [61, 62]. It will transfer into a different crystallographic phase during Li^+ insertion/deinsertion process. However, the inactive Li_2Se leads to irreversible capacity loss. And the practical application of SnSe also is limited by its large volume change about like the majority of Sn-based compounds. Here, we will introduce relative progress in SnS-based SnSe-based anode materials fabricated via electrospinning method in LIBs.

Xia et al. [63] successfully developed the self-supported SnS/carbon nanofibers (SnS/CNFs) anode, whose graphical representation of manufacture is displayed within Fig. 13.9a. The TEM images (Fig. 13.9b) show that SnS/CNFs-650 °C possesses a fiber diameter of around 130 nm and the four different elements have a

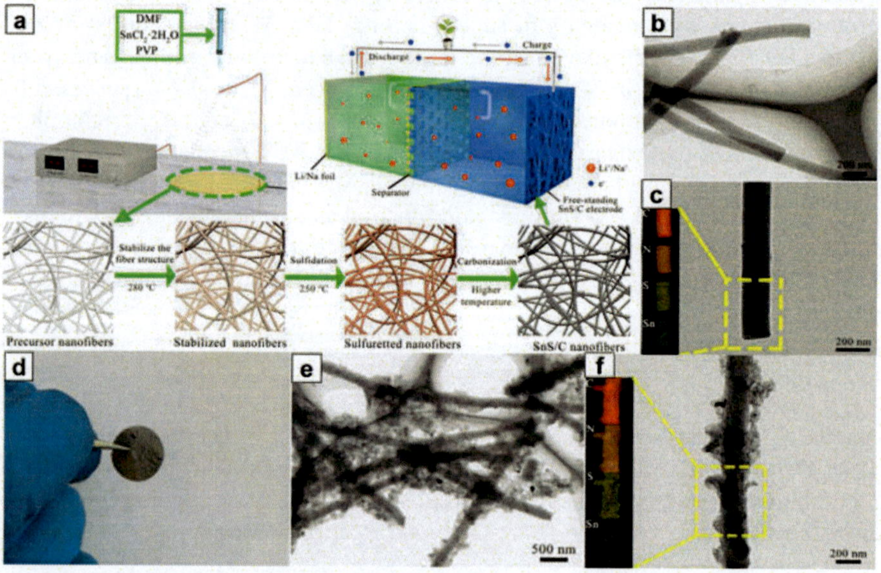

Fig. 13.9 **a** Graphical representation preparation of self-supported SnS/CNFs; **b** and **c** TEM images of SnS/CNFs-650 °C; **d** Optical photograph, **e, f** TEM images of the SnS/CNFs-650 °C anode after 500th cycles at 0.5 A g^{-1} in LIBs. Adapted and reproduced by Xia et al. [63]. Copyright 2019 Elsevier

homogeneous distribution in the CNFs (Fig. 13.9c). For the sake of examining the structural steadiness and completeness of SnS/CNFs-650 °C self-supported anode, the lithium half-cell after the cyclic test was disassembled and the micromorphology of the SnS/CNFs-650 °C was detected by TEM. As seen from the optical photograph, SnS/CNFs anode can maintain the integrity of the whole (Fig. 13.9d). Moreover, the diameter of the fibers (Fig. 13.9e) was approximate to as-prepared fibers in Fig. 13.9b, indicating that the 3D CNFs network could availably adjust the change in volume of SnS in the time of cyclic process. Besides, as shown by its inset EDX element mapping (Fig. 13.9f), the C, N, Sn, and S element none the less evenly dispersed within CNFs, revealing outstanding structure steadiness of composite. Zhang et al. [64] firstly prepared nanoscale SnSe powders by ball milling commercial SnSe and then fabricated SnSe@CNFs nanocomposite by electrospinning method and subsequent carbonization process. The optimized SnSe/CNFs-750 anode showed a steady invertible capacity output of 840 mAh g^{-1} behind 100 times of cycle at 0.2 A g^{-1}. The CNFs not only offered a buffering matrix for Li^+ alloying/ dealloying process but also improved the electrical conductivity of the anode. Some researchers also tried to introduce SnS or SnSe into SnO_x/CNFs system to enhance the electrical conductivity of generated irreversible Li_2O. For instance, Yuan et al. [65] successfully made SnSe nanorods attached to SnO_x@CNFs in situ via electrospinning method and subsequent carbonization process. The optimal SnSe/SnO_x@CNFs system displayed a higher ICE (86%) and higher reversible specific capacity output (741 mAh g^{-1} at 0.2 A g^{-1}) than contrast sample SnO_x@CNFs (corresponding value: 68%, 550 mAh g^{-1}).

13.4 Sn-Based Intermetallic Compound Anodes

Many Sn-based intermetallic compounds (M-Sn) have been tested as anode materials. These Sn-based intermetallics include Cu–Sn [66, 67], Sn–Sb [68–72], Co–Sn [73, 74, 76], Ni–Sn [75, 76], Sn–Mn [77], Ge–Sn [78, 79], and Sn–Fe [80]. The introduced heterogeneous metallic elements can compactly unite with Sn by metallic binding, offering a supple buffer skeleton. These intermetallic compounds possess a relatively high specific capacity output of 500–800 mAh g^{-1}. General speaking, they have shown enhanced charge/discharge rates and long-term cycling steadiness compared to metallic Sn [31, 81]. However, large initial irreversible capacity is a problem for Sn-based intermetallic compounds that cannot be ignored. The majority of metals which can form M-Sn intermetallic compounds also form LixMy during Li^+ insertion stage. The structure of M-Sn will experience a refactor accompanying large volume change for that there is a discrepancy between Sn and different metals. In the meantime, uneven volume change will easily exist in the phase boundary region of M-Sn, thus leading to breakdown or pulverization of the active material. The appearance of electrospinning technique can introduce elastic carbon material into M-Sn system and then the buffering role of carbonaceous material is capable of adjusting large volume change of Sn-based intermetallic compounds. The SnSb, CoSn, and CuSn electrospinning system are investigated relatively more.

13.4.1 SnSb

Shiva et al. [69] successfully developed a one-dimensional assembly where crystalline single-phase SnSb alloy NPs were encapsulated within porous CNFs (namely SnSb–C). And the SnSb NPs were not agglomerated but predominantly encapsulated within hollow CNFs. Thanks to the nanometer size markedly facilitating both the Li^+ flux and Li^+ insertion/deinsertion, the SnSb-C anode displayed excellent rate performance. Profiting from the CNFs shell functioning as a "buffer zone" to adjust large internal strain/stress of SnSb NPs, an outstanding cyclic performance was achieved. Niu et al. [70] first made $Sn_{0.92}Sb_{0.08}O_{2.04}$ NPs (Fig. 13.10a) coated by a glucose-derived carbon (Fig. 13.10b) and obtained a carbon-coated SnSb NPs encapsulated within CNFs (namely SnSb@C/C, Fig. 13.10c) via electrospinning method and subsequent carbonization process. The morphology of $Sn_{0.92}Sb_{0.08}O_{2.04}$ NPs is displayed in Fig. 13.10d and e, respectively. The Sn0.92Sb0.08O2.04 NPs have a diameter distribution (10–20 nm). The carbon-coated Sn0.92Sb0.08O2.04 NPs have a clear core/shell structure and no distinct clusters could be observed between NPs. The SnSb@C NPs are encapsulated in CNFs and many small pores exist around SnSb@C NPs (Fig. 13.10f). Byproduct CO_2 gas effused during carbonation process of PAN, thus leading to the generation of the voids nearby SnSb@C NPs. [82] The initial discharge specific capacity and charge specific capacity of SnSb@C/C composite is 984 and 674 mAh g^{-1}, respectively. The pores around SnSb@C NPs

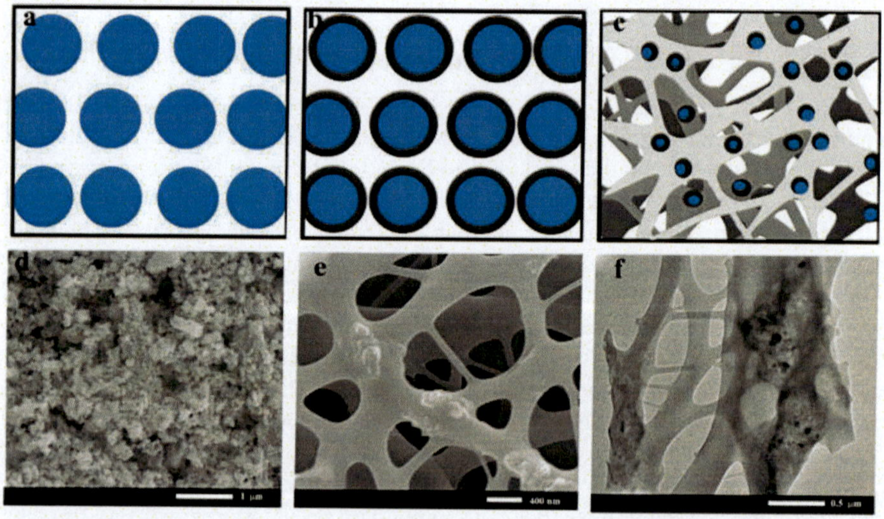

Fig. 13.10 Graphical representation preparation of SnSb@C/C composite: **a** $Sn_{0.92}Sb_{0.08}O_{2.04}$ NPs, **b** Carbon-coated $Sn_{0.92}Sb_{0.08}O_{2.04}$ NPs, **c** Carbon-coated SnSb NPs encapsulated within CNFs (namely SnSb@C/C), **d** SEM **e** TEM image of precursor NPs, **f** TEM image of final product. Adapted and reproduced by Niu et al. [70]. Copyright 2015 Elsevier

are in favor of improving the Li-alloying/dealloying reversibility of SnSb, resulting in a relatively high initial coulombic efficiency.

13.4.2 Co–Sn and Cu–Sn

Jang et al. [83] successfully developed CNFs composites embedded with Co–Sn NPs. They investigated different carbonization temperature influences on Co–Sn/CNFs composite, obtaining three samples namely Co–Sn/CNFs-700, Co–Sn/CNFs-800, and Co–Sn/CNFs-900. And the corresponding Co–Sn NPs have a size range of ~15–20, 30–35, and 40–45 nm in CNFs, respectively. Overall speaking, sample Co–Sn/CNF-800 anode showed the best cyclic performance for its centered carbonization temperature as well as NPs size. Lu et al. [19] also investigated carbonization temperature influence on CoSn/CNFs composite. They found that the sample carbonized at 850 °C showed the best specific capacity output and cyclic performance.

Shen et al. [84] successfully developed a three-dimensional system where Sn–Cu NPs were equably dispersed within CNFs (Sn–Cu–CNFs) via electrospinning technique and subsequent one-step carbonization-alloying reactions. And they found that Sn–Cu alloy was made up of activated Cu_6Sn_5 as well as inert Cu_3Sn components. The inactive Cu_3Sn component could function as a 'buffer zone' to adjust large volumetric change of active Cu_6Sn_5.

13.5 Conclusion

We have made a summary of the significant progress about electrospinning Sn-based anode materials in this chapter. We come to the conclusion that the dominating modification methods of electrospinning Sn-based anode materials including nanocrystallization and synchronously attached on CNFs matrix, internal buffer void structure design, heteroatom/second active substance composition introduction, and so on. Nanotechnology can availably minish the absolute volume swelling and buffer the internal stress/strain, thus delaying particle smash. The CNFs can suppress the NPs uniting as well as growing during material production and repetitive discharge/charge process. The reserved plentiful internal void among nanoscale particles supplies buffer function for volume change, thus protecting the structural integrity of the anode. The introduction of heteroatom or second active substance composition is not only in favor of enhancing the capacity output of anode materials but also likely to acquire synergistic effect with Sn or Sn-based compound.

Although abundant works have been done to enhance specific capacity output and cyclic steadiness the with the help of electrospinning technique, few researchers set their sights on the areal/volumetric mass loading of Sn or Sn-based compound and corresponding areal/volumetric capacities. The rapidly growing consumer market for portable electronics and vehicles desires the LIBs possessing high energy density which is strongly associated with areal/volumetric capacities of electrodes. So the path of developing energy-storage devices based on Sn-based composites is still full of difficulties. In future studies, more effort should be put on developing practically feasible technologies capable of manufacturing Sn-based anodes possessing high tap/packing density on a large scale at low cost in order that electrospinning Sn-based anode materials enable to be widely put into use in the energy-storage fields.

References

1. Foster MS, Crouthamel CE, Wood SE (1966) Thermodynamics of binary alloys. II. The lithium—tin system[1]. J Phys Chem 70(10):3042–3045. https://doi.org/10.1021/j100882a004
2. Idota Y, Kubota T, Matsufuji A, Maekawa Y, Miyasaka T (1997) Tin-based amorphous oxide: a high-capacity lithium-ion-storage material. Science 276(5317):1395–1397. https://doi.org/10.1126/science.276.5317.1395
3. http://www.bytecamera.com/content/view/372/2
4. http://www.sony.net/SonyInfo/News/Press/200502/05-006E
5. http://neasia.nikkeibp.com/neasiarrchivedtail/001100
6. http://techon.nikkeibp.co.jp/english/NEWS_EN/20110720/193457/
7. http://www.sony.com.cn/newscenter/techonology/118.htm
8. Kim S, Choi JH, Lim DS, Lee JH, Kim ID (2014) Phase and microstructural evolution of Sn particles embedded in amorphous carbon nanofibers and their anode properties in Li-ion batteries. J Electroceram 32(4):261–268. https://doi.org/10.1007/s10832-014-9941-1
9. Zheng J, Yu X, Wang C, Cao Z, Yang H, Ma D, Xu X (2016) Facile synthesis of three-dimensional reinforced Sn@polyaniline/sodium alginate nanofiber hydrogel network for high performance lithium-ion battery. J Mater Sci Mater Electron 27(5):4457–4464. https://doi.org/10.1007/s10854-016-4317-8
10. Agubra VA, Zuniga L, Garza DDL, Gallegos L, Pokhrel M, Alcoutlabi M (2016) Forcespinning: a new method for the mass production of Sn/C composite nanofiber anodes for lithium ion batteries. Solid State Ionics 286:72–82. https://doi.org/10.1016/j.ssi.2015.12.020
11. Wang Y, Aponte M, Leon N, Ramos I, Furlan R, Pinto N, Evoy S, Santiago-Aviles S (2005) Synthesis and characterization of ultra-fine tin oxide fibers using electrospinning. J Am Ceram Soc 88(8):2059–2063. https://doi.org/10.1111/j.1551-2916.2005.00409.x
12. Luo L, Xu WZ, Xia ZK, Fei YQ, Zhu JD, Chen C, Lu Y, Wei QF, Qiao H, Electrospun Zhang X G (2015) ZnO–SnO_2 composite nanofibers with enhanced electrochemical performance as lithium-ion anodes. Ceram Int 42(9):10826–10832. https://doi.org/10.1016/j.ceramint.2016.03.211
13. Lian QW, Zhou G, Zeng XH, Wu C, Wei YH, Cui C, Wei WF, Chen LB, Li CC (2016) Carbon Coated SnS/SnO_2 heterostructures wrapping on CNFs as an improved-performance anode for Li-Ion batteries: lithiation-induced structural optimization upon cycling. ACS Appl Mater Interf 8(44):30256–30263. https://doi.org/10.1021/acsami.6b10391
14. Hu YM, Yang QR, Ma JM, Chou SL, Zhu MY, Li Y (2015) Sn/SnO_2@C composite nanofibers as advanced anode for lithium-ion batteries. Electrochim Acta 186:271–276. https://doi.org/10.1016/j.electacta.2015.10.185
15. Narsimulu D, Vadnala S, Srinadhu ES, Satyanarayana N (2018) Electrospun Sn–SnO_2/C composite nanofibers as an anode material for lithium battery applications. J Mater Sci Mater Electron 29(13):11117–11123. https://doi.org/10.1007/s10854-018-9195-9
16. Kim JC, Kim DW (2014) Synthesis of multiphase SnSb nanoparticles-on-SnO_2/Sn/C nanofibers for use in Li and Na ion battery electrodes. Electrochem Commun 46:124–127. https://doi.org/10.1016/j.elecom.2014.07.005

17. Winter M, Besenhard JO (1999) Electrochemical lithiation of tin and tin-based intermetallics and composites. Electrochim Acta 45(1–2):31–50. https://doi.org/10.1016/S0013-4686(99)00191-7
18. Tirado JL (2003) Inorganic materials for the negative electrode of lithium-ion batteries: state-of-the-art and future prospects. Mater Sci Eng R Rep 40(3):103–136. https://doi.org/10.1016/S0927-796X(02)00125-0
19. Zhu ZQ, Wang SW, Du J, Zhang TR, Cheng FY, Chen J (2013) Ultrasmall Sn nanoparticles embedded in nitrogen-doped porous carbon as high-performance anode for lithium-ion batteries. Nano let. 14(1):153–157. https://doi.org/10.1021/nl403631h
20. Wang YG, Li HQ, He P, Hosono E, Zhou HS (2010) Nano active materials for lithium-ion batteries. Nanoscale 2(8):1294–1305. https://doi.org/10.1039/C0NR00068J
21. Guo YG, Hu JS, Wan LJ (2008) Nanostructured materials for electrochemical energy conversion and storage devices. Adv Mater 20(15):2878–2887. https://doi.org/10.1002/adma.200800627
22. Yu YH, Yang Q, Teng DH, Yang XP, Ryu S (2010) Reticular Sn nanoparticle-dispersed PAN-based carbon nanofibers for anode material in rechargeable lithium-ion batteries. Electrochem Commun 12(9):1187–1190. https://doi.org/10.1016/j.elecom.2010.06.015
23. Wang HY, Gao P, Lu SF, Liu HD, Yang G, Pinto J, Jiang XF (2011) The effect of tin content to the morphology of Sn/carbon nanofiber and the electrochemical performance as anode material for lithium batteries. Electrochim Acta 58:44–51. https://doi.org/10.1016/j.electacta.2011.08.075
24. Wang J, Song WL, Wang ZY, Fan LZ, Zhang YF (2015) Facile fabrication of binder-free metallic tin nanoparticle/carbon nanofiber hybrid electrodes for lithium-ion batteries. Electrochim Acta 153:468–475. https://doi.org/10.1016/j.electacta.2014.12.026
25. Wang XL, Feygenson M, Aronson MC, Han WQ (2010) Sn/SnO_x core−shell nanospheres: synthesis, anode performance in Li ion batteries, and superconductivity. J Phys Chem C 114(35):14697–14703. https://doi.org/10.1021/jp101852y
26. Yu Y, Gu L, Wang CL, Dhanabalan A, Maier J (2009) Encapsulation of Sn@ carbon nanoparticles in bamboo-like hollow carbon nanofibers as an anode material in lithium-based batteries. Angew Chem Int Ed 48(35):6485–6489. https://doi.org/10.1002/anie.200901723
27. Shen Z, Hu Y, Chen YL, Zhang XW, Wang KH, Chen RZ (2015) Tin nanoparticle-loaded porous carbon nanofiber composite anodes for high current lithium-ion batteries. J Power Sources 278:660–667. https://doi.org/10.1016/j.jpowsour.2014.12.106
28. Yang ZX, Meng Q, Yan WH, Lv J, Guo ZP, Yu XB, Chen ZX, Guo TL, Zeng R (2015) Novel three-dimensional tin/carbon hybrid core/shell architecture with large amount of solid cross-linked micro/nanochannels for lithium ion battery application. Energy 82:960–967. https://doi.org/10.1016/j.energy.2015.01.105
29. Zhang GH, Zhu J, Zeng W, Hou SC, Gong FL, Li F, Li CC, Duan HG (2014) Tin quantum dots embedded in nitrogen-doped carbon nanofibers as excellent anode for lithium-ion batteries. Nano Energy 9:61–70. https://doi.org/10.1016/j.nanoen.2014.06.030
30. Qie L, Chen WM, Wang ZH, Shao QG, Li X, Yuan LX, Hu LX, Zhang WX, Huang YH (2012) Nitrogen-doped porous carbon nanofiber webs as anodes for lithium ion batteries with a superhigh capacity and rate capability. Adv Mater 24(15):2047–2050. https://doi.org/10.1002/adma.201104634
31. Ying H, Han WQ (2017) Metallic Sn-based anode materials: application in high-performance lithium-ion and sodium-ion batteries. Adv Sci 4(11):1700298. https://doi.org/10.1002/advs.201700298
32. Wang H, Huang H, Chen L, Wang CG, Yan B, Yu YT, Yang Y, Yang G (2014) Preparation of Si/Sn-based nanoparticles composited with carbon fibers and improved electrochemical performance as anode materials. ACS Sustain Chem Eng 2(10):2310–2317. https://doi.org/10.1021/sc500290x
33. Lou XW, Chen JS, Chen P, Archer LA (2009) One-pot synthesis of carbon-coated SnO_2 nanocolloids with improved reversible lithium storage properties. Chem Mater 21(13):2868–2874. https://doi.org/10.1021/cm900613d

34. Chen JS, Lou XW (2013) SnO_2-based nanomaterials: synthesis and application in lithium-ion batteries. Small 9(11):1877–1893. https://doi.org/10.1002/smll.201202601
35. Larcher D, Beattie S, Morcrette M, Edstrom K, Jumas JC, Tarascon JM (2007) Recent findings and prospects in the field of pure metals as negative electrodes for Li-ion batteries. J Mater Chem 17(36):3759–3772. https://doi.org/10.1039/B705421C
36. Huang JY, Zhong L, Wang CM, Sullivan JP, Xu W, Zhang LQ, Mao SX, Hudak NS, Liu XH, Subramanian A, Fan HY, Qi L, Kushima A, Li J (2010) In situ observation of the electrochemical lithiation of a single SnO_2 nanowire electrode. Science 330(6010):1515–1520. https://doi.org/10.1126/science.1195628
37. Fu Z, Li X, Xu G (2014) Novel electrospun SnO_2@carbon nanofibers as high performance anodes for lithium-ion batteries. Cryst Res Technol 49(7):441–445. https://doi.org/10.1002/crat.201300211
38. Bonino CA, Ji LW, Lin Z, Toprakci O, Khan SA (2011) Electrospun carbon-tin oxide composite nanofibers for use as lithium ion battery anodes. ACS Appl Mater Interfaces 3(7):2534–2542. https://doi.org/10.1021/am2004015
39. Liu Y, Lan JL, Cai Q, Yu YH, Lin YH, Yang YP (2015) Encapsulating tin Dioxide@Porous carbon in carbon tubes: a fiber-in-tube hierarchical nanostructure for superior capacity and long-life lithium storage. Part Part Syst Charact 32(10):952–961. https://doi.org/10.1002/ppsc.201500073
40. Xie WH, Gu LL, Xia FY, Hou XY, Wang Q, Liu DQ, He DY (2016) Fabrication of voids-involved SnO_2@C nanofibers electrodes with highly reversible Sn/SnO_2 conversion and much enhanced coulombic efficiency for lithium-ion batteries. J Power Sources 327:21–28. https://doi.org/10.1016/j.jpowsour.2016.07.030
41. Ji LW, Lin Z, Guo BK, Medford AJ, Zhang XW (2010) Assembly of Carbon–SnO_2 core-sheath composite nanofibers for superior lithium storage. Chem Eur J 16(38):11543–11548. https://doi.org/10.1002/chem.201001564
42. Wang W, Liang YH, Kang YF, Liu LS, Xu ZW, Tian X, Mai W, Fu HJ, Lv HM, Teng KY, Jiao XN, Li FY (2019) Carbon-coated SnO_2@carbon nanofibers produced by electrospinning-electrospraying method for anode materials of lithium-ion batteries. Mater Chem Phys 223:762–770. https://doi.org/10.1016/j.matchemphys.2018.11.066
43. Xia L, Wang SQ, Liu GX, Ding LX, Li DD, Wang HH, Qiao SZ (2016) Flexible SnO_2/N-doped carbon nanofiber films as integrated electrodes for lithium-ion batteries with superior rate capacity and long cycle life. Small 12(7):853–859. https://doi.org/10.1002/smll.201503315
44. Kim D, Lee D, Kim J, Moon J (2012) Electrospun Ni-added SnO_2–carbon nanofiber composite anode for high-performance lithium-ion batteries. ACS Appl Mater Interfaces 4(10):5408–5415. https://doi.org/10.1021/am301328u
45. Tran T, McCormac K, Li JL, Bi ZG, Wu J (2014) Electrospun SnO_2 and TiO_2 composite nanofibers for lithium ion batteries. Electrochim Acta 117:68–75. https://doi.org/10.1016/j.electacta.2013.11.101
46. Zhao Y, Li X, Dong L, Yan B, Shan H, Li DJ, Sun XL (2015) Electrospun SnO_2–ZnO nanofibers with improved electrochemical performance as anode materials for lithium-ion batteries. Int J Hydrog Energy 40(41):14338–14344. https://doi.org/10.1016/j.ijhydene.2015.06.054
47. Xie WH, Li SY, Wang SY, Xue S, Liu ZJ, Jiang XY, He DY (2014) N-doped amorphous carbon coated Fe_3O_4/SnO_2 coaxial nanofibers as a binder-free self-supported electrode for lithium ion batteries. ACS Appl Mater Interfaces 6(22):20334–20339. https://doi.org/10.1021/am505829v
48. Qiao L, Wang XH, Qiao L, Sun XW, Li XW, Zheng YX, He DY (2013) Single electrospun porous NiO–ZnO hybrid nanofibers as anode materials for advanced lithium-ion batteries. Nanoscale 5(7):3037–3042. https://doi.org/10.1039/C3NR34103H
49. Sun YK, Myung ST, Park BC, Prakash J, Belharouak I, Amine K (2009) High-energy cathode material for long-life and safe lithium batteries. Nat Mater 8(4):320–324. https://doi.org/10.1038/nmat2418
50. Liu Y, Yan X, Lan JL, Teng DH, Yu YH, Yang XP (2014) Ti-doped SnO_x encapsulated in Carbon nanofibers with enhanced lithium storage properties. Electrochim Acta 137:9–16. https://doi.org/10.1016/j.electacta.2014.05.052

51. Liu XW, Teng DH, Li T, Yu YH, Shao XH, Yang XP (2014) Phosphorus-doped tin oxides/carbon nanofibers webs as lithium-ion battery anodes with enhanced reversible capacity. J Power Sources 272:614–621. https://doi.org/10.1016/j.jpowsour.2014.08.084
52. Chi C, Lan JL, Sun JM, Liu Y, Yu YH, Yang XP (2015) Amorphous Cu-added/SnO_x/CNFs composite webs as anode materials with superior lithium-ion storage capability. RSC Adv. 5(51):41210–41217. https://doi.org/10.1039/C5RA03658E
53. Li Q, Lan JL, Liu Y, Yu YH, Yang XP (2015) Carbon nanofiber-supported B_2O_3–SnO_x glasses as anode materials for high-performance lithium-ion batteries. RSC Adv. 5(108):89099–89104. https://doi.org/10.1039/C5RA19387G
54. Liu Y, Yan XD, Yu YH, Yang XP (2016) Eco-friendly fabricated porous carbon nanofibers decorated with nanosized SnO_x as high-performance lithium-ion battery anodes. ACS Sustain Chem Eng 4(6):2951–2959. https://doi.org/10.1021/acssuschemeng.5b01236
55. Liu Y, Yan XD, Lan JL, Yu YH, Yang XP, Lin YH (2017) Phase-separation induced hollow/porous carbon nanofibers containing in situ generated ultrafine SnO_x as anode materials for lithium-ion batteries. Mater Chem Front 1(7):1331–1337. https://doi.org/10.1039/C6QM00377J
56. Jung HR, Lee WJ (2011) Electrochemical characterization of electrospun SnO_x-embedded carbon nanofibers anode for lithium ion battery with EXAFS analysis. J Electroanal Chem 662(2):334–342. https://doi.org/10.1016/j.jelechem.2011.09.006
57. Zhou XS, Dai ZH, Liu SH, Bao JC, Guo YG (2014) Ultra-Uniform SnO_x/carbon nanohybrids toward advanced lithium-ion battery anodes. Adv Mater 26(23):3943–3949. https://doi.org/10.1002/adma.201400173
58. Joshi BN, An S, Jo HS, Song KY, Park HG, Hwang SW, Al-Deyab SS, Yoon WY, Yoon SS (2016) Flexible, freestanding, and binder-free SnO_x–ZnO/carbon nanofiber composites for lithium ion battery anodes. ACS Appl Mater Interfaces 8(14):9446–9453. https://doi.org/10.1021/acsami.6b01093
59. Xiong XH, Yang CH, Wang GH, Lin YW, Qu X, Wang JH, Zhao BT, Liu ML, Lin Z, Huang K (2017) SnS nanoparticles electrostatically anchored on three-dimensional N-doped graphene as an active and durable anode for sodium-ion batteries. Energy Environ Sci 10(8):1757–1763. https://doi.org/10.1039/C7EE01628J
60. Wei ZX, Wang L, Zhuo M, Ni W, Wang HX, Ma JM (2018) Layered tin sulfide and selenide anode materials for Li- and Na-ion batteries. J Mater Chem A 6(26):12185–12214. https://doi.org/10.1039/C8TA02695E
61. Lee DH, Park CM (2017) Tin selenides with layered crystal structures for Li-Ion batteries: interesting phase change mechanisms and outstanding electrochemical behaviors. ACS Appl Mater Interfaces 9(18):15439–15448. https://doi.org/10.1021/acsami.7b01829
62. Wang W, Li PH, Zheng H, Liu Q, Lv F, Wu JD, Wang S, Guo SJ (2017) Ultrathin layered SnSe nanoplates for low voltage, high-rate, and long-life Alkali-Ion batteries. Small 13(46):1702228. https://doi.org/10.1002/smll.201702228
63. Xia J, Liu L, Jamil S, Xie JJ, Yan HX, Yuan YT, Zhang Y, Nie S, Pan J, Wang XY, Cao GZ (2019) Free-standing SnS/C nanofiber anodes for ultralong cycle-life lithium-ion batteries and sodium-ion batteries. Energy Storage Mater. 17:1–11. https://doi.org/10.1016/j.ensm.2018.08.005
64. Zhang L, Lu L, Zhang DC, Hu WT, Wang N, Xu B, Li YM, Zeng H (2016) Dual-buffered SnSe@CNFs as negative electrode with outstanding lithium storage performance. Electrochim Acta 209:423–429. https://doi.org/10.1016/j.electacta.2016.05.106
65. Yuan HC, Jin YQ, Lan JL, Liu Y, Yu YH, Yang XP (2018) In situ synthesized SnSe nanorods in a SnO_x@CNF membrane toward high-performance freestanding and binder-free lithium-ion batteries. Inorg. Chem. Front. 5(4):932–938. https://doi.org/10.1039/C7QI00762K
66. Naille S, Dedryvere R, Martinez H, Leroy S, Lippens PE, Jumas JC, Gonbeau D (2007) XPS study of electrode/electrolyte interfaces of η-Cu6Sn5 electrodes in Li-ion batteries. J Power Sources 174(2):1086–1090. https://doi.org/10.1016/j.jpowsour.2007.06.043
67. Hu RZ, Zhang Y, Zhu M (2008) Microstructure and electrochemical properties of electron-beam deposited Sn–Cu thin film anodes for thin film lithium ion batteries. Electrochim Acta 53(8):3377–3385. https://doi.org/10.1016/j.electacta.2007.11.064

68. Hassoun J, Derrien G, Panero S, Scrosati B (2009) A SnSb–C nanocomposite as high performance electrode for lithium ion batteries. Electrochim Acta 54(19):4441–4444. https://doi.org/10.1016/j.electacta.2009.03.027
69. Shiva K, Rajendra HB, Bhattacharyya AJ (2015) Electrospun SnSb crystalline nanoparticles inside porous carbon fibers as a high stability and rate capability anode for rechargeable batteries. ChemPlusChem 80(3):516–521. https://doi.org/10.1002/cplu.201402291
70. Niu X, Zhou HM, Li ZY, Shan XH, Xia X (2015) Carbon-coated SnSb nanoparticles dispersed in reticular structured nanofibers for lithium-ion battery anodes. J Alloy Compd 620:308–314. https://doi.org/10.1016/j.jallcom.2014.09.150
71. Xia X, Li ZY, Xue LG, Qiu YP, Zhang CY, Zhang XW (2017) The electrochemical performance of SnSb/C nanofibers with different morphologies and underlying mechanism. J Mater Res 32(6):1184–1193. https://doi.org/10.1557/jmr.2016.508
72. Yuan ZX, Dong LX, Gao QL, Huang ZG, Wang LW, Wang GF, Yu XB (2019) SnSb alloy nanoparticles embedded in N-doped porous carbon nanofibers as a high-capacity anode material for lithium-ion batteries. J Alloy Compd 777:775–783. https://doi.org/10.1016/j.jallcom.2018.10.295
73. Xue LJ, Xu YF, Huang L, Ke FS, He Y, Wang YX, Wei GZ, Li JT, Sun SG (2011) Lithium storage performance and interfacial processes of three dimensional porous Sn–Co alloy electrodes for lithium-ion batteries. Electrochim Acta 56(17):5979–5987. https://doi.org/10.1016/j.electacta.2011.04.103
74. Jang BO, Park SH, Lee WJ (2013) Electrospun Co–Sn alloy/carbon nanofibers composite anode for lithium ion batteries. J Alloy Compd 574:325–330. https://doi.org/10.1016/j.jallcom.2013.05.063
75. Barakat NAM, Amen MT, Al-Mubaddel FS, Karim MR, Alrashed M (2019) NiSn nanoparticle-incorporated carbon nanofibers as efficient electrocatalysts for urea oxidation and working anodes in direct urea fuel cells. J Adv Res 16:43–53. https://doi.org/10.1016/j.jare.2018.12.003
76. Milanova V, Atanasova S, Avdeev G, Markova I (2017) Morphology of intermetallic (Co–Sn, Ni–Sn) nanoparticles, electrochemicaly tested as electrodes in li-ion battery. J Chem Technol Metal 52(3):542–556. https://doi.org/10.1515/rams-2018-0031
77. Arbizzani C, Beninati S, Lazzari M, Mastragostino M (2006) On the lithiation–delithiation of tin and tin-based intermetallic compounds on carbon paper current collector-substrate. J Power Sources 158(1):635–640. https://doi.org/10.1016/j.jpowsour.2005.06.042
78. Cho YJ, Kim CH, Im HS, Myung Y, Kim HS, Back SH, Lim YR, Jung CS, Jang DM, Park J, Lim SH, Cha EH, Bae KY, Song MS, Cho Wl (2013) Germanium–tin alloy nanocrystals for high-performance lithium ion batteries. Phys Chem Chem Phys PCCP 15(28):11691–11695. https://doi.org/10.1039/C3CP51366A
79. Lee H, Cho J (2007) $Sn_{78}Ge_{22}$@ carbon core—shell nanowires as fast and high-capacity lithium storage media. Nano Lett 7(9):2638–2641. https://doi.org/10.1021/nl071022n
80. Mao O, Dunlap R A, Dahn J R (1999) Mechanically alloyed Sn–Fe (–C) powders as anode materials for Li-Ion batteries: I. the Sn2Fe-C system. J Electrochem Soc 146(2):405–413. https://doi.org/10.1149/1.1391622
81. Zhao MM, Zhao QX, Qiu JQ, Xue HG, Pang H (2016) Tin-based nanomaterials for electrochemical energy storage. RSC Adv 6(98):95449–95468. https://doi.org/10.1039/C6RA19877E
82. Xue LG, Xia X, Tucker T, Kun F, Zhang S, Zhang XW (2013) A simple method to encapsulate SnSb nanoparticles into hollow carbon nanofibers with superior lithium-ion storage capability. J Mater Chem A 1(44):13807–13813. https://doi.org/10.1039/C3TA12921G
83. Lu WL, Luo CH, Li Y, Feng YY, Zhao YY, Yuan XY (2013) CoSn/carbon composite nanofibers for applications as anode in lithium-ion batteries. J Nanopart Res 15(9):1736. https://doi.org/10.1007/s11051-013-1736-9
84. Shen Z, Hu Y, Chen RZ, He X, Chen YL, Shao HF, Zhang XW, Wu KS (2017) Split Sn–Cu alloys on carbon nanofibers by one-step heat treatment for long-lifespan lithium-ion batteries. Electrochim Acta 225:350–357. https://doi.org/10.1016/j.electacta.2016.12.143

Chapter 14
Electrospun Carbon-Based Nanocomposites as Anodes for Lithium Ion Batteries

Apurva Patrike, M. Thripuranthaka, Golu Parte, Indrapal Karbhal, and Manjusha Shelke

14.1 Introduction

With increasing population, demand for energy sources is increasing globally because of insufficient fossil fuel. In the context of increasing energy demand, clean and efficient energy conversion and high capacity energy storage are becoming an urgent task. Rechargeable lithium ion batteries (LIB) are one of the promising electrochemical storage devices due to its high energy density and long working life. Hence, LIB has gained intense attention from academic community as well as industry in commercialization point of view. LIB technology has dominated in portable electronic market in past two decades. Nowadays, these are intensively used in transportation application like hybrid electric vehicles (HEV), electric vehicles (EV) and in stationary energy storage application like grid storage technology. Considering huge demand for energy and power density, tremendous research has been devoted on LIB to improve its electrochemical performance. The electrochemical performance of LIB including high capacity, high energy density, operation potential and current densities with longer working life is mainly depending on electrode materials [1, 2]. State-of-the-art LIB utilizes Li containing metal oxides/phosphates as cathode and graphite as anode material. Electronic and ionic conductivity are the two crucial factors for any material which is considered to be electrode material for LIB. These two parameters can be improved through doping, nanostructure engineering and carbonaceous material addition [3].

A. Patrike · M. Thripuranthaka · G. Parte · I. Karbhal · M. Shelke (✉)
B101, Polymer and Advanced Materials Laboratory Physical & Material's Chemistry Division, CSIR-National Chemical Laboratory, Pune, India
e-mail: mv.shelke@ncl.res.in

Academy of Scientific and Innovative Research (AcSIR), Ghaziabad 201002, UP, India

N. T. M. Balakrishnan and R. Prasanth (eds.), *Electrospinning for Advanced Energy Storage Applications*, Materials Horizons: From Nature to Nanomaterials,
https://doi.org/10.1007/978-981-15-8844-0_14

At present, graphite is used as anode in commercial LIB with limited capacity (372 mAh g^{-1}) and long Li^{+}-ion diffusion path. Tremendous efforts have been taken to explore new materials as an anode for LIB. Out of which the number for carbon-based materials is more as carbon has promising properties like low cost, safety, easy availability and also it has been conventionally utilized in commercial LIBs. However, the development of new carbon materials is at most important in order to meet requirement of ever increasing energy demand, growing number of electronic devices and large-scale applications for stable electrochemical performance in LIB.

Recently, the number of carbon materials with different nanostructures has been explored as an anode for LIB like graphene [4], carbon nanotubes [5], carbon nanofibers [6], etc. Out of which carbon nanofibers attracted more attention due to its one-dimensional (1D) nature which provides favorable properties for electrochemical storage of Li^{+}-ions in LIB. CNF provides good access for electrolyte due to enhanced surface-volume ratio and conductivity along the length. In high surface area, CNF provides shorter path for Li insertion, efficient 1D electron transport along length compared to powder materials which ultimately gives lower resistance and high capacity [7, 8].

Several methods have been employed for CNF synthesis like catalytic synthesis [9], electrospinning technique [10], chemical vapor deposition growth [11] and template-based methods [12]. Out of which the electrospinning technique is simple, versatile and cost effective. It can be used from laboratory level synthesis to industry level large-scale synthesis to prepare various CNFs and its composites. Electrospinning provides an advantage of synthesis of continuous CNFs with controllable morphology, porous CNFs, core-shell CNFs and metal loading into CNFs [13]. Diameter of CNFs produced from electrospinning technique varies from few nanometers to few hundred nanometers.

14.1.1 Working Principle of Electrospinning

This technique uses precursors dissolved in suitable solvent as starting material to draw nanofibers through application of high electric potential (several kV range). The high voltage is applied between the needle of syringe comprised of precursor solvent and metallic plate which is used to collect nanofibers (collector) and is grounded. With instrumental parameters like voltage, distance between syringe and metal collector, flow rate of precursor solvent and viscosity of solvent, one can adjust diameter and length of nanofibers. The shape of collector also plays an important role and there are two types of collectors, viz., plate collector and drum collector. Nanofibers collected on plate collector are not orientated in a particular direction, whereas drum collector provides orientated nanofibers aligned in a particular direction. Hence, the synthesis of nanofibers with desired physical properties is achieved in electrospinning technique [10]. Schematic of electrospinning technique with different collector is shown in Fig. 14.1.

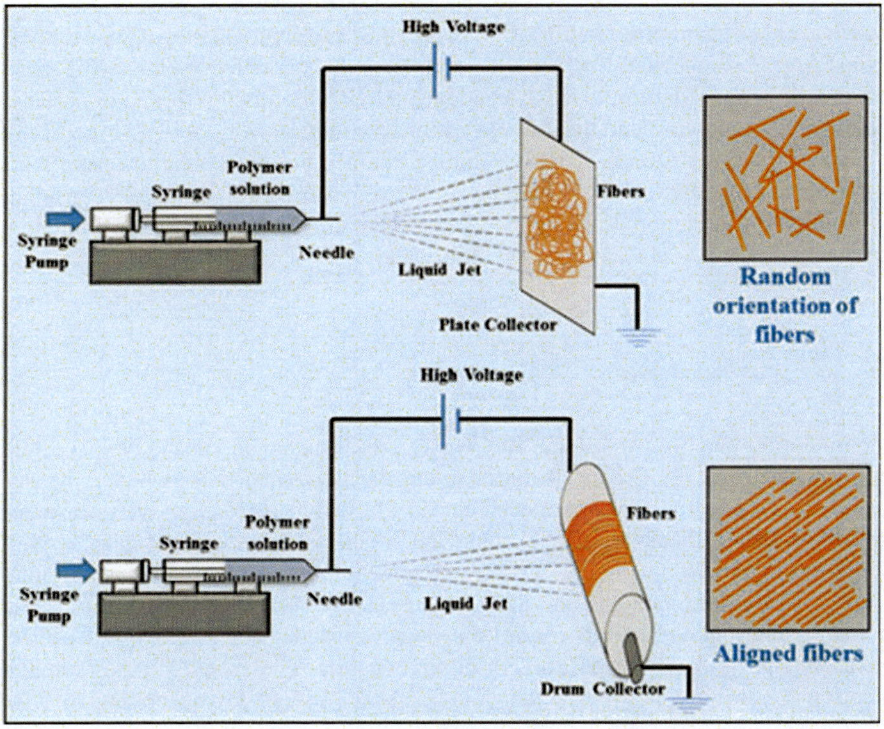

Fig. 14.1 Schematic of electrospinning technique with plate and drum collector

The current chapter emphasizes on electrospun carbon nanofibers as anode materials for LIB including recent trends, scientific developments and strategies to improve electrochemical performance of CNF and detailed synthesis protocols. Moreover, it covers the importance of composites of CNF for better performance.

14.2 Electrospun Fibers

14.2.1 Solid Fibers

Various synthetic and natural polymers have been utilized for synthesis of CNFs via electrospinning technique. The commonly used polymers are polyvinyl alcohol (PVA) [14], polyvinyl pyrrolidone (PVP) [15], polyacrylonitrile (PAN) [16] and cellulose [17]. The physical and chemical properties of CNF are determined by the precursor. Out of these PAN has been extensively used as carbon precursor as it has good spinnability property and can yield good amount of carbon after stabilization and carbonization process. Synthesis of PAN-based CNFs is performed in three stages which are electrospinning, stabilization of polymer and carbonization. CNFs derived from PAN possess robust structure and good electrical conductivity. Hence, the direct use of CNF web can be employed as an anode for LIB which avoids the use of polymer binder and conducting additives which ultimately reduces weight of LIB cell [18]. Yang et al. [18] has synthesized N, P and Si tridoped nonporous carbon nanofibers with increased N content in carbon matrix to use it as electrode for supercapacitor. This nonporous structure has demonstrated to find out heteroatom doping amount in carbon matrix with enhanced rate capability and specific capacitance.

14.2.2 Porous Fibers

Incorporation of porous structure in CNFs increases specific capacitance and rate performance for LIB. This is attributed to interconnected particles which enhances surface for charge transfer and better electrode–electrolyte interaction. Various strategies have been utilized to make porous CNFs. Chan Kim et al. [19] have synthesized porous carbon fibers via electrospinning technique with two stable and immiscible polymer solutions and thermal treated at 1000 °C. Use of two immiscible polymer solutions utilized to control the pore size and surface area of nanofibers by changing the blend proportions in different ratios. The schematic diagram and electron microscopy images are shown in Figs. 14.2 and 14.3, respectively.

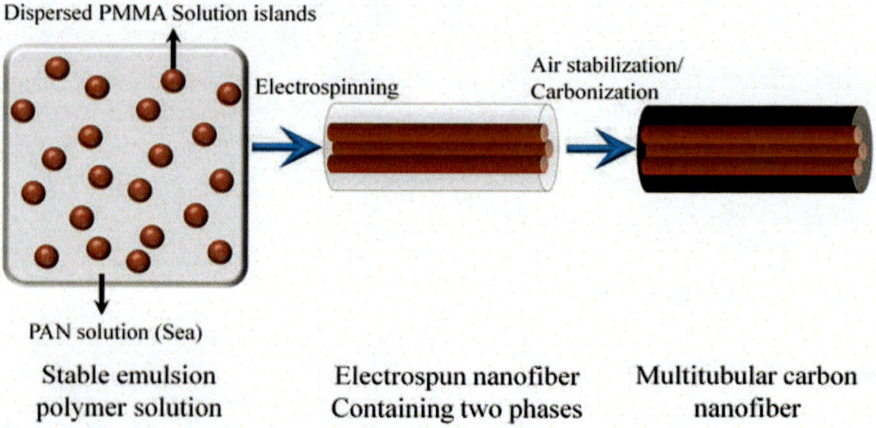

Fig. 14.2 Schematic diagram of procedural steps involved in synthesis of porous nanofibers. Adapted and reproduced from Ref. [19], Copyright 2007 John Wiley and Sons

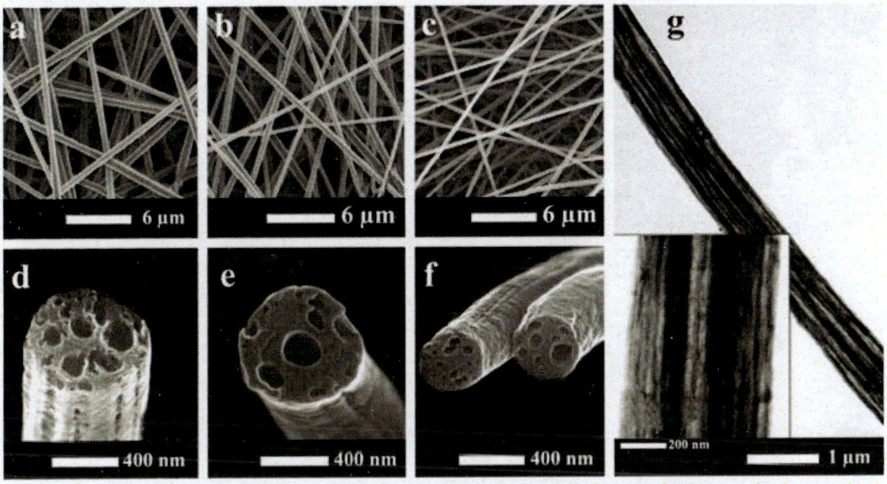

Fig. 14.3 **a**, **b**, **c** Morphology of electrospun nanofibers comprising of two polymer phases. PAN:PMMA **a** 5:5, **b** 7:3 and **c** 9:1. **d**, **e**, **f** Cross-sectional FESEM images of thermally treated nanofibers at 1000 °C with similar PAA:PMMA ratios 5:5, 7:3 and 9:1, respectively, **g** TEM images of 5:5 PAN:PMMA sample. The inset is a magnified TEM image. Adapted and reproduced from Ref. [19], Copyright 2007 John Wiley and Sons

14.2.3 Core-Shell/Co-axial Fibers

In certain cases, the functionalization of nanofibers over surface through functional coatings or particles is required to fulfill the requirements. Electrospinning technique is a versatile technique for fabrication of such fibers. This involves spinning of two different polymers precursor solutions via concentrically aligned nozzles. Same potential is applied to spin the fibers and the thickness of shell can be maintained by flow rate of precursor solution. Depending on requirements, materials for shell and core are decided [20]. Yang et al. [21] fabricated hollow CNF from PAN with styrene-co-acrylonitrile (SAN) polymer as sacrificial core materials and later it is filled with Si nanoparticles. These Si nanoparticles stuffed CNFs help to control volume expansion of Si during lithiation/de-lithiation process, and CNFs matrix over the surface of Si provides mechanical support path lengths for charge transfer. TEM images of hollow CNF and Si encapsulated hollow CNFs are shown in Fig. 14.4.

14.3 Synthesis of CNF from Different Precursors

Carbon nanofibers can be synthesized by different methods such as vapor growth, chemical vapor deposition, laser ablation, arc discharge and electrospinning. Among these, electrospinning provides a simple and scalable process for the production

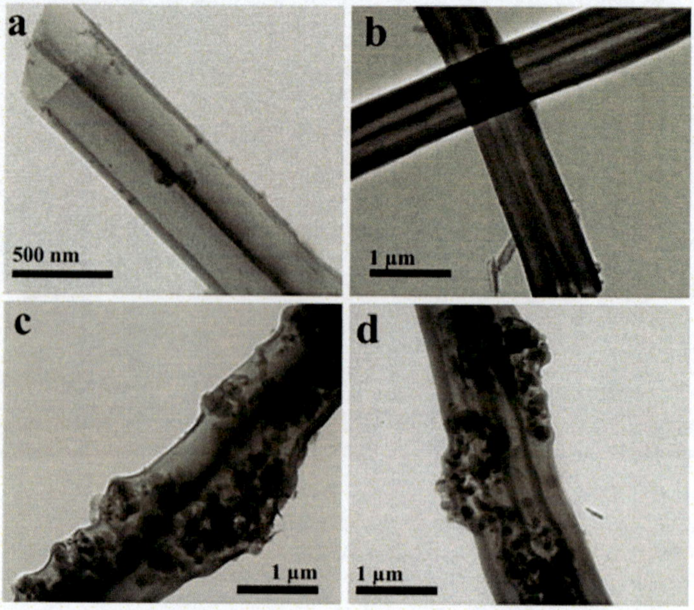

Fig. 14.4 Transmission electron microscopy (TEM) images of **a** two-channeled hollow CNF, **b** four-channeled hollow CNF, **c** Si encapsulated two-channeled hollow CNF and **d** Si encapsulated four-channeled hollow CNF. Adapted and reproduced from Ref. [21], Copyright 2014 Royal Society of Chemistry

of carbon nanofibers. In typical CNF synthesis, electrospinning of polymer solution followed by stabilization and carbonization of electrospun polymer fibers into CNFs. Various factors determine the structure and properties of the electrospun CNFs which includes (1) solution properties like polymer, concentration, electrical conductivity and surface tension; (2) processing conditions such as strength of applied electric field, flow rate of polymer solution and distance between the spinneret and the collector. Along with these, surrounding temperature and humidity also affect the morphology and properties of CNFs. CNFs can be prepared by any material having carbon chain. Although the variety of polymer precursors have been electrospun for the CNF synthesis, commonly used polymers are PAN, PVA, polyimide, PVdF, PVP and phenolic resins have been used for the synthesis of electrospun CNF. The electrospinning parameter and polymers employed for the synthesis of carbon nanofibers by various research groups have been described here.

PAN has been extensively used to get the electrospun CNFs due to its high carbon yield, easy processing and mechanical stability of synthesized carbon nanofibers. Kim et al. developed carbon nanofibers web (CNFW) from PAN as a carbon precursor. Electrospinning solution was prepared by dissolving PAN in DMF. Electrospun carbon fibers were stabilized at 280 °C in air for 1 h and further carbonized at 1000 °C in inert atmosphere to obtain carbon nanofibers web. The structural variations of CNFW with different heat treatment temperatures (700–2800 °C) were studied [22–24].

Despite the enormous success of PAN as carbon precursor for the CNF synthesis, its solubility in many solvent is relatively low. DMF is most used as solvent to make electrospinning solution of PAN. To counter this problem, P. Wang and coworker used PVP as carbon precursor for the CNF synthesis as its solubility is better in solvent, including water and ethanol, cheaper than PAN and environmentally benign. PVP was dissolved in ethanol to prepare precursor solution for electrospinning. The electrospun nanofibers were stabilized at 150 °C for 24 h in air and pre-oxidized for 4 h at 360 °C and annealed in N_2 atmosphere at 800 °C for 4 h to produce CNFs and investigated as an electrode material for Li-ion battery [15]. Figure 14.5 shows the SEM images of PVP synthesized CNFs. Dong et al. [25] used PVP as a carbon precursor in DMF to synthesize carbon nanofibers via electrospinning. The electrospun PVP nanofibers were stabilized at 300 °C for 3 h and then annealed at 500 °C in argon atmosphere for 3 h to obtain CNF. The PVP synthesized CNFs was doped with cobalt and used as an electrode material for Li-ion battery.

PVA, another water soluble polymer also studied as a carbon precursor for the synthesis of CNFs. Although PVA does not withstand high temperature and gives low carbon yield, Ding et al. [26] synthesized PVA cross-linked carbon nanofibers to study the factors that affect the morphology and porosity of PVA CNFs. The morphology mostly influenced by solution concentration and applied voltage. Some

Fig. 14.5 SEM images of PVP synthesized CNFs. Adapted and reproduced from Ref. [15], Copyright 2012 Royal Society of Chemistry

other reports are on PVA as carbon precursor for the CNFs synthesis. W. K. Sun studied the effect of pH on electrospinning of poly(vinyl alcohol). They found that in acidic condition CNFs are not continuous [14, 27].

Hong et al. produced porous CNFs prepared from PVdF for CO_2 capture. They dissolved PVdF in mixture of acetone and N, N-dimethylacetamide to form 11% PVdF solution. This solution was electrospun to get nanofibrous mat, which dehydrofluorinated to stabilize the CNFs and then carbonized for 1 h in the temperature range of 300–1000 °C to obtain porous carbon nanofibers. Various other papers are also available for the synthesis of PVdF-based CNFs for Li-ion batteries. Figures 14.6 and 14.7 show the schematic and representative SEM images of PVdF synthesized CNFs, respectively [28–31].

For the manipulation of morphology and porosity in CNFs, mainly two strategies have been employed sacrificial and activation methods. Different nanostructures such as SiO_2, nano-$CaCO_3$ or polymers such as PVA, polymethyl methacrylate (PMMA), PVP, poly (ethylene oxide), Nafion, polysulfone, polystyrene, poly-L-lactic acid have been introduced during polymer precursor solution preparation as a sacrificial component. By thermal or chemical treatments, these sacrificial components (i.e., nanostructures or polymer) extracted from the electrospun fiber that helps in creating porosity.

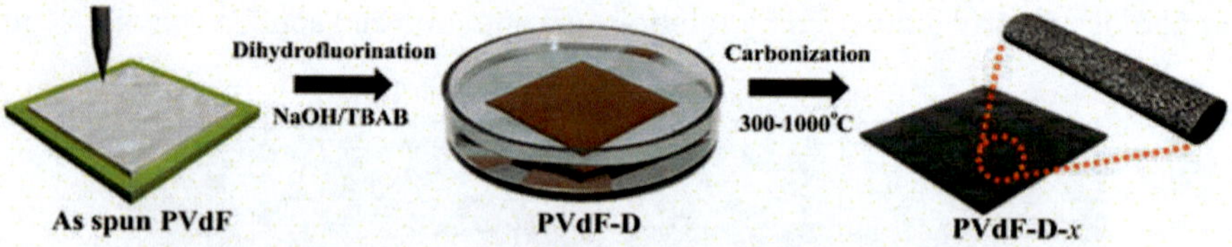

Fig. 14.6 Schematic for synthesis of PVdF-based CNFs. Adapted and reproduced from Ref. [28], Copyright 2014 Royal Society of Chemistry

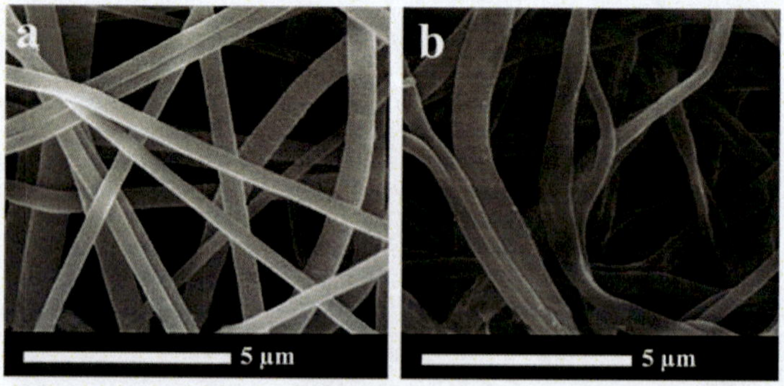

Fig. 14.7 SEM images of synthesized PVdF-based CNFs. Adapted and reproduced from Ref. [28], Copyright 2014 Royal Society of Chemistry

L. Ji and coworker prepared porous carbon nanofibers by adding SiO_2 as porogen in PAN/DMF precursor solution. The electrospun PAN/SiO_2 composite nanofibers were then stabilized at 280 °C and then carbonized at 700–1000 °C in N_2 atmosphere for an hour to get carbon/SiO_2 composite nanofibers. By HF treatment, SiO_2 nanoparticles were removed to create porous carbon nanofibers [32, 33]. Zhang et al. prepared hierarchical porous carbon nanofibers by electrospining a polymer solution composed of polyacrylonitrile (PAN) and nano-$CaCO_3$ in DMF. They first dispersed nano-$CaCO_3$ in DMF/THF and then added PAN to make polymer precursor solution [34].

Numerous papers have been reported on the preparation of bicomponent polymer fibers comprising PAN as primary carbon precursor and integration of sacrificial polymers in PAN/DMF solution. These composite nanofibers get stabilized during carbonization by molecular interaction between the both the polymers.

Peng et al. reported electrospinning of 12 wt% PAN/PMMA bicomponent polymer precursor in DMF. The stabilization was carried out at 250 °C in air for 6 h followed by carbonization at 800 °C for 1 h in N_2 atmosphere to obtain porous CNF. The PAN without PMMA showed long and bread-free morphology while with PMMA it forms interconnected network. This is due to melting of PMMA during carbonization which leads to fibers-fiber connection. Figure 14.8 shows the representative SEM images of PAN and morphology development with addition of PMMA in PAN synthesized CNFs. Several other papers have been reported to improve the fiber structure and morphology by adding thermally liable polymer [35–39].

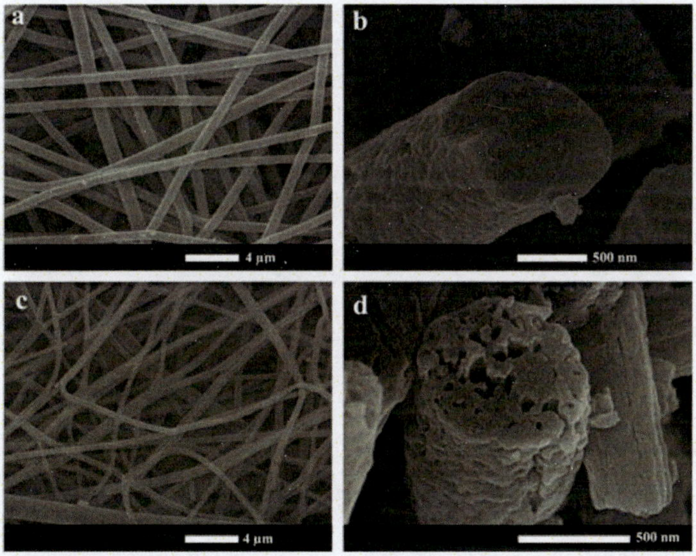

Fig. 14.8 SEM images of PAN/PMMA-based CNFs. Adapted and reproduced from Ref. [35]. Copyright 2015 Springer Nature

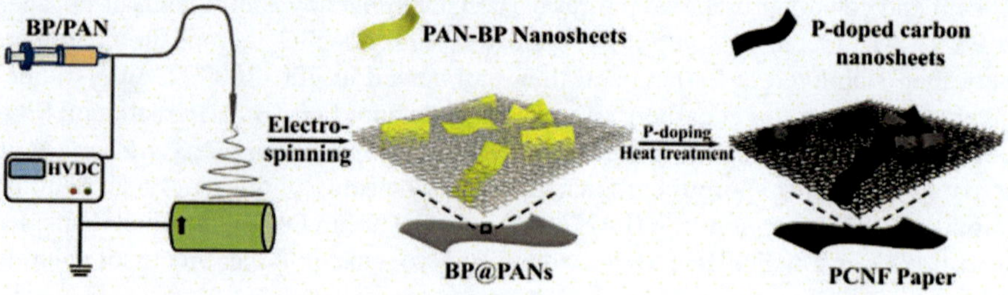

Fig. 14.9 Schematic illustration of preparation of PCNF paper. Adapted and reproduced from Ref. [43], Copyright 2018 Elsevier

Heteroatom (e.g., boron, nitrogen, sulfur and phosphorus) doping in carbon material is an effective way to tailor their electronic and chemical properties. These doping enhances the electrochemical performance of the materials. Nan et al. synthesized N-doped carbon nanofibers using PAN and melamine. In this, the solution concentration of 10 wt% PAN in DMF was prepared and melamine was added in the 2:1 ratio of PAN/DMF: melamine. This solution was electrospun and stabilized at 250 °C for an hour and carbonized at 850 °C for 1 h to obtain nitrogen-doped carbon nanofibers networks. These fibers further activated by 20% ammonia in N_2 atmosphere to get N_2-rich carbon nanofibers [40–42].

Li et al. reported the phosphorus-doped carbon nanosheets/nanofibers free-standing paper from PAN as carbon precursor in DMF, and subsequently black phosphorus/red phosphorus was added to form electrospun solution. The obtained electrospun fibers annealed at 800 and 900 °C in N_2 atmosphere for 1 h to form carbon nanosheets/nanofibers. Phosphorus enhances the electrocatalytic activity of carbon fibers. Figure 14.9 shows the schematic of synthesis of P-doped carbon nanosheets/nanofibers [43].

From the given synthesis protocols, it is evident that the electrospinning has emerged as a promising technique for the synthesis of carbon nanofibers from variety of polymers. The morphology and structure properties of electrospun nanofibers can be modified and controlled in number of different chemical or physical methods. These nanofibers material have been utilized in large number of applications such as catalysis, sensors, biomedical, adsorption, energy conversion and storage applications.

14.4 CNF-Based Nanomaterials as Anode for LIB

Electrospun carbon nanofibers are attracted toward energy storage application due to unique 1D physical and chemical property. Long fiber length of CNF provides easy access for Li^+-ions to the innermost area of anode. CNFs are extensively researched as an anode for lithium ion batteries. Some of the selected research works are discussed in this section of chapter.

14.4.1 CNF-Based Anode

C. Kim et al. have synthesized CNF from PAN polymer from electrospinning followed by stabilization at 260 °C in air and thermally treated at 1000 °C in inert atmosphere. Further, this CNF has shown high capacity of 435 mAh g^{-1} at 30 mA g^{-1} after two cycles. The authors elaborated the detail study of temperature (700, 1000 and 2800 °C). The study shows that CNF prepared at 1000 °C exhibited highest electrochemical performance as interlinked CNF provides high electrical conductivity [44]. Utilization of electrospinning parameters comes with improved capacity by tuning CNF in different morphology and incorporation of other active materials. One of such strategy was incorporated to make porous hollow CNFs (pHCNFs) via co-axial electrospinning by Lee et al. Improved electrochemical performance for pHCNFs is elaborated by authors where styrene-co-acrylonitrile (SAN) was used as sacrificial material for core and for porosity generation on shell of PAN-based HCNF. pHCNFs were obtained after subsequent heat treatments where SAN get decomposed and burnt. Improved LIB performance was obtained for porous pHCNFs as porous nature of CNFs improves intecalation/de-intercalation mechanism. Initial discharge capacity for pHCNFs and HCNFs was 1003 mAh g^{-1} and 653 mAh g^{-1}, respectively, at 50 mA g^{-1}. The initial capacity at 200 mA g^{-1} was reduced to 827 mAh g^{-1} for pHCNFs. This is because at high current rates, intercalation gets disturbed. The schematic and battery performance for the same is shown in Fig. 14.10(a) and Fig. 14.10(b) respectively [45].

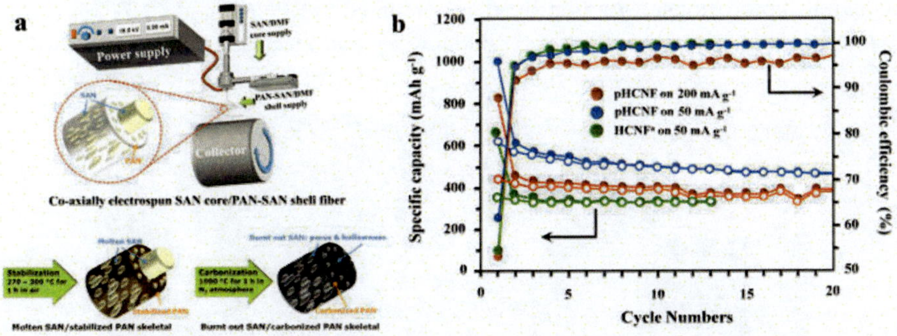

Fig. 14.10 (**a**) Schematic for synthesis of pHCNFs. (**b**) Battery performance. Adapted and reproduced from Ref. [45], Copyright 2012 American Chemical Society

To tailor electronic and chemical properties of CNFs, chemical doping into carbon with heteroatom is an effective and beneficial strategy. Many dopants have been used for improvement in conductivities like nitrogen [46], phosphorous [47], boron [48] and sulfur [49]. Among these, the nitrogen is an effective dopant as it has atomic radius (56 pm) similar to carbon (67 pm) and has more electronegativity (3.04) as compared to that of carbon (2.55). N doping gives following benefits to carbon: (1) As N has lone pair of electrons, it can introduces donor states which gives n-type conducting nature, (2) N can easily bond with carbon and this helps for easy Li^+-ion insertion [50], (3) N introduces defect sites in carbon which improves Li^+-ion storage reservoir sites [40]. Based on these, in order to achieve high performance for LIB with good rate capability, high capacity and long cycling stability, N doping to porous carbonaceous material is highly desirable. D. Nan et al. developed N enriched porous CNF through electrospinning method as free-standing anode for LIB application as shown in Fig. 14.11a. Typical synthesis procedure involves PAN and melamine as precursors for carbon and nitrogen. Porous N enriched CNFs (NPCNFs) were achieved after stabilization and NH_3 treatments of electrospun fibers. TEM image of NPCNFs is shown in Fig. 14.11b. NPCNFs showed high initial capacity of 1323 and 1150 mAh g^{-1} after 50 cycles at 50 mA g^{-1} current density value as shown in Fig. 14.12a. It indicates that N doping in porous CNFs provided much improved Li^+-ion storage performance with good rate capability (Fig. 14.12b) as compared to nonporous and without doped CNFs [39].

Moreover, the phosphorous doping also contributes to improve electronic and chemical properties of carbon. Li et al. developed phosphorus-doped carbon nanosheets/nanofibers via electrospinning and subsequent heat treatments to use it as anode for LIB. The electrode was fabricated in flexible and free-standing form which allows complete utilization of active material in electrochemical tastings. These types of free-standing flexible electrodes are need for the next generation flexible batteries. Black phosphorus (BPCNF) and red phosphorus (RPCNF) doping in carbon nanosheets/nanofibers are demonstrated in this work. With BP doping electrochemical performance got improved as compared to RPCNF and CNF, this could be due to enhancement in electrical conductivity and electrochemical reactivity of

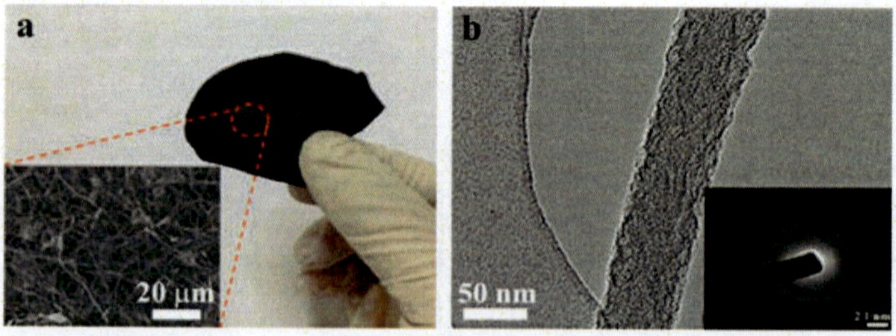

Fig. 14.11 **a** Free-standing electrode images with microscope image in inset, **b** TEM of NPCNF with SAED pattern in inset. Adapted and reproduced from Ref. [40], Copyright 2014 Royal Society of Chemistry

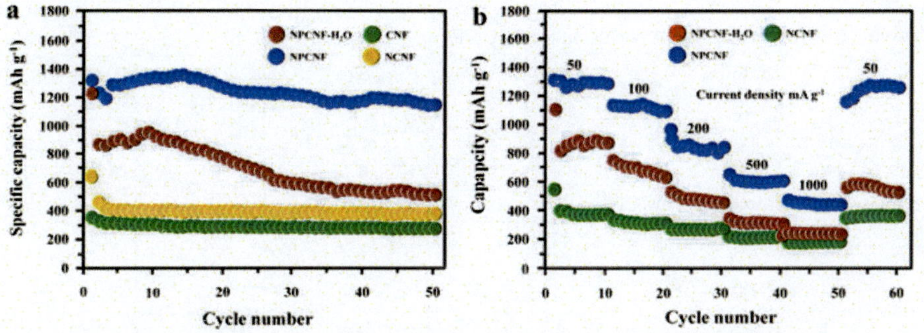

Fig. 14.12 **a** Cycling stability at 50 mA g^{-1} current density and **b** rate performance of different samples. Adapted and reproduced from Ref. [40], Copyright 2014 Royal Society of Chemistry

carbon upon doping with BP. Among all allotropes of phosphorus, BP is thermodynamically stable one which provides high P doping during thermal treatment. This is attributed to improved capacity value for BPCNF as compared to RPCNF and CNF as shown in Fig. 14.13. At 1 A g^{-1} current density, after 700 cycles capacity values for BPCNF-800, RPCNF-800 and CNFs are 607 mAh g^{-1}, 356 mAh g^{-1} and 192 mAh g^{-1}, respectively, where 800 indicates annealing temperature of electrospun nanofibers. The self-supporting flexible CNF electrode accommodated volume changes during lithiation/de-lithiation as well as provided conductivity to electrode. This works direct research in the field of flexible electrodes and BP material as an anode for LIB [43].

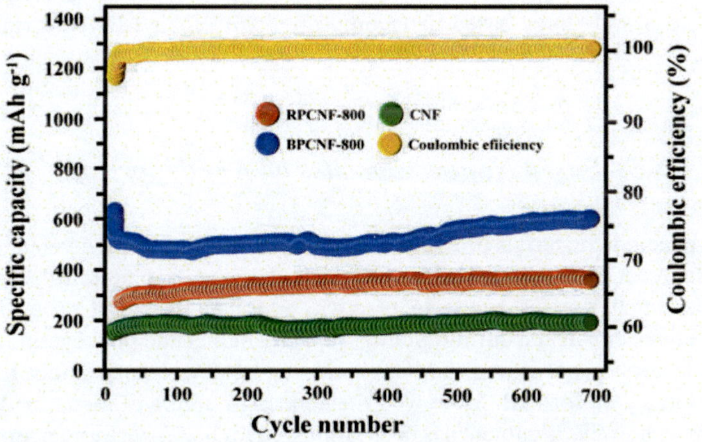

Fig. 14.13 Cycling performance of BPCNF-800, RPCNF-800 and CNFs at 1 A g^{-1} current density. Adapted and reproduced from Ref. [43], Copyright 2018 Elsevier

Table 14.1 Electrochemical performance of CNFs as anode for LIB

S. No.	Material	Surface area ($m^2\ g^{-1}$)	Capacity ($mAh\ g^{-1}$), cycle number	Current density ($mA\ g^{-1}$)	Reference
1	CNFs	NA	450, 2	30	[44]
2	Porous hollow CNF	13	501, 10	50	[45]
3	Free-standing N enriched porous CNF	1198	1150, 50	50	[40]
4	Black phosphorus-doped CNF (BPCNFs)	NA	607, 700	1000	[43]
5	N-doped porous CNF web	2381	943, 600	2000	[51]

Besides these, the porosity incorporation to CNF is also equally important to improve electrochemical performance of battery. Pores in CNF provide high surface area, more surface exposure for electrode/electrolyte interface and more charge transfer. In Table 14.1, papers reporting CNFs for LIB application are summarized with electrochemical performance. Though carbon is low cost, good cycling stability and lower electrochemical potential, its performance as an anode gives less reversible capacity value which turns in lower energy and power density. Hence, the development of nanocomposite of carbon with other high capacity materials like transition metal oxides/sulfides is a next step to achieve high capacity value with good cycling stability and high energy density. In the next section of chapter, transition metal oxides/sulfides and CNFs-based nanocomposites with their LIB performance are discussed.

14.4.2 CNFs-Based Nanocomposites as Anode for LIB

Metal-based anode materials provide more number of Li^+-ions involvement in electrochemical reaction as compared to commercially used graphite. Still, the utilization of these materials as anode in practical applications is not promising as it involves volume expansion during cycling, capacity fading and poor cycling stability. Various transition metal oxides (TMOs) including binary, ternary oxides have been researched as anode for LIB. Moreover, the nanoenginnering of metal oxides to get different morphology, alignment has been employed for better performance of oxides. However, the hurdle of low diffusion rate of Li^+-ion and poor electrical conductivity of metal oxides lowers the electrochemical performance and end in poor Li^+-ion storage [52]. To solve this issue, carbon may serve as support electrical conductivity to TMOs. Out of different carbon morphologies, CNFs are most suitable candidate as 1D CNFs may control volume expansion caused in TMOs during electrochemical reactions.

High energy density, long cycle life as well as high rate capability is an essential parameter of Li^+-ion battery for the next generation electric vehicle and renewable energy storage device [53–55]. Current state-of-the-art graphite anode-based material is not able to meet the require energy density with their limited theoretical capacity of 372 mAh g^{-1}, and on the other hand, metal oxide has high theoretical capacity of >800 mAh g^{-1}. These metal oxides have one of the greatest problems of volume expansion and deformation of structural morphology resulting capacity fading [53, 56, 57]. To overcome of this problem and to get require essential energy density, one-dimensional carbon nanofiber and metal oxide-based composite are the possible and cheap way to improve the energy density. Also, it have several advantages including distinct electrical contact, high electron and Li^+-ion transport, unique electronic conduction, strain relaxation, outstanding durability and short diffusion pathways. Electrospun-based carbon nanofiber can prevent the integrating of metal oxide NPs, deterioration as well from volume expansion [58].

There are several electrospun-based syntheses have done for the CNF/metal oxide composite to meet the required amount of energy and power density. Fe_2O_3-based composite is one of the highly studied materials with CNF due to it has high theoretical capacity of 1005 mAh g^{-1}, natural abundance, low cost and environmental-friendly nature. Figure 14.14a shows the design and synthesis of bubbled-nanorod-structured Fe_2O_3-carbon nanofiber, which has been synthesized by Chao et al. [59] Another metal oxide which has been widely used as composite with CNF is SnO_2 because of the high theoretical capacity of 1494 mAh g^{-1}. Liu et al. has been synthesized SnO_x decorated CNF for the high performance LIBs [60]. There are several metal oxides, like Co_2O_3, CuO, SiO_2/Sb and TiO_2 have also been synthesized by different groups [59, 61–63]. One of the most excellent ideas to make CNF-based is to protect the volume expansion, cracking and crumbling of the electrode material resulting continual formation of unstable and insulating solid electrolyte interphase (SEI) layer, which are one of the reason for the loss of electric contact between electrode material and current collector as well as drastically capacity fading.

The electrochemical LIBs performance of Fe_2O_3 CNF has been shown in Fig. 14.15. The cyclic voltammogram of bubbled-nanorod-structure Fe_2O_3-C composite nanofiber is at 0.01 mV s^{-1}. The reduction peak observed at 0.7 V versus Li^+/Li which is due to reduction of Fe(III) to Fe (0), formation of Li_2O as well as irreversible reduction of electrolyte and formation of SEI layers. The appearance of the two anodic peaks at 1.5 and 1.8 V is because of the reversible oxidation of Fe (0) to Fe (II) and Fe (II) to Fe (III), respectively. Moreover, the peak at 0.5 V could be the partial decomposition of SEI layer. The hollow Fe_2O_3 and bubbled-nanorod-structured Fe_2O_3-C composite nanofiber are able to achieve initial first discharge and charge capacity of 1406 and 1145 mAh g^{-1}, 1335 and 1957 mAh g^{-1}, respectively.

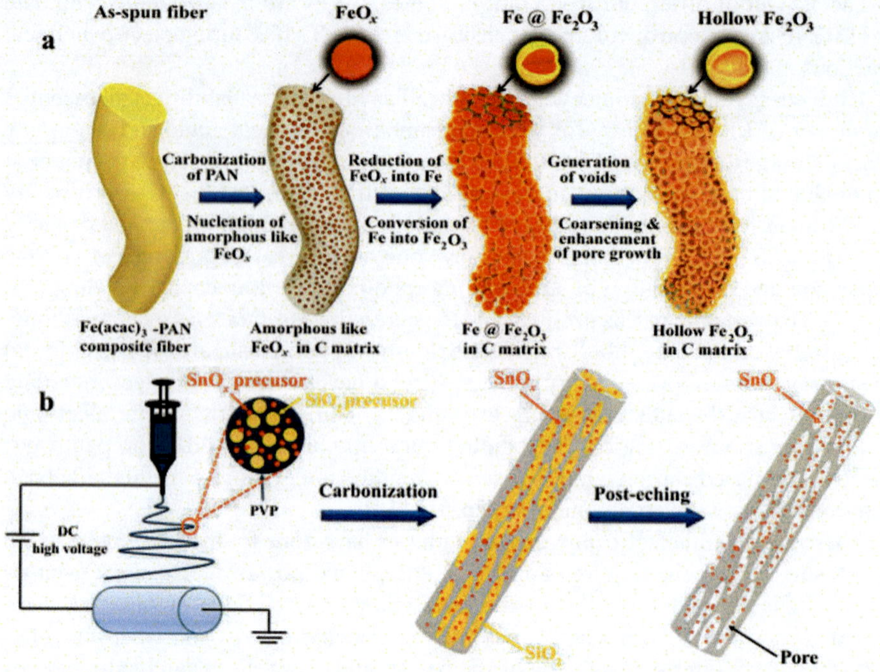

Fig. 14.14 **a** Synthesis of hollow Fe_2O_3 in C matrix, [59]. **b** SnO_x decorated porous carbon nanofibers. Adapted and reproduced from Ref. [60], Copyright 2016 American Chemical Society

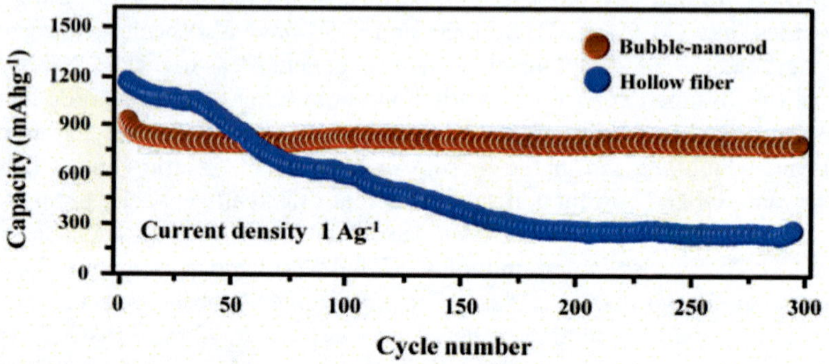

Fig. 14.15 Cyclic performance of bubbled-nanorod-structure Fe_2O_3-C composite nanofiber at 0.1 A g^{-1} Adapted and reproduced from Ref. [59], Copyright 2015 American Chemical Society

Figure 14.15 shows the stability of bubbled-nanorod-structured Fe_2O_3-C composite nanofiber at current density of 1 A g^{-1} up to 300 cycles. The initial capacity loss of first few cycles can be attributed to the partial destruction of the internal structure and decomposition of electrolyte as well as formation of the SEI layers.

The cyclic performance of three different compositions of SnO_x@PCNF and PCNF at 0.5 A g^{-1} is evaluated. Among all the samples, SnO_x@PCNF-2 has highest discharge capacity of 684 mAh g^{-1} and after 100 cycles retention capacity is 57.7%, and it exhibits rate performance with a reversible capacity of 819, 639, 468 and 323 mAh g^{-1} at a different current density of 0.2, 0.5, 1 and 2 A g^{-1}, respectively. Figure 14.16a, b shows charge and discharge capacity respectively for long cycle stability of SnO_x@PCNF-2 and even after 900 cycles the reversible capacity is 511 mA g^{-1}.

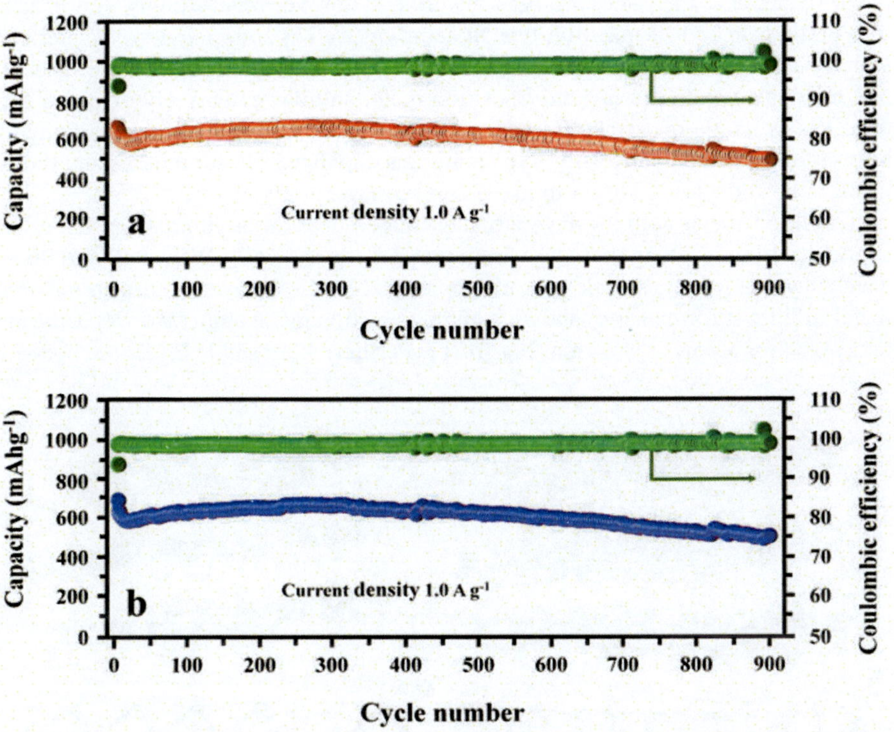

Fig. 14.16 **a** Charge capacity, **b** discharge capacity of SnO_x@PCNFs-2 for long term cycling at current density 1 A g^{-1} up to 900 cycles. Adapted and reproduced from Ref. [60], Copyright 2016 American Chemical Society

14.4.3 Transition Metal Sulfides Composites with Electrospun Carbon Nanofibers as Anodes for Lithium Ion Batteries

As discussed in the previous sections, there are several materials used as anodes for lithium ion batteries due to their interesting properties. As known, there are different types of anode materials such as insertion-type materials which are layered in structures and conversion-type anode materials which are generally transition metal oxides and then alloying type materials. However, interestingly, few materials participate both as insertion-type materials as well as conversion-type materials which has the advantage of gaining enhanced specific capacity. Here, we discuss few materials which are layered transition metal sulfides extensively explored for the LIBs as anode materials. Even though the transition metal sulfides participate as insertion-type and conversion-type materials they have few limitations and those needs to be addressed in order to make them commercially viable.

Transition metal sulfides have attracted huge attention as lithium ion batteries anode materials. Among these transition metal sulfides, MoS_2, WS_2, SnS and SnS_2 have a layered structure analogous to graphene. The single layer of these transition metal sulfides is formed as a sandwich of the transition metal atom (Mo, W) between the two sulfur atoms as shown in Fig. 14.17. Further, these single layers are bonded

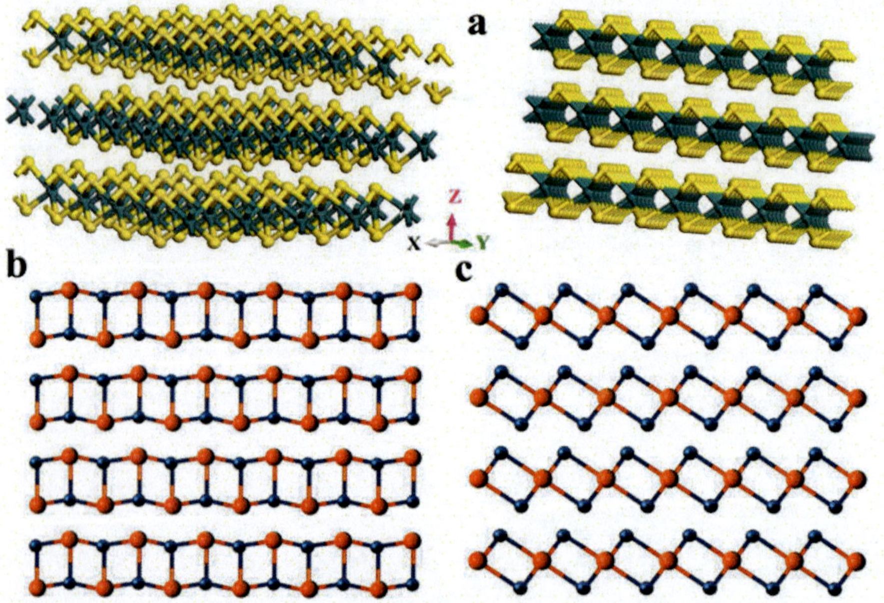

Fig. 14.17 Layered structure of **a** MoS_2. Adapted and reproduced from Ref. [67], Copyright 2020 Elsevier. **b** SnS_2 and SnS. **b** and **c** The blue balls are tin atoms, whereas the yellow balls are sulfur atoms. Adapted and reproduced from Ref. [68], Copyright 2017 Springer

to their adjacent layer through weak van der Waal's forces between the sulfur atoms [64]. And the atoms within the layer are bonded through a strong covalent bond. The layered nature of SnS and SnS_2 are also found to be akin to that of MoS_2 and WS_2 [65]. Since the layers of transition metal sulfides are bonded through week van der Waal's force, the Li^+-ions can easily undergo intercalation and de-intercalation to and from the layered structures, which is advantageous for the fast charging and discharging of the batteries based on these materials [66].

In the case of layered MoS_2, after intercalation of Li^+-ions into the MoS_2 layers, it undergoes conversion reaction to produce Li_2S and Mo [64] through the reaction-resulting in an increase in capacity. However, MoS_2 has poor electrical conductivity for the transfer of electrons; hence, for the fabrication of electrodes, one needs to add conductive carbons and polymeric binders. The polymeric binders being insulating in nature hinder the diffusion of ions and hence electron transfer. Therefore, it is very important to develop novel methods to design materials with good electrical conductivity as anodes for lithium ion batteries. Researchers have reported two methods to increase the specific capacity of MoS_2 and similar metal sulfides-based Li^+-ion batteries. One is to synthesize few-layered and even single-layered nanostructures of MoS_2, thereby reducing the strain developing during the intercalation of Li^+-ions. The second method is to synthesize MoS_2 coated with conductive carbon materials or composites of MoS_2 with conductive carbon nanostructures such as carbon nanotubes, carbon nanofibers, graphene, etc. [64]. In this expedition, there have been reports investigating the effect of making composites of MoS_2 and its nanostructures with carbon nanofibers synthesized by different methods. Here, we focus particularly on the composites of transition metal sulfides and its nanostructures with electrospun carbon nanofibers.

Electrospinning is a very useful technique to produce carbon nanofibers of varying aspect ratio and doping heteroatoms such as nitrogen, sulfur by using proper precursors. Electrospinning has been extensively used to synthesize carbon nanofibers and their composites with transition metal sulfides such as MoS_2, WS_2, SnS and SnS_2. We will discuss those works and try to understand the effect and advantages of making composites of electrospun carbon nanofibers with transition metal sulfides.

MoS_2 has a theoretical specific capacity of 670 mAh g^{-1} [66] which is much higher than that of graphite which is about 372 mAh g^{-1}. Hence, it is of very high interest to explore the possibilities to improve and use these materials from a commercialization point of view. However, in spite of having a high specific capacity MoS_2 does suffer from a few issues, such as poor electrical conductivity. Zhu et al. [64] reported the synthesis of nanoplates of MoS_2 embedded in carbon nanofibers using the precursors $(NH_4)_2MoS_4$ and polyvinylpyrrolidone in the required stoichiometry. They also point out that MoS_2 can undergo significant volume change during cycling of the cells. Hence, it is important to come up with a solution. Hence, they proposed and synthesized MoS_2 embedded in carbon nanofiber through electrospinning method which formed like a 3D network with the nanofiber diameter of around

50 nm. The thickness of the MoS_2 nanoplate thus obtained was 0.4 nm which shows the single-layered nature of the MoS_2 with a lateral dimension of 4 nm.

The TEM images of the MoS_2-nanoplate-CNF composite are shown in Fig. 14.18. For first cycle, with 1712 mAh g^{-1} and 1267 mAh g^{-1} discharge and charge capacity values, the coulombic efficiency is 74%. In the second cycle, the Coulombic efficiency increases to 95.5% and which increases to 99.1% after ten cycles. The

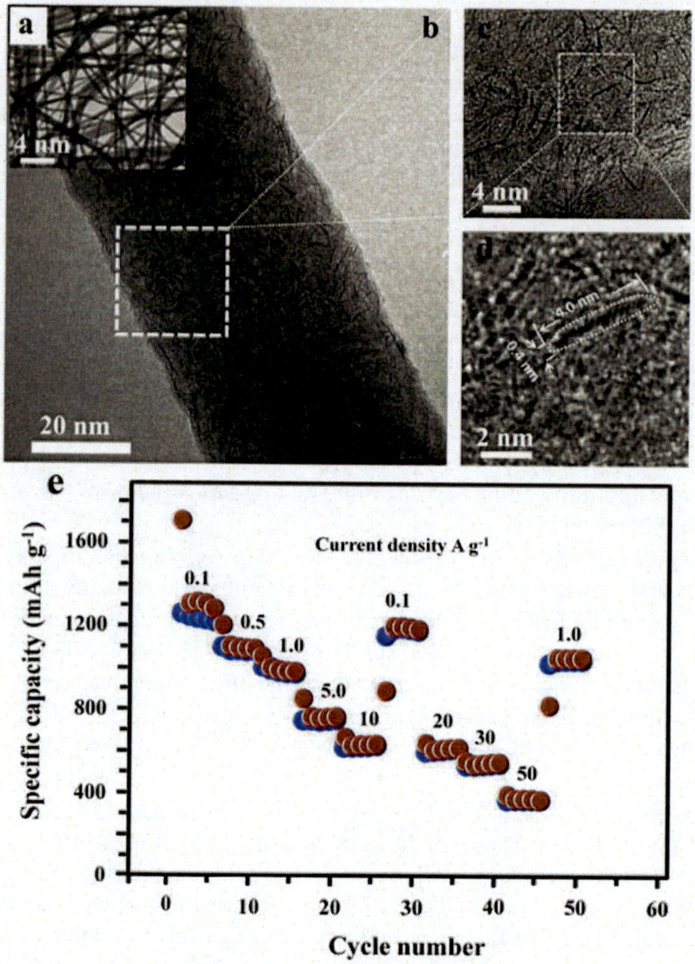

Fig. 14.18 **a** Bright-field TEM images of the MoS_2-CNF composites, **b** HRTEM image of the ultrathin MoS_2 on the surface of the CNF, **c, d** HRTEM images of marked region, **e** rate performance of MoS_2-CNF composites. Adapted and reproduced from Ref. [64], Copyright 2014 Wiley

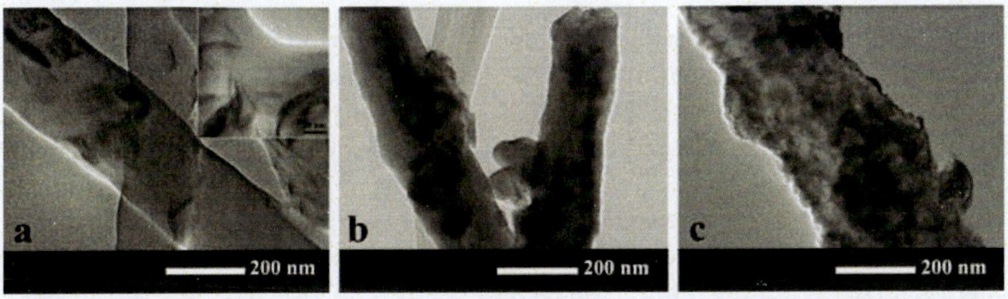

Fig. 14.19 **a**, **b**, **c** TEM images of MoS_2/C-47, MoS_2/C-73 and MoS_2/C-95, respectively. Adapted and reproduced from Ref. [69], Copyright 2014 American Chemical Society

composite also shows a remarkable rate capability of 374 mAh g^{-1} at 50 A g^{-1} together with excellent cycling stability of 661 mAh g^{-1} after 1000 cycles at a current density of 10 A g^{-1}. The authors ascribe this outstanding performance to the following; (1) the single layer nature of MoS_2 allows for the faster transport of the charge and storage through intercalation, conversion and alloying reactions. (2) The 1D nature of the carbon nanofibers helps easy access of ions.

Zhao et al. [69] synthesized MoS_2 nanoflakes encapsulated in carbon nanofibers prepared by electrospinning. They first synthesized MoS_2 nanoflakes by a solvothermal method and then dispersed different amounts of MoS_2 flakes 150, 300 and 900 mg in the precursor (PAN in DMF) for the synthesis of carbon nanofibers using single spinneret electrospinning technique at a voltage of 9.5–12.5 kV. Then, the obtained fibers were subjected to thermal treatment at steps to obtain carbon nanofibers. Thus, obtained nanofibers with MoS_2 nanoflakes were characterized with TEM which is shown in Fig. 14.19a–c. The authors then carried out electrochemical characterizations for the lithium ion batteries fabricated using the composite materials and the data is shown in Fig. 14.20a, b. In the cyclic voltammetry curve, the peak at 1.6 V in the cathodic part is attributed to the reduction of the oxygen and nitrogen-containing carbon obtained from the PAN. The slope appearing at 1.1 V indicates the

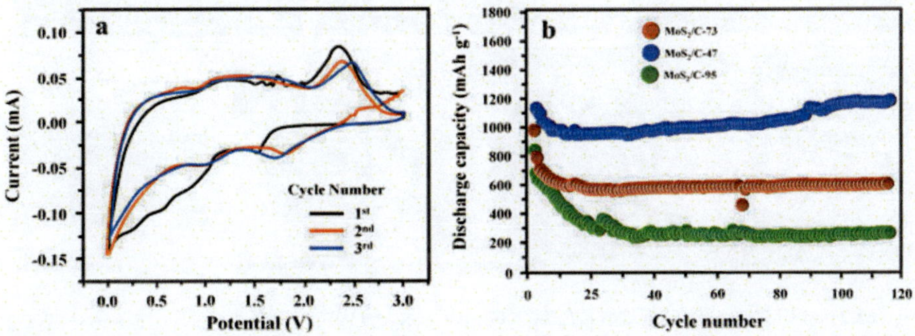

Fig. 14.20 **a** Cyclic voltammogram and (**b**) Cycling performance at 50 mA g^{-1} of prepared samples. Adapted and reproduced from Ref. [69], Copyright 2014 American Chemical Society

formation of the Li_xMoS_2 which then decomposes to Mo metal nanoparticles and Li_2S resulting in a cathodic peak at 0.5 V. Of the composites prepared with different loading of MoS_2, the one with 47% showed good cycling stability and rate capability. The sample delivered a capacity of 1133 mAh g^{-1} with a Coulombic efficiency of 73% at a current density of 50 mA g^{-1}. The excellent stability of the materials is resulting because of the confinement of the active material in the carbon matrix and the structural stability. Hence, the MoS_2 nanoflakes show very good performance as an anode material for the lithium ion batteries.

Yu et al. [70] reported the use of another transition metal sulfide tungsten disulfide (WS_2) single layer as an anode material in composite with nitrogen-doped carbon nanofiber (WS_2@NCNF) prepared by electrospinning. The composite materials were synthesized through a one-step synthetic route followed by heat treatment in two steps. WS_2 has a theoretical specific capacity of 433 mAh g^{-1} and the charge storage mechanism in this material is similar to MoS_2. That is, the lithium ions are stored as Li_2S and the electrons in the W in the metallic form. The material delivered a specific capacity of 590.4 mAh g^{-1} at a current density of 0.1 A g^{-1} with a Coulombic efficiency of 81.7% and exhibits capacity retention of 437.5 mAh g^{-1} after 200

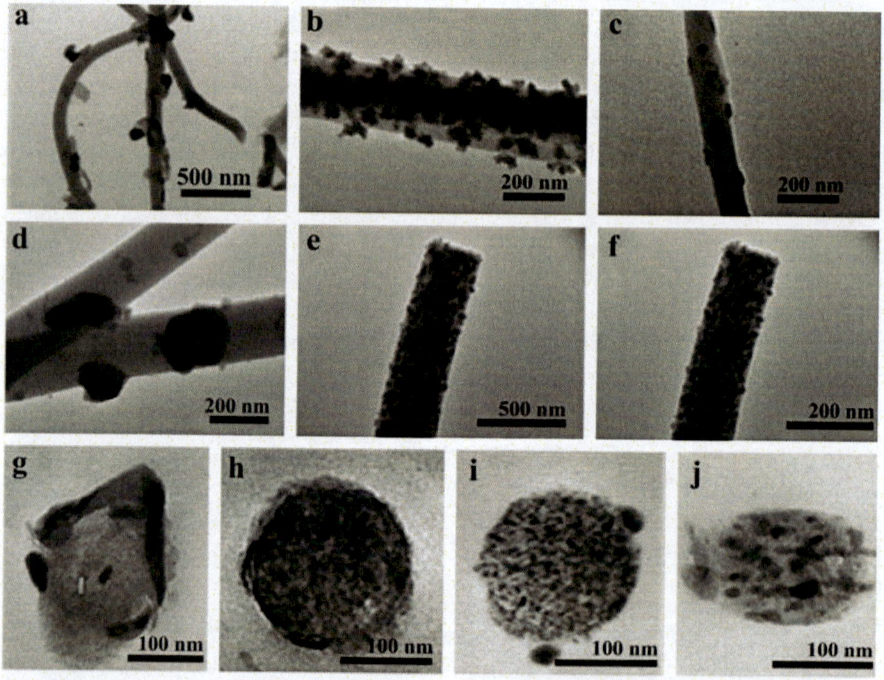

Fig. 14.21 TEM images of **a** $Cu_{1.96}S$@CNFs; **b** Co_9S_8@CNFs; **c** MnS@CNFs; **d** FeS@CNFs; **e** NiS@CNFs; **f** SnS@CNFs and microtomed cross section of **g** $Cu_{1.96}S$@CNFs; **h** MnS@CNFs; **i** NiS@CNFs; **j** SnS@CNFs. Adapted and reproduced from Ref. [71], Copyright 2016 Royal Society of Chemistry

cycles at 0.5 A g^{-1} current density and delivering a capacity of 367.1 mAh g^{-1} at 2 A g^{-1} which shows a very good rate capability. The cyclic voltammogram shows the characteristic peaks for the oxidation and reduction reactions. The reduction peak at 1.27 V represents the insertion of Li-ions into the WS_2, and the formation of Li_xWS_2 and the peak at 0.57 V has been ascribed to the conversion of Li^+-ions with the WS_2 and subsequent decomposition of the electrolyte. The peak at 0.57 V disappears after the first cycle and the peak at 1.27 V shifts to 1.83 V. The oxidation peaks appearing at 1.43 and 2.18 V represent the extraction of lithium from the WS_2. The reversible specific capacity of 590.4 mAh g^{-1} is higher than the theoretical specific capacity of the WS_2, this additional increase in capacity is ascribed to the contribution from the lithium storage at the interfaces of WS_2 nanoplates and formation of the W nanoparticles after the conversion reaction. The nitrogen-doped carbon nanofibers with improved electrical conductivity provide additional pathways for the electrons.

Fei et al. [71] reported the synthesis of different metal sulfides with carbon nanofibers using electrospinning by dispersing the sulfur into the solution of polymer and metal acetates and subjecting the obtained fibers to thermal treatment in the nitrogen gas environment This way the authors have synthesized SnS@CNFs, MnS@CNFs, FeS@CNFs, $Cu_{1.96}S$@CNFs, and Co_9S_8@CNFs, NiS@CNFs, and tested their lithium ion batteries performance as anode materials and the results are shown in Fig. 14.22. Their morphological studies have been depicted in Fig. 14.21. Out of all these composite samples, SnS@CNFs showed high capacity owing to the highest theoretical capacity of SnS together with high rate capability. The high performance and the excellent rate capability are attributed to the embedding of nanoparticles in the structure of the fiber which not only accommodates volume expansion but also prevents the aggregation of nanoparticles. The presence of a large number of micropores on the nanofibers allows for the fast diffusion of electrolyte and fast approach of the Li^+-ions to the metal sulfide particles.

Zhang et al. [72] designed graphene-wrapped carbon nanofibers and grew WS_2 nanosheets on them to improve the electrical conductivity and accommodate the large volume expansion and contraction during lithiation and de-lithiation. Graphene

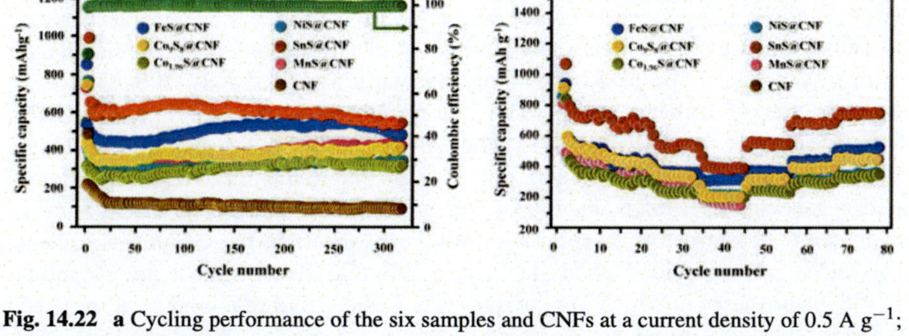

Fig. 14.22 **a** Cycling performance of the six samples and CNFs at a current density of 0.5 A g^{-1}; **b** rate performance of the six samples. Adapted and reproduced from Ref. [71], Copyright 2016 Royal Society of Chemistry

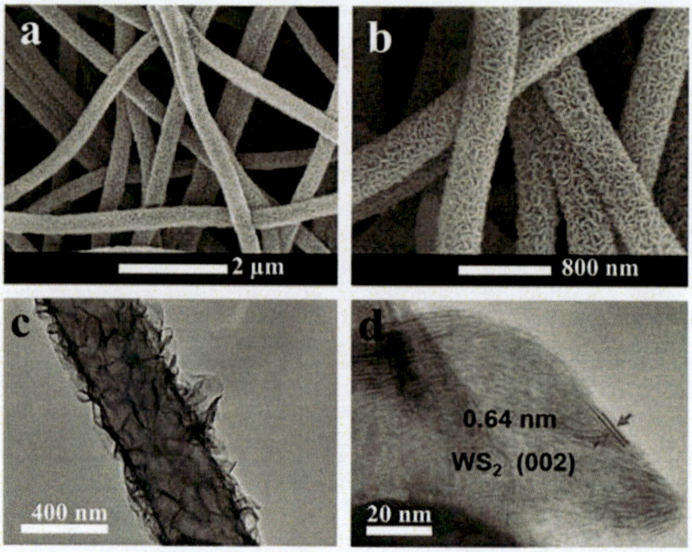

Fig. 14.23 Microstructures of WS_2/GCNF-2 hybrid membrane. **a**, **b** FESEM images, **c** TEM and **d** HRTEM images of WS_2/GCNF-2. Adapted and reproduced from Ref. [72], Copyright 2016 Royal Society of Chemistry

nanosheets were separately synthesized by the modified Hummer's method and the pre-oxidized polyacrylonitrile (PAN) nanofibers were synthesized by electrospinning. Later, the PAN were added to graphene oxide dispersion in water for 24 h and then rinsed, dried in a vacuum and carbonizing at 950 °C in nitrogen ambiance to get graphene-wrapped carbon nanofibers. WS_2/GCNF composite was synthesized by a solvothermal method. Figure 14.23a–d shows FESEM images and HRTEM images of the WS_2 nanoflakes grown on the graphene-wrapped CNF. The authors prepared WS_2/GCNF composites with varying amount of WS_2 loading on the fibers by varying the amount of the WS_2 precursor ($(NH_4)_2WS_4$) during synthesis and then carried out electrochemical tests on the lithium ion batteries fabricated using the same materials without adding any polymeric binder or conductive carbon to make the electrodes.

The electrochemical data is shown in Fig. 14.24a, b. Among the different weight ratios of WS_2/GCNF prepared, the composites with 74.6% of WS_2 (WS_2/GCNF-2) showed better performance compared to other composites with 54.8% (WS_2/GCNF-1) and 88.3% (WS_2/GCNF-3) of WS_2. The WS_2/GCNF-2 exhibits an initial discharge capacity of 1624.3 and 1128.2 mAh g^{-1} as compared to the only WS_2 which delivered a capacity of 1305.6 and 693.4 Ah g^{-1}. This enhancement in capacity is attributed to the effective dispersion of WS_2 nanosheets in the GCNF nanofibers. The GCNF also provided good stability to WS_2/GCNF as against poor stability of WS_2 nanosheets in which conductive carbon and polymeric binders were used. WS_2/GCNF-2 anode delivers high initial charge capacity of 1128.2 mAh g^{-1} and retention of 1068.5 mAh g^{-1} after 100 cycles which is quite a good result.

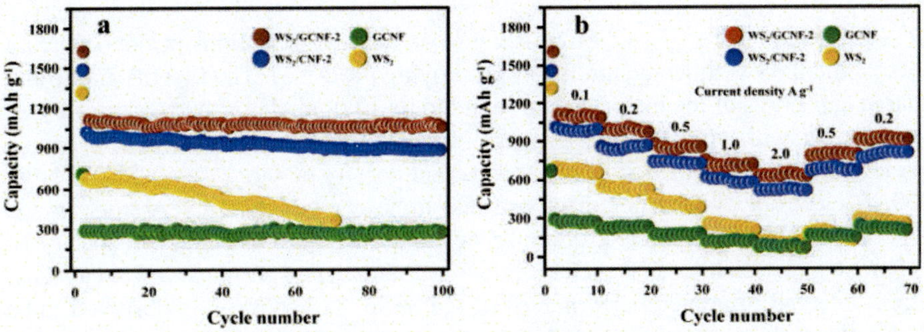

Fig. 14.24 **a** Cycling stability at 0.1 A g^{-1} and **b** rate performance of prepared samples. Adapted and reproduced from Ref. [72], Copyright 2016 Royal Society of Chemistry

Xia et al. [73] produced flexible and free-standing SnS/carbon nanofibers by dissolving suitable amounts of $SnCl_2$ and polyvinylpyrrolidone (PVP) in DMF and carrying out electrospinning. The electrospun fibers were then heated in a tube furnace in the presence of Ar/H_2 gas keeping thiourea in the upstream of the gas flow. The fibers obtained were again subjected to carbonization at different temperatures (550, 650 and 750 °C). Thus, the produced materials were used as anodes to fabricate binder and conductive carbon-free lithium ion batteries. The SnS has a layer spacing of 4.04 Å which is favorable for both Li^+ and Na^+-ion intercalation. The electrochemical characterization of the lithium ion batteries is shown in Fig. 14.25.

For the SnS/CNF prepared at 650 °C, in the cyclic voltammogram the peak at 0.98 V in the first cathodic curve is attributed to the conversion reaction of SnS with Li^+ to produce Li_2S and Sn which is given as $SnS + 2Li^+ + 2e^- \rightarrow Sn + Li_2S$. The peak at 0.54 V appears due to the formation of the solid electrolyte interface film. The peak appearing below 0.3 V may be representing the alloying reaction of Sn and

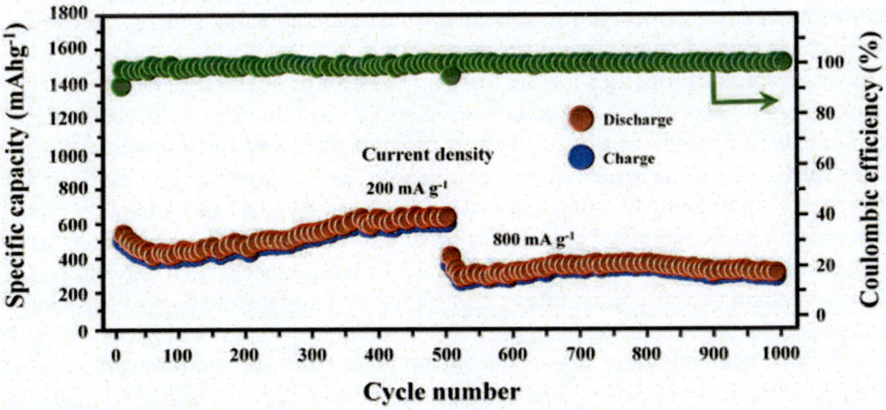

Fig. 14.25 Long-term cycling performance for SnS/C NFs-650 °C. Adapted and reproduced from Ref. [73], Copyright 2019 Elsevier

Li^+ ($Sn + xLi^+ + xe^- \rightarrow Li_xSn$ ($0 < x \leq 4.4$), and the oxidations peaks occurring at 0.52 and 0.68 V show the multistep de-alloying processes. The first discharge and charge capacities of the materials are found to be 1278 mAh g^{-1} and 898 mAh g^{-1}, respectively, with a Coulombic efficiency of 70.3%, the loss in capacity could be due to electrolyte decomposition. Even at a current density of 4 A g^{-1}, the SnS/C NFs-650 °C electrodes deliver a capacity of 206 mAh g^{-1} which is a very good value. The SnS/C NFs-650 °C electrode also delivers an impressive higher capacity of 548 mAh g^{-1} even after 500 cycles while the other two deliver much lower capacity. The stable and improved performance of the SnS/C NFs-650 °C is observed to be due to the surface defect formed due to carbonization at 650 °C which favor more storage of Li^+-ions in the electrode. From these results, the authors conclude that the concentration of defects, electrical conductivity, and the amount of SnS present in the composite decides the capacity values obtained.

From all the above-discussed reports, we can understand that for the best performance of a given transition metal sulfide one needs to design proper composite which can improve the electrical conductivity, porosity, defects and the optimized content of the metal sulfide. With this in mind, one can design various possible metal sulfide electrospun carbon nanofiber composites for excellent performance as anode materials for lithium ion batteries.

14.5 Conclusion

In this chapter, we have discussed the synthesis of CNFs using electrospinning method. Wide varieties of material such as polymers, nanoparticles and composites have been employed to electrospinning to obtain nanofibers. Depending upon the need and current prospective, CNFs have been designed and functionalized by various materials. Although LIBs have been explored worldwide using graphite as anode material, yet to meet the demand of the next generation high energy and power density due to its limited theoretical capacity of 372 mAh g^{-1}. Hence, 1D CNF provides favorable properties for electrochemical storage of Li^+-ions in LIB through good access for electrolyte due to enhanced surface-volume ratio and conductivity along the length. High capacity and lower resistance are key electrochemical properties for CNFs as it provides shorter path length for Li^+ insertion and efficient 1D electron transport along length compared to powder materials. The varieties of nanomaterials such as CNF/metal oxides, CNF/metal sulfides, conducting polymers have been discussed along with their electrochemical performances. Also, advantages of high theoretical capacity materials as anodes for LIB and the challenges associated with it for composite making have been discussed with their Li^+-ion battery performance. The materials have been controllably, designed and synthesized as need of electrochemical view of points. Overall, the chapter summarized, designed and synthesis with unique strategies for the application of LIBs anode.

References

1. Marom R, Amalraj SF, Leifer N, Jacob D, Aurbach D (2011) A review of advanced and practical lithium batteries materials. J Mater Chem 21(27):9938–9954. https://doi.org/10.1039/c0jm04225k
2. Kim SW, Seo DH, Ma X, Ceder G, Kang K (2012) Electrode materials for rechargeable sodium-ion batteries: potential alternatives to current lithium ion batteries. Adv Energy Mater 2(7):710–721. https://doi.org/10.1002/aenm.201200026
3. Li W, Li M, Adair KR, Sun X, Yu Y (2017) Carbon nanofiber-based nanostructures for lithium-ion and sodium-ion batteries. J Mater Chem A 5(27):13882–13906. https://doi.org/10.1039/c7ta02153d
4. Chen XC, Wei W, Lv W, Su FY, He YB, Li B, Kang F, Yang QH (2012) A graphene-based nanostructure with expanded ion transport channels for high rate Li-ion batteries. Chem Commun 48(47):5904–5906. https://doi.org/10.1039/c2cc32276e
5. Reddy ALM, Shaijumon MM, Gowda SR, Ajayan PM (2009) Coaxial MnO_2/carbon nanotube array electrodes for high-performance lithium batteries. Nano Lett 9(3):1002–1006. https://doi.org/10.1021/nl803081j
6. Cui LF, Yang Y, Hsu CM, Cui Y (2009) Carbon–silicon core–shell nanowires as high capacity electrode for lithium ion batteries. Nano Lett 9(9):3370–3374. https://doi.org/10.1021/nl901670t
7. Weng W, Kurihara R, Wang J, Shiratori S (2019) Electrospun carbon nanofiber-based composites for lithium ion batteries: structure optimization towards high performance. Compos Commun 15:135–148. https://doi.org/10.1016/j.coco.2019.07.005
8. Zhang L, Aboagye A, Kelkar A, Lai C, Fong H (2014) A review: carbon nanofibers from electrospun polyacrylonitrile and their applications. J Mater Sci 49(2):463–480. https://doi.org/10.1007/s10853-013-7705-y
9. De Jong KP, Geus JW (2000) Carbon nanofibers: catalytic synthesis and applications. Catal Rev 42(4):481–510. https://doi.org/10.1081/CR-100101954
10. Inagaki M, Yang Y, Kang F (2012) Carbon nanofibers prepared via electrospinning. Adv Mater 24(19):2547–2566. https://doi.org/10.1002/adma.201104940
11. Zheng R, Zhao Y, Liu H, Liang C, Cheng G (2006) Preparation, characterization and growth mechanism of platelet carbon nanofibers. Carbon 44(4):742–746. https://doi.org/10.1016/j.carbon.2005.09.015
12. Tu JP, Zhu LP, Hou K, Guo SY (2003) Synthesis and frictional properties of array film of amorphous carbon nanofibers on anodic aluminum oxide. Carbon 41(6):1257–1263. https://doi.org/10.1016/S0008-6223(03)00047-2
13. Yu Y, Liu Y, Yang X (2016) Carbon nanofiber-based materials as anode materials for lithium ion batteries. InTech, 1, 978-953-51-2396-5, (3-30)
14. Kim MJ, Lee J, Jung D, Shim SE (2010) Electrospun poly (vinyl alcohol) nanofibers incorporating PEGylated multi-wall carbon nanotube. Synth Met 160(13–14):1410–1414. https://doi.org/10.1016/j.synthmet.2010.04.020
15. Wang P, Zhang D, Ma F, Ou Y, Chen QN, Xie S, Li J (2012) Mesoporous carbon nanofibers with a high surface area electrospun from thermoplastic polyvinylpyrrolidone. Nanoscale 4(22):7199–7204. https://doi.org/10.1039/c2nr32249h
16. Arshad SN, Naraghi M, Chasiotis I (2011) Strong carbon nanofibers from electrospun polyacrylonitrile. Carbon 49(5):1710–1719. https://doi.org/10.1016/j.carbon.2010.12.056
17. Ko Y, Kim J, Kim D, Kwon G, Yamauchi Y, You J (2019) Fabrication of highly conductive porous cellulose/PEDOT: PSS nanocomposite paper via post-treatment. Nanomaterials 9(4):612. https://doi.org/10.3390/nano9040612
18. Yang Y, Hou X, Ding C, Lan JL, Yu Y, Yang X (2017) Eco-friendly fabricated nonporous carbon nanofibers with high volumetric capacitance: improving rate performance by tri-dopants of nitrogen, phosphorus, and silicon. Inorg Chem Front 4(12):2024–2032. https://doi.org/10.1039/C7QI00517B

19. Kim C, Jeong YI, Ngoc BTN, Yang KS, Kojima M, Kim YA, Endo M, Lee JW (2007) Synthesis and characterization of porous carbon nanofibers with hollow cores through the thermal treatment of electrospun copolymeric nanofiber webs. Small 3(1):91–95. https://doi.org/10.1002/smll.200600243
20. McCann JT, Marquez M, Xia Y (2006) Melt coaxial electrospinning: a versatile method for the encapsulation of solid materials and fabrication of phase change nanofibers. Nano Lett 6(12):2868–2872. https://doi.org/10.1021/nl0620839
21. Yang HS, Lee BS, You BC, Sohn HJ, Yu WR (2014) Fabrication of carbon nanofibers with Si nanoparticle-stuffed cylindrical multi-channels via coaxial electrospinning and their anodic performance. RSC Adv 4(88):47389–47395. https://doi.org/10.1039/c4ra10031j
22. Kim C, Yang KS (2003) Electrochemical properties of carbon nanofiber web as an electrode for supercapacitor prepared by electrospinning. Appl Phys Lett 83(6):1216–1218. https://doi.org/10.1063/1.1599963
23. Wang Y, Serrano S, Santiago-Aviles JJ (2002) Conductivity measurement of electrospun PAN-based carbon nanofiber. J Mater Sci Lett 21(13):1055–1057. https://doi.org/10.1023/A:1016081212346
24. Zussman E, Chen X, Ding W, Calabri L, Dikin DA, Quintana JP, Ruoff RS (2005) Mechanical and structural characterization of electrospun PAN-derived carbon nanofibers. Carbon 43(10):2175–2185. https://doi.org/10.1016/j.carbon.2005.03.031
25. Dong L, Wang G, Li X, Xiong D, Yan B, Chen B, Li D, Cui Y (2016) PVP-derived carbon nanofibers harvesting enhanced anode performance for lithium ion batteries. RSC Adv 6(5):4193–4199. https://doi.org/10.1039/c5ra23924a
26. Ding B, Kim HY, Lee SC, Lee DR, Choi KJ (2002) Preparation and characterization of nanoscaled poly (vinyl alcohol) fibers via electrospinning. Fibers Polym 3(2):73–79. https://doi.org/10.1007/BF02875403
27. Son WK, Youk JH, Lee TS, Park WH (2005) Effect of pH on electrospinning of poly (vinyl alcohol). Mater Lett 59(12):1571–1575. https://doi.org/10.1016/j.matlet.2005.01.025
28. Hong SM, Kim SH, Jeong BG, Jo SM, Lee KB (2014) Development of porous carbon nanofibers from electrospun polyvinylidene fluoride for CO_2 capture. RSC Adv 4(103):58956–58963. https://doi.org/10.1039/c4ra11290c
29. Kim JR, Choi SW, Jo SM, Lee WS, Kim BC (2004) Electrospun PVdF-based fibrous polymer electrolytes for lithium ion polymer batteries. Electrochim Acta 50(1):69–75. https://doi.org/10.1016/j.electacta.2004.07.014
30. Zhao Z, Li J, Yuan X, Li X, Zhang Y, Sheng J (2005) Preparation and properties of electrospun poly (vinylidene fluoride) membranes. J Appl Polym Sci 97(2):466–474. https://doi.org/10.1002/app.21762
31. Yang Y, Centrone A, Chen L, Simeon F, Hatton TA, Rutledge GC (2011) Highly porous electrospun polyvinylidene fluoride (PVDF)-based carbon fiber. Carbon 49(11):3395–3403. https://doi.org/10.1016/j.carbon.2011.04.015
32. Ji L, Lin Z, Medford AJ, Zhang X (2009) Porous carbon nanofibers from electrospun polyacrylonitrile/SiO_2 composites as an energy storage material. Carbon 47(14):3346–3354. https://doi.org/10.1016/j.carbon.2009.08.002
33. Fan L, Yang L, Ni X, Han J, Guo R, Zhang CJ (2016) Nitrogen-enriched meso-macroporous carbon fiber network as a binder-free flexible electrode for supercapacitors. Carbon 107:629–637. https://doi.org/10.1016/j.carbon.2016.06.067
34. Zhang L, Jiang Y, Wang L, Zhang C, Liu S (2016) Hierarchical porous carbon nanofibers as binder-free electrode for high-performance supercapacitor. Electrochim Acta 196:189–196. https://doi.org/10.1016/j.electacta.2016.02.050
35. Peng YT, Lo CT (2015) Electrospun porous carbon nanofibers as lithium ion battery anodes. J Solid State Electrochem 19(11):3401–3410. https://doi.org/10.1007/s10008-015-2976-7
36. Niu H, Zhang J, Xie Z, Wang X, Lin T (2011) Preparation, structure and supercapacitance of bonded carbon nanofiber electrode materials. Carbon 49(7):2380–2388. https://doi.org/10.1016/j.carbon.2011.02.005

37. Zhang Z, Li X, Wang C, Fu S, Liu Y, Shao C (2009) Polyacrylonitrile and carbon nanofibers with controllable nanoporous structures by electrospinning. Macromol Mater Eng 294(10):673–678. https://doi.org/10.1002/mame.200900076
38. Heo YJ, Zhang Y, Rhee KY, Park SJ (2019) Synthesis of PAN/PVDF nanofiber composites-based carbon adsorbents for CO2 capture. Compos B Eng 156:95–99. https://doi.org/10.1016/j.compositesb.2018.08.057
39. Ju YW, Park SH, Jung HR, Lee WJ (2009) Electrospun activated carbon nanofibers electrodes based on polymer blends. J Electrochem Soc 156(6):A489–A494. https://doi.org/10.1149/1.3116245
40. Nan D, Huang ZH, Lv R, Yang L, Wang JG, Shen W, Lin Y, Yu X, Ye L, Sun H, Kang F (2014) Nitrogen-enriched electrospun porous carbon nanofiber networks as high-performance free-standing electrode materials. J Mater Chem A 2(46):19678–19684. https://doi.org/10.1039/c4ta03868a
41. Zhou T, Jiang Q, Wang L, Qiu Z, Liu Y, Zhou J, Liu B (2018) Facile preparation of nitrogen-enriched hierarchical porous carbon nanofibers by $Mg(OAc)_2$-assisted electrospinning for flexible supercapacitors. Appl Surf Sci 456:827–834. https://doi.org/10.1016/j.apsusc.2018.06.214
42. Chen LF, Lu Y, Yu L, Lou XWD (2017) Designed formation of hollow particle-based nitrogen-doped carbon nanofibers for high-performance supercapacitors. Energy Environ Sci 10(8):1777–1783. https://doi.org/10.1039/c7ee00488e
43. Li D, Wang D, Rui K, Ma Z, Xie L, Liu J, Zhang Y, Chen R, Yan Y, Lin H, Xie X (2018) Flexible phosphorus doped carbon nanosheets/nanofibers: electrospun preparation and enhanced Li-storage properties as free-standing anodes for lithium ion batteries. J Power Sources 384:27–33. https://doi.org/10.1016/j.jpowsour.2018.02.069
44. Kim C, Yang KS, Kojima M, Yoshida K, Kim YJ, Kim YA, Endo M (2006) Fabrication of electrospinning-derived carbon nanofiber webs for the anode material of lithium-ion secondary batteries. Adv Funct Mater 16(18):2393–2397. https://doi.org/10.1002/adfm.200500911
45. Lee BS, Son SB, Park KM, Lee G, Oh KH, Lee SH, Yu WR (2012) Effect of pores in hollow carbon nanofibers on their negative electrode properties for a lithium rechargeable battery. ACS Appl Mater Interfaces 4(12):6702–6710. https://doi.org/10.1021/am301873d
46. Hu C, Xiao Y, Zhao Y, Chen N, Zhang Z, Cao M, Qu L (2013) Highly nitrogen-doped carbon capsules: scalable preparation and high-performance applications in fuel cells and lithium ion batteries. Nanoscale 5(7):2726–2733. https://doi.org/10.1039/c3nr34002c
47. Wu Y, Fang S, Jiang Y (1998) Carbon anode materials based on melamine resin. J Mater Chem 8(10):2223–2227. https://doi.org/10.1039/A805080E
48. Endo M, Kim C, Karaki T, Nishimura Y, Matthews MJ, Brown SDM, Dresselhaus MS (1999) Anode performance of a Li ion battery based on graphitized and B-doped milled mesophase pitch-based carbon fibers. Carbon 37(4):561–568. https://doi.org/10.1016/S0008-6223(98)00222-X
49. Wu YP, Fang S, Jiang Y, Holze R (2002) Effects of doped sulfur on electrochemical performance of carbon anode. J Power Sources 108(1–2):245–249. https://doi.org/10.1016/S0378-7753(02)00013-7
50. Mao Y, Duan H, Xu B, Zhang L, Hu Y, Zhao C, Wang Z, Chen L, Yang Y (2012) Lithium storage in nitrogen-rich mesoporous carbon materials. Energy Environ Sci 5(7):7950–7955. https://doi.org/10.1039/c2ee21817h
51. Qie L, Chen WM, Wang ZH, Shao QG, Li X, Yuan LX, Hu XL, Zhang WX, Huang YH (2012) Nitrogen-doped porous carbon nanofiber webs as anodes for lithium ion batteries with a superhigh capacity and rate capability. Adv Mater 24(15):2047–2050. https://doi.org/10.1002/adma.201104634
52. Reddy MV, Subba Rao GV, Chowdari BVR (2013) Metal oxides and oxysalts as anode materials for Li ion batteries. Chem Rev 113(7):5364–5457. https://doi.org/10.1021/cr3001884
53. Zhang X, Liu H, Petnikota S, Ramakrishna S, Fan HJ (2014) Electrospun Fe_2O_3–carbon composite nanofibers as durable anode materials for lithium ion batteries. J Mater Chem A 2(28):10835–10841. https://doi.org/10.1039/c3ta15123a

54. Basavaraja BM, Majumder SB, Sharma A (2015) Electrospun hollow glassy carbon–reduced graphene oxide nanofibers with encapsulated ZnO nanoparticles: a free standing anode for Li-ion batteries. J Mater Chem A 3(10):5344–5351. https://doi.org/10.1039/c4ta07220k
55. Kim D, Lee D, Kim J, Moon J (2012) Electrospun Ni-added SnO_2–carbon nanofiber composite anode for high-performance lithium ion batteries. ACS Appl Mater Interfaces 4(10):5408–5415. https://doi.org/10.1021/am301328u
56. Zhang B, Huang J, Kim JK (2015) Ultrafine amorphous SnO_x embedded in carbon nanofiber/carbon nanotube composites for Li-ion and Na-ion batteries. Adv Funct Mater 25(32):5222–5228. https://doi.org/10.1002/adfm.201501498
57. Zhou X, Dai Z, Liu S, Bao J, Guo YG (2014) Ultra-uniform SnO_x/carbon nanohybrids toward advanced lithium ion batteries anodes. Adv Mater 26(23):3943–3949. https://doi.org/10.1002/adma.201400173
58. Wu Y, Zhu P, Reddy MV, Chowdari BVR, Ramakrishna S (2014) Maghemite nanoparticles on electrospun CNFs template as prospective lithium ion batteries anode. ACS Appl Mater Interfaces 6(3):1951–1958. https://doi.org/10.1021/am404939q
59. Cho JS, Hong YJ, Kang YC (2015) Design and synthesis of bubble-nanorod-structured Fe_2O_3–carbon nanofibers as advanced anode material for Li-ion batteries. ACS Nano 9(4):4026–4035. https://doi.org/10.1021/acsnano.5b00088
60. Liu Y, Yan X, Yu Y, Yang X (2016) Eco-friendly fabricated porous carbon nanofibers decorated with nanosized SnO_x as high-performance lithium ion batteries anodes. ACS Sustain Chem Eng 4(6):2951–2959. https://doi.org/10.1021/acssuschemeng.5b01236
61. Sahay R, Suresh Kumar P, Aravindan V, Sundaramurthy J, Chui Ling W, Mhaisalkar SG, Ramakrishna S, Madhavi S (2012) High aspect ratio electrospun CuO nanofibers as anode material for lithium ion batteries with superior cycleability. J Phys Chem C 116(34):18087–18092. https://doi.org/10.1021/jp3053949
62. Wang H, Yang X, Wu Q, Zhang Q, Chen H, Jing H, Wang J, Mi SB, Rogach AL, Niu C (2018) Encapsulating silica/antimony into porous electrospun carbon nanofibers with robust structure stability for high-efficiency lithium storage. ACS Nano 12(4):3406–3416. https://doi.org/10.1021/acsnano.7b09092
63. Zhang X, Suresh Kumar P, Aravindan V, Liu HH, Sundaramurthy J, Mhaisalkar SG, Duong HM, Ramakrishna S, Madhavi S (2012) Electrospun TiO_2–graphene composite nanofibers as a highly durable insertion anode for lithium ion batteries. J Phys Chem C 116(28):14780–14788. https://doi.org/10.1021/jp302574g
64. Zhu C, Mu X, van Aken PA, Yu Y, Maier J (2014) Single-layered ultrasmall nanoplates of MoS_2 embedded in carbon nanofibers with excellent electrochemical performance for lithium and sodium storage. Angew Chem Int Ed 53(8):2152–2156. https://doi.org/10.1002/anie.201308354
65. Wang L, Zhuo L, Yu Y, Zhao F (2013) High-rate performance of SnS_2 nanoplates without carbon-coating as anode material for lithium ion batteries. Electrochim Acta 112:439–447. https://doi.org/10.1016/j.electacta.2013.08.154
66. Xu X, Liu W, Kim Y, Cho J (2014) Nanostructured transition metal sulfides for lithium ion batteries: progress and challenges. Nano Today 9(5):604–630. https://doi.org/10.1016/j.nantod.2014.09.005
67. Nandi DK, Sen UK, Choudhury D, Mitra S, Sarkar SK (2014) Atomic layer deposited MoS_2 as a carbon and binder free anode in Li-ion battery. Electrochim Acta 146:706–713. https://doi.org/10.1016/j.electacta.2014.09.077
68. Xia C, Zhang F, Liang H, Alshareef HN (2017) Layered SnS sodium ion battery anodes synthesized near room temperature. Nano Research 10(12):4368–4377. https://doi.org/10.1007/s12274-017-1722-0
69. Zhao C, Kong J, Yao X, Tang X, Dong Y, Phua SL, Lu X (2014) Thin MoS_2 nanoflakes encapsulated in carbon nanofibers as high-performance anodes for lithium ion batteries. ACS Appl Mater Interfaces 6(9):6392–6398. https://doi.org/10.1021/am4058088
70. Yu S, Jung JW, Kim ID (2015) Single layers of WS_2 nanoplates embedded in nitrogen-doped carbon nanofibers as anode materials for lithium ion batteries. Nanoscale 7(28):11945–11950. https://doi.org/10.1039/c5nr02425k

71. Fei L, Williams BP, Yoo SH, Carlin JM, Joo YL (2016) A general approach to fabricate free-standing metal sulfide@carbon nanofiber networks as lithium ion batteries anodes. Chem Commun 52(7):1501–1504. https://doi.org/10.1039/c5cc06957b
72. Zhang L, Fan W, Liu T (2016) Flexible hierarchical membranes of WS_2 nanosheets grown on graphene-wrapped electrospun carbon nanofibers as advanced anodes for highly reversible lithium storage. Nanoscale 8(36):16387–16394. https://doi.org/10.1039/c6nr04241d
73. Xia J, Liu L, Jamil S, Xie J, Yan H, Yuan Y, Zhang Y, Nie S, Pan J, Wang X, Cao G (2019) Free-standing SnS/C nanofiber anodes for ultralong cycle-life lithium ion batteries and sodium-ion batteries. Energy Storage Mater 17:1–11. https://doi.org/10.1016/j.ensm.2018.08.005

Chapter 15
Electrospun Silicon-Based Nanocomposite Anodes for Lithium-Ion Batteries

Indrapal Karbhal, Golu Parte, Apurva Patrike, and Manjusha Shelke

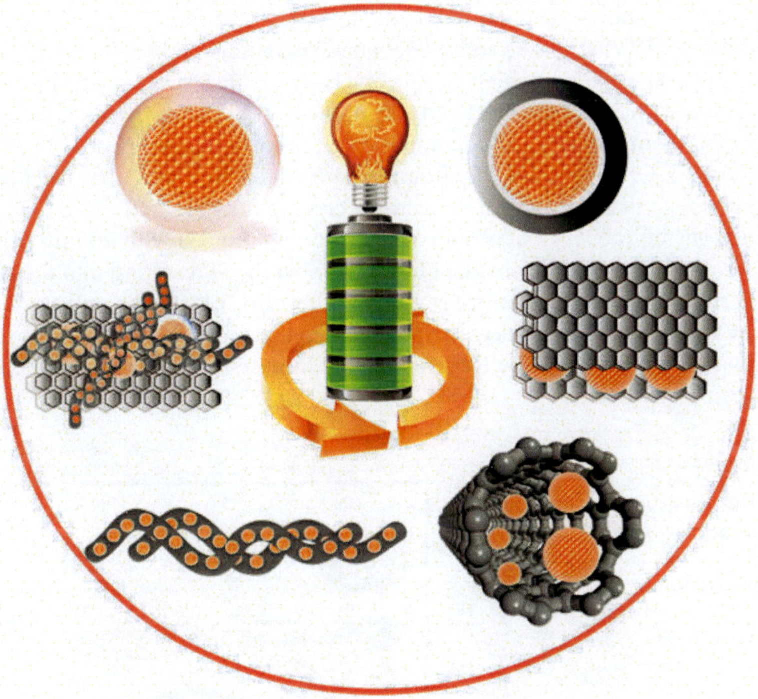

I. Karbhal · G. Parte · A. Patrike · M. Shelke (✉)
B101, Polymer and Advanced Materials Laboratory, Physical & Material's Chemistry Division, CSIR-National Chemical Laboratory, Pune, India
e-mail: mv.shelke@ncl.res.in

Academy of Scientific and Innovative Research (AcSIR), Ghaziabad, Uttar Pradesh 201002, India

N. T. M. Balakrishnan and R. Prasanth (eds.), *Electrospinning for Advanced Energy Storage Applications*, Materials Horizons: From Nature to Nanomaterials,
https://doi.org/10.1007/978-981-15-8844-0_15

15.1 Introduction

Li-ion batteries (LIBs) have been very well developed and commercialized due to its excellent high energy, power density, and long cycle life [1]. The Ragone plot of various energy storage devices has been given in Fig. 15.1 [2]. Commercial LIBs have layer-type cathode $LiCoO_2$ and graphite as anode, the lithiation and delithiation occur as reaction which is $LiCoO_2 + 6\,C \leftrightarrow Li_{1-x}CoO_2 + Li_xC_6$. LIBs are very lightweight with operating voltage of 3.6 V with deliverable capacity ranging from 700 to 2400 mAh for a single cell (battery) and extensively used in many electronic gadgets such as cell phones, laptops, medical equipments, video camcorders, electric vehicles (EV), hybrid electric vehicles (HEV) [1, 3].

The state-of-the-art graphite anode has limited theoretical capacity (372 $mAhg^{-1}$) which is not able to meet the requirement of next-generation energy and power density. Therefore, it is needed to improve the energy density and power density of LIBs. Table 15.1 shows the common anode materials and their theoretical capacity with respect to LIBs [4]. There are several anode materials such as Si, Sn, Ge, metal oxide, and metal sulfides, which have very high theoretical capacity but among them Si is one of the best material and earth-abundant, having very high theoretical capacity of 4200 $mAhg^{-1}$ ($Li_{22}Si_5$ composition), low discharge potential of 0–0.4 V, and low cost [5]. Unfortunately, the commercialization of Si-based LIBs faces several challenges due to their low intrinsic electrical conductivity (1×10^{-5} S cm^{-1}), large volume expansion (400%) that occurs during alloying process with lithium ($Li_{4.4}$ Si) and Li^+ drastically break the covalent bond between Si–Si resulting into the change in crystal structure either by amorphization or alloy, this leads to the pulverization of the electrolyte resulting low electrical contact with the current collector, furthermore rapid capacity loss [6–11]. The lithiation and delithiation process, as well as phase change, can be written as:

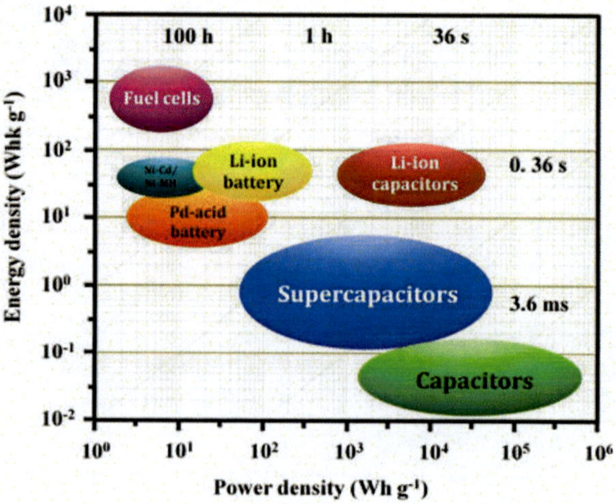

Fig. 15.1 Ragone plot shows the demands of high energy/power for electrochemical energy storage devices for the next generation. Adapted and reproduced from Ref. [2], Copyright 2014 American Chemical Society

Table 15.1 Common anode material and theoretical capacity with respect to LIBs [4]

Active anode materials			Theoretical capacity (mAh g^{-1})	Advantages	Common issue
Insertion/de-insertion materials					
A.	Carbonaceous				
	a.	Hard Carbons	200–600	• Good working potential • Low cost • Good safety	• Low coulombic efficiency • High voltage hysteresis • High irreversible capacity
	b.	CNTs	1116		
	c.	Graphene	780		
	d.	Graphite	372		
B	Titanium Oxides				
	a.	$LiTi_4O_5$	175	• Extreme safety • Good cycle life • Low cost • High power capability	• Very low capacity • Low energy density
	b.	TiO_2	330		
Alloy/de-alloy materials					
a.	Silicon		4212	• Higher specific capacity • High energy density • Good safety	• Large irreversible capacity • Huge capacity fading • Poor cycling
b.	Germanium		1624		
c.	Tin		993		
d.	Antimony		660		
e.	Tin Oxide		790		
f.	SiO		1600		
Conversion materials					
a.	Metal oxides (Fe_2O_3, Fe_3O_4, CoO, Co_3O_4, Mn_xO_y, Cu_2O/CuO, NiO, Cr_2O_3, RuO_2, MoO_2/MoO_3, etc.)		500–1200	• High capacity • High energy • Low cost • Environmentally compatibility	Low coulombic efficiency • Unstable SEI formation • Large potential hysteresis • Poor cycle life
b.	Metal phosphides/sulfides/nitrides (MX_y; M ¼ Fe, Mn, Ni, Cu, Co, etc., and X ¼ P, S, N)		500–1800	• High specific capacity • Low operation potential and Low polarization than counter oxides	• Poor capacity retention • Short cycle life • High cost of production

$$\text{lithiation: } x - \text{Si} \rightarrow a - \text{Li}_y\text{Si} \rightarrow a - \text{Li}_x\text{Si} \rightarrow x - \text{Li}_{15}\text{Si}_4$$

$$\text{delithaition: } x - \text{Li}_{15}\text{Si}_4 \overset{-\text{Li}}{\rightarrow} a - \text{Li}_z\text{Si} \overset{-\text{Li}}{\rightarrow} a - \text{Si}$$

$$\text{subsequent cycling: } a - \text{Si} \overset{-\text{Li}}{\rightarrow} a - \text{Li}_z\text{Si} \overset{-\text{Li}}{\rightarrow} x - \text{Li}_{15}\text{Si}_4$$

where "x" refers to a crystalline phase and "a" refers to an amorphous phase [12]. The fundamental challenges are discussed ahead based on the Si and Li chemistry.

15.2 Challenges in Si-Based LIBs

There is no wonder that Si-based anode has high demand due to its high theoretical capacity, abundant, very good anodic characteristics with potential range of 0–0.4 V. But there are three main challenges which limit their use as commercial application [9, 13]. The fundamental challenges have been discussed below, and Fig. 15.2 is the detail about the challenges in Si-based LIBs.

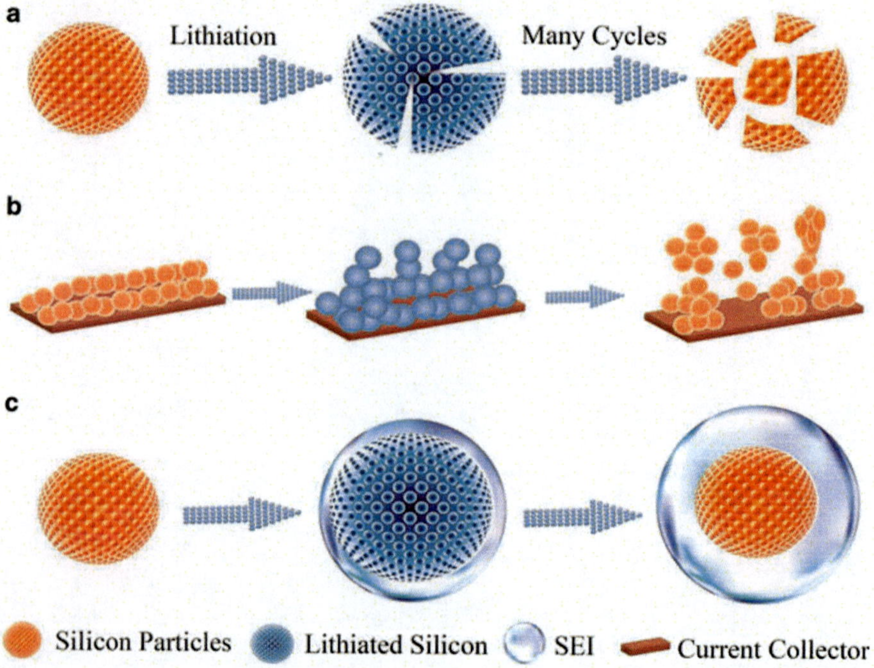

Fig. 15.2 Challenge in Si-based LIBs **a** Pulverization **b** Morphology and volume change **c** Continuous SEI growth. (Figure is made by using adobe illustrator CS5 software)

15.2.1 Pulverization

During the lithiation and delithiation process due to the high degree of stress at Si, it expands about 400% volume change with alloying $Li_{4.4}Si$. The high degree of stress causes the cracking as well as pulverization of material resulting in the loss of electrical contact. This phenomenon shows the drastic capacity fading, such kind of result was observed at early practical studies when Si used as bulk, film, and large particles [9, 14].

15.2.2 Morphology and Volume Changes

The large volume changes of Si during the lithiation and delithiation process are key challenges. During lithiation process, Si makes alloy with Li and expands up to 400% and while delithiation Si contract resulting detachment of particles as well as current collectors. This all process offer for the loss of electrical contact and capacity fading. Moreover, the total volume of Si increases and decreases in lithiation and delithiation, leading to material peel out, failure, and electrical contact with the current collector [13].

15.2.3 Solid Electrolyte Interphase (SEI)

The decomposition of organic electrolyte is thermodynamically favorable at the electrode surface when the potential of the anode is below—1 V. The formation of the layer at the electrode surface due to the decomposition of the organic electrolyte is known as "solid electrolyte interphase" (SEI). SEI layer should be stable and dense for the no further side reaction, ionically conducting, and electrically insulating. The SEI layer mainly consists of Li_2CO_3, LIF, Li_2O, lithium alkyl carbonates ($ROCO_2Li$), and nonconducting polymers. The formation of a stable SEI layer with Si is very challenging due to the large volume expansion and cracking. This volume expansion and cracking can further form the SEI layer and goes upon cycle by cycle, resulting in thicker and thicker SEI layer which is responsible for the loss of electrical contact among particles [9, 15].

To solve this fundamental problem, there are several engineered tailored method which have been designed and synthesized with carbon material such as Si nanotubes, [16], 3D porous Si (three-dimensional porous silicon) [17], Si/graphene [18, 19] SiNWs/graphene (Si nanowires/graphene) [20], Si@carbon core/shell [21], 3DCNT (three-dimensional carbon nanotube) mesoporous Si [22], silicon oxycarbide-CNT shell/core [23], polymer capped Si [24], Si/CNF composites [25], SiC-C core/shell [26], sandwich graphene protected Si/CNFs (Si/carbon nanofibers) [27], Si encapsulated hollow CNFs [28], flexible binder-free Si/silica/carbon nanofibers [29], and also with different porous and non-porous structural morphologies have been designed.

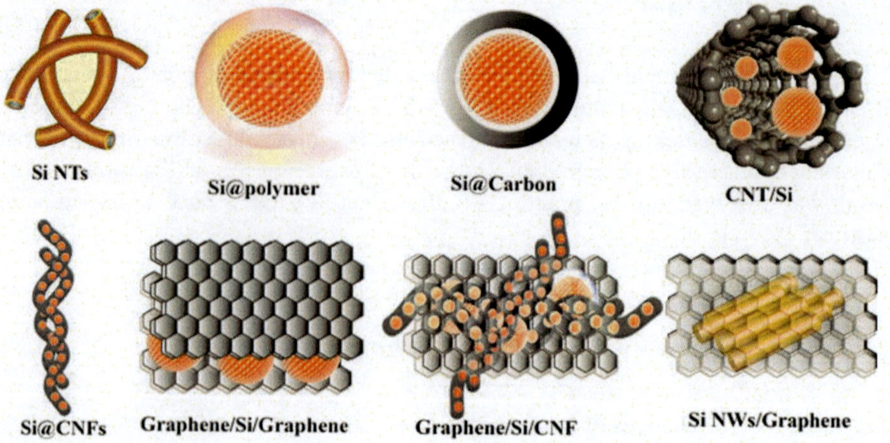

Fig. 15.3 Schematic of various Si and carbon-based morphologies to overcome the problem of Si-based LIBs. (Figure is made by using adobe illustrator CS5 software and BIOVIA Materials Studio)

Figure 15.3 shows the schematics of different morphological (1D, 2D, and 3D) approach for Si-based nonmaterial. Among all the method, electrospinning method is more favorable for the synthesis of Si-based carbon morphology due to its cost-effective, industry-viable, simplicity, high efficiency, and high yield. The synthesis of Si-based nonmaterial via electrospinning method has been discussed.

15.3 Electrospinning Method for Synthesis

Due to the three different controlled parameters: (i) operating parameters such as electric potential, needle gauge, flow rate, tip-to-collector, and nozzle type; (ii) solution parameters such as solubility and viscosity of the precursor, polymer templates, dielectric constant, and surface tension of the solvents; (iii) ambient parameters such as temperature and humidity along with advanced versatility, high efficiency and high yield, electrospinning method has been taken great attention for the synthesis of desirable nanomaterials. By controlling the electrospinning and thermal treatment, the 1D NFs (one-dimensional nanofibers), controlled phase and morphology of desired materials (NFs and composites) can be easily obtained. Moreover, we can control the electrospun NFs with (i) surface-to-volume ratio and high surface area with the desired active site (ii) porous and cavity structure can buffer the large volume change while electrochemical reactions. Figure 15.4a shows the electrospinning apparatus with their parts, and Fig. 15.4b shows the various type of controlled and desirable morphology can be obtained by electrospinning methods.

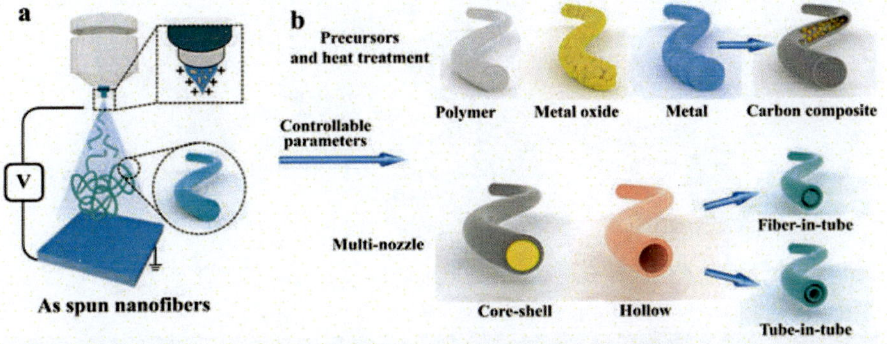

Fig. 15.4 **a** Electrospinning apparatus with parts (i) stainless needle-equipped plastic syringe with solution (ii) Voltage supplier (iii) grounded conducting collector. **b** Various types of electrospun NFs obtained by controlling the parameter such as precursor, calcination, and type of nozzle Adapted and reproduced from Ref. [30], Copyright 2016 Royal Society of Chemistry

There is a growing demand for high capacity and high energy Li-ion batteries for large-scale applications. Currently, graphite is used as anode material for the commercial battery. However, its low specific capacity hinders its large-scale application. So there is growing research interest to find the electrode materials with high specific capacity and energy density. Among various materials, Si has been extensively investigated because of its high theoretical capacity of ~4200 mAh g^{-1}. However, during cycling Si particles suffer from large volume change leading to pulverization of Si resulting in loss of electrical contact and fast capacity fading. Various strategies were used to overcome these issues by constructing nanostructures and nanocomposites of silicon with carbon. Electrospinning-based synthesis of Si composites has been discussed. Various 1D nanostructures of Si with various forms such as nanotubes [31], hollow nanospheres [32], nanowires [6], hollow porous nanoparticles [33], and nanofibers [34] as well as its composite with variety of carbon nanostructures [35] especially electrospun carbon nanofibers have been synthesized to counter the problem and improve the rate capability and cycling stability of the Li-ion batteries. Several synthesis methods have been utilized to get the nanostructures of Si as well as carbon composites with Si. However, all these methods are complex and have low yield which prevents its practical application. Electrospinning is low cost, simple, and scalable process for the Si nanofibers and its composites with carbon nanofibers.

D. J. Lee et al. reported mesoporous silicon nanofibers (m-SiNFs) via electrospinning and reduction with magnesium. Electrospinning solution was prepared in the ratio of 1:1 wt% by adding colloidal SiO_2 in aqueous polyacrylic acid (PAA). This solution is electrospun to get PAA/SiO_2 fibers. As PAA acts as a sacrificial template, it is removed by carbonizing electrospun fibers at 500 °C for 1.5 h. Then, the synthesized SiO_2 nanofibers were reduced by Mg powder in Ar atmosphere at 650 °C for 2 h, producing Si and MgO. Further, HCl etching was carried out to remove the MgO formed during reduction. These primary Si nanoparticles interconnect to form 3D m-SiNFs. Although a primary Si nanoparticle provides fast electronic and

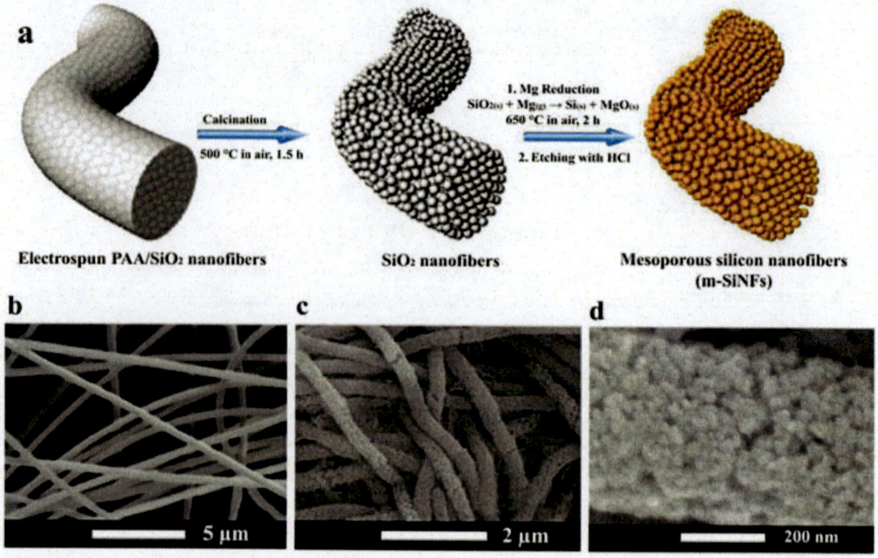

Fig. 15.5 **a** Schematic of synthesis of mesoporous Si nanofibers SEM images of the **b** electrospun PAA/SiO_2 fibers **c** and **d** SiNFs after Mg reduction and HCl etching. Adapted and reproduced from Ref. [36], Copyright 2013 American Chemical Society

ionic diffusion, the 3D m-SiNFs gives uniform distribution of nanoparticles which binds each other during disintegration during cycling. These m-SiNFs shows remarkable electrochemical performance as an anode material for Li-ion battery. Figure 15.5 shows the schematic of the synthesis of m-SiNFs. [36]. Cho and co-workers have also used Mg reduction procedure to prepare the electrospun SiNFs and made composite with graphene by plasma-enhanced chemical vapor deposition to improve the cycling performance of SiNFs. The graphene covers the open surface and absorbs the volume change and hence suppresses the pulverization [37].

X. Zhou et al. developed Si/porous carbon nanofibers (PCNFs) by electrospinning to study electrochemical performance of Si as anode material and schematic illustration as well scanning electron microscopy (SEM), as well as transmission electron microscopy (TEM) images, have been given in 15.6 b, c respectively. The SiO_x was coated on the Si nanoparticles by heating at 800 °C for 1 h in air to form a uniform suspension of Si@SiO_x in PAN/DMF solution. Then electrospinning solution was prepared by dissolving Si@SiO_x in PAN/DMF solution. Electrospun Si@SiO_x/PAN nanofibers were stabilized at 250 °C in air for 1 h and carbonized at 1000 °C in Argon atmosphere to obtain Si@SiOx@CNFs. Finally, HF treatment was done to remove the SiO_x and create Si@PCNFs. Removal of SiO_x leaves the space between the carbon and Si NPs (Si nanoparticles). Figure 15.6a shows the schematic illustration of the electrospinning-based synthesis of Si@PCNFs and scanning electron microscopy (SEM), as well as transmission electron microscopy (TEM) images,

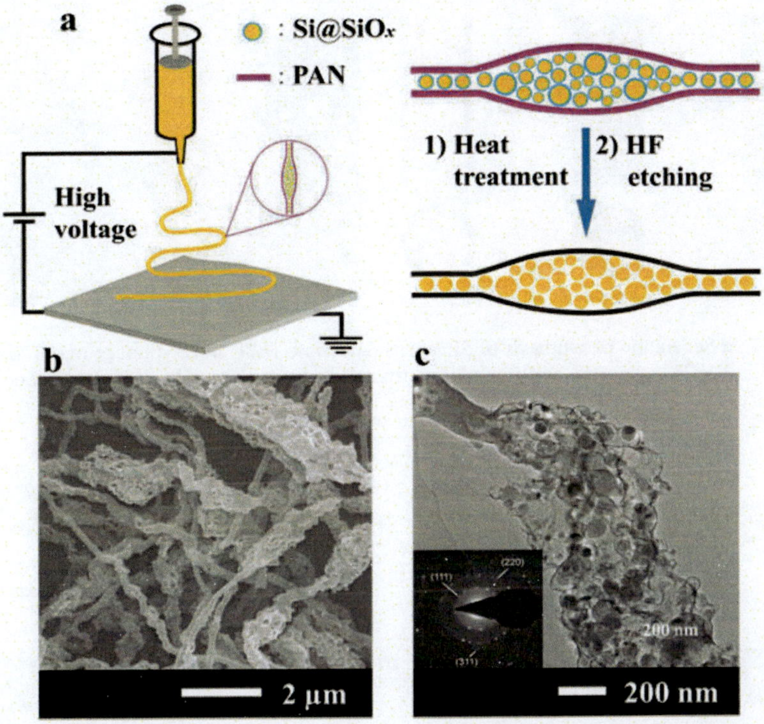

Fig. 15.6 **a** Schematic illustration of electrospinning-based synthesis of Si@PCNFs. **b** SEM images of the Si@PCNFs and **c** Tem image of Si@PCNFs. Adapted and reproduced from Ref. [38], Copyright 2013 John Wiley and Sons

have been given in 15.6 b, c respectively. Si@PCNFs composites show excellent rate capability and cycling stability [38].

H. Wu et al. demonstrated Si nanoparticles encapsulated inside the hollow carbon tubes (HCT) via electrospinning method and schematic of synthesis protocol, and TEM images have been given in Fig. 15.7. Electrospinning solution was prepared by addition of Si nanoparticles with Tetraethoxysilane (TEOS) and PVP in ethanol/HCl. This solution was electrospun to produce Si nanoparticles embedded with SiO_2 nanofibers. Further, polystyrene (PS) was carbonized for coating a layer of carbon onto composite nanofibers. The carbon-coated composite nanofibers were treated with HF to remove the SiO_2 and generate Si encapsulated hollow carbon tubes [39].

K. Fu and co-workers also coated carbon by chemical vapor deposition (CVD) on to the electrospun Si@CNF to compensate for the defects created during HF etching. This helps in encapsulating Si within the CNF and restricts the pulverization of Si NPs. They demonstrated the high electrochemical performance of carbon-coated Si@CNF compares to the Si@CNF. Figure 15.8a shows the schematics of the carbon coating of Si@CNFs, and Fig. 15.8b–d shows the TEM and SEM images of Si@CNF@C composite [25]. Similar synthesis protocols for the electrospinning

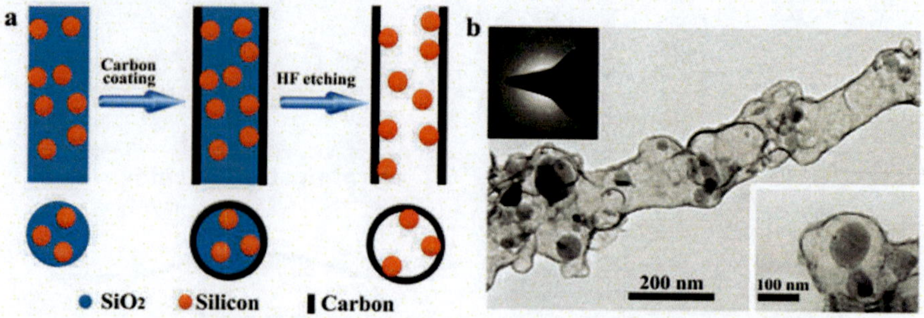

Fig. 15.7 **a** Schematic of synthesis of Si encapsulated into HCT, **b** TEM image of the Si encapsulated into HCT. Adapted and reproduced from Ref. [39], Copyright 2012 Royal Society of Chemistry

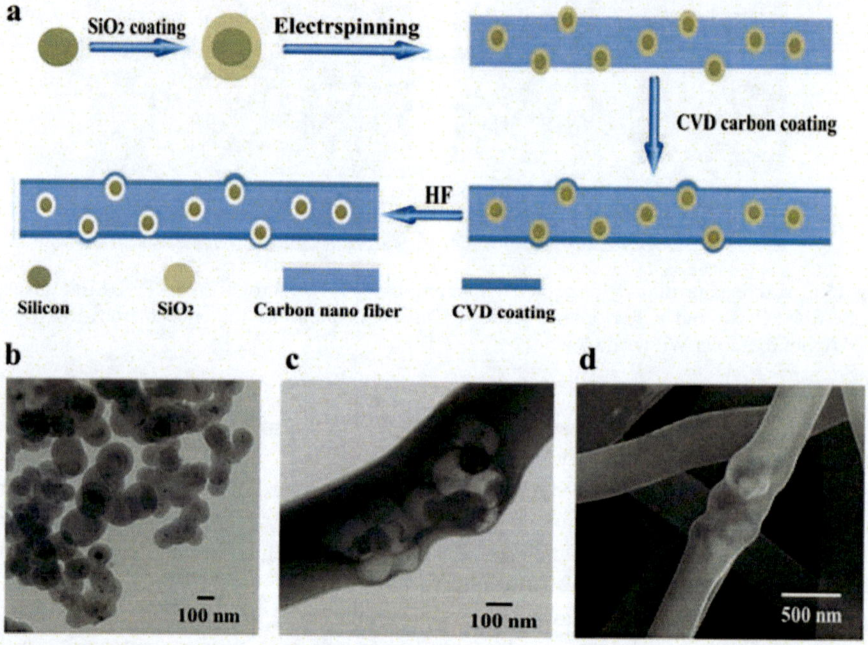

Fig. 15.8 **a** Schematic of synthesis of Si@CNF@C composite, **b** TEM image of Si@SiO_2 NPs, **c** TEM image of Si@CNF@C; **d** SEM images of Si@CNF@C at low and high magnification. Adapted and reproduced from Ref. [25], Copyright 2014 American Chemical Society

of Si and its composites with carbon nanofibers have been utilized by various other research groups which are listed in Table 15.2 to overcome the pulverization and volume expansion during the cycling. The capacities of the prepared materials as an anode for the Li-ion battery are shown in Table 15.3.

Table 15.2 Preparation of Si-NFs and Si-carbon nanofibers composite via electrospinning

Precursor	Electrospinning conditions[a]	Stabilization, carbonization, and activation	Ref.
PAA + aq. SiO_2 → m-SiNFs	12 kV/1 cm^3h^{-1}/-	Carbonization at 500 °C for 1.5 h, Reduction by Mg at 650 °C in Ar for 2 h	[36]
Sitetraacetate/H_2O/AcOH + PVA/H_2O → SiNFs → graphene coated SiNFs	18 kV/1.2 mLh^{-1}/15	Carbonization at 600 °C for 1 h, Reduction by Mg at 650 °C in Ar for 30 min, HCl treatment	[37]
Si + SiO_x + PAN/DMF → Si@PCNF	25 kV/1 mLh^{-1}/-	Stabilization at 250 °C for 30 min in air, Carbonization for 1 h at 1000 °C in Ar, HF treatment	[38]
Si + TEOS + PAN +/EtOH/HCl + PS/DMF → SiNP@HCT	15 kV/-	Stabilization at 600 °C for 5 h in air, carbonization at 850 °C for 10 min in N_2, HF treatment	[39]
Si NPs + TEOS/EtOH/H_2O/NH_4OH + PAN/DMF → Si@CNF + C_2H_2	15 kV/0.75 mL h^{-1}/-	Stabilization at 280 °C for 5.5 h in air, carbonization at 760 °C for 1 h in Ar; thermal heating at 760 °C for 30 min in C_2H_2 gas	[25]
PAN + Si NPs/DMF → Si/C	17 kV/0.75 mL h^{-1}/15	Stabilization at 280 °C for 5.5 h in air, carbonization at 700 °C for 1 h in Ar	[40]
PAN/DMF + PMMA +Si NPs/Acetone/DMF → SiNP@C core–shell	10.5 kV/1.5 and 2.0 mL h^{-1}/9 cm	Stabilization at 280 °C for 1 h in air, carbonization at 1000 °C for 5 h in Ar	[41]

(continued)

Table 15.2 (continued)

Precursor	Electrospinning conditions[a]	Stabilization, carbonization, and activation	Ref.
PVP/EtOH + Titanium (IV)Isopropoxide + Si/C/HOAc/EtOH → Si/C/TiO_2	25 kV/3 mL h^{-1}/12 cm	Annealed in air at 540 °C for 24 h followed by further 2 h annealing at 700 °C	[42]
Si NPs + PAN/DMF + PAN/DMF → Si/C–C core–shell composite NFs	20 kV/0.3 and 0.6 mL h^{-1}/15 cm	Stabilization at 280 °C for 5.5 h in air, carbonization at 800 °C for 2 h in Ar	[26]
Si + TEOS/EtOH/H_2O + PAN/DMF → Si@HC/CNFs	15 kV/0.7 mL h^{-1}/15	Carbonization at 700 °C and HF treatment	[43]
Si NPs/THF/EtOH + Phosphonitrilic chloride trimer + 4,4-sulfonyldiphenol + triethylamine PAN/DMF + Si@C → Si/C (Si@C/CNFs)	15 kV/0.75 mL h^{-1}/15 cm/-	Calcination at 900 °C for 2 h in Argon atmosphere to form Si@C Stabilization at 280 °C for 5 h in air, carbonization at 700 °C for 2 h in Ar	[44]
Si NPs + PAN/DMF	18 kV/0.75 mL h^{-1}/15 cm	Stabilization at 280 °C for 5 h in air, carbonization at 700, 800, 900 °C for 2 h in Ar	[45]
Si NPs + PAN/DMF	1.5 kV/0.75 mL h^{-1}	Stabilization at 280 °C for 5.5 h in air, carbonization at 760 °C for 1 h in Ar; thermal heating at 760 °C in C_2H_2 gas	[46]
PAN + Si/Mineral Oil → Si–C composite NFs	–/20 and 8 μL min^{-1}/10 cm	electrospun fibers were soaked in *n*-octane for 12 h, carbonization at 800 °C for 4 h in Ar/H_2	[47]

[a]Voltage/flow rate/distance between nozzle and collector

PAA Polyacrylic acid, *PAN* Polyacrylonitrile, *TEOS* Tetraethoxyorthosilicate, *HCT* Hollow carbon tubes, *DMF* N,N-dimethylformamide, *PVP* Polyvinylpyrrolidone, *PMMA* Polymethyl methacrylate, *THF* Tetrahydrofurane, *EtOH* Ethanol

Table 15.3 Capacities of Si-NFs and Si-carbon NFs composites

Materials	Electrochemical performance			Ref.
	Capacity (mAh g^{-1})	Current density (A g^{-1})	Cycle	
Si nanoparticles	1125.4	2	100	[36]
m-SiNFs	1363.4	2	300	[36]
Graphene-coated SiNFs	760	0.5	50	[37]
Si@PCNFs	1104	0.5	100	[38]
Si encapsulated HCT	870	1	200	[39]
Si@CNF@C composite	620	0.1	200	[25]
Si/carbon nanofibers	773	0.1	20	[40]
SiNP@C core–shell	721	2.75	300	[41]
Si/C/TiO_2 composite nanofibers	677	0.048	55	[42]
Si/C–C core–shell	886	0.05	50	[26]
Hollow core–shell structured silicon@carbon nanoparticles	1020	0.2	100	[43]
Si/C composite nanofibers	670	0.05	40	[44]
Si/C nanofibers	647	0.05	50	[45]
carbon coatings on Si@CNF (1) 10 wt% Si@CNF-C (2) 30 wt% Si@CNF-C	700 1000			[46]
Si (core)–hollow carbon nanofibers	1300	3	80	[47]

15.4 Advantages in 1D CNF/Si Composite for LIBs

As one of the most promising substituent to commercialized graphite which has 372 mAh g^{-1} theoretical capacity value, Si has attracted tremendous attention as an anode due to its very high theoretical capacity value (4200 mAh g^{-1}). With this holly nature, Si suffers poor electronic and ionic transport properties if it contains oxygen in it. Also, as Si is an alloying/dealloying anode type, it undergoes a huge volume expansion around 400% during lithiation/delithiation process as it forms an alloy with Li [48]. This volume expansion causes capacity fading during subsequent cycling. Many approaches have been taken to get the benefit of high capacity Si. The most commonly employed approaches are the nanoengineering of Si [49–51] and composite making with carbon [52–55]. Out of these Si-C composite is a promising strategy to control volume expansion of Si as carbon provides electrical conductivity and acts as robust mechanical support to volume expansion of Si which avoids deterioration of electrode. The super combination of Si/C composite provides high stability due to the presence of carbon with the advantage of high capacity associated with Si.

So overall electrochemical performance gets improved [56]. The selection of carbon material is the next important task. As there are extensively researched carbon forms at present, selecting a proper carbon is utmost required. Among different carbon forms, CNF is a promising candidate that provides electrical conductivity, provides a 1D path for Li^+-ion diffusion, more surface for electrode/electrolyte interphase, and can control volume expansion of Si by encapsulating Si in it [57].

J. Wang et al. demonstrated Si-hollow CNFs as anode for LIB. The authors have synthesized Si-hollow CNFs via simple coaxial electrospinning technique. The said nanocomposite shows improved electrochemical performance as compared to only Si. Si-hollow CNFs shows 1300 mAh g^{-1} specific capacity after 80 cycles at 0.5 °C current density (1500 mAg^{-1}). On the other hand, only Si shows a rapid decrease in capacity. The electron microscopy images are shown in Fig. 15.9. Outer coverage of CNFs accommodates volume change and carbon prevents direct contact of Si to electrolyte so the formation of SEI is also controlled. This makes Si-hollow CNFs composite material a promising anode material [47].

Ho-Sung Yang et al. encapsulated Si nanoparticles in multi-channeled hollow CNF and employed it as anode for LIB through coaxial electrospinning method. The multi-channels in CNFs provide more surface area to encapsulate Si. The authors have synthesized single channeled (1cHCNF), two channeled (2cHCNF), and 4

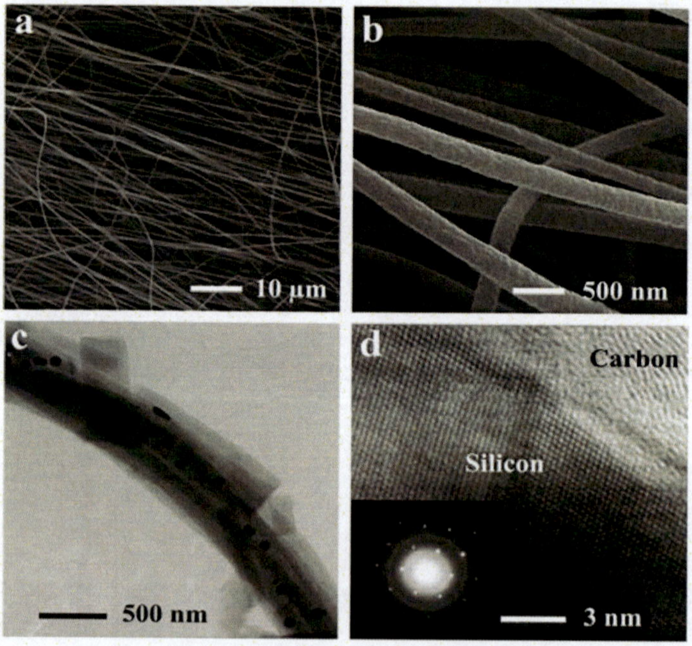

Fig. 15.9 **a**, **b** SEM images of PAN CNFs, **c** TEM image of Si nanoparticles encapsulated in CNFs, **d** HRTEM (high-resolution transmission electron microscopy) and SAED pattern of Si-hollow CNFs. Adapted and reproduced from Ref. [47], Copyright 2013 Royal Society of Chemistry

channeled hollow CNFs (4cHCNF) and encapsulated Si in it named as Si1cHCNF, Si2cHCNF, and Si4cHCNF, respectively. The electron microscope images are shown in Fig. 15.10. The charge and discharge capacity of Si1cHCNF, Si2cHCNF, and Si4cHCNF are displayed in Fig. 15.11. The specific capacities of Si1cHCNF, Si2cHCNF, and Si4cHCNF were 2615 mAh g^{-1}, 2506 mAh g^{-1}, and 2276 mAh g^{-1}, respectively. The highest capacity for Si4cHCNF is attributed to more number of contact points in between Si and CNFs. Also, capacity retention for Si4cHCNF is highest (72.6%) as compared to Si1cHCNF and Si2cHCNF after 50 cycles (Fig. 15.11a, b).

These multi-channeled CNFs provided improved conductive path and accommodated volume change of Si. This strategy improves contact surface in between Si and CNFs which helps to improve electrochemical performance [57]. Hence, CNFs is a promising carbon form for making composter with Si nanoparticles in order to achieve high capacity and cycling stability life. The 1D nature of CNFs is the key parameter to improve electrochemical performance. Also, the synthesis of CNF via electrospinning provides an advantage to get desired diameter CNFs and in situ synthesis of composite making with other nanomaterials.

Due to the several advantages of electrospun-based CNF/Si toward the LIBs, it has been performed and able to meet the high energy and power density demands. The electrochemical performance of the electrospun Si-based composites with different strategies has been discussed in detail.

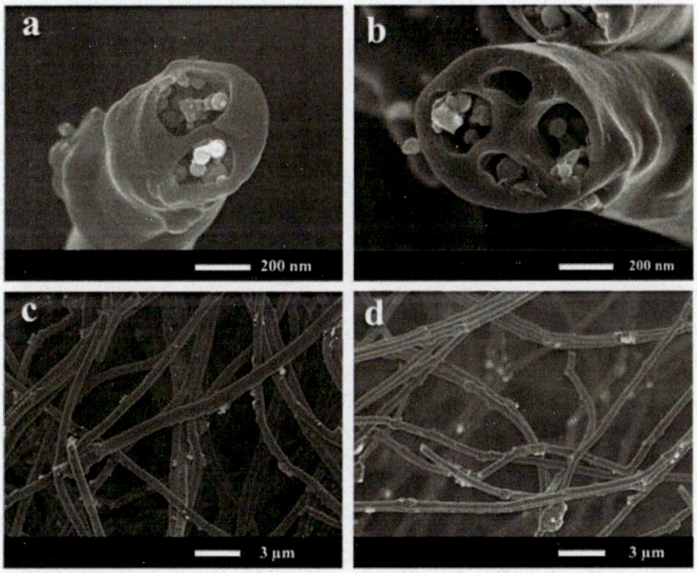

Fig. 15.10 **a** and **c** are FE-SEM images of Si2cHCNF and **b** and **d** are FE-SEM images of Si4cHCNF at high and low resolution, respectively. Adapted and reproduced from Ref. [57], Copyright 2014 Royal Society of Chemistry

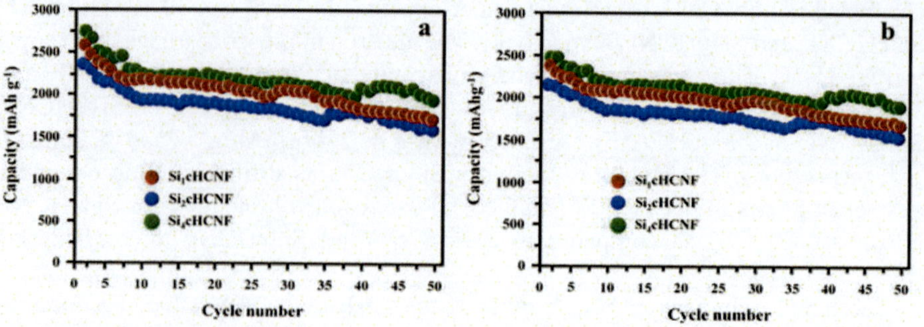

Fig. 15.11 **a** Charge and **b** discharge capacity over different cycles of Si1cHCNF, Si2cHCNF and Si4cHCNF samples with normalized Si content. Adapted and reproduced from Ref. [57], Copyright 2014 Royal Society of Chemistry

15.5 LIBs Performance of Electrospun-Based Synthesized Si Composites

There are several strategies that have been applied to overcome the challenge in Si-based anode for LIBs. Electrospun is one of the excellent methods to overcome the fundamental problem. Zhu et al. reported that atomic-scale control of silicon expansion as ultrastable LIBs anode. They have made binder-free three-dimensional flexible Si and graphene/carbon nanofibers (FSiGCNF) [58].

Figure 15.12a: Schematic design of synthesis of 3DFSiGCNFs, Fig. 15.13a: SEM images of 3DFSiGCNFs (top view) Fig. 15.13b: SEM images of 3DFSiGCNFs (side-view) Fig. 15.13c TEM images of SEM images of 3DFSiGCNFs and clearly showing the hollow space between Si and carbon layer (d–g) TEM elemental mapping of 3DFSiGCNFs showing the very well distribution of C, O, and Si and also empty space. Figure 15.14a, b shows electrochemical performance different samples while charging and discharging; in the case of Si NPs, the fast capacity fading is observed; due to huge volume change, SiCNF has slightly improved capacity and due to the coating of carbon which enhances the electrical conductivity and stability. It is notable that the FSiGCNF has a very high retention capacity of 2002 mAh g^{-1} at a current density of 700 mAg^{-1} also after 1050 cycles able to achieve very high capacity 3840 mAh g^{-1} for silicon alone. The high reversible capacity of Si alone is due to the voids present in the graphene carbon matrix which provide sufficient space for intercalation–deintercalation resulting in very high storage capacity with 99% coulombic efficiency. Figure 15.15a, b shows the rate performance of FSiGCNF during charging and discharging at current density range from 280 to 28,000 mAg^{-1}. 3D FSiGCNF delivers the reversible capacity of 1862, 1583, 1240, 1028, 886, and 761 mAh g^{-1} with current densities of 280, 700, 1400, 2800, 7000, and 14,000 mAg^{-1}, respectively.

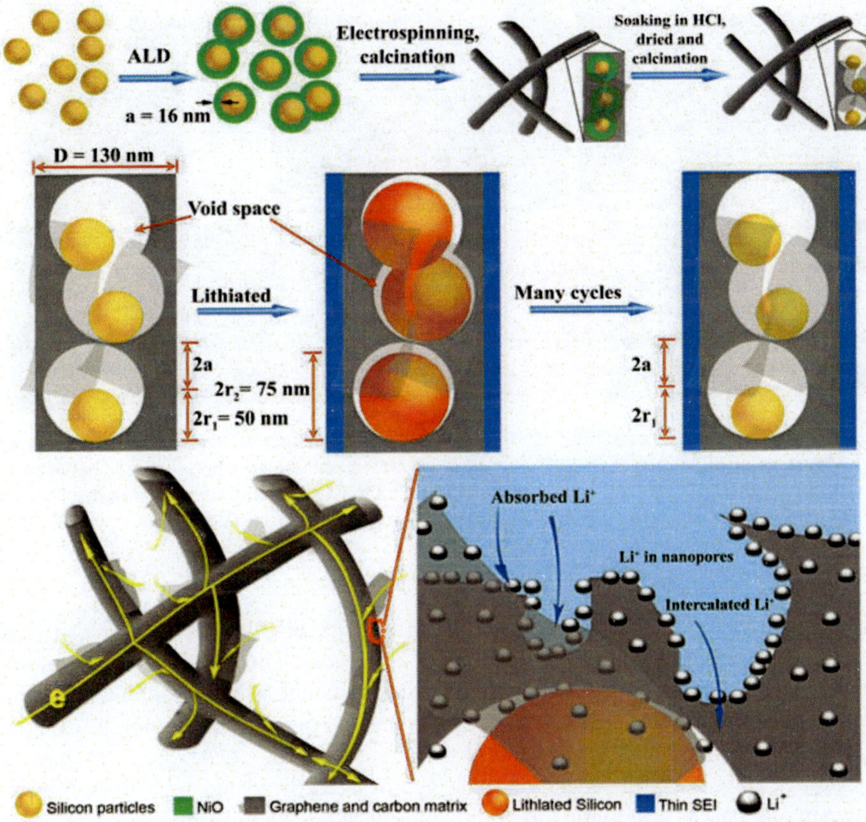

Fig. 15.12 **a** Schematic of synthesis of 3DFSiGCNFs. Adapted and reproduced from Ref. [58], Copyright 2016 American Chemical Society

These all data confirm that the electrospun-based method can control the fundamental challenges and able to deliver very high energy density and power density.

Wu et al. have been designed as engineered empty space between Si and CNTs by low-cost electrospinning methods to overcome the fundamental challenges. The electrochemical LIBs performance has been given in the figure. Figure 15.16a shows the schematic of conformal carbon coating on Si NPs and disintegration of Si NPs due to volume changes Fig. 15.16b shows the schematic of the design of carbon coating on Si NPs with empty space between Si NPs and Carbon coating. This empty space will help for the volume expansion and contraction of Si NPs during the electrochemical process. Figure 15.17a shows the stability of SiNPs@CT and shows very high capacity 969 mAh g^{-1} at current density of 1 A g^{-1} which is 3 time higher than state-of-the-art graphite anode at 1C. SiNPs@CTs shows the superior cyclability up to 200 cycles with 90% retention capacity. Figure 15.17 (b) shows the cyclic performance of bare Si, carbon-coated Si and SiNPs@CT. Si NPs show the drastically capacity fading due to large volume expansion, while carbon coated Si has

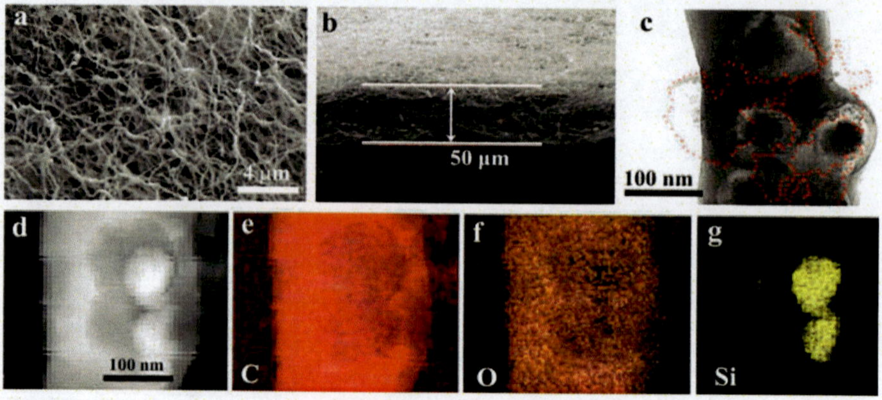

Fig. 15.13 **a** SEM images of 3DFSiGCNFs (top view) **b** SEM images of 3DFSiGCNFs (side-view) **c** TEM images of SEM images of 3DFSiGCNFs and clearly showing the hollow space between Si and carbon layer **d–g** TEM elemental mapping of 3DFSiGCNFs showing the very well distribution of C, O, and Si and also empty space. Adapted and reproduced from Ref. [58], Copyright 2016 American Chemical Society

somewhat improved capacity but SiNPs@CTs show excellent capacity retention even after 50 cycle negligible capacity fading (95% retention). The high capacity retention is due to the space available between SiNPs and CTs which help for suitable lithiation and delithiation process. Figure 15.17c shows the rate performance ranging from the current densities of 0.8 to 8 A g^{-1} [39].

Chen et al. successfully designed and synthesized Si coated with the CNFs as shown in Fig. 15.18a schematic illustration of Si@MC-CNF with void space between Si and carbon layer and Fig. 15.18b SEM image of Si@MC-CNF and Fig. 15.18c TEM image of Si@MC-CNF. They have used this composite with a fascinating strategy for the LIBs application. The electrochemical performance of Si@MC-CNF (Si@multichannel carbon fibers) has been given in Fig. 15.19a, b. Figure 15.19a shows the cyclic performance of Si@MC-CNF at a current density of 100 mAg^{-1} and able to achieve the discharge and charge capacity of 1400 and 1213 mAh g^{-1} which is very high as compared to graphite anode. Figure 15.19b shows the rate performance of Si@MC-CNF in the potential window of 0.01–2 V. Si@MC-CNF has been performed with different current densities 100, 200, 500, 1000, 1500 mAg^{-1} and achieved the capacity of 1342, 1331, 1215, 1007, and 784 mAh g^{-1}, respectively [59].

Hawang et al. Figure 15.20a, b has been designed core–shell SiNPs@C by electrospun using duel node. Figure 15.20a shows the electrospun duel node method for the synthesis of SiNPs@C to overcome the problem of volume expansion. Figure 15.20b shows the cross-sectional TEM images of SiNPs@C that also clearly indicates the core–shell morphology of SiNPs and carbon. [41]. Li et al. also designed the binder-free silicon/carbon fabric using double-nozzle electrospinning and Fig. 15.20c shows the schematic of the synthesis of silicon/carbon binder-free fabrics. Figure 15.20d shows the TEM images of C_{PAN}-Si/C_{PAN} [60].

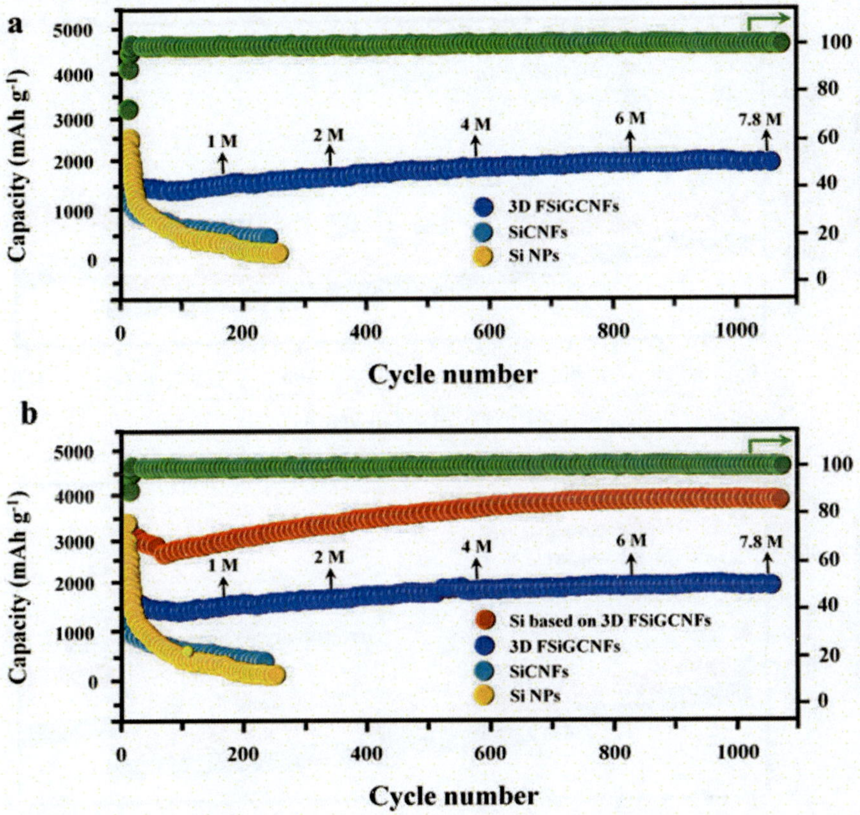

Fig. 15.14 Cyclic performance of 3DFSiGCNFs, SiCNFs and pure Si NPs during **a** charging, **b** discharging at current density of 700 mAg^{-1}. Adapted and reproduced from Ref. [58], Copyright 2016 American Chemical Society

Zhange et al. designed and synthesized electrospun core–shell Si/carbon fibers. Figure 15.21a is the schematic illustration of structural design of Si/Po-C@C (Si in porous carbon@C) composite and 3D sketch of overall structure, Fig. 15.21b SEM image of Si/Po-C@C, Also, Ma et al. designed as shown in Fig. 15.21c schematic illustration for the synthesis process of G/Si@CFs (carbon nanofibers interwined

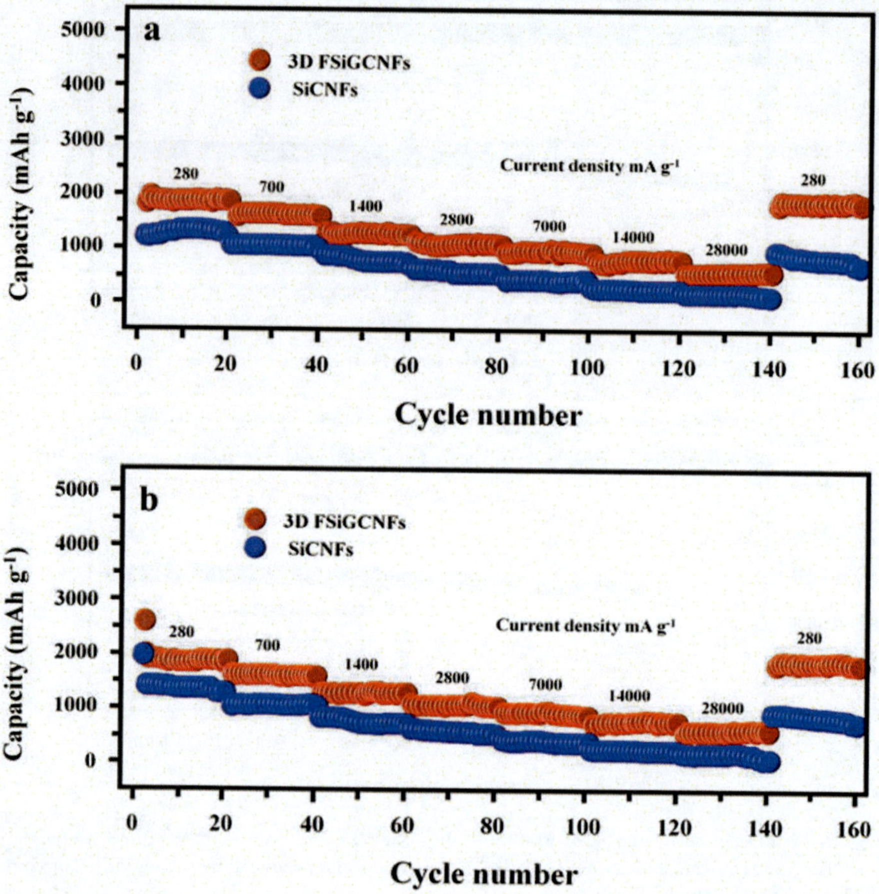

Fig. 15.15 Rate performance of 3DFSiGCNFs as well as SiCNFs during **a** charging, **b** discharging at different current densities. Adapted and reproduced from Ref. [58], Copyright 2016 American Chemical Society

graphene/silicon) composite, Fig. 15.21d SEM image of G/Si@CFs, Fig. 15.22a cyclic performance of Si/Po-C@C, Si NPs only, and Po-C@C at current density of 0.2 Ag^{-1} Fig. 15.22b rate capability of Si/Po-C@C at current densities of 0.1–1 A g^{-1} [61]. Figure 15.23a cycling performance with coulombic efficiency of G/Si@CFs, G/Si, and Si NPs@CFs, Fig. 15.23b rate capability of G/Si@CFs and Si NPs @ CFs [62].

Table 15.3 shows the capacities of Si NPs, Si-NFS, and Si-carbon NFs composites for the performance of LIBs.

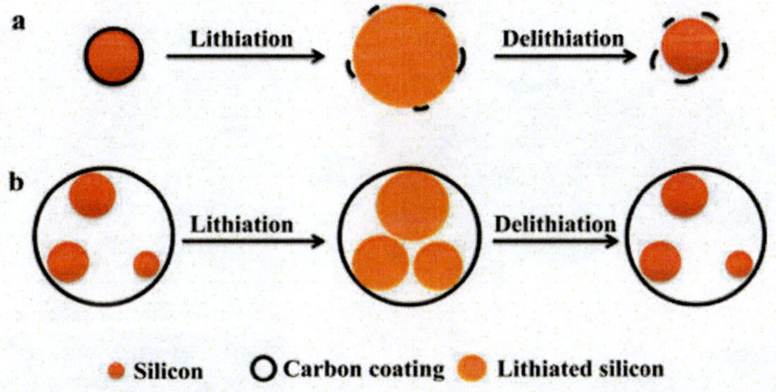

Fig. 15.16 **a** Schematic of carbon-coated silicon NPs **b** schematic of carbon-coated silicon with empty space between Si NPs and carbon. Adapted and reproduced from Ref. [39], Copyright 2012 American Chemical Society

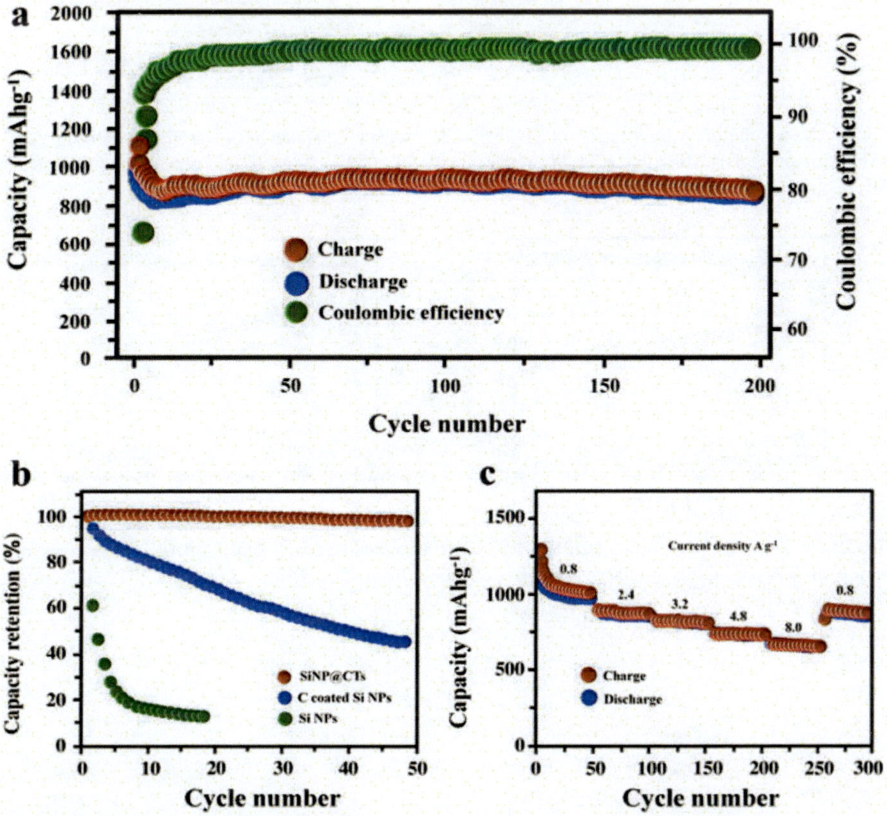

Fig. 15.17 **a** Cyclic stability of SiNPs@CT at 1C **b** cyclic performance of bare Si, carbon-coated Si and SiNPs@CT **c** rate performance of SiNPs@CT. Adapted and reproduced from Ref. [39], Copyright 2012 American Chemical Society

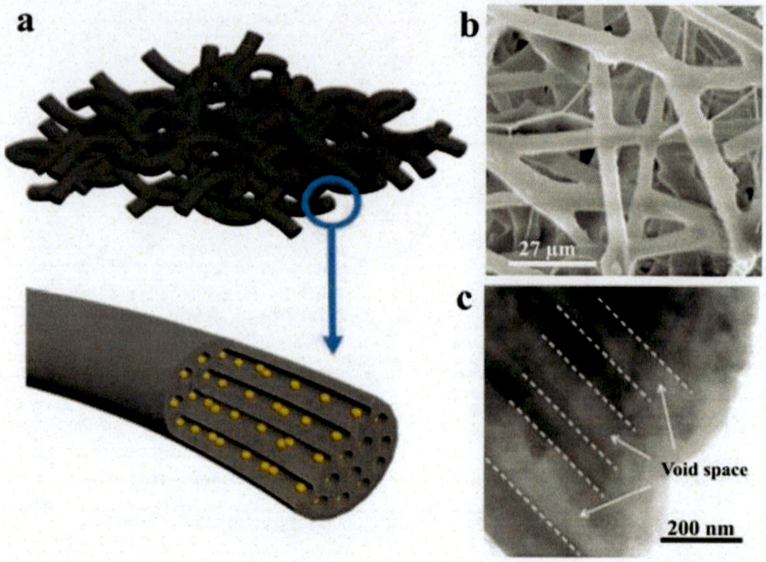

Fig. 15.18 **a** Schematic illustration of Si@MC-CNF with void space between Si and carbon layer. **b** SEM image of Si@MC-CNF **c** TEM image of Si@MC-CNF. Adapted and reproduced from Ref. [59], Copyright 2019 American Chemical Society

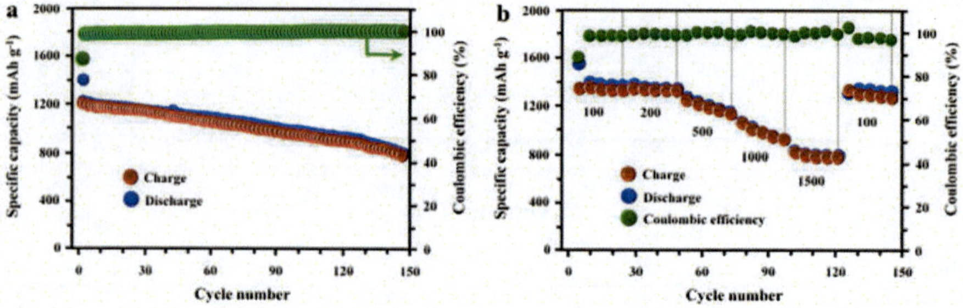

Fig. 15.19 **a** Cyclic performance of Si@MC-CNF at current density of 100 mAg^{-1} **b** rate performance of Si@MC-CNF. Adapted and reproduced from Ref. [59], Copyright 2019 American Chemical Society

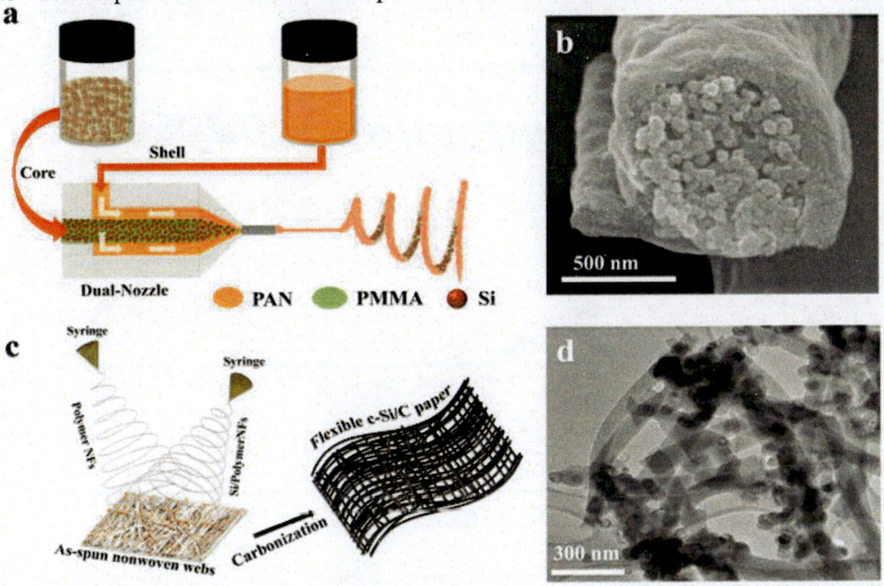

Fig. 15.20 **a** Synthesis of SiNPs@C by electrospun. **b** Cross-sectional TEM images for SiNPs@C [41]. **c** Synthesis of C-Si/C using double nozzle electrospinning. **d** TEM images of C_{PAN}-Si/C_{PAN}. Adapted and reproduced from Ref. [60], Copyright 2016 American Chemical Society

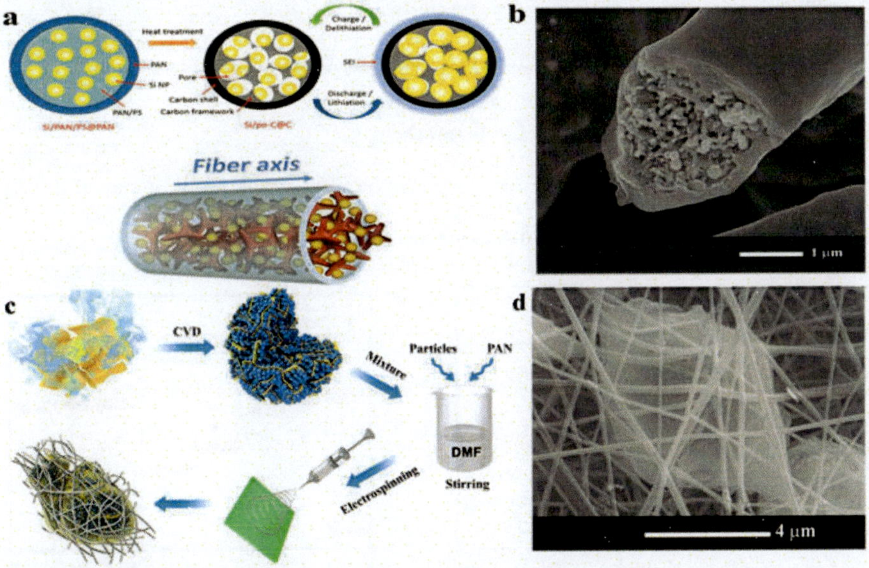

Fig. 15.21 **a** Schematic illustration of structural design of SiPo-C@C composite and 3D sketch of overall structure; **b** SEM image of SiPo-C@C. Adapted and reproduced from Ref. [61], Copyright 2015 Royal Society of Chemistry **c** Schematic illustration for the synthesis process of G/Si@CFs composite, **d** SEM image of G/Si@CFs. Adapted and reproduced from Ref. [62], Copyright 2017 Springer Nature

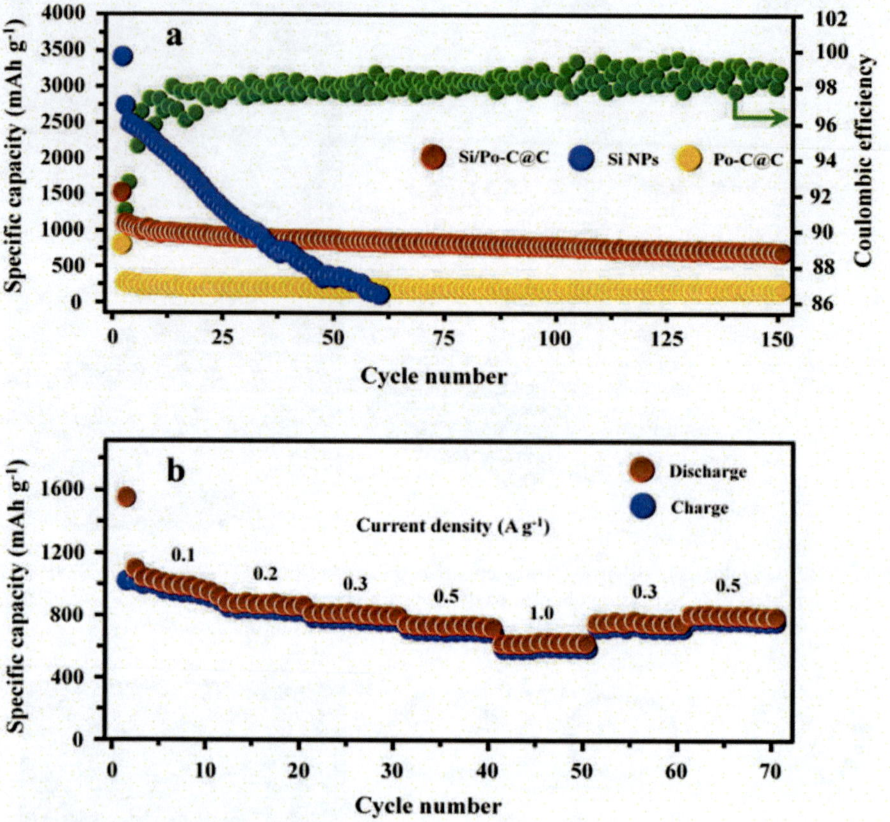

Fig. 15.22 **a** Cyclic performance of Si/Po-C@C, Si NPs only and Po-C@C at current density of 0.2 A g^{-1}; **b** rate capability of Si/Po-C@C at current densities of 0.1–1 A g^{-1}. Adapted and reproduced from Ref. [61], Copyright 2015 Royal Society of Chemistry

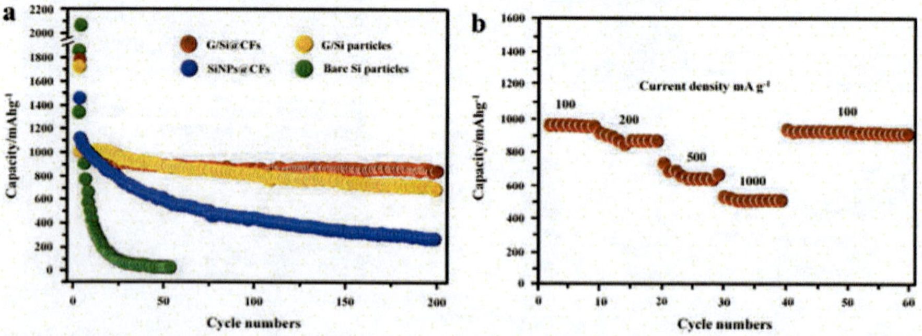

Fig. 15.23 **a** Cycling performance with coulombic efficiency of G/Si@CFs, G/Si, and Si NPs@ CFs; **b** rate capability of G/Si@CFs and Si NPs @ CFs. Adapted and reproduced from Ref. [62], Copyright 2017 Springer Nature

15.6 Conclusions

In this chapter, we have summarized the need for high energy density, high power density, and long cycle life LIBs. As Si has high theoretical capacity among various anode materials, low operating potential, earth abandon makes Si as promising candidate. But due to some challenges such as pulverization, morphology changes, volume expansion and SEI layers formation with cycle by cycles. But to overcome these problems, many engineered tailored strategies have been employed to meet the next generation energy density and power density demands. Therefore, electrospun-based method with tuned space sized between Si NPs core and carbon shell has fulfilled the requirements of the above three challenges. All these strategies are able to meet the requirement, and the performances of the materials have been given in Table 15.3.

References

1. Reddy MV, Subba Rao GV, Chowdari BVR (2013) Metal oxides and oxysalts as anode materials for Li ion batteries. Chem Rev 113(7):5364–5457. https://doi.org/10.1021/cr3001884
2. Aravindan V, Gnanaraj J, Lee YS, Madhavi S (2014) Insertion-type electrodes for nonaqueous Li-ion capacitors. Chem Rev 114(23):11619–11635. https://doi.org/10.1021/cr5000915
3. Deng J, Ji H, Yan C, Zhang J, Si W, Baunack S, Oswald S, Mei Y, Schmidt OG (2013) Naturally rolled-up C/Si/C trilayer nanomembranes as stable anodes for lithium-ion batteries with remarkable cycling performance. Angew Chem Int Ed 52(8):2326–2330. https://doi.org/10.1002/anie.201208357
4. Goriparti S, Miele E, De Angelis F, Di Fabrizio E, Zaccaria RP, Capiglia C (2014) Review on recent progress of nanostructured anode materials for Li-ion batteries. J Power Sources 257:421–443. https://doi.org/10.1016/j.jpowsour.2013.11.103
5. Shelke MV, Gullapalli H, Kalaga K, Rodrigues MTF, Devarapalli RR, Vajtai R, Ajayan PM (2017) Facile synthesis of 3D anode assembly with Si nanoparticles sealed in highly pure few layer graphene deposited on porous current collector for long life Li-ion battery. Adv Mater Interfaces 4(10):1601043. https://doi.org/10.1002/admi.201601043
6. Chan CK, Peng H, Liu G, McIlwrath K, Zhang XF, Huggins RA, Cui Y (2008) High-performance lithium battery anodes using silicon nanowires. Nat Nanotechnol 3(1):31. https://doi.org/10.1038/nnano.2007.411
7. Hu YS, Demir-Cakan R, Titirici MM, Müller JO, Schlögl R, Antonietti M, Maier J (2008) Superior storage performance of a Si@ SiOx/C nanocomposite as anode material for lithium-ion batteries. Angew Chem Int Ed 47(9):1645–1649. https://doi.org/10.1002/anie.200704287
8. Zhou X, Yin YX, Wan LJ, Guo YG (2012) Facile synthesis of silicon nanoparticles inserted into graphene sheets as improved anode materials for lithium-ion batteries. Chem Commun 48(16):2198–2200. https://doi.org/10.1039/C2CC17061B
9. Wu H, Cui Y (2012) Designing nanostructured Si anodes for high energy lithium ion batteries. Nano today 7(5):414–429. https://doi.org/10.1016/j.nantod.2012.08.004
10. Tardif S, Pavlenko E, Quazuguel L, Boniface M, Maréchal M, Micha JS, Gonon L, Mareau V, Gebel G, Bayle-Guillemaud P, Rieutord F (2017) Operando Raman spectroscopy and synchrotron X-ray diffraction of lithiation/delithiation in silicon nanoparticle anodes. ACS Nano 11(11):11306–11316. https://doi.org/10.1021/acsnano.7b05796
11. Su X, Wu Q, Li J, Xiao X, Lott A, Lu W, Sheldon BW, Wu J (2014) Silicon-based nanomaterials for lithium-ion batteries: a review. Adv Energy Mater 4(1):1300882. https://doi.org/10.1002/aenm.201300882

12. Obrovac MN, Chevrier VL (2014) Alloy negative electrodes for Li-ion batteries. Chem Rev 114(23):11444–11502. https://doi.org/10.1021/cr500207g
13. Luo W, Chen X, Xia Y, Chen M, Wang L, Wang Q, Li W, Yang J (2017) Surface and Interface engineering of silicon-based anode materials for lithium-ion batteries. Adv Energy Mater 7(24):1701083. https://doi.org/10.1002/aenm.201701083
14. Rahman MA, Song G, Bhatt AI, Wong YC, Wen C (2016) Nanostructured silicon anodes for high-performance lithium-ion batteries. Adv Func Mater 26(5):647–678. https://doi.org/10.1002/adfm.201502959
15. Lee PK, Tan T, Wang S, Kang W, Lee CS, Yu DY (2018) Robust micron-sized silicon secondary particles anchored by polyimide as high-capacity, high-stability li-ion battery anode. ACS Appl Mater Interfaces 10(40):34132–34139. https://doi.org/10.1021/acsami.8b09566
16. Park MH, Kim MG, Joo J, Kim K, Kim J, Ahn S, Cui Y, Cho J (2009) Silicon nanotube battery anodes. Nano Lett 9(11):3844–3847. https://doi.org/10.1021/nl902058c
17. Ge M, Lu Y, Ercius P, Rong J, Fang X, Mecklenburg M, Zhou C (2013) Large-scale fabrication, 3D tomography, and lithium-ion battery application of porous silicon. Nano Lett 14(1):261–268. https://doi.org/10.1021/nl403923s
18. Ko M, Chae S, Jeong S, Oh P, Cho J (2014) Elastic a-silicon nanoparticle backboned graphene hybrid as a self-compacting anode for high-rate lithium ion batteries. ACS Nano 8(8):8591–8599. https://doi.org/10.1021/nn503294z
19. Park JH, Moon J, Han S, Park S, Lim JW, Yun DJ, Kim DY, Park K, Son IH (2017) Formation of stable solid–electrolyte interphase layer on few-layer graphene-coated silicon nanoparticles for high-capacity Li-ion battery anodes. J Phys Chem C 121(47):26155–26162. https://doi.org/10.1021/acs.jpcc.7b05876
20. Hassan FM, Elsayed AR, Chabot V, Batmaz R, Xiao X, Chen Z (2014) Subeutectic growth of single-crystal silicon nanowires grown on and wrapped with graphene nanosheets: high-performance anode material for lithium-ion battery. ACS Appl Mater Interfaces 6(16):13757–13764. https://doi.org/10.1021/am5032067
21. Guo S, Hu X, Hou Y, Wen Z (2017) Tunable synthesis of yolk–shell porous silicon@ carbon for optimizing Si/C-based anode of lithium-ion batteries. ACS Appl Mater Interfaces 9(48):42084–42092. https://doi.org/10.1021/acsami.7b13035
22. Yang Y, Yang X, Chen S, Zou M, Li Z, Cao A, Yuan Q (2017) Rational design of hierarchical carbon/mesoporous silicon composite sponges as high-performance flexible energy storage electrodes. ACS Appl Mater Interfaces 9(27):22819–22825. https://doi.org/10.1021/acsami.7b05032
23. Bhandavat R, Singh G (2013) Stable and efficient Li-ion battery anodes prepared from polymer-derived silicon oxycarbide–carbon nanotube shell/core composites. J Phys Chem C 117(23):11899–11905. https://doi.org/10.1021/jp310733b
24. Zhang J, Fan S, Wang H, Qian J, Yang H, Ai X, Liu J (2019) Surface-bound silicon nanoparticles with a planar-oriented n-type polymer for cycle-stable li-ion battery anode. ACS Appl Mater Interfaces 11(14):13251–13256. https://doi.org/10.1021/acsami.9b00939
25. Fu K, Lu Y, Dirican M, Chen C, Yanilmaz M, Shi Q, Bradford PD, Zhang X (2014) Chamber-confined silicon–carbon nanofiber composites for prolonged cycling life of Li-ion batteries. Nanoscale 6(13):7489–7495. https://doi.org/10.1039/C4NR00518J
26. Li Y, Xu G, Yao Y, Xue L, Yanilmaz M, Lee H, Zhang X (2014) Coaxial electrospun Si/C–C core–shell composite nanofibers as binder-free anodes for lithium-ion batteries. Solid State Ionics 258:67–73. https://doi.org/10.1016/j.ssi.2014.02.003
27. Chen Y, Hu Y, Shen Z, Chen R, He X, Zhang X, Zhang Y, Wu K (2016) Sandwich structure of graphene-protected silicon/carbon nanofibers for lithium-ion battery anodes. Electrochim Acta 210:53–60. https://doi.org/10.1016/j.electacta.2016.05.086
28. Nan D, Huang ZH, Lv R, Lin Y, Yang L, Yu X, Ye L, Shen W, Sun H, Kang F (2014) Silicon-encapsulated hollow carbon nanofiber networks as binder-free anodes for lithium ion battery. J Nanomaterials 2014:9. https://doi.org/10.1155/2014/139639

29. Dirican M, Yildiz O, Lu Y, Fang X, Jiang H, Kizil H, Zhang X (2015) Flexible binder-free silicon/silica/carbon nanofiber composites as anode for lithium–ion batteries. Electrochim Acta 169:52–60. https://doi.org/10.1016/j.electacta.2015.04.035
30. Jung JW, Lee CL, Yu S, Kim ID (2016) Electrospun nanofibers as a platform for advanced secondary batteries: a comprehensive review. J Mater Chem A 4(3):703–750. https://doi.org/10.1039/C5TA06844D
31. Song T, Xia J, Lee JH, Lee DH, Kwon MS, Choi JM, Wu J, Doo SK, Chang H, Park WI, Zang DS (2010) Arrays of sealed silicon nanotubes as anodes for lithium ion batteries. Nano Lett 10(5):1710–1716. https://doi.org/10.1021/nl100086e
32. Yao Y, McDowell MT, Ryu I, Wu H, Liu N, Hu L, Nix WD, Cui Y (2011) Interconnected silicon hollow nanospheres for lithium-ion battery anodes with long cycle life. Nano Lett 11(7):2949–2954. https://doi.org/10.1021/nl201470j
33. Chen D, Mei X, Ji G, Lu M, Xie J, Lu J, Lee JY (2012) Reversible lithium-ion storage in silver-treated nanoscale hollow porous silicon particles. Angew Chem Int Ed 51(10):2409–2413. https://doi.org/10.1002/anie.201107885
34. Xu D, Huang Z, Miao R, Bie Y, Yang J, Yao Y, Che S (2014) Rigid bolaform surfactant templated mesoporous silicon nanofibers as anode materials for lithium-ion batteries. J Mater Chcm A 2(46):19855–19860. https://doi.org/10.1039/C4TA04088K
35. Song T, Lee DH, Kwon MS, Choi JM, Han H, Doo SG, Chang H, Park WI, Sigmund W, Kim H, Paik U (2011) Silicon nanowires with a carbon nanofiber branch as lithium-ion anode material. J Mater Chem 21(34):12619–12621. https://doi.org/10.1039/C1JM12511G
36. Lee DJ, Lee H, Ryou MH, Han GB, Lee JN, Song J, Choi J, Cho KY, Lee YM, Park JK (2013) Electrospun three-dimensional mesoporous silicon nanofibers as an anode material for high-performance lithium secondary batteries. ACS Appl Mater Interfaces 5(22):12005–12010. https://doi.org/10.1021/am403798a
37. Cho D, Kim M, Hwang J, Park JH, Joo YL, Jeong Y (2015) Facile synthesis of porous silicon nanofibers by magnesium reduction for application in lithium ion batteries. Nanoscale Res Lett 10(1):424. https://doi.org/10.1186/s11671-015-1132-8
38. Zhou X, Wan LJ, Guo YG (2013) Electrospun silicon nanoparticle/porous carbon hybrid nanofibers for lithium-ion batteries. Small 9(16):2684–2688. https://doi.org/10.1002/smll.201202071
39. Wu H, Zheng G, Liu N, Carney TJ, Yang Y, Cui Y (2012) Engineering empty space between Si nanoparticles for lithium-ion battery anodes. Nano Lett 12(2):904–909. https://doi.org/10.1021/nl203967r
40. Ji L, Jung KH, Medford AJ, Zhang X (2009) Electrospun polyacrylonitrile fibers with dispersed Si nanoparticles and their electrochemical behaviors after carbonization. J Mater Chem 19(28):4992–4997. https://doi.org/10.1039/B903165K
41. Hwang TH, Lee YM, Kong BS, Seo JS, Choi JW (2012) Electrospun core–shell fibers for robust silicon nanoparticle-based lithium ion battery anodes. Nano Lett 12(2):802–807. https://doi.org/10.1021/nl203817r
42. Wu Q, Tran T, Lu W, Wu J (2014) Electrospun silicon/carbon/titanium oxide composite nanofibers for lithium ion batteries. J Power Sources 258:39–45. https://doi.org/10.1016/j.jpowsour.2014.02.047
43. Chen Y, Hu Y, Shen Z, Chen R, He X, Zhang X, Li Y, Wu K (2017) Hollow core–shell structured silicon@ carbon nanoparticles embed in carbon nanofibers as binder-free anodes for lithium-ion batteries. J Power Sources 342:467–475. https://doi.org/10.1016/j.jpowsour.2016.12.089
44. Xue L, Fu K, Li Y, Xu G, Lu Y, Zhang S, Toprakci O, Zhang X (2013) Si/C composite nanofibers with stable electric conductive network for use as durable lithium-ion battery anode. Nano Energy 2(3):361–367. https://doi.org/10.1016/j.nanoen.2012.11.001

45. Li Y, Guo B, Ji L, Lin Z, Xu G, Liang Y, Zhang S, Toprakci O, Hu Y, Alcoutlabi M, Zhang X (2013) Structure control and performance improvement of carbon nanofibers containing a dispersion of silicon nanoparticles for energy storage. Carbon 51:185–194. https://doi.org/10.1016/j.carbon.2012.08.027
46. Fu K, Xue L, Yildiz O, Li S, Lee H, Li Y, Xu G, Zhou L, Bradford PD, Zhang X (2013) Effect of CVD carbon coatings on Si@ CNF composite as anode for lithium-ion batteries. Nano Energy 2(5):976–986. https://doi.org/10.1016/j.nanoen.2013.03.019
47. Wang J, Yu Y, Gu L, Wang C, Tang K, Maier J (2013) Highly reversible lithium storage in Si (core)–hollow carbon nanofibers (sheath) nanocomposites. Nanoscale 5(7):2647–2650. https://doi.org/10.1039/C3NR00322A
48. Jerliu B, Hüger E, Dorrer L, Seidlhofer BK, Steitz R, Oberst VV, Geckle U, Bruns M, Schmidt H (2014). Volume expansion during lithiation of amorphous silicon thin film electrodes studied by in-operando neutron reflectometry. J Phys Chem C 118(18):9395–9399. https://doi.org/10.1021/jp502261t
49. Sun L, Su T, Xu L, Du HB (2016) Preparation of uniform Si nanoparticles for high-performance Li-ion battery anodes. Phys Chem Chem Phys 18(3):1521–1525. https://doi.org/10.1039/C5CP06585B
50. Wang Y, Xie K, Guo X, Zhou W, Song G, Cheng S (2016) Mesoporous silica nanoparticles as high performance anode materials for lithium-ion batteries. New J Chem 40(10):8202–8205. https://doi.org/10.1039/C6NJ01698G
51. Jiang S, Hu B, Sahore R, Zhang L, Liu H, Zhang L, Lu W, Zhao B, Zhang Z (2018) Surface-Functionalized silicon nanoparticles as anode material for lithium-ion battery. ACS Appl Mater Interfaces 10(51):44924–44931. https://doi.org/10.1021/acsami.8b17729
52. Palumbo S, Silvestri L, Ansaldo A, Brescia R, Bonaccorso F, Pellegrini V (2019) Silicon few-layer graphene nanocomposite as high-capacity and high-rate anode in lithium-ion batteries. ACS Applied Energy Mater 2(3):1793–1802. https://doi.org/10.1021/acsaem.8b01927
53. Ng SH, Wang J, Wexler D, Konstantinov K, Guo ZP, Liu HK (2006) Highly reversible lithium storage in spheroidal carbon-coated silicon nanocomposites as anodes for lithium-ion batteries. Angew Chem Int Ed 45(41):6896–6899. https://doi.org/10.1002/anie.200601676
54. Zhang R, Du Y, Li D, Shen D, Yang J, Guo Z, Liu HK, Elzatahry AA, Zhao D (2014) Highly reversible and large lithium storage in mesoporous Si/C nanocomposite anodes with silicon nanoparticles embedded in a carbon framework. Adv Mater 26(39):6749–6755. https://doi.org/10.1002/adma.201402813
55. Hu YS, Adelhelm P, Smarsly BM, Maier J. (2010) Highly stable lithium storage performance in a porous carbon/silicon nanocomposite. Chem Sus Chem: Chem Sustain Energy Mater 3(2):231–235. https://doi.org/10.1002/cssc.200900191
56. Magasinski A, Dixon P, Hertzberg B, Kvit A, Ayala J, Yushin G (2010) High-performance lithium-ion anodes using a hierarchical bottom-up approach. Nat Mater 9(4):353. https://doi.org/10.1038/nmat2725
57. Yang HS, Lee BS, You BC, Sohn HJ, Yu WR (2014) Fabrication of carbon nanofibers with Si nanoparticle-stuffed cylindrical multi-channels via coaxial electrospinning and their anodic performance. RSC advances 4(88):47389–47395. https://doi.org/10.1039/C4RA10031J
58. Zhu J, Wang T, Fan F, Mei L, Lu B (2016) Atomic-scale control of silicon expansion space as ultrastable battery anodes. ACS Nano 10(9):8243–8251. https://doi.org/10.1021/acsnano.6b04522
59. Chen X, Hu P, Xiang J, Zhang R, Huang Y (2019) Confining silicon nanoparticles within free-standing multichannel carbon fibers for high-performance li-ion batteries. ACS Appl Energy Mater. https://doi.org/10.1021/acsaem.9b00898
60. Li C, Liu C, Wang W, Bell J, Mutlu Z, Ahmed K, Ye R, Ozkan M, Ozkan CS (2016) Towards flexible binderless anodes: silicon/carbon fabrics via double-nozzle electrospinning. Chem Commun 52(76):11398–11401. https://doi.org/10.1039/C6CC04074H
61. Zhang H, Qin X, Wu J, He YB, Du H, Li B, Kang F (2015) Electrospun core–shell silicon/carbon fibers with an internal honeycomb-like conductive carbon framework as an anode for lithium ion batteries. J Mater Chem A 3(13):7112–7120. https://doi.org/10.1039/C4TA06044J

62. Ma X, Hou G, Ai Q, Zhang L, Si P, Feng J, Ci L (2017) A heart-coronary arteries structure of carbon nanofibers/graphene/silicon composite anode for high performance lithium ion batteries. Scientific reports 7(1):9642. https://doi.org/10.1038/s41598-017-09658-4

Chapter 16
Lithium Cobalt Oxide ($LiCoO_2$): A Potential Cathode Material for Advanced Lithium-Ion Batteries

Anjumole P. Thomas, Akhila Das, Leya Rose Raphael, Neethu T. M. Balakrishnan, Jou-Hyeon Ahn, M. J. Jabeen Fatima, and Raghavan Prasanth

16.1 Introduction

Lithium-ion batteries (LIB) have long been considered as a power source for portable electronics due to its high energy density and very low self-discharge of all rechargeable batteries. Today, lithium-ion batteries dominating the energy storage device market at least by a factor of 2.5 to any competing technology because of its high value of energy density, *i.e.*, 150 Wh kg^{-1} [1]. The performance of a battery is a measure of its cell potential, capacity, and energy density which is directly related to the properties of the material that forms positive and negative electrodes. Cycle time and stability are also a function of the electrode material. Compared to other battery technology such as lead–acid and Ni–Cd, lithium batteries are in the state of infancy, and there is much hope for innovations in battery chemistry and cell engineering. Fabricating efficient batteries with suitable properties is a key challenge that will lead to developments of novel materials for battery such as cathode, anode, and electrolyte [1]. The components of lithium-ion batteries are electrodes, positive and negative electrodes, and electrolytes. The schematic of a typical lithium-ion battery is displayed in Fig. 16.1.

Inorganic materials that reversibly react with lithium were later identified as intercalation compounds that are crucial in the development of promising lithium-ion

A. P. Thomas · A. Das · L. R. Raphael · N. T. M. Balakrishnan · M. J. Jabeen Fatima (✉) ·
R. Prasanth (✉)
Department of Polymer Science and Rubber Technology (PSRT), Cochin University of Science and Technology (CUSAT), Cochin, Kerala 682022, India
e-mail: jabeen@cusat.ac.in

R. Prasanth
e-mail: prasanth@cusat.ac.in

J.-H. Ahn · R. Prasanth
Department of Materials Engineering and Convergence Technology, Gyeongsang National University, 501 Jinju-Daero, Jinju 52828, Republic of Korea

N. T. M. Balakrishnan and R. Prasanth (eds.), *Electrospinning for Advanced Energy Storage Applications*, Materials Horizons: From Nature to Nanomaterials,
https://doi.org/10.1007/978-981-15-8844-0_16

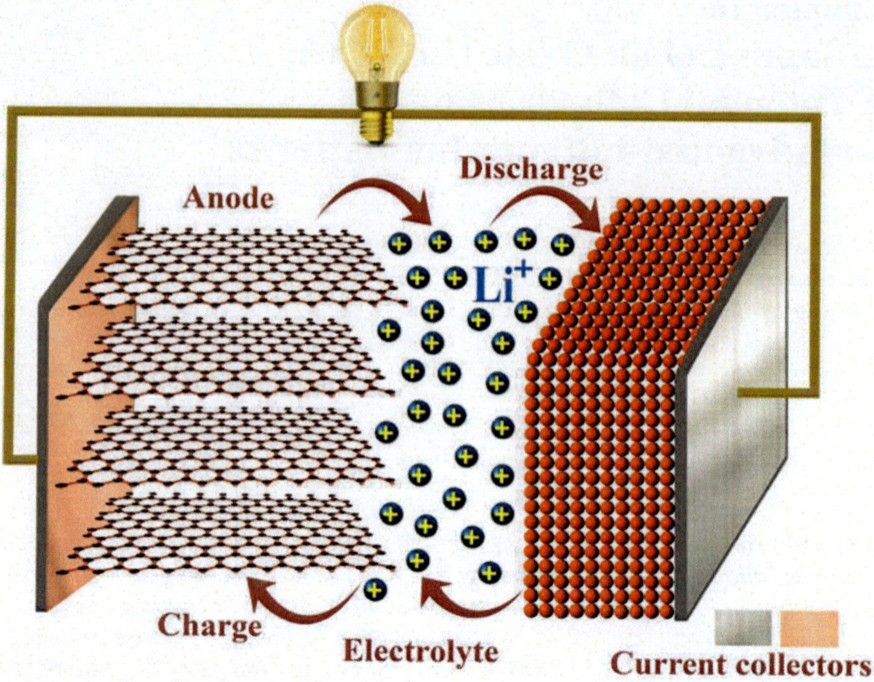

Fig. 16.1 Schematic illustration of basic mechanism in LIBs

batteries. In 1972, titanium disulfide (TiS_2) was the best intercalation compound available at that time and was started using as the positive electrode in Li-ion batteries were Li metal as the anode. But its impeccable operation strategies made it non-viable for battery application [2]. Later, higher chalcogenides were introduced as intercalation electrode materials such as sulfides and selenides. In 1983, Goodenough proposed lithium metal oxide insertion electrodes that are successfully used in today's battery technologies [3]. Different combinations of lithium-ion electrode materials are discovered out of which metal oxide/sulfide electrodes for non-aqueous secondary batteries stand a major role due to their high specific energy and power. Metal oxides and sulfides are extensively investigated for electrode material and many of them have been synthesized either by lithium extraction or by insertion. Oxides include vanadium pentoxide (V_2O_5), molybdenum oxide (MoO_3), tungsten oxide (WO_2), iron (III) phosphate ($FePO_4$), cobalt oxide (CoO_2), etc., and sulfides such as molybdenum sulfide (MoS_2), chromium sulfide (CrS_2), and nickel phosphorus trisulfide ($NiPS_3$), etc. [4]. A high voltage and material stability make lithium/metal oxide electrode more preferable over metal sulfides.

Lithium cobalt oxide ($LiCoO_2$) is one of the important metal oxide cathode materials in lithium battery evolution and its electrochemical properties are well investigated. The hexagonal structure of $LiCoO_2$ consists of a close-packed network of oxygen atoms with Li^+ and Co^{3+} ions on alternating (111) planes of cubic rock-salt sub-lattice [5]. Goodenough et al. reported that lithium-ion can be reversibly extracted from the structure by a corresponding change of cobalt oxidation and over the range $0.5 \leq x \leq 1$ in Li_xCoO_2 exhibit low overpotential and good reversibility [4]. The first commercial LIBs are introduced by SONY, but relatively low capacity and stability of $LiCoO_2$ make it less desirable for cathode material. Later, several modifications are reported on $LiCoO_2$ by coating with metal oxides and phosphates in order to improve the capacity of batteries [6, 7]. Availability, low thermal stability, and high price are the main challenges faced in $LiCoO_2$ market. To overcome these problems, many synthetic approaches are experimented, and out of these, the electrospinning method offers low-cost and simple procedure for the preparation of one-dimensional (1D) nanostructures with excellent properties such as high specific surface area, short ionic and electron diffusion pathways, and mechanical stability [8]. And hence it also improves electrochemical properties and battery performance.

Electrospinning is a widely used fabrication technique for the processing of novel 1D nanostructure such as nanofibers, nanotubes, and nanowires for a wide range of applications such as tissue engineering, energy conversion and storage, and electronic devices. The reduction of the size of the particle from micro to nanoscale improves the structural properties such as porosity, surface area to volume ratio, and electrical and mechanical properties. The electrospinning technique has a wider application in battery technology for the preparation of efficient electrode material for longer applications. In typical LIBs, powder electrodes are used such as a carbon-based anode and lithium metal oxide cathode. During lithium-ion intercalation and de-intercalation cycles, ions having long diffusion pathways that diminish the kinetics of electrochemical reactions and result in poor battery performance [9, 10]. It is reported that electrospun nanofibers show better intercalation–de-intercalation mechanism as well as good electrochemical reaction kinetics at the electrode-electrolyte interface due to high surface area, porosity, and high stability [11]. Nanostructured $LiCoO_2$ cathode material can be prepared by sol-gel-assisted electrospinning method were metal acetate and citric acid as precursors with applied voltage 25 kV and distance of 30 cm were reported. Electrochemical studies showed that nanostructured electrode shows high initial discharge capacity of 182 mAh g^{-1} compared with capacity 140 mAh g^{-1} of powder $LiCoO_2$ electrodes [12], promises a new model micro-lithium battery for future energy storage device market.

Higher performance Li-ion batteries for advanced transportation applications required batteries with high energy and power density. Enhancement of rate capability of batteries can be achieved by structural improvements in electrode materials and which leads to overcoming the barriers such as low cycle time, low-temperature tolerance, and poor life. Mizuno et al. proposed nanostructured wire-like $LiCoO_2$ which could reduce crystal growth at high temperature and show better rate capability [13]. Technological developments in electrospun $LiCoO_2$ electrode material merits structural stability, cyclic performance, and safety. Also, doping, surface engineering, and incorporation of carbon materials are the key research methodologies in this area for further developments in electrospun $LiCoO_2$ materials.

16.2 $LiCoO_2$ Nanostructures of Different Morphology

To fabricate micro-scale lithium batteries, effective techniques are required for the fabrication of micro-scale anode, cathode, and electrolytes [1, 14]. There are lots of investigations carried out in the field of electrode materials, especially $LiCoO_2$ for improving its electrochemical properties. Most of the preparation methods are focused on high-temperature solid-state reactions which lead to the formation of $LiCoO_2$ with abnormal grain growth, inhomogeneity, and poor control on stoichiometry [15]. Solution-assisted preparation techniques are also employed for preparing $LiCoO_2$ fibers, out of these, electrospinning is a simple and versatile method to generate uniaxially aligned arrays or mats with well-controlled chemical compositions as well as precision in diameter.

There are several reports on inorganic hollow nanofibers of TiO_2 and SiO_2 [16, 17], however, the first hollow nanofibers of $LiCoO_2$ prepared by Zhan et al. [18] with higher surface area and fancy nanostructure. In his experiment, precursor solution contains lithium acetate, cobalt acetate, citric acid, and water were used. In the current experiment, the precursor solution was heated up to 60 °C the pellucid solution would be gradually polymerized to form the sol with part of the acetic acid and water evaporating slowly. While electrospinning, the sol was elongated to form the core-shell structured fibers accompanied by the solvent volatilizing. After the fibers were put into cyclohexane to eliminate the machine oil (core) and the following drying, the volatile components rapidly vaporized, and the gel hollow nanofibers would be transformed to the xerogel nanofibers. The typical xerogel hollow nanofibers were continuous as long as 10–20 cm with the diameter of 0.5–4 mm, which overlapped each other and seem as a piece of brown nonwoven cotton agglomerate (Fig. 16.2).

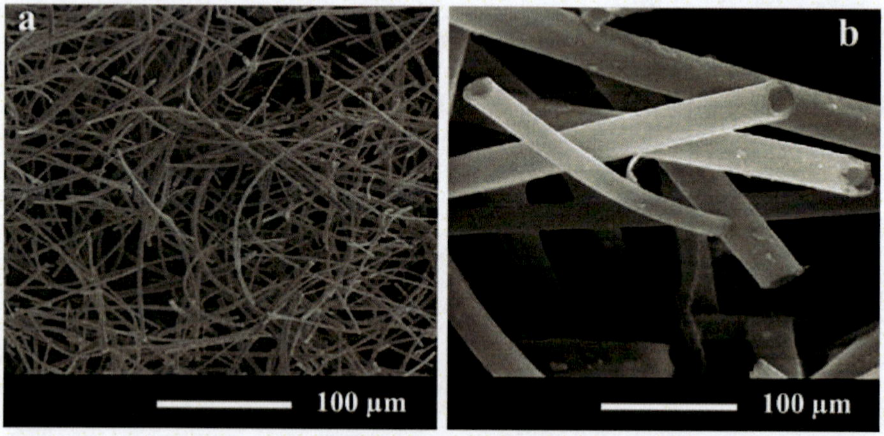

Fig. 16.2 SEM images on the surface morphology of the **a** xerogels fibers, **b** $LiCoO_2$ hollow fibers calcined at 400 °C for 6 h in air Adapted and reproduced from [18], Copyright 2007 Taylor and Francis

Morphology of xerogel nanofibers by SEM showed smooth and featureless appearance. After calcination, nanofibers become brickle due to the decomposition of citric acid and reaction between lithium acetate and cobalt acetate. Increasing of calcination temperature to 700 °C, there are no structural deformations and the properties are well maintained. Further, observations by TEM technique indicate that wall of xerogel nanofibers was made up of polycrystalline $LiCoO_2$ nanoparticles.

In advanced transportation system such as hybrid electric vehicle (HEV) and electric vehicle (EV), Li batteries are prior choice due to its high performance [13]. Li batteries show better charge–discharge capability as well as faster lithium-ion transfer in the cathode. Morphology of cathode material is a determining factor in battery performance. Honma et al. [19] reported nanosized electrode materials that are expected to possess much higher power output due to short diffusion length. Nanosized $LiCoO_2$ as suggested as electrode material for Li-ion rechargeable batteries has anisotropic surface properties affecting electronic structures, which is evidenced by electron energy loss spectroscopy (EELS). To obtain nanostructured cathode material, there are several methods reported yet such as post-templating method [20], sol-gel method [21], spray-drying method [22], co-precipitation method [23], and hydrothermal method [24–26]. Out of these, sol-gel method is more preferred due to its high control on chemical composition of the obtained nanostructured material. But the disadvantage of these processes is large size of the product and this is due to crystal growth during heat treatment. In contrast, electrospun wire structure could suppress the crystal growth even at high temperature because of its unwoven fabric morphology [24].

Mizuno et al. [13] fabricated wire-structured $LiCoO_2$ material from wire-structured precursor by applying high voltage to the precursor solution by electrospinning method. This method has attracted much attention recently, since this method can easily fabricate mass production of wire-structured electrode material than conventional hydrothermal method. By using electrospinning method, various wire-structured materials such as metal–carbon composites, electrode material, such as $LiFePO_4$, have been synthesized successfully [27–29].

Mizuno et al. fabricated as-spun fiber of $LiCoO_2$ with/without vapor grown carbon fiber (VGCF) from the precursor aqueous solutions of Li, Co, polyvinyl alcohol (PVA) and polyoxyethylene octyl phenyl ether. SEM images showed that the diameter of obtained as-spun fiber with/without VGCF is 0.5 and 1.2 μm, respectively. It is clear that the precursor solution with VGCF could control the diameter of the wire-structured $LiCoO_2$. The charge–discharge and rate capability experiments revealed that both resulting materials show the reversible Li^+-ion insertion/extraction reaction. However, due to the existence of a small irreversible capacity at the initial cycles, the interfacial resistance increases, resulting in the poor cyclability and lower charge–discharge rate capability, especially for nanowire $LiCoO_2$ fabricated with VGCF.

Larger the size of electrode material smaller will be the surface area and lead to poor electrochemical performance [30]. Smaller surface area also leads to slower charge–discharge rate and lower power density [19, 31]. Since sol-gel synthetic approach is used for smaller particle but still have relatively lager size over than micron meters [32]. Xu et al. [33] developed a new synthetic approach for producing ultrafine $LiCoO_2$ powders via a combined sol-gel and electrospinning method. In this experiment, acrylic acid was used as chelating agent to avoid pH change in gel formation during electrospinning. The obtained electrospun nanofiber is grinded to ultrafine powder of sub-micron particle size under lower calcination temperature and shorter calcination time compared to regular sol-gel process. In this method, precursor solution of $LiCoO_2$ is prepared from lithium acetate and cobalt acetate in the molar ration of 1.05:1 (Li: Co). Acrylic acid (AA) was added to the above solution to form 1:1 molar AA: $LiCoO_2$ sol-gel solution at 70–80 °C. Ethanol solution containing polyvinyl pyrrolidone (PVP) was added to the above solution with continues stirring. This composite solution was loaded in 1 mL syringe equipped with a stainless steel needle and electrospun at DC voltage of 40 kV under controlled rate of 0.4 mL/min. For comparing the properties of ultrafine phase, pure $LiCoO_2$ powders and sol-gel derived nanofibers are also prepared. Lower calcination temperature and calcination time give smaller particle size, which is evident from SEM images of electrospun $LiCoO_2$ are shown in Fig. 16.3. Powders derived from nanofiber calcined at 650 °C shows diameter less than 300 nm and calcined at temperature above 700 °C shows diameter around 0.3–1 μm. Sol-gel powders show substantially larger diameter of 0.3–2.6 μm diameter were calcined above 700 °C. Crystalline structure of $LiCoO_2$

Fig. 16.3 SEM images on the surface morphology of $LiCoO_2$, **a** $LiCoO_2$ nanofibers calcined at 650 °C, **b** nanofibers derived powders calcined at 650 °C, **c** nanofibers derived powders calcined at 700 °C, **d** sol-gel powders calcined at 700 °C Adapted [33], Copyright 2013 Elsevier

calcined at 700 °C was examined by TEM shows nanocrystalline grains with size in the range of 300–500 nm. In contrast, sol-gel derived $LiCoO_2$ powder shows larger sizes much above than 500 nm. These results show that electrospinning is indeed a very good synthetic route for producing fine $LiCoO_2$ powders as compared to conventional methods. The smaller size of $LiCoO_2$ powder shows enhanced surface chemistry and lithium-ion intercalation, which is evident from the results of cyclic voltammetry (CV) carried out with calcined powder at 650 and 700 °C.

The charge–discharge curve versus number of cycles for the $LiCoO_2$ powder derived from nanofiber shows higher discharge capacity and cyclic stability than sol-gel derived powders. Surface modification by lanthanum oxide (La_2O_3) coating on $LiCoO_2$ powder also shows improved electrochemical properties. Phase pure ultra-fine $LiCoO_2$ powders derived from sol-gel-based electrospinning method is suitable for high efficiency LIBs. Compared to other conventional method, electrospinning is found as effective approach for producing fine-sized electrode material without losing chemical composition. This technique is also a facile method for other electrode materials in LIBs.

16.3 Effect of Coating on $LiCoO_2$ Nanostructures

$LiCoO_2$ is widely accepted electrode material in LIBs, and many researches are developed to improve its performance in advanced applications by preparing $LiCoO_2$ in different sizes and shapes also in surface modifications by coatings [34–36]. Surface modification of $LiCoO_2$ with metal oxides such as zinc oxide (ZnO), magnesium oxide (MgO), and aluminum oxide (Al_2O_3) showed better structural stability as well as improved cyclability. Wang et al. proposed molten-salt treatment on $LiCoO_2$ and the result showed enhancement of structural stability of electrode material due to surface solid solution, also it suppresses Co^{4+} dissolution at charged state [34]. Coatings on $LiCoO_2$ make it less vulnerable to electrolyte electrode reactions and hence improve performance. Myung et al. proposed that amphoteric metal oxide (Al_2O_3 and ZnO) coating on $Li[Li_{0.05}Ni_{0.4}Co_{0.15}Mn_{0.4}]O_2$ show improved cycling performance at 60 °C. Studies revealed that metal oxide coating layers gradually transformed to metal fluoride film due to the scavenging of F^- from hydrogen fluoride (HF), which is one of the by-products from electrolytic salt decomposition into electrolyte. This will form an extra metal fluoride layer on metal oxide, which prevents acid attack. Particle morphology is well maintained, and hence it showed better cyclability even at 60 °C.

Structural stability of electrospun $LiCoO_2$ is also enhanced by coating active material on its surface. Lu et al. proposed coating of lithium phosphorous oxinitride (LiPON) on the surface of 3D nanostructured fibrous $LiCoO_2$. In his report, $LiCoO_2$ fibers are prepared from Li-Co-PVP sol, which is used as precursor for electrospinning with stoichiometric ratio of Li: Co: PVP = 1:1:2. Radio frequency sputtering method was used to coat $LiCoO_2$ fibers with 2.4 μm LiPON film [37]. The length of the as-spun fibers with a smooth surface could extend up to 1–2 cm. The diameters of the fibers are relatively uniform with an average value of 80–100 nm. The 3D network structure is well reserved after being calcined. Electrochemical properties of $LiCoO_2$ fibers as cathode material in lithium-ion batteries with and without LiPON coating are investigated; results poor cycle performance and loss of capacity in consecutive cycles. Hence, LiPON layer coated $LiCoO_2$ fibers show better structural stability during examination and exhibited the discharge capacity of 120.4 mAh g^{-1} with the loss of 0.11% per cycle. During the 100th cycle, it showed discharge rate of 0.05 mA cm^{-2}.

Morphological studies revealed mechanical collapse of 3D architectures of $LiCoO_2$ fibers after CV and discharge–charge measurements. Figure 16.4d shows the SEM images of fibers after the charging and discharging 10th cycles. As shown by the arrow, many fibers have been interrupted, and the compact clusters of fibers collected on the substrate as shown in Fig. 16.4c become sparse. These changes indicate that the 3D net architecture has been destroyed, indicating a poor character. To prevent the fibers from breaking down, one LiPON layer was coated onto the surface of the fibers. As shown in Fig. 16.4e, the fibers collected on the substrate, as shown in Fig. 16.4c, become the smooth surface after the deposition of LiPON film, indicating that the new layer of LiPON film is densely coated onto the surface of these fibers. The solid-state electrolyte of LiPON film could provide a very fast Li^+-ion diffusion inside the fibers electrode. Figure 16.4f shows, as the typical 3D network architecture as cathode was assembled with lithium into the cell, the SEM images of its surface after the first electrochemical reaction of 3D architecture with Li^+. It can be seen that, after electrochemical cycling, the LiPON layer is also well coated on the surface of the 3D structure of the fibers as shown in Fig. 16.4e. The smooth surface as shown in Fig. 16.4f supports the 3D network structure stability of the $LiCoO_2$ fibers with a LiPON coating layer after cycling.

Li et al. [33] developed a synthetic route for electrospun phase pure $LiCoO_2$ nanofibers with ultrafine particle sizes by sol-gel-based electrospinning method. In electrospinning process, pH of the precursor solution is a determining factor. Here, results showed that using acrylic acid as chelating agent, which made electrospinning easier and obtained nanofibers of $LiCoO_2$ is grinded to ultrafine powders of sub-micron size. This sub-micron sized ultrafine powder derived from electrospun nanofibers possess discharge capacity of 148–153 mAh g^{-1}, 84% of initial capacity is retained after 30 cycles. Surface modification of this powdered $LiCoO_2$ with La_2O_3 coating the retention of initial capacity is increased further to 91% at 30th cycle and 83% at 50th cycle without decreasing in its initial capacity, making it attractive for Li-ion batteries. The discharge capacities and cyclic stability of various $LiCoO_2$ nanopowders (coated with and without 3% La_2O_3) derived from electrospun nanofibers and sol-gel synthesis as shown in Fig. 16.5. Calcined $LiCoO_2$ powders derived from nanofibers at 650 °C retain 76% of initial capacity at 30th cycle, while those calcined at 700 °C retain 84% of it, though their initial capacity is slightly lower. Furthermore, the nanofiber derived $LiCoO_2$ powders with La_2O_3 coating retains 91% of initial capacity at 30th cycle, and 83% of it at 50th cycle, showing much improved cyclic stability without decreasing its initial capacity. Such improved stability is

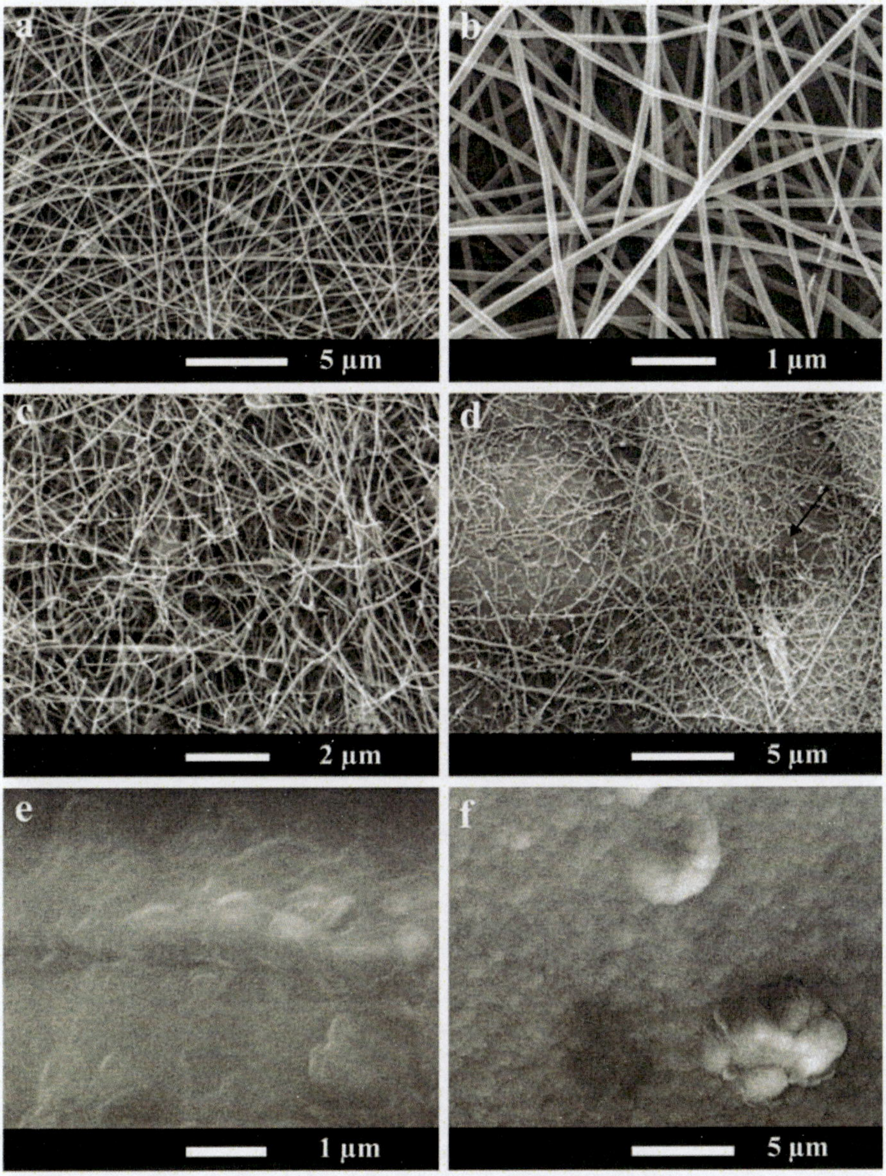

Fig. 16.4 SEM images on the surface morphology of the as-spun $LiCoO_2$ fibers at magnification of **a** 5000 and **b** 20,000. SEM images of the 600 °C annealed fibers **c** before and **d** after charge and discharge cycles; SEM images of the LiPON coated fibers **e** before and **f** after charge and discharge cycles Adapted and reproduced from [37], Copyright 2008 Elsevier

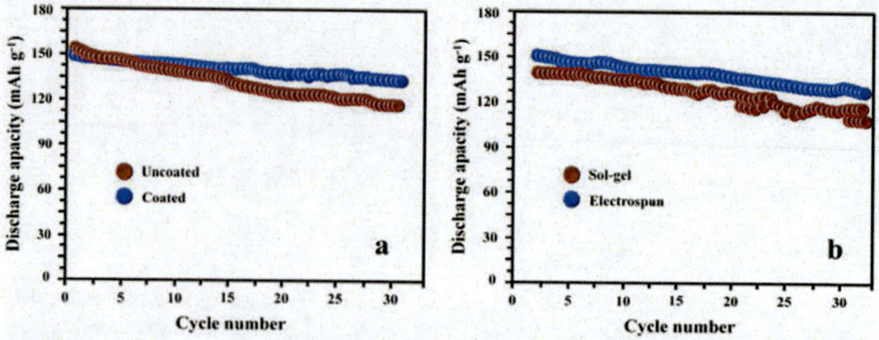

Fig. 16.5 Discharge capacity versus number of cycles for different $LiCoO_2$ nanoparticles; **a** nanopowders (with and without La_2O_3 coating) derived from electrospun nanofibers calcined at 650 °C, **b** nanopowders derived from electrospun nanofibers calcined at 700 °C and sol-gel powders calcined at 750 °C Adapted and reproduced [33], Copyright 2012 Elsevier

believed to be caused by the higher resistivity of La_2O_3 coating that prevents HF from destroying the structure of $LiCoO_2$ powder.

Chen et al. [38] successfully prepared PVP/$LiCoO_2$ nanofibers using typical electrospinning route. The obtained nanofibers are composed of small crystalline grains, which are found to be uniformly linked. PVP is employed in synthesis of nanofiber, which shows better aqueous solubility, less toxicity, and better stability [39]. In this preparation method, 0.001 mol lithium acetate and 0.001 mol cobalt acetate were dissolved in 2 mL methanol is used as precursor solution. 0.06 g of PVP in methanol is added to the above solution of 0.025 mol $LiCoO_2$ is electrospun under DC voltage of 8 kV. The obtained nanofiber was calcined at temperature of 700 °C for 12 h. Electrospinning is continued with different applied voltages of 10 and 12 kV. Figure 16.6 shows the SEM images of $LiCoO_2$ nanofibers obtained from electrospinning at 12 kV voltage with an annealing temperatures of 300, 500, and 700 °C. After annealing at 300 and 500 °C, the diameter of the fiber reduced due to removal of acetate group

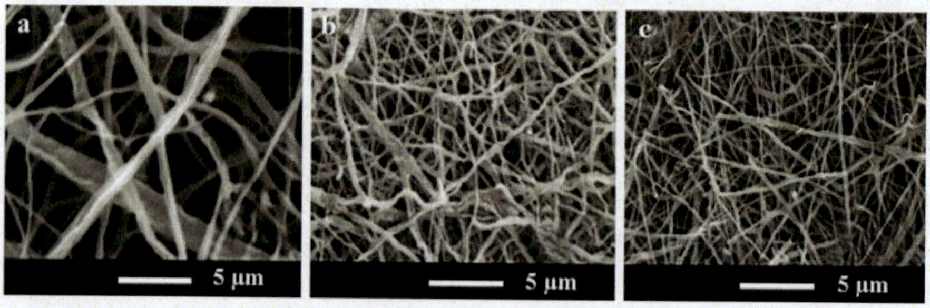

Fig. 16.6 SEM images on the surface morphology of $LiCoO_2$ fibers (bar 5 μm) electrospinning (12 kV) various annealing temperature at **a** 300 °C, **b** 500 °C, and **c** 700 °C for 12 h Adapted and reproduced from [38], Copyright 2011 John Wiley and Sons

from cobalt acetate and lithium acetate molecules. Annealing at 700 °C reduces the size even smaller by the complete removal of organic residues and lead to the formation of crystalline lithium oxide and cobalt oxide.

With a strong interaction of PVP as capping agent, it is assembled on the surface of $LiCoO_2$ as long chains. $LiCoO_2$ nanofibers itself show strong interaction due to high Gibbs free surface energy of small sized nanofiber. These two interactions make cluster type nanofibers and with few surface defects. Low aggregation of nanofiber can be induced by lowering the viscosity of precursor solution. Gu et al. [40] reported a core-shell $LiCoO_2$–MgO coaxial fiber for lithium-ion batteries were prepared by combining electrospinning with sol-gel process. Co-electrospinning is considered as an effective method for the preparation coaxial fibers as compared to self-assembly approach, laser ablation, etc. In this work, core-shell $LiCoO_2$–MgO were coaxial fibers are prepared from one step process and thickness of the fiber is controlled by adjusting the pressure of sol during electrospinning process. In this method, $LiCoO_2$ spinnable sol and MgO spinnable sol were prepared and its viscosity is adjusted to 4.0 Pa s. Electrospinning is carried out under 25 kV voltage and pressure of $LiCoO_2$ and MgO sols were controlled at 0.3 MPa and 0.02 MPa, respectively.

SEM images of the obtained nanofiber are shown in Fig. 16.7, diameter of the fiber is 2–3 μm, and flaked portion clearly shows the core-shell morphology of the fiber calcined core-shell fiber have length of 10–15 cm and diameter of 1–2 μm after the removal of organic residue. TEM images of coaxial fiber show the clear boundary of core-shell structure. The composite fiber has a shell thickness of 50–100 nm and shows excellent flexibility under optical microscope when one end is bend with a copper wire. This indicates that the presence of MgO coatings on $LiCoO_2$ makes it more flexible as compared to bare $LiCoO_2$. CV experiments are applied to investigate the reversibility of obtained core-shell fiber electrode during charge–discharge process. CV curve shape is similar to $LiCoO_2$ fibers, and the two small redox peaks are corresponding to first-order phase transformation. It indicates that MgO coating does not suppress the phase transformation from hexagonal to monoclinic, also it improves the reversibility of phase transformation by the migration of Mg^{2+} into Li_2O layers. Cyclic stability of core-shell fiber is found to be higher due to maximum diffusion of

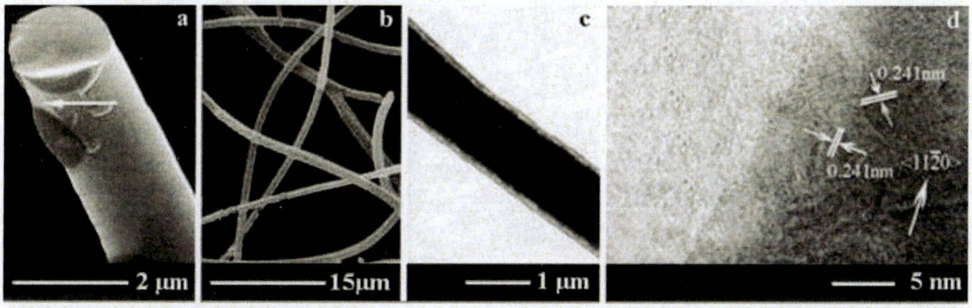

Fig. 16.7 Four SEM images on the surface morphology of the xerogels fibers **a,** the calcined fibers **b**, TEM **c,** and HR-TEM **d** images of the calcined fibers Adapted and reproduced from [40], Copyright 2007 Royal Society of Chemistry

Mg^{2+} and Li^+ having similar ionic radii. The core-shell fiber electrode shows initial charge capacity of 192 mAh g^{-1} which is found to be lower than bare $LiCoO_2$ fibers, which implies MgO layer changes lithium diffusion kinetics. Cycle performance of the core-shell fiber after 40th cycle shows discharge capacity of 163.0 mAh g^{-1} with a capacity retention of 90.0%. Bare $LiCoO_2$ fiber shows discharge capacity of 90.0 mAh g^{-1} exhibiting only 52.0% of the initial discharge capacity. This improved electrochemical performance is attributed by the stabilizing effect of MgO coating on $LiCoO_2$ fiber. Impedance spectra of core-shell $LiCoO_2$–MgO fiber suggest that MgO coating on $LiCoO_2$ effectively avoids impedance growth by protecting surface from passive surface film formation on cycling. Coating the surface of $LiCoO_2$ with inert material is an effective method for improving the electrochemical properties.

16.4 Electrospun $LiCoO_2$ Composites

16.4.1 Metal/$LiCoO_2$ Composites

Lithium transition metal oxides such as lithium cobalt oxide ($LiCoO_2$), lithium vanadium oxide (LiV_2O_5), lithium titanium oxide ($Li_4Ti_5O_{12}$), lithium manganese oxide ($LiMn_2O_2$), lithium copper oxide ($LiCuO_2$), lithium manganese chromium oxide (LiMnCrO), lithium iron phosphate ($LiFePO_4$), and lithium nickel oxide ($LiNiO_2$) are used as cathode material in lithium-ion rechargeable batteries [12, 41–43]. Among these, $LiCoO_2$ is widely used as cathode material in lithium-ion batteries due to its layered crystalline structure, good capacity, energy density, high cell voltage, high specific energy density, high power rate, low self-discharge, and excellent cycle life [43]. Studies showed that doping of $LiCoO_2$ with some metals such as manganese, chromium, nickel, titanium, and boron improves its electrochemical properties [44–48]. Alcantara et al. [44] reported that boron-doped $LiCoO_2$ exhibited higher capacity at first discharge and enhancement in capacity retention. Step potential electrochemical spectroscopy and galvanostastic cycling revealed that boron dopants improved the reversibility of lithium de-intercalation–intercalation process [44]. Waki et al. [45] reported that manganese substitution up to 20% increase in the cyclic stability of $LiCoO_2$ material. From potential step experiment, $LiCo_{0.8}Mn_{0.2}O_2$ is recommended as one of the promising cathode materials for a 4 V class secondary lithium battery [45]. On the other hand, doping with non-substituting material like silver also enhances electrochemical properties of $LiCoO_2$ [49].

Aykut et al. [49] developed silver/lithium cobalt oxide ($Ag/LiCoO_2$) nanofiber via sol-gel electrospinning. 1D nanofibers of $LiCoO_2$ with high surface to volume ratio have been produced by combination of sol-gel-assisted electrospinning. He adopted a new method for the preparation of ($Ag/LiCoO_2$) nanofiber because the atomic ratio of Ag^+ is larger than Co and Li cause loss of nanofiber uniformity during the production. In this preparation method, homogeneous electrospinning solution of 8 wt% of PVA/water in which lithium acetate and cobalt acetate were added in

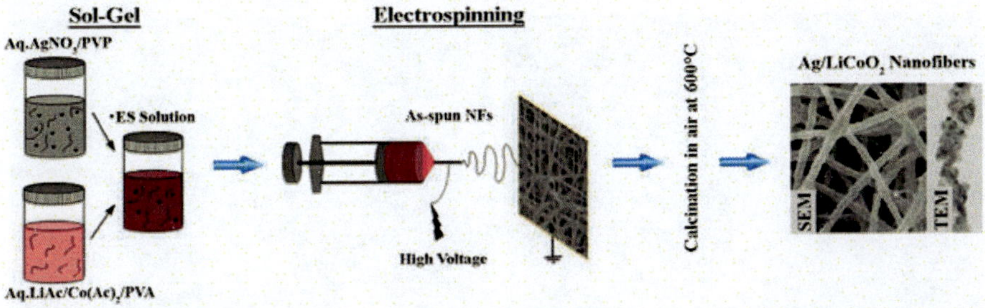

Fig. 16.8 Schematic illustration of sol-gel electrospinning of Ag/$LiCoO_2$ Adapted and reproduced from [49], Copyright 2013 Elsevier

1:1 ratio. The above mixture is magnetically stirred under 60 °C and this homogeneous solution is electrospun under 15 kV voltage and 15 cm from the collector. The obtained nanofiber is calcined at temperature of 600 °C with increasing rate of 5 °C/min which allow proper oxidation and removal of the combustion gas and organic residue. After calcination process, a nonporous structure of Ag/$LiCoO_2$ is obtained. Figure 16.8 shows the sol-gel electrospinning of Ag/$LiCoO_2$.

In order to obtain smooth nanofibers from electrospinning, different precursor solutions were prepared and the obtained nanofibers were shown in Fig. 16.9. Addition of silver nanoparticles makes nanofiber rougher and is visible from the SEM images as shown in figure denoted by red arrows. Diameter of the fiber is in range from 100 nm to 2 μm and increase with the addition of silver nanoparticle. On comparing, the XRD pattern of $LiCoO_2$ and Ag/$LiCoO_2$ shows that the presence of silver does not make any shift in diffraction angles of $LiCoO_2$ peaks. This indicates that highly mobile Ag^+ is entered into interlayer space of $LiCoO_2$ during the calcination process and does not replace lithium or cobalt atoms. X-ray photoelectron spectroscopy analysis of Ag/$LiCoO_2$ nanofibers is also reported in which clear peaks of silver nanoparticle with binding energy of 367.3 and 373.3 eV indicate the presence of silver in electrospun Ag/$LiCoO_2$ nanofiber.

In this study, Aykut et al. [49] reported a facile preparation of Ag-doped $LiCoO_2$ NFs via sol-gel electrospinning technique Ag nanoparticles. Well-distributed Ag nanoparticles in the nanofibers were produced and these were calcined at 600 °C in air atmosphere for 3 h. Self-alignment of Ag nanoparticles in the nanofibers was observed in the morphological studies. Ag/$LiCoO_2$ nanofibers with nano-porous structure were obtained after the calcination process. The porous structure was possibly caused by the Ag nanoparticles leaving the fibers during the calcination process in which atomic radius of Ag ions is bigger than Li and Co ions. The Ag phases thus exist only at the interlayers of $LiCoO_2$ crystals in the nanofibers. The prepared Ag-doped $LiCoO_2$ nanofibers can be used as the cathode material in Li^+-ion batteries.

Kim et al. [50] developed ruthenium oxide/cobalt oxide (RuO_2/Co_3O_4) nanowires in a simple electrospinning method. This composite RuO_2/CoO_4 is used as bi-functional electro-catalysts in rechargeable lithium–oxygen battery cathodes. Among

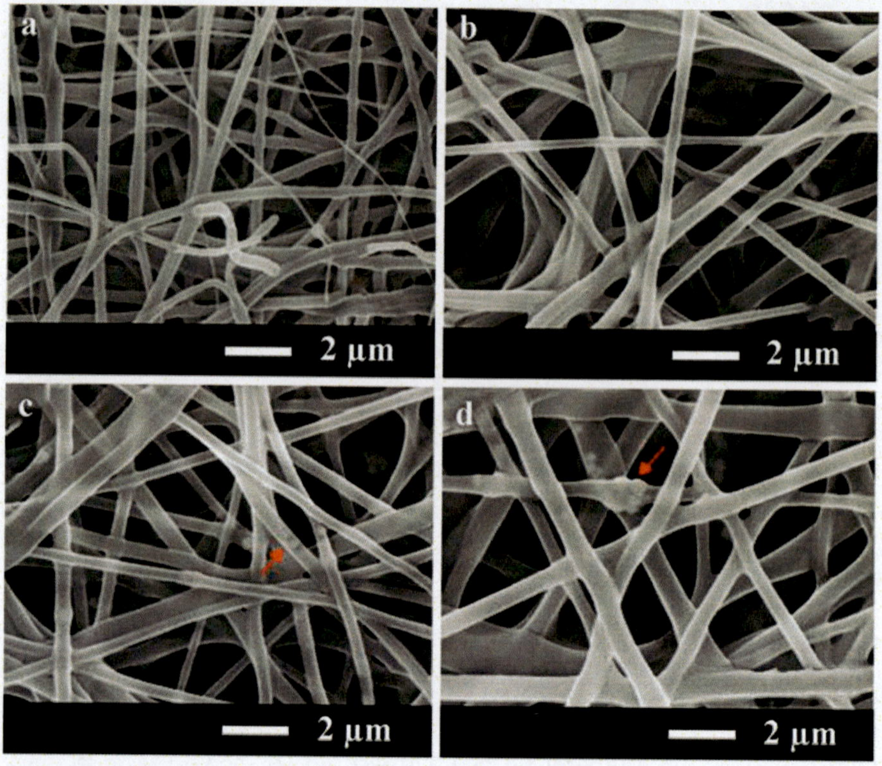

Fig. 16.9 FE-SEM images on the surface morphology of as-spun **a** Li(Ac)/Co(Ac)$_2$/PVA NFs, and Ag/Li(Ac)/Co(Ac)$_2$/PVA/PVP NFs with different Ag contents: **b** 1 wt% Ag, **c** 2 wt% Ag, and **d** 3 wt% Ag in ES solution Adapted and reproduced from [49], Copyright 2013 Elsevier

other transition metal oxides, Co_3O_4 is common choice for Li-O_2 batteries as a catalyst because of its favorable catalytic activity with oxygen [51–54]. But early studies show that the initial discharge capacity of O_2 electrode faded after some cycles due to the presence of Co_3O_4 [55]. Noble-metal oxides such as RuO_2 is used to fabricate composite with Co_3O_4, for improving its electronic conductivity and also reduce the charging overpotentials of Li-O_2 batteries [56–58]. Therefore, a composite of RuO_2 and Co_3O_4 is of great importance in Li-O_2 batteries.

In this work, nanowire RuO_2/Co_3O_4 composite with Co: Ru ratio of 9:1 is prepared by facile electrospinning method. Both Co_3O_4 and RuO_2/Co_3O_4 nanowires are prepared. Structural characterization Co_3O_4 and RuO_2/Co_3O_4 by SEM and TEM shows 1D nanowires composed of many crystalline particles of small size. The crystalline phase of both nanowires Co_3O_4 and RuO_2/Co_3O_4 is examined by XRD experiment in which both shows characteristic crystalline peaks of Co_3O_4. Due to the small amount of ruthenium in RuO_2/Co_3O_4, the diffraction peak is obscured. The electrochemical properties of both Co_3O_4 and RuO_2/Co_3O_4 are examined. The discharge–charge profiles for the first cycle in the Li-O_2 cell fabricated with the

Co_3O_4 nanowire, and RuO_2/Co_3O_4 nanowire cathode catalysts are presented. The initial discharge capacity of the LiO_2 cells RuO_2/Co_3O_4 nanowire catalysts was approximately 10,850 mAh g^{-1} which is much higher than those of cells with Co_3O_4 nanowires. LiO_2 cells with RuO_2/Co_3O_4 nanowires show lower overpotential for the discharge and charge functions than Co_3O_4 nanowires.

The cyclic stability of the Li-O_2 cells with the Co_3O_4 NWs, and RuO_2/Co_3O_4 NW cathode catalysts were also examined under a limited capacity operation of 500 mAh g^{-1} at a current density of 200 mA g^{-1}. The initial potentials of Li-O_2 cells with the Co_3O_4 NWs, and RuO_2/Co_3O_4 NWs are plotted. The Co_3O_4 NWs and RuO_2/Co_3O_4 NWs catalysts showed sequentially decreased overpotentials of 1.44 and 1.14 V, respectively. The lowest charge potential of the RuO_2/Co_3O_4 NWs indicates that the reaction products of Li_2O_2 are decomposed more easily, whereas the highest discharge potential means that the reaction products of Li_2O_2 are formed more readily.

Incorporating the RuO_2/Co_3O_4 NWs as the bi-functional catalysts can clearly facilitate the oxygen evolution reaction (OER) and oxygen reduction reaction (ORR) processes by reducing the overpotentials at the Li-O_2 battery cathode. The discharge voltage profile of the cathode with the RuO_2/Co_3O_4 NWs was maintained for 300 h, whereas the discharge voltage dropped to below 2.3 V after 225 and 175 h for the cathodes made with the Co_3O_4 NWs and KB respectively. The high initial discharge capacity and good cycle stability of the RuO_2/Co_3O_4 NW electrode can be attributed to the high bi-functional catalytic activity of the RuO_2/Co_3O_4 nanowires from its 1D structural features and from the interaction between the composite metal oxide components.

16.4.2 Carbon/$LiCoO_2$ Composite

Carbon is considered as an ideal material for air cathode in Li-O_2 batteries [59–62]. Mitchell et al. [63] developed the carbon nanofibers electrode for lithium–oxygen batteries and achieved a discharge capacity of 7200 mAh g^{-1} and of higher gravimetric energy density, which is almost four times higher compared with $LiCoO_2$ cathode for LIBs. But the evolution of CO_2 from the electrode surface diminishes battery performance. To protect the carbon from degradation, the high LiO_2 adsorption energy material which activates the growth of Li_2O_2 on its surface rather than carbon surface should be utilized [64, 65]. Recently, Zhang et al. [66] reported that LiO_2 shows more stable adsorption on Co/CoO surface rather than pure carbon electrode leads to the formation of Li_2CO_3 which enhanced cycling performance of LiO_2 batteries. In order to minimize the degradation of carbon materials, many researchers have developed composite cathodes.

Hyun et al. [67] developed a bi-functional air cathode Co-CoO nanoparticles embedded in the nitrogen-doped carbon nanorods (Co-CoO/N-CNR) for lithium–oxygen battery. In this report, a highly efficient air cathode made up of cobalt-cobalt oxide nanoparticles which are exposed on the surface also encapsulated in the

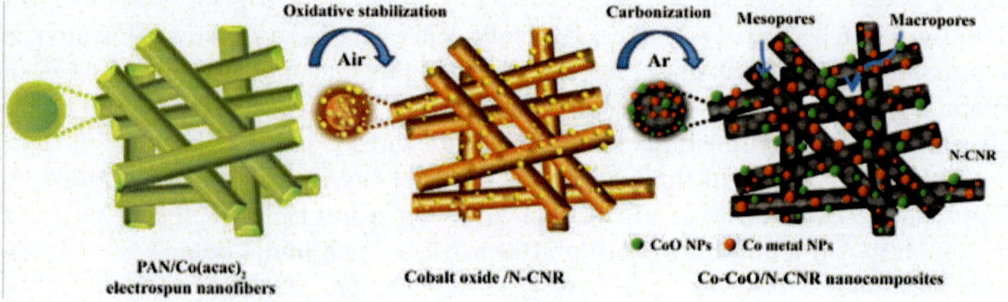

Fig. 16.10 Preparation process of the Co-CoO/N-CNR composites Adapted and reproduced from [67], Copyright 2017 Elsevier

nitrogen-doped carbon nanorod (Co-CoO/N-CNR) with superior specific capacity and cyclability based on Li-O_2 batteries. Precursor solution was prepared from poly(acrylonitrile) PAN as nitrogen source, cobalt actylactonate Co(acac)$_2$ as carbon source. This precursor solution was electrospun on an aluminum foil by applying a voltage of 15 kV voltage and 10 cm distance between needle tip (23 gauge) and collector at room temperature condition. The electrospun mat was collected and dried before calcination. After that the PAN/Co(acac)$_2$ electrospun nanofibers were stabilized in air at 275 °C for 4.5 h with a slow heating rate of 3 °C min^{-1}. After that the mat was carbonized at 800 °C for 30 min in argon with a 5 °C min^{-1} heating rate. PAN/Co(acac)$_2$ nanofibers obtained from electrospinning having an average diameter of 400 nm. After that it is exposed to thermal treatment, Co(acac)$_2$ oxidizes and transforms into cobalt oxide nanoparticles while parts of PAN decompose, thereby generating some pores, producing cobalt oxide/N-CNR composites as shown in Fig. 16.10. During the subsequent carbonization process under Argon gas, it reduces and forms Co metal/CoO which are embedded in the electrode and also developed onto the N-CNR surface. PAN residue converts into the graphitized structure of carbon with large mesopores, formed Co-CoO/N-CNR composites. The mesopores found in Co-CoO/N-CNR composites which are seen in figure can handle reversible formation and decomposition of discharge product.

Electrochemical performance of the Co-CoO/N-CNR composite electrode is evaluated. It is found that, co-existence of Co-CoO nanoparticles directly introduced in the Co-CoO/N-CNR composite cathode contributes to decrease in overall overpotential and enhance the stability of battery performance. Co-CoO nanoparticles also avoid the direct contact of Li_2O_2 on the CNR surface, so that can prevent air cathode from undesirable side reactions during cycling. The porous Co-CoO/N-CNR nanostructure with rod-like morphology accommodates continuous formation and decomposition of solid discharge product enables efficient transport of electrons and ions, and thereby improves the rate performance and cycling ability. The Li-O_2 battery with Co-CoO/N-CNR cathode exhibits high discharge capacity of 10,555 mAh g^{-1} at a current density of 100 mA g^{-1}. The well-organized design of Co-CoO/N-CNR composite as an air cathode bi-functional catalyst by the in situ electrospinning approach shows the importance of not only high porous structures but also the efficient catalyst for high performance Li-O_2 batteries which can also protect the electrode well.

Wang et al. [68] reported composite of Co_xO_y nanoparticles decorated nitrogen-doped carbon nanofibers synthesized by coaxial electrospinning method. It is found that, zinc precursor used in this method undergo self-evaporation from inside to outside can form the pore channels inside the carbon composites. These pore channels facilitate the migration of ions and the pore channels density can be controlled by varying the amount of zinc nitrate. Electrochemical measurements of Co_xO_y nanoparticles decorated nitrogen-doped carbon nanofibers show an extraordinary high reversible capacity of 983.5 mAh g^{-1} after 400 cycles at 1000 mA g^{-1} and excellent rate capability as the result of both metal oxide nanoparticles and special porous channels structure over those capacities of Co_xO_y—carbon composite materials.

Electrospinning is considered as a facile, low-cost technique to produce continuous carbon nanofibers (CNFs) in the form of thin film by applying a strong electric field between the nozzle and collector while injecting a polymer solution [69]. The carbon nanofiber (CNF) film produced can be directly acts as electrodes without adding polymer binders or conductive additives. They possess many advantages over the other forms of carbon used as electrodes, such as easy and controllable creation of an extremely large surface area and hierarchical pores as well as tuneable compositions of the CNF composites by incorporation of active materials [70, 71].

Abouli et al. [69] reported a facile route to synthesize porous carbon nanofibers containing cobalt and cobalt oxide nanoparticles (CoO_x/PCNF) as anodes for Li^+-ion batteries. This is a one-pot strategy to produce Co and Co_3O_4 nanoparticles embedded in electrospun porous carbon nanofibers (CoO_x/PCNF). The composite electrodes showed excellent cyclic and rate performance resulting from the high capacity spinel cobalt oxides and improved conducting networks provided by the graphitic carbon layers and CNFs. All these advantages make this hybrid material a good candidate as electrodes in high performance LIBs. Figure 16.11 shows the schematic illustration of synthesis of Co_3O_4/PCNF. Polyacrylonitrile and polymethyl

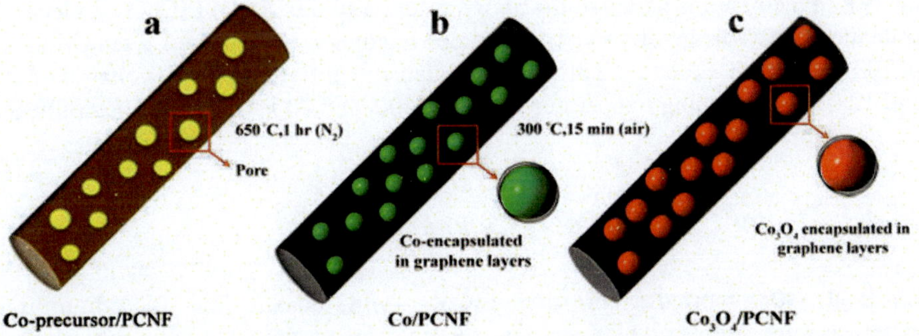

Fig. 16.11 Schematic illustration of the synthesis process: **a** co-precursor/PCNF after stabilization; **b** Co/PCNF after carbonization, and **c** Co_3O_4/PCNF after post-heat treatment Adapted and reproduced from [69], Copyright 2014 Royal Society of Chemistry

methacrylate were used as precursors and N,N-dimethyl formamide (DMF) as the solvent to produce the porous carbon nanofiber (PCNF) matrix. Cobalt (II) acetate tetrahydrate was used as the source of Co was dissolved in 10 ml of DMF at 80 °C for 5 h under magnetic stirring. The mixture was transferred to a plastic capillary and electrospun on a spinning machine at 16 kV. The film obtained after electrospinning was carbonized at 650 °C for 1 h in a nitrogen atmosphere using a tube furnace. Carbonization at temperatures higher than 650 °C was beneficial to improving the electrical conductivity of the carbon matrix. A lower temperature caused incomplete crystallization of the nanoparticles. The resultant CNFs consisted of Co nanoparticles uniformly dispersed in the porous carbon matrix (Co/PCNF). The carbonized films were post-heat treated at 300 °C for 15 min in air to precipitate Co_3O_4 nanoparticles, designating the final product as Co_3O_4/PCNF.

The electrochemical properties of the Co_3O_4/PCNF electrode measured in the voltage window of 0.01–3.0 V are compared with those of the neat PCNF and Co/PCNF electrodes are studied. The discharge–charge profiles of the Co_3O_4/PCNF electrode presented a long plateau at about 1.03 V, followed by a slope down to 0.01 V upon the first discharge. In the first cycle of the CV curves, a large cathodic peak appeared at 0.8 V corresponding to lithiation of Co_3O_4 to Co, followed by a small peak at around 0.36 V corresponding to the formation of a solid electrolyte interphase (SEI) film at below 0.8 V. The main anodic peak observed at about 2.1 V is attributed to delithiation of Co to Co_3O_4. The very small anodic peak at 1.35 V suggests that the oxidation took place in a multistep process. The Co_3O_4/PCNF electrode delivered discharge capacity of 952 mA g^{-1} after 100 cycles at a current density of 100 mA g^{-1}.

Chen et al. [38] successfully prepared PVP/$LiCoO_2$ nanofibers using typical electrospinning route. The obtained nanofibers are composed of small crystalline grains, which are found to be uniformly linked. Polyvinyl pyrrolidone is employed in synthesis of nanofiber, which shows better aqueous solubility, less toxicity, and better stability. In this preparation method, 0.001 mol lithium acetate and 0.001 mol cobalt acetate were dissolved in 2 mL methanol is used as precursor solution. 0.06 g of PVP in methanol is added to the above solution of 0.025 mol $LiCoO_2$ is electrospun under DC voltage of 8 kV. The obtained nanofiber was calcined at temperature of 700 °C for 12 h. Electrospinning is continued with different applied voltages of 10 and 12 kV an annealing temperatures of 300, 500, and 700 °C to obtain nanofibers.

16.4.3 Metal Oxide/$LiCoO_2$ Composite

Gu et al. [40] reported a core-shell $LiCoO_2$–MgO coaxial fiber for lithium-ion batteries were prepared by combing electrospinning with sol-gel process. Co-electrospinning is considered as an effective method for the preparation coaxial fibers as compared to self-assembly approach, laser ablation, etc. In this work, core-shell $LiCoO_2$–MgO were coaxial fibers are prepared from one step process and thickness of the fiber is controlled by adjusting the pressure of sol during electrospinning process.

In this method, $LiCoO_2$ spinnable sol and MgO spinnable sol were prepared and its viscosity is adjusted to 4.0 Pa s. Electrospinning is carried out under 25 kV voltage and pressure of $LiCoO_2$ and MgO sols were controlled at 0.3 MPa and 0.02 MPa, respectively.

XRD and HR-TEM results revealed that the $LiCoO_2$ core was composed of partially oriented $LiCoO_2$ nanoparticles, 10 nm in size, growing in the direction of axis of the fiber. The electrochemical properties of the coaxial fibers were investigated using bare $LiCoO_2$ fibers as a reference. Impedance spectra of the coaxial fiber electrodes suggest that the MgO coating on the $LiCoO_2$ effectively avoids impedance growth by protecting the surface from passive surface film formation during cycling. The CV experiment indicates an improved reversibility of the core-shell fiber electrode, and a preferable cyclability was also revealed by the charge–discharge experiment. After 40 cycles, the discharge capacity of the core-shell fiber electrode still remained at 90.0% of the initial value, compared to 52.0% retention for the uncoated $LiCoO_2$ fiber electrode. Coating the surface of $LiCoO_2$ with an appropriate inert substance is an effective approach to improve its electrochemical properties.

Yuan et al. [72] reported a facile and general electrospinning technique followed by controlled annealing in air to fabricate Co_3O_4–CeO_2 composite oxide nanotubes (CCONs). The effect of the annealing temperature on the lithium-ion battery performance and catalytic activity toward CO oxidation was investigated in this report. Advantages of 1D hollow nanostructure and the synergistic effect of Co_3O_4 and CeO_2, the as-obtained Co_3O_4–CeO_2 composite oxide nanotubes exhibit improved lithium-ion battery performance and high catalytic activity toward CO oxidation. In this report, he also predicted the design and synthesis of novel metal oxides composite as advanced multifunctional materials for more potential applications. Figure 16.12 shows the schematic illustration of the formation of Co_3O_4–CeO_2 composite oxide nanotubes.

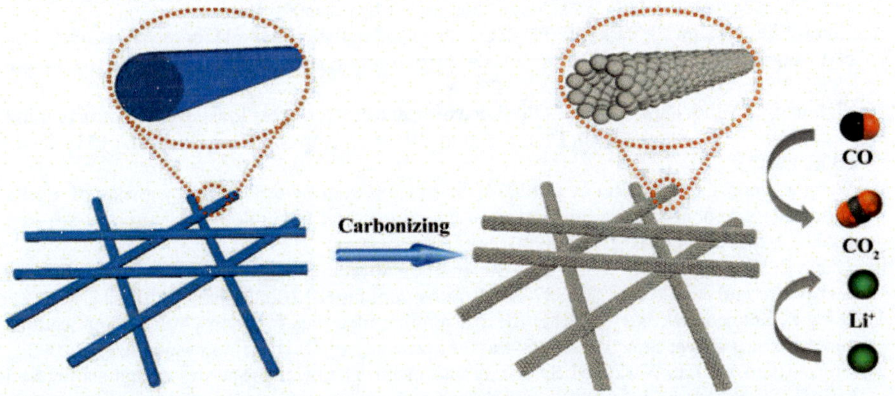

Fig. 16.12 Schematic illustration of the formation of Co_3O_4–CeO_2 composite oxide nanotubes Adapted and reproduced from [72], Copyright 2017 Elsevier

16.5 Conclusion

Many cathode materials were explored for the development of lithium-ion batteries. Among these developments, lithium cobalt oxide plays a vital role in the effective performance of lithium-ion batteries. Electrospun LCO attained its popularity owing to its higher porosity, larger surface to volume ratio, better efficiency, etc. Earlier, pristine electrospun LCO was used as such and later on, coating materials on the surface of electrospun LCO materials, the addition of metal oxide-based composites, the addition of carbon-based material composites, etc., was synthesized for the achievement of higher ionic conductivity, cycling stability, etc. Electrospun materials are always preferable compared to other techniques and there are many other techniques such as sol-gel method handled along with the electrospinning method. Nanosized fibers, wires, tubes, etc., were also incorporated for the effective implementation of lithium-ion batteries. The structural stability, safer and cycling stabilities are the major highlights of using electrospun LCO cathodes.

References

1. Armand MJ-MT (2001) Issues and challenges facing rechargeable lithium batteries. Nat Commun 414:359–367
2. Whittingham MS (1976) Electrical energy storage and intercalation chemistry 192
3. Thackeray MM, David WIF, Bruce PG, Goodenough JB (1983) Lithium insertion into magnesium spinels. Mater Res Bull 18:461–472. https://doi.org/10.1016/0025-5408(83)90138-1
4. Mizushima K, Jones PC, Wiseman PJ, Goodenough JB (1981) Li_xCoO_2 ($0 < x < -1$): A new cathode material for batteries of high energy density. Solid State Ionics 4:171–174
5. Orman HJ, Wiseman PJ (1984) Cobalt (III) lithium oxide, $CoLiO_2$: structure refinement by powder 12–14
6. Cho J, Kim YJ, Park B (2000) Novel $LiCoO_2$ cathode material with Al_2O_3 coating for a Li ion cell. Chem Mater 12:3788–3791. https://doi.org/10.1021/cm000511k
7. Cho J, Kim YJ, Kim TJ, Park B (2001) Zero-strain intercalation cathode for rechargeable Li-ion cell. Angew Chemie Int Ed 40:3367–3369. https://doi.org/10.1002/1521-3773(20010917)40:18%3c3367:AID-ANIE3367%3e3.0.CO;2-A
8. Sigmund W, Yuh J, Park H et al (2006) Processing and structure relationships in electrospinning of ceramic fiber systems. J Am Ceram Soc 89:395–407. https://doi.org/10.1111/j.1551-2916.2005.00807.x
9. Wang J, Feng CQ, Sun ZQ et al (2014) In-situ one-step hydrothermal synthesis of a lead germanate-graphene composite as a novel anode material for lithium-ion batteries. Sci Rep 4:7030. https://doi.org/10.1038/srep07030
10. Zhang R, Wen Q, Qian W et al (2011) Superstrong ultralong carbon nanotubes for mechanical energy storage. Adv Mater 23:3387–3391. https://doi.org/10.1002/adma.201100344
11. Dong Z, Kennedy SJ, Wu Y (2011) Electrospinning materials for energy-related applications and devices. J Power Sour 196:4886–4904. https://doi.org/10.1016/j.jpowsour.2011.01.090
12. Gu Y, Chen D, Jiao X (2005) Synthesis and electrochemical properties of nanostructured $LiCoO_2$ fibers as cathode materials for lithium-ion batteries. J Phys Chem B 109:17901–17906. https://doi.org/10.1021/jp0521813
13. Mizuno Y, Hosono E, Saito T, et al (2012) Electrospinning synthesis of wire-structured $LiCoO_2$ for electrode materials of high-power Li-ion batteries. https://doi.org/10.1021/jp2123482

14. Sun YK, Oh IH, Hong SA (1996) Synthesis of ultrafine $LiCoO_2$ powders by the sol-gel method. J Mater Sci 31:3617–3621. https://doi.org/10.1007/BF00352769
15. Reimers JN, Dahn JR (1992) Electrochemical and in situ x-ray diffraction studies of lithium intercalation in Li_xCoO_2. J Electrochem Soc 139:2091–2097. https://doi.org/10.1149/1.2221184
16. Dersch R, Steinhart M, Boudriot U et al (2005) Nanoprocessing of polymers: applications in medicine, sensors, catalysis, photonics. Polym Adv Technol 16:276–282. https://doi.org/10.1002/pat.568
17. Li D, Xia Y (2004) Electrospinning of nanofibers: reinventing the wheel? Adv Mater 16:1151–1170. https://doi.org/10.1002/adma.200400719
18. Zhan S, Li Y, Yu H (2008) $LiCoO_2$ hollow nanofibers by co-electrospinning sol-gel precursor. J Dispers Sci Technol 29:702–705. https://doi.org/10.1080/01932690701758517
19. Honma I, Okubo M, Kim J et al (2009) Anisotropic surface effect on electronic structures and electrochemical properties of $LiCoO_2$. J Phys Chem C 113:15337–15342. https://doi.org/10.1021/jp904877d
20. Shaju J, Bruce P (2005) Synthesis of nanowire and mesoporous low-temperature $LiCoO_2$ by a post templating reaction. Angew Chem 15:6553
21. Peng ZS (1998) Synthesis by sol-gel process and characterization of $LiCoO_2$ cathode materials 215–220
22. Vertruyen B, Eshraghi N, Piffet C, et al (2018) Spray-drying of electrode materials for lithium-ion and sodium-ion batteries. Materials (Basel) 11. https://doi.org/10.3390/ma11071076
23. Chen H, Qiu X, Zhu W, Hagenmuller P (2002) Synthesis and high rate properties of nanoparticled lithium cobalt oxides as the cathode material for lithium-ion battery. Electrochem Commun 4:488–491. https://doi.org/10.1016/S1388-2481(02)00357-0
24. Hosono E, Kudo T, Honma I et al (2009) Synthesis of single crystalline spinel $LiMn_2O_4$ nanowires for a lithium ion battery with high power density. Nano Lett 9:1045–1051. https://doi.org/10.1021/nl803394v
25. Dai X, Wang L, Xu J et al (2014) Improved electrochemical performance of $LiCoO_2$ electrodes with ZnO coating by radio frequency magnetron sputtering. ACS Appl Mater Interfaces 6:15853–15859. https://doi.org/10.1021/am503260s
26. Burukhin A, Brylev O, Hany P, Churagulov BR (2002) Hydrothermal synthesis of $LiCoO_2$ for lithium rechargeable batteries 151:259–263
27. Cohen J, Perl JM (1988) Atmospheric noise implementation in an HF-channel simulator 355
28. Li D, Wang Y, Xia Y (2003) Electrospinning of polymeric and ceramic nanofibers as uniaxially aligned arrays. Nano Lett 3:1167–1171. https://doi.org/10.1021/nl0344256
29. Cavaliere S, Subianto S, Savych I et al (2011) Electrospinning: designed architectures for energy conversion and storage devices. Energ Environ Sci 4:4761–4785. https://doi.org/10.1039/c1ee02201f
30. Zhang W, Hu Y, Tao X et al (2010) Synthesis of spherical $LiFePO_4$/C via Ni doping. J Phys Chem Solids 71:1196–1200. https://doi.org/10.1016/j.jpcs.2010.04.015
31. Kawamura T, Makidera M, Okada S et al (2005) Effect of nano-size $LiCoO_2$ cathode powders on Li-ion cells. J Power Sour 146:27–32. https://doi.org/10.1016/j.jpowsour.2005.03.012
32. Ding N, Ge XW, Chen CH (2005) A new gel route to synthesize $LiCoO_2$ for lithium-ion batteries. Mater Res Bull 40:1451–1459. https://doi.org/10.1016/j.materresbull.2005.04.022
33. Ou Y, Wen J, Xu H et al (2013) Ultrafine $LiCoO_2$ powders derived from electrospun nanofibers for Li-ion batteries. J Phys Chem Solids 74:322–327. https://doi.org/10.1016/j.jpcs.2012.10.007
34. Bai Y, Shi H, Wang Z, Chen L (2007) Performance improvement of $LiCoO_2$ by molten salt surface modification. J Power Sour 167:504–509. https://doi.org/10.1016/j.jpowsour.2007.02.036
35. Kim YJ, Kim TJ, Shin JW et al (2002) The effect of Al_2O_3 coating on the cycle life performance in thin-film $LiCoO_2$ cathodes. J Electrochem Soc 149:1337–1341. https://doi.org/10.1149/1.1505634

36. Yu Y, Shui JL, Jin Y, Chen CH (2006) Electrochemical performance of nano-SiO_2 modified $LiCoO_2$ thin films fabricated by electrostatic spray deposition (ESD). Electrochim Acta 51:3292–3296. https://doi.org/10.1016/j.electacta.2005.09.021
37. Lu HW, Yu L, Zeng W et al (2008) Fabrication and electrochemical properties of three-dimensional structure of $LiCoO_2$ fibers. Electrochem Solid State Lett 11:140–144. https://doi.org/10.1149/1.2932054
38. Chen L-J, Liao J-D, Chuang Y-J, Hsu K-C, Chiang Y-F, Fu Y-S (2010) Synthesis and characterisation of PVP/$LiCoO_2$ nanofibers by electrospinning route. J Appl Polym Sci 116:2658–2667. https://doi.org/10.1002/app
39. Zhang J, Liu J, Wang S et al (2004) Facile methods to coat polystyrene and silica colloids with metal. Adv Funct Mater 14:1089–1096. https://doi.org/10.1002/adfm.200400119
40. Gu Y, Chen D, Jiao X, Liu F (2007) $LiCoO_2$–MgO coaxial fibers: co-electrospun fabrication, characterization and electrochemical properties. J Mater Chem 1769–1776. https://doi.org/10.1039/b614205b
41. Nakai I, Takahashi K, Shiraishi Y et al (1998) Study of the Jahn-Teller distortion in $LiNiO_2$, a cathode material in a rechargeable lithium battery, by in situ X-ray absorption fine structure analysis. J Solid State Chem 140:145–148. https://doi.org/10.1006/jssc.1998.7943
42. Zhu GN, Wang YG, Xia YY (2012) Ti-based compounds as anode materials for Li-ion batteries. Energ Environ Sci 5:6652–6667. https://doi.org/10.1039/c2ee03410g
43. Tao H, Feng Z, Liu H et al (2011) Reality and future of rechargeable lithium batteries. Open Mater Sci J 5:204–214. https://doi.org/10.2174/1874088X01105010204
44. Alcantara R, Lavela P, Tirado JL (1998) Structure and electrochemical properties of boron-doped $LiCoO_2$. J Solid State Chem 23:781–786
45. Waki S, Dokko K, Itoh T et al (2000) High-speed voltammetry of Mn-doped $LiCoO_2$ using a microelectrode technique. J Solid State Electrochem 4:205–209. https://doi.org/10.1007/s100080050196
46. Stoyanova R, Zhecheva E, Zarkova L (1994) Effect of Mn-substitution for Co on the crystal structure and acid delithiation of $LiMn_yCo_{1-y}O_2$ solid solutions. Solid State Ionics 73:233–240. https://doi.org/10.1016/0167-2738(94)90039-6
47. Mladenov M, Stoyanova R, Zheheva E, Vassilev S (2001) Effect of Mg doping and MgO-surface modification on the cycling. Electrochem Commun 3:410–416
48. Madhavi S, Subba Rao GV, Chowdari BVR, Li SFY (2002) Effect of Cr dopant on the cathodic behavior of $LiCoO_2$. Electrochim Acta 48:219–226. https://doi.org/10.1016/S0013-4686(02)00594-7
49. Aykut Y, Pourdeyhimi B, Khan SA (2013) Synthesis and characterization of silver/lithium cobalt oxide (Ag/$LiCoO_2$) nanofibers via sol-gel electrospinning. J Phys Chem Solids 74:1538–1545. https://doi.org/10.1016/j.jpcs.2013.05.021
50. Kim Y, Park JH, Kim JG et al (2017) Ruthenium oxide incorporated one-dimensional cobalt oxide composite nanowires as lithium-oxygen battery cathode catalysts. ChemCatChem 9:3554–3562. https://doi.org/10.1002/cctc.201700560
51. Zhang J, Li P, Wang Z et al (2015) Three-dimensional graphene-Co_3O_4 cathodes for rechargeable Li-O_2 batteries. J Mater Chem A 3:1504–1510. https://doi.org/10.1039/c4ta05573j
52. Wang F, Wen Z, Shen C et al (2015) Open mesoporous spherical shell structured Co_3O_4 with highly efficient catalytic performance in Li-O_2 batteries. J Mater Chem A 3:7600–7606. https://doi.org/10.1039/c5ta00295h
53. Zeng J, Francia C, Amici J et al (2014) Mesoporous Co_3O_4 nanocrystals as an effective electrocatalyst for highly reversible Li-O_2 batteries. J Power Sour 272:1003–1009. https://doi.org/10.1016/j.jpowsour.2014.09.055
54. Hou Y, Li J, Wen Z et al (2015) Co_3O_4 nanoparticles embedded in nitrogen-doped porous carbon dodecahedrons with enhanced electrochemical properties for lithium storage and water splitting. Nano Energ 12:1–8. https://doi.org/10.1016/j.nanoen.2014.11.043
55. Débart A, Bao J, Armstrong G, Bruce PG (2007) An O_2 cathode for rechargeable lithium batteries: the effect of a catalyst. J Power Sour 174:1177–1182. https://doi.org/10.1016/j.jpowsour.2007.06.180

56. Zhang X, Shi F, Yu X et al (2004) Polyelectrolyte multilayer as matrix for electrochemical deposition of gold clusters: toward super-hydrophobic surface. J Am Chem Soc 126:3064–3065. https://doi.org/10.1021/ja0398722
57. Wittmaier D, Danner T, Wagner N, Friedrich KA (2014) Screening and further investigations on promising bi-functional catalysts for metal-air batteries with an aqueous alkaline electrolyte. J Appl Electrochem 44:73–85. https://doi.org/10.1007/s10800-013-0602-x
58. Lee J, Jeong B, Ocon JD (2013) Oxygen electrocatalysis in chemical energy conversion and storage technologies. Curr Appl Phys 13:309–321. https://doi.org/10.1016/j.cap.2012.08.008
59. Xiao J, Mei D, Li X et al (2011) Hierarchically porous graphene as a lithium-air battery electrode. Nano Lett 11:5071–5078
60. Xu JJ, Wang ZL, Xu D et al (2013) Tailoring deposition and morphology of discharge products towards high-rate and long-life lithium–oxygen batteries. Nat Commun 4:1–10. https://doi.org/10.1038/ncomms3438
61. Lim HD, Park KY, Song H et al (2013) Enhanced power and rechargeability of a Li-O_2 battery based on a hierarchical-fibril CNT electrode. Adv Mater 25:1348–1352. https://doi.org/10.1002/adma.201204018
62. Sun B, Chen S, Liu H, Wang G (2015) Mesoporous carbon nanocube architecture for high-performance lithium–oxygen batteries. Adv Funct Mater 25:4436–4444. https://doi.org/10.1002/adfm.201500863
63. Mitchell RR, Gallant BM, Thompson CV, Shao-Horn Y (2011) All-carbon-nanofiber electrodes for high-energy rechargeable Li-O_2 batteries. Energ Environ Sci 4:2952–2958. https://doi.org/10.1039/c1ee01496j
64. Shang C, Dong S, Hu P et al (2015) Compatible interface design of CoO-based Li-O_2 battery cathodes with long-cycling stability. Sci Rep 5:1–7. https://doi.org/10.1038/srep08335
65. Wang X, Wang L, Zhao F, Fu C-G (2015) Monoatomic-thick graphitic carbon nitride dots on graphene sheets as an efficient catalyst in oxygen reduction reaction 1–48. https://doi.org/10.1039/b000000x
66. Zhang P, Wang R, He M et al (2016) 3D hierarchical Co/CoO-graphene-carbonized melamine foam as a superior cathode toward long-life lithium oxygen batteries. Adv Funct Mater 26:1354–1364. https://doi.org/10.1002/adfm.201503907
67. Hyun S, Shanmugam S (2017) Mesoporous Co-CoO/N-CNR nanostructures as high-performance air cathode for lithium-oxygen batteries. J Power Sour 354:48–56. https://doi.org/10.1016/j.jpowsour.2017.04.029
68. Wang X, Tang Y (2017) Self-evaporating from inside to outside to construct cobalt oxide nanoparticles-embedded nitrogen-doped porous carbon nanofibers for high-performance lithium ion batteries. Chem Eng J. https://doi.org/10.1016/j.cej.2017.11.155
69. Abouali S, Garakani MA, Zhang B et al (2014) Co_3O_4/porous electrospun carbon nanofiber as anodes for high performance Li-ion batteries. J Mater Chem A Mater Energ Sustain 2:16939–16944. https://doi.org/10.1039/C4TA03206C
70. Agarwal S, Greiner A, Wendorff JH (2013) Functional materials by electrospinning of polymers. Prog Polym Sci 38:963–991. https://doi.org/10.1016/j.progpolymsci.2013.02.001
71. Thavasi V, Singh G, Ramakrishna S (2011) Electrospun nanofibers for energy and environmental applications. In: ICAFPM 2011—proceedings of 2011 international conference on advanced fibers and polymer materials, pp 413–420. https://doi.org/10.1039/b809074m
72. Yuan C, Wang H, Liu J et al (2017) Journal of colloid and interface science facile synthesis of Co_3O_4–CeO_2 composite oxide nanotubes and their multifunctional applications for lithium ion batteries and CO oxidation. J Colloid Interface Sci 494:274–281. https://doi.org/10.1016/j.jcis.2017.01.074

Chapter 17
Electrospun Lithium Iron Phosphate ($LiFePO_4$) Electrodes for Lithium-Ion Batteries

Neethu T. M. Balakrishnan, M. A. Krishnan, Akhila Das, Nikhil Medhavi, Jou-Hyeon Ahn, M. J. Jabeen Fatima, and Raghavan Prasanth

17.1 Introduction

The use of renewable systems at its best hand was made possible by the use of energy storage devices. Among the different energy storage systems, lithium-ion batteries are seeking much attention since it can offer high energy and power density [1, 2]. Eventhough, there was certain limits in its practical application due to the safety hazards and performance degradation [3]. The performances of batteries are primarily determined by the type of integral components used in it; the electrodes and the electrolytes. Even though electrolytes are more crucial for the batteries, electrodes have a vital role in determining the performance LIBs. Among the anodes and cathodes used, cathodes which serve as the positive electrodes in LIBs are considered to be more important to determine the electrochemical properties of the device. The capacity and specific power of LIBs are greatly depends on the choice of cathodes materials [4]. The morphology and compositional characteristics of electrodes are important to determine its electrochemical performance. Different types of cathodes such as spinel [5], olivine [6], and conversion [7] types are employed for the enhancement in

N. T. M. Balakrishnan · A. Das · N. Medhavi · M. J. Jabeen Fatima (✉) · R. Prasanth (✉)
Department of Polymer Science and Rubber Technology, Cochin University of Science and Technology (CUSAT), Cochin 682022, India
e-mail: jabeen@cusat.ac.in

R. Prasanth
e-mail: prasanth@cusat.ac.in

M. A. Krishnan
Department of Mechanical Engineering, Amrita Vishwa Vidyapeetham, Amritapuri, India

Department of Electrical Engineering, Pennsylvania State University, 311, Electrical Engineering East Building, State College, PA 16802, USA

J.-H. Ahn · R. Prasanth
Department of Materials Engineering and Convergence Technology, Gyeongsang National University, 501 Jinju-daero, Jinju 52828, Republic of Korea

N. T. M. Balakrishnan and R. Prasanth (eds.), *Electrospinning for Advanced Energy Storage Applications*, Materials Horizons: From Nature to Nanomaterials,
https://doi.org/10.1007/978-981-15-8844-0_17

performance. When the lithium-ion batteries was commercialized in 1991, cathode that used was lithium cobalt oxide ($LiCoO_2$ Simply LCO) [8] which was introduced by John Good enough. It was considered as the most commercially successful form of layered cathodes. LCO exhibits high theoretical capacity (274 mAh g^{-1}) and excellent volumetric density (1363 mAh cm^{-3}). The self-discharge of these cathodes is very low and they are enough to provide good cycling performance. However, the toxic nature of cobalt makes severe problems including environmental issues while using LCO in LIBs. Apart from this, expensive nature and low thermal stability made it to hunt for the new cathode materials in LIBs. For LCO, when it heated above certain temperature results, the release of oxygen that leads to the thermal runway which finally results the burst out of cells. This issue is the major concern of this material for the commercial application in LIBs, for example, the incident that was happened in Boeing 787 airplanes in 2013 [9]. Similar thermal issues where reported recently with Samsung Note 7 [10] and Tesla cars [11], all these incidents catalyzed the research over the development of thermally stable as well as electrochemically best performing electrode-electrolyte combination. After LCO, lithium nickel oxide ($LiNiO_2$ simply LNO) [12] introduced that possess similar crystal structure of LCO, but could not reach at its performance level due to the tendency of Ni^{2+} ions to substitute Li^+-ions. Compared to cobalt and nickel, manganese is considered to be cheaper and less toxic so, lithium manganese oxide and lithium nickel manganese oxide [13] are considered to be effective than LCO and LNO. During the lithium extraction process, the tendency of these materials to change its spinal structure by Jahn-Teller distortion made its performance unsatisfactory [14]. On the continuous effort of developing a thermally stable and well performing cathode material for LIBs, the research over the polyanions was extensively made and as a result, in 1998, John Goodenough developed an olivine-structured cathode material, lithium iron phosphate ($LiFePO_4$ simply LFP) [15] that exhibits high thermal stability as well as high power capability. Compared to other batteries, LIBs with LFP cathodes are safer and can be used without any safety hazards even at abused conditions and this material is environmentally benign. LFP batteries can work at a wide range of temperature in between −20 and 150 °C [16]. Moreover, the environmental friendliness, good shelf life along with long cycle life that ensured by this cathode material open up LIBs to a new era of safer and reliable lithium-ion technology [17]. Eventhough, they have a deft to use as cathode material in LIBs, is that the lower practical capacity compared to its theoretical capacity (170 mAh g^{-1}) due to its low electronic conductivity. Inorder to tackle this issue and thereby improve the electrochemical performance of LFP, different methods such as composite as well as physical modifications of LFP were introduced. Carbon coating [18] and doping [19] are the two most effective methods and even the commercially available LFP cathodes used are coated with carbon for ensuring good electrochemical performance. Through these methods, the lithium-ion kinetics can enhanced by providing a convenient lithium diffusion pathway and thereby the final electrochemical activity. Similarly, flexible LFP cathodes are observed to be exhibit best functioning in LIBs which can be prepared by electrospinning [20], sol–gel method [21], hydrothermal method [22], and spray pyrolysis method [23].

Electrospinning [24] is considered as the excellent method for the preparation of flexible electrodes. The versatility and scalability of electrospinning technique make it important method for the synthesis of nanostructured electrode materials, especially nanofibrous morphology than any other methods used. Electrospinning can be effectively used for making nanostructured cathodes [25], anodes [26], and separators [27]. Usage of electrospun material provides economical as well as functional advantages in LIB performance. Electrospun fibrous electrodes can easily use as free standing and do not require any binders for support, which greatly enhance the energy density of LIBs. Moreover, they have a two-dimensional (2D) morphology, high surface area with a tunable density, and most importantly, a highly porous structure that provides easier pathway for lithium diffusion and accommodate huge volume expansion during the lithiation and delithiation process thereby preventing the cracking and pulverization of electrodes and loosing of contact with current collector [28]. Anodic nanofibers are effective to reduce the dendrite growth which is severe issue with anodes like silicon [29]. Separators, that prepared by the electrospinning process produce porous structures that can effectively control the motion of ions through it. Comparing with commercially using separators such as polypropylene (PP) [30] and polyethylene (PE) [31], electrospunned separators are observed to be exhibiting superior electrochemical properties due to the porous structure that it results. So the fibrous membranes of separators are extensively studied for LIBs. For cathodes such as LFP, the major factor which limits its application is its degradation that happens during the cycling as well as its high internal resistance. Electrospinning can be effectively utilized in order to overcome this issue with the LFP [32, 33]. The porous structural morphology which enhances the lithium diffusion kinetics and tackle other issues related to the cathode materials. Several studies revealed that electrospun LFP can exhibit improved electrochemical performance than normal LFP that will discuss detail in this chapter.

17.2 Structure of Lithium Iron Phosphate (LFP)

LFP cathodes possess an olivine structure that shown in Fig. 17.1 generally represented as M_2XO_4. In the structure, M atoms are occupied in half of the octahedral sites, whereas one eighth of tetrahedral sites of the hexagonally closed packed oxygen arrays are occupied by the X atoms.

Compare to that of spinal-structured cathodes such as $LiMnO_4$ [5], olivine structures are distinct by the presence of crystallographically different octahedral sites. Normally those structures composed of ‘M’ as Fe, Mn, Co or Ni are observed to be exhibiting an ordered olivine structure. The structure of cathodes has significant over the electronic conductivity it poss. For olivine-structured cathode materials different lithium-ion diffusion pathways are proposed such as along the *b*-axis, *c*-axis and between the *bc*-layers. Among these different pathways, diffusion through the *c*-axis is observed to the best (Fig. 17.2).

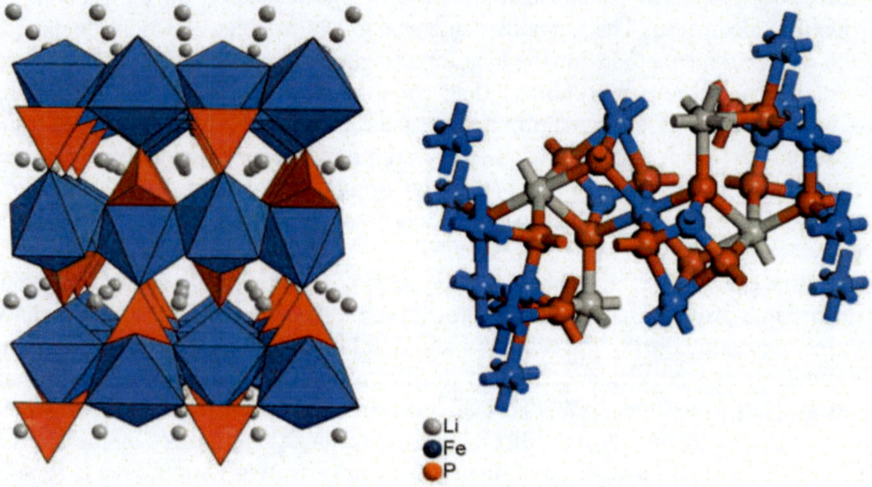

Fig. 17.1 Olivine structures of lithium iron phosphate

Fig. 17.2 Movement of lithium-ions through lithium iron phosphate (LFP) cathode

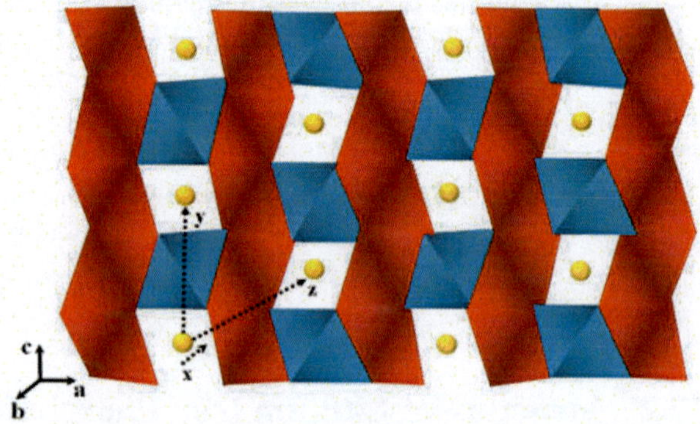

Low electronic conductivity of LFP leads to research over the development of nano-engineered structures, that can results high conductivity and there by good electrochemical properties. Electrospinning forms leaf-like structures with highly interconnected polymer chains. The comparison of electrospunned LFP/C and conventional LFP was made by Bachtin et al. [28] reveals that, for electrospunned LFP, the specific capacity increases with cycling. This is attributed by the activation of LFP particles by cycling. The reason for this may be the volume change associated with the lithiation and delithiation process that will make the accessibility of more LFP particles with the electrolytes. In this aspect, electrospun LFP can be considered as best for the development of better cathodes in LIBs than the conventional cathodes.

17.3 Precursors for Electropsun Lithium Iron Phosphate Cathodes

The morphology of LFP has a significant impact in determining its electrochemical properties. LFP microspheres with nanoparticles having a three-dimensional structure appears to be exhibiting high rate capability and tape density [34]. Similarly, nanospheres and hierarchical spindle-like architectures produced via hydrothermal method of synthesis exhibits good electrochemical properties [35]; however, nanofibers of LFP exhibit superior electrochemical properties than any of these nanostructures [36]. Also the fibrous LFP is capable to exhibit excellent rate capability even at 10 C-rate [36]. LFP nanofibers can be synthesized using different precursor solutions. The solution that employed for the electrospinning often contains lithium (Li) and iron (Fe) containing compounds along with a polymer. The presence of polymer is important, because at high temperature and at an inert atmosphere, they can easily get converted into amorphous carbon, hence that could forms a uniform layer of carbon coating over the electrode active material. In most of the studies, electrospun LFP, polyvinyl pyrrolidone (PVP) is used as the polymer and it has a significant role in the formation of porous structures. PVP is highly polar polymer and water soluble. The ability of PVP to form hydrogen bond with the citrate group influences the textural properties such as pore size, porousity, and surface area of the solution of LFP [37]. Qiu et al. [38] synthesized LFP/C nanofibers using lithium dihydrogen phosphate (LiH_2PO_4), ferric citrate ($FeC_6H_5O_7.5H_2O$), polyvinyl pyrrolidone (PVP), and polyethylene oxide in water. After electrospinning, the as-spun fibers were treated at a temperature of 450 °C under nitrogen atmosphere. Depending on the concentration of precursors used, the morphology of resulting fibers also varied. Different morphologies of spunned fibers with varying concentration of $LiH_2PO_4/FeC_6H_5O_7$ (0.8 to 1.4) were shown in Fig. 17.3. The best morphology of fiber with smooth surface and uniform diameter is observed for higher concentration of $LiH_2PO_4/FeC_6H_5O_7$ [38].

LFP/C nanobelts synthesized with lithium hydroxide monohydrate ($LiOH.H_2O$), ferric nitrate ($FeNO_3.9H_2O$) and phosphoric acid (H_3PO_4) in dimethyl formamide (DMF) at a stoichiometric ratio 1:1:1. PVP was employed as the spinning template as well as the carbon source. The resulting fibers exhibit a reduction in diameter with the decomposition of PVP [20]. Kang et al. [39] synthesized LFP by using the solution containing PVP, deionized water, methanol (CH_3OH), nitric acid (HNO_3), lithium nitrate ($LiNO_3$), iron(II) nitrate nonahydrate [$Fe(NO_3)_3.9H_2O$], and ammonium dihydrogen phosphate ($NH_4H_2PO_4$). The resulting hollow tubular structures were compared with LFP-CNT tubes and no significant difference in morphology as well as electrochemical performance was found. So PVP is one of the best choices for the preparation of porous LFP electrode. The XRD pattern of prepared LFP exhibits their orthorhombic structure and the videoscopic image and surface morphology were analyzed. The as-spun fiber exhibits a diameter 3–5 μm with very smooth surface (Fig. 17.4). On calcined at high temperature, the prepared LFP/C nanotubes

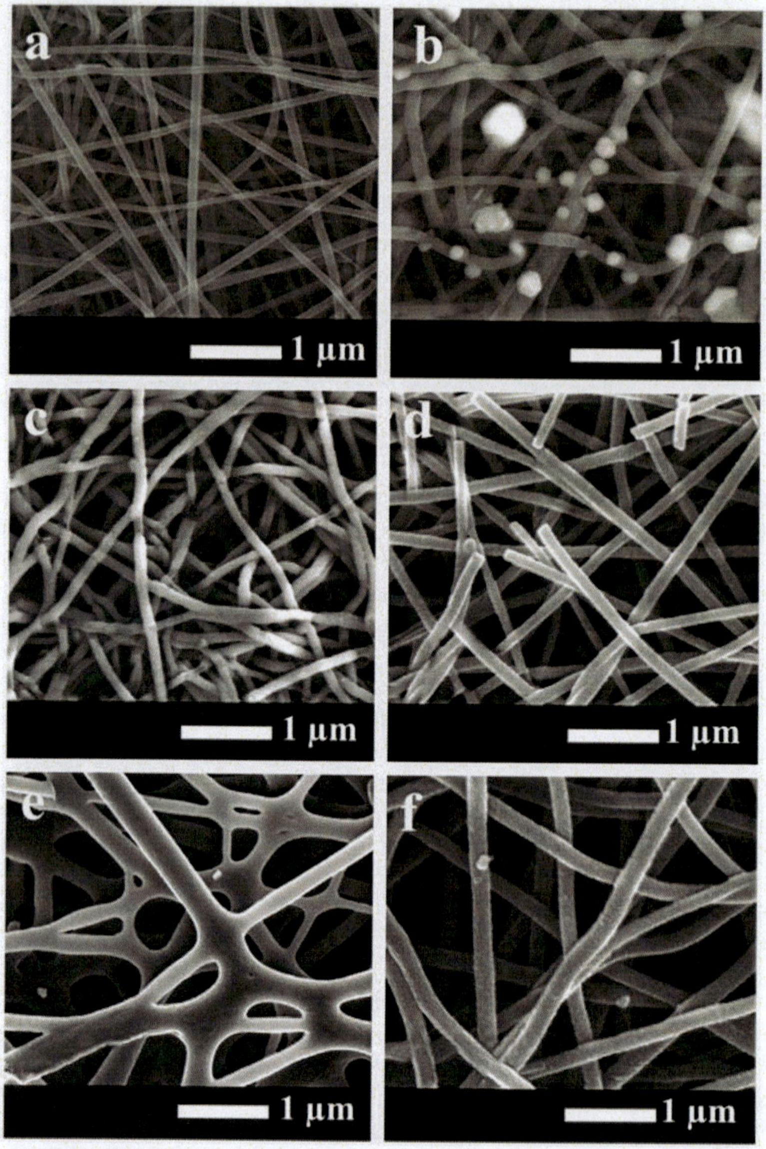

Fig. 17.3 SEM images on the surface morphology of **a** as-spun LiH_2PO_4-$FeC_6H_5O_7$/PVP-PEO precursor fibers and $LiFePO_4$/C nanofibers prepared using different mole ratio of LiH_2PO_4/$FeC_6H_5O_7$: **b** 0.8; **c** 1.0; **d** 1.1; **e** 1.2; **f** 1.3. Adapted and reproduced from Ref. [38]. Copyright 2014 Elsevier

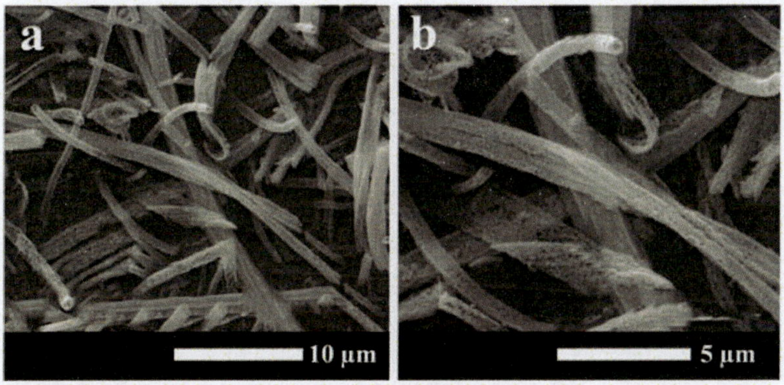

Fig. 17.4 SEM images on the surface morphology of porous $LiFePO_4$ nanotubes **a, b** calcination at 800 °C for 10 h. Adapted and reproduced from Ref. [39], Copyright 2012, The Chemical Society of Japan

exhibit a diameter 150–500 nm but even at high temperature, they maintained the original morphology [39].

Using the solvent (DMF), Lin et al. [40] synthesized LFP fibers using the precursors $Fe(NO_3)_3 \cdot 9H_2O$, $LiNO_3$ and H_3PO_4. Polyacrylonitrile (PAN) is used as the polymer in this synthesis method. The precursors, PAN, and DMF were taken in a concentration 13, 3, and 84% [40]. The lower precursor concentration yields best fiber morphology with good chemical stability. Nanofibers are synthesized with coaxial and uniaxial electrospinning method using the solution of LiH_2PO_4, $Fe(NO_3)_3 \cdot 9H_2O$ and 1.0 g of polyethylene oxide (PEO) in water, followed by sintering. An amorphous layer of carbon which act as the conducting carbon layer over the surface of the LFP was formed that can reduce the lithium-ion diffusion path and there by enhances the electrochemical properties of the cathode material [41]. Carbon-coated single crystalline LFP nanowires (SCNW-LFP) prepared using an aqueous precursor solution containing LiH_2PO_4, $Fe(NO_3)_3$, and polyethylene oxide result fibrous structure with a uniform diameter of about 100 nm. This cathode material retained about 98% of its theoretical capacity even after 100 Cycles at a temperature of 60 °C (Fig. 17.5). This superior capacity is due to the presence of single crystalline nanowires as well the carbon coating which can act as the protective layer over the single crystalline nanowire and there by improve the capacity [42].

Structurally stable LFP electrode was synthesized using the solution containing precursors such as PAN and commercially available LFP particle in DMF. The raw LFP (rLFP) appears in the form of clusters with independent particle. But, in LFP/C complexes, a uniform dispersion of the LFP particles that linked with carbon is normally observed. Electrospinning results high density fiber structure that can efficiently supply electrons to the LFP particles (Fig. 17.6). This can be seen from the fibers synthesized from PAN and LFP particles. This nanofibers heat treated at a temperature of 800 °C exhibits an improved electronic conductivity of about 2.23×10^{-2} S cm^{-1}. In addition to this, they are sufficient to exhibit a discharge capacity

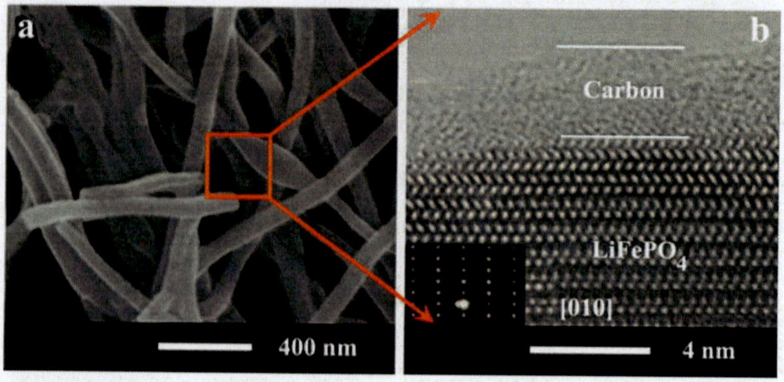

Fig. 17.5 **a** SEM image on the surface morphology of SCNW-LFP heated at 600 °C for 2 h, **b** overview TEM micrograph of SCNW-LFP showing nanowires with a diameter of around 100 nm. Adapted and reproduced from Ref. [42], Copyright 2011 John Wiley and Sons

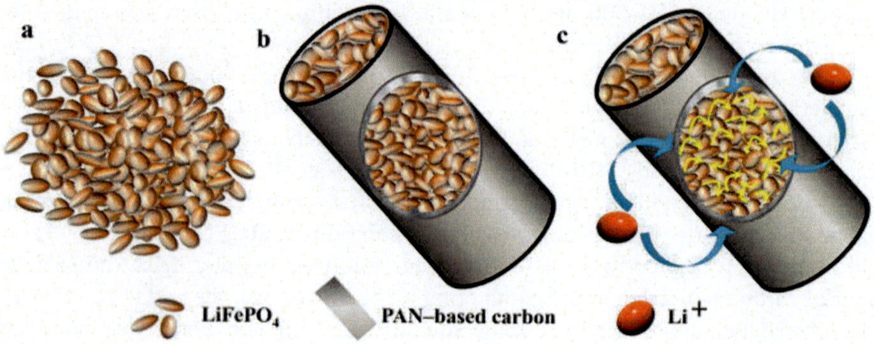

Fig. 17.6 Schematic diagrams of **a** r-LFP, **b** prepared $LiFePO_4$/carbon complex, **c** electron transfer pathway of the $LiFePO_4$/carbon complex. Adapted and reproduced from Ref. [44], Copyright 2011 Springer

of about 153 and 61 mAh g^{-1} at current density of 0.1 and 10 C rate, respectively [43]. Taprakci et al. [44] Synthesized LFP/C fibers by using PAN as the carbon source and LFP precursor solution in DMF. LFP precursors such as lithium acetate ($C_2H_3LiO_2$), phosphoric acid (H_3PO_4), and iron acetate (Fe $(C_2H_3O_2)_2$) were used at a stoichiometric ratio of 1:1:1. Resulting fibers exhibit a homogeneously distributed carbonaceous layer. The calcination/carbonization temperature could highly influence over the electrochemical property of the resulting fibers. The average discharge capacity exhibited by the fibers calcined at temperature of 600, 700, and 800 °C is about 125, 141, and 131 mAh g^{-1}, respectively, at a discharge rate of 0.1 C. Moreover, on evaluating the cycling performance of the resulting fiber, best capacity retention is observed up to 50 cycles. Both the fibers calcined at higher temperature (700 and 800 °C) displayed low capacity fading per cycle but it is higher for the fiber that calcined at temperature of 600 °C. Hence, the best charge–discharge and cycling performance of the electrospun composite electrode is observed for the sample calcined at a temperature of about 700 °C within the nanofibers prepared using 8% LFP precursor and 4% PAN. The fiber exhibits a capacity of about 166 mAh g^{-1} at a current rate of 0.05 C. This is achieved by the good penetration of electrolyte within the membrane [44].

17.4 Enhancing the Performance of Electrospun Lithium Iron Phosphate Cathode by Structural Modification

Eventhough carbon coating can enhance the electronic conductivity of LFP, the resulting conducting nature is not satisfactory. So inorder to improve it further, different conducting additives are incorporated with the LFP particles. Carbon nanotubes (CNT) [45, 46] graphite [47] and graphene [48] are the major carbon allotropes that employed as conducting additives in most of the electrodes due to its ability to form three dimensional (3D) conducting network channels that can ensure large surface area and high aspect ratio [49, 50].

17.4.1 Composites of Lithium Iron Phosphate with Conducting Materials

Mixing the electronically conducting materials with LFP is a very common method to prepare composite electrode having enhanced electrochemical properties and long-term charge–discharge cycling stability. In the simplest form, a sufficient amount of electronically conducting materials, mostly nanostructured materials such as CNT, graphene, or conducting graphitic carbon uniformly mixed by mechanical mixing in a mortar and pestle and transferred to a glass vial containing sufficient amount of NMP solvent, then air tightly lose it. The material is then subjected to ultrasound sonication flowed by magnetic stirring typically for overnight. The uniformly mixed and grinded composite slurry then coated on a pre-cleaned and flattened aluminum foil using the doctor blade technique and dried at 80 °C under vacuum for 12 h and punched out the circular disc of it used as the cathodes in LIBs. However, for preparing the fibrous morphology, electrospinning is used for the fabrication of electrodes. For that, a nanocomposite slurry mentioned above could directly electrospun on the current collector or the precursor LFP solution having electrically conducting nanostructured materials is electrospun first and the resulting fibers are calcined [20, 28] at high temperature (typically 600 to 800 °C). Electrospinning of precursor solution of LFP with sufficient amount of PAN is also a widely reported method [51]. LFP/CNT/C composite nanofibers were prepared by Topraskci et al. [51] by the combination of electrospinning process with sol–gel method. The precursor solution

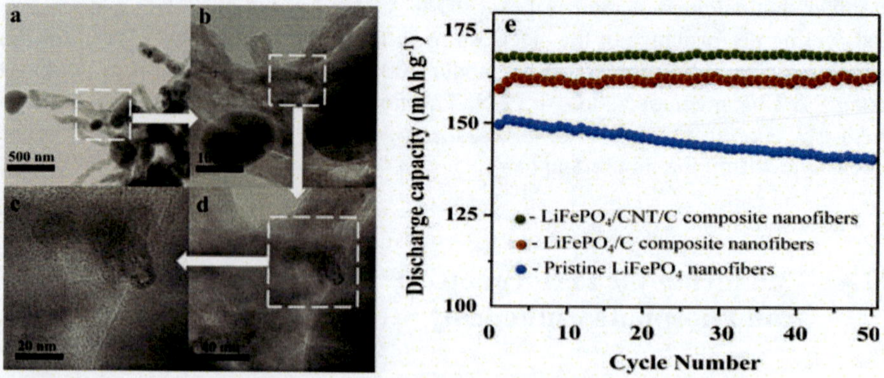

Fig. 17.7 TEM images of $LiFePO_4$/CNT/C composite nanofibers at various magnifications: **a** 20,000×, **b** 80,000×, **c** 200,000×, and **d** 400,000×, **e** cycling performance, **f** initial voltage versus capacity curves of pristine $LiFePO_4$ powder, $LiFePO_4$/C composite nanofibers, and $LiFePO_4$/CNT/C composite nanofibers at 0.05 C. Adapted and reproduced from Ref. [51]. Copyright 2012 American Chemical Society

that contains LFP precursors lithium acetate ($LiCOOCH_3$) phosphoric acid (H_3PO_4) and iron(II) acetate ($Fe(COOCH)_2$ along with CNT and polyacrylonitrile (PAN) can be converted into LFP/CNT/C composite simply by heat treatment. In addition to this, adding functionalized CNT can further enhance the electrochemical performance of the composite. The functionalized CNT forms bridges between the LFP particles (Fig. 17.7a, b) and forms a distribution throughout the matrix (Fig. 17.7c, d). Comparing this composite structure formed with normal LFP/C, a rapid and continuous capacity fading is observed for pristine LFP. The irreversible capacities of pristine LFP, LFP/C, and LFP/CNT/C are observed to be 150, 162 and 169 mAh g^{-1} (Fig. 17.7e). Compared to carbon modified LFP, the composite LFP/CNT/C electrode exhibited superior discharge capacity corresponds to 99% of the theoretical value (170 mAh g^{-1}) and retained up to 50 cycles (Fig. 17.7f). The exceptionally improved electrochemical performance of these cathodes (fibrous LFP/C and LFP/CNT/C nanocomposite electrode) are attributed by the one-dimensional fiber structure that results best lithium-ion transport [51].

This same research group [51] developed LFP/r-GO/C composite combining both electrospinning and sol–gel method, where the LFP particles are encapsulated inside the graphene containing carbon nanofiber matrix. Spinning solution consists of PAN, LFP precursors and functionalized CNT dissolved in DMF was electrospunned and the as-spunned fibers were heat treated at 700 °C. PAN functions as the carbon source were as the graphene helps to improve the electronic conductivity of as-spun fibrous electrode material. The electrochemical studies on electrospunned composite nanofiber displayed a reversible capacity of about 166 mAh g^{-1} at a discharge rate of 0.05 C. Comparing with the discharge capacity of LFP/C, the nanocomposite LFP/r-GO/C composite, shows significant enhancement in specific capacity retention, which exhibited a stable discharge capacity of 166 mAh g^{-1} up to 50 cycles without any considerable amount of capacity fading which is corresponds to the 98% theoretical capacity of LFP. The enhanced capacity in the LFP/rGO/C composite was attributed to the presence of highly conducting and electrochemically active graphene [52] and very stable cycling stability resulted from the role of graphene nanosheets as the mechanical buffer which prevent the cracking and pulverization of electrode due to the continues volume expansion and contraction of the nanofibers during the lithiation and delithiation process. The main limiting factors of using transition metal oxides as cathode is that, they will only utilize one of their oxidation states during the intercalation and de-intercalation process [53]. Hence, inorder to achieve high energy density and specific capacity, one of the important aspects that can be made easy to possible is the utilization of all the oxidation state of the active material. From this idea of developing high capacity cathodes, Zhang et al. [54] developed LiF/ferrocene/polyacrylonitrile (PAN) composite fibrous electrodes by electrospinning and heat treatment of LiF/F/C precursors. The electrochemical performance of the electrode is evaluated by preparing precursor solution having different concentration of LiF + ferrocene + PAN. During the first charge–discharge cycle, the nanofibers prepared with precursor concentration (LiF + ferrocene + PAN) 11, 16, 21, and 26% exhibit a discharge capacity about 262, 435, 472, and 294 mAh g^{-1}, respectively (Fig. 17.8). Comparing the discharge capacity of this cathode material with currently using cathodes, it is evident that the cathode with 21% precursor solution exhibits an enhanced electrochemical performance with high capacity retention (472 mAh g^{-1}) even at 50th cycle [54].

Hosono et al. [55] developed a triaxial LFP nanowires with multiwall carbon nanotubes (VGCF) core column along with an outer amorphous carbon shell. Through the electrospinning technique CNT core that gets oriented along the direction of the wire enhances the electronic conduction during the charge–discharge cycling. At the same time, the amorphous carbon present at the outer surface will suppress the oxidation of Fe^{2+}. The FE-SEM image of the fibers formed by the precursor solution without VGCF is shown in Fig. 17.9 which shows structure similar

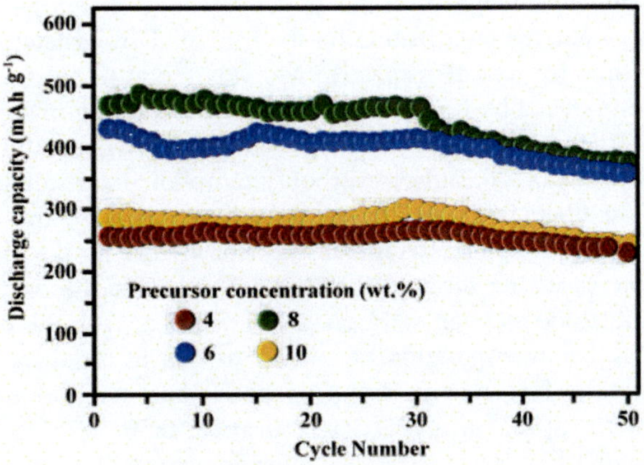

Fig. 17.8 Cycling performance of the LiF/Fe/C nanofiber composites prepared from precursor solutions with different concentrations. Adapted and reproduced from Ref. [54], Copyright 2012 IOP Science

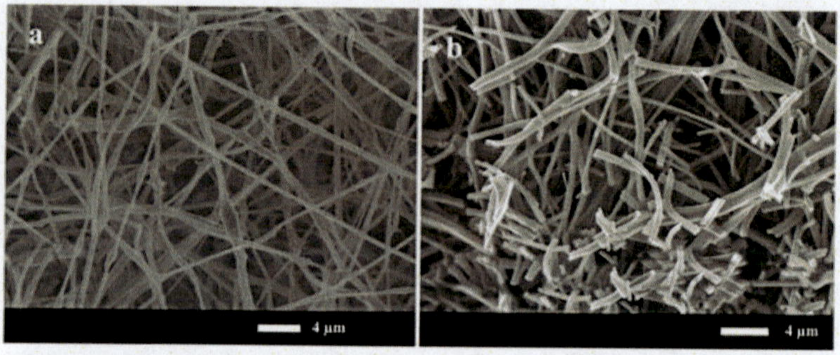

Fig. 17.9 FE-SEM images on the surface morphology of the **a** dried, **b** materials heated at 800 °C for 10 h. The precursor solution includes the poly acryl acid, and the sources of Li, Fe, and P without VGCF. Adapted and reproduced from Ref. [55], Copyright 2010 American Chemical Society

to the fibers formed by the precursor solution containing VGCF. The discharge capacity of triaxial nanowires of LFP with VGCF core column (a), triaxial nanowires of LFP with acetylene black (b), and LFP nanowires without VGCF (c) was compared [55].

17.4.2 Insertion Compounds of Lithium Iron Phosphate and Their Modifications

Compared to other olivine cathodes such as lithium cobalt phosphate ($LiCoPO_4$), lithium nickel phosphate ($LiNiPO_4$), LFP exhibits a low redox potential of 3.5 V that will lead to lower working voltage. This is due to the fact that, chemical potential of Fe^{2+}/Fe^{3+} is lower than that of Mn^{2+}/Mn^{3+}, Co^{2+}/Co^{3+} and Ni^{2+}/Ni^{3+} (4.1, 4.8, and 5.1, respectively) [56, 57]. Incorporation of manganese results an increment in the cell voltage from 3.5 to 4.1 V [58, 59]. Furthermore, on decreasing the iron content, the polarization for the iron oxidation/reduction was found to be decreased while for that of respective Mn was found to be increased. Structural change was introduced by Kang et al. [60] by incorporating manganese to the olivine structure. $Li[Fe_{1-x}Mn_x]PO_4$ nanofibers with varying Mn and Fe concentrations were prepared ($x = 0$, 0.1, 0.3); however, only morphological studies were reported in this study. The fibers prepared possess a diameter ranging from 100 to 500 nm and found that surface roughness is increased with increasing the manganese content. Roughness increases the surface area that will improve the electrochemical properties of the electrode and can set a good interface between the electrode and electrolyte [60]. Further incorporation of carbon to this material can enhance the battery performance. Hagen et al. [61] proposed in situ incorporation of carbon into the electrospun $Li[Fe_{1-x}Mn_x]PO_4$ ($x = 0$, 0.25, 0.50, 0.75, and 1) Fig. 17.10. The electrospun LFP was prepared using the commercially available precursors such as lithium hydroxide, iron Sulfate, manganese Sulfate, phosphoric acid, and PVP. After that,

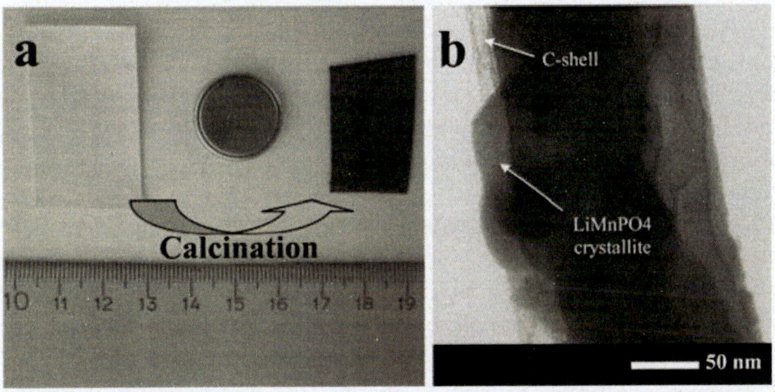

Fig. 17.10 **a** Image showing a nanofiber electrode before and after calcination and a CR2025 bottom cell, **b** at higher magnifications the homogeneous carbon coating of the $LiMnPO_4$ crystallites is resolved. Adapted and reproduced from Ref. [61]. Copyright 2012 John Wiley and Sons

these nanofibers were calcined at a temperature of about 850 °C under vacuum, which results in the formation of graphitic carbon to the final fibrous nanostructure. On incorporation of carbon, the resulting fiber formed consists of a porous conducting matrix that can secure the fibrous morphology as well as improve the electronic conductivity, specific capacity, and cyclability. The porous carbon matrix consists of uniformly dispersed $LiMnPO_4$ and the as-prepared nanofibers are capable to exhibit a high charge–discharge capacity 125 mAh g^{-1} even after 50 cycles [61] Moreover, the sample exhibits very low capacity fades at a current rate up to 4 C. Even at a discharge rate of C/2, charge–discharge capacity of system exhibits an increase in trend which is attributed by the structural stability of the nanofiber electrode and also there an increase in contact area of the conduction channel that result by the fusion of fiber connection during cycling which favors the electrochemical performance of the system. The significantly enhanced electrochemical performance is attributed to the highly porous conducting matrix formed around the fibers which consist of homogeneously distributed $LiMnPO_4$ crystallites.

A123 Systems Inc. and Sony introduced LFP batteries for the powering of electric vehicles due to its good electrochemical performance which made it significant for meeting the advanced application of lithium-ion batteries such as in electric vehicles. However, the system exhibits a low energy density which is about 578 Wh Kg^{-1} due to its low operating voltage. Insertion LFP cathodes are satisfactory for fabrication of battery modules to power the electric vehicles, if the researchers could find a solution to improve its cell voltage. Because of the effort taken in this direction by the battery technologists, a new system of cathode materials was extensively studied such as $LiMnPO_4$, which is capable to exhibit a high energy density of 700 Wh Kg^{-1} with an operating voltage of about 4.1 V vs. Li [62]. Again, lithium vanadium phosphate (LVP) are considered to be the similar class of material that exhibits a monoclinic structure with three-dimensional network and exhibit a good specific capacity, excellent thermal, and cyclic stability. The three-dimensional (3D) network structures provide an easy lithium-ion diffusion pathway and there by exhibit a theoretical reversible capacity of about 197 mAh g^{-1}. Similar to many other cathode materials like LFP, LMO, etc., the electrochemical performance of these cathodes are limited by its low electronic conductivity [63, 64]. Similar to LFP materials, in LVP also, for the enhancement of electrochemical properties, particle reduction and carbon coating are considered as the effective method. The LVP/C composite cathodes are observed to be exhibiting high electronic and ionic conductivity due to the presence of nanosized LVP particles that will enhance the lithium-ion kinetics and the carbon coating will provide good electronic conductivity [65]. For the LVP cathodes, the electrospunned fibrous structures are observed to be exhibiting enhanced electrochemical performance because the percolation behavior, which is observed to be much better in fibrous structures than the particles [66]. From these results, it is expected that the fibrous LVP/C composite can provide excellent electrochemical performance for LIBs. Chen et al. [66] developed LVP/C nanofiber composite and decorated as high performance cathode for LIBs. Using a typical precursor solution containing NH_4VO_3, $NH_4H_2PO_4$, and $CH_3COOLi.2H_2O$ in citric acid along with PVP was electrospun and prepared nanosized fibers having uniform morphology. The resulting composite fiber exhibited good cyclic stability and rate capability in a voltage range of 3.0–4.8 V. These composite fibers exhibited a discharge capacity 190 mAh g^{-1} at 0.1 C which corresponds to 99% theoretical capacity of LVP. It is interesting to note that the composite fiber is capable to exhibit high discharge capacity of about 132 mAh g^{-1} at a high discharge rate of 20 C along with better cycling performance. This superior electrochemical properties are attributed to the exceptionally high surface area offered by the nanosized fibers and enhanced electronic conductivity provided by carbon fibers [66]. Study over this fibrous LVP, Wei et al. [67] prepared similar one-dimensional and highly porous-structured LVP cathode material by adopting the simple electrospinning-assisted solid-state synthesis method. The precursor solution containing PVP was electrospunned first and the as-spun fibers are calcined at a temperature of 350 °C and then treated at a temperature of 800 °C under vacuum. This composite fibers exhibited an excellent electrochemical performance and can be used in LIBs at a potential range of 3.0–4.8 V. The cell studies on this porous composite fibrous electrode showed high reversible capacity of about 191 mAh g^{-1} at a discharge rate of 0.1 C and 110 mAh g^{-1} at 20 C. Eventhough the discharge capacity exhibited by this porous fibers are inferior compared to that of the LVP fibers developed by Chen et al. [66] or currently using cathode materials in commercial LIBs, the performance is excellent and promising [67]. Contrast to the usual methods of preparing electrospun LVP, ionic liquids were firstly introduced in electrospinning process by Peng et al. [68] that finally results to the formation of nanocubes embedded N-doped carbon nanofibers. The role of ionic liquids (ILs) in LIBs as an additive in electrolytes is extensively studied to improve the safety of the batteries [69–71], but it very rarely used in electrodes. ILs are considered to be important in this process because they enhance the feasibility of PAN based electrospinning process and facilitate the formation of self-supporting conducting network moreover

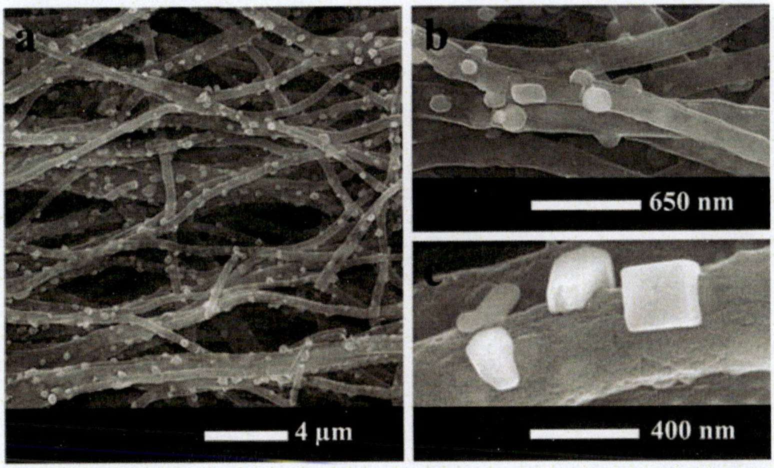

Fig. 17.11 Representative morphologies and structures of the as-electrospun LVP-NC/NCNF, **a** low, **b, c** high magnification. Adapted and reproduced from Ref. [68]. Copyright 2019 Royal Society of Chemistry

they can facilitate the formation of (100) facet oriented nanocubes by acting a structure directing agent. The presence of ILs can further enhance the electronic conductivity by forming a dual phase carbon coating layer. The precursor solution containing with PVP and PAN along with IL, 1-n-Butyl-3-methylimidazolium dihydrogen phosphate ([Bmim] H_2PO_4) was electrospun and annealed at a temperature of 800 °C to obtain LVP nanocubes embedded in N-doped carbon fibers (LVP-NC/NCNF). The electrochemical activity and electrochemical performance of the obtained nanofibers were compared with LVP nanoparticles/carbon nanofibers (LVP-NP/CNF) and LVP nanoparticles (LVP-NC). Investigating the morphology of the as-spunned fibers it is evident that the as-spunned fibers possess high aspect ratio and a diameter of about 500 nm and they are intertwined to form a mechanically stable free standing electrode (Fig. 17.11a, b). Apart from this, the LVP particles with a cubic structure having a side length of 200 nm are connected by the carbon fibers as presented in Fig. 17.12 LVP-NC/NCNF is capable to exhibit high electrochemical performance and is exhibited a discharge capacity of 155 mAh g^{-1} at 0.1 C which is significantly higher (>40–50 mAh g^{-1}) than that of LVP-NP/CNF and LVP-NC, which exhibits a discharge capacity of about 114.8 and 105.2 mAh g^{-1}, respectively. Also these LVP-NC/NCNF electrodes displayed excellent cycling stability and retained the discharge capacity even after 60 cycles [68]. Hence, in summary, it can be say that among the phosphate-based cathodes, LVP is one of the promising groups of cathodes that can ensure high electrochemical performance. Similar to LFP cathodes even it suffers low electronic conductivity metal ion doping, coating with conducting materials is considered to be effective methods to reach at the high performance. Moreover, electrospun LVP materials are considered to be the best due to the shorter diffusion

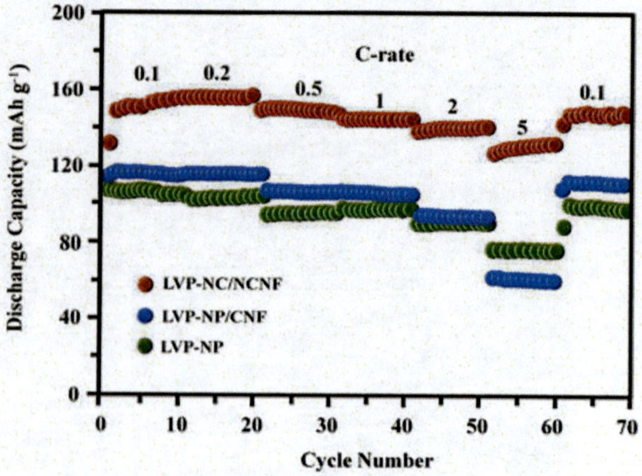

Fig. 17.12 Cycling performance at 1C, of the as-prepared LVP-NC/NCNF, LVP-NP/CNF, and LVP-NP, respectively. Adapted and reproduced from Ref. [68]. Copyright 2019 Royal Society of Chemistry

pathways, faster Li^+-ion intercalation that favors the electrochemical performance of the system.

17.5 Conclusion

LFP-based cathodes are considered to be the best for LIBs especially for the high temperature applications due to the high thermal stability and good electrochemical performance that it can deliver. The factor that limits the application of LFP is its low electronic conductivity that can be enhanced by coating with conducting materials such as carbon which will inhibit the oxidation of iron, metal doping, and controlling the size of particles LFP that will enhance the electronic conductivity there by improves the electrochemical performance. Electrospinning is opted as the best method for the preparation of best performing LFP cathodes because, it results the formation of nanofibrous structures that will have shorter lithium diffusion path, fast lithium intercalation kinetics, good mechanical strength, high surface area and porousity. During the electrospinning process, different structural modifications were adopted for LFP by incorporating different conducting additives such as CNTs and graphene. This lead to the formation of cathode material having good electronic property and it helps to overcome the issue related with the LFP cathode used.

References

1. Whittingham MS (2012) History, evolution, and future status of energy storage. Proc IEEE 100:1518–1534. https://doi.org/10.1109/JPROC.2012.2190170
2. Scrosati B, Hassoun J, Sun YK (2011) Lithium-ion batteries. A look into the future. Energy Environ Sci 4:3287–3295. https://doi.org/10.1039/c1ee01388b

3. Balakrishnan PG, Ramesh R, Prem Kumar T (2006) Safety mechanisms in lithium-ion batteries. J Power Sources 155:401–414. https://doi.org/10.1016/j.jpowsour.2005.12.002
4. Fergus JW (2010) Recent developments in cathode materials for lithium ion batteries. J Power Sour 195:939–954. https://doi.org/10.1016/j.jpowsour.2009.08.089
5. Liu D, Zhu W, Trottier J et al (2014) Spinel materials for high-voltage cathodes in Li-ion batteries. RSC Adv 4:154–167. https://doi.org/10.1039/c3ra45706k
6. Korthauer R (2018) Lithium-ion batteries: basics and applications. Lithium-Ion Batter Basics Appl 1–413. https://doi.org/10.1007/978-3-662-53071-9
7. Ates MN, Jia Q, Shah A et al (2014) Mitigation of layered to spinel conversion of a li-rich layered metal oxide cathode material for li-ion batteries. J Electrochem Soc 161:290–301. https://doi.org/10.1149/2.040403jes
8. Mizushima K, Jones PC, Wiseman PJ, Goodenough JB (1981) Li_xCoO_2 (0 < x ~ l): a new cathode material for batteries of high energy density k. Solid State Ionics 3–4:171–174. https://doi.org/10.1016/0167-2738(81)90077-1
9. News B Dreamliner: boeing 787 planes grounded on safety fears
10. News B Samsung confirms battery faults as cause of note 7 fires
11. Buisness Insier I Life, death and spontaneous combustion—here's why the debate about Tesla fires just got more fierce
12. Van Hassel BA, Montross CS (1996) Optimization of the composition of the $Li_{1-}Ni_{1+}O_2$ electrode materials: sfructural, magnetic, and electrochemical studies. Journl Electrochem Soc 143:1168–1175
13. Rougier A, Gravereau P, Delmas C (1993) Optimization of the composition of the LNO electrode materials: structural, magnetic, and electrochemical studies. Electrochem Soc Proc Ser Pennington, NJ 143:1168–1175. https://doi.org/10.1002/1521-4095(200107)13:12/13%3c943:AID-ADMA943%3e3.0.CO;2-J
14. Gowda SR, Gallagher KG, Croy JR et al (2014) Oxidation state of cross-over manganese species on the graphite electrode of lithium-ion cells. Phys Chem Chem Phys 16:6898–6902. https://doi.org/10.1039/c4cp00764f
15. Padhi AK (1997) Phospho-olivines as positive-electrode materials for rechargeable lithium batteries. J Electrochem Soc 144:1188. https://doi.org/10.1149/1.1837571
16. Yang S, Song Y, Zavalij PY, Stanley Whittingham M (2002) Reactivity, stability and electrochemical behavior of lithium iron phosphates. Electrochem Commun 4:239–244. https://doi.org/10.1016/S1388-2481(01)00298-3
17. Eftekhari A (2017) $LiFePO_4$/C nanocomposites for lithium-ion batteries. J Power Sour 343:395–411. https://doi.org/10.1016/j.jpowsour.2017.01.080
18. Huang X, Du Y, Qu P et al (2017) Effect of carbon coating on the properties and electrochemical performance of $LiFePO_4$/C composites by vacuum decomposition method. Int J Electrochem Sci 12:7183–7196. https://doi.org/10.20964/2017.08.77
19. Yang CC, Jang JH, Jiang JR (2015) Preparation of carbon and oxide co-modified $LiFePO_4$ cathode material for high performance lithium-ion battery. Mater Chem Phys 165:196–206. https://doi.org/10.1016/j.matchemphys.2015.09.018
20. Shao D, Wang J, Dong X et al (2013) Electrospinning fabrication and electrochemical properties of $LiFePO_4$/C composite nanofibers. J Mater Sci: Mater Electron 24:4263–4269. https://doi.org/10.1007/s10854-013-1395-8
21. Zhang Y, Xin P, Yao Q (2018) Electrochemical performance of $LiFePO_4$/C synthesized by sol-gel method as cathode for aqueous lithium ion batteries. J Alloys Compd 741:404–408. https://doi.org/10.1016/j.jallcom.2018.01.083
22. Wu G, Liu N, Gao X et al (2018) A hydrothermally synthesized $LiFePO_4$/C composite with superior low-temperature performance and cycle life. Appl Surf Sci 435:1329–1336. https://doi.org/10.1016/j.apsusc.2017.11.276
23. Guan XM, Li GJ, Li CY, Ren RM (2017) Synthesis of porous nano/micro structured $LiFePO_4$/C cathode materials for lithium-ion batteries by spray-drying method. Trans Nonferrous Met Soc China (English Ed) 27:141–147. https://doi.org/10.1016/S1003-6326(17)60016-5

24. Greiner A, Wendorff JH (2007) Electrospinning: a fascinating method for the preparation of ultrathin fibers. Angew Chemie Int Ed 46:5670–5703. https://doi.org/10.1002/anie.200604646
25. Kalluri S, Seng KH, Guo Z et al (2013) Electrospun lithium metal oxide cathode materials for lithium-ion batteries. RSC Adv 3:25576–25601. https://doi.org/10.1039/c3ra45414b
26. Wu H, Cui Y (2012) Designing nanostructured Si anodes for high energy lithium ion batteries. Nano Today 7:414–429
27. Liang X, Yang Y, Jin X et al (2015) The high performances of SiO2/Al_2O_3-coated electrospun polyimide fibrous separator for lithium-ion battery. J Memb Sci 493:1–7. https://doi.org/10.1016/j.memsci.2015.06.016
28. Bachtin K, Kaus M, Pfaffmann L et al (2016) Comparison of electrospun and conventional $LiFePO_4$/C composite cathodes for Li-ion batteries. Mater Sci Eng B Solid State Mater Adv Technol 213:98–104. https://doi.org/10.1016/j.mseb.2016.04.006
29. Wang L, Ding CX, Zhang LC et al (2010) A novel carbon-silicon composite nanofiber prepared via electrospinning as anode material for high energy-density lithium ion batteries. J Power Sour 195:5052–5056. https://doi.org/10.1016/j.jpowsour.2010.01.088
30. Gao H, Chen Y, Sun H, et al (2018) Silicone modified polypropylene separator for high temperature lithium ion battery applications. Mater Res Exp 5. https://doi.org/10.1088/2053-1591/aacb10
31. Shi C, Zhang P, Chen L et al (2014) Effect of a thin ceramic-coating layer on thermal and electrochemical properties of polyethylene separator for lithium-ion batteries. J Power Sour 270:547–553. https://doi.org/10.1016/j.jpowsour.2014.07.142
32. Miao J, Miyauchi M, Simmons TJ et al (2010) Electrospinning of nanomaterials and applications in electronic components and devices. J Nanosci Nanotechnol 10:5507–5519. https://doi.org/10.1166/jnn.2010.3073
33. Zhang X, Ji L, Toprakci O et al (2011) Electrospun nanofiber-based anodes, cathodes, and separators for advanced lithium-ion batteries. Polym Rev 51:239–264. https://doi.org/10.1080/15583724.2011.593390
34. Nan C, Lu J, Li L et al (2013) Size and shape control of $LiFePO_4$ nanocrystals for better lithium ion battery cathode materials. Nano Res 6:469–477. https://doi.org/10.1007/s12274-013-0324-8
35. Park Y, Shin W, Lee JW (2012) Synthesis of hollow spherical $LiFePO_4$ by a novel route using organic phosphate. Cryst Eng Comm 14:4612–4617. https://doi.org/10.1039/c2ce25159k
36. Lim S, Yoon CS, Cho J (2008) Synthesis of nanowire and hollow $LiFePO_4$ cathodes for high-performance lithium batteries. Chem Mater 20:4560–4564. https://doi.org/10.1021/cm8006364
37. Gu Y, Chen D, Jiao X (2007) Synthesis and characterization of hollow $LiNiO_2$ fibers via sol-electrospinning method. J Sol-Gel Sci Technol 43:245–249. https://doi.org/10.1007/s10971-007-1572-4
38. Qiu Y, Geng Y, Li N et al (2014) Nonstoichiometric $LiFePO_4$/C nano fibers by electrospinning as cathode materials for lithium-ion battery. Mater Chem Phys 144:226–229. https://doi.org/10.1016/j.matchemphys.2013.12.027
39. Kang CS, Kim C, Park TJ, Son JT (2012) Morphology study of new shaped porous nanotubes of $LiFePO_4$ by electrospinning. Chem Lett 41:1428–1429. https://doi.org/10.1246/cl.2012.1428
40. Lin HY, Yeh SM, Chen JS (2014) Physical and electrochemical properties of $LiFePO_4$/C nanofibers synthesized by electrospinning. Int J Electrochem Sci 9:6936–6948
41. Hu Y, Gu D, Jiang H et al (2016) Electrochemical performance of $LiFePO_4$/C via coaxial and uniaxial electrospinning method. Adv Chem Eng Sci 06:149–157. https://doi.org/10.4236/aces.2016.62017
42. Zhu C, Yu Y, Gu L et al (2011) Electrospinning of highly electroactive carbon-coated single-crystalline $LiFePO_4$ nanowires. Angew Chemie Int Ed 50:6278–6282. https://doi.org/10.1002/anie.201005428
43. Lee SH, Jung MJ, Im JS et al (2010) Preparation and characterization of electrospun $LiFePO_4$/carbon complex improving rate performance at high C-rate. In: Research on chemical intermediates, pp 591–602

44. Toprakci O, Ji L, Lin Z et al (2011) Fabrication and electrochemical characteristics of electrospun $LiFePO_4$/carbon composite fibers for lithium-ion batteries. J Power Sour 196:7692–7699. https://doi.org/10.1016/j.jpowsour.2011.04.031
45. Bhuvaneswari MS, Bramnik NN, Ensling D et al (2008) Synthesis and characterization of carbon nano fiber/$LiFePO_4$ composites for Li-ion batteries. J Power Sour 180:553–560. https://doi.org/10.1016/j.jpowsour.2008.01.090
46. Zhou Y, Wang J, Hu Y et al (2010) A porous $LiFePO_4$ and carbon nanotube composite. Chem Commun 46:7151–7153. https://doi.org/10.1039/c0cc01721c
47. Divakar D, Manikandan D, Sivakumar T (2008) Vapor-phase selective hydrogenation of citral over Pd/bentonite : effect. Society 1478:1472–1478. https://doi.org/10.1002/jctb
48. Katsnelson MI (2007) Graphene: carbon in two dimensions. Mater Today 10:20–27. https://doi.org/10.1016/S1369-7021(06)71788-6
49. Ebbsen TW, Lezec HJ, Hiura H, Bennett JW, Ghaqemi HF Thio T (1996) Electrical conductivity of individual carbon nanotubes. Lett to Nat 382
50. Baughman RH, Zakhidov AA, De Heer WA (2002) Carbon nanotubes—the route toward applications. Science 297(80):787–792. https://doi.org/10.1126/science.1060928
51. Toprakci O, Toprakci HAK, Ji L et al (2012) Carbon nanotube-loaded electrospun $LiFePO_4$ carbon composite nanofibers as stable and binder-free cathodes for rechargeable lithium-ion batteries. ACS Appl Mater Interfaces 4:1273–1280. https://doi.org/10.1021/am201527r
52. Toprakci O, Toprakci HAK, Ji L et al (2012) $LiFePO_4$ nanoparticles encapsulated in graphene-containing carbon nanofibers for use as energy storage materials. J Renew Sustain Energ 4. https://doi.org/10.1063/1.3690936
53. Badway F, Pereira N, Cosandey F, Amatucci GG (2003) Carbon-metal fluoride nanocomposites. J Electrochem Soc 150:A1209. https://doi.org/10.1149/1.1596162
54. Zhang S, Lu Y, Xu G, et al (2012) LiF/Fe/C nanofibres as a high-capacity cathode material for Li-ion batteries. J Phys D Appl Phys 45. https://doi.org/10.1088/0022-3727/45/39/395301
55. Hosono E, Wang Y, Kida N et al (2010) Synthesis of triaxial $LiFePO_4$ nanowire with a VGCF core column and a carbon shell through the electrospinning method. ACS Appl Mater Interfaces 2:212–218. https://doi.org/10.1021/am900656y
56. Zhou F, Cococcioni M, Kang K, Ceder G (2004) The Li intercalation potential of $LiMPO_4$ and $LiMSiO_4$ olivines with M = Fe, Mn Co, Ni. Electrochem Commun 6:1144–1148. https://doi.org/10.1016/j.elecom.2004.09.007
57. Wolfenstine J, Allen J (2005) Ni^{3+}/Ni^{2+} redox potential in $LiNiPO_4$. J Power Sour 142:389–390. https://doi.org/10.1016/j.jpowsour.2004.11.024
58. Jiang Y, Liu R, Xu W et al (2013) A novel graphene modified $LiMnPO_4$ as a performance-improved cathode material for lithium-ion batteries. J Mater Res 28:2584–2589. https://doi.org/10.1557/jmr.2013.235
59. Hong J, Wang F, Wang X, Graetz J (2011) $LiFe_xMn_{1-x}PO_4$: a cathode for lithium-ion batteries. J Power Sour 196:3659–3663. https://doi.org/10.1016/j.jpowsour.2010.12.045
60. Kang CS, Kim C, Kim JE et al (2013) New observation of morphology of $Li[Fe_{1-x}Mn_x]PO_4$ nano-fibers (x = 0, 0.1, 0.3) as a cathode for lithium secondary batteries by electrospinning process. J Phys Chem Solids 74:536–540. https://doi.org/10.1016/j.jpcs.2012.11.010
61. Von Hagen R, Lorrmann H, Möller KC, Mathur S (2012) Electrospun $LiFe_{1-y}Mn_yPO_4$/C nanofiber composites as self-supporting cathodes in Li-ion batteries. Adv Energ Mater 2:553–559. https://doi.org/10.1002/aenm.201100534
62. Aravindan V, Gnanaraj J, Lee YS, Madhavi S (2013) $LiMnPO_4$—a next generation cathode material for lithium-ion batteries. J Mater Chem A 1:3518–3539. https://doi.org/10.1039/c2ta01393b
63. Xu J, Chou SL, Zhou C et al (2014) Three-dimensional-network $Li_3V_2(PO_4)_3$/C composite as high rate lithium ion battery cathode material and its compatibility with ionic liquid electrolytes. J Power Sour 246:124–131. https://doi.org/10.1016/j.jpowsour.2013.07.055
64. Saïdi MY, Barker AJ, Huang H, Swoyer J (2002) Properties of lithium vanadium phosphate as a cathode material for lithium-ion batteries. Transition 2:A149–A151

65. Mao WF, Yan J, Xie H et al (2013) Short communication an in-situ copolymerization synthesis of $Li_3V_2(PO_4)_3/C$ nanocomposite with ultrahigh discharge capacity. J Power Sour 237:167–171. https://doi.org/10.1016/j.jpowsour.2013.03.039
66. Chen Q, Zhang T, Qiao X et al (2013) $Li_3V_2(PO_4)_3/C$ nanofibers composite as a high performance cathode material for lithium-ion battery. J Power Sour 234:197–200. https://doi.org/10.1016/j.jpowsour.2013.01.164
67. Wei S, Yao J, Shi B (2017) 1D highly porous $Li_3V_2(PO_4)_3/C$ nanofibers as superior high-rate and ultralong cycle-life cathode material for electrochemical energy storage. Solid State Ionics 305:36–42. https://doi.org/10.1016/j.ssi.2017.04.019
68. Peng Y, Tan R, Ma J et al (2019) Electrospun $Li_3V_2(PO_4)_3$ nanocubes/carbon nanofibers as free-standing cathodes for high-performance lithium-ion batteries. J Mater Chem A 7:14681–14688. https://doi.org/10.1039/c9ta02740h
69. Yamagata M, Tanaka K, Tsuruda Y et al (2015) The first lithium-ion battery with ionic liquid electrolyte demonstrated in extreme environment of space. In: Electrochemistry, pp 918–924
70. Ishikawa M, Sugimoto T, Kikuta M et al (2006) Pure ionic liquid electrolytes compatible with a graphitized carbon negative electrode in rechargeable lithium-ion batteries. 162:658–662. https://doi.org/10.1016/j.jpowsour.2006.02.077
71. Liu H, Yu H (2019) Ionic liquids for electrochemical energy storage devices applications. J Mater Sci Technol 35:674–686

Chapter 18
Electrospun Fibrous Vanadium Pentoxide Cathodes for Lithium-Ion Batteries

N. S. Jishnu, Neethu T. M. Balakrishnan, Akhila Das, Jou-Hyeon Ahn, M. J. Jabeen Fatima, and Raghavan Prasanth

18.1 Introduction

The main concern with the renewable energy is their proper storage that is essential to establish a proper energy supply. An efficient and cost-effective storage is the next hurdle that has to be overcome to meet the effective utilization and storage of renewable energy such as solar, wind and tidal energy. Among different energy storage systems such as capacitors, supercapacitors, fuel cells and batteries, batteries are important and play a crucial role. They are the essential components of most of the electronic devices, and portability is the fact that makes batteries important over other energy storage devices. Most of the portable electronic devices are making use of the lithium-ion battery; it is prime important due to high energy density as well as power density. The minimum self-discharge, lack of memory effect and excellent storage capacity are the important factors that make LIBs important over the other battery technology. Even though perfect energy storage is always a serious concern, nanostructured materials especially nanofibers are promising candidate for the development of high-performance energy storage devices, especially in LIB systems. Different methods such as wet chemical synthesis [1], sol–gel templating

N. S. Jishnu
Rubber Technology Centre, Indian Institute of Technology-Kharagpur (IIT-KGP), Kharagpur, West Bengal 721302, India
e-mail: jishnuns1995@gmail.com

N. T. M. Balakrishnan · A. Das · M. J. Jabeen Fatima · R. Prasanth (✉)
Department of Polymer Science and Rubber Technology, Cochin University of Science and Technology (CUSAT), Cochin 682022, India
e-mail: prasanth@cusat.ac.in

J.-H. Ahn · R. Prasanth
Department of Materials Engineering and Convergence Technology, Gyeongsang National University, 501 Jinju-daero, Jinju 52828, Republic of Korea

N. T. M. Balakrishnan and R. Prasanth (eds.), *Electrospinning for Advanced Energy Storage Applications*, Materials Horizons: From Nature to Nanomaterials,
https://doi.org/10.1007/978-981-15-8844-0_18

[2], self-assembly [3], molten salt method [4], precursor methods [5] and electrospinning [6] are introduced for the synthesis of nanofibrous structures. Among these different methods of preparation, electrospinning is one of the versatile, simple and cost-effective techniques that can produce fibrous structures with different morphologies like core shell [7], hollow [8], multilayer [9] and porous [10] structures. Apart from the capability to fabricate the nano- to submicron-sized fibers having unique morphologies, electrospinning leads to the development of fibers having high surface area to volume ratio, ease of using material combinations and ease of fiber functionalization as well a mass production capability [11, 12]. In energy storage devices, electrospunned electrode and electrolyte materials are considered as the ideal candidate for high-performance batteries since it possesses large surface area, tunable properties, different material compositions and high aspect ratio.

The type of electrode materials plays a crucial role on the electrochemical performance such as output voltage, charge–discharge capacity and cycling stability rate capability of the LIBs. Transition metal oxides seek greater interest in recent studies in both their fundamental and technological aspects. The advances of lithium-ion battery come from the choice of suitable intercalation material as cathode. The electrochemical reaction that involves insertion and extraction of Li^+-ions from the cathode matrix without collapsing its structure determines the performance of the battery. The intercalation anodes such as graphite ensure the proper intercalation of ions without the formation of dendrite, which was the severe problem faced while using lithium metal anodes. While in cathode materials, transition metal oxide-based cathodes that possess layered, spinal and olivine structures are superior. In the long journey on hunting of high-capacity cathode materials, transition metal oxide gained great attention due to its ability to react with more than one lithium-ion per redox ion. Initially, the commercialized lithium-ion battery used $LiCoO_2$ (LCO) as cathode [13, 14], but due to the toxicity of cobalt as well as its low thermal stability leads to the usage of MnO_2-based cathodes [15, 16]. Still, the problem was there within this cathode material that causes structural changes during the cycling that result by the Jahn–Teller distortion effect. For thermally stable LIBs, $LiFePO_4$ (LFP) was a promising candidate since it can stable even at high temperature. But the low electronic conductivity makes issues with the use of LFP [17]. Recently, layered structured vanadium pentoxide-based cathodes are the most studied cathode materials due to its ability to form variety of possible structures. It was firstly reported by Wittingham in 1976 [18, 19]. The low electronic conductivity and low lithium-ion diffusion restricted its application in LIBs, but the issues and challenges associated with V_2O_5 were successfully resolved by fabricating nanostructured materials via different methods such as sol–gel synthesis [20], hydrothermal synthesis [21], electrospraying [22] and electrospinning [23]. Among these methods, electrospinning takes up a significant impact that comes from the porous nanostructures. Because of the unique structure and high surface area, electrospun nanofibers exhibit excellent electrochemical properties. Electrospunned nano-entities exhibit one-dimensional (1D) fibrous, rode or tube-like structures and thereby enhance the Li^+-ions diffusion rate. Different nanostructures 1D structure such as V_2O_5 and Nb_2O_5 [24] were extensively studied that can effectively replace the current cathodes in LIBs in order to ensure high electrochemical properties [25]. Among these, V_2O_5 nanofibers seek major attention.

18.2 Synthesis of Electropsun Vanadium Pentoxide Cathode by Using Different Precursors

The tunable oxidation state and intercalation property make V_2O_5 a viable alternative active material for cathodes in lithium-ion battery. The layered structure of V_2O_5 results the reversible intercalation of the Li^+-ions, and the tunable oxidation state results the structural variation in the cathode which is beneficial for the battery performance. But the problem with this cathode material was its low structural stability and cyclic stability that limited its practical application in LIBs. Electrospinning is proposed as an alternative method to produce the V_2O_5-based cathode with improved electrochemical performance. Electrospinning results the formation of nanostructures, especially very long seamless nanofibrous structures. It can employed for the fabrication of polymer metal oxide, ceramic or carbon fibers. In the method of electrospinning, suitable precursor solution is continuously drawn as a charged thread from a spinneret (Fig. 18.1) by means of an electrostatic force which possesses the combinational characteristics of electrospraying and dry spinning or dry jet wet spinning.

For the electrospinning of V_2O_5, different precursor solutions such as vanadyl acetyl acetonate or simple vanadium pentoxide powder dissolved or uniformly dispersed in a suitable polymer solution like polymethyl methacrylate (PMMA), polyvinyl alcohol (PVA), polyvinyl pyrrolidone (PVP), etc., can be used. Among different spinning parameters such as solution concentration, applied voltage and feed rate that determine the properties of resultant nanofiber structure, the precursors used are also impart vital role in determining the membrane morphology. Vanadium sol prepared by using vanadium oxide isoproxide as precursor followed by mixing with polyvinyl acetate was electrospunned, and heat treatment results in the formation of highly crystalline V_2O_5fibrous nanostructures [26]. Vanadium oxitriisopropoxide precursor dissolved in PMMA solution was electrospunned, and hydrothermal reaction over this as-spun fibers results in the nanoscale vanadium oxide having a layered structure with interlayer distance of 11.4 Å [27]. Same vanadium oxitripropoxide is used as the precursor to synthesize the one-dimensional (1D) nanofibers consisting of nanostructured particle from the sol–gel solution of the precursor [28]. Commercially available V_2O_5 powders can also be successfully employed as the precursor for the development of V_2O_5-based fibrous cathodes. This system shows an enhanced rate capability and superior long-term charge–discharge cycling stability. The advantage of using commercially available V_2O_5 is fact of cost-effectiveness. The synthesis results in the formation of the nanofibers made up of nanoplates [29]. Synthesis of

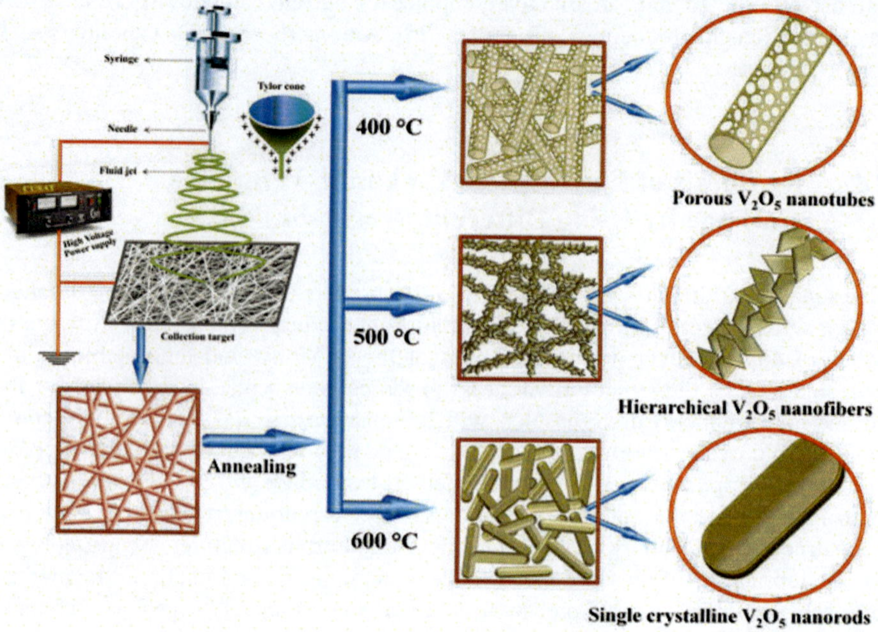

Fig. 18.1 Schematic representation of electrospinning and effect of annealing temperature over V_2O_5 nanostructures

electrospun nanofibers using vanadium acetyl acetonate precursor followed by calcination results in fibrous cathode material in which the effect of doping substantially increased the electrochemical performance [30]. Among different polymers, polymethyl methacrylate is successfully used for the synthesis of V_2O_5 nanofibers along with suitable vanadium oxides or other vanadium-based compounds [9]. Polyvinyl acetate (PVAc) [31] is another polymer-based precursor that is used along with vanadium oxitripropoxide in order to synthesis V_2O_5 cathode for LIBs. Polyvinyl alcohol [32] also used along with the vanadium compounds. The polymer PVP is also used along with the commercially available V_2O_5 in order to form a uniform precursor for the production of V_2O_5 nanofibers by sol–gel and electrospinning method. The type of precursor solution and method of preparation along with synthesis procedures and annealing temperature plays a vital role in the properties of resulting V_2O_5 nanofibers. The use of different precursors forms different fiber morphologies (Fig. 18.2), and this is the beauty and versatility of electrospinning [33].

18.3 Structure and Property of Electrospun V_2O_5

Vanadium pentoxide (V_2O_5), yellow/brown solid (Fig. 18.4b) with a layered crystalline structure, is a promising cathode material for lithium-ion batteries (LIBs).

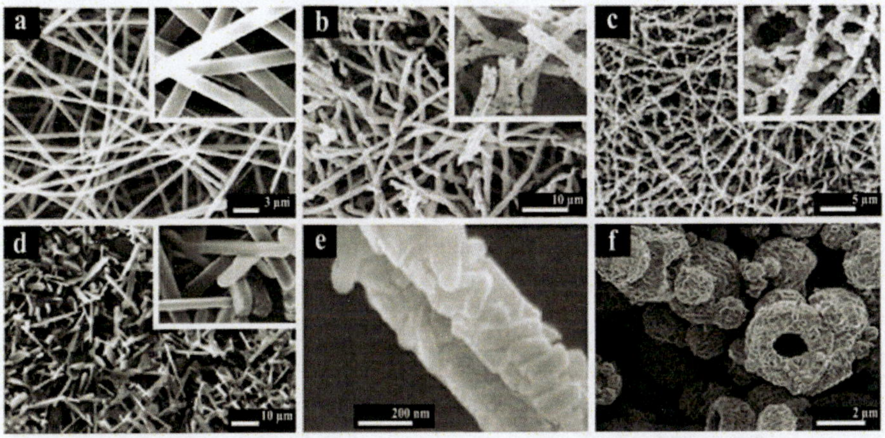

Fig. 18.2 SEM images on the surface morphology with inset scales of 500 nm of **a** as-spun fibers, **b** porous nanofibers calcined at 400 °C, **c** hierarchical nanofibres calcined at 500 °C, **d** single-crystalline nanofibres calcined at 600 °C, **e** electrospun hierarchical fibers composed of V_2O_5 nanorods, **f** Spherical hollow hierarchical V_2O_5 structures. Adapted and reproduced from Ref. [33]. Copyright 2018 Elsevier

Vanadium was discovered by Andrés Manuel del Río, a Spanish-Mexican mineralogist, in 1801 and is the 20th most abundant element in the earth's crust. Vanadium-oxygen system is a widely studied material that exhibits 13 distinct phases, and each one of the systems is similar in their lattice structure and spacing. Different oxides of vanadium exhibit different properties and many of vanadium-oxygen phases are non-stoichiometric. Different types of vanadium oxides are vanadium (II) oxide (vanadium monoxide, VO), vanadium (III) oxide (vanadium sesquioxide or trioxide, V_2O_3), vanadium (IV) oxide (vanadium dioxide VO_2) and vanadium (V) oxide (vanadium pentoxide, V_2O_5). In addition to these principal oxides of vanadium, various other distinct phases also exist:

(i) Phases with the general formula V_nO_{2n+1} exist between V_2O_5 and VO_2; examples of these phases include V_3O_7, V_4O_9 and V_6O_{13}.
(ii) Phases with the general formula V_nO_{2n-1} exist between VO_2 and V_2O_3; examples of these phases include V_4O_7, V_5O_9, V_6O_{11}, V_7O_{13} and V_8O_{15}. Called Magnéli phases and are examples of crystallographic shear compounds based on the rutile structure.
(iii) V_3O_5 appears as the mineral oxyvanite

Among these principal oxides of vanadium, V_2O_5 attracted much attention due to its large spacing and open layered crystal structure (Figs. 18.3 and 18.4a) [26, 34]. V_2O_5 consist of both V^{4+} and V^{5+} ions. It has an orthorhombic cubic crystal structure with the lattice parameters of $a = 11.54$ Å, $b = 3.571$ Å and $c = 4.383$ Å with space group Pmmn [35]. The structure of V_2O_5 is made up of pyramidal structures that are formed by oxygen and vanadium atoms which form the most stable oxide structure of vanadium. The stacked pyramidal structured V_2O_5 shares edge to form

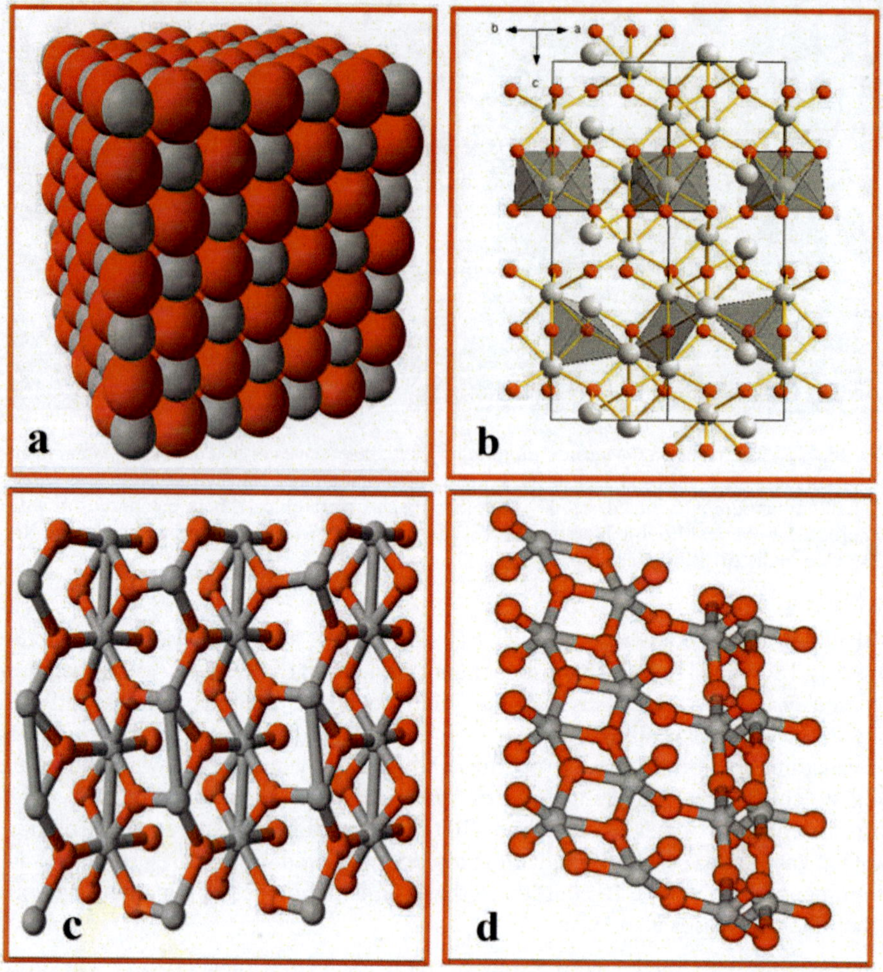

Fig. 18.3 Structural representation of V_2O_5 (where white = vanadium and red = oxygen), **a** close packed structure, **b** 3D structure, **c** crystal structure, **d** monolayer

zigzag double chains. These layers get bonded over the z-direction or along the (001) direction. The bonding over the layer is through the weak van der Waals bond that is formed between the oxygen and vanadium atoms of adjacent layered structure [36, 37].

When vanadium pentoxide intercalated by water, it forms a V_2O_5.n H_2O xerogels that consist of bilayer V_2O_5 with water in between them. This structure results a change in vanadyl position that allows the layers to get close to each other [38].

Currently using cathode materials such as $LiCoO_2$ (LCO), $LiFePO_4$ (LFP), $LiNiMnCoO_2$ (NMC) and $LiMn_2O_4$ (LMO) deliver a capacity value which is less than 200 mAh g^{-1}; hence, the development of superior cathode materials which

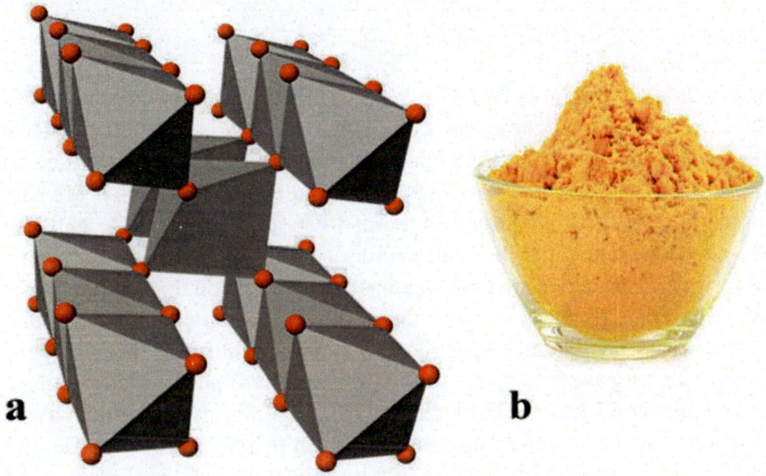

Fig. 18.4 **a** 3D structural representation of V_2O_5, **b** V_2O_5 Powder

have higher practical capacity is the prime importance to fabricate LIBs that can possess high energy density and power density. The choice of metal oxide that can provide multi-electrons redox reaction is important, since they can insert and extract multiple electrons which can enhance the reaction rate and thereby improve the specific capacity of the LIBs. By using such lithium-free oxide structure, e.g., V_2O_5, the problem faced was the compulsory usage of these metal oxides along with lithium metal anode in order to provide the required Li^+-ions. But there was a serious problem with the use of lithium metal anode called dendrite formation. The formation and growth of finger-like dendrite over the lithium anode cause short circuit that can lead to the battery explosion [39, 40]. Hence, the safety of the lithium cells is the main concern while making use of lithium metal as anode in LIBs. Latter, this problem was rectified with the use of ionic liquids as well as the use of lithium alloy such as Li–Al alloy [41]. V_2O_5 is a best choice due to its open layered structure that can facilitate the easy intercalation and deintercalation of Li^+-ions as well all that can accommodate more Li^+-ions, which provide a high specific capacity value in the potential range of 4.0–2.0 V (vs. Li/Li^+). V_2O_5 possesses theoretical capacities of 442 mAh g^{-1} for three Li^+ intercalations, or 294 mAh g^{-1} for two Li^+ intercalations per the molecular formula [42] which are much larger than those of traditional cathode materials, such as $LiCoO_2$ (140 mAh g^{-1}) [43] $LiMn_2O_4$ (148 mAh g^{-1}) [44] and $LiFePO_4$ (170 mAh g^{-1}) [45]. The natural abundance (20th most abundant element in the earth's crust), low cost, ease of synthesis and processing further make it suitable as a cathode material in LIBs [46–48]. As mentioned earlier, V_2O_5 as a cathode material exhibits a multiple electron intercalation and easy to intercalate and deintercalation of Li^+-ions into their layered structure. The lithium intercalation and deintercalation reaction in Li/V_2O_5 half-cell is expressed as:

$$V_2O_5 + xLi^+ + xe^- \rightarrow Li_xV_2O_5$$

However, several problems largely restrict the battery performance of V_2O_5, such as small Li^+-ion diffusion coefficient, poor lithium diffusion kinetics, low electronic conductivity (1×10^{-2} to 1×10^{-3} S cm^{-1}), irreversible phase transitions, instability in structure and dissolution of vanadium into the electrolyte which are the main issues associated with this cathode material [49]. Therefore, V_2O_5 exhibits poor discharge capacity, rate capability and cycling stability. In order to overcome these issues, a variety of strategies are employed and demonstrated that the nanostructured V_2O_5 materials can effectively enhance the electrochemical performance of the V_2O_5 as a cathode material. Nanostructured V_2O_5 can shorten the diffusion length of Li^+-ions by reducing particle sizes, facilitate electrolyte penetration by generating porous structures and buffer volume changes during lithium insertion/extraction by free voids. One of the novel strategies adopted is to integrate cation doping with the V_2O_5 nanostructures. This design strategy not only has the advantages of the V_2O_5 nanomaterials, but also increases the intrinsic electric conductivity of V_2O_5. In addition, it is significant to modify the nanostructured V_2O_5 with conductive additives, such as carbon materials and conductive polymers (coating on V_2O_5 or mixing with carbon nanomaterials or conducting polymers). Both conducting carbon and polymers can enhance the electric conductivity on the V_2O_5 surface and prevent particle aggregation. The polymer coating also works as a protective surface layer which prevents the direct contact of V_2O_5 with the electrolyte.

A variety of nanostructured V_2O_5 cathode materials was synthesized to improve the cycling performance and the rate capability of LIBs. The novel nanostructures of V_2O_5 cover zero-dimensional (0D) nanoparticles [50], one-dimensional (1D) nanorods [51], nanoribbons, [52], nanowires [35], nanotubes [49], two-dimensional (2D) nanosheets [42] and three-dimensional (3D) hierarchical architectures [52]. Porous carbon [53], carbon nanotube [54–57] and graphene [58–63] are the widely used carbonaceous materials with V_2O_5. Nanobelts [61, 64–66] nanowires [35, 55, 67, 68] and nanotubes [69] are the most studied 1D nanostructure for V_2O_5. 3D nanostructures such as nanoflowers [70], nanofibrous membrane [32] and 3D porous V_2O_5 microspheres [71] are also reported. Other 1D nanostructures, such as nanorods [72], nanotubes [69] and porous V_2O_5 nanofibers [73], were also investigated.

V_2O_5 nanoparticles (size ranging from 30 to 60 nm) were prepared by a one-step flame spray pyrolysis process [50], hollow nanospheres (diameter of around 28 nm) by a sol–gel method with polymeric micelles as a soft template [74] having good cycling performance, and rate capability is reported. The V_2O_5 nanoparticles showed excellent cycling performance by keeping a specific capacity of 110 mAh g^{-1} after 100 cycles at a current rate of 10 C and a superior high rate capacity of 100 mAh g^{-1} at 20 C. Hybrid nanocomposites can offer advanced performance that is not possessed by an individual component. Hence, the V_2O_5 nanostructures composited with CNTs by filling the V_2O_5 nanoparticles in the interior of CNTs [57], coating on the surface of CNTs [75], or intertwining ultra-long CNTs and V_2O_5 nanowires [55] having improved electrochemical properties and charge–discharge cycling property along with good rate capability were studied. The intertwining ultra-long CNTs and V_2O_5

nanowires were prepared as freestanding and highly robust electrode materials for flexible LIBs, which are promising in wearable devices and roll-up displays [55]. The intertwining electrode structure showed high rate capability and excellent cycling stability. The MWCNT/V_2O_5 core/shell sponge [75] carbon tube-in-tube (CTIT) (Fig. 18.5) [56] structure also reported. The areal capacity of the MWCNT/V_2O_5 sponge was as high as 816 μAh cm^{-2} at 1 C rate (1.1 mA cm^{-2}) in a voltage range of 4.0–2.1 V, which was 450 times that of a planar V_2O_5 thin film. At a high

Fig. 18.5 TEM image of the V_2O_5/carbon tube-in-tube composite, where most of the V_2O_5 nanoparticles were encapsulated within carbon tube-in-tube. Adapted and reproduced from Ref. [56]. Copyright 2009 John Wiley and Sons

current rate of 50 C (55.9 mA cm^{-2}), this composite preserved the areal capacity of 155 μAh cm^{-2}, leading to a high power density of 21.7 mW cm^{-2}.

V_2O_5 nanowires were homogeneously integrated with graphene sheets to form a robust V_2O_5 nanowire–graphene composite paper via a simple vacuum filtration process [60] and rGO-supported porous V_2O_5 spheres (V_2O_5/rGO) with diameters of 10–50 nm were grown on the surfaces of rGO sheets in a facile solvothermal process [61], hierarchical nanocomposite of V_2O_5 nanosheets self-assembled on rGO sheets was prepared by a solvothermal method [58] as a high-power cathode material for LIBs are also reported. V_2O_5 nanowires prepared by a hydrothermal treatment process and graphene nanosheets exfoliated by a modified Hummer's method is used to form a robust V_2O_5 freestanding electrode. The uniform mixture of graphene and V_2O_5 nanowires was filtered to obtain a V_2O_5 nanowire–GO composite paper, in which V_2O_5 nanowires homogeneously integrated with GO. This free-standing nanocomposite electrode exhibited extremely high cycling stability and high rate performance. When measured at a current density of 10,000 mAg^{-1}, the V_2O_5 nanowire–graphene composite cycled up to 100,000 cycles with a stable specific capacity of 94.4 mAh g^{-1}. The extremely long cycle numbers are two or three orders of magnitude larger than those of conventional electrode materials, suggesting that highly organized nanostructured electrodes can address the poor cycling issues [60]. In this entire hybrid V_2O_5 composite with carbon nanotubes or graphene, the voids within the interpenetrative network facilitate the easy penetration of electrolyte and facilitate the space for volume expansion and contraction of the electrode during lithium intercalation and deintercalation, thereby improving the cycling stability and rate capability of LIBs. There are three conducting polymers widely used in developing V_2O_5-based cathode materials, such as polypyrrole (PPy) [76–78], polyaniline (PANI) [79–81] and polythiophene [82–85]. The electrochemical performance of these hybrid materials is summarized in Table 18.1. However, the methods to prepare these nanostructured materials are not simple, easy or cost effective, and many of the cases need high pressure or vacuum. For example, the hydrothermal or solvothermal method required expensive autoclaves which have serious safety issues and the impossibility of observing the crystal as it grows if a steel tube is used [86, 87].On the other hand, the sol–gel method is a very popular fabrication technique, but the purity of the materials in this technique is greatly substrate dependent [88, 89]. Again, this technique has little control over porosity of the gel which in turn affects the rate of solvent removal from the gel in order to form the final powder. Moreover, formation of gel is a slow process, which makes sol–gel a time-consuming fabrication technique, and post-preparation heating is required as compared to other methods. Also, it needs very high-temperature furnace or heating device, and metal alkoxides are the most preferred precursor, which make this technique more expensive. To get the material completely in crystalline form, we have to do the synthesis at very high temperature; otherwise, the resulting material in amorphous form along with considerable amount of hydroxides, defects and phase composition may vary from batch to batch. Moreover, this technique requires the use of significant quantities of citric acid and ethylene glycol, but produces only a small volume of product relative to the precursor sol–gel [82, 83].

Table 18.1 Electrochemical performance of V_2O_5 hybrid nanocomposite cathode in LIBs

Strategies	Material	Voltage range (V)	Cycling stability			Rate capability		Reference
			Current density (mA g^{-1})	Cycles No	Capacity (mAh g^{-1})	C-rate	Capacity	
Carbon coating	Carbon-coated V_2O_5 nanocrystals	2.0–4.0	1000	50	288	10	130	[90]
	Carbon-coated V_2O_5 nanoparticles	2.0–4.0	15	50	270	–	–	[91]
	V_2O_5/mesoporous carbon composite	2.0–4.0	500	100	163	5	70	[92]
CNT composition	V_2O_5 nanoparticles/carbon tube-in-tube	2.0–4.0	59	20	265	58.8	90	[56]
	V_2O_5 nanowire/MWCNT composite	1.8–4.0	70	50	260	2.80	169	[55]
	CNTs encapsulating V_2O_5 nanosheets	2.0–4.0	150	200	211	30	90	[93]
	V_2O_5 nanosheets assembled on CNTs	2.0–4.0	294	100	190	8. 82	116	[94]
	V_2O_5nanospheres/MWCNT layer-by-layer composite	2.0–4.0	294	100	200	1.18	130	[95]
Graphene composition	rGO-supported porous V_2O_5 spheres	2.0–4.0	90	50	202	5.70	102	[93]
	V_2O_5 nanowires/graphene	1.7–3.8	1000	3000	192	10	256	[60]
	V_2O_5 nanosheets self-assembled on rGO	2.0–4.0	600	160	102	15	76	[58]
	3D V_2O_5/rGO/CNT composite	2.0 – 4.0	294	80	220	5.88	100	[59]

(continued)

Table 18.1 (continued)

Strategies	Material	Voltage range (V)	Cycling stability			Rate capability		Reference
			Current density (mA g^{-1})	Cycles No	Capacity (mAh g^{-1})	C-rate	Capacity	
	V_2O_5 nanoparticles embedded in rGO balls	2.0–4.0	1000	100	214	1.50	200	[62]
	V_2O_5/rGO xerogels	1.5–3.7	443	150	252	4.43	201	[63]
Metal doping	Self-doped $V^{4+}-V_2O_5$ nanoflakes	2.0–4.0	100	100	260	4	139	[96]
	Al-doped V_2O_5/rGO nanocomposite	2.0 – 4.0	300	50	247	3	122	[97]
	ZrO_2-doped V_2O_5 particles	1.5–4.0	294	50	222	–	–	[98]
	Mn-doped V_2O_5 flakes	2.0–4.0	300	50	201	1.50	171	[99]
	Sn-doped V_2O_5 microspheres	2.0–4.0	200	50	212	2	163	[100]
Polymer coating	PEDOT/V_2O_5nanobelt/graphite foam	2.0–4.0	1500	500	265	24	115	[82]
	PANI-coated V_2O_5 nanowires	1.5–4.0	44	50	95	0.44	125	[101]
	PPy-coated V_2O_5 nanowires	1.5–4.0	40	100	188	–	–	[102]
	PPy-coated V_2O_5 rods	2.0–4.0	7	100	170	–	–	[76]
Electro spinning	V_2O_5 nanofibers	1.75–4	35	50	150	35	525	[9]
	V_2O_5 nanotubes	2.5–4	2000	250	105.6	0.2Ag^{-1}	15Ag^{-1}	[103]
	Porous V_2O_5 nanofibers	2–4	800	100	133.9	100	3000	[29]
	$Al_{(0.5)}$-doped V_2O_5	2–4	35	50	158	–	–	[104]
	$Al_{(1.0)}$-doped V_2O_5	2–4	35	50	298	–	–	[104]

(continued)

Table 18.1 (continued)

Strategies	Material	Voltage range (V)	Cycling stability			Rate capability		Reference
			Current density (mA g^{-1})	Cycles No	Capacity (mAh g^{-1})	C-rate	Capacity	
	Mesoporous V_2O_5 nanofibers	2–4	800	10	370	–	–	[69]
	Silver-doped V_2O_5	2–3.6	20	30	180			[105]
	V_2O_5/graphitic nanotubes	2–4	150	200	211	3	3000	[57]
	rGO-V_2O_5 nanowires	2–4	50	–	225	–	–	[106]

Electrospinning is considered as an effective method to prepare V_2O_5 cathodic materials having long nanofibrous structure, exceptionally high surface area, high aspect ratio, large surface to volume ratio and easy to prepare even a non-woven fabric-type freestanding electrodes without any binder. The morphology of electrospun V_2O_5 fiber can be turned by varying the electrospinning parameters, which is a unique advantage of electrospinning [18, 103]. Different morphologies such as core shell [107], hollow [108], porous [109], aligned [110], random [9], beaded [111], single phase [9], multi-axial fibers composed of three or more phases [112], etc., are easily possibility with the electrospinning technique. Many studies reported that electrospinning results in the formation of one-dimensional V_2O_5 nanofiber that exhibits improved performance in LIBs [32, 113]. Electrospunned 1D V_2O_5 fibers result in the superior properties due to the reduced size that improves the contact area between the electrodes and electrolyte. Moreover, these nanostructures allow reduction in the diffusion distance that enhances the properties of the resulting cathode material. For example, the carbon nanotubes encapsulating V_2O_5 nanosheets (V_2O_5@G) have diameters of around 150 nm and lengths of several microns, which were interwoven to form a flexible freestanding web (Fig. 18.6b, c). The preparation of V_2O_5@G started through a in situ strategy with producing precursor nanowires by the electrospinning of polyvinyl pyrrolidone solution containing a vanadium source and tetraethylorthosilicate. Layered graphitic carbon with average thickness of 5 nm was then deposited on the precursor nanowires in the presence of methane at 1070 °C in a CVD process (Fig. 18.6a) [93]. Due to the good electric conductivity and mechanical strength, the film of V_2O_5@G web was directly used as a freestanding electrode for flexible LIBs. This freestanding electrode showed ultra-fast and stable Li^+-ion storage performance. It retained a capacity of over 90 mAh g^{-1} at a high rate of 100 C and also showed a small capacity fading of 0.04% per cycle over 200 cycles. Noteworthy, an energy density of ca. 360 Wh kg^{-1} at a power rate of 15.2 kW kg^{-1} is one of the best results ever reported for V_2O_5-based cathode materials [113]. Eventhough the simply electrospun V_2O_5nanofiberous electrode shows superior electrochemical properties and improved lithium-ion battery performance, still there are several strategies such as doping with charged metal ions [114] and composite with carbon nanomaterials [106] which are adopted for the further improvement in electrochemical properties of V_2O_5-based cathode.

18.4 Methods to Improve the Electrochemical Properties of Electrospunned V_2O_5

In order to improve the electrochemical properties of electrodes in LIBs, different strategies have been introduced like coating, doping, composite preparation, etc. Through these approaches, the crystalline structure of the active material will get modified. In vanadium pentoxide cathodes as well, these methods are successfully in cooperated in order to improve its intercalation properties. The modification of

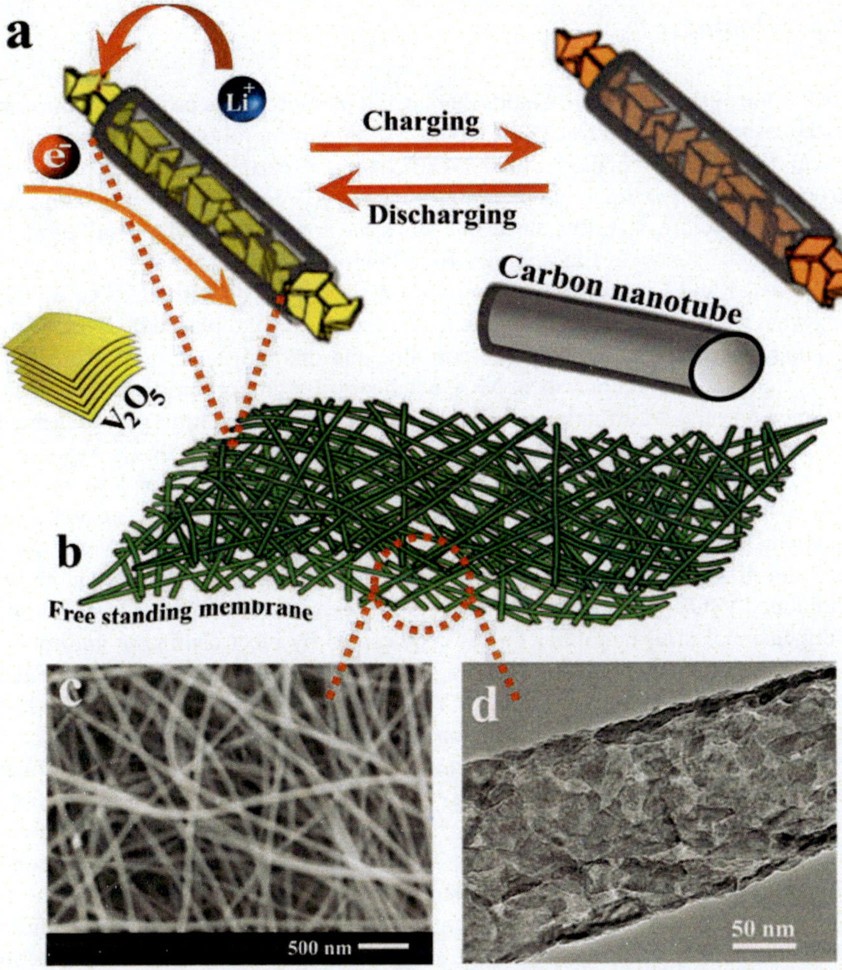

Fig. 18.6 **a** Schematic diagram of a single nanocable where V_2O_5 uniformly encapsulated into the graphitic nanotube. **b** Schematic of the configuration of self- supported flexible V_2O_5@G membrane interwoven by nanocables. **c** SEM images of V_2O_5@G membrane. **d** TEM image of V_2O_5@G. Adapted and reproduced from Ref. [93]. Copyright 2010 Royal Society of Chemistry

crystalline structure enhances the interlayer attraction, and it will result in an increase in the interlayer distance that will further results in the enhancement of intercalation property. A composite structure is prepared that results in the formation of V_2O_5-TiO_2 nanorods that can exhibit 1.5 times discharge capacity than the vanadium pentoxide cathode [115].

18.4.1 Doping

Doping is one of the effective methods that is used to improve the property of most of the cathodes material used in LIBs. Doping will impart morphological and electrochemical differences in cathode. In V_2O_5, inclusion of certain cations increases its electrochemical and extreme temperature properties. The presence of ionic dopants improves its structure, stability, electronic and ionic conductivity which will proportionally enhance its cycle life and electrode kinetics. Different charged ions such as Ag, Mn, Al and Cu are used as the doping agent in V_2O_5cathode. During the doping what happens inside this cathode is, V^{5+} ions will get replaced by this doping ions. The replacement will take place partially, and this will lead to the formation of V_2O_5 framework that consists of MO_6 octahedra. By the formation of this modified framework, the deformation caused during the lithium insertion gets prevented and shows an enhanced structural stability as cathode. Apart from this, a high electronic conduction can result in the material due to the formation of low valence state in vanadium by the insertion of doping ions [116–118]. Doping is also used in V_2O_5 which is synthesized by methods other than electrospinning [119]. V_2O_5 films doped with different dopants [114] as well as silver show a superior conductivity than undoped V_2O_5 [46]. The same effect can be seen in electrospunned V_2O_5 also. One-dimensional (1D) nanofiber that is synthesized by electrospinning and inclusion of various amounts of aluminum in V_2O_5 interlayer shows an enhanced elevated temperature property [104]. Chromium (Cr)-doped vanadium pentoxide produced by sol–gel method exhibits a lithium diffusion coefficient between 10^{-11} and 10^{-12} cm^2 s^{-1} that is greater than the crystalline vanadium pentoxide [36]. Same doping effect is studied by Zhang et al. [114] in V_2O_5 using Cr, and this system exhibits a reversible capacities of 200, 170 and 120 mAh g^{-1} at cyclic rate of C/10, C/2 and 1C rate, respectively [120]. Similar to Cr, copper (Cu), terbium (Tb) [121], niobium (Nb) and thallium (Ta) [122] are also used as an effective dopant in V_2O_5 [123]. Yan L. Cheah et al. [104] investigated the effect of aluminum over the electrospunned vanadium pentoxide and observed that, by the addition of even 0.5% aluminum, it exhibits an initial discharge capacity of 360 mAh g^{-1} with a capacity retention of about 50% even after 50 cycles (Fig. 18.7), and the best result was observed in the 1% aluminum-doped V_2O_5 that exhibits enhanced high rate and elevated temperature performance. The reason for this improved property was the retention of fibrous morphology as well as the stable inclusion of aluminum ions in the crystal structure [104].

18.4.2 Composite with Carbon Materials

Like doping, one of the efficient methods that is used for the enhancement of properties of vanadium pentoxide is making composite with carbon materials such a carbon coating, mixing with CNT, graphene, bucky ball and quantum dots. Introduction of carbon sources such as carbon nanotubes, graphene, active carbon or any other

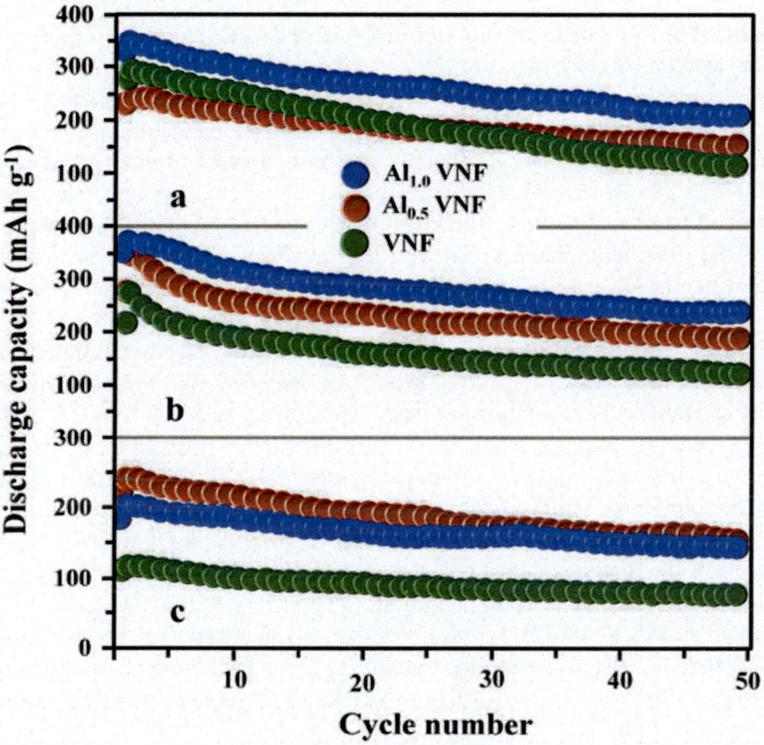

Fig. 18.7 Plots of specific discharge capacity versus cycle number of VNF, $Al_{0.5}$VNF and $Al_{1.0}$VNF cycled at **a** 0.1 C rate at room temperature, **b** 0.1 C rate at 55 °C and **c** 1 C rate at room temperature. Adapted and reproduced from Ref. [104]. Copyright 2012 American Chemical Society

carbon sources can effectively improve the structural stability of vanadium pentoxide nanostructures. Different allotropes of carbon such as graphite, carbon nanotubes, graphene and bucky ball are considered to be an effective material for electrodes in LIBs. By incorporating such conducting materials, the electronic conductivity and lithium diffusion property of V_2O_5, which have great control over the electrode properties, further get improved [124–126]. The carbon nanolayer coated over the V_2O_5 can act as a protective layer over the active material [127, 128]. Carbon coating by different methods was observed in nanostructured vanadium pentoxide cathode material other than that prepared by electrospinning [122]. Vanadium oxide nanostructures that are produced by the hydrothermal method are modified with carbon by making use of sucrose as carbon source which exhibits an enhanced surface area along with improved conductivity [129]. The resulting electrochemical properties greatly depend on the source of carbon choosed. Multiwalled carbon nanotubes (MWCNTs) were also used for the modification of vanadium oxide nanostructures produced by the hydrothermal method [130, 131]. ElectrospunV_2O_5 can be also modified by using different carbon sources that can ensure the advantageous property of both V_2O_5 and graphitic carbon. Graphene-modified electrospunned V_2O_5 shows an improved property that resulted by the capability to exhibit both the properties of electrospunned V_2O_5 structure as well as that of graphene. By the incorporation of graphitic carbon, a composite was produced with the vanadium pentoxide nanostructure that exhibits high Li^+-ions storage and high discharge capacity of about 224 mAh g^{-1} with a scan rate of 0.2 C. The electrode also showed good cyclic stability and rate capability that resulted from the increased kinetics of the material by composite preparation route of the GVO composite [106]. Kong et al. [93] synthesized electrospunned vanadium pentoxide nanosheets that are encapsulated with vapor phase deposited graphitic carbon layers. It results in the formation of a flexible web film in which the vanadium pentoxide nanoparticles are encapsulated in carbon nanotubes. Along with excellent flexibility, the device level analysis shows that the resulted cathode structure shows excellent lithium-ion storage performance as well. At 0.5 C, the cell exhibits a reversible capacity value of about 224 mAh g^{-1} (Fig. 18.8). Even over 200 cycles, the cell showed only 0.04% capacity decay which demonstrates the excellent cyclic stability of the V_2O_5-carbon composite cathode. Also, the cell shows an energy density of 360 Wh kg^{-1} at a power rate of 15.2 kW kg^{-1} which is observed. This is one of the works which shows the excellent performance that is attributed by the modification with the carbon nanotubes [93].

Hyoung An et al. [132] synthesized carbon encapsulated porous vanadium pentoxide (C/HPV_2O_5) by using electrospinning along with the post-calcination technique using PAN and VO(acac)$_2$ (Fig. 18.9a); this formed fiber was then carbonized

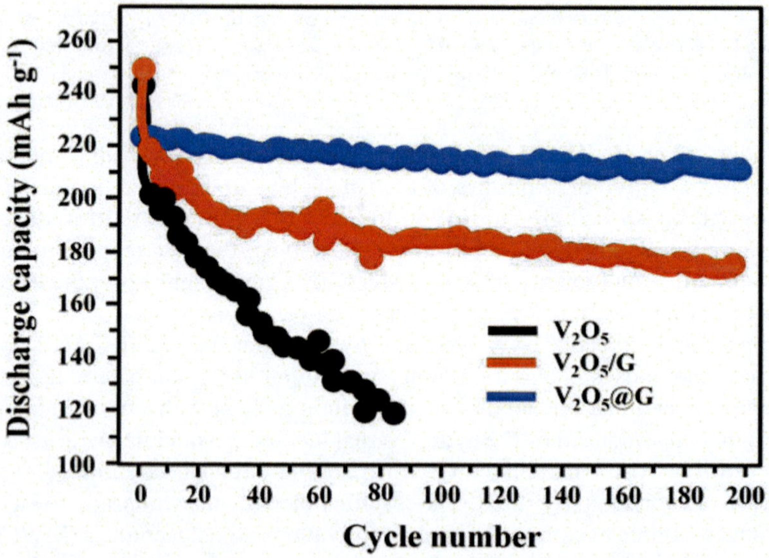

Fig. 18.8 Electrochemical characteristics of half-cell: different cycle performance of V_2O_5@G, V_2O_5/G and pure V_2O_5 electrodes at current density of 0.5 C. Adapted and reproduced from Ref. [93]. Copyright 2010 Royal Society of Chemistry

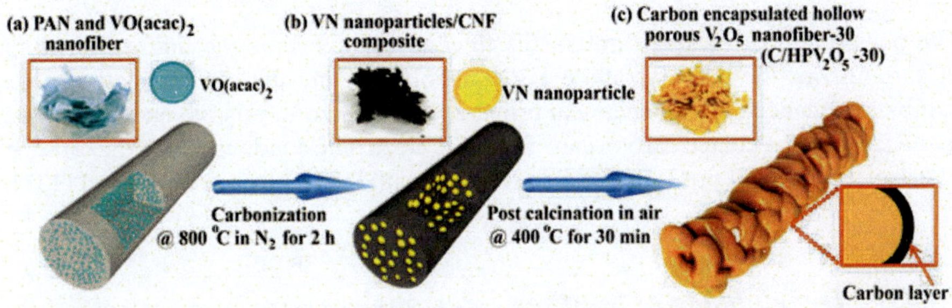

Fig. 18.9 Schematic illustration of the ideal C/HPV_2O_5-30 synthesis process. **a** PAN and VO$(acac)_2$ nanofiber obtained by electrospinning, **b** VN nanoparticles/CNF composite obtained after a carbonization process, **c** C/HPV_2O_5-30 with interior void spaces and well-defined pores as well as a uniform carbon layer on the V_2O_5 surface prepared by post-calcination. Adapted and reproduced from Ref. [132]. Copyright 2016 American Chemical Society

under nitrogen atmosphere to form VF/CNF (Fig. 18.9b) which is then finally get converted into C/HPV_2O_5—30 (Fig. 18.9c). The possible formation mechanism of this system was proposed based on the SEM and TEM results (Fig. 18.10). The hollow porous structure consists of the carbon encapsulation with well-defined pores. Compared to commercially available V_2O_5, this modified cathode material exhibits improved Li^+-ion storage with a discharge capacity of 241 mAh g^{-1} at 100 cycles. Also, the cell exhibits an improved rate performance of about 155 mAh g^{-1} at 1000 mA g^{-1} [132].

Plasma-enhanced carbon coating was successfully introduced by Cheah et al. [133]. In this study, a nanothick layer of carbon was uniformly coated over the electrospunned vanadium pentoxide using plasma-enhanced chemical vapor deposition.

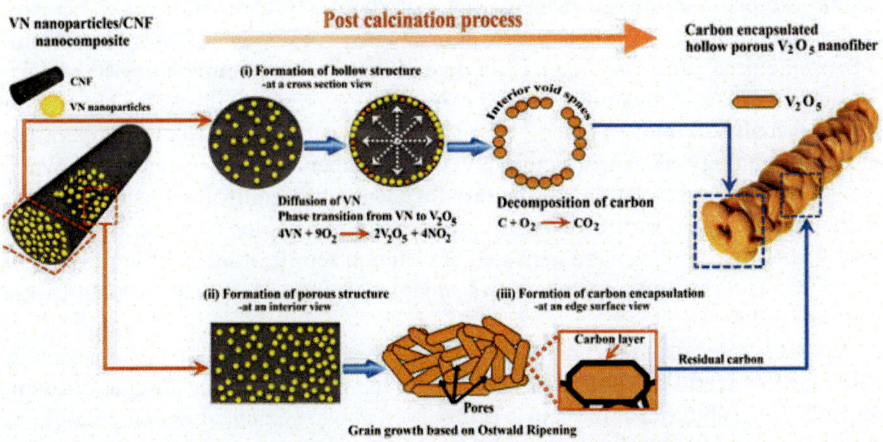

Fig. 18.10 Schematic illustration of the formation mechanism of C/HPV_2O_5-30. Adapted and reproduced from Ref. [132]. Copyright 2016 American Chemical Society

The penetration of plasma through the fiber meshes ensures the uniform coating of carbon over the polycrystalline V_2O_5 nanofibers, and the carbon-coated fibers exhibited enhanced electrochemical properties compared to pristine nanofibers due to the fact that the coated carbon can ensure the electronic conductivity to the cathode material as well as it prevents the side reaction with the electrolyte employed [133].

18.4.3 Dependence of Synthesis Parameters

In the method of electrospinning for the synthesis of vanadium pentoxide nanofibers, the annealing temperature is one of the important parameters which determines the surface morphology as well as the electronic and electrochemical properties of the nanofibers. For the determination of structural morphology of the resulting fiber, annealing atmosphere also has a great influence. By investigating the structure of the V_2O_5 nanofiber, annealing of the as-spun fiber under air is appeared to exhibit nanowire morphology, while under the N_2 atmosphere is favorable for the nanorod structures [134]. By combining the sol–gel method along with electrospinning, a cathode material with high lithium-ion storage was synthesized by Yu et al. [128], and he observed a storage capacity above 370 mAh g^{-1} and charge/discharge at a higher rate up to 800 mA g^{-1} with low cyclic degradation. The starting materials used are V_2O_5, hydrogen peroxide (H_2O_2) along with PVP. The resulting cathode material has a mesoporous structure with average fiber diameter of 350 nm and specific surface area of ~97 m^2 g^{-1} [73]. One-step electrospinning process followed by sintering at a temperature of 400 °C for 15 min produced highly porous structure due to the shrinking of fibers caused by the thermal degradation of PVP. The precursors used for this process are vanadium vanadyl acetylacetonate along with PVP. The resulting cathode exhibits a discharge capacity value of about 316 and 158 mAh g^{-1} during the first and 50th cycles, respectively, at 0.1 C rate (only 50% capacity retention). At the same time, electrospunned cathode that sintered at a temperature 400 °C for 2 h shows a capacity retention value of about 72% even after 100 cycles along with improved high rate performance [19]. The dependence of annealing temperature on the morphology of nanostructured V_2O_5 nanofibers was investigated by Wang et al. [103]. The investigation was mainly focused on the controlling parameters for producing V_2O_5 nanostructures with controllable morphology by electrospinning method, and the samples were annealed at different temperatures varying from 350 to 600 °C. The resulting morphologies obtained for the V_2O_5 nanostructures are displayed in Fig. 18.11.

The V_2O_5 electrospunned nanofibers were prepared from the precursors such as vanadium(IV) acetyl acetone and PVP. During the process of annealing at different temperatures, a significant mass loss is observed above an annealing temperature of 400 °C (Fig. 18.12) [103] that is primarily due to the decomposition of PVP. At an annealing temperature of 500 C, the disappearance of nanograins was observed, and further increase in annealing temperature results in the formation of nanobelts.

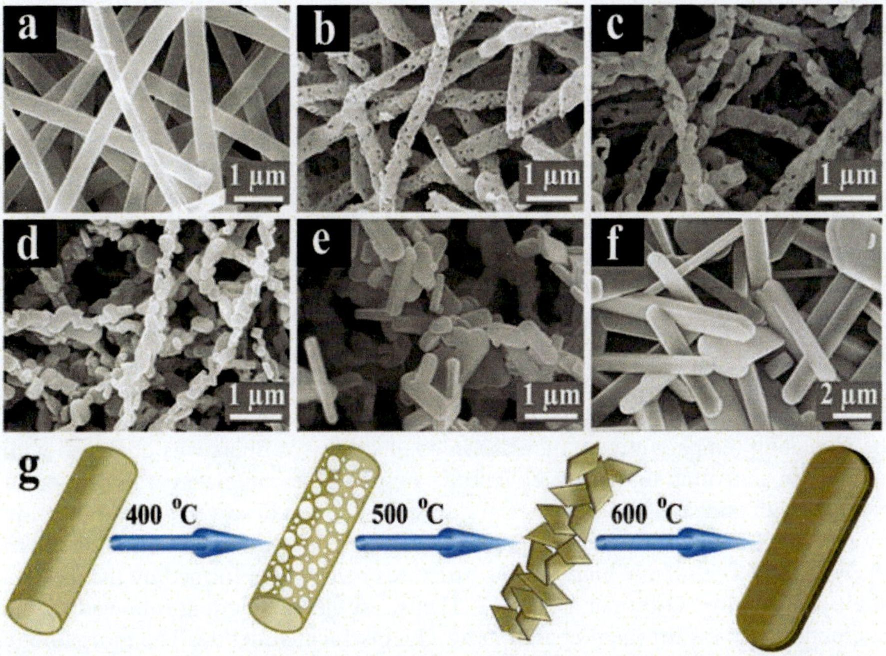

Fig. 18.11 SEM images on the surface morphology of V_2O_5 nanostructures that were annealed at: **a** 350 °C, **b** 400 °C, **c** 450 °C, **d** 500 °C, **e** 550 °C, **f** 600 °C, **g** formation of V_2O_5 nanostructures. Adapted and reproduced from Ref. [103]. Copyright 2012 John Wiley and Sons

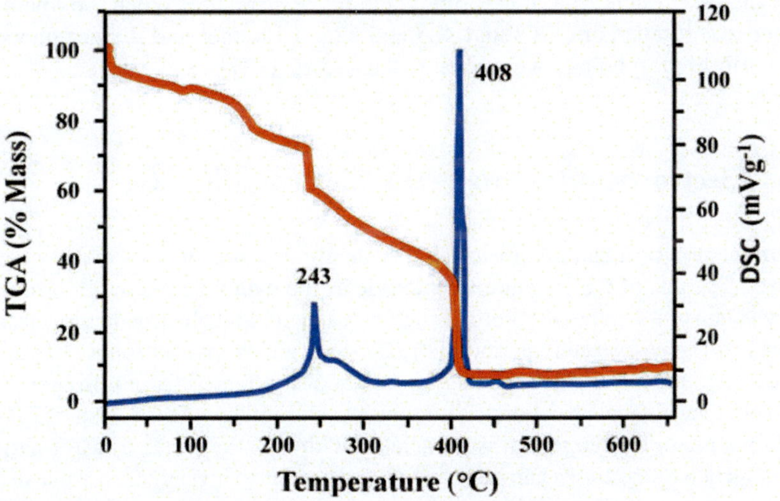

Fig. 18.12 Thermogravimetric analysis (TGA) of the as-prepared vanadium (IV)-acetylacetone/PVP precursor nanofibers. Adapted and reproduced from Ref. [103]. Copyright 2012 John Wiley and Sons

Liqiang Mai et al. [129] synthesized ultra-long vanadium pentoxide nanowires having a diameter 100–200 nm using a single spinneret for the first time. The synthesis has been done by controlling the precursor solution properties. Vanadium oxytrihydroxide solution was prepared by the hydrolysis of vanadyl acetyl acetonate in PVP, and the resulting solution was partially miscible. The core of the electrospunned polymer was produced by the phase-separated polymer during the spinning, and the $VO(OH)_3$ forms the shell of the fiber. The electrospunned fibers were further annealed at temperatures of 350, 400 and 500 °C. The sintering process results in the burning of PVP fibers completely which will finally result in the crystallization of V_2O_5 crystals. High-temperature annealing results in the destruction of complete continuous morphology of fiber that is obtained at a temperature of 350 °C (Fig. 18.13) [135].

Berezina et al. [130] proposed electrospinning method to develop both micro- and nanostructured V_2O_5 using vanadyl acetyl acetonate as the precursor. On studying the annealing temperature, an increase in conductivity of fibers was observed with increase in annealing temperature, and this suggests the complete carbonization of polymer with annealing at high temperatures. The best is observed at a temperature between 400 and 450 °C, and the longer annealing will result in the solidification of V_2O_5 [136]. Controllable hierarchical structures can be also formed by the method of electrospinning. Ostermann et al. [131] suggest electrospinning followed by the calcination process for the preparation of a hierarchical structure. The morphology and phase structures were modified through this method of preparation. Single-crystal V_2O_5 nanorods were allowed to form on the V_2O_5/TiO_2 composite nanofibers using the process of calcination. Calcination has a major role in determining the fiber morphology (Fig. 18.14). In addition, post-spinning treatment of the electrospunned fiber can also influence the morphology and phase structure. When the fibers were calcined at a temperature of about 425 and 525 °C, longer and thinner fibers with more uniform morphology were obtained at 425 °C [137].

18.5 Electrochemical Properties Electrospun V_2O_5

The choice of a best cathode material completely depends over its electrochemical property. The use of the vanadium pentoxide as the cathode material in lithium-ion battery is due to its higher theoretical specific capacity and electrochemical property that leads to the easier intercalation of the Li^+-ions which make it superior over other cathode materials. The electrospun V_2O_5 cathodes prepared from different precursors exhibit varying electrochemical properties. During the intercalation of Li^+-ions, the main structural change that is associated with it is the phase transformation of V_2O_5. Cyclic voltagramms can effectively use in order to investigate the phase transformation during the intercalation and deintercalation of Li^+-ions. Cheah et al. [28] studied the cyclic voltammogram of electrospunned V_2O_5 that is produced by using the precursor solution of vanadium sol–gel. The first cycling displays several oxidation–reduction peaks, whereas the cycling below 2 V shows a reduction peak at 1.95 which results in the accommodation of more than two Li^+-ion per V_2O_5 [9]. Ban

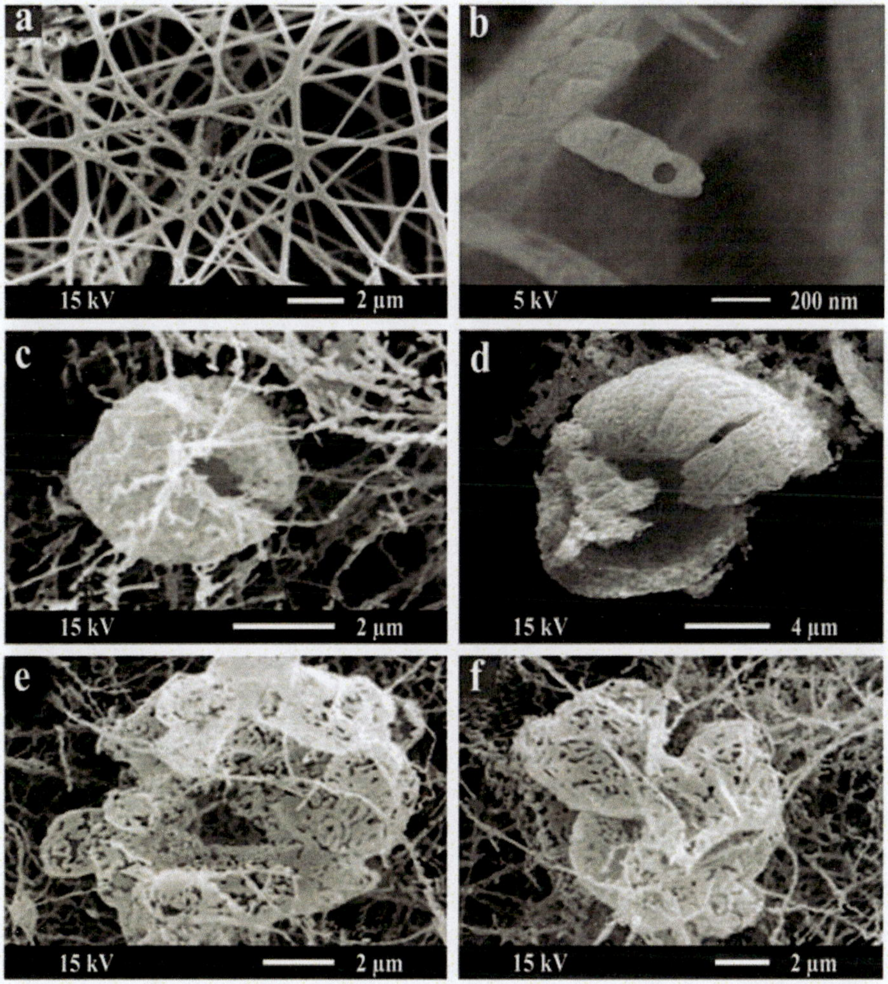

Fig. 18.13 **a–f** FE-SEM images on the surface morphology of fibers heated at 350, 450 and 500 °C for $t = 1$ and 3 h at the heating/cooling rate of 2 degrees/min to obtain TNFs, spheres, flakes and nanowires. Adapted and reproduced from Ref. [135]. Copyright 2012 Springer Nature

et al. [28] increased the calcination temperature after the hydrothermal treatment of electrospunned fibers from 420 to 500 °C. When discharged to 1.75 V versus Li/Li^+, it exhibits a low initial capacity of about 110 mAh g^{-1} that gets decayed after few cycling. When annealing the as-spun fibers between the temperature ranges of 450–500 °C, V_2O_5 nanobelts were formed, and then, it showed an improved discharged capacity of 350 mAh g^{-1} at current density of about 0.1 mA cm^{-2} within 1.75–3.75 V versus Li/Li^+ (Fig. 18.15) [28]. The charge–discharge cyclic stability of electrospunned V_2O_5 is highly influenced by the phase changes that occurred by the intercalation of Li^+-ion into the nanostructure which further determines its

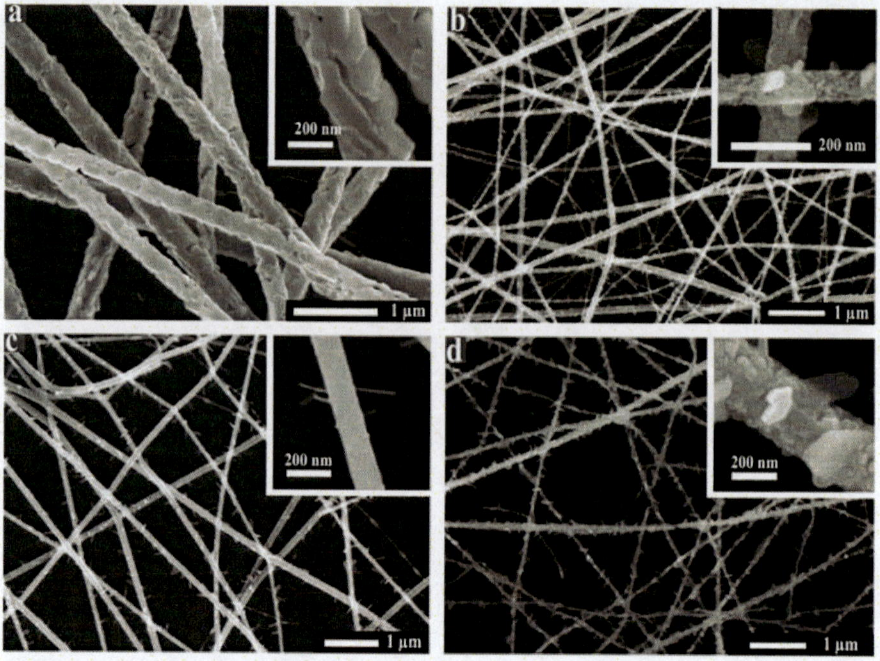

Fig. 18.14 SEM images on the surface morphology of V_2O_5-TiO_2 nanofibers that were electrospun from a 2-propanol solution containing different ratios (r) of $VO(OiPr)_3$ to $Ti(OiPr)_4$ and calcinating at different temperature: **a** r = 1:0, T = 475°C; **b** r = 4:1, T = 475 °C; **c** r = 1:1, T = 425 °C; **d** r = 1:1, T = 525 °C. Adapted and reproduced from Ref. [137]. Copyright 2006 American Chemical Society

electrochemical potential window [136–138]. Ban et al. [27] synthesized nanoscale single-crystal vanadium oxide by making use of electrospinning followed by the hydrothermal method. The resulting structures form a fibrous morphology with a capacity of 120 mAh g^{-1} in the first cycle, which is about 5% less than that of V_2O_5 nanobelts [27].

The porous fiber structures produced by Yu et al. [23] using electrospunned PMMA fibers as the template will get removed by the high-temperature annealing process at about 400 °C, and it results in structure with tube-like pore size of one micrometers in diameters (T.V_2O_5). Following the similar procedure, but without making use of electrospunned PMMA fibers, another sample is made (S.V_2O_5). The cycling performance of both the electrodes was investigated at different C-rates at a potential range of 4.0-2.0 V (vs. Li/Li^+) at room temperature. The discharge capacity is well maintained for T-V_2O_5 up to 300 cycles, but it is gradually decreasing in the case of S-V_2O_5. About 85% of capacity retention is observed for T.V_2O_5 which only 56% for S.V_2O_5 up to 300 cycles [23]. A phase pure V_2O_5 nanofibers produced by electrospinning followed by annealing resulted in the formation of fibers having nanoscale diameter [26]. Compared to other nanostructured V_2O_5 electrodes, these

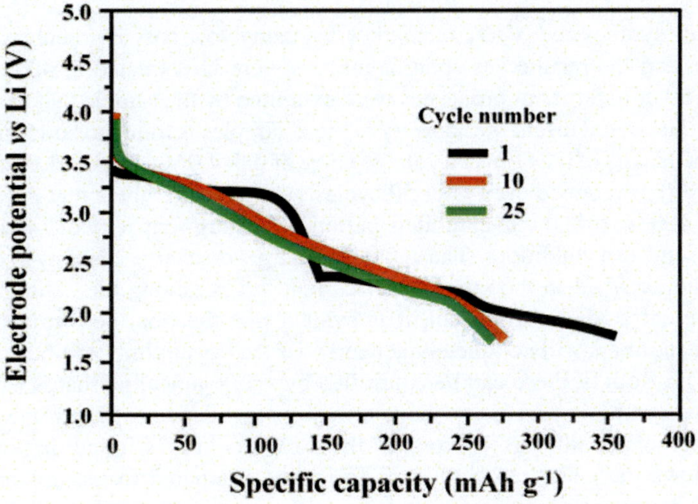

Fig. 18.15 Discharge curves during 1st, 10th and 25th cycle of nano-V_2O_5 at 0.1 mA cm^{-2} between 4.0 and 1.75 V. Adapted and reproduced from Ref. [28]. Copyright 2009 Elsevier

fibers exhibit superior lithium storage capacity. It is also noteworthy that the as-spun V_2O_5 nanofibers exhibits a high charge and discharge capacity compared to that of commercially available V_2O_5 cathodes. The resulting nanofibrous structure provide a specific capacity of 133.9 mAh g^{-1} at a current density of 800 mA g^{-1} even after 100 cycles which shows a capacity retention of 96%, while for the commercial V_2O_5 the discharge capacity was only about 40.6 mAh g^{-1} after 100 cycles with a capacity retention of 71.85% (Fig. 18.16) [29].

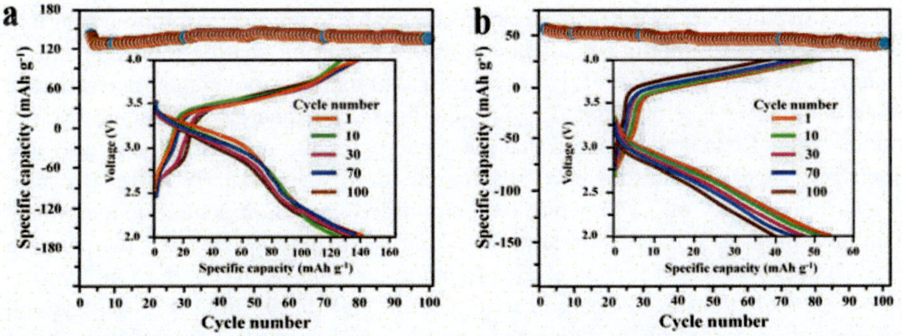

Fig. 18.16 Electrochemical properties of the prepared materials: Specific capacity of the **a** commercial V_2O_5 and **b** V_2O_5 nanofibers for 100 cycles at a current density of 800 mA g^{-1}, Inset shows the charge/discharge curves correspond to different cycles. Adapted and reproduced from Ref. [29]. Copyright 2015 Elsevier

Li et al. synthesized V_2O_5 nanotubes by using low-cost inorganic vanadium pentoxide, and the resulted as-spun nanofibers were heat treated at 400 °C either 15 min or 60 min. Superior properties were exhibited by the sample that heat treated at 15 min. The initial discharge capacity of these samples is found to be 150 mAh g^{-1} at a current density of 0.2 C, over a voltage range of 2.5–4.0 V versus Li/Li^+, and about 86% capacity is retained even after 50 cycles, while for the fibers that produced by annealing 400 °C for 60 min exhibited an initial discharge capacity of 144 mAh g^{-1} under the same test conditions. The enhanced charge–discharge capacity and cycling stability are ascribed to the retention of carbon traces among the nanotubes [18]. From different studies carried out, it is evident that the final performance of the resulting nanofibers will completely depend over the crystallinity of fibers and fiber morphology. Both of these can be controlled by using annealing that is carried out finally. The synthesis parameters have a major dependence to determine the final performance of the fiber [33]. From all these reports, it was found that the doped system shows the enhanced battery performance compared to its undoped counterpart. In summary, it can be concluded that the electrochemical properties of the electrospunned V_2O_5 nanofibers can be tuned by doping with other electroactive materials. Doped V_2O_5 nanostructures will exhibit enhanced electrochemical properties due to the presence of charged cations, which results in structural variations in the materials capable to influence the electrochemical characteristics.

18.6 Electrochemical Properties of Doped Electrospun V_2O_5

Doping is one of the methods that is effectively used in most of the cathode material in order to enhance its electrochemical properties. The dopants can easily incorporate to the electrospun fibers by the incorporation of the same to the electrospinning solution. The same effect of doping can also be observed in carbon coating-induced cathode materials as well. Cheah et al. [133] investigated the electrochemical performance of carbon-coated electrospunned V_2O_5 nanofibers. The charge–discharged cycling studies in the LIBs demonstrated that the capacity retention in carbon-coated V_2O_5 fibers gets improved and showed a capacity retention greater than 65% after 50 cycles at a current density of 0.1 C, while the unmodified vanadium pentoxide nanofibers exhibited only 40% capacity retention. Along with this, the elevated temperature (55 °C) performance of the material is found to be better than that in room temperature. An increment in discharge capacity from 280 to 298 mAh g^{-1} is observed for the carbon-coated fibers during the first cycle at 500 °C. But for the unmodified fibers, a discharge capacity 284 mAh g^{-1} is observed during the second cycle, and it is observed to be 32 mAh g^{-1} less than that of room temperature capacity [133]. Sodium (Na)-, potassium (K)- and molybdenum (Mo)-based vanadium oxides prepared by the method other than electrospinning are also reported as high-performance cathodes in LIBs [139–141]. Sodium vanadium oxides are attractive cathode materials that exhibit excellent electrochemical properties. Hydrothermally synthesized material exhibits a mesoporous flake-like structure that enhances the lithium-ion diffusion with β-$Na_{0.33}V_2O_5$ phase. The resulting structure exhibits discharge capacity of 339 and 226 mAh g^{-1} at a current density of 20 and 300 mAg^{-1} which is higher than undoped V_2O_5 cathodes [142]. A novel silver-doped vanadium oxide is synthesized by combining both electrospinnings along with hydrothermal method. The resulting fibrous structure is composed of single-crystalline nanorods. The electrochemical studies over this cathode material show that there is an improved cyclic performance. At a current density of 20 and 100 mA g^{-1}, these electrodes are delivered and initial discharge capacity of 250 and 175 mAh g^{-1} with a cyclic efficiency of about 72 and 71% over 30 cycles, respectively, in the voltage range of 2.0–3.6 V versus Li/Li^+ (Fig. 18.17) [105]. By the introduction of dopants, the phase change effect of V_2O_5 cathodes during the lithiation and delithiation can be nullified. Among different dopants, Ti^{4+} is considered as the best that can prevent the phase change effect in electrospunned V_2O_5-based cathode materials. Hence, it can offer an enhanced cyclic stability throughout the electrochemical reaction process [31].

Vanadium-based nanomaterials are employed as negative electrodes (anodes) in LIBs as well. LiV_3O_8 nanofibers are prepared by electrospinning using 2D nanosheets with an exposed (001) facet which is proposed as the best anode for LIBs. The hierarchical LiV_3O_8 nanofibers produced by the electrospinning show improved electrochemical performance than the normal LiV_3O_8 nanofibers because of its unique structure. LIBs performed with liquid electrolytes, LiV_3O_8 nanofibers and $LiMnO_4$ cathode, can exhibit a reversible capacity of about 72 mAh g^{-1} even

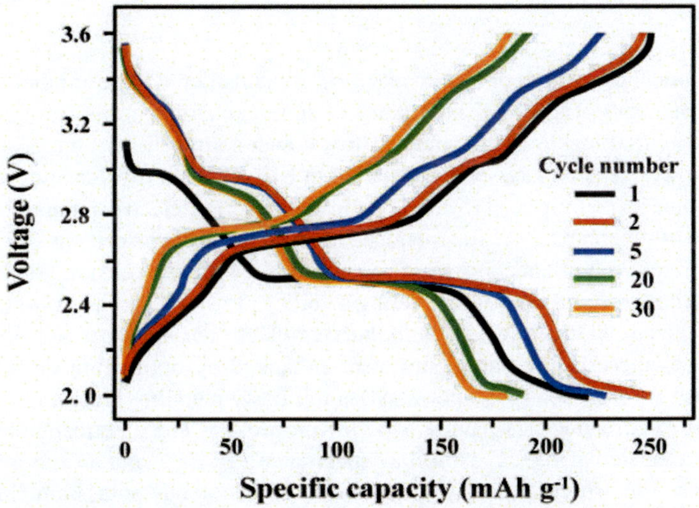

Fig. 18.17 Electrochemical performance of β-$Ag_{0.33}V_2O_5$: Galvanostatic cycling profiles at 1, 2, 5, 20 and 30 cycles in the voltage range 2–3.6 V at 20 mA g^{-1}. Adapted and reproduced from Ref. [105]. Copyright 2013 Royal Society of Chemistry

after 50 cycles which show 69.7% retention of capacity obtained in the first cycle at a current density of 60 mA g^{-1} [143]. Similarly, different mixed oxides such as hollow nanofibers synthesized with different mixed oxides ($M_xV_3O_8$ where M = Cu, Ni and Fe) also exhibit remarkable electrochemical properties. Xiang et al. [33] synthesized metal oxide vanadium-based nanofiber $M_xV_3O_8$ using different metals such as copper (Cu), nickel (Ni) and iron (Fe) which were prepared by electrospinning followed by calcination process. Compared to other metal oxide structures, copper-based V_2O_5 exhibits superior electrochemical properties. The electrospinning followed by the calcination process results in the transformation from electrospunned dense fiber to core shell fibers of $Co_3V_2O_8$ that exhibit 900 mAh g^{-1} reversible capacity over 2000 cycles at a current rate of 5 A g^{-1} [144].

Similar to LIBs, electrospunned V_2O_5 cathodes are employed for the other metal-ion batteries. There are few number of studies, which investigate the electrochemical property of V_2O_5 electrodes in sodium [145, 146]- and aluminum [147, 148]-based batteries; however, there is only limited investigation over electrospunned materials for this metal-ion batteries yet. Vanadium oxide-based cathodes prepared by the low-temperature hydrothermal method exhibit a capacity of 116 mAh g^{-1} even after 100 cycles at a charging rate of 50 mA g^{-1} in aluminum ion batteries. This cathode material synthesized is capable to exhibit good intercalation toward lithium as well as aluminum ions [149].

18.7 Electrospun V_2O_5 for Energy Storage Solutions other than LIBs

Electrochemical capacitors or supercapacitors are considered to be another potential source for energy storage. The importance of supercapacitors arises from their high power density that makes them an alternative power source against batteries, which have low power density. However, compared to LIBs, the energy densities exhibited by super capacitors are poor [150, 151]. Similar to LIBs, electrospunned materials can be used in supercapacitors for enhanced electrochemical performance [152, 153]. Among different materials choosed, electrospunned V_2O_5 can ensure good electrochemical properties and long-term cycling stability [154]. Wee et al. [148] proposed the use of electrospunned V_2O_5 nanofibers as electrodes in supercapacitors. The pseudocapacitances of the nanofibers were enhanced by optimizing the annealing temperature. About 400 °C is optimized as the best annealing temperature for the electrode in supercapacitors having superior properties. The electrospunned V_2O_5 nanofibers annealed at 400 °C exhibit a capacitance 190 Fg^{-1} and an energy density of 5 Wh g^{-1}. Electrolyte possessing neutral pH like aqueous potassium chloride is optimized for the enhanced electrochemical properties, and further enhancement of the capacity is done by the usage of lithium containing salt such as $LiClO_4$ [155]. Similar to LIBs cathode, structural modification can impart in supercapacitors electrode in order to improve its electrochemical activity. V_2O_5 coated over multiwalled carbon nanotube is reported as an excellent choice for energy storage device since it possesses both high energy density and power density [156]. Modifying the material with graphitic carbon, carbon nanotube or graphene is reported as one of the methods to improve its electrochemical properties of electrodes in pseudocapacitors [56, 157–160]. Huang et al. [155] studied the electrochemical properties of a composite material prepared by embedding vanadium oxide with different valencies over the carbon fibers by electrospinning followed by calcination. The resulting fibrous structures were investigated for high performance in supercapacitors, and the best performance is observed for the V_2O_5-embedded carbon nanofibers. A stable gravimetric charge–discharge capacitance of about 606 Fg^{-1} is observed for this electrode at a charging rate of 0.5 A g^{-1} over 5000 cycles due to the superior electrochemical property exhibited by V^{5+} than that of V^{4+} and V^{3+} ions [161]. Pasquier et al. [162] synthesized V_2O_5/carbon nanofiber composite produced by electrospinning of polyacrylonitrile/V_2O_5 in N,N dimethylformamide (DMF). On varying the loading of V_2O_5, there was a change in nanometer size diameter, specific surface area, porosity, pore size and pore distribution. These factors determine the electrochemical behavior of resulting electrode. The electrode with maximum amount of V_2O_5 (20 wt%) exhibits large interfacial area. The resultant electrode exhibits 150 Fg^{-1} capacitance with an energy density 18.8 Wh kg^{-1} over a power density of 400–20,000 Wh kg^{-1} [162]. α-Fe_2O_3-based composite is prepared by doping with V_2O_5, and the composite nanofibers were produced by electrospinning. The resulting material exhibits magnetic properties as well that completely depends over the amount of V_2O_5 in the composite [163]. A similar material selection is made by Jiang et al. [164] and investigated the electrochemical property of flexible and highly porous Fe_2O_3 and V_2O_5 nanofibers by electrospinning followed by high-temperature calcination. The nanofibers of these two distinct materials interconnected to form a binder-free 3D hierarchical structure. The resulting fibrous structure provides specific capacitance about 255 Fg^{-1} at a current rate of 2 mV s^{-1} [164].

Methods other than electrospinning also employed for the facile synthesis of vanadium oxide-based electrodes for supercapacitors. Polypyrrole, a conducting polymer that allows to grow over V_2O_5 nanosheets using anionic dodecylbenzenesufonate as surfactant, provides excellent electrochemical properties like long-term cyclic stability and rate capability [47]. The interconnected porous vanadium oxide synthesized exhibits a highest specific capacitance of 316 Fg^{-1} with an energy density of 43.8 Wh kg^{-1} [165]. Dependence of morphology over the electrochemical behavior is investigated by Zhang et al. [166], and hydrothermally synthesized microspheres show largest specific area and a capacitance of 308 Fg^{-1} [166].

18.8 Conclusion and Future Outlook

The future of LIBs lies in the development of materials that can ensure the complete battery performance even at extreme conditions. The choice of material as well as the method of preparation plays a major role in determining the electrochemical performance of the nanostructured electrodes. This chapter summarizes the significance of vanadium pentoxide-based cathodes in LIBs and the methods adopted for the enhanced electrochemical performance for the same. The ability of vanadium pentoxide to be stable at extreme conditions, capability to exhibit multiple oxidation state and ability to intercalate Li^+-ions readily due to the open layered structure make it most attractive cathode in LIBs. Among different methods such a sol–gel or hydrothermal synthesis, etc., electrospinning is economically viable. Most simple and effective technique for the fabrication of nanofibers has hierarchal structure and a range of morphologies such as nanorods, nanotube, nanoribbon, nanobelt, nanofibers, core shell or hollow structures, etc., with enhanced electrochemical properties. The process of spinning can perform at ambient conditions with simple equipment without any risk. Further improvement in electrochemical performance of V_2O_5 electrospun fibrous electrodes was made by structural modifications of the active material. Coating and doping are considered as the most effective method for the development of advanced vanadium pentoxide-based cathodes. Different conducting materials are employed for this purpose that can result in higher material properties and that will finally result in outstanding functioning of the rechargeable system. The application of electrospunned V_2O_5 and its composite nanofibers is one of the efficient and finest options for the development of advanced LIBs. However, their applications are not limited to only lithium-ion technology, but extended to other battery systems such as magnesium-ion batteries, aluminum-ion batteries or fluoride-ion batteries. Besides battery systems, for supercapacitors and fuel cells as well, these electrospunned V_2O_5 or other metal oxides and its composite nanofibrous materials including carbon materials are found to be better in performing the charge storage than the other normally using active materials. In nutshell, all the aforementioned studies suggest that the future of LIBs lies in the development of electrode material other than lithium and among which transition metal oxide-based cathodes such as vanadium pentoxide are going to be the important candidate in LIBs.

References

1. Zhou J, Liu X, Cai W et al (2017) Wet-chemical synthesis of hollow red-phosphorus nanospheres with porous shells as anodes for high-performance lithium-ion and sodium-ion batteries. Adv Mater 29:1–7. https://doi.org/10.1002/adma.201700214
2. Li N, Martin CR (2001) A high-rate, high-capacity, nanostructured Sn-based anode prepared using sol-gel template synthesis. J Electrochem Soc 148:A164–A170. https://doi.org/10.1149/1.1342167
3. Yin XM, Li CC, Zhang M et al (2010) One-step synthesis of hierarchical SnO_2 hollow nanostructures via self-assembly for high power lithium ion batteries. J Phys Chem C 114:8084–8088. https://doi.org/10.1021/jp100224x
4. Wang Y, Lee JY (2004) Molten salt synthesis of tin oxide nanorods: morphological and electrochemical features. J Phys Chem B 108:17832–17837. https://doi.org/10.1021/jp0467447
5. Grosjean R, Delacroix S, Gouget G et al (2017) High pressures pathway toward boron-based nanostructured solids. Dalt Trans 47:7634–7639. https://doi.org/10.1039/x0xx00000x

6. Zhao LM, Shi LE, Zhang ZL et al (2011) Preparation and application of chitosan nanoparticles and nanofibers. Braz J Chem Eng 28:353–362. https://doi.org/10.1590/S0104-66322011000300001
7. Hwang TH, Lee YM, Kong B et al (2012) Electrospun core–shell fibers for robust silicon nanoparticle-based lithium ion battery anodes. Nano Lett 12:802–807. https://doi.org/10.1021/nl203817r
8. Chen Y, Lu Z, Zhou L et al (2012) Triple-coaxial electrospun amorphous carbon nanotubes with hollow graphitic carbon nanospheres for high-performance Li ion batteries. Energy Environ Sci 5:7898–7902. https://doi.org/10.1039/c2ee22085g
9. Cheah YL, Gupta N, Pramana SS et al (2011) Morphology, structure and electrochemical properties of single phase electrospun vanadium pentoxide nanofibers for lithium ion batteries. J Power Sources 196:6465–6472. https://doi.org/10.1016/j.jpowsour.2011.03.039
10. Li L, Yin X, Liu S et al (2010) Electrospun porous SnO_2nanotubes as high capacity anode materials for lithium ion batteries. Electrochem Commun 12:1383–1386. https://doi.org/10.1016/j.elecom.2010.07.026
11. Jung JW, Lee CL, Yu S, Kim ID (2016) Electrospun nanofibers as a platform for advanced secondary batteries: a comprehensive review. J Mater Chem A 4:703–750. https://doi.org/10.1039/c5ta06844d
12. Sun G, Sun L, Xie H, Liu J (2016) Electrospinning of nanofibers for energy applications. Nanomaterials 6:129. https://doi.org/10.3390/nano6070129
13. Julien C, Mauger A, Zaghib K, Groult H (2014) Comparative issues of cathode materials for Li-ion batteries. Inorganics 2:132–154. https://doi.org/10.3390/inorganics2010132
14. Li M, Lu J, Chen Z, Amine K (2018) 30 years of lithium-ion batteries. Adv Mater 30:1800561. https://doi.org/10.1002/adma.201800561
15. Padhi AK (1997) Phospho-olivines as positive-electrode materials for rechargeable lithium batteries. J Electrochem Soc 144:1188. https://doi.org/10.1149/1.1837571
16. Schipper F, Erickson EM, Erk C et al (2017) Review—recent advances and remaining challenges for lithium ion battery cathodes. J Electrochem Soc 164:A6220–A6228. https://doi.org/10.1149/2.0351701jes
17. Nitta N, Wu F, Lee JT, Yushin G (2015) Li-ion battery materials: present and future. Mater Today 18:252–264. https://doi.org/10.1016/j.mattod.2014.10.040
18. Li Z, Liu G, Guo M et al (2015) Electrospun porous vanadium pentoxide nanotubes as a high-performance cathode material for lithium-ion batteries. Electrochim Acta 173:131–138. https://doi.org/10.1016/j.electacta.2015.05.057
19. Liu Y, Guan D, Gao G et al (2017) Enhanced electrochemical performance of electrospun V_2O_5 nanotubes as cathodes for lithium ion batteries. J Alloys Compd 726:922–929. https://doi.org/10.1016/j.jallcom.2017.07.214
20. Karami H, Mohammadi A (2015) Poly vinyl alcohol-based sol-gel synthesis of V_2O_5 nanoflakes as positive electrodes of Li-ion batteries. Int J Electrochem Sci 10:7392–7408
21. Livage J (2010) Hydrothermal synthesis of nanostructured vanadium oxides. Materials (Basel) 3:4175–4195. https://doi.org/10.3390/ma3084175
22. Wang S, Li S, Sun Y et al (2011) Three-dimensional porous V_2O_5 cathode with ultra high rate capability. Energy Environ Sci 4:2854–2857. https://doi.org/10.1039/c1ee01172c
23. Yu JJ, Yang J, Nie WB et al (2013) A porous vanadium pentoxide nanomaterial as cathode material for rechargeable lithium batteries. Electrochim Acta 89:292–299. https://doi.org/10.1016/j.electacta.2012.11.032
24. Viet AL, Reddy MV, Jose R, Chowdari BVR, Ramakrishna S (2010) Nanostructured Nb_2O_5 polymorphs by electrospinning for rechargeable lithium batteries. J Phys Chem C 114:664–671. https://doi.org/10.1021/jp9088589
25. Ramaseshan R, Sundarrajan S, Jose R, Ramakrishna S (2007) Nanostructured ceramics by electrospinning. J Appl Phys 102:7. https://doi.org/10.1063/1.2815499
26. Viswanathamurthi P, Bhattarai N, Kim HY, Lee DR (2003) Vanadium pentoxide nanofibers by electrospinning. Scripta Mater 49:577–581. https://doi.org/10.1016/S1359-6462(03)00333-6

27. Ban C, Whittingham MS (2008) Nanoscale single-crystal vanadium oxides with layered structure by electrospinning and hydrothermal methods. Solid State Ionics 179:1721–1724. https://doi.org/10.1016/j.ssi.2008.01.037
28. Ban C, Chernova NA, Whittingham MS (2009) Electrospun nano-vanadium pentoxide cathode. Electrochem Commun 11:522–525. https://doi.org/10.1016/j.elecom.2008.11.051
29. Yan B, Li X, Bai Z et al (2015) Superior lithium storage performance of hierarchical porous vanadium pentoxide nanofibers for lithium ion battery cathodes. J Alloys Compd 634:50–57. https://doi.org/10.1016/j.jallcom.2015.01.292
30. Zhang Z, Shao C, Zhang L et al (2010) Electrospun nanofibers of V-doped TiO_2 with high photocatalytic activity. J Colloid Interface Sci 351:57–62. https://doi.org/10.1016/j.jcis.2010.05.067
31. Armer CF, Lowe A, Lübke M et al (2017) Phase change effect on the structural and electrochemical behaviour of pure and doped vanadium pentoxide as positive electrodes for lithium ion batteries. J Power Sources 353:40–50. https://doi.org/10.1016/j.jpowsour.2017.03.121
32. Mai L, Xu L, Han C et al (2010) Electrospun ultralong hierarchical vanadium oxide nanowires with high performance for lithium ion batteries. Nano Lett 10:4750–4755. https://doi.org/10.1021/nl103343w
33. Armer CF, Yeoh JS, Li X, Lowe A (2018) Electrospun vanadium-based oxides as electrode materials. J Power Sources 395:414–429. https://doi.org/10.1016/j.jpowsour.2018.05.076
34. Liu P, Zhu K, Bian K et al (2018) 3D hierarchical porous sponge-like V_2O_5 micro/nanostructures for high-performance Li-ion batteries. J Alloys Compd 765:901–906. https://doi.org/10.1016/j.jallcom.2018.06.314
35. Zhai T, Liu H, Li H et al (2010) Centimeter-long V_2O_5 nanowires: from synthesis to field-emission, electrochemical, electrical transport, and photoconductive properties. Adv Mater 22:2547–2552. https://doi.org/10.1002/adma.200903586
36. Zhan SY, Wang CZ, Nikolowski K et al (2009) Electrochemical properties of Cr doped V_2O_5 between 3.8 V and 2.0 V. Solid State Ionics 180:1198–1203. https://doi.org/10.1016/j.ssi.2009.05.020
37. Zhang Z, Henrich VE (1994) Surface electronic structure of V_2O_5(001): defect states and chemisorption. Surf Sci 321:133–144. https://doi.org/10.1016/0039-6028(94)90034-5
38. Kristoffersen HH, Metiu H (2016) Structure of $V_2O_5{\cdot}nH_2O$ Xerogels. J Phys Chem C 120:3986–3992. https://doi.org/10.1021/acs.jpcc.5b12418
39. Takeda Y, Yamamoto O, Imanishi N (2016) Lithium dendrite formation on a lithium metal anode from liquid, polymer and solid electrolytes. Electrochemistry 84:210–218. https://doi.org/10.5796/electrochemistry.84.210
40. Fang C, Wang X, Meng YS (2019) Key issues hindering a practical lithium-metal anode. Trends Chem 1:152–158. https://doi.org/10.1016/j.trechm.2019.02.015
41. Li H, Shi L, Wang Q et al (2002) Nano-alloy anode for lithium ion batteries. Solid State Ionics 148:247–258
42. Pan A, Zhang JG, Nie Z et al (2010) Facile synthesized nanorod structured vanadium pentoxide for high-rate lithium batteries. J Mater Chem 20:9193–9199. https://doi.org/10.1039/c0jm01306d
43. Ji Y, Li S, Zhong G et al (2015) Synergistic effects of suberonitrile-LiBOB binary additives on the electrochemical performance of high-voltage $LiCoO_2$ electrodes. J Electrochem Soc 162:A7015–A7023. https://doi.org/10.1149/2.0041513jes
44. Kim DK, Muralidharan P, Lee HW et al (2008) Spinel $LiMn_2O_4$ nanorods as lithium ion battery cathodes. Nano Lett 8:3948–3952. https://doi.org/10.1021/nl8024328
45. Wu XL, Jiang LY, Cao FF et al (2009) $LiFePO_4$ nanoparticles embedded in a nanoporous carbon matrix: superior cathode material for electrochemical energy-storage devices. Adv Mater 21:2710–2714. https://doi.org/10.1002/adma.200802998
46. Coustier F, Passerini S, Hill J, Smyrl WH (1997) Silver-doped vanadium oxides as host materials for lithium intercalation. MRS Proc 496:1355–1360. https://doi.org/10.1557/proc-496-353

47. Takahashi K, Wang Y, Cao G (2005) Growth and electrochromic properties of single-crystal V_2O_5 nanorod arrays. Appl Phys Lett 86:1–3. https://doi.org/10.1063/1.1857087
48. Liu F, Song S, Xue D, Zhang H (2012) Selective crystallization with preferred lithium-ion storage capability of inorganic materials. Nanoscale Res Lett 7:1–17. https://doi.org/10.1186/1556-276X-7-149
49. Wang Y, Takahashi K, Lee K, Cao G (2006) Nanostructured vanadium oxide electrodes for enhanced lithium-ion intercalation. Adv Funct Mater 16:1133–1144. https://doi.org/10.1002/adfm.200500662
50. Pinna N, Hochepied JF, Niederberger M, Gregg M (2009) Chemistry and physics of metal oxide nanostructures. Phys Chem Chem Phys 11:3607. https://doi.org/10.1039/b905768d
51. Wang Y, Cao G (2006) Synthesis and enhanced intercalation properties of nanostructured vanadium oxides. Chem Mater 18:2787–2804. https://doi.org/10.1021/cm052765h
52. Chou SL, Wang JZ, Sun JZ et al (2008) High capacity, safety, and enhanced cyclability of lithium metal battery using a V_2O_5 nanomaterial cathode and room temperature ionic liquid electrolyte. Chem Mater 20:7044–7051. https://doi.org/10.1021/cm801468q
53. Zhang XF, Wang KX, Wei X, Chen JS (2011) Carbon-coated V_2O_5 nanocrystals as high performance cathode material for lithium ion batteries. Chem Mater 23:5290–5292. https://doi.org/10.1021/cm202812z
54. Sathiya M, Prakash AS, Ramesha K et al (2011) V_2O_5-anchored carbon nanotubes for enhanced electrochemical energy storage. J Am Chem Soc 133:16291–16299. https://doi.org/10.1021/ja207285b
55. Jia X, Chen Z, Suwarnasarn A et al (2012) High-performance flexible lithium-ion electrodes based on robust network architecture. Energy Environ Sci 5:6845–6849. https://doi.org/10.1039/c2ee03110h
56. Hu YS, Liu X, Müller JO et al (2009) Synthesis and electrode performance of nanostructured V_2O_5 by using a carbon tube-in-tube as a nanoreactor and an efficient mixed-conducting network. Angew Chem Int Ed 48:210–214. https://doi.org/10.1002/anie.200802988
57. Kong D, Li X, Zhang Y et al (2016) Encapsulating V_2O_5 into carbon nanotubes enables the synthesis of flexible high-performance lithium ion batteries. Energy Environ Sci 9:906–911. https://doi.org/10.1039/c5ee03345d
58. Cheng J, Wang B, Xin HL et al (2013) Self-assembled V_2O_5 nanosheets/reduced graphene oxide hierarchical nanocomposite as a high-performance cathode material for lithium ion batteries. J Mater Chem A 1:10814–10820. https://doi.org/10.1039/c3ta12066j
59. Palanisamy K, Um JH, Jeong M, Yoon WS (2016) Porous V_2O_5/RGO/CNT hierarchical architecture as a cathode material: emphasis on the contribution of surface lithium storage. Sci Rep 6:1–12. https://doi.org/10.1038/srep31275
60. Lee JW, Lim SY, Jeong HM et al (2012) Extremely stable cycling of ultra-thin V_2O_5 nanowire-graphene electrodes for lithium rechargeable battery cathodes. Energy Environ Sci 5:9889–9894. https://doi.org/10.1039/c2ee22004k
61. Rui X, Zhu J, Sim D et al (2011) Reduced graphene oxide supported highly porous V_2O_5 spheres as a high-power cathode material for lithium ion batteries. Nanoscale 3:4752–4758. https://doi.org/10.1039/c1nr10879d
62. Choi SH, Kang YC (2014) Uniform decoration of vanadium oxide nanocrystals on reduced graphene-oxide balls by an aerosol process for lithium-ion battery cathode material. Chem A Eur J 20:6294–6299. https://doi.org/10.1002/chem.201400134
63. Liu Q, Li Z, Liu Y et al (2015) Electrochemical performance for Li-ion batteries. 1–10. https://doi.org/10.1038/ncomms7127
64. Li G, Pang S, Jiang L et al (2006) Environmentally friendly chemical route to vanadium oxide single-crystalline nanobelts as a cathode material for lithium-ion batteries. J Phys Chem B 110:9383–9386. https://doi.org/10.1021/jp060904s
65. Rui X, Tang Y, Malyi OI et al (2016) Ambient dissolution-recrystallization towards large-scale preparation of V_2O_5 nanobelts for high-energy battery applications. Nano Energy 22:583–593. https://doi.org/10.1016/j.nanoen.2016.03.001

66. Aravindan V, Gnanaraj J, Lee YS, Madhavi S (2013) $LiMnPO_4$—a next generation cathode material for lithium-ion batteries. J Mater Chem A 1:3518–3539. https://doi.org/10.1039/c2ta01393b
67. Lazzari M, Scrosati B (1980) A cyclable lithium organic electrolyte cell based on two intercalation electrodes. J Electrochem Soc 127:773–774. https://doi.org/10.1149/1.2129753
68. Chan CK, Peng H, Twesten RD et al (2007) Fast, completely reversible Li insertion in vanadium pentoxide nanoribbons. Nano Lett 7:490–495. https://doi.org/10.1021/nl062883j
69. Wang Y, Takahashi K, Shang H, Cao G (2005) Synthesis and electrochemical properties of vanadium pentoxide nanotube arrays. J Phys Chem B 109:3085–3088. https://doi.org/10.1021/jp044286w
70. Tang Y, Rui X, Zhang Y et al (2013) Vanadium pentoxide cathode materials for high-performance lithium-ion batteries enabled by a hierarchical nanoflower structure via an electrochemical process. J Mater Chem A 1:82–88. https://doi.org/10.1039/c2ta00351a
71. Zhang C, Chen Z, Guo Z, Lou XW (2013) Additive-free synthesis of 3D porous V_2O_5 hierarchical microspheres with enhanced lithium storage properties. Energy Environ Sci 6:974–978. https://doi.org/10.1039/c3ee24134c
72. Pan KY, Wei DH (2018) Enhanced electronic and electrochemical properties of core-shelled V_2O_5-Pt nanowires. Appl Surf Sci 427:1064–1070. https://doi.org/10.1016/j.apsusc.2017.09.033
73. Yu D, Chen C, Xie S et al (2011) Mesoporous vanadium pentoxide nanofibers with significantly enhanced Li-ion storage properties by electrospinning. Energy Environ Sci 4:858–861. https://doi.org/10.1039/c0ee00313a
74. Sasidharan M, Gunawardhana N, Yoshio M, Nakashima K (2012) V_2O_5 hollow nanospheres: a lithium intercalation host with good rate capability and capacity retention. J Electrochem Soc 159:2012–2015. https://doi.org/10.1149/2.082205jes
75. Chen X, Zhu H, Chen YC et al (2012) MWCNT/V_2O_5 core/shell sponge for high areal capacity and power density Li-ion cathodes. ACS Nano 6:7948–7955. https://doi.org/10.1021/nn302417x
76. Zhao H, Yuan A, Liu B et al (2012) High cyclic performance of V_2O_5@PPy composite as cathode of recharged lithium batteries. J Appl Electrochem 42:139–144. https://doi.org/10.1007/s10800-012-0380-x
77. Wang YH, Liu H, Zhu D et al (2011) Preparation and electrochemical performance of hollow-spherical polypyrrole/V_2O_5 composite. Trans Nonferrous Met Soc China (Engl Ed) 21:1303–1308. https://doi.org/10.1016/S1003-6326(11)60857-1
78. Li L, Peng S, Chen H et al (2016) Polypyrrole-coated hierarchical porous composites nanoarchitectures for advanced solid-state flexible hybrid devices. Nano Energy 19:307–317. https://doi.org/10.1016/j.nanoen.2015.11.026
79. Gittleson FS, Hwang J, Sekol RC, Taylor AD (2013) Polymer coating of vanadium oxide nanowires to improve cathodic capacity in lithium batteries. RSC J Mater Chem A 1:7979–7984. https://doi.org/10.1039/c3ta11049d
80. Park KI, Song HM, Kim Y et al (2010) Electrochemical preparation and characterization of V_2O_5/polyaniline composite film cathodes for Li battery. Electrochim Acta 55:8023–8029
81. Chen Y, Yang G, Zhang Z et al (2010) Polyaniline-intercalated layered vanadium oxide nanocomposites—one-pot hydrothermal synthesis and application in lithium battery. Nanoscale 2:2131–2138. https://doi.org/10.1039/c0nr00246a
82. Chao D, Xia X, Liu J et al (2014) A V_2O_5/conductive-polymer core/shell nanobelt array on three-dimensional graphite foam: a high-rate, ultrastable, and freestanding cathode for lithium-ion batteries. Adv Mater 26:5794–5800. https://doi.org/10.1002/adma.201400719
83. Song HM, Yoo DY, Hong SK et al (2011) Electrochemical impedance analysis of V_2O_5 and PEDOT composite film cathodes. Electroanalysis 23:2094–2102. https://doi.org/10.1002/elan.201100177
84. Kwon CW, Murugan AV, Campet G (2003) Preparation, characterization and electrochemical lithium insertion into the new organic-inorganic poly(3,4-ethylene dioxythiophene)/V_2O_5 hybrid. Act Passive Electron Compon 26:171–183. https://doi.org/10.1080/1042015031000073904

85. Murugan AV, Kwon CW, Campet G et al (2002) Electrochemical lithium insertion into a poly(3,4-ethylenedioxythiophene)PEDOT/V_2O_5 nanocomposite. J Power Sources 105:1–5. https://doi.org/10.1016/S0378-7753(01)00992-2
86. Mahajan S (2001) Hydrothermal synthesis reactor. Encycl Mater Sci Technol 3989–3992
87. Byrappa K, Yoshimura M (2013) Hydrothermal technology for nanotechnology—a technology for processing of advanced materials. https://doi.org/10.1016/B978-0-12-375090-7.00010-4
88. Livage J, Beteille F, Roux C et al (1998) Sol-gel synthesis of oxide materials. Acta Mater 46:743–750. https://doi.org/10.1016/S1359-6454(97)00255-3
89. Chen D-H, He X-R (2001) Synthesis of nickel ferrite nanoparticles by sol-gel method. Mater Res Bull 36:1369–1377. https://doi.org/10.1016/S0025-5408(01)00620-1
90. Amatucci G (1996) Cobalt dissolution in $LiCoO_2$-based non-aqueous rechargeable batteries. Solid State Ionics 83:167–173. https://doi.org/10.1016/0167-2738(95)00231-6
91. Odani A, Pol VG, Pol SV et al (2006) Testing carbon-coated VO_x prepared via reaction under autogenic pressure at elevated temperature as li-insertion materials. Adv Mater 18:1431–1436. https://doi.org/10.1002/adma.200501611
92. Ihsan M, Meng Q, Li L et al (2015) V_2O_5/mesoporous carbon composite as a cathode material for lithium-ion batteries. Electrochim Acta 173:172–177. https://doi.org/10.1016/j.electacta.2015.05.060
93. Kong D, Li X, Zhang Y, Hai X, Wang B, Qiu X, Song Q, Yang Q-H, Zhi L (2010) Encapsulating V_2O_5 into carbon nanotube enables flexible high-performance lithium ion batteries. Energy Environ Sci 2058–2061. https://doi.org/10.1039/b000000x
94. Yu R, Zhang C, Meng Q et al (2013) Facile synthesis of hierarchical networks composed of highly interconnected V_2O_5 nanosheets assembled on carbon nanotubes and their superior lithium storage properties. ACS Appl Mater Interfaces 5:12394–12399. https://doi.org/10.1021/am4033444
95. Sun B, Huang K, Qi X et al (2015) Rational construction of a functionalized V_2O_5 nanosphere/MWCNT layer-by-layer nanoarchitecture as cathode for enhanced performance of lithium-ion batteries. Adv Funct Mater 25:5633–5639. https://doi.org/10.1002/adfm.201502382
96. Song H, Liu C, Zhang C, Cao G (2016) Self-doped V^{4+}-V_2O_5 nanoflake for 2 Li-ion intercalation with enhanced rate and cycling performance. Nano Energy 22:1–10. https://doi.org/10.1016/j.nanoen.2016.02.004
97. Zhu K, Qiu H, Zhang Y et al (2015) Electrochemical properties of ZrO_2-doped V_2O_5 amorphous powders with spherical shape and fine size. Chemsuschem 8:1017–1025. https://doi.org/10.1002/cssc.201500027
98. Ko YN, Choi SH, Kang YC, Bin PS (2013) Electrochemical properties of ZrO_2-doped V_2O_5 amorphous powders with spherical shape and fine size. ACS Appl Mater Interfaces 5:3234–3240. https://doi.org/10.1021/am303150v
99. Zeng H, Liu D, Zhang Y et al (2015) Nanostructured Mn-doped V_2O_5 cathode material fabricated from layered vanadium jarosite. Chem Mater 27:7331–7336. https://doi.org/10.1021/acs.chemmater.5b02840
100. Li Z, Zhang C, Liu C et al (2016) Enhanced electrochemical properties of Sn-doped V_2O_5as a cathode material for lithium ion batteries. Electrochim Acta 222:1831–1838. https://doi.org/10.1016/j.electacta.2016.11.174
101. Gittleson FS, Hwang J, Sekol RC, Taylor AD (2013) Polymer coating of vanadium oxide nanowires to improve cathodic capacity in lithium batteries. J Mater Chem A 1:7979–7984. https://doi.org/10.1039/c3ta11049d
102. Noerochim L, Wang JZ, Wexler D et al (2012) Impact of mechanical bending on the electrochemical performance of bendable lithium batteries with paper-like free-standing V_2O_5-polypyrrole cathodes. J Mater Chem 22:11159–11165. https://doi.org/10.1039/c2jm16470a
103. Wang HG, Ma DL, Huang Y, Zhang XB (2012) Electrospun V_2O_5 nanostructures with controllable morphology as high-performance cathode materials for lithium-ion batteries. Chem A Eur J 18:8987–8993. https://doi.org/10.1002/chem.201200434

104. Cheah YL, Aravindan V, Madhavi S (2012) Improved elevated temperature performance of al-intercalated V_2O_5 electrospun nanofibers for lithium-ion batteries. ACS Appl Mater Interfaces 4:3270–3277. https://doi.org/10.1021/am300616k
105. Wu Y, Zhu P, Zhao X et al (2013) Highly improved rechargeable stability for lithium/silver vanadium oxide battery induced via electrospinning technique. J Mater Chem A 1:852–858. https://doi.org/10.1039/c2ta00042c
106. Pham-Cong D, Ahn K, Hong SW et al (2014) Cathodic performance of V_2O_5 nanowires and reduced graphene oxide composites for lithium ion batteries. Curr Appl Phys 14:215–221. https://doi.org/10.1016/j.cap.2013.10.022
107. Yarin AL (2011) Coaxial electrospinning and emulsion electrospinning of core–shell fibers. https://doi.org/10.1002/pat.1781
108. Yang G, Xu X, Yan W et al (2014) Single-spinneret electrospinning fabrication of $CoMn_2O_4$ hollow nanofibers with excellent performance in lithium-ion batteries. Electrochim Acta 137:462–469. https://doi.org/10.1016/j.electacta.2014.05.167
109. Wang B, Cheng JL, Wu YP et al (2012) Porous NiO fi bers prepared by electrospinning as high performance anode materials for lithium ion batteries. Electrochem Commun 23:5–8. https://doi.org/10.1016/j.elecom.2012.07.003
110. Zhu J, Chen L, Xu Z, Lu B (2015) Electrospinning preparation of ultra-long aligned nano fi bers thin films for high performance fully flexible lithium-ion batteries. Nano Energy 12:339–346. https://doi.org/10.1016/j.nanoen.2014.10.026
111. Raghavan P, Lim D, Ahn J et al (2012) Electrospun polymer nanofibers: the booming cutting edge technology. React Funct Polym 72:915–930. https://doi.org/10.1016/j.reactfunctpolym.2012.08.018
112. Shi X, Zhou W, Ma D et al (2015) Electrospinning of Nanofibers and Their Applications for Energy Devices. J Nanomater 2015. https://doi.org/10.1155/2015/140716
113. Online VA, Afrooz M (2015) RSC Adv. https://doi.org/10.1039/b000000x
114. Iida Y, Kanno Y (2009) Doping effect of M (M=Nb, Ce, Nd, Dy, Sm, Ag, and/or Na) on the growth of pulsed-laser deposited V_2O_5 thin films. J Mater Process Technol 209:2421–2427. https://doi.org/10.1016/j.jmatprotec.2008.05.033
115. Takahashi K, Wang Y, Lee K, Cao G (2006) Fabrication and Li^+-intercalation properties of V_2O_5-TiO_2 composite nanorod arrays. Appl Phys A Mater Sci Process 82:27–31. https://doi.org/10.1007/s00339-005-3375-1
116. Yu DM, Zhang ST, Liu DW et al (2010) Effect of manganese doping on Li-ion intercalation properties of V_2O_5 films. J Mater Chem 20:10841–10846. https://doi.org/10.1039/c0jm01252a
117. Hurst SL (1978) X-Ray powder diffraction study of a new vanadium oxide $Cr_{0.11}$ V_2O_5 synthesized by a sol–gel process. J Mater Chem C 985–997
118. Pradeep I, Ranjith Kumar E, Suriyanarayanan N et al (2018) Effect of Al doping concentration on the structural, optical, morphological and electrical properties of V_2O_5 nanostructures. New J Chem 42:4278–4288. https://doi.org/10.1039/c7nj03607h
119. Nam SC, Lim YC, Park HY et al (2001) The effects of Cu-doping in V_2O_5 thin film cathode for microbattery. Korean J Chem Eng 18:673–678. https://doi.org/10.1007/BF02706385
120. Zhan S, Chen G, Liu D et al (2009) Effects of Cr doping on the structural and electrochemical properties of V_2O_5. J Alloys Compd 479:652–656. https://doi.org/10.1016/j.jallcom.2009.01.023
121. Yiwata N, Watada A, Yamamuro T et al (1998) X-ray absorption fine structure of V_2O_5 and $Li_xV_2O_5$. J Synchrotron Radiat 5:1146–1148. https://doi.org/10.1107/s0909049597018670
122. Sakunthala A, Reddy MV, Selvasekarapandian S et al (2011) Energy storage studies of bare and doped vanadium pentoxide, $(V_{1.95}M_{0.05})O_5$, M=Nb, Ta, for lithium ion batteries. Energy Environ Sci 4:1712–1721. https://doi.org/10.1039/c0ee00513d
123. Wei Y, Ryu CW, Kim KB (2007) Improvement in electrochemical performance of V_2O_5 by Cu doping. J Power Sources 165:386–392. https://doi.org/10.1016/j.jpowsour.2006.12.016
124. Liu H, Zou Y, Huang L et al (2018) Enhanced electrochemical performance of sandwich-structured polyaniline-wrapped silicon oxide/carbon nanotubes for lithium-ion batteries. Appl Surf Sci 442:204–212. https://doi.org/10.1016/j.apsusc.2018.02.023

125. Hu L-H, Wu FY, Lin CT et al (2013) Graphene-modified $LiFePO_4$ cathode for lithium ion battery beyond theoretical capacity. Nat Commun 4:1–7. https://doi.org/10.1038/ncomms2705
126. Bin Luo W, Wen L, Luo HZ et al (2014) Carbon nanotube-modified $LiFePO_4$ for high rate lithium ion batteries. Xinxing Tan Cailiao/New Carbon Mater 29:287–294. https://doi.org/10.1016/S1872-5805(14)60138-4
127. Zhou G, Ma J, Chen L (2014) Selective carbon coating techniques for improving electrochemical properties of NiO Uanosheets. Electrochim Acta 133:93–99. https://doi.org/10.1016/j.electacta.2014.03.161
128. Li H, Zhou H (2012) Enhancing the performances of Li-ion batteries by carbon-coating: Present and future. Chem Commun 48:1201–1217. https://doi.org/10.1039/c1cc14764a
129. Reddy Channu VS, Ravichandran D, Rambabu B, Holze R (2014) Carbon and functionalized graphene oxide coated vanadium oxide electrodes for lithium ion batteries. Appl Surf Sci 305:596–602. https://doi.org/10.1016/j.apsusc.2014.03.140
130. Jia X, Zhang L, Zhang R et al (2014) Carbon nanotube-penetrated mesoporous V_2O_5microspheres as high-performance cathode materials for lithium-ion batteries. RSC Adv 4:21018–21022. https://doi.org/10.1039/c4ra01316f
131. Qin M, Liu J, Liang S ct al (2014) Facile synthesis of multiwalled carbon nanotube–V_2O_5 nanocomposites as cathode materials for Li-ion batteries. J Solid State Electrochem 18:2841–2846. https://doi.org/10.1007/s10008-014-2543-7
132. An GH, Lee DY, Ahn HJ (2016) Carbon-encapsulated hollow porous vanadium-oxide nanofibers for improved lithium storage properties. ACS Appl Mater Interfaces 8:19466–19474. https://doi.org/10.1021/acsami.6b05307
133. Cheah YL, von Hagen R, Aravindan V et al (2013) High-rate and elevated temperature performance of electrospun V_2O_5 nanofibers carbon-coated by plasma enhanced chemical vapour deposition. Nano Energy 2:57–64. https://doi.org/10.1016/j.nanoen.2012.07.012
134. Huang S, Cai Y, Jin J, Li Y (2014) Annealed vanadium oxide nanowires and nanotubes as high performance cathode materials for lithium ion battery. J Mater Chem A 2:14099–14108. https://doi.org/10.1039/C4TA02339K
135. Lala L, Yusoff M, Ramakrishna S (2012) Continuous tubular nanofibers of vanadium pentoxide by electrospinning for energy storage devices. J Nanoparticle Res 14:1201. https://doi.org/10.1007/s11051-012-1201-1
136. Berezina OY, Kirienko DA, Markova NP, Pergament AL (2015) Synthesis of vanadium pentoxide micro and nanofibers by electrospinning. Tech Phys 60:1361–1366. https://doi.org/10.1134/S1063784215090054
137. Ostermann R, Li D, Yin Y et al (2006) V_2O_5 nanorods on TiO_2 nanofibers: a new class of hierarchical nanostructures enabled by electrospinning and calcination. Nano Lett 6:1297–1302. https://doi.org/10.1021/nl060928a
138. Cheah YL, Aravindan V, Madhavi S (2013) Synthesis and enhanced lithium storage properties of electrospun V_2O_5 nanofibers in full-cell assembly with a spinel $Li_4Ti_5O_{12}$ anode. ACS Appl Mater Interfaces 5:3475–3480. https://doi.org/10.1021/am400666n
139. Ding T, Xu J, Chen C et al (2017) Controlled synthesis of NaV_6O_{15} nanorods with high reversible capacity and excellent cycling stability. J Mater Sci Technol 33:271–275. https://doi.org/10.1016/j.jmst.2016.04.012
140. Bach S, Boudaoud A, Emery N et al (2014) $K_{0.5}V_2O_5$: a novel Li intercalation compound as positive electrode material for rechargeable lithium batteries. Electrochim Acta 119:38–42. https://doi.org/10.1016/j.electacta.2013.12.039
141. Yu H, Zeng J, Hao W et al (2018) Mo-doped V_2O_5 hierarchical nanorod/nanoparticle core/shell porous microspheres with improved performance for cathode of lithium-ion battery. J Nanoparticle Res 20:135. https://doi.org/10.1007/s11051-018-4242-2
142. Liang S, Zhou J, Fang G et al (2014) Synthesis of mesoporous β-$Na_{0.33}V_2O_5$ with enhanced electrochemical performance for lithium ion batteries. Electrochim Acta 130:119–126. https://doi.org/10.1016/j.electacta.2014.02.131

143. Liang L, Zhou M, Xie Y (2012) Electrospun hierarchical LiV_3O_8 nanofibers assembled from nanosheets with exposed {100} facets and their enhanced performance in aqueous lithium-ion batteries. Chem Asian J 7:565–571. https://doi.org/10.1002/asia.201100757
144. Xiang J, Yu XY, Paik U (2016) General synthesis of vanadium-based mixed metal oxides hollow nanofibers for high performance lithium-ion batteries. J Power Sources 329:190–196. https://doi.org/10.1016/j.jpowsour.2016.08.079
145. Renard MS, Emery N, Baddour-Hadjean R, Pereira-Ramos JP (2017) γ'-V_2O_5: a new high voltage cathode material for sodium-ion battery. Electrochim Acta 252:4–11. https://doi.org/10.1016/j.electacta.2017.08.175
146. Van Nghia N, Long PD, Tan TA et al (2017) Electrochemical performance of a V_2O_5 cathode for a sodium ion battery. J Electron Mater 46:3689–3694. https://doi.org/10.1007/s11664-017-5298-y
147. Wang H, Bai Y, Chen S et al (2015) Binder-free V_2O_5 cathode for greener rechargeable aluminum battery. ACS Appl Mater Interfaces 7:80–84. https://doi.org/10.1021/am508001h
148. Reed LD, Menke E (2013) The roles of V_2O_5 and stainless steel in rechargeable Al-ion batteries. J Electrochem Soc 160:915–917. https://doi.org/10.1149/2.114306jes
149. Wang W, Jiang B, Xiong W et al (2013) A new cathode material for super-valent battery based on aluminium ion intercalation and deintercalation. Sci Rep 3:2–7. https://doi.org/10.1038/srep03383
150. Conway BE (1991) Transition from "Supercapacitor" to "Battery" behavior in electrochemical energy storage. J Electrochem Soc 138:1539. https://doi.org/10.1149/1.2085829
151. Du Pasquier A, Plitz I, Menocal S, Amatucci G (2003) A comparative study of Li-ion battery, supercapacitor and nonaqueous asymmetric hybrid devices for automotive applications. J Power Sources 115:171–178. https://doi.org/10.1016/S0378-7753(02)00718-8
152. Chen S, He S, Hou H (2013) Electrospinning technology for applications in supercapacitors. Curr Org Chem 17:1402–1410. https://doi.org/10.2174/1385272811317130007
153. Desbrières J, Babak VG (2006) Interfacial properties of amphiphilic natural. Polymer 55:1177–1183. https://doi.org/10.1002/pi.2055
154. Jabbarnia A, Khan WS, Ghazinezami A, Asmatulu R (2016) Tuning the ionic and dielectric properties of electrospun nanocomposite fibers for supercapacitor applications. J Eng Res Appl 6:65–73. www.ijera.com
155. Wee G, Soh HZ, Cheah YL et al (2010) Synthesis and electrochemical properties of electrospun V_2O_5 nanofibers as supercapacitor electrodes. J Mater Chem 20:6720–6725. https://doi.org/10.1039/c0jm00059k
156. Sathiya M, Prakash AS, Ramesha K et al (2011) Ja207285B.Pdf. 16291–16299
157. Liu Z, Fu K, Wang Z et al (2016) Cut-and-stack nanofiber paper toward fast transient energy storage. Inorg Chem Front 3:681–688. https://doi.org/10.1039/c5qi00288e
158. Lee M, Hong WG, Jeong HY et al (2014) Graphene oxide assisted spontaneous growth of V_2O_5 nanowires at room temperature. Nanoscale 6:11066–11071. https://doi.org/10.1039/c4nr01780c
159. Zhu C, Hu D, Liu Z (2017) Interconnected three-dimensionally hierarchical heterostructures with homogeneously-dispersed V_2O_5 nanocrystals and carbon for high performance supercapacitor electrodes. Electrochim Acta 229:155–165. https://doi.org/10.1016/j.electacta.2017.01.144
160. Chen Z, Augustyn V, Wen J et al (2011) High-performance supercapacitors based on intertwined CNT/V_2O_5 nanowire nanocomposites. Adv Mater 23:791–795. https://doi.org/10.1002/adma.201003658
161. Huang G, Li C, Sun X, Bai J (2017) Fabrication of vanadium oxide, with different valences of vanadium, -embedded carbon fibers and their electrochemical performance for supercapacitor. New J Chem 41:8977–8984. https://doi.org/10.1039/c7nj01482a
162. Kim BH, Kim CH, Yang KS et al (2012) Electrospun vanadium pentoxide/carbon nanofiber composites for supercapacitor electrodes. Electrochim Acta 83:335–340. https://doi.org/10.1016/j.electacta.2012.07.093

163. Nie G, Lu X, Lei J et al (2014) Electrospun V_2O_5-doped α-Fe_2O_3 composite nanotubes with tunable ferromagnetism for high-performance supercapacitor electrodes. J Mater Chem A 2:15495–15501. https://doi.org/10.1039/c4ta01732c
164. Jiang H, Niu H, Yang X et al (2018) Flexible Fe_2O_3 and V_2O_5 nanofibers as binder-free electrodes for high-performance all-solid-state asymmetric supercapacitors. Chem A Eur J 24:10683–10688. https://doi.org/10.1002/chem.201800461
165. Saravanakumar B, Purushothaman KK, Muralidharan G (2012) Interconnected V_2O_5 nanoporous network for high-performance supercapacitors. ACS Appl Mater Interfaces 4:4484–4490. https://doi.org/10.1021/am301162p
166. Zhang Y, Zheng J, Zhao Y et al (2016) Fabrication of V_2O_5 with various morphologies for high-performance electrochemical capacitor. Appl Surf Sci 377:385–393. https://doi.org/10.1016/j.apsusc.2016.03.180

Chapter 19
Electrospun Manganese Oxide-Based Composites as Cathodes for Lithium-Ion Batteries

Aiswarya Bhaskar, Ditty Dixon, Ammu Surendran, and Harsha Enale

19.1 Introduction

The state-of-the-art energy storage systems mainly support the use of lithium-ion batteries due to its high energy density and moderate power density. The careful optimization of lithium-ion batteries (LIBs) can be a milestone for stationary as well as electromobility application. In a lithium-ion battery, positive electrode (cathode) acts as lithium source and hence, the capacity limiting electrode and negative electrode (anode) is the lithium accepting electrode (e.g. Graphite, $Li_4Ti_5O_{12}$, etc.). Hence, it is important to choose carefully the cathode materials for efficient and prolonged use of a lithium-ion battery.

The cathode materials are divided mainly into three classes based on their crystal structures, such as olivines (Fig. 19.1), layered (Fig. 19.2) and spinels (Fig. 19.3) [1–3]. Other than these, miscellaneous materials such as silicates are also under investigation by the researchers [4]. The olivine family has a nominal composition LiMPO$_4$, M=Fe, Mn, Ni, Co, etc. [1]. However, this class of materials suffers from a lower energy density. Among those, $LiFePO_4$ is intensively investigated by the researchers as a high rate and safe cathode material for lithium-ion battery. It has a theoretical specific capacity of 170 mAh g^{-1} and a flat charge-discharge plateau around ~3.45 V versus Li^+/Li. Nevertheless, $LiFePO_4$ has a poor electronic conductivity and hence is used with a carbon coating to ensure good electrochemical properties. Moreover, due to the one-dimensional movement of Li^+ ions in $LiFePO_4$, any defect that can hinder the lithium channels can adversely affect the electrochemical performance.

The layered class of materials, having a composition LiMO$_2$ (M=Co, Mn, Ni, Fe, etc.), are attractive cathode materials facilitating a two-dimensional Li^+ diffusion. Most of these materials have a crystal structure belonging to $R\overline{3}m$ space group, where

A. Bhaskar (✉) · D. Dixon · A. Surendran · H. Enale
Electrochemical Power Sources Division, CSIR-Central Electrochemical Research Institute, Karaikudi, Tamil Nadu 630003, India
e-mail: aiswarya@cecri.res.in

N. T. M. Balakrishnan and R. Prasanth (eds.), *Electrospinning for Advanced Energy Storage Applications*, Materials Horizons: From Nature to Nanomaterials,
https://doi.org/10.1007/978-981-15-8844-0_19

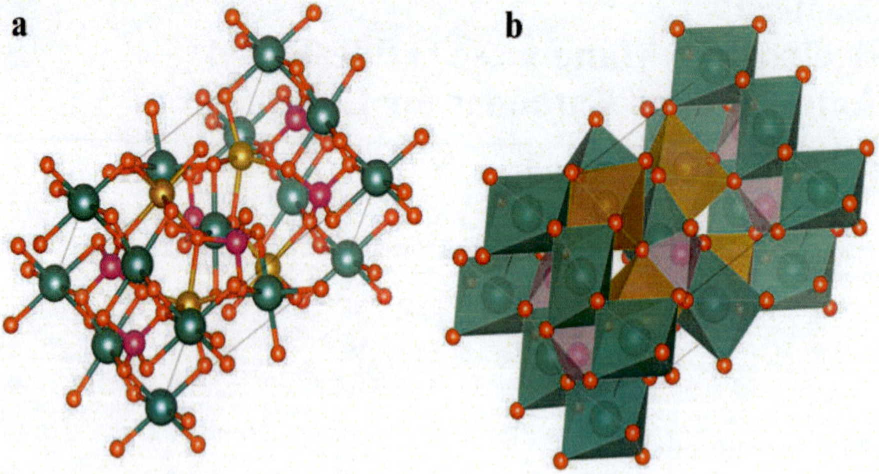

Fig. 19.1 Olivine structure of $LiFePO_4$, **a** ball and stick model, **b** polyhedron model

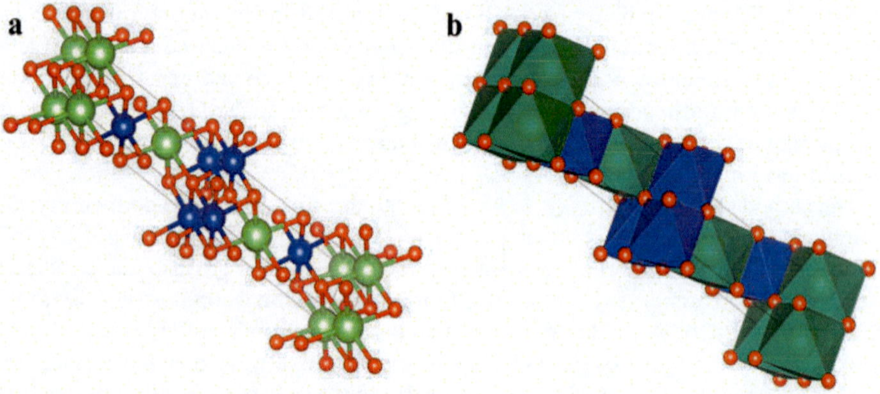

Fig. 19.2 Layered structure of $LiCoO_2$, **a** ball and stick model, **b** polyhedron model

there exist alternate layers of transition metal-oxygen and lithium [5]. In the transition metal-oxygen layer (MO_2 layers), the metal occupies octahedral sites whose corners are made up of oxygen. $LiCoO_2$ is the widely known commercial cathode material in this class, and the $LiCoO_2$/C lithium-ion technology was commercialized by Sony in 1991 [2]. This material exhibits a stable electrochemical performance for several hundreds of cycles. However, only 0.5 mol of Li^+ out of 1 mol of total Li^+ ions present (corresponds to a capacity of ~140 mAh g^{-1}) can be extracted from this material, which means only half of the theoretical capacity (275 mAh g^{-1}) can be reached. Moreover, due to the presence of toxic and expensive Cobalt, this material is not environmentally friendly and cost effective [5]. By substituting a part (or complete) of Cobalt with Manganese and/or Nickel, this material can be modified in such a

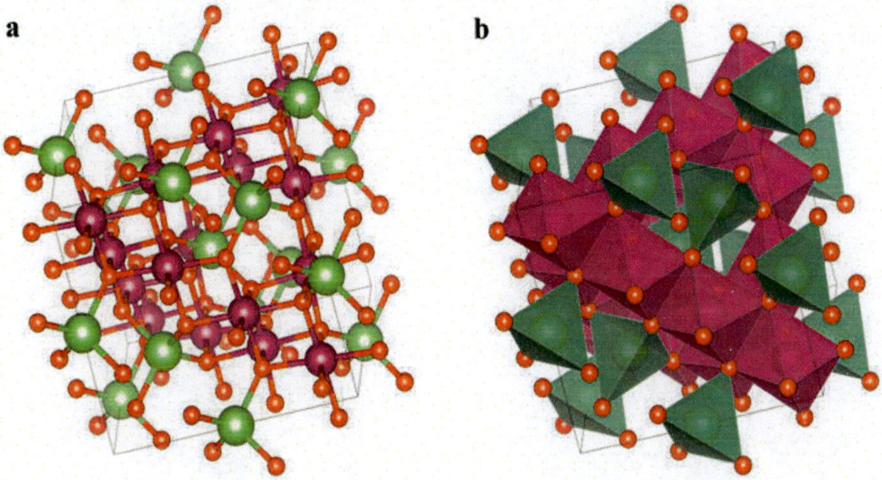

Fig. 19.3 Normal spinel structure of $LiMn_2O_4$, **a** ball and stick model, **b** polyhedron model

way that more than 50% of Li^+ can be extracted. Mn ions will exist in tetravalent form, and Ni ions will exist in divalent form in the material. The Mn^{4+} ions remain in octahedral coordination, ensuring the structural stability of the material [6].

In order to extract complete lithium from the layered compounds while keeping the structure intact, methods such as integrating it with Li_2MnO_3 having a monoclinic structure were adopted [6, 7]. This class of cathode materials is named "Lithium-rich layered" as the overall lithium amount exceeds 1 mol and the excess lithium resides in the transition metal layers (Fig. 19.4). The nominal composition can be written as $xLi_2MnO_3 \cdot (1-x)LiMO_2$ (M=Co, Mn, Ni, Fe, etc.). This class of material shows a specific capacity, exceeding 280 mAh g^{-1} [8]. However, this material still suffers from poor first cycle coulombic efficiency as well as voltage decay during cycling [9]. Moreover, the material exhibits poor rate capability due to poor Li^+ transport kinetics [10]. Initially, Li_2MnO_3 was considered as electrochemically inactive in view of the fact that the tetravalent Mn in octahedral coordination cannot be oxidized to higher valencies. However, extending the upper cut off voltage to >4.5 V versus Li^+/Li during charging, it was observed that Li and oxygen can be simultaneously removed from the material with a nominal composition of "Li_2O", which leads to the formation of MnO_2-like species [11]. In the subsequent discharge, Li^+-ions can be intercalated into the MnO_2-like species, forming $LiMnO_2$. It was observed that Li_2MnO_3 as the starting material shows a better electrochemical performance than $LiMnO_2$. The reason for this superior electrochemical performance was attributed to the existence of a composite structure in Li_2MnO_3 after the first "Li_2O" removal, in which Li_2MnO_3 domains are distributed in a $LiMnO_2$ matrix [12]. Several Li-rich layered compositions were researched thereafter which contain Co, Ni, Fe, etc., in addition to Manganese [13]. In the first charge against Li anode in a half cell, a plateau is visible in the electrochemical profile of these materials, corresponds to the

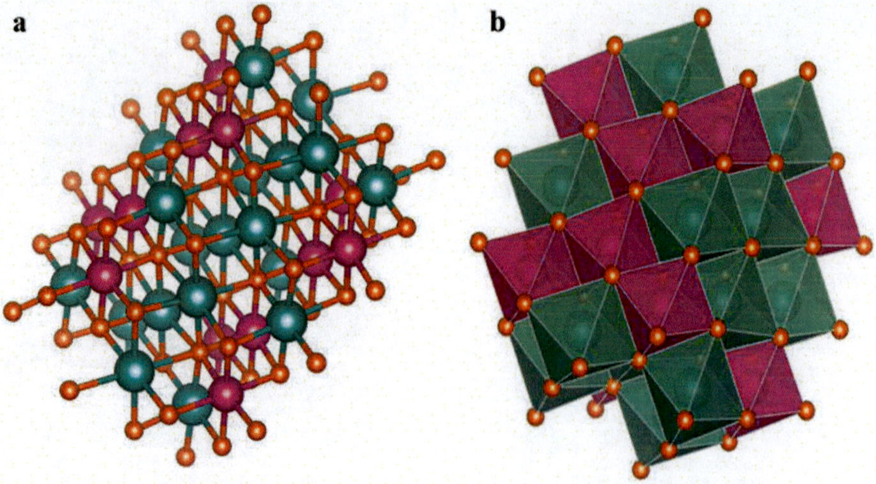

Fig. 19.4 Li-rich layered structure of Li_2MnO_3, **a** ball and stick model, **b** polyhedron model

simultaneous removal of Li^+ and stoichiometric amounts of oxygen. In the subsequent discharge as well as charge, when performed at slow charge–discharge rates, the plateau disappears, and the profile becomes sloppy indicating the completion of the "Li_2O" extraction reaction. However, these materials are yet to be commercialized mainly due to the issue of voltage decay upon continuous cycling. In order to overcome these issues, several methods are suggested in the literature such as surface coating using carbon, fluorides and phosphates. [2].

Manganese-based spinel-type cathode material $LiMn_2O_4$ and its transition metal (Ni, Co, Cr, Fe, Zn, etc.) doped derivatives have attracted huge attention as high voltage cathodes for lithium-ion batteries since 90s [14]. $LiMn_2O_4$, the 4 V spinel cathode, suffers from issues like (i) disproportionation of Mn^{4+} to form trivalent and divalent Mn ions, leading to dissolution of Mn in electrolyte and further forming oxygen vacancies, which if further aggravated at high temperatures and (ii) Jahn–Teller distortion, resulting in cubic to tetragonal phase transition during deep discharge [14, 15]. It was observed that substituting a part of Mn with other transition metals can increase the average oxidation state of Mn and solve these issues to a certain extent [3]. Hence, its transition metal derivatives have also gained huge attention as cathode materials (>4 V) for LIBs. $LiMn_2O_4$ was commercialized by NEC in 1996 [16]. The crystal structure of these materials mainly belongs to normal spinels with the space group $Fd\bar{3}m$ (disordered), where Li^+ ions occupy tetrahedral 8*a* sites, manganese ions and the other transition metal, if doped, occupy octahedral 16*d* sites, and oxygen atoms occupy the 32*e* sites, forming a face-centred cubic close packing [17]. In this material, Li^+ extraction (charging step) from 8*a* sites of the crystal lattice takes place via the formation of two intermediate cubic spinel phases, resulting in the appearance of two peaks around 4 V. Both these phases could be assigned to a space group of $Fd\bar{3}m$. In addition, another initial crystal structure belonging to

$P4_332$ (ordered) exists for some of the transition metal substituted $LiMn_2O_4$, where the substituted transition metal as well as manganese do not share same Wyckoff site 16*d*. In the cation-ordered spinel, Mn and the substituted transition metal (for, e.g. Mn and Ni in $LiNi_{0.5}Mn_{1.5}O_4$) occupy distinct octahedral sites such as 4*b* and 12*d*, respectively [17, 18]. These materials possess high rate capability as a result of the three-dimensional Li^+ diffusion pathway, formed by the 8*a* tetrahedral sites and vacant 16*c* octahedral sites. Among these, the Ni-substituted spinel with composition $LiNi_{0.5}Mn_{1.5}O_4$ is highly attractive as this material contains all the Mn in tetravalent form and Ni in divalent form. As a result, complete capacity is obtained around ~4.7 V for this material where the Ni^{4+}/Ni^{2+} electrochemical activity occurs.

Among these different classes of cathode materials, Mn-based layered, Li-rich layered and spinel materials are particularly attractive as cathodes in lithium-ion batteries for electric vehicles [19]. While the layered and Li-rich layered materials deliver a high capacity, which translates into a high energy density in a lithium-ion battery, the spinel class of materials exhibits a high rate capability that translates into a high power density in a lithium-ion battery. As noted here, it is still necessary to improve the rate capability of layered and Li-rich layered materials in order to increase their penetration as cathodes in lithium-ion batteries with high power density and energy density [20]. To achieve that, the materials should be modified in such a way to enhance the lithium diffusion kinetics. Several methods are adopted for shortening the Li^+ diffusion pathway such as going to nano-sized particles, synthesizing nanoarchitectures, creating composites with carbon, etc. Among these, electrodes synthesized using electrospinning technique are very interesting due to the improved Li^+ transport kinetics, provided via 1D nanostructures formed through facile and low-cost synthesis routes that are industrially viable [21–23]. One of the advantages is that the electrospun electrodes can be prepared sometimes binder free and that way the electrochemically inactive components of the electrode can be reduced. In addition, the quicker intercalation/deintercalation of Li^+ ions, minimized dissolution throughout cycling arising from stability of the morphology and increased surface area will result in enhanced electrochemical performance of the cathode in the cell. At the same time, low packing density is the major disadvantage of electrospun electrodes, which can be overcome by using needleless electrospinning technology [24].

19.2 Electrospun Mn-Based Layered Cathode Materials

The investigations performed on electrospun electrodes of layered and Li-rich layered materials containing manganese will be described here. In 2009, Ding et al. reported the synthesis of aluminium-doped $LiNi_{1/3}Co_{1/3}Mn_{1/3}O_2$ nanofibres via electrospinning [25]. For this, initially an electrospinning precursor solution was prepared by dissolving appropriate amounts of metal nitrate precursors in deionized water. To the prepared solution, polyvinylpyrrolidone (PVP) was added and stirred intensively. The obtained sol was then electrospun at a voltage of 25 kV, keeping a distance of

15 cm between the needle and the collector. The collector loaded with the electrospun fibres was then heated under vacuum at 100 °C, followed by peeling off the fibres and annealing them at high temperatures such as 500 and 800 °C. The obtained nanofibres were found to exist in layered structure (space group $R\overline{3}m$), as revealed by X-ray diffraction (XRD) analysis. Atomic force microscopic (AFM) studies before calcination confirmed the cylindrical structure of the electrospun nanofibres, with diameter in the range of 200–250 nm and with smooth surface. After calcination at 500 °C, the surface of the fibre was found to be rougher with decrease of the diameter. Increasing the calcination temperature furthermore to 800 °C resulted in the complete shrinking of the fibres with decrease of diameter to 100 nm. For comparison, nanopowder of the same material was prepared through a sol-gel process. Electrochemical investigations in lithium half cells (note that the electrochemical experiments mentioned in this chapter are always done in half cells against Li metal as counter and reference electrodes unless otherwise specified) reveal superior rate capability for the electrospun nanofibres of aluminium-doped $LiNi_{1/3}Co_{1/3}Mn_{1/3}O_2$ in comparison with the sol-gel synthesized nanopowder [25].

Another work came out in 2012 by Kang et al. added support to the results reported by Ding et al. on layered $LiNi_{1/3}Co_{1/3}Mn_{1/3}O_2$ with an only difference that the sample was not doped with aluminium [26]. Here, the synthesis of electrospun nanofibres was carried out using the same method adopted by Ding et al. The final annealing temperatures used were 700 °C after a pre-annealing at 500 °C, and the calcination was performed under air as well as oxygen for comparison. XRD patterns obtained for the electrospun nanofibres show that the synthesized materials belong to hexagonal α-$NaFeO_2$-type structure. However, the degree of cation mixing was found to be influenced by the annealing atmosphere. Morphology studies conducted using scanning electron microscope (SEM) on air- and oxygen-annealed samples indicated the presence of nanofibres with ~100 nm and 800 nm diameters, respectively. Additionally, the sample annealed under air displayed more agglomeration. Cyclic voltammetry studies indicated better reversibility of cathodic and anodic peaks for the oxygen-annealed nanofibres. Furthermore, higher absolute capacity was delivered by the oxygen-annealed electrospun $LiNi_{1/3}Co_{1/3}Mn_{1/3}O_2$ nanofibres, which could be attributed to the better uniformity and smaller diameter of these fibres [26].

19.3 Electrospun Mn-Based Li-Rich Layered Cathode Materials

Hosono et al. reported the synthesis of $0.5Li_2MnO_3$–$0.5LiNi_{1/3}Co_{1/3}Mn_{1/3}O_2$ hollow wires as positive electrodes for high energy Li-ion batteries [27]. Synthesis of $0.5Li_2MnO_3$–$0.5LiNi_{1/3}Co_{1/3}Mn_{1/3}O_2$ precursor solution was carried out by dissolving stoichiometric amounts of metal acetates and 1.5 g of polyvinyl alcohol (PVA) in a mixture of water, methanol and acetic acid by stirring at 90 °C for 1 h. The electrospinning of the solution was then carried out at a voltage of 20 kV applied across the metal needle and the collector, which here is an aluminium foil. After a preheating of the precursors with the collector under vacuum, the obtained nanowires were removed from the collector and annealed at a higher temperature of 800 °C for 3 h under air. The prepared samples were subjected to XRD, and the obtained patterns confirm the crystallization of $0.5Li_2MnO_3$–$0.5LiNi_{1/3}Co_{1/3}Mn_{1/3}O_2$ in a layered-layered solid-solution. High grade of crystallinity was revealed by the existence of sharp reflections in the XRD pattern. The obtained material was found to possess a hollow nanowire morphology using field emission SEM and transmission electron microscopy investigations. Moreover, the hollow structure morphology of the wires was found to change even with hand crushing. According to the authors, the idea of electrospinning here is to suppress three-dimensional grain growth at high temperature annealing and to obtain smaller particles with high crystallinity and high tap density than getting hollow structures. In fact, the hollow structures would result in low volumetric energy density in a battery. Voltage profiles of electrospun $0.5Li_2MnO_3$–$0.5LiNi_{1/3}Co_{1/3}Mn_{1/3}O_2$ hollow wires in LIBs showed fingerprint behaviour of the Li-rich layered material during the first charging with the plateau around 4.5 V which could be attributed to the activation of Li_2MnO_3 component. The material delivered a charge capacity of around 328 mAh g^{-1} in the first cycle. The first discharge capacity delivered was 273 mAh g^{-1}. In addition, a coulombic efficiency of 83% was obtained in the first cycle, which is quite good for layered Li-rich materials. The authors attribute this high capacity and first cycle coulombic efficiency to the volume relaxation of the hollow wires during cycling due to its peculiar morphological features. However, a capacity decrease was observed after 20 cycles [27].

Electrospun $Li_{1.2}Ni_{0.17}Co_{0.17}Mn_{0.5}O_2$ nanofibres were synthesized by Min et al., as high rate cathode materials for LIBs [22]. For this, an electrospinning solution consisting of polyvinylpyrrolidone (PVP), N,N-dimethylformamide (DMF) and stoichiometric amounts of lithium acetate dihydrate, manganese acetate tetrahydrate, nickel acetate tetrahydrate and cobalt acetate tetrahydrate was prepared. The prepared mixture was subjected to intense stirring for a day at room temperature and subjected to electrospinning at a high voltage of ~15–20 kV. The formed nanofibres were then collected on an aluminium collector and annealed for 12 h at 600 °C in air. For comparison, powder form of the $Li_{1.2}Ni_{0.17}Co_{0.17}Mn_{0.5}O_2$ material also synthesized via a co-precipitation process using the same starting materials. Morphological investigations conducted using SEM reveal the existence of nanofibre morphology with 100–200 nm diameter in the electrospun synthesized samples and aggregated particles in the case of sample synthesized via co-precipitation (Fig. 19.5a–d). TEM images of electrospun $Li_{1.2}Ni_{0.17}Co_{0.17}Mn_{0.5}O_2$ nanofibres confirm the uniform distribution of $Li_{1.2}Ni_{0.17}Co_{0.17}Mn_{0.5}O_2$ nanoparticles inside the nanofibre matrix and also indicate the crystalline nature of the sample (Fig. 19.5e, f).

Structural analysis using XRD confirmed the existence of $Li_{1.2}Ni_{0.17}Co_{0.17}Mn_{0.5}O_2$ in a single-phase α-$NaFeO_2$-type structure belonging to the space group $R\overline{3}m$ and was consistent with the observation from TEM analysis. The crystallization of $Li_{1.2}Ni_{0.17}Co_{0.17}Mn_{0.5}O_2$ in the layered-layered solid-solution was further confirmed by the presence of superlattice reflections around 20−23°

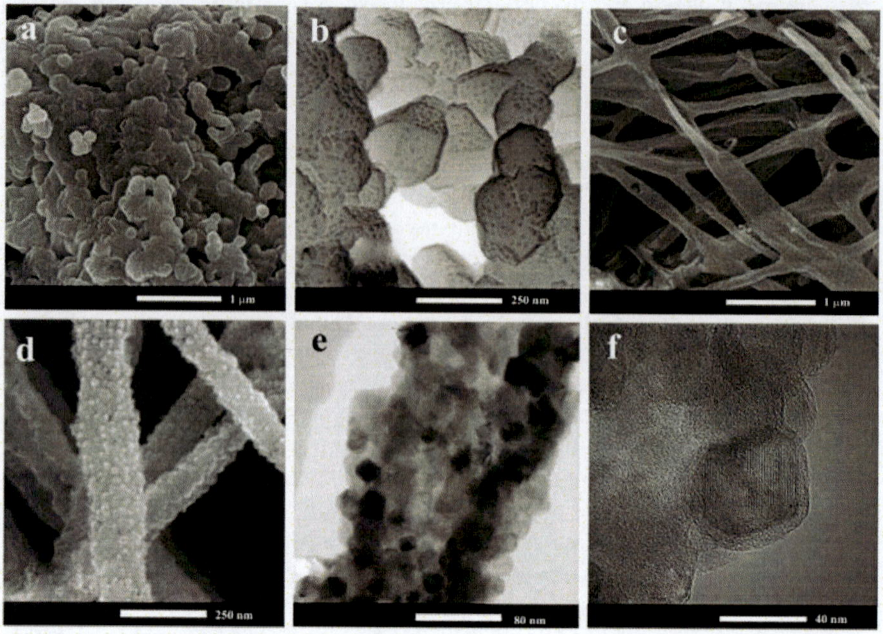

Fig. 19.5 SEM images of (**a**, **b**) $Li_{1.2}Ni_{0.17}Co_{0.17}Mn_{0.5}O_2$ nanoparticles, (**c**, **d**) $Li_{1.2}Ni_{0.17}Co_{0.17}Mn_{0.5}O_2$ nanofibres and TEM images of (**e**, **f**) $Li_{1.2}Ni_{0.17}Co_{0.17}Mn_{0.5}O_2$ nanofibres. Adapted and reproduced from Ref. [22]. Copyright 2013 American chemical Society

2θ. Additionally, Fourier transform infrared (FT-IR) spectroscopy analysis of $Li_{1.2}Ni_{0.17}Co_{0.17}Mn_{0.5}O_2$ samples was carried out in order to examine the effect of the nature of the PVP participating in the chemical reactions. Characteristic peaks for the formation of inorganic material were observed in the range of wavenumbers 500–700 cm^{-1} for the electrospun $Li_{1.2}Ni_{0.17}Co_{0.17}Mn_{0.5}O_2$ nanofibres. In comparison with the peaks observed for the co-precipitation via synthesized sample, the electrospun nanofibre peaks appeared smoother, indicating poor crystallinity of the compound. Electrochemical performance of the $Li_{1.2}Ni_{0.17}Co_{0.17}Mn_{0.5}O_2$ samples was studied by conducting cycling stability investigations at room temperature, in the voltage range 2.0–4.8 V in galvanostatic mode. The electrospun $Li_{1.2}Ni_{0.17}Co_{0.17}Mn_{0.5}O_2$ nanofibres delivered a high charge capacity as well as discharge capacity of 331 mAh g^{-1} and 256 mAh g^{-1}, respectively, in the first cycle. However, the charge as well as discharge capacities of $Li_{1.2}Ni_{0.17}Co_{0.17}Mn_{0.5}O_2$ nanopowder obtained via co-precipitation were observed to be 269 and 193 mAh g^{-1}. The $Li_{1.2}Ni_{0.17}Co_{0.17}Mn_{0.5}O_2$ nanofibres displayed a capacity retention of 66% at the 60th cycle, whereas the nanoparticles synthesized via co-precipitation exhibited 79%, which is slightly higher. Large surface area could be the reason for large capacity delivered by electrospun $Li_{1.2}Ni_{0.17}Co_{0.17}Mn_{0.5}O_2$ nanofibres electrode at first cycle. The nanofibre structure enhances the electrode–electrolyte contact in the cell. Even though the authors attribute the poor capacity retention of electrospun

$Li_{1.2}Ni_{0.17}Co_{0.17}Mn_{0.5}O_2$ nanofibres to its poor crystallinity, the enhanced parasitic reactions with increased surface area could also contribute to capacity decay. As revealed from the rate capability studies, nanofibre $Li_{1.2}Ni_{0.17}Co_{0.17}Mn_{0.5}O_2$ electrodes prepared via electrospinning method showed superior high rate performance. Meanwhile, an inferior rate capability was observed for the nanopowder $Li_{1.2}Ni_{0.17}Co_{0.17}Mn_{0.5}O_2$ (Fig. 19.6). This limited rate capability could be attributed to the sluggish solid-state Li^+ ion diffusion in the sample. The authors suggest that the diffusion distance of Li^+-ion could be further shortened by reducing the nanofibre diameter to 10–30 nm. A further improvement of intercalation kinetics can be achieved by using effective conducting nanofibres that enhance the electrical contact between the electrode particles. In summary, this investigation reveals that the electrospun $Li_{1.2}Ni_{0.17}Co_{0.17}Mn_{0.5}O_2$ nanofibre electrode has superior electrochemical performance than the nanoparticle electrode mainly due to the improved Li^+ diffusion kinetics provided by the peculiar fibre morphology [22].

$Li_{1.2}Mn_{0.54}Ni_{0.13}Co_{0.13}O_2$ encapsulated carbon nanofibres, synthesized by combination of electrospinning method and heat treatment were reported by Ma et al. [28]. For comparison, pristine $Li_{1.2}Mn_{0.54}Ni_{0.13}Co_{0.13}O_2$ particles were synthesized via sol-gel method. In this work, PVP-assisted sol-gel method was used for the preparation of $Li_{1.2}Mn_{0.54}Ni_{0.13}Co_{0.13}O_2$ particles. In order to do this, firstly, lithium acetate dihydrate, manganese acetate tetrahydrate, nickel acetate tetrahydrate and cobalt acetate tetrahydrate were dissolved in 10 wt% polyvinylpyrrolidone (PVP) solution (in water) under intense stirring for 6 h. A viscous violet gel was obtained by evaporating the solution at 85 °C. The obtained gel was then dried under vacuum at 120 °C for 18 h. The powder was annealed at 500 °C for 6 h to decompose PVP completely

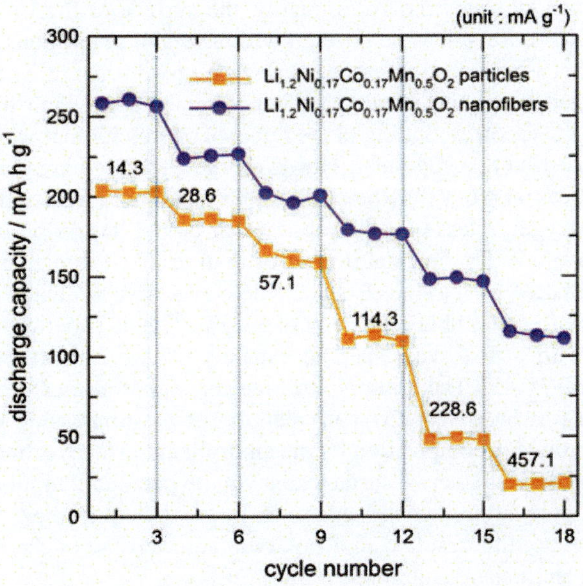

Fig. 19.6 Rate capabilities of $Li_{1.2}Ni_{0.17}Co_{0.17}Mn_{0.5}O_2$ nanoparticles and $Li_{1.2}Ni_{0.17}Co_{0.17}Mn_{0.5}O_2$ nanofibres (voltage 2.0–4.8 V, charging current was 14.3 mA g^{-1}). Adapted and reproduced from reference [22]. Copyright 2013 American chemical Society

and further calcined at 950 °C for 8 h in air to get the final product. After this step, to prepare electrospinning suspension, dedicated amounts of $Li_{1.2}Mn_{0.54}Ni_{0.13}Co_{0.13}O_2$ and carbon black were added into polyacrylonitrile (PAN) in dimethyl formamide (DMF) solution (8 wt%) and dispersed ultrasonically. The prepared suspension was then subjected to electrospinning at a high voltage of 18 kV with a feeding rate of 1 ml/h and with a 20 cm needle-to-collector distance. The obtained electrospun nanofibres were collected on the aluminium foil and dried in a vacuum oven for 12 h at 120 °C. The electrospun nanofibres obtained thereafter were subjected to different subsequent heating steps in order to get the final $Li_{1.2}Mn_{0.54}Ni_{0.13}Co_{0.13}O_2$ encapsulated carbon nanofibres. Pristine $Li_{1.2}Mn_{0.54}Ni_{0.13}Co_{0.13}O_2$ was also synthesized via the same heating process for comparison. From the XRD analysis, the pristine sample was found to crystallize in α-$NaFeO_2$ structure without any additional reflections indicating the absence of crystalline admixtures. Superstructure reflections correspond to the existence of Li_2MnO_3 phase (Space group *C2/m*) was visible confirming the presence of layered-layered solid-solution. Similar XRD patterns with no additional reflections corresponding to the presence of crystalline form of carbon were observed in case of the electrospun samples, indicating the amorphous nature of the carbon present. Morphology investigations using SEM reveal a nanofibre nature for the electrospun nanofibres, whereas slight agglomeration with particle size distribution varying between 300 and 500 nm was observed for the pristine sample. Additionally, it was observed that the electrospun sample is comprised of $Li_{1.2}Mn_{0.54}Ni_{0.13}Co_{0.13}O_2$ particles, uniformly distributed with carbon black, which after heat treatment resulted in shrinking and breaking of the nanofibres together with squeezing out of particles from inside the fibres. High resolution TEM investigations reveal the dispersion of $Li_{1.2}Mn_{0.54}Ni_{0.13}Co_{0.13}O_2$ and carbon black inside the carbon nanofibre matrix. Additionally, the surface of the $Li_{1.2}Mn_{0.54}Ni_{0.13}Co_{0.13}O_2$ fibres was observed to be covered with layers of amorphous carbon which is supposed to act as a protection barrier during charge–discharge. The cyclic voltammetric investigations on both pristine and nanofibre materials show fingerprint anodic peak around 4.6 V, corresponding to the removal of stoichiometric amounts of Li and O from the Li_2MnO_3 component. However, the decreased area and intensity of the oxidation peak at 4.6 V for the electrospun nanofibres were attributed to the suppression of oxygen release by the presence of carbon layers on the $Li_{1.2}Mn_{0.54}Ni_{0.13}Co_{0.13}O_2$ surface. Cycling stability investigations have shown that the electrospun nanofibres of $Li_{1.2}Mn_{0.54}Ni_{0.13}Co_{0.13}O_2$ deliver a first discharge capacity of 263.7 mAh g^{-1} with coulombic efficiency of 83.5%. This is superior to that of the pristine sample, with a first cycle specific capacity of 247.2 mAh g^{-1} and coulombic efficiency of 75.4%. The authors attribute this superior electrochemical performance of the $Li_{1.2}Mn_{0.54}Ni_{0.13}Co_{0.13}O_2$ nanofibres to suppression of oxygen release and electrolyte decomposition by the amorphous surface carbon layer and enhanced electrochemically active surface area via the peculiar nanofibre morphology with minimal aggregation, which leads to an improvement in the Li^+ transport kinetics. Furthermore, this resulted in a predominant increase of the rate capability in case of the electrospun nanofibre samples [28].

Recently in 2019, Al_2O_3 coated $Li_{1.2}Mn_{0.54}Ni_{0.13}Co_{0.13}O_2$ nanotubes were prepared by electrospinning by Chen et al. for the application as cathodes in LIBs [29]. The amount of Al_2O_3 was varied and optimized according to the electrochemical performance. In order to prepare the electrospinning solution, stoichiometric amounts of metal acetates were mixed with a pre-prepared solution of polyacrylonitrile (PAN) in DMF under vigorous stirring for 12 h at 60 °C. The solution was then subjected to electrospinning with a voltage of 15 kV and a flow rate of 1 ml h^{-1}. A distance of 200 mm was used between the spinneret and collector (copper foil). The nanofibres were then calcined at 800 °C under air to obtain final product. To coat with Al_2O_3, the obtained fibres were mixed with aluminium nitrate in distilled water under stirring. After washing and filtering, the obtained material was freeze dried and subjected to annealing at 500 °C for 10 h. XRD analysis confirms the formation of a layered structure for all the materials. The superlattice reflections indicate the presence of monoclinic domains of Li_2MnO_3. No considerable differences in the unit cell parameters were observed before and after Al_2O_3 coating. Morphology examination using SEM reveals a nanofibre morphology for the uncoated and coated samples after electrospinning which is changed to a nanotube morphology after annealing, as resulted from the decomposition of the polymer. In addition, a shrinking in the nanofibre diameter was observed after annealing at 500 °C with increase in surface roughness. Al_2O_3 coating was confirmed by X-ray photoelectron spectroscopy (XPS) investigations on the samples. A peak around 73.8 eV was observed, which could be attributed to the Al $2p_{3/2}$ binding energy in Al_2O_3. An amorphous layer of Al_2O_3 with $\leq$10 nm thickness was observed in HRTEM images for the medium Al_2O_3 coated $Li_{1.2}Mn_{0.54}Ni_{0.13}Co_{0.13}O_2$ nanotubes. Furthermore, spinel domains were observed using HRTEM for the Al_2O_3 coated samples. Cyclic voltammetric studies further confirm this observation, as revealed by the oxidation/reduction peaks around 3.0 V. For the coated and uncoated nanotubes, a high intensity peak was observed around 4.6 V in the charge, associated with the Li_2MnO_3 activation. Additionally, peaks correspond to the electrochemical activity of Ni and Co ions were observed in the 4 V region similar to that of the Li-rich layered samples containing Ni and Co as mentioned in the former text. The observations from CV curves were further supported by the voltage profiles obtained from galvanostatic cycling experiments performed on the samples between 2.0 and 4.8 V. Among the coated and uncoated nanotubes, sample with the highest percentage of Al_2O_3 coating delivered the highest first cycle charge capacity which is about 359 mAh g^{-1}. The discharge capacity was found to be 272 mAh g^{-1}, and the coulombic efficiency was observed to be ~76% in the first cycle for this sample. However, taking into account the first cycle coulombic efficiency (CE) and specific capacities, the medium Al_2O_3 coated $Li_{1.2}Mn_{0.54}Ni_{0.13}Co_{0.13}O_2$ nanotubes sample is the optimum one as it delivers a charge capacity of ~331 mAh g^{-1} and a CE of 85%, respectively. Additionally, the Al_2O_3 coated $Li_{1.2}Mn_{0.54}Ni_{0.13}Co_{0.13}O_2$ nanotubes prepared through electrospinning followed by calcination were found to exhibit superior cycling stability, compared with the Al_2O_3 coated nanoparticles of the same Li-rich material. High rate electrochemical investigations reveal excellent performance for the nanotubes synthesized via electrospinning owing to the enhanced

Li^+ diffusion and electronic conductivity in the one-dimensional nanostructures. Further, among those, the coated samples exhibited better rate capability and high rate performance than the uncoated $Li_{1.2}Mn_{0.54}Ni_{0.13}Co_{0.13}O_2$ nanotubes. In summary, the diameter and calcination temperatures were found to influence the electrochemical performance of the nanostructured electrodes synthesized via electrospinning [29].

19.4 Electrospun Mn-Based Spinel Cathode Materials

As mentioned in the former text, Mn-based spinels are promising cathode materials for LIBs, owing to their normal spinel structure, facilitating a three-dimensional Li^+ diffusion, resulting in high power density in a LIB. However, the rate capability of these high voltage cathode materials is still not sufficient in order to meet the requirements for high power applications. Hence, several research groups have adopted electrospinning to prepare nanofibres, which facilities faster Li^+ diffusion due to shortening of diffusion length.

Electrospun nanofibres of $LiMn_2O_4$ were synthesized by several research groups. Liu et al. synthesized $LiMn_2O_4$ nanofibres via electrospinning for application as LIB cathode materials in 2012 [30]. The precursor solution was prepared by dissolving manganese acetate, lithium carbonate and PVP in a mixture of solvents such as ethanol, distilled water and acetic acid. The solution was then taken in a syringe, fitted with a needle and electrospun at an applied voltage of 20 kV on an aluminium collector. A distance of 12 cm was maintained between the needle and collector. The obtained fibres were then subjected to a high temperature annealing at 800 °C for 5 h. Calcination was found to decrease the fibre diameter from 350 nm to 180 nm. Furthermore, it resulted in the transformation of fibres to compact grain structures. XRD analysis revealed the presence of single-phase cubic spinel $LiMn_2O_4$. The first discharge capacity for the electrospun, and 800 °C calcined fibres were observed to be 80 mAh g^{-1}, which showed a pronounced capacity fading even after 5 cycles. This shows that for good electrochemical performance, the electrospinning parameters and annealing temperatures need to be carefully optimized [30].

Zhao et al., in 2012, reported the fabrication of electrospun $LiMn_2O_4$ nanofibres as LIB cathode material [31]. Initially, metal acetates in stoichiometric ratios were added to a PVP in ethanol solution and stirred vigorously. The solution was then delivered into a syringe. A voltage of 15 kV was applied across the spinneret and the collector (here, Aluminium foil). After a precalcination at 450 °C, the fibres were annealed at 600 °C for 3 h at a rate of 5 °C min^{-1}. After confirming the phase purity with XRD, the prepared fibres were subjected to morphology investigations. The obtained fibres after calcination displayed diameters in the range 150–350 nm and a length of 3–5 μm. Each fibre was made of particles with face size ranging from 50 to 150 nm. The electrospun $LiMn_2O_4$ material after calcination displays two redox peaks in the cyclic voltammogram (3.5–4.3 V). These two peaks correspond to the stepwise Li^+ extraction from the 8*a* wyckoff sites in the crystal lattice of the cubic spinel structure. A higher hysteresis was observed for the electrospun $LiMn_2O_4$ fibres in comparison with $LiMn_2O_4$ synthesized via other methods. Cycling stability results reveal a discharge capacity retention of 94% after 50 cycles. However, the coulombic efficiency in the first cycle was found to be quite low (79%) for the $LiMn_2O_4$ spinels [31].

Zhou et al. have prepared $LiMn_2O_4$ with a porous "network-like" morphology and compared its electrochemical performance with the material prepared through a conventional sol-gel synthesis [32]. For this, solution of manganese acetate tetrahydrate and lithium nitrate in distilled water was prepared. Dedicated amount of PVP was added to this solution to prepare the electrospinning precursor. The solution was then taken in a syringe, fitted with a metal needle (diameter 1.4 mm). A voltage of 15 kV was applied across the needle and the collector, which is an aluminium foil. The distance between the needle and collector was maintained at 12 cm. A flow rate of 0.5 ml h^{-1} was used. The final nanofibres of $LiMn_2O_4$ were acquired by annealing the electrospun fibres at 700 °C for 4 h. The obtained material was found to crystallize in $Fd\bar{3}m$ space group using XRD analysis (Fig. 19.7d) [32]. SEM images display a diameter of 150 nm for the electrospun $LiMn_2O_4$ samples (Fig. 19.7a, b)

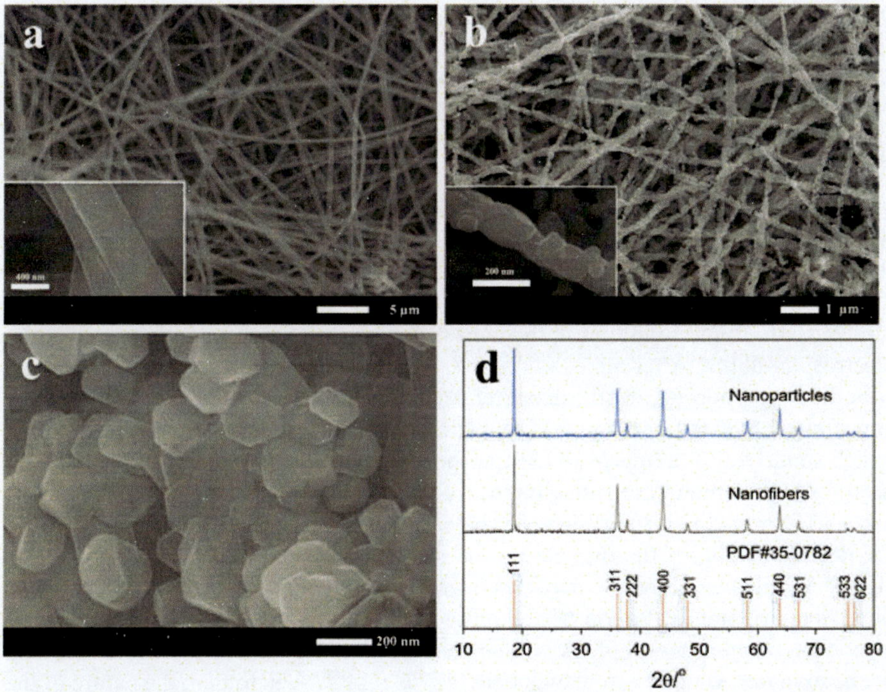

Fig. 19.7 **a** SEM images of electrospun $LiMn_2O_4$ precursor, **b** $LiMn_2O_4$ nanofibre cathode after annealing at 700 °C, **c** $LiMn_2O_4$ nanoparticles (sol-gel synthesized) after annealing at 700 °C, **d** XRD patterns of both synthesized samples. Adapted and reproduced from reference [32]. Copyright 2014 Elsevier

[32]. The sol-gel synthesized sample was found to exist in a particle size of 200 nm and exhibited strong agglomeration (Fig. 19.7c) [32]. Even though voltage profiles (3.0 V to 4.5 V) show similar behaviour for the electrospun and conventional sol-gel synthesized electrodes, the cycling stability investigations, conducted at room temperature, reveal a superior behaviour for the electrospun nanofibres. The authors attribute this enhanced capacity retention (92% at 60th cycle) to the suppression of Mn dissolution. This observation is further supported by the fact that the electrospun $LiMn_2O_4$ samples exhibit a better cycling stability than the sol-gel synthesized sample even at 55 °C. For the cycling stability experiments conducted at 10 C (1C = 148 mAh g^{-1}), the electrospun nanofibres delivered a higher specific capacity (90 and 78 mAh g^{-1} for the electrospun and sol-gel samples, respectively) and capacity retention (85% and 58% for the electrospun and sol-gel samples, respectively) than the sol-gel synthesized nanoparticles. Additionally, enhanced rate capability was also observed for the electrospun samples. This could be attributed to the enhanced surface area as well as shorter Li^+ diffusion path due to the typical nanofibre morphology, resulted from electrospinning [32].

Porous hollow nanofibres of $LiMn_2O_4$ were synthesized by Duan et al., in 2017, by electrospinning the fibres on a fluorine-doped tin oxide glass collector [33]. For the preparation of electrospinning precursor solution, firstly, metal acetates in stoichiometric amounts were dissolved into a pre-prepared solution of PVP in ethanol, followed by addition of acetic acid. The precursor solution was then delivered into a syringe with a metal needle. A voltage of 20 kV was applied across the needle and metal collector, and the distance between needle and collector was kept as 20 cm. After a heating step at 60 °C, for 1 h, the obtained fibres were subjected to various annealing temperatures for comparison. The 700 °C annealed $LiMn_2O_4$ nanofibres were phase pure (space group $Fd\bar{3}m$) as revealed by XRD analysis. Meanwhile, the diffraction pattern obtained for 650 °C annealed samples exhibited reflections corresponding to the presence of Mn_2O_3 impurity phase. The annealing temperature was also found to influence the nanostructure morphology and diameter of the electrospun fibres. After annealing at 700 °C, the smooth surface of the precursor nanofibres gets porous, and the diameter decreases from 500 to 200 nm. Additionally, annealing at high temperature was found to create a network structure for the fibres, which would result in improved Li^+ transport kinetics. Furthermore, the heating rate used for high temperature annealing was also found to influence the morphology of the nanofibres. This is dependent on the speed of release of CO_2 and H_2O (during the decomposition of the organic moieties). When the gas and water are released slowly during annealing, the particles have a tendency to grow, which will result in increased size and disruption of nanofibre morphology. Whereas, when the release is too fast, the fibre morphology will be disturbed. Hence, it is important to maintain the heating rate as optimum, which here is 5 °C min^{-1}. Add to that, the nature of the collector was found to influence the fibre morphology. The fluorine-doped tin oxide glass collector was found to provide the nanofibres more uniformity and integrity than that of an aluminium collector, the use of which in comparison results in more agglomeration. Electrochemical performance analysis of the $LiMn_2O_4$ nanofibres, after 700 °C annealing at a heating rate of 5 °C min^{-1}, shows charge and discharge capacities of 132.8 and 125.9 mAh g^{-1}, respectively, at a rate of C/10 (1C = 148 mAh g^{-1}). The voltage range used for this galvanostatic cycling measurement was 3.0–4.4 V. A capacity retention of 91% was observed after 200 cycles and 83% after 400 cycles. Moreover, it exhibited satisfactory rate capability. The cycling stability was found to be highest for this sample in comparison with the samples annealed to 700 °C with heating rates 3 and 7 °C min^{-1} [33].

Hosono et al. have reported the fabrication of $LiNi_{0.5}Mn_{1.5}O_4$ hollow wires as cathodes for high energy Li-ion batteries [27]. The synthesis was performed using metal acetate precursors and PVA. The precursors were dissolved in a mixture of distilled water, methanol and acetic acid. After dissolving at 90 °C, the solution was delivered to a syringe having a metal needle. At a voltage of 20 kV, the precursor solution was ejected, and the obtained nanofibres were collected in an aluminium collector. After a pre-drying step at 100 °C under vacuum, the fibres were removed from the collector and annealed at 800 °C for 3 h to get the final nanofibres. In this work, the authors assign the crystal structure of electrospun $LiNi_{0.5}Mn_{1.5}O_4$ hollow wires to $P4_332$ space group even though no neutron diffraction data are provided. Normally, it is not possible to distinguish between the ordering of Ni and Mn in using XRD. Sharp reflections in the XRD pattern indicate a high degree of crystallinity for the hollow wire sample. TEM images show that the thin walls of hollow wires consist of faceted nanoparticles interconnected via a two-dimensional manner. Cycling stability investigations were conducted in galvanostatic mode. The first 20 cycles were conducted in a voltage range 3.5–4.9 V, whereas the later cycles were carried out between 2.0 and 4.9 V. In the broader voltage range, where the second Li^+ intercalation/deintercalation takes place, the capacity fade was more prominent, owing to the huge volume change associated with spinel to tetragonal phase transition. However, the authors claim an accommodation of this volume change by the smaller particles as revealed by its superior electrochemical performance in comparison with that of the $LiMn_2O_4$ spinels reported in the literature [27].

One-dimensional nanoarchitecture of $LiNi_{0.5}Mn_{1.5}O_4$ was prepared via electrospinning by Arun et al. [34]. Firstly, required amounts of metal acetates were dissolved in methanol, and acetic acid was added to the solution. Subsequently, the solution was mixed with a PVP solution in ethanol and subjected to vigorous stirring. This precursor solution was then taken in a syringe and electrospun at a voltage of 18 kV and a flow speed of 0.8 ml h^{-1}. Aluminium foil was used as a collector, and a distance of 15 cm was maintained between the collector and needle. After a precalcination at 300 °C for 3 h, the fibres were subjected to high temperature annealing at 600 °C for 1 h in a box furnace at a heating rate of 5 °C min^{-1}. The synthesized $LiNi_{0.5}Mn_{1.5}O_4$ electrospun fibres were then electrochemically investigated in half cells and full cells (against anatase TiO_2 anode) after basic characterization. Structural analysis using XRD shows the crystallization of $LiNi_{0.5}Mn_{1.5}O_4$ in cubic spinel phase with slight amount of $Li_xNi_{1-x}O$ impurity phase. The electrospun $LiNi_{0.5}Mn_{1.5}O_4$ nanofibres were of about 100–200 nm diameter as observed from FESEM analysis. After final calcination, the diameter of nanofibres reduced to ~50–100 nm, and TEM analysis showed the existence of intergrown crystalline nanoparticles with diameters in the range 10–20 nm on the fibres. Prepared $LiNi_{0.5}Mn_{1.5}O_4$

nanofibre electrodes, when investigated in Li half cells, exhibited typical Ni^{4+}/Ni^{2+} redox peaks above 4.7 V. An initial specific discharge capacity of ~118 mAh g^{-1} was delivered by the electrospun fibres (at a charge–discharge rate of 1 C = 150 mA g^{-1}) with a coulombic efficiency of 84.3%. Moreover, the material exhibited ~93% capacity retention after 50 cycles. The full cell assembly of electrospun $LiNi_{0.5}Mn_{1.5}O_4$ nanofibres was performed using electrospun anatase TiO_2 as counter electrode and electrospun PVdF–HFP membrane as separator, soaked and gelled with 1 M $LiPF_6$ in EC (ethylene carbonate): DMC (dimethyl carbonate) electrolyte. The full cell delivered a capacity of ~102 mAh g^{-1} at 0.1 C. The specific capacity for the full cell was calculated by taking the active mass of cathode into account. An energy density of 285 W h kg^{-1} was obtained for the prepared full cell assembly. However, the rate capability of the full cell was not as good as the reported half cell values for electrospun cathodes and hence needs further optimization [34].

Excellent high rate performance was delivered by the electrospun $LiNi_{0.5}Mn_{1.5}O_4$ hierarchical nanofibres, synthesized by Liu et al., in 2013 [35]. For the synthesis of nanofibres, lithium acetate, manganese nitrate, nickel nitrate and PVP were dissolved in stoichiometric amounts, and the precursor solution was then subjected to electrospinning, followed by pre-annealing at 450 °C and then annealing at 800 °C. The obtained product, when subjected to SEM investigations, revealed the presence of nanofibres (diameter 300-400 nm) formed with $LiNi_{0.5}Mn_{1.5}O_4$ nanooctahedrons. These one-dimensional nanostructures were intergrown to form a network-like morphology. The hierarchical nanofibres were found to exist in normal spinel structure (space group $Fd\overline{3}m$). Galvanostatic cycling in the voltage range 3.5–5.0 V at a charge–discharge rate of C/2 (1C = 150 mAh g^{-1}) shows a specific discharge capacity of 133 mAh g^{-1} and a coulombic efficiency of 93% in the first cycle. The material delivered an excellent rate capability (discharge capacities of 90 mAh g^{-1} and 80 mAh g^{-1} at 10 C and 15 C, respectively), owing to the peculiar 1D hierarchical nanofibre morphology [35]. Caterpillar-like $LiNi_{0.5}Mn_{1.5}O_4$ was fabricated by Haridas et al. as high voltage cathodes for LIBs in 2016 [36].

For the preparation, initially lithium acetate, manganese acetate and nickel nitrate were dissolved in ethanol, and the solution was added to a premixed solution of polyethylene oxide (PEO) in distilled water. Electrospinning was performed by transferring the precursor solution into a syringe fitted with a needle. The spinning was carried out at a voltage of 15 kV while maintaining a distance of 22 cm between the electrodes. The submicron fibres collected on the aluminium foil collector were subjected to heat treatment at 850 °C in air for 2 h to obtain the caterpillar-like $LiNi_{0.5}Mn_{1.5}O_4$ nanostructures (Fig. 19.8). When other annealing temperatures were used, the morphology was found to be different. At 550 and 650 °C, a fibre morphology was observed, and at 750 °C, spiral morphology was seen (Fig. 19.8). The caterpillar-like morphology mentioned in this work could also be related to intergrown fibre morphology. The electrospun material was found to exist in cubic spinel structure as revealed by XRD analysis.

Raman spectroscopy analysis confirmed the existence of caterpillar-like $LiNi_{0.5}Mn_{1.5}O_4$ in the disordered spinel structure (space group $Fd\overline{3}m$). Meanwhile, the conventional sol-gel synthesized $LiNi_{0.5}Mn_{1.5}O_4$ showed peaks in Raman spectra,

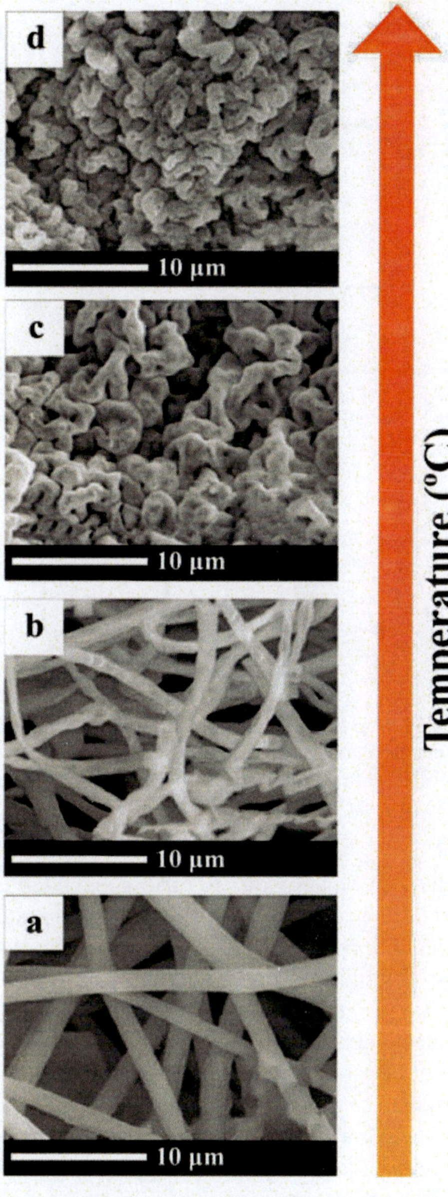

Fig. 19.8 FESEM images of electrospun $LiNi_{0.5}Mn_{1.5}O_4$/PEO precursor fibres at different calcination temperatures. **a** As synthesized electrospun $LiNi_{0.5}Mn_{1.5}O_4$/PEO precursor fibres, **b**, **c** fibre morphology obtained at 550 °C and 650 °C, respectively, **d** spiral-like morphology obtained at 750 °C, **e** caterpillar-like morphology obtained at 750 °C. Adapted and reproduced from reference [36]. Copyright 2016 Elsevier

corresponding to ordered $P4_332$ structure. Confirming this observation, the cyclic voltammetry investigations (3.5–5.0 V) show the presence of Mn^{4+}/Mn^{3+} electrochemical activity for the caterpillar-like fibres. Results for the cycling stability and rate capability are given in (Fig. 19.9). Rate capability investigations conducted in galvanostatic mode show that at 1C (150 mA g^{-1}), the caterpillar-like $LiNi_{0.5}Mn_{1.5}O_4$

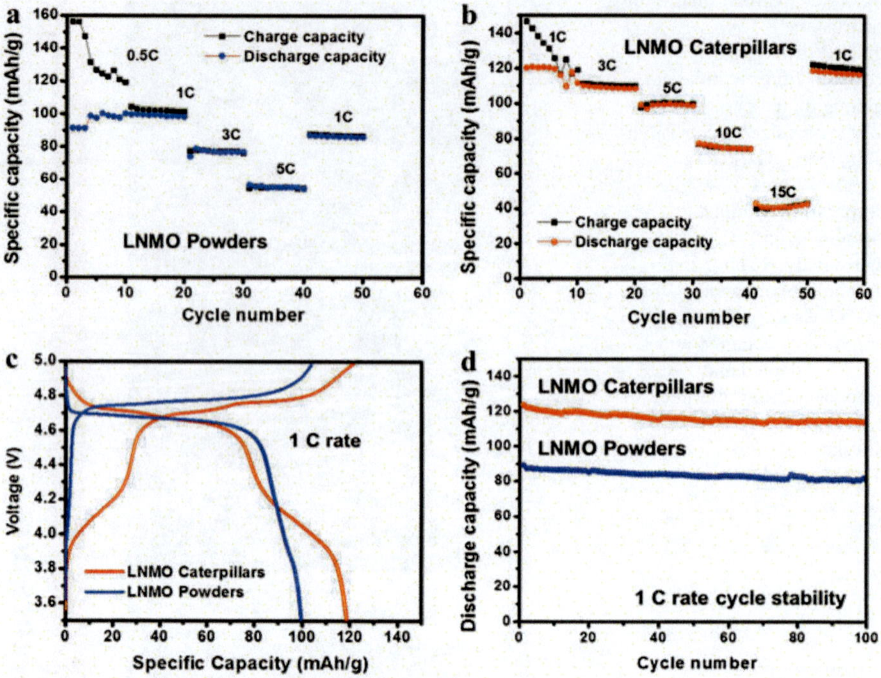

Fig. 19.9 **a**, **b** Rate capability studies of $LiNi_{0.5}Mn_{1.5}O_4$ powder and caterpillar, **c** first charge–discharge plot of $LiNi_{0.5}Mn_{1.5}O_4$ powder and caterpillar, **d** cycling stability studies performed at 1C for $LiNi_{0.5}Mn_{1.5}O_4$ powder and caterpillar. Adapted and reproduced from reference [36]. Copyright 2016 Elsevier

delivers a capacity of 120 mAh g^{-1}, which is higher than the conventionally synthesized $LiNi_{0.5}Mn_{1.5}O_4$ powder. Very good rate capability was observed for the electrospun $LiNi_{0.5}Mn_{1.5}O_4$ (~98 mAh g^{-1} and 40 mAh g^{-1} at 5 C and 15 C, respectively) than the conventional powder (~47 mAh g^{-1} at 5 C). The capacity losses observed for the electrospun and powder $LiNi_{0.5}Mn_{1.5}O_4$ were found to be 18% and 1.7%, respectively, after 100 cycles. Apart from the morphology influence, this could also be an effect of the cation ordering/disordering in the samples as disordered crystal structure ($Fd\overline{3}m$) was found to be advantageous for smoother Li^+ ion diffusion [36].

$Li_{1.5}MnTiO_{4+\delta}$ ultralong nanofibres were synthesized by Vu et al. using electrospinning as cathode materials for LIBs [37]. As a first step, stoichiometric amounts of Li acetate, manganese acetate and titanium (IV) isopropoxide were mixed separately in DMF and acetic acid, respectively. Both solutions were then mixed with each other, stirred, and PVP was added. PVP concentration was varied and optimized to control the diameter of the electrospun fibres. The precursor solution was vigorously stirred and subjected to electrospinning at voltages of 15–20 kV and collected on an aluminium foil mat, followed by annealing at 600 °C for 12 h in air after preheating at 120 °C for 10 h in vacuum. Analysis of XRD patterns of the prepared electrospun fibres shows the existence of sample in $Fd\bar{3}m$ space group. However, superlattice reflections correspond to Li-rich layered structure (monoclinic phase with *C2/c* space group) were also visible. This phase coexistence was further confirmed by HRTEM investigations. SEM images imply that the $Li_{1.5}MnTiO_{4+\delta}$ nanofibres are formed by octahedral particles intergrown/arranged in a uniform manner. The fibres formed were found to have a length of 30 μm and diameter 80 nm. Cycling stability studies were conducted in galvanostatic mode at a voltage range 2.0–4.8 V. The $Li_{1.5}MnTiO_{4+\delta}$ nanofibres were found to exhibit less polarization, high absolute capacity (235 mAh g^{-1}) as well as high cycling stability (94% retention at 1C = 154 mA g^{-1} after 100 cycles) in comparison with the same material synthesized via solid-state (184 mAh g^{-1}) and sol-gel (235 mAh g^{-1}) routes. However, at lower C-rates (C/5), the electrospun nanofibres exhibited much faster capacity fading than the samples synthesized via other routes. This could be a result of the enhanced parasitic reactions of the nanofibres with electrolyte due to the large surface area. The electrospun nanofibres of $Li_{1.5}MnTiO_{4+\delta}$ also exhibited enhanced rate capability than the samples obtained through other synthesis routes, owing to the enhanced Li^+ diffusion kinetics facilitated by the 1D nanostructure. PVA was also used in this study instead of PVP. However, only shorter nanofibres were obtained, which could be due to the low bonding ability with titanium (IV) isopropoxide and hence not investigated electrochemically [37].

19.5 Conclusions

Electrospun nanofibres of several Mn-oxide-based cathode materials were synthesized by various research groups worldwide. In general, the synthesis method consists of mainly three steps such as (i) preparation of precursor solution for electrospinning comprised of metal salts and a polymer, (ii) electrospinning the solution on a collector, which could be metal foils like aluminium, copper, etc., or special mats like fluorinated tin oxide, by applying a voltage across the needle (spinneret) and the collector, resulting in the formation of smooth fibres (iii) drying the collected fibres, followed by annealing at high temperatures to obtain shrinked fibres or hollow structures with rough surfaces. The list of electrospun Mn-oxide-based materials investigated as cathodes in LIBs is summarized in Table 19.1.

The electrospun fibres exhibited superior electrochemical performance in terms of high rate cycling stability and rate capability. The capacity retention was found to be enhanced in certain cases, revealing that the synthesis route and morphology have a strong influence on cycling stability. However, only very limited lithium-ion full cell investigations have been conducted to analyse the electrochemical behaviour of electrospun fibres in real systems. Furthermore, volumetric energy density of these electrospun nanofibres is still in question, mainly due to their special nanoarchitectures resulting in low tap densities.

Table 19.1 List of electrospun Mn-oxide-based materials investigated as cathodes in LIBs

Material class	Material	Carbon precursor used for electrospinning (polymer)	Metal precursors used	References
Layered	Al-doped $LiNi_{1/3}Co_{1/3}Mn_{1/3}O_2$	PVP	Metal nitrates	25
	$LiNi_{1/3}Co_{1/3}Mn_{1/3}O_2$	PVP	Metal nitrates	26
Layered Li-rich	$0.5Li_2MnO_3$–$0.5LiNi_{1/3}Co_{1/3}Mn_{1/3}O_2$	PVA	Metal acetates	27
	$Li_{1.2}Ni_{0.17}Co_{0.17}Mn_{0.5}O_2$	PVP	Metal acetates	22
	$Li_{1.2}Mn_{0.54}Ni_{0.13}Co_{0.13}O_2$	PVP	Metal acetates	28
	Al_2O_3-coated $Li_{1.2}Mn_{0.54}Ni_{0.13}Co_{0.13}O_2$	PAN	Metal acetates	29
Spinel	$LiMn_2O_4$	PVP	Mn acetate, Li carbonate	30
	$LiMn_2O_4$	PVP	Metal acetates	31
	$LiMn_2O_4$	PVP	Mn acetate, Li nitrate	32
	$LiMn_2O_4$	PVP	Metal acetates	33
	$LiNi_{0.5}Mn_{1.5}O_4$	PVA	Metal acetates	27
	$LiNi_{0.5}Mn_{1.5}O_4$	PVP	Metal acetates	34
	$LiNi_{0.5}Mn_{1.5}O_4$	PVP	Li acetate, Mn acetate, Ni nitrate	35
	$LiNi_{0.5}Mn_{1.5}O_4$	PEO	Li acetate, Mn acetate, Ni nitrate	36
	$Li_{1.5}MnTiO_{4+\delta}$	PVP	Li acetate, Mn acetate, Ti isopropoxide	37

References

1. Mauger A, Julien CM (2018) Olivine positive electrodes for Li-ion batteries: status and perspectives. Batteries 4:1–32. https://doi.org/10.3390/batteries4030039
2. Rozier P, Tarascon JM (2015) Review-Li-rich layered oxide cathodes for next-generation Li-ion batteries: chances and challenges. j electrochem Soc 162:A2490–A2499. https://doi.org/10.1149/2.0111514jes
3. Bhaskar A, Mikhailova D, Kiziltas-Yavuz N, Nikolowski K, Oswald S, Bramnik NN, Ehrenberg H (2014) 3d-Transition metal doped spinels as high-voltage cathode materials for rechargeable lithium-ion batteries. Prog Solid State Chem 42:128–148. https://doi.org/10.1016/j.progsolidstchem.2014.04.007
4. Goriparti S, Miele E, De Angelis F, Di Fabrizio E, Proietti Zaccaria R, Capiglia C (2014) Review on recent progress of nanostructured anode materials for Li-ion batteries. J Power Sour 257:421–443. https://doi.org/10.1016/j.jpowsour.2013.11.103
5. Jung YS, Cavanagh AS, Dillon AC, Groner MD, George SM, Lee S-H (2010) Enhanced stability of $LiCoO_2$ cathodes in lithium-ion batteries using surface modification by atomic layer deposition. J Electrochem Soc 157:A75. https://doi.org/10.1149/1.3258274
6. Deshpande RD, Ridgway P, Fu Y, Zhang W, Cai J, Battaglia V (2014) The limited effect of VC in graphite/NMC cells. J Electrochem Soc 162:A330–A338. https://doi.org/10.1149/2.0221503jes
7. Amalraj F, Kovacheva D, Talianker M, Zeiri L, Grinblat J, Leifer N, Goobes G, Markovsky B, Aurbach D (2010) Synthesis of integrated cathode materials $x Li_2MnO_3$ $(1 - x)LiMn_{1/3}Ni_{1/3}Co_{1/3}O_2$ ($x = 0.3, 0.5, 0.7$) and studies of their electrochemical behavior. J Electrochem Soc 157:A1121–A1130. https://doi.org/10.1149/1.3463782
8. Yan J, Liu X, Li B (2014) Recent progress in Li-rich layered oxides as cathode materials for Li-ion batteries. RSC Adv 4:63268–63284. https://doi.org/10.1039/C4RA12454E
9. He X, Wang J, Wang R, Qiu B, Frielinghaus H, Niehoff P, Liu H, Stan MC, Paillard E, Winter M, Li J (2016) A 3D porous Li-rich d *in situ* modified surface for high performance lithium ion batteries with reduced voltage decay. J Mater Chem A 4:7230–7237. https://doi.org/10.1039/C6TA01448H
10. Li J, Klöpsch R, Stan MC, Nowak S, Kunze M, Winter M, Passerini S (2011) Synthesis and electrochemical performance of the high voltage cathode material $Li[Li_{0.2}Mn_{0.56}Ni_{0.16}Co_{0.08}]O_2$ with improved rate capability. J Power Sour 196:4821–4825. https://doi.org/10.1016/j.jpowsour.2011.01.006
11. Kalyani P, Chitra S, Mohan T, Gopukumar S (1999) Lithium metal rechargeable cells Using Li_2MnO_3 as the positive electrode. J Power Sour 80:103–106. https://doi.org/10.1016/S0378-7753(99)00066-X
12. Doan TNL, Yoo K, Hoang TKA, Chen P (2014) Recent developments in synthesis of $x Li_2MnO_3 \cdot (1 - x)LiMO_2$ (M=Ni Co, Mn) cathode powders for high-energy lithium rechargeable batteries. Front Energy Res 2:1–7. https://doi.org/10.3389/fenrg.2014.00036
13. Liu GB, Liu H, Shi YF (2013) The synthesis and electrochemical properties of $x Li_2MnO_3$–$(1 - x)MO_2$($M=Mn_{1/3}Ni_{1/3}Fe_{1/3}$) via co-precipitation method. Electrochim Acta 88:112–116. http://dx.doi.org/10.1016/j.electacta.2012.10.054
14. Thackeray MM, de Picciotto LA, de Kock A, Johnson PJ, Nicholas VA, Adendorff KT (1987) Spinel electrodes for lithium batteries—a review. J Power Sour 21:1–8. https://doi.org/10.1016/0378-7753(87)80071-X
15. Thackeray MM (1997) manganese oxides for lithium batteries. Prog Solid State Chem 25:1–71. https://doi.org/10.1016/S0079-6786(97)81003-5
16. Normakhmedov OO, Brylev OA, Petukhov DI, Kurilenko KA, Kulova TL, Tuseeva EK, Skundin AM (2018) Cryochemically processed $Li_{1+y}Mn_{1.95}Ni_{0.025}Co_{0.025}O_4$ ($y = 0, 0.1$) cathode materials for Li-ion batteries. Materials 11:1–14. https://doi.org/10.3390/ma11071162
17. Bhaskar A, Bramnik NN, Senyshyn A, Fuess H, Ehrenberg H (2010) Synthesis, characterization, and comparison of electrochemical properties of LiM0.5Mn1.5O4 (M=Fe Co, Ni)

at different temperatures. J Electrochem Soc 157:A689–A695. https://doi.org/10.1149/1.3365025

18. Amdouni N, Zaghib K, Gendron F, Mauger A, Julien CM (2006) Structure and insertion properties of disordered and ordered $LiNi_{0.5}Mn_{1.5}O_4$ spinels prepared by wet chemistry. Ionics 12:117–126. https://doi.org/10.1007/s11581-006-0021-7
19. Manthiram A, Knight JC, Myung S-T, Oh S-M, Sun Y-K (2015) Nickel-rich and lithium-rich layered oxide cathodes: progress and perspectives. Adv Energy Mater 6:1501010. https://doi.org/10.1002/aenm.201501010
20. Bhaskar A, Krueger S, Siozios V, Li J, Nowak S, Winter M (2015) Synthesis and characterization of high-energy, high-power spinel-layered composite cathode materials for lithium-ion batteries. Adv Energy Mater 5:1401156. https://doi.org/10.1002/aenm.201401156
21. Zhang X, Ji L, Toprakci O, Liang Y, Alcoutlabi M (2011) Electrospun nanofiber-based anodes, cathodes, and separators for advanced lithium-ion batteries. Polym Rev 51:239–264. https://doi.org/10.1080/15583724.2011.593390
22. Min JW, Yim CJ, Im WB (2013) Facile synthesis of electrospun $Li_{1.2}Ni_{0.17}Co_{0.17}Mn_{0.5}O_2$ nanofiber and its enhanced high-rate performance for lithium-ion battery applications. ACS Appl Mater Interfaces 5:7765–7769. https://doi.org/10.1021/am402484f
23. Aravindan V, Sundaramurthy J, Suresh Kumar P, Lee YS, Ramakrishna S, Madhavi S (2015) Electrospun nanofibers: a prospective electro-active material for constructing high performance Li-ion batteries. Chem Commun 51:2225–2234. https://doi.org/10.1039/c4cc07824a
24. Kalluri S, Seng KH, Guo Z, Liu HK, Dou SX (2013) Electrospun lithium metal oxide cathode materials for lithium-ion batteries. RSC Adv 3:25576–25601. https://doi.org/10.1039/c3ra45414b
25. Ding Y, Zhang P, Long Z, Jiang Y, Xu F (2009) Morphology and electrochemical properties of Al doped $LiNi_{1/3}Co_{1/3}Mn_{1/3}O_2$ nanofibers prepared by electrospinning. J Alloys Compd 487:507–510. https://doi.org/10.1016/j.jallcom.2009.08.002
26. Kang CS, Son JT (2012) Synthesis and electrochemical properties of $LiNi_{1/3}Co_{1/3}Mn_{1/3}O_2$ cathode materials by electrospinning process. J Electroceramics 29:235–239. https://doi.org/10.1007/s10832-012-9762-z
27. Hosono E, Saito T, Hoshino J, Mizuno Y, Okubo M, Asakura D, Kagesawa K, Nishio-Hamane D, Kudo T, Zhou H (2013) Synthesis of $LiNi_{0.5}Mn_{1.5}O_4$ and $0.5Li_2MnO_3$-$0.5LiNi_{1/3}Co_{1/3}Mn_{1/3}O_2$ hollow nanowires by electrospinning. Cryst Eng Comm 15:2592–2597. https://doi.org/10.1039/c3ce26972h
28. Ma D, Zhang P, Li Y, Ren X (2015) $Li_{1.2}Mn_{0.54}Ni_{0.13}Co_{0.13}O_2$-encapsulated carbon nanofiber network cathodes with improved stability and rate capability for Li-ion batteries. Sci Rep 5:1–8. https://doi.org/10.1038/srep11257
29. Chen Y, Wang X, Zhang J, Chen B, Xu J, Zhang S, Zhang L (2019) Al_2O_3-coated $Li_{1.2}Mn_{0.54}Ni_{0.13}Co_{0.13}O_2$ nanotubes as cathode materials for high-performance lithium-ion batteries. RSC Adv 9:2172–2179. https://doi.org/10.1039/c8ra09428d
30. Liu S, Long YZ, Di Zhang H, Sun B, Tang CC, Li HL, Kan GW, Wang C (2012) Preparation and electrochemical properties of $LiMn_2O_4$ nanofibers via electrospinning for lithium ion batteries. Adv Mater Res 562–564:799–802. https://doi.org/10.4028/www.scientific.net/AMR.562-564.799
31. Zhao X, Reddy MV, Ramakrishna S, Mahishalkar S, Chowdari BVR (2012) Electrospun $LiMn_2O_4$ nanofibers As cathode for lithium ion batteries. Solid State ionics: ionics for sustainable world 1:3–9. https://doi.org/10.1142/9789814415040_0001
32. Zhou H, Ding X, Yin Z, Xu G, Xue Q, Li J, Jiao S, Wang X (2014) Fabrication and electrochemical characteristics of electrospun $LiMn_2O_4$ nanofiber cathode for Li-ion batteries. Mater Lett 117:175–178. https://doi.org/10.1016/j.matlet.2013.11.086
33. Duan L, Zhang X, Yue K, Wu Y, Zhuang J, Lü W (2017) Synthesis and electrochemical property of $LiMn_2O_4$ porous hollow nanofiber as cathode for lithium-ion batteries. Nanoscale Res Lett 12:2–9. https://doi.org/10.1186/s11671-017-1879-1

34. Arun N, Aravindan V, Jayaraman S, Shubha N, Ling WC, Ramakrishna S, Madhavi S (2014) Exceptional performance of a high voltage spinel $LiNi_{0.5}Mn_{1.5}O_4$ cathode in all one dimensional architectures with an anatase TiO_2 anode by electrospinning. Nanoscale 6:8926–8934. https://doi.org/10.1039/c4nr01892c
35. Liu J, Liu W, Ji S, Zhou Y, Hodgson P, Li Y (2013) Electrospun spinel $LiNi_{0.5}Mn_{1.5}O_4$ hierarchical nanofibers as 5 V cathode materials for lithium-ion batteries. Chem Plus Chem 78:636–641. https://doi.org/10.1002/cplu.201300180
36. Haridas AK, Sharma CS, Rao TN (2016) Caterpillar-like sub-micron $LiNi_{0.5}Mn_{1.5}O_4$ structures with site disorder and excess Mn^{3+} as high performance cathode material for lithium ion batteries. Electrochim Acta 212:500–509. https://doi.org/10.1016/j.electacta.2016.07.039
37. Vu NH, Arunkumar P, Im WB (2017) High-performance spinel-rich $Li_{1.5}$ $MnTiO_{4+\delta}$ ultralong nanofibers as cathode materials for Li-ion batteries. Sci Rep 7:1–12. https://doi.org/10.1038/srep45579

Chapter 20
Electrospun Mixed Oxide-Based Composites as Cathodes for Lithium-Ion Batteries

Satishkumar R. Naik, Anand I. Torvi, and Mahadevappa Y. Kariduraganavar

20.1 Electrospinning Technique for Lithium-Ion Batteries—An Overview

Energy conversion and storage devices have drawn significant attention owing to depletion of fossil fuels, climate change and environmental deterioration. The commercialization of lithium-ion batteries (LIBs) was achieved in the early 1990s by employing an intercalated lithium metal oxide compound as a cathode and graphite as an anode material. However, the rapidly growing market for electric vehicles has created a demand for high energy storage capacity and fast charge/discharge capability with high energy density and durability. To meet these requirements, much research has been focused on improving the performance of pre-existing electrodes and/or developing new electrode materials. Novel nanofiber technologies particularly created the opportunity to design new materials for advanced rechargeable lithium-ion batteries. Currently, much effort has been devoted in developing nanostructure materials by employing various techniques, such as sol-gel method [1], wet chemical synthesis [2], chemical vapor deposition [3], the molten salt method [4–9], polymer precursor method [10–15] and electrospinning. Among these methods, electrospinning is a versatile, efficient and low-cost method, and has been used widely to synthesize nanofibers (NFs) with different morphologies. The electrospun NFs play a crucial role in various energy storage devices because of high surface area, controllable porosity and ease of accessibility.

S. R. Naik · A. I. Torvi · M. Y. Kariduraganavar (✉)
Department of Chemistry, Karnatak University, Dharwad 580003, India
e-mail: mahadevappayk@gmail.com

N. T. M. Balakrishnan and R. Prasanth (eds.), *Electrospinning for Advanced Energy Storage Applications*, Materials Horizons: From Nature to Nanomaterials,
https://doi.org/10.1007/978-981-15-8844-0_20

20.2 Electrospinning

Electrospinning is a widely used fabrication technique for developing continuous electro-active nanofibers from a wide range of materials with diameters in the range of several nanometers to the micrometer regime. Among the different materials, nanofibers can be successfully obtained by electrospinning from polymers, semiconductors, ceramics, and their composites. Because of the decrease in the fiber diameter to nanoscale, these materials show enhanced properties of high surface area to volume ratio, porosity, flexibility, electrochemical and mechanical properties, which make them excellent candidates for a wide range of applications, such as energy conversion and storage devices, electronic devices and tissue engineering, etc.

20.2.1 Principle of Electrospinning

The basic principle behind the processing of nanofibers by electrospinning is based on the unidirectional elongation of a spinnable visco-elastic solution by considering various parameters involved in the process of electrospinning. Continuous nanofibers are formed due to the electrostatic Coulombic repulsive forces applied during elongation of the visco-elastic solution. The schematic representation of the electrospinning set-up is shown in Fig. 20.1. During the process of electrospinning, a visco-elastic solution is first loaded into a syringe and fed through a flow meter pump. When a high electric potential is applied to the spinnable solution at a threshold voltage of ~ 6 kV, the repulsive force developed in the electro-active solution is greater than its surface tension and a droplet, namely a Taylor cone, is formed at the tip of the syringe [16, 17]. This droplet is further elongated owing to electrostatic forces, which results in evaporation of the solvent and formation of solidified nanofibers, which are usually collected randomly on the grounded static/rotating mandrel collector substrate.

20.2.2 Parameters of Electrospinning

Electrospinning of fibers is influenced by various parameters. Some of the important parameters are briefly discussed below:

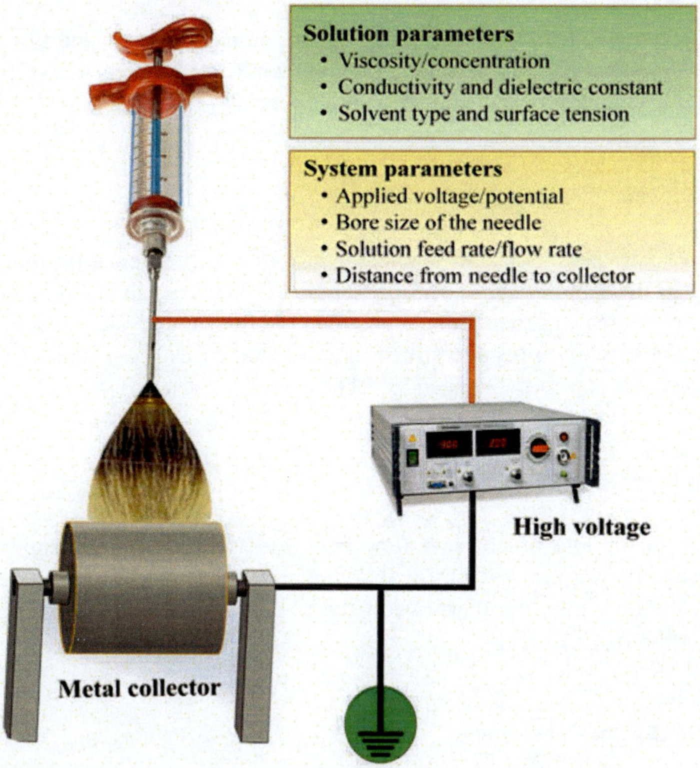

Fig. 20.1 Schematic representation of the electrospinning process

20.2.2.1 Process Parameters

Collector

The physical properties of electrospun fibers such as crystal morphology and molecular orientation are affected by the nature of collectors. Generally, a rotating drum collector is used as a collector. The diameter of the fibers can be controlled by monitoring the drum rotating speed. In some cases, the rotating disk is also used to develop uniaxially aligned fibers. The speed of collector could improve the crystal orientation of fibers due to polymer molecular chains' alignment in the direction of the fiber axis.

Applied Voltage

The applied voltage plays an important role in fiber development. An increase or a decrease in applied voltage can cause a change of morphology and structure of fibers. Increasing the voltage can cause an increase in the spinning current [18]. Increasing spinning current can lead to an incidence of beaded morphology, and this structure can reduce the surface area. The studies have shown that the increase in voltage leads to an increase in fiber length and decrease in fiber size [19–21].

Distance Between Nozzle to Collector

The distance between nozzle and collector has a direct influence on morphology, structure, physical and chemical properties of the electrospun fibers. The studies show that by decreasing the distance between the nozzle and the collector, results in electrospun fiber with beaded structure and some cases, the shape of the fibers changed from circular to flat shape [19, 21].

Dispersion Flow Rate

The change in the rate of polymer flow from the syringe of electrospinning unit causes change in morphology, physical and chemical properties of the fibers. With increasing the flow rate, the diameter of the fiber could be increased and leads to beaded morphology [22–24].

20.2.2.2 Solution Parameters

Solution Concentration

The variation in the viscosity and surface tension of the polymer solution influences the spinning of fibers and controls the fiber structure and morphology. Low concentration solution generally forms droplets due to the influence of surface tension, while higher concentration prohibits fiber formation due to higher viscosity.

Volatility of Solvent

Electrospinning technique involves phase separation and rapid solvent evaporation. The solvent vapor pressure critically determines the evaporation rate and drying time. Thus, the solvent volatility plays a major role in the formation of nanostructures by influencing the phase separation process.

20.2.2.3 Ambient Parameters

The surrounding conditions such as temperature, humidity, pressure, and air velocity in the chamber may have a direct influence on the fiber diameter and morphology. For instance, pores are formed on the surface of the electrospun fibers because of huge variations in humidity level [25, 26]. Further, humidity in the surroundings of the electrospinning determines the rate of evaporation of the solvents in the electrospinning solution [27, 28].

20.3 Lithium-Ion Batteries

Among the various existing energy storage technologies, rechargeable lithium-ion batteries are considered an effective solution to the increasing need for high-energy electrochemical power sources. As one of the most important energy storage systems, the Li-ion battery (LIB) has been used not only in portable electronics but also in power batteries for electric vehicles. To meet the demands of all-electric vehicles in the long term, researchers have been devoted to developing other battery systems with lithium metal as the anode material to improve the energy density, such as lithium-sulfur battery and lithium-oxygen battery.

Primary Li batteries have become commercial during the 1970s. Attempts to develop rechargeable Li batteries with Li-metal anodes have accompanied the R&D of Li batteries from their early stages. However, a few nearly commercial products of secondary Li (metal) batteries appeared during the early 1990s. From the early stages of R&D of Li-ion batteries, it was clear that transition metal oxides and sulfides can serve as excellent reversible cathode materials for rechargeable Li batteries [29, 30]. The Li-ion battery technology evolution, which enabled the commercialization of the rechargeable, high-energy density batteries that are conquering the market, emerged due to introduction of graphite as the anode material instead of Li-metal, and the use of lithiated transition metal oxide as cathode materials; $LiMO_2$ as the source of lithium in the cell [31]. Graphite-$LiCoO_2$ became the leading Li-ion battery systems that are being used in most of the portable electronic devices, like laptops, cellular phone, digital cameras, etc.

Lithium-ion battery consists of a cathode and an anode separated by an electrolyte containing dissociated lithium salts, which enable the transfer of lithium ions between the two electrodes, as illustrated in Fig. 20.2. The electrolyte is typically contained in a porous separator membrane that prevents the physical contact between the cathode and anode. When the battery is being charged, an external electrical power source injects electrons into the anode. At the same time, the cathode gives up some of its lithium ions, which move through the electrolyte to the anode. During this process, electricity is stored in the battery in the form of chemical energy. During discharging, the lithium ions move back to the cathode, enabling the release of electrons to the outer circuit to do the electrical work.

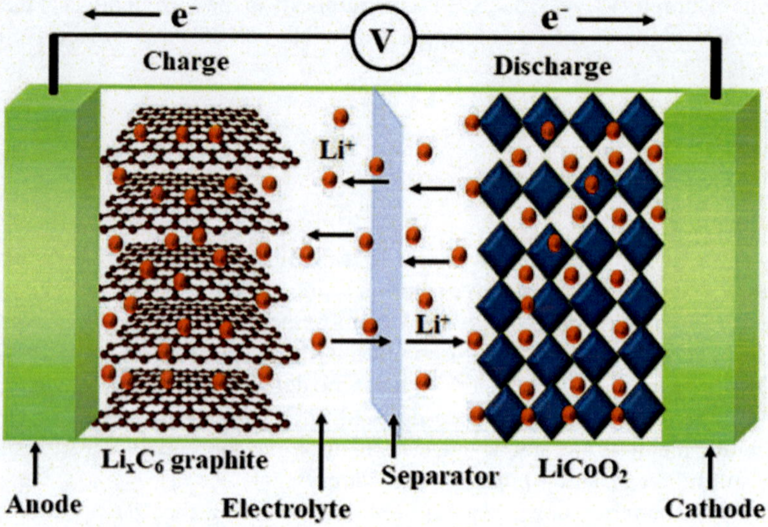

Fig. 20.2 Schematic representation of a lithium-ion battery

20.4 Electrospun Nanofiber-Based Lithium-Ion Batteries

Among the various existing energy storage technologies, lithium-ion batteries (LIBs) have become one of the most advanced rechargeable batteries for portable devices and more recently, for electric vehicles, because of their high energy density, negligible memory effect, and low self-discharge. Rechargeable lithium-ion batteries offer energy densities 2–3 times and power densities 5–6 times higher than the conventional nickel-cadmium (Ni-Cd) and nickel-metal hydride (Ni-MH) batteries, and as a result, they weigh less, take less space and deliver more energy [32–34]. In addition to high energy and power densities, lithium-ion batteries also have other advantages, such as high Coulombic efficiency, low self-discharge and high operating voltage [35].

At present, lithium-ion batteries depend on active powder materials such as graphite powder in the anode and $LiFePO_4$ powder in the cathode to store energy. However, powder materials have a long diffusion path for lithium ions and slow electrode reaction kinetics. The long migration pathways for the Li^+ of powder materials may lead to the large expansion volume during cycling, resulting in the low rate of performance and poor cyclability. Thus, electrodes with advanced nanostructured materials must be developed to obtain potential lithium-ion batteries that outperform current technologies and can be used in large-scale systems. Recently, the application of electrospun nanofibers in lithium-ion batteries has attracted much research attention. The electrospunned nanofiber-based electrodes exhibited the shorter diffusion pathways for the Li^+-ions in comparison with the powder materials. These electrodes also showed an efficient electrochemical reaction kinetic due to a faster intercalation-deintercalation mechanism at the electrode-electrolyte interface. Besides, electrospun based materials are promising materials for LIBs because of their good electrochemical activity, high mechanical strength, and large specific surface area. Results also demonstrate that lithium-ion batteries using electrospun nanofiber cathodes have excellent overall performance including large capacity, high charge/discharge rate capability and extended cycle life. In this section, mixed oxide-based composites as cathodes for lithium-ion batteries are briefly discussed.

20.4.1 Mixed Nickel-Cobalt Dioxide, $LiNi_{1-Y}Co_YO_2$

In a series of research papers, Delmas group [36–39] and Zhecheva et al. [40] reported the structural details and physical properties of $LiNi_{1-y}Co_yO_2$ system and showed that there is an increased ordering as the cobalt concentration increases. The cobalt suppresses the migration of nickel to the lithium site in the mixed Li nickel/cobalt compounds. The same behavior was found in the system like Li nickel/manganese/cobalt oxides. Other ions, such as iron, do not have the same positive effect as cobalt. For example, in the case of the compound $LiNi_{1-y}Fe_yO_2$, the capacity is reduced with increasing iron [41]. The electronic conductivity is an important parameter which influences the properties of such layered oxides. Thus, cobalt substitution in $LiNiO_2$, as in $LiNi_{0.8}Co_{0.2}O_2$, reduces the conductivity [39]. Also, as lithium is removed from the phase $Li_xNi_{0.1}Co_{0.9}O_2$ or from Li_xCoO_2 the dramatic increase in conductivity was observed [42].

The studies have also shown that the addition of a little of a redox-inactive element such as magnesium as in case of $LiNi_{1-y}Mg_yO_2$ the capacity was reduced [43]. Similarly, in the case of substituted nickel oxides, such as $LiNi_{1-y-z}Co_yAl_zO_2$, the nickel is oxidized first to Ni^{4+} then the cobalt to Co^{4+} during charging these mixed oxides [44]. These are the prime candidates for the cathode of advanced lithium batteries for large-scale applications.

20.4.2 Lithium Manganese Dioxide, $LiMnO_2$

Lithium manganese dioxide ($LiMnO_2$) is a low-cost and environmentally friendly cathode material [45–47]. This can be prepared from three different approaches reported in the literature. The first approach includes ion exchange mechanism with the sodium compound, giving $LiMnO_2$, which was accomplished independently by Bruce and Armstrong [48] and Delmas et al. [49].

The second synthetic approach includes low temperatures preparation method. For example, the hydrothermal synthesis/decomposition of alkali permanganates, in the presence of lithium results in the composition $Li_{0.5}MnO_2.nH_2O$ [46, 49–51].

Mild warming causes the loss of water to give the desired layered Li_xMnO_2 and overheating to 150 °C leads to the formation of the spinel $LiMn_2O_4$.

The third approach includes electronic stabilization method. The idea behind this method is to make the electronic properties of Mn to be more cobalt-like by substitution of Mn with more electron-rich elements such as Ni [52]. The successful substitution of Mn by Ni and Co has been reported [53–55]. Spahr et al. [56] demonstrated a high capacity and reversibility for $LiNi_{0.5}Mn_{0.5}O_2$. More recently, the compounds $LiNi_{1-y-z}Mn_yCo_zO_2$ have been extensively investigated and found to have properties that qualify them as possible candidates for the replacement of $LiCoO_2$ [57–59]. In addition to their high lithiation capacities and reversibility, these compounds show higher thermal stabilities compared to the cobalt-free compounds.

20.4.3 Mixed Manganese-Cobalt Dioxide, $LiMn_{1-Y}Co_yO_2$

Bruce et al. [60] reported the synthesis and electrochemical performance of cobalt-substituted $LiMn_{1-y}Co_yO_2$. In such compounds, the partial substitution of manganese ions by cobalt, iron, or nickel was found to significantly increase the electronic conductivity of the manganese oxide. To obtain sufficiently dense material for conductivity measurement, the potassium analogs were synthesized at elevated temperatures with 10% of the manganese substitution. The study showed the enhanced conductivity by almost 2 orders of magnitude upon adding cobalt. These cobalt-substituted materials can also be prepared hydrothermally, and their cycling behavior is much improved over the cobalt-free compounds [61].

Doping elements other than cobalt have also been investigated, but substitution by nickel leads to a system where the manganese becomes the structure stabilizer and nickel is the electrochemically active ion. These compounds are thus best described as substituted nickel oxides in which the manganese remains in the tetravalent state and the nickel is redox-active between the +2 and +4 oxidation states. The manganese helps to stabilize the lattice and reduces the cost as well. Cobalt plays a critical role in controlling the ordering of the 3d ions in the structure.

20.4.4 Mixed Nickel-Manganese Dioxide, $LiNi_{1-Y}Mn_yO_2$, Multi-electron Redox Systems

Ammundsen and Davidson research groups studied the mixed metal compounds like Li–Mn–Cr–O_2 system [62, 63]. The lithium ions in the transition metal layer were found clustered around the manganese ions as in Li_2MnO_3, and the system can be considered as a solid solution of Li_2MnO_3 and $LiCrO_2$. Considering the toxicity of Cr(VI), this pioneering research provoked much thinking about other multi-electron redox reaction and is discussed here. The Dahn research group [64] studied $LiNi_{1-y}Mn_yO_2$ phase system and reported a solid solution for $y \leq 0.5$ but a deterioration of the electrochemical behavior with increasing the manganese content. Spahr et al. [56] repeated the optimum electrochemical behavior for the composition $LiNi_{0.5}Mn_{0.5}O_2$. This compound is called as 550 material (0.5 Ni, 0.5 Mn, 0.0 Co).

Spahr et al. [56] prepared 550 compounds at 700 °C and reported the capacity of 150 mAh g^{-1} falling to 125 mAh g^{-1} after 25 cycles and to 75 mAh g^{-1} after 50 cycles. They also showed that the capacity and its retention increased as the synthesis temperature was increased from 450 to 700 °C. Similarly, Ohzuku et al. [65] prepared the 550 material at 1000 °C and reported a constant capacity of 150 mAh g^{-1} at 0.1 mA cm^{-2}. The 550 material, synthesized at 900 °C and quenched to room temperature, also showed a capacity exceeding 150 mAh g^{-1} for over 50 cycles in thin-film configuration [66]. A material formed at 1000 °C showed a lower capacity of ~ 120 mAh g^{-1} at 0.1 mA cm^{-2}; the capacity was increased to about 140 mAh g^{-1} by addition of 5% cobalt, aluminum or titanium [67]. This suggests that the synthesis temperature of 1000 °C may be too high, leading possibly to excess nickel in the lithium layer. This 550 compound can intercalate second lithium, particularly when some titanium is added, forms $y LiNi_{0.5}Mn_{0.5}O_2.(1-y)Li_2TiO_3$, which results from the reduction of Mn(IV) to Mn(II) [68].

In conclusion, the mixed nickel-manganese dioxides have the following cathode characteristics:

i. It has a capacity of ~ 180 mAh g^{-1} for at least 50 cycles under mild cycling condition,
ii. The synthesis temperature should be over 700 °C and less than 1000 °C, probably optimally ~ 900 °C,
iii. Additions of cobalt can reduce the level of nickel in the lithium layer,
iv. The lithium in the transition metal layer may be a necessary structural component,
v. Nickel is the electrochemically active ion, and
vi. The electronic conductivity needs to be increased.

20.4.5 *Mixed Nickel-Manganese-Cobalt Dioxide, $LiNi_{1-Y-Z}Mn_yCo_zO_2$*

The synthesis of mixed nickel-manganese-cobalt dioxides was first reported in 1999 by Liu et al. [69] and in 2000 by Yoshio et al. [52]. The addition of cobalt to $LiMn_{1-y}Ni_yO_2$ would stabilize the structure in a strictly two-dimensional pattern. Ohzuku et al. [57] synthesized the symmetric compound $LiNi_{0.33}Mn_{0.33}Co_{0.33}O_2$ at 1000 °C and reported a capacity of ~150 mAh g^{-1}. This compound will be hereafter referred to as 333 materials. The synthesis of these $LiNi_{1-y-z}Mn_yCo_zO_2$ compounds is typically accomplished using a modified mixed-hydroxide approach by reacting $Ni_{1-y-z}Mn_yCo_z(OH)_2$ with a lithium salt in air or oxygen as described in Liu's first synthesis at 750 °C [69].

The synthesis temperature has a profound effect on the composition. For example, for the composition $LiNi_{0.4}Mn_{0.4}Co_{0.2}O_2$, the sample prepared at 1000 °C when cooled rapidly to ambient temperatures has almost 10% Ni occupancy in the lithium layer. Kim et al. [70] also reported a high Ni content of 5.9% on the Li site for samples of the 333 compositions prepared at 950 °C. At 900 °C even with more cobalt than nickel in the material, there is still considerable nickel disorder, almost 2% more Ni in the lithium layer at 900 °C than at 800 °C for all compositions. This suggests that high temperature increases the disorder of the nickel ions and this could be reduced by slow cooling of the sample in an oxidizing environment. This will allow the partial reordering of the ions [71].

Although these materials show good electrochemical behavior, their electronic conductivity is still low for a high-rate cathode. The conductivity needs to be improved without the addition of excessive amounts of a conductor such as carbon black, which will reduce the volumetric energy storage capacity. There has also been a report [72] on the low packing density of powders which will also severely reduce the volumetric energy density. The conductivity of $LiNi_{0.5}Mn_{0.5}O_2$ was 6.2×10^{-5} S cm^{-1}; this increased on cobalt addition to 1.4×10^{-4} S cm^{-1} for $LiNi_{0.4}Mn_{0.4}Co_{0.2}O_2$ at 21 °C and 6.8×10^{-4} S cm^{-1} at 100 °C [73].

For the four oxides $Li_xNi_{1.02}O_2$, $Li_xNi_{0.89}Al_{0.16}O_2$, $Li_xNi_{0.70}Co_{0.15}O_2$, and $Li_xNi_{0.90}Mn_{0.10}O_2$, a structural transformation first to a spinel phase and then to a rock salt phase was found [74, 75] for lithium x values of 0.5 or less. The second transformation is accompanied by a loss of oxygen, and the first may be depending on the composition but usually when x is less than 0.5; the latter oxygen release occurs at a lower temperature as the lithium content decreases and as low as 190 °C for $Li_{0.3}Ni_{1.02}O_2$. The stability is improved on aluminum or cobalt substitution. The compound $Li_{0.1}NiO_2$ is reported [76] to lose weight at 200 °C forming a rock salt structure. The substitution of manganese for nickel appears to move the transition to the spinel to higher temperatures; thus, $Li_{0.5}Ni_{0.5}Mn_{0.5}O_2$ even after 3 days at 200 °C is still layered [77], but a spinel phase is formed above 400 °C and is stable to much higher temperatures for the 1:1 Ni:Mn lithium-free compound, eventually giving a mixture of spinel and nickel oxide in air and $NiO + Mn_3O_4$ in nitrogen [78]. The compounds $Li_{0.5}Ni_{0.4}Mn_{0.4}Co_{0.2}O_2$ and $Li_{0.5}Ni_{0.33}Mn_{0.33}Co_{0.33}O_2$ both begin to lose weight above 300 °C with major weight loss, 7–8%, only above 450 °C, which corresponds to reduction of Co(III) to Co(II) and any Ni(IV) to Ni(II); the manganese remains Mn(IV), and the structure begins to change to spinel by 350 °C, and the spinel phase is still present at 600 °C.

In conclusion, the mixed nickel-manganese-cobalt dioxides have the following cathode characteristics:

i. The synthesis temperature should be over 700 °C and less than 1000 °C, probably optimally ~ 900 °C,
ii. The cobalt reduces the number of nickel ions in the lithium layer,
iii. Nickel is the electrochemically active ion at low potentials,
iv. Cobalt is only active at the higher potentials,
v. The electronic conductivity need to be increased, and

vi. The optimum composition is still to be determined for energy storage, power capability, life-time, and cost considerations.

20.4.6 *Lithium-Rich Mixed Metal Dioxides, $Li_{1+X}M_{1-X}O_2$*

As explained in the case of chromium [79, 80] and cobalt [81] systems, excess lithium can be incorporated into the layered structure through a solid solution of Li_2MnO_3 and LMO_2, where M = Cr or Co. Yoshio et al. [82], Thackeray et al. [83, 84] and Dahn et al. [66] reported that the transition metal cation can also be nickel or manganese including mixtures such as $LiNi_{1-y}Co_yO_2$ and Li_2MnO_3 can be replaced by related materials such as Li_2ZrO_3 and Li_2TiO_3. Li_2MnO_3 can be represented in the normally layered notation as $Li[Li_{1/3}Mn_{2/3}]O_2$. These solid solutions can thus be represented as $LiM_{1-y}[Li_{1/3}Mn_{2/3}]_yO_2$, where M can be, for example, Cr, Mn, Fe, Co, Ni, or mixtures thereof. Addition of extra lithium will tend to push the manganese away from trivalent to tetravalent, and thus minimizing the impact of any Jahn-Teller distortion coming from Mn^{3+}. The end-member Li_2MnO_3 which has been shown [85, 86] to exhibit unexpected electrochemical activity on charging as the manganese is already in the 4^+ oxidation state. This overcharging can be associated with two phenomena, removal of lithium with the concomitant loss of oxygen giving a defective oxygen lattice and the removal of lithium by decomposition of the electrolyte giving protons which can ion exchange for the lithium. The predominance of the mechanism depends on the temperature and chemical composition of the oxide lattice [87]. In both cases, the manganese oxidation state remains unchanged. Acid leaching of Li_2MnO_3 also results in the removal of lithium, and here again, both mechanisms of Li_2O removal [88, 89] and proton exchange [90] are believed to be operative. Acid leaching of the lithium stoichiometric compounds, such as $LiNi_{0.4}Mn_{0.4}Co_{0.2}O_2$, also results in the removal of lithium and a small amount of proton exchange. Thackeray et al. [91] showed that Li_2TiO_3 forms a solid solution with $LiNi_{0.5}Mn_{0.5}O_2$ and that the titanium helped allow the intercalation of second lithium into the structure [68]. Addition of some cobalt to these manganese-rich compounds was reported to help retain the capacity at higher discharge rates [92]. Magnesium has also been proposed as a stabilizing agent for manganese-rich materials [93]. The lithium excess content is one of the important parameters to be considered in addition to the nickel, cobalt and manganese ratios in designing the optimum composition for the ideal cathode. Each of these elements has a role to play. Manganese helps to stabilize the lattice; nickel acts as the electrochemically active member; cobalt helps in ordering the transition metals and thus increasing conductivity and the rate capability, and the lithium improves the capacity.

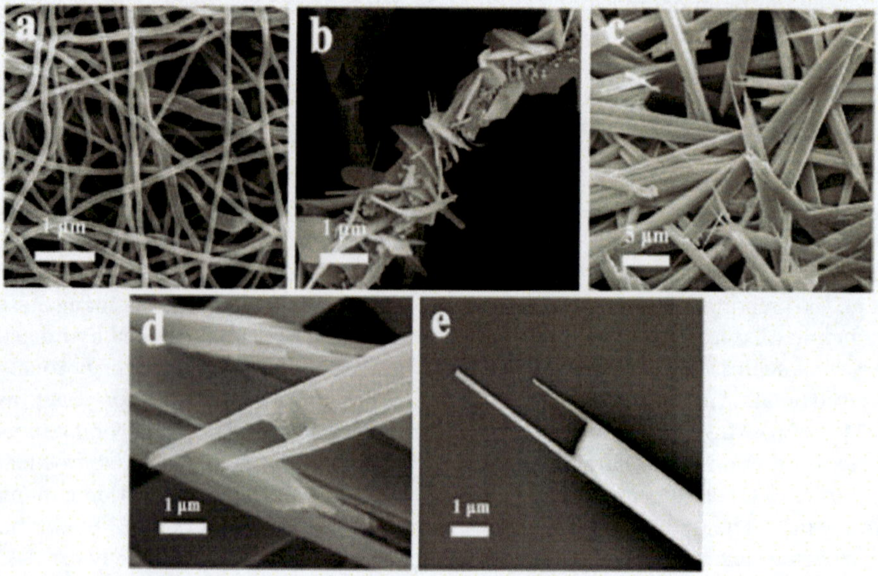

Fig. 20.3 Different magnifications SEM images of the precursor nanofibers (**a**), $K_2V_8O_{21}$ nanostructures annealed at 350 °C (**b**), and 500 °C (**c–e**). Adapted and reproduced from Ref. [94]

20.4.7 Electrospun Single-Crystalline Fork-like $K_2V_8O_{21}$

Anqiang et al. [94] reported the synthesis of single-crystalline fork-like potassium vanadate ($K_2V_8O_{21}$) by electrospinning method and a subsequent annealing process. A unique layer-by-layer stacked structure with fork-like morphology was seen in the SEM images as shown in Fig. 20.3. The prepared materials exhibit high specific discharge capacity and excellent cyclic stability with high specific discharge capacities of 200.2 and 131.5 mAh g^{-1} can be delivered at the current densities of 50 and 500 mA g^{-1}, respectively. Furthermore, the $K_2V_8O_{21}$ electrode exhibits excellent long-term cycling stability which maintains a capacity of 108.3 mAh g^{-1} after 300 cycles at 500 mA g^{-1}. These excellent results demonstrate their potential applications in next-generation high-performance lithium-ion batteries.

20.4.8 Lithium Iron Phosphate ($LiFePO_4$)—3D Carbon Nanofiber Composites

Dimesso et al. [95] reported the composite consist of $LiFePO_4$ as cathode material and deposited on carbon nanofiber 3D nonwovens prepared via electrospinning technique. The $LiFePO_4$ was prepared by Pechini-assisted sol-gel process. The cathode material was then coated on carbon nanofiber 3D nonwovens by soaking in

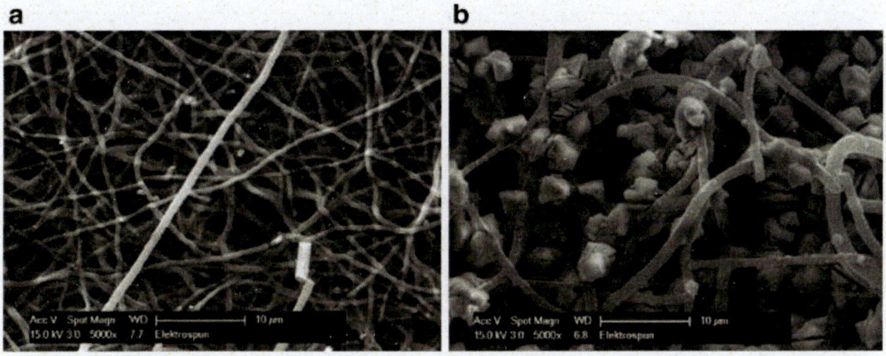

Fig. 20.4 The SEM images of **a** carbon nanofibers, **b** $LiFePO_4$/C nanofiber composites Adapted and reproduced from Ref. [95]. Copyright 2012 AIP Publishing

an aqueous solution containing lithium, iron salts and phosphates for 2–4 h at 70 °C. The composites were then annealed at 600 °C for 5 h under an inert atmosphere. The SEM images of the prepared composites are shown in Fig. 20.4 indicated a uniform coating of the carbon nanofibers and the uniform distribution of cauliflower-like crystalline structures all over the surface area of the carbon nanofibers. The electrochemical measurements on the composites showed good performances delivering a discharge specific capacity of 156 mAh g^{-1} at a discharging rate of C/25 and 152 mAh g^{-1} at a discharging rate of C/10 at ambient temperature.

20.5 Conclusion

Lithium batteries have made a substantial and significant contribution in dominant rechargeable battery for consumer portable applications. Electrospinning is a versatile technique presents many opportunities in the enhancement of Li-ion battery performance by easy modifications of components. The literature explicitly showed the clear advantage of increased charge retention and achievable energy densities approaching the theoretical capacity of the active materials. The electrospun NMC/C fibers exhibited the higher capacities of ~ 200–250 mAh g^{-1} with energy densities of ~200 Wh kg^{-1}. Recent studies also demonstrated the advantages of nanofiber usage because of increased gravimetric storage capacities. A further potential advantage, in addition to higher gravimetric energy densities, is capacity retention. This is also an essential factor for short-term energy storage during high peak loads. There are notable potential gains in using fibrous cathodes, while some drawbacks require solutions. The next market opportunities mostly demand higher power capabilities at lower costs with enhanced safety. The developments on the area layered oxides and mixed transition metals replacing the cobalt system and proving enhanced safety with improved electrochemical performance. Though the material made of lithium iron phosphate is a potentially low-cost cathode, the costs of the other cell components including the electrolyte, anode and the separators, need to be reduced.

In addition, the balance of the active material/C ratio is extremely important to achieve a proper volumetric and gravimetric energy density of the product with a suitable nanofibrous structure. The large volumetric energy density, resulting from the high porosity of nanofiber webs, restricts the application of fibrous cathodes in vehicles and some portable devices, and instead suggests the potential application of fibrous cathodes in stationary storage systems. Another drawback observed in laboratory experiment is the *in situ* synthesis of active material containing CNFs. The active material is often more brittle than fibers containing already-synthetized nanoparticles. These two challenges would impede industrial production.

References

1. Clark MD, Walker LS, Hadjiev VG, Khabashesku V, Corral EL, Krishnamoorti R, Cinibulk M (2015) Fast sol-gel preparation of silicon carbide-silicon oxycarbide nanocomposites. J Am Ceram Soc 94:4444–4452. https://doi.org/10.1111/j.1551-2916.2011.04707.x
2. Balde CP, Hereijgers BP, Bitter JH, Jong KP (2008) Sodium alanate nanoparticles—linking size to hydrogen storage properties. J Am Chem Soc 130:6761–6765. https://doi.org/10.1021/ja710667v
3. Li M, Wu X, Zeng J, Hou Z, Liao S (2015) Heteroatom doped carbon nanofibers synthesized by chemical vapor deposition as platinum electrocatalyst supports for polymer electrolyte membrane fuel cells. Electrochim Acta 182:351–360. https://doi.org/10.1016/j.electacta.2015.09.122
4. Reddy M, Beichen Z, Nicholette JLE, Kaimeng Z, Chowdari B (2011) Molten salt synthesis and its electrochemical characterization of Co_3O_4 for lithium batteries. Electrochem Solid State Lett 14:A79–A82. https://doi.org/10.1149/1.3556984
5. Reddy M, Yu C, Jiahuan F, Loh KP, Chowdari B (2012) Molten salt synthesis and energy storage studies on $CuCo_2O_4$ and $CuO{\cdot}Co_3O_4$. RSC Adv. 2:9619–9625. https://doi.org/10.1039/C2RA21033A
6. Reddy M, Khai V, Chowdari B (2015) Facile one pot molten salt synthesis of nano $(M_{1/2}Sb_{1/2}Sn)O_4$ (M = V, Fe, In). Mater Lett 140:115–118. https://doi.org/10.1016/j.matlet.2014.10.145
7. Wu Y, Reddy M, Chowdari B, Ramakrishna S (2013) Long-term cycling studies on electrospun carbon nanofibers as anode material for lithium ion batteries. ACS Appl Mater Interf 5:12175–12184. https://doi.org/10.1021/am404216j
8. Reddy M, Quan CY, Teo KW, Ho LJ, Chowdari B (2015) Mixed oxides, $(Ni_{1-x}Zn_x)Fe_2O_4$ ($x = 0, 0.25, 0.5, 0.75, 1$): Molten salt synthesis, characterization and its lithium-storage performance for lithium ion batteries. J Phys Chem C 119:4709–4718. https://doi.org/10.1021/jp5121178
9. Reddy M, Cherian CT, Ramanathan K, Jie KCW, Daryl TYW, Hao TY, Adams S, Loh KP, Chowdari BVR (2014) Molten synthesis of $ZnO.Fe_3O_4$ and Fe_2O_3 and its electrochemical performance. Electrochim Acta 118:75–80. https://doi.org/10.1016/j.electacta.2013.11.125
10. Sakunthala A, Reddy M, Selvasekarapandian S, Chowdari B, Selvin PC (2010) Synthesis of compounds, $Li(MMn_{11/6})O_4$ ($M = Mn_{1/6}, Co_{1/6}, (Co_{1/12}Cr_{1/12}), (Co_{1/12}Al_{1/12}), (Cr_{1/12}Al_{1/12})$) by polymer precursor method and its electrochemical performance for lithium-ion batteries. Electrochim Acta 55:4441–4450. https://doi.org/10.1016/j.electacta.2010.02.080
11. Reddy M, Sakunthala A, Selvashekara Pandian S, Chowdari B (2013) Preparation, comparative energy storage properties, and impedance spectroscopy studies of environmentally friendly

cathode, Li($MMn_{11/6}$)O_4 (M = $Mn_{1/6}$, $Co_{1/6}$, ($Co_{1/12}Cr_{1/12}$)). J Phy Chem C 117:9056–9064. https://doi.org/10.1021/jp309180k

12. Prabu M, Reddy M, Selvasekarapandian S, Rao GS, Chowdari B (2013) (Li, Al)-co-doped spinel, Li($Li_{0.1}Al_{0.1}Mn_{1.8}$)O_4 as high performance cathode for lithium ion batteries. Electrochim Acta 88:745–755. https://doi.org/10.1016/j.electacta.2012.10.011
13. Zhao X, Reddy M, Liu H, Ramakrishna S, Rao GS, Chowdari B (2012) Nano $LiMn_2O_4$ with spherical morphology synthesized by a molten salt method as cathodes for lithium ion batteries. RSC Adv 2:7462–7469. https://doi.org/10.1039/C2RA01110G
14. Reddy M, Raju MS, Sharma N, Quan P, Nowshad SH, Emmanuel H-C, Peterson V, Chowdari B (2011) Preparation of $Li_{1.03}Mn_{1.97}O_4$ and $Li_{1.06}Mn_{1.94}O_4$ by the polymer precursor method and X-ray, neutron diffraction and electrochemical studies. J Electrochem Soc 158:A1231–A1236. https://doi.org/10.1149/2.074111jes
15. Reddy M, Cheng H, Tham J, Yuan C, Goh H, Chowdari B (2012) Preparation of Li($Ni_{0.5}Mn_{1.5}$)O_4 by polymer precursor method and its electrochemical properties. Electrochim Acta 62:269–275. https://doi.org/10.1016/j.electacta.2011.12.029
16. Ashammakhi N, Ndreu A, Piras AM, Nikkola L, Sindelar T, Ylikauppila H, Harlin A, Gomes ME, Neves NM, Chiellini E, Chiellini F, Hasirci V, Redl H, Reis RL (2006) Biodegradable nanomats produced by electrospinning: Expanding multifunctionality and potential for tissue engineering. J Nanosci Nanotechnol 6:2693–2711. https://doi.org/10.1166/jnn.2007.485
17. Reneker DH, Yarin AL (2008) Bending instability of electrically charged liquid jets of polymer solutions in electrospinning. Polymer 49:2387–2425. https://doi.org/10.1063/1.373532
18. Deitzel JM, Kleinmeyer J, Harris D, Tan NCB (2001) The effect of processing variables on the morphology of electrospun nanofibers and textiles. Polymer 42(1):261–272. https://doi.org/10.1016/S0032-3861(00)00250-0
19. Megelski S, Stephens JS, Chase DB, Rabolt JF (2002) Micro- and nanostructured surface morphology on electrospun polymer fibers. Macromolecules 35(22):8456–8466. https://doi.org/10.1021/ma020444a
20. Baumgarten PK (1971) Electrostatic spinning of acrylic microfibers. J Colloid Interf Sci 36(1):71–79. https://doi.org/10.1016/0021-9797(71)90241-4
21. Buchko CJ, Chen LC, Shen Y, Martin DC (1999) Processing and microstructural characterization of porous biocompatible protein polymer thin films. Polymer 40(26):7397–7407. https://doi.org/10.1016/S0032-3861(98)00866-0
22. Eda G, Shivkumar S (2006) Bead structure variations during electrospinning of polystyrene. J Mater Sci 41(17):5704–5708. https://doi.org/10.1007/s10853-006-0069-9
23. Tao J, Shivkumar S (2007) Molecular weight dependent structural regimes during the electrospinning of PVA. Mater Lett 61(11–12):2325–2328. https://doi.org/10.1016/j.matlet.2006.09.004
24. Fridrikh SV, Yu JH, Brenner MP, Rutledge GC (2003) Controlling the fiber diameter during electrospinning. Phys Rev Lett 90(14):144502(1–4). https://doi.org/10.1103/PhysRevLett.90.144502
25. Nezarati RM, Eifert MB, Hernandez EC (2012) Effects of humidity and solution viscosity on electrospun fiber morphology. Tissue Eng Part C 19:810–819. https://doi.org/10.1089/ten.tec.2012.0671
26. Casper CL, Stephens JS, Tassi NG, Chase DB, Rabolt JF (2004) Controlling surface morphology of electrospun polystyrene fibers: effect of humidity and molecular weight in the electrospinning process. Macromolecules 37:573–578. https://doi.org/10.1021/ma0351975
27. Raghavan BK, Coffin DW (2011) Control of inter-fiber fusing for nanofiber webs via electrospinning. J Eng Fibers Fabr 6:1–5. https://www.jeffjournal.org/papers/Volume6/6.4.1Coffin.pdf
28. Htike HH, Chen L, Sachiko S (2012) The effect of relative humidity on electrospinning of poly(vinyl alcohol) with soluble eggshell membrane. J Textil Eng 58:9–12. https://doi.org/10.4188/jte.58.9
29. Gabano JP (1983) Lithium batteries. Academic Press, London and New York, 467. https://scholar.google.co.il/citations?user=EO-B1RMAAAAJ&hl=iw

30. Nazri GA, Pistoia G (2009) Lithium batteries: Science and technology. Springer. https://www.springer.com/gp/book/9781402076282
31. Scrosati B (1993) Insertion compounds for lithium rocking chair batteries: the electrochemistry of novel materials. VCH Publishers 3:111–137. https://d-nb.info/941450163/04
32. Tarascon M, Armand M (2001) Issues and challenges facing rechargeable lithium batteries. Nature 414:359–367. https://doi.org/10.1038/35104644
33. Long JW, Dunn B, Rolison DR, White HS (2004) Three-dimensional battery architectures. Chem Rev 104:4463–4492. https://doi.org/10.1021/cr020740l
34. Ji L, Zhang X (2008) Ultrafine polyacrylonitrile/silica composite fibers via electrospinning. Mater Lett 62:2165–2168. https://doi.org/10.1016/j.matlet.2007.11.051
35. Guo YG, Hu JS, Wan LJ (2008) Nanostructured materials for electrochemical energy conversion and storage devices. Adv Mater 20:2878–2887. https://doi.org/10.1002/adma.200800627
36. Rougier A, Saadouane I, Gravereau P, Willmann P, Delmas C (1996) Effect of cobalt substitution on cationic distribution in $LiNi_{1-y}$ Co_yO_2 electrode materials. Solid State Ionics 90:83–90. https://doi.org/10.1016/S0167-2738(96)00370-0
37. Saardoune I, Delmas C (1996) On the $Li_xNi_{0.8}Co_{0.2}O_2$ System. J Solid State Chem 136:8–15. https://doi.org/10.1006/jssc.1997.7599
38. Saadoune I, Menetrier M, Delmas C (1997) Redox processes in $Li_xNi_{1-y}Co_yO_2$ cobalt-rich phases. J Mater Chem 7:2505–2511. https://doi.org/10.1039/A703368K
39. Saadoune I, Delmas C (1996) $LiNi_{1-y}Co_yO_2$ positive electrode materials: relationships between the structure, physical properties and electrochemical behaviour. J Mater Chem 6:193–199. https://doi.org/10.1039/JM9960600193
40. Zhecheva E, Stoyanova R (1993) Stabilization of the layered crystal structure of $LiNiO_2$ by co-substitution. Solid State Ionics 66:143–149. https://doi.org/10.1016/0167-2738(93)90037-4
41. Prado G, Rougier A, Fournes L, Delmas C (2000) Electrochemical behavior of iron-substituted lithium nickelate. J Electrochem Soc 147:2880–2887. https://doi.org/10.1149/1.1393620
42. Imanishi N, Fujiyoshi M, Takeda Y, Yamamoto O, Tabuchi M (1999) Preparation and Li-NMR study of chemically delithiated $Li_{1-x}CoO_2$ $(0 < x < 0.5)$. Solid State Ionics 118:121–128. https://doi.org/10.1016/S0167-2738(98)00441-X
43. Pouillerie C, Croguennec L, Delmas C (2000) The $Li_xNi_{1-y}Mg_yO_2$ ($y = 0.05, 0.10$) system: structural modifications observed upon cycling. Solid State Ionics 132:15–29. https://doi.org/10.1016/S0167-2738(00)00699-8
44. Nakai I, Nakagome T (1998) In-situ transmission X-ray absorption fine structure analysis of the Li deintercalation process in Li(Ni0.5Co0.5)O2. Electrochem Solid State Lett 1:259–261. https://doi.org/10.1149/1.1390705
45. Delmas C, Capitaine F (1996) Abstracts of the 8th international meeting, lithium batteries. Electrochemical Society, Pennington, NJ
46. Chen R, Whittingham MS (1997) Cathodic behavior of alkali manganese oxides from permanganate. J Electrochem Soc 144:L64–L67. https://doi.org/10.1149/1.1837554
47. Armstrong AR, Bruce PG (1996) Synthesis of layered $LiMnO_2$ as an electrode for rechargeable lithium batteries. Nature 381:499–500. https://doi.org/10.1149/1.1837554
48. Capitaine F, Gravereau P, Delmas C (1996) A new variety of $LiMnO_2$ with a layered structure. Solid State Ionics 89:197–202. https://doi.org/10.1016/0167-2738(96)00369-4
49. Chen R, Zavalij PY, Whittingham MS (1997) New manganese oxides by hydrothermal reaction of permanganates. Mater Res Soc Proc 453:653. https://doi.org/10.1557/PROC-453-653
50. Chen R, Zavalij P, Whittingham MS (1996) Hydrothermal synthesis and characterization of $K_xMnO_2{\cdot}yH_2O$. Chem Mater 8:1275. https://doi.org/10.1021/cm950550+
51. Chen R, Chirayil T, Whittingham MS (1996) The hydrothermal synthesis of sodium manganese oxide and a lithium vanadium oxide. Solid State Ionics 1:86–88. https://doi.org/10.1016/0167-2738(96)00086-0
52. Yoshio M, Yamato K, Itoh J, Noguchi H, Okada M, Mouri T (1994) Preparation and properties of $LiCo_yMn_xNi_{1-x-y}O_2$ as a cathode for lithium ion batteries. Electrochem Soc Proc 251:28–94. https://doi.org/10.1016/S0378-7753(00)00407-9

53. Nitta Y, Okamura K, Haraguchi K, Kobayashi S, Ohta A (1995) Crystal structure study of $LiNi_{1-x}Mn_xO_2$. J Power Sour 54:511–515. https://doi.org/10.1016/0378-7753(94)02137-R
54. Armstrong A, Gitzendanner R (1998) The intercalation compound $Li(Mn_{0.9}Co_{0.1})O_2$ as a positive electrode for rechargeable lithium batteries. Chem Commun 17:1833–1834. https://doi.org/10.1039/A803741H
55. Numata K, Yamanaka S (1999) Preparation and electrochemical properties of layered lithium-cobalt-manganese oxides. Solid State Ionics 118:117–120. https://doi.org/10.1016/S0167-2738(98)00425-1
56. Spahr ME, Novak P, Schnyder B, Haas O, Nesper R (1998) Characterization of layered lithium nickel manganese oxides synthesized by a novel oxidative coprecipitation method and their electrochemical performance as lithium insertion electrode materials. J Electrochem Soc 145:1113–1121. https://doi.org/10.1149/1.1838425
57. Ohzuku T, Makimura Y (2001) Layered lithium insertion material of $LiCo_{1/3}Ni_{1/3}Mn_{1/3}O_2$ for lithium-ion batteries. Chem Lett 30:642–643. https://doi.org/10.1246/cl.2001.642
58. Lu Z, MacNeil DD, Dahn JR (2001) Layered Li [Ni_xCo_1—$2x$Mnx]O_2 cathode materials for lithium-ion batteries. Electrochem Solid State Lett 4:A200–A203. https://doi.org/10.1149/1.1413182
59. Wang Z, Sun Y, Chen L, Huang X (2004) Electrochemical characterization of positive electrode material LiNi1/3Co1/3Mn1/3O2 and compatibility with electrolyte for Lithium-ion batteries. J Electrochem Soc 151:A914–A921. https://doi.org/10.1149/1.1740781
60. Armstrong AR, Paterson AJ, Robertson AD, Bruce PG (2002) Nonstoichiometric layered $Li_xMn_yO_2$ with a high capacity for lithium intercalation/deintercalation. Chem Mater 14:710–719. https://doi.org/10.1021/cm010382n
61. Zhang F, Whittingham MS (2000) Electrochemistry of the layered manganese dioxides: a xMn1—y(Co, Ni, Fe)yO2 (A = Li, K) rate effects. Electrochem Solid State Lett 3:309–311. https://doi.org/10.1149/1.1391132
62. Grincourt Y, Storey C, Davidson IJ (2001) Lithium ion cells using a new high capacity cathode. J Power Source 97–98:711–713. https://doi.org/10.1016/S0378-7753(01)00750-9
63. Storey C, Kargina I, Grincourt Y, Davidson IJ, Yoo YC, Seung DY (2001) Electrochemical characterization of a new high capacity cathode. J Power Sour 97–98:541–544. https://doi.org/10.1016/S0378-7753(01)00692-9
64. Rossen E, Jones CDW, Dahn JR (1992) Structure and electrochemistry of $Li_xMn_yNi_{1-y}O_2$. Solid State Ionics 57:311–318. https://doi.org/10.1016/0167-2738(92)90164-K
65. Ohzuku T, Makimura Y (2001) Layered lithium insertion material of $LiNi_{1/2}Mn_{1/2}O_2$: a possible alternative to $LiCoO_2$ for advanced lithium-ion batteries Chem. Lett. 30:744–745. https://doi.org/10.1246/cl.2001.744
66. Lu Z, MacNeil DD, Dahn JR (2001) Layered cathode materials Li [Ni_xLi (1/3–2x/3)Mn(2/3–x/3)]O2 for lithium-ion batteries. Electrochem Solid State Lett 4:A191–A194. https://doi.org/10.1149/1.1407994
67. Kang S-H, Kim J, Stoll ME, Abraham D, Sun YK, Amine K (2002) Layered $Li(Ni_{0.5-x}Mn_{0.5-x}M_{2x}{}')O_2$ ($M' = $ Co, Al, Ti; $x = 0, 0.025$) cathode materials for Li-ion rechargeable batteries. J Power Sour 112:41–48. https://doi.org/10.1016/S0378-7753(02)00360-9
68. Johnson CS, Kim JS, Kropf AJ, Kahaian AJ, Vaughey JT, Thackeray MM (2002) The role of Li_2MO_2 structures (M=metal ion) in the electrochemistry of $(x)LiMn_{0.5}Ni_{0.5}O_2\cdot(1-x)Li_2TiO_3$ electrodes for lithium-ion batteries. Electrochem Commun 4:492–498. https://doi.org/10.1016/S1388-2481(02)00346-6
69. Liu Z, Yu A, Lee JY (1999) Synthesis and characterization of $LiNi_{1-x-y}Co_xMn_yO_2$ as the cathode materials of secondary lithium batteries. J Power Sour 81–82:416–419. https://doi.org/10.1016/S0378-7753(99)00221-9
70. Kim JM, Chung HT (2004) The first cycle characteristics of $Li[Ni_{1/3}Co_{1/3}Mn_{1/3}]O_2$ charged up to 4.7 V. Electrochim Acta 49:937–944. https://doi.org/10.1016/j.electacta.2003.10.005
71. Jouanneau S, Eberman KW, Krause LJ, Dahn JR (2003) Synthesis, characterization and electrochemical behavior of improved Li [NixCo1 $-$ $2_x$$Mn_x$] O2 ($0.1 \leq x \leq 0.5$). J Electrochem Soc 150:A1637–A1642. https://doi.org/10.1149/1.1622956

72. Jouanneau S, Macneil DD, Lu Z, Beattie SD, Murphy G, Dahn JR (2003) Morphology and safety of Li $[Ni_xCo1{-}2_xMn_x]O2(0 \leq x \leq 1/2)$. J Electrochem Soc 150:A1299–A1304. https://doi.org/10.1149/1.1602077
73. Ngala JK, Chernova NA, Ma M, Mamak M, Zavalij PY, Whittingham MS (2004) The synthesis, characterization and electrochemical behavior of the layered $LiNi_{0.4}Mn_{0.4}Co_{0.2}O_2$ compound. J Mater Chem 14:214–220. https://doi.org/10.1039/B309834F
74. Guilmard M, Croguennec L, Delmas C (2003) Thermal stability of lithium nickel oxide derivatives. Part I: $Li_xNi_{1.02}O_2$ and $Li_xNi_{0.89}Al_{0.16}O_2$ ($x = 0.50$ and 0.30). Chem Mater 15:4476–4483. https://doi.org/10.1021/cm030059f
75. Guilmard M, Croguennec L, Delmas C (2003) Thermal stability of lithium nickel oxide derivatives Part II: $Li_xNi_{0.70}Co_{0.15}Al_{0.15}O_2$ and $Li_xNi_{0.90}Mn_{0.10}O_2$ ($x = 0.50$ and 0.30). Comparison with $Li_xNi_{1.02}O_2$ and $Li_xNi_{0.89}Al_{0.16}O_2$. Chem Mater 15:4484–4493. https://doi.org/10.1021/cm030340u
76. Arai H, Sakurai Y (1999) Characteristics of $LixNiO_2$ obtained by chemical delithiation. J Power Sour 80–81:401–405. https://doi.org/10.1016/S0378-7753(99)00115-9
77. Choi S, Manthiram A (2000) Synthesis and electrode properties of metastable $Li_2Mn_4O_9$- δ spinel oxides. J Electrochem Soc 147(5):1623–1629. https://scholar.google.com/citations?hl=en&user=DGy5QdUAAAAJ&view_op=list_works&sortby=pubdate
78. Chen R, Zavalij PY, Whittingham MS, Greedan JE, Raju NP, Bieringer M (1999) The hydrothermal synthesis of the new manganese and vanadium oxides, $NiMnO_3H$, MAV_3O_7 and $MA_{0.75}V_4O_{10}.0.67H_2O$ ($MA{=}CH_3NH_3$). J Mater Chem 9:93–100. https://doi.org/10.1039/a805312j
79. Ammundsen B, Desilvestro H, Paulson JM, Steiner R, Pickering PJ (2000) 10th international meeting on lithium batteries, Italy, 2000. Electrochemical Society, Pennington, NJ
80. Storey C, Kargina I, Grincourt Y, Davidson IJ, Yoo Y, Seung DY (2000) 10th International meeting on lithium batteries, Italy, Electrochemical Society: Pennington, NJ, vol 10, abstract 234
81. Numata K, Sakaki C, Yamanaka S (1999) Synthesis and characterization of layer structured solid solutions in the system of $LiCoO_2$-Li_2MnO_3. Solid State Ionics 117:257–263. https://doi.org/10.1016/S0167-2738(98)00417-2
82. Yoshio M, Noguchi H, Itoh J, Okada M, Mouri T (2000) Preparation and properties of $LiCo_yMn_xNi_{1-x-y}O_2$ as a cathode for lithium ion batteries. J Power Sour 90:176–181. https://doi.org/10.1016/S0378-7753(00)00407-9
83. Thackeray MM, Johnson CS, Amine K, Kim J (2004) Lithium metal oxide electrodes for lithium cells and batteries. U.S. Patent 6,677,082
84. Thackeray MM, Johnson CS, Amine K, Kim J (2004) Lithium metal oxide electrodes for lithium cells and batteries. U.S. Patent 6,680,143
85. Robertson AD, Bruce PG (2002) The origin of electrochemical activity in Li_2MnO_3. Chem Commun 2790–2791. http://doi.org/10.1039/b207945c
86. Robertson AD, Bruce PG (2003) Mechanism of electrochemical activity in Li_2MnO_3. Chem Mater 15:1984–1992. https://doi.org/10.1021/cm030047u
87. Armstrong AR, Bruce PG (2004) Electrochemistry beyond Mn4 + in LixMn1 − yLiy O2 Electrochem. Solid State Lett. 7:A1–A4. https://doi.org/10.1149/1.1625591
88. Russouw MH, Thackeray MM (1991) Lithium manganese oxides from Li_2MnO_3 for rechargeable lithium battery applications. Mater Res Bull 26:463–473. https://doi.org/10.1016/0025-5408(91)90186-P
89. Russouw MH, Liles DC, Thackeray MM (1993) Synthesis and structural characterization of a novel layered lithium manganese oxide, $Li_{0.36}Mn_{0.91}O_2$, and its lithiated derivative, $Li_{1.09}Mn_{0.91}O_2$. J Solid State Chem 104:464–466. https://doi.org/10.1006/jssc.1993.1182
90. Paik Y, Grey CP, Johnson CS, Kim JS, Thackeray MM (2002) Lithium and deuterium NMR studies of acid-leached layered lithium manganese oxides. Chem Mater 14:5109–5115. https://doi.org/10.1021/cm0206385
91. Kim JS, Johnson CS, Thackeray MM (2002) Layered $xLiMO_2{\cdot}(1-x)Li_2M'O_3$ electrodes for lithium batteries: a study of $0.95LiMn_{0.5}Ni_{0.5}O_2{\cdot}0.05Li_2TiO_3$. Electrochem Commun 4:205–209. https://doi.org/10.1016/S1388-2481(02)00251-5

92. Kim JH, Park CW, Sun YK (2003) Synthesis and electrochemical behavior of $Li[Li_{0.1}Ni_{0.35-x/2}Co_xMn_{0.55-x/2}]O_2$ cathode materials. Solid State Ionics 164:43–49. https://doi.org/10.1016/j.ssi.2003.08.003
93. Lee CW, Sun YK, Prakash J (2004) A novel layered Li $[Li_{0.12}Ni_zMg_{0.32-z}Mn_{0.56}]O_2$ cathode material for lithium-ion batteries. Electrochim Acta 49:4425–4432. https://doi.org/10.1016/j.electacta.2004.04.033
94. Hao P, Zhu T, Su Q, Lin J, Cui R, Cao X, Wang Y, Pan A (2018) Electrospun single crystalline fork-like $K_2V_8O_{21}$ as high-performance cathode materials for lithium-ion batteries. Front Chem 6:195(1–9). https://doi.org/10.3389/fchem.2018.00195
95. Dimesso L, Spanheimer C, Jaegermann W, Zhang Y, Yarin AL (2012) $LiFePO_4$-3D carbon nanofiber composites as cathode materials for Li-ions batteries. J Appl Phys 111:064307. https://doi.org/10.1063/1.3693575